Major Geological, Climatic and Biological Events

Culmination of mountain-building followed by erosion and moderate, short-lived invasions of continental margins by the sea. Early warming trends were reversed by the middle of the period to cooler and finally to glacial conditions. Subtropical forests gave way to temperate forests and finally to extensive grasslands. Transition from archaic mammals to modern orders and eventually families. Evolution of humans during the last 5-8 million years.

Last great spread of epicontinental seas and shoreline swamps. At the end of the period extensive mountain-building cooled the climate worldwide. Angiosperm dominance began. Extinction of archaic birds and many reptiles by the end of the period.

Climate was warm and stable with little latitudinal or seasonal variation. Modern genera of many gymnosperms and advanced angiosperms appeared. Reptilian diversity was high in all habitats. First birds appeared.

Continents were relatively high with few shallow seas. The climate was warm; deserts were extensive. Gymnosperms dominated; angiosperms first appeared. Mammal-like reptiles were replaced by precursors of dinosaurs and the earliest true mammals appeared.

Land was generally higher than at any previous time. The climate was cold at the beginning of the period but warmed progressively. Glossopterid forests developed with the decline of the coal swamps. Mammal-like reptiles were diverse; widespread extinctions at the end of the period.

Generally warm and humid, but some glaciation in the Southern Hemisphere. Extensive coal-producing swamps with large anthropod faunas. Many specialized amphibians and the first appearance of reptiles.

Mountain-building produced locally arid conditions, but extensive lowland forests and swamps were the beginning of the great coal deposits. Extensive radiation of amphibians; extinction of some fish lineages and expansion of others.

The land was higher and climates cooler. Freshwater basins developed in addition to shallow seas. The first forests appeared and the first winged insects. There was an explosive radiation of fishes, followed by the disappearance of many jawless forms. The earliest tetrapods appeared.

The land was slowly being uplifted, but shallow seas were extensive. The climate was warm and terrestrial plants radiated. Eurypterid arthropods were at their maximum abundance in aquatic habitats and the first terrestrial arthropods appeared. The first gnathostomes appeared among a diverse group of marine and freshwater jawless fishes.

The maximum recorded extent of shallow seas was reached and the warming of the climate continued. Algae became more complex, vascular plants may have been present, and there was a variety of large invertebrates. Jawless fish fossils from this period are fragmentary but more widespread.

There were extensive shallow seas in equatorial regions. The climate was warm. Algae were abundant and there are records of trilobites and brachiopods. The first remains of vertebrates are found at the end of this period.

Changes in the lithosphere produced major land masses and areas of shallow seas. Multicellular organisms appeared and flourished—algae, fungi, and many invertebrates.

Formation of the earth and slow development of the lithosphere, hydrosphere, and atmosphere. Development of life in the hydrosphere.

ECOLOGY
of a Changing Planet
SECOND EDITION

Mark B. Bush

Florida Institute of Technology

PRENTICE HALL
Upper Saddle River, New Jersey 07458

Library of Congress Cataloging-in-Publication Data
Bush, Mark B.
    Ecology of a changing planet / Mark B. Bush. — 2nd ed.
        p.   cm.
    Includes bibliographical references.
    ISBN 0-13-011202-X
    1. Ecology.   2. Nature—Effect of human beings on.
    3. Environmental sciences.   I. Title
    QH541.B88   2000
    577—dc21   99-24607
                                                    CIP

Senior Editor: Teresa K. Ryu
Executive Editor: Sheri L. Snavely
Editor in Chief: Paul F. Corey
Production Editor: Joanne Hakim
Assistant Vice President of Production & Manufacturing: David W. Riccardi
Executive Managing Editor: Kathleen Schiaparelli
Assistant Managing Editor: Lisa Kinne
Marketing Manager: Jennifer Welchans
Creative Director: Paula Maylahn
Art Director: Ann France
Interior and Cover Design: Jill Little
Cover Photo: Gabriela Staebler Wildlife Photography
Art Manager: Karen Branson
Manufacturing Manager: Trudy Pisciotti
Manufacturing Buyer: Benjamin Smith
Associate Editor: Mary Hornby
Photo Editor: Beth Boyd
Photo Research: Linda Sykes
Image Coordinator: Charles Morris
Illustrations: Kandis Elliot
Part Illustrations: Tamara Newnam-Cavallo
Icons: Batelman Illustration
Copyeditor: Jane Loftus
Editorial Assistants: Lisa Tarabokjia and Nancy Bauer

Printed in the United States of America

10 9 8 7 6 5 4 3 2 1

ISBN 0-13-011202-X

Prentice-Hall International (UK) Limited, *London*
Prentice-Hall of Australia Pty. Limited, *Sydney*
Prentice-Hall Canada, Inc., *Toronto*
Prentice-Hall Hispanoamericana, S.A., *Mexico*
Prentice-Hall of India Private Limited, *New Delhi*
Prentice-Hall of Japan, Inc., *Tokyo*
Prentice-Hall (*Singapore*) Pte Ltd
Editora Prentice-Hall do Brasil, Ltda., *Rio de Janeiro*

Printed on Recycled Paper

Brief Contents

Contents

FOUR   **ECOLOGY AND SOCIETY**

Preface

My goal has been to write an introductory applied ecology text that centers on how ecology relates to environmental issues. This edition contains two entirely new chapters that reflect some of the fastest growing fields in applied ecology: wetland science and fishery management. A number of chapters, such as those on climate, habitat fragmentation, conservation biology, disease, and the future have been completely rewritten. In many areas of the book, ideas have been rearranged to improve the flow of information and make the book more usable. Inevitably the book has grown, though this has been minimized by eliminating some text (always a hard decision). The end of each chapter has some recommended readings. These are not meant to be exhaustive, but they contain a mix of classic papers, conflicting viewpoints, and accessible summaries of related ideas. Many students may have missed the availability of a Web-site that accompanied the first edition, as the only reference to it was buried in the Preface. In this edition, each chapter ends with a reminder to visit the Web-site to find further information. My Web-site is organized around the chapters in the book. You will find self-grading quizzes and a set of hotlinks to Web pages that I have found useful and accurate. The hotlinks access the Web sites of select governments, universities, and reputable organizations. If you find more that should be included (or ones that no longer exist), please let me know at mbush@fit.edu. Please remember that these sites are not refereed and should not be given the same credence as the primary literature. Teachers will also have access to 75 overhead transparencies of key figures from the text. I have learned much as I researched these chapters, frequently heading from the keyboard into the classroom to try out a new line of thinking on my students. I hope that some of my enthusiasm shines through in the text. A textbook is no substitute for an effective teacher, but it should be a valuable ally to teacher and student alike, providing a common baseline of knowledge. I believe that a textbook should be topical, relevant, and interesting to read. Although scientists must learn facts, I have tried to emphasize the concepts and examples of applied ecology, rather than produce an encyclopedia of indigestible statistics.

Although many teaching variations are possible, I have written the material in the sequence that I teach it. The first half of the book establishes a strong framework of applied ecology. The second half of the book develops environmental issues such as habitat fragmentation, acid deposition, and the emergence of new human diseases. Each topic is investigated from a scientific rather than an activist viewpoint and is written to stand alone. This book is meant for students who have had a college-level introductory biology class and want to pursue biology, ecology, or environmental science to the next level.

To the Student

Science requires precision of measurement, careful thought, and recall. It is often said that a little knowledge is a dangerous thing. It is not enough to grasp only the "big picture." Facts are used to support an argument, and it is essential that you understand how the details relate to larger concepts. Misinformation about the environment occurs when these details, conveniently or accidentally, are ignored. The natural world is incredibly complex and full of detail. In this book I have distilled, from all this information, what I believe to be the essentials for an introduction to applied ecology. To get the most from this book or a course in ecology, try to think like an ecologist. Nature is all around us as we walk between buildings, drive down the road, or stare out of a window. Can you explain what you see? What does the news have to do with ecology? New disease outbreaks, wars, refugees, famines, an oil spill, a new medicine to combat malaria, a hydroelectric project stalled by concern for an endangered species are all going to have ecological and environmental impacts. In writing this book, I have concentrated on presenting scientific arguments relating to environmental issues rather than encouraging environmental activism. Make your own judgment. Use your knowledge of ecology to determine what needs to be done to the environment, and then talk to others, including your instructor, about how to achieve it. It's your Earth! Be informed!

To the Instructor

I have found the content of this book to be a successful blend of material for teaching an introduction to applied ecology course and for heightening environmental awareness. In my class, I try to maintain some emotional distance from the issues, but I encourage my students to become active. The students can monitor Internet home pages of such organizations as the Sierra Club, Envirolink Network, Defenders of Wildlife, and federal and state governments. Links to these addresses can be found from my Web-site: http://www.prenhall.com/bush. Information about the home page can be obtained from your local Prentice Hall representative.

I hope that the users of this book—both teachers and students—will find it to be a valuable tool. My publisher and I have made every effort to ensure that it is completely accurate and error-free. Should you find an error—or perhaps even a way to make the book better—I would be delighted to hear from you.

Acknowledgments

Writing a book represents a tremendous output of personal energy and, in my case, could only be achieved with the support of family, friends, and colleagues.

I owe a special debt of gratitude to my longtime friend, field companion, colleague, and mentor Paul Colinvaux. Without his example and inspiration I would never have tried to write a book. I wish to thank all those who have collaborated on past field projects, especially John Flenley and John Pethick, for the training they gave me as a graduate student. My later projects owe much of their success to Robert Whittaker and Tukurin Partomihardjo (expeditions to Krakatau, Indonesia), Paulo De Oliveira, Melanie Reidinger, Michael Miller, Miriam Steinitz-Kannan, Eduardo Asanza, Ana-Cristina Asanza-Sosa and Fausto Sarmiento (Amazonian and Andean studies), and Robert Rivera and George DeBusk (Central America). Many of my colleagues have provided valuable insights, and I particularly wish to thank James Clark, Richard Cook, Randy Kramer, Daniel Livingstone, Carol Mansfield, Dolores Piperno, Curtis Richardson, William Schlesinger, and Peter Thrall.

I thank my colleagues at Florida Institute of Technology and my students in "Ecology of a Changing Planet" for being my sounding board for new ideas; they have been integral to the development of this text. My publishing house has provided me with wonderful associates who have converted a pile of manuscript pages into a polished text. My special thanks go to my editor Teresa Ryu for her steadfast belief in this project, and to my production editor, Joanne Hakim, who has ensured that the project moved swiftly and smoothly through the production phase. The manuscript benefited greatly from the attentions of Ann France and Karen Horton. To this whole group I extend a heartfelt thanks.

Reviewers

Scott Anderson
Northern Arizona University

Royce Ballinger
Boston University

Scott Brady
Central Washington University

Peter Busher
Boston University

Spencer Courtright
Indiana University Northwest

Craig Davis
Ohio State University

Susan Foster
Clark University

Margaret Fusari
University of California, Santa Cruz

Heather Gallacher
Cleveland State University

Grant Gerrish
University of Hawaii-Hilo

Teresa Horton
Northwestern University

Linda Margaret Hunt
University of Notre Dame

Matthew Kelty
University of Massachusetts

Steven Malcolm
Western Michigan University

Margaret Reisinger
Northeastern Illinois University

Steve Stephenson
Michigan State University

Richard Vance
University of California, Los Angeles

Paul Weihe
Davison & Elkins College

About the Author

Mark B. Bush is an Associate Professor of conservation biology at the Florida Institute of Technology. His BS and Ph.D. degrees were earned at the University of Hull in England. Between undergraduate and graduate school, Bush spent several years working for the British Trust for Conservation Volunteers as a specialist in conservation education. During this time, he designed and implemented the city of Hull's first inner-city nature area, a project that involved over 300 schoolchildren and adult volunteers. He used his garden as a tree nursery to provide more than 2000 native trees to be planted each year. Between 1985 and 1987, he also managed a small wetland nature reserve owned by the Yorkshire Wildlife Trust.

Professor Bush has spent more than 18 years in ecological research and has worked in some of the world's most remote locations. His field sites include Amazonia, Panama, Costa Rica, Ecuador, and the islands of Krakatau, Indonesia. He is an authority on the history of South and Central American tropical ecosystems and on island biogeography. He has lived in the United States since 1987, spending four years as a researcher at the Ohio State University, a year as a Mellon Fellow at the Smithsonian Tropical Research Institute in Panama, and four years teaching ecology and environmental science at Duke University, prior to moving to the Florida Institute of Technology.

Professor Bush is a member of Sigma Xi, the Society of Wetland Scientists, and numerous environmental organizations. His hobbies include scuba diving, kayaking, and hiking.

Mark Bush
Associate Professor
Department of Biological Sciences
Florida Institute of Technology
150 W. University Blvd.
Melbourne, FL 32901
Phone (407) 674-7166
Fax   (407) 674-7238

To Virginia, my
love and thanks.

P A R T

ONE

Diversity

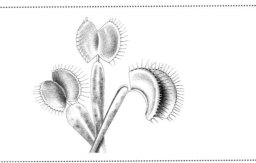

1

Ecology, Environmentalism, and the First Polluters

2

Evolution and Natural Selection: The Heart of Ecology

3

The Ecological Efficiency of Living Things

4

Climate

5

Biomes: The Great Vegetation Types

6

Ecosystems, Nutrient Cycles, and Soil

7

Aquatic Ecosystems

8

Why Wetlands Aren't Worthless

1

Ecology, Environmentalism, and the First Polluters

Ecology is the ideal science for the naturally curious. If you have ever wondered why leaves turn color before they fall from the tree, or why insects are small, or why there are two sexes of humans rather than three or four, you are pondering ecological questions. The term **ecology** comes from the Greek *oikos* meaning "house," or our environment. *Ecology* was first used as a scientific term in the late nineteenth century; it eventually came into popular use in the 1960s. Ecologists are scientists who study the distribution and abundance of species and their relationship to the environment, but more than that they always ask Why?

1.1 Developing and Testing Hypotheses

Good ecology, like all science, is investigative. Accurate observations and attention to details are the basis of science and are crucial to forming a clearly defined **hypothesis**. A hypothesis is an explanation of events or observations, and it usually accounts for all the available data on a subject. Above all, it must be testable. For it to be testable, the hypothesis must lead directly to a prediction, and it is the accuracy of the prediction that we can test. It is important to note that *a scientific hypothesis can never be proven*; it can only be disproven. The testing may take the form of running an experiment in a laboratory, constructing a computer simulation, or spending long hours in the field observing, measuring, and quantifying. Data collection and analysis must avoid bias or preconceived ideas of what the result should be. Repetitions of the whole procedure must produce the same results for a hypothesis to be viable. The results of the analyses are then formulated into a logical scientific conclusion that is based on the results of the tests that confirmed the prediction. Despite gathering all this information, we have not proved the hypothesis. For example, an observation might be that all the swans that I have ever seen are white. I can then formulate the hypothesis that all swans are white. To test my hypothesis, I search throughout North America, perhaps finding another thousand white swans. I have added to my data set and strengthened my argument, but I have not proved the hypothesis. The search continues abroad. After traveling through Europe, Africa, and Asia, I have still seen only white swans; my hypothesis remains intact but unproved. I then go to Australia and find a black swan. This single new datum refutes my hypothesis. It does not matter how many white swans I find: A single black swan is enough to disprove the hypothesis. Science demands that I set up a new hypothesis to accommodate the new data. One hypothesis would be that all American swans are white. But this hypothesis is less useful than the original one because it contains the qualifier "American." Excluding a portion of the observed data weakens the hypothesis, especially when no factor is identified to justify or explain the exclusion. Furthermore, although the more limited hypothesis is testable, it does not lead the investigator into any new avenue of understanding. A better hypothesis—one that might lead to an improved understanding of the situation—would be that all white swans are more closely related to each other than a white swan is to a black swan. This prediction could be tested by comparing the genetic heritage of black and white swans.

When numerous hypotheses have been tested and found to be viable, and when those hypotheses all deal with the same basic subject matter, a more general and more powerful statement can be made to encompass all the hypotheses. Such a statement is a scientific **theory**. A theory represents the highest level of scientific understanding (it cannot be a fact because, by definition, it cannot be proved but only disproved). For example, we have the theory of gravity, the theory of light, and the theory of evolution, all of which have withstood every scientific test; they are powerful tools to explain how our universe works. Theories are not carved in stone and can be modified by improved understanding.

As do all other disciplines of learning, ecology deals with a relative truth, not an absolute one. As we gain knowledge, our science and our hypotheses are continually being modified to take into account new information. Our scientific forebears founded their hypotheses on a contemporaneous understanding of what was true. Even if their argument had flawless logic, they would have reached faulty conclusions if their starting assumptions were false. Just because earlier workers did not always arrive at the right answer does not make them poor scientists. The faltering steps, or sometimes huge leaps, of these investigators sometimes advanced knowledge and sometimes took them down a blind alley. But even blind alleys and negative results form an important part of accumulated scientific knowledge. Later workers can use those results to help shape their hypotheses and arrive at a perception of nature. It is the willingness to be bound by fact and to strive to disprove theories that makes the scientific method unique.

1.2 Science and Society

Because so many of our scientific discoveries have direct implications for future generations, the practice of science is tied intimately to the future of society. Scientific innovations have brought much good to our species. Science has helped us feed the hungry and prevent disease, and it has given us heat and light in our homes. However, some scientific discoveries can be unpopular, even disturbing to our social fabric. For example, the sixteenth-century assertion of Copernicus that Earth revolved around the sun was a massive heresy in European Christianity. For centuries it had been taught that the sun and all other heavenly bodies revolved around Earth. Orthodoxy placed man (I use man deliberately) at the center of the universe, and to relegate him to an out-of-the-way planet in a backwater of the Milky Way was sacrilege. This simple scientific postulate threatened the supremacy of the Catholic Church, the infallibility of the Pope, and the structure of power in Renaissance Europe. In 1633, Galileo was able to verify many of the calculations of Copernicus through detailed observation of the movement of the stars and their moons using his own invention, the telescope. Galileo, the father of modern astronomy, was placed under house arrest for eight years because he "taught and held" Copernican views (the Vatican officially apologized for this injustice in 1979).

The technological improvements and discoveries of science continue to present society with mighty moral dilemmas. One such area of concern surrounds the morality of eugenics (the deliberate improvement of a species through manipulation of genes). The idea of altering the genetic code of an organism and thus improving a crop to produce a greater yield, or a cow to produce more milk, has obvious appeal. A simple genetic operation that alters the configuration of a parent's chromosomes might mean the difference between the birth of a healthy child and one afflicted with Down's syndrome or epilepsy. These examples would seem to be laudable improvements. However, we have entered a moral, ethical, and even technological twilight zone. To what extent can we ethically and safely change the genes of an organism (and by extension, its evolution)? And who will decide? Is it right to alter chromosomes to prevent susceptibility to midlife breast cancer or the senility caused by Alzheimer's disease? Perhaps our genes could be manipulated to prevent dwarfism. But how tall should people be? Should we allow the cloning of humans? The capacity for genetic engineering is growing rapidly, and a new moral code will be needed, because no rules exist to cover the contingencies that are emerging.

A parallel discussion is whether euthanasia, sometimes known as the "Right to Die," should be allowed. Battles have already begun over the legitimacy of living wills, the preapproval of a victim of disease or accident to have life support withdrawn, and the ethics of a doctor being involved in assisted suicide. The capacity of medicine to keep people "alive," despite failure of most of their natural functions, will increase. The malfunction of heart, lung, and kidney can all be offset by bedside machinery, and in the future, perhaps, the brain may be kept active through electrical stimulation. At what point should the machinery be turned off? And, again, who should decide?

Science and technology will also bring change to developing nations in the form of increased use of refrigerators, more cars, more air travel, new pesticides, and more food. Costs in energy, raw materials, and human labor must be paid for this development. Before long, we will be forced to reassess our use of natural resources. Conventional energy sources are finite, and eventually we will run short of coal, oil, and gas. The loss of these resources conjures powerful images of darkened cities in which artificial light is a luxury, cars are a distant memory, and industry has ground to a halt.

No one can deny that running out of manufactured energy would radically change society and that, unless something changes, it will happen eventually. Most environmental concerns may be less dramatic or less haunting, but they are no less real. Our development of the planet is causing natural resources and natural systems to become degraded. Some of these depletions, for example the loss of fertile soil, may cause natural systems to collapse long before we run out of coal. Pressing environmental problems include overpopulation, desertification, threatened and unclean water supplies (Figure 1.1), the destabilization of climate patterns, and the loss of potential medicinal chemicals as genetic biodiversity is reduced. Ecology cannot solve these problems. Solutions will require societal change, but ecology is the appropriate science to determine the extent of each problem and to suggest potential remedies. It is also the best science to point out the potential cost of ignoring evidence that we are changing our environment.

In their quest to unravel and explain the workings of natural systems, ecologists have tended to devote most of their attention to pristine habitats that have been untouched by humans. It seems eminently rea-

Figure 1.1   An open sewer in Jakarta, Indonesia. The water of the sewer is covered by trash, but the area beneath the bridge is kept clear as a bathing area. Water is drawn from this sewer for sale as drinking water. Unfit for drinking or bathing, but used for both, dirty water such as this is a major source of disease among the poor of less-developed countries.

sonable to remove the messy influence of people—our pollution, our cutting and burning of forests, our hunting—to reach an understanding of the truly "natural" state. However, the more we find out about the history of almost any area, the more we find the influence of our forebearers. Hominids (members of the genus *Homo* and their immediate evolutionary ancestors), including the latest form, *Homo sapiens*, have been a feature of East African landscapes for more than 4 million years. Throughout that period, hominids have exerted a hunting pressure on animal populations.

The entire ecology of North America may have been changed by the first humans to enter the land at the end of the last ice age, just 14,000 years ago. These hunters exterminated large mammals such as mammoths, ground sloths, and cave lions (Figure 1.2). Humans as hunters, fishers, and farmers have been shap-

Figure 1.2 **A scene from North America 12,000 years ago.** Humans hunted many large mammal species, such as this mammoth, to extinction.

ing the landscape for millennia, and to exclude them from ecological studies as "unnatural" is highly artificial. Humans are part of nature. However, the consequences of human actions have been accelerating through time, as both our population and technological capacity increase. From an evolutionary perspective, the changes we effect are taking place at lightning speed. Although evolution is the natural response of nature to change, evolutionary processes cannot bring about change in natural populations of species fast enough to keep pace with human-induced changes in the environment.

In this book we will visit the areas of ecology needed to comprehend the environmental effects that modern humans are exerting on the landscape. With that foundation of knowledge, we can investigate the realities and fallacies of the environmental movement and try to understand the underlying ecological cause of environmental problems.

1.3 Ecology Is Not Environmentalism

Perhaps it is surprising to you that I am separating ecology from environmentalism. Despite their almost interchangeable use by the media and in common parlance, a true distinction does exist between

them. Ecology is a science; environmentalism is a concern.

Tracing the roots of environmental concern is almost impossible. For as long as there have been humans, it is likely that someone was concerned over the changing state of the planet. In 1798, the English scientist Thomas Malthus expressed doubts about the ability of agriculture to feed a rapidly growing human population. In the early years of the twentieth century, John Muir (founder of the Sierra Club) worried that all of the United States would be covered by industry, urbanization, or agriculture. He crusaded to set aside and protect the California Sierra as a wilderness area and to secure a portion of the landscape where humans were secondary to nature. Muir succeeded in his quest and persuaded President Theodore Roosevelt to establish Yellowstone as the first national park in the United States.

The true environmental movement began in the 1960s, however, when environmental issues became part of the general public consciousness. In 1963, Rachel Carson published *Silent Spring*, a book that described the dangers of pesticides and the consequences of chemical pollution to our environment. Her message was stark and troubling: We are poisoning our planet. It was a message that changed how the public viewed agriculture, industry, and science. For the first time, the American public became concerned for the fate of their environment.

The scientific study of the influence of human actions on natural processes is Environmental Science. These scientists attempt to integrate information from many traditionally separate areas of learning such as ecology, economics, sociology, politics, agronomy, anthropology, archaeology, and the law. However, more vocal and influential than either ecologists or environmental scientists are those members of the general public who are environmentalists. Recent polls in the United States show that 70% of the public are concerned that human actions are leading to the degradation of the planet. Environmentalists object to this degradation on aesthetic, moral, and pragmatic grounds. Their arguments may be drawn on scientific data, but they are just as likely to be based on emotional appeal or on ethical or moral criteria. Both science and environmentalism have their place in modern society, but a distinction should be maintained. It is important that testable fact as determined by scientific investigation is kept discrete from heartfelt opinion or speculation.

The greatest value of the environmental movement is that it has increasingly sensitized our society to the relationship that exists between ourselves, nature, industry, and development. The danger of increased cancer rates from living close to nuclear waste sites or from the degradation of Earth's ozone shield is now common knowledge. The repeated message that pollution not only damages nature but also harms human health is now widely accepted. These messages have been delivered through the environmental movement. Ecologists may reach the same conclusions, but they frequently fail to communicate effectively with the general public.

On the other hand, not all environmental arguments have been grounded in good science. Some alarming scenarios have been put forward that are unfounded; they are based on emotion and a half-understanding of ecology. For example, the suggestion that cutting down the tropical rain forests will cause us to run out of oxygen is untrue. In the course of this text, we will explore some of the major environmental issues that confront society today, but we will always view them from an ecological standpoint.

Although A. A. Milne described Pooh as a bear of very little brain, Pooh was not mistaken in declaring that the best place to start a story is at the beginning. For us to understand modern environmental issues we need some background in ecology, and to understand ecology we need to look at the formation of our planet and its atmosphere, for that is where our story begins.

1.4 A Brief History of Earth: The First Billion Years

From space, Planet Earth appears blue, because 71% of the surface is ocean. The blue is decorated with irregular drifts of white. Vast white masses of ice sit astride both poles, and water-laden clouds swirl in the atmosphere. A closer look reveals that the blue of the ocean is broken by irregular green masses. The brown and gray land is shrouded by a veil of photosynthetic plants composed largely of water. Although the lasting image gained by a visitor from another planet would be that Earth is an exceedingly wet place, our planet has not always looked that way.

As Galileo realized, our solar system, with its one sun and circling planets, lies toward the edge of our galaxy. On a clear night, we can see the pale blur of the Milky Way as a band that stretches from one horizon to the other. Staring at the glow from these remote stars, we are looking from the edge into the center of our discus-shaped galaxy. Current scientific hypothe-

ses for the formation of Earth start with the explosion of a sun (not ours) in the Milky Way.

About 5 billion years ago, one of myriad suns in the Milky Way exploded (Figure 1.3). The debris of that explosion formed a loose cloud in space. Gradually the cloud coalesced, chunks of rock and dust fused together to form miniplanets, or planetesimals, which in turn collided and fused to form the planets and moons of our solar system. About 4.5 billion years ago, Earth had attained its present size.

Shortly after Earth was formed, there was no free water and the atmosphere was a cloud of volcanic gases. The planet would have appeared as a gray-brown orb. The surface was pitted with craters from meteorite impact and puckered with volcanoes spitting fire, dust, and lava. Meteorites may have added

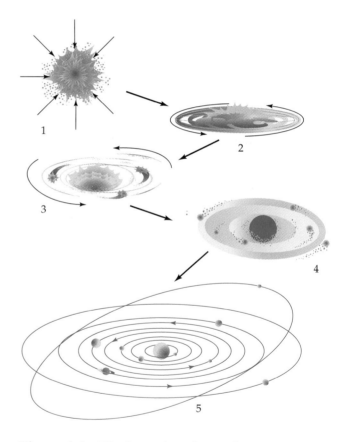

Figure 1.3 The formation of our solar system.
(1) Diffuse gas cloud, or nebula, begins to contract inward. (2) Cloud flattens into nebular disk as it spins about its central axis. (3) Particles in the outer parts of the disk collide with each other to form planetesimals. (4) Planetesimals coalesce into planets and settle into orbit around the hot center. (5) The outcome: our solar system with nine planets orbiting a central sun. (After McKnight, Tom, *Physical Geography: A Landscape Appreciation.* 5th ed., 1996. Adapted by permission of Prentice-Hall, Inc., Upper Saddle River, NJ.)

the chemical element carbon, previously missing from Earth, to the chemical composition of the planet. In the overheated, volcanic environment there was no recognizable life, and for a billion years (almost a quarter of Earth's history) the planet was lifeless, but not unchanging.

In this early phase of Earth's history a protoatmosphere was formed from water vapor and gases, such as carbon dioxide (CO_2) and sulfur dioxide (SO_2), released by volcanic activity. Because of the volcanic upheaval within the planet, the surface of Earth was so hot that water could not exist in liquid form. This early atmosphere was rich in nitrogen but had no free oxygen. A little oxygen would have been produced as ultraviolet light from the sun reacted with water vapor, but this oxygen would have been absorbed almost instantly by chemical reactions with oxygen-hungry gases or rocks that made up the surface of Earth.

1.5 Trying to Create Life in a Test Tube

How could life ever evolve under these conditions? Evolutionary explanations of the origin of life hinge on the belief that it is possible for the gulf that divides the living from the nonliving to be crossed. Living things are built of carbon-based (organic) molecules that, when strung together, can form amino acids. These acids are the basic building blocks of proteins. If it can be demonstrated that simple organic molecules can form in the absence of living organisms, one of the fundamental obstacles to an evolutionary explanation of the origin of life can be overcome. Laboratory experiments run in the 1950s by Stanley Miller, a graduate student, and his adviser, Harold Urey, of the University of Chicago, found that simple organic molecules could be manufactured when a lifeless protoatmosphere and ocean were charged with an electrical shock. In nature, the electrical energy input could have come from lightning. Later experiments showed that ultraviolet light could also provide the requisite energy to manufacture these complex organic chemical compounds. With the addition of phosphorus to the laboratory experiment, some organic molecules joined together to form simple amino acids. It is relatively easy to produce colloids, accumulations of molecules that stay clumped together under experimental conditions. Such droplets of organic chemicals are measurably different from their surrounding medium and, in that respect, are remarkably similar to the simplest living unicellular organisms. However, experimenters have yet to succeed in creating "life" in the sense of a set of organic chemicals, bound by a membrane, and capable of reproduction and respiration. The creation of life still remains one of the unsolved, but probably solvable, mysteries of science.

The experiments yielding amino acids could be replicated only when the atmosphere contained no oxygen. If oxygen was present, it broke down the organic molecules as soon as they formed. Although we think of oxygen as a health-giving gas and essential to life, it is in fact a highly reactive and destructive gas that stops many chemical reactions dead in their tracks. It is probable that if oxygen (O_2) had existed in the early atmosphere, life as we know it could not have evolved. Space is oxygen-free, and some meteorites have been found to contain amino acids, suggesting that the chemical processes that were the precursors to the evolution of life on Earth have also taken place at other times and places in the galaxy. There is a long step between amino acids and recognizable life, however, and the presence of these protein precursors on a meteorite does not signify successful extraterrestrial evolution. No evidence exists to suggest that life itself arrived from outer space.

1.6 Oceans and Life

For the first billion years of Earth's history, the planet was a lifeless chunk of volcanically active rock spinning in space. About 3.8 billion years ago, Earth's volcanic activity began to stabilize and the planet began to cool (Figure 1.4). As the cooling passed below the critical threshold that keeps water as vapor, a deluge of rain fell and the oceans were filled. Flowing water cutting across a bare landscape of lava and ash began to shape the surface of Earth. The process of **erosion** is a combination of dissolution and mechanical damage done by moving water and by the scouring action of rock fragments carried in the water or moved by wind. The water amassing in the oceans carried the dissolved minerals from the surface rocks and so formed the brew of chemicals that we call seawater. With the formation of the oceans, the environment existed in which life would evolve.

The thin soup of chemicals in the oceans provided the raw materials for life. Consequently, organisms are largely composed of water-soluble chemicals that were abundant in these ancient oceans. Because the chemistry of life is remarkably consistent from one organism to the next, our makeup still reflects that initial availability of basic ingredients. In contrast, many

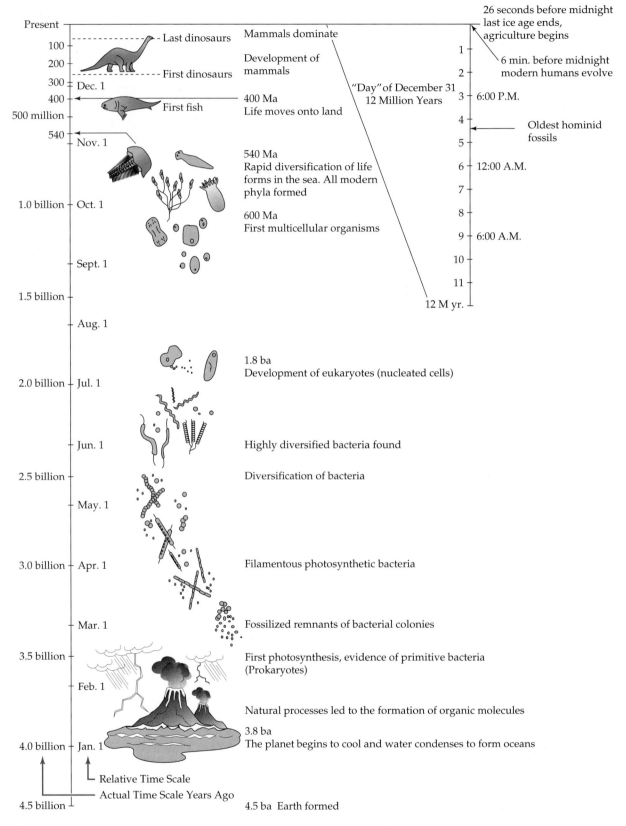

Present
100
200
300 — Dec. 1
400
500 million
540
— Nov. 1
1.0 billion — Oct. 1
— Sept. 1
1.5 billion
— Aug. 1
2.0 billion — Jul. 1
— Jun. 1
2.5 billion
— May. 1
3.0 billion — Apr. 1
— Mar. 1
3.5 billion
— Feb. 1
4.0 billion — Jan. 1
Relative Time Scale
Actual Time Scale Years Ago
4.5 billion

Last dinosaurs
First dinosaurs

First fish

Mammals dominate

Development of
mammals

400 Ma
Life moves onto land

540 Ma
Rapid diversification of life
forms in the sea. All modern
phyla formed

600 Ma
First multicellular organisms

1.8 ba
Development of eukaryotes (nucleated cells)

Highly diversified bacteria found

Diversification of bacteria

Filamentous photosynthetic bacteria

Fossilized remnants of bacterial colonies

First photosynthesis, evidence of primitive bacteria
(Prokaryotes)

Natural processes led to the formation of organic molecules

3.8 ba
The planet begins to cool and water condenses to form oceans

4.5 ba Earth formed

"Day" of December 31
12 Million Years

26 seconds before midnight
last ice age ends,
agriculture begins

6 min. before midnight
modern humans evolve

1
2
3 — 6:00 P.M.
4
5
6 — 12:00 A.M.
7
8
9 — 6:00 A.M.
10
11
12 M yr.

Oldest hominid
fossils

Figure 1.4 A condensed history of the development of life on Earth. The last 12 million years of history are enlarged, and the analogy of a 24-hour period is used to demonstrate the brevity of human history. However disappointing it may be to Hollywood film producers, dinosaurs had died out more than 65 million years before the first human evolved.

of the chemicals most toxic to life forms are those that are rarely found in seawater, such as mercury, arsenic, lead, and beryllium.

Miller and Urey had copied the probable chemical makeup of the early oceans in their test-tube attempt to replicate the formation of the first living organisms (Figure 1.5). Instead of the experimenter's flask of chemicals and a few zaps of electricity in the laboratory, the real experiment covered two-thirds of the planetís surface and lasted for 300 million years. The laws of probability suggest that what seems to be an impossibly unlikely event becomes a near certainty over time. For example, it seems unlikely that a person without a parachute could survive a fall from an airplane. But the more people who fall, the more likely it becomes that conditions will be right for at least one person to survive. During World War II, when many people were jumping from planes, one crewman falling onto a fir forest in Europe survived with broken legs, but no worse damage. Similarly, the apparently improbable set of chemical circumstances that would lead to life became more probable as more and more

reactions occurred. Over the course of 300 million years—between 3.8 and 3.5 billion years ago—and an ocean of potentially life-forming chemical reactions, the unlikely happened: Life evolved.

Biochemical clues suggest that there may have been life on the planet as early as 3.8 billion years ago. However, no fossils have been found to support this, and the first certain evidence of life is found in oceanic sediments that are about 3.5 billion years old. These first life forms resemble modern bacteria and may have been able to obtain energy by breaking apart simple organic molecules, such as acetate, that could have formed through nonbiological processes.

The general assumption that life evolved near the surface of the ocean has been challenged with new biological data from the deep ocean. On the seabed, thousands of meters below the surface, are hot springs known as **hydrothermal vents**. Where the ocean crust is cracked by volcanic activity, seawater seeps into the crust and is heated by the underlying molten mantle layer. The heated water is chemically altered by this process and enriched with sulfide. The hot sulfide-rich water squirting from the vents can provide the basis of a food chain in which sunlight is not needed. Perhaps the high pressure experienced at depth and the geothermal heat could have offered a suitable environment for the chemical reactions that led to life. This hypothesis gained a lot of supporters when it was discovered that modern deep-sea vents support huge worms and clams, pale crabs and multicolored bacteria. These creatures have evolved to live in total darkness where photosynthesis is impossible. The food chains at these vents draw their initial energy from the geothermal heat, not from sunlight. The surprising discovery of a whole ecosystem that does not rely on sunlight for energy has caused many scientists to consider the possibility that life may have evolved in the depths of the ocean rather than in the surface layers.

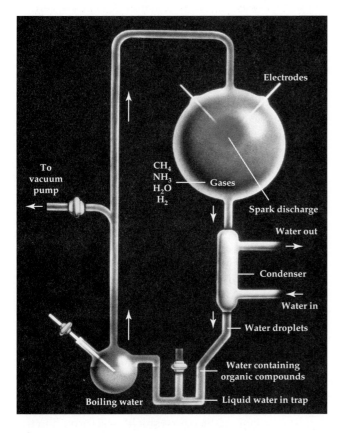

Figure 1.5   The apparatus Miller and Urey used to try to create life in the laboratory.

1.7   The Evolution of Photosynthesis

As organisms evolved, there followed a progression of ever more complex chemical pathways for liberating the energy needed to obtain body-building carbon from carbon dioxide. The earliest form of photosynthesis was conducted by bacteria-like organisms that could split hydrogen sulfide (H_2S) and carbon dioxide using energy from sunlight. The carbon dioxide and hydrogen sulfide molecules were reassembled to form carbohydrate (CH_2O).

$$\text{sunlight} + CO_2 + 2\,H_2S \rightarrow CH_2O + 2\,S + H_2O$$

carbon dioxide   hydrogen sulfide   carbohydrate   sulfur   water

Carbohydrates can later be "burned" by the body to gain energy. An important group of carbohydrates are sugars such as glucose, which have the formula $C_6H_{12}O_6$. Note how glucose and the general carbohydrate CH_2O have the same ratio of carbon to hydrogen to oxygen (1:2:1). The carbohydrate formed by photosynthesis was an energy source that fueled the metabolism (basic life-supporting processes going on within an organism) of the bacterium.

Photosynthetic bacteria appear in the fossil record about 3.5 billion years ago, a date that appears to be the same as for the earliest nonphotosynthetic organisms. Of course, at this distance in time the dating error could easily be a few tens of millions of years. It is almost certain that photosynthetic bacteria are a later evolutionary form than the nonphotosynthetic organisms, although they probably shared the oceans for millions of years. The evolution of bacteria capable of using water as an input for photosynthesis was the next major evolutionary leap. This form of photosynthesis yielded real advantages over the form that used sulfur. Water is much more readily available than hydrogen sulfide, and splitting the water yields four times as much energy per reaction as splitting the sulfide. The photosynthetic reaction of modern green plants had its origin in the first water-based photosynthesis.

$$\text{sunlight} + CO_2 + H_2O \rightarrow CH_2O + O_2$$

This single chemical reaction would bring about the most dramatic environmental change in the known galaxy: It would convert a planet that originally had a reducing (containing no free oxygen) atmosphere to one having a unique, oxidizing (containing free oxygen) atmosphere.

Although oxygen was being produced by photosynthesis, it was all absorbed by dissolved chemicals in the oceans and the surface layers of rocks. The clue to the timing of the first photosynthetic production of oxygen comes from iron occurring as iron oxides (Fe_2O_3) in ancient sedimentary rocks. For these oxides to form, the iron must have come into contact with free oxygen at the time of rock formation. The formation of these oxides, and, by inference, the age of the first oxygen-producing life forms, dates to about 3.5 billion years ago. As before, this date is only approximate and does not imply that acetate-metabolizers, sulfide-photosynthetic, and water-photosynthetic organisms all evolved at the same time. However, it

suggests that the oceans were probably shared by these various unicellular organisms.

1.8 Oxygen Producers Pollute the Planet

The remains of generation upon generation of these oxygen-producing organisms drifted down into the ocean depths, where their bodies failed to decay and were incorporated into the ocean sediments. Each body was composed of thousands of carbon molecules. For each carbon molecule incorporated into a body part, a molecule of oxygen had been released during photosynthesis. Countless billions of these single-celled organisms living, dying, and failing to decompose led to a net accumulation of oxygen in the ocean. The release of oxygen started to affect the marine environment. The rocks were oxidized, basically the same process as rust forming on iron, and the geology of the oceanbed began to change. The biology of the oceans was also subject to change. Oxygen is a highly reactive gas and potentially destructive to cell chemistry. These early photosynthetic bacteria vented the oxygen as a waste gas. Free oxygen would have been a noxious poison to the sulfur- and acetate-metabolizing organisms. As oxygen gradually accumulated in the ocean, organisms that could not tolerate oxygen would have gone extinct or survived only in very restricted habitats where oxygen was absent, such as in ocean and lake sediment, around volcanic vents, or in the intestines of animals. Thus, while we may look upon this process of oxygen liberation as paving the way for life as we know it, at the time the oxygen was a severe pollutant that made many areas uninhabitable for the acetate- and sulfur-metabolizing bacteria.

Over the next 3 billion years, the oceans were gradually saturated with oxygen, and tolerance to this chemical would have become a prime requirement for successful evolution in most habitats. Once the reduced minerals of the ocean, the seabed, and the land surface were sated with oxygen and stopped soaking up the oxygen produced by photosynthesis, free oxygen could accumulate in the atmosphere (Figure 1.6).

About 800 million years ago, the atmosphere of Earth gradually started to gain free oxygen. By about 400 million years ago, the concentration of atmospheric oxygen approximated modern levels. With abundant oxygen, the ozone shield was formed in the upper atmosphere and blocked most ultraviolet light from the sun. Water protects organisms from many of the worst effects of ultraviolet light, thus until the ozone layer was formed, evolution had taken place in the sea. With the ozone shield in place, life could invade the land. The fossil record reveals that this was the time when life spread from aquatic environments onto dry land. After 4 billion years of Earth history and 3 billion years of life, all the surfaces of Earth were now accessible to the biological invasion.

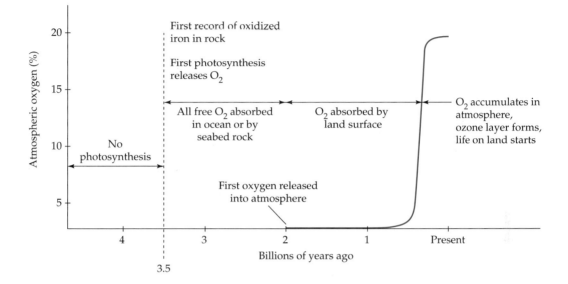

Figure 1.6 **The accumulation of atmospheric oxygen through geologic time.** The first oxygen produced was absorbed first by oxygen-hungry minerals that produced a banded-iron formation and later by "Red-Beds" (terrestrial rocks that absorbed atmospheric oxygen). Once these minerals had been saturated, oxygen could accumulate in the atmosphere. Modern oxygen levels were attained about 400 million years ago. (After M. Schidlowski, 1980.)

Summary

- Science is a process of logical reasoning and hypothesis testing. A hypothesis cannot be proved.
- A theory must be consistent with all substantiated hypotheses and is the highest certainty offered by the natural sciences.
- A new integration of science, philosophy, religion, and ethics will be needed as we improve our ability to manipulate genes.
- Ecology is a science; environmentalism is a concern.
- Earth is about 4.5 billion years old and life may have evolved as early as 3.8 billion years ago.
- The earliest fossils are found in sediments that are 3.5 billion years old.
- The initial atmosphere of Earth contained no oxygen.

- Photosynthesis used by modern plants requires carbon dioxide, sunlight, and water as inputs. Energy in the form of carbohydrate is the output, and oxygen is a waste gas.
- The oxidation of the marine environment took more than 1.5 billion years. Oxygen released by photosynthetic organisms living in the oceans started to be released into the atmosphere about 2 billion years ago.
- A rapid increase in atmospheric oxygen concentrations took place about 400 million years ago and led to the formation of the ozone shield.
- The establishment of an ozone shield reduced input of ultraviolet light and allowed organisms to colonize the land 400 million years ago.

Further Readings

De Duve, C. 1995. The beginnings of life on Earth. *American Scientist* 83:428–437.

Haines-Young, R. and J. Petch. 1986. *Physical Geography: Its Nature and Methods*. London: Harper and Row.

Sagan, C. 1980. *Cosmos*. New York: Ballantine Books.

Schlesinger, W. H. 1996. *Biogeochemistry: An Analysis of Global Change*. 2nd ed. San Diego: Academic Press.

Zimmerman, M. 1995. *Science, Nonscience and Nonsense: Approaching Environmental Literacy*. Baltimore: Johns Hopkins University Press.

Web Connections

On-line resources for this chapter are on the World Wide Web at:
http://www.prenhall.com/bush/
(From this web-site, select Chapter 1.)

Evolution and Natural Selection: The Heart of Ecology

2

To an ecologist, evolution and natural selection are not just theories or abstract concepts but rather powerful explanatory tools that act like a sixth sense. Every sight in a walk through a woodlot or beside a lake is explicable through evolution and natural selection. Why do dogs have four legs and not five? Why do birds make themselves so obvious by squawking and cackling? Why do ants appear to bump heads when they meet before proceeding on their way? An answer can be found to each of these questions: A fifth leg adds little stability when standing and would increase the chance of tripping when running; the birds are defending their territories against an invader (you); the ants are passing chemical messages about feeding locations. Evolution and natural selection are different, and the concepts should be kept distinct. Evolution is the genetic change in a lineage through time that leads to the formation of a new species. Whereas natural selection is the most important process determining which of a variety of competing forms is the one that survives.

2.1 Change, Evolution, and Chance

In Chapter 1, we thought about evolution of life in terms of the chemical impact it had on Earth. We should now consider the biological aspects of evolution that led to the full diversity of life, from single-celled organisms to humans and giant redwoods. The first step is to think about time.

When the leaves turn color in the forests of Carolina it is a sure sign that winter lies just ahead. The cycle of seasons, spring through winter in the high latitudes or wet season and dry season in the tropics, seems predictable. Nevertheless, the environment and climate of Earth are anything but uniform, and once we start to ponder questions of evolution, we must think of changes in the environment over long time scales. For most of us, the scale of a human life, 70–100 years, is easily comprehensible, but it takes some mental effort to contemplate the millions of years over which most evolutionary change occurs. However, this is the scale of thought we must adopt. That is not to say that evolution cannot take place rapidly; it is just that the events that trigger major bouts of evolutionary change are likely to be separated by many millions of years.

Most of our evidence for evolutionary history is drawn from fossils, the skeletal remains and impressions of organisms preserved in rocks. Such fossils provide an imperfect record, containing tantalizing gaps, but each year more of the puzzle falls into place as new fossil discoveries are made. We know that the early phase of evolution took place in the oceans. Fortunately for us, through the course of millions of years some of the oceanic rocks that contain fossils have been uplifted to become part of the continental landmasses. Many of the rocks that we see in our modern terrestrial landscape were once the soft sediments of ancient seabeds. Corpses were incorporated into the sediments and preserved to form fossils. Limestones, for example, are composed almost entirely of the bodies of calcium-rich plants and animals. Through the course of time, these sediments were hardened by the pressure of overlying sediment or by geothermal (originating from within Earth) heat to become solid rock. Because the oldest fossils are overlain by more recent ones, starting at the bottom of a cliff and working our way up allows us to travel forward through time. The changes in the structure and physiology of fossilized organisms trapped in successive layers of rock provide the history of the evolution of life.

The first traces of life—single-celled, bacteria-like organisms—are found in rocks about 3.5 billion years old. These organisms, **prokaryotes**, had DNA but lacked a cell nucleus and lived by metabolizing simple organic chemicals drifting in the oceans. Later evolutionary lines started to build their own organic compounds, drawing energy from photosynthesis. About 1.8 billion years ago, a major evolutionary revolution occurred as the first **eukaryotes**, organisms in which each cell contains a nucleus, evolved. Because eukaryotic cells have a nucleus that contains genetic information, the code for living, it was possible for the cells to become specialized, leading to the evolution of multiple-celled organisms.

By 1 billion years ago the first multiple-celled algae were drifting in the oceans. Recent analyses of DNA suggest that some multicellular organisms may have existed about 1.2 billion years ago. However, the earliest known fossils of multicellular animals are about 600 million years old. Microscopic sponges found in phosphate deposits in China indicate that the phylum to which modern sponges belong, the Porifera, had evolved between 600 million and 580 million years

ago. Multiple-celled organisms offer myriad body plan possibilities compared with unicellular structures. In a unicellular organism, all body functions must rely on rates of diffusion, which are slow, and this constraint will set a practical limit to the size of the organism. With a eukaryotic structure, a vascular system, a digestive system, appendages for feeding, and improved locomotion all become possible. Size is no longer limited by diffusion, so larger bodies can evolve. Recent discoveries of fossils of soft-bodied creatures suggest that around 565 million years ago there was a substantial increase in the size and complexity of organisms. But this diversification is overshadowed by a major evolutionary change; one that appears to have happened fairly rapidly between 530 and 525 million years ago. Suddenly the fossil record is full of creatures with limbs, specialized mouthparts, spines, armored shells, and even skeletons (Figure 2.1). Among these more complex life forms, there were few organisms that we would recognize, and none lived on the land.

This explosion of life-form diversity took place early in the Cambrian geologic period, which lasted from 540 million years ago to about 505 million years ago. The ocean-dwelling organisms of the Cambrian demonstrate a bewildering array of body plans that can be divided into basic taxonomic groups called **phyla**. **Taxonomy**, which is the classification of life forms, makes a first basic division into six categories called *Kingdoms*: Metaphyta (plants), Metazoa (animals), and Mycota (fungi) are the kingdoms of multiple-celled organisms, and there are three further kingdoms of primarily unicellular organisms (protista, eubacteria, and archaea). The modern animal kingdom is then divided into 20–30 phyla (singular, *phylum*); the number varies from one school of taxonomic classification to another. Phyla are large, inclusive units that have a defining component to their body plan. For example, the defining component of the chordate body plan is the presence of a notochord, a central nervous system that is gathered into a bundle running from the head down the central axis of the animal. The phylum Chordata includes all vertebrates, such as fish, birds, reptiles, amphibians, and mammals, plus invertebrates that have a notochord, such as sea squirts. Clearly, within this overall body plan, there can be a huge range of life forms, which are divided into ever-smaller taxonomic groups. Classes, for example the class *mammalia*, are divided into orders (for example, primates), which are divided into families (hominids), which are divided into genus and species (*Homo sapiens*).

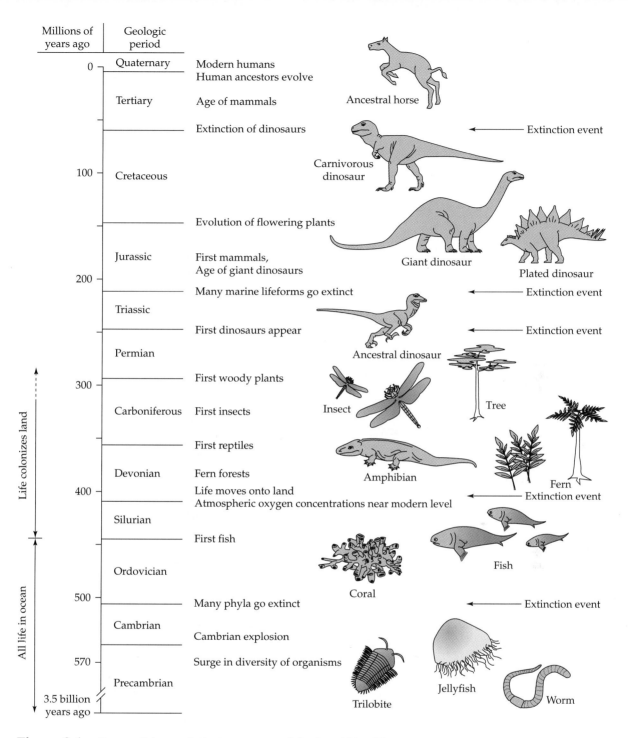

Millions of years ago | **Geologic period**

- 0 — Quaternary — Modern humans / Human ancestors evolve
- Tertiary — Age of mammals
- Extinction of dinosaurs ← Extinction event
- 100 — Cretaceous
- Evolution of flowering plants
- Jurassic — First mammals, Age of giant dinosaurs
- 200 — Many marine lifeforms go extinct ← Extinction event
- Triassic — First dinosaurs appear ← Extinction event
- Permian
- 300 — First woody plants
- Carboniferous — First insects
- First reptiles
- Devonian — Fern forests
- 400 — Life moves onto land / Atmospheric oxygen concentrations near modern level ← Extinction event
- Silurian
- First fish
- Ordovician
- 500 — Many phyla go extinct ← Extinction event
- Cambrian — Cambrian explosion
- 570 — Surge in diversity of organisms
- Precambrian
- 3.5 billion years ago

Life colonizes land

All life in ocean

Ancestral horse · Carnivorous dinosaur · Giant dinosaur · Plated dinosaur · Ancestral dinosaur · Insect · Tree · Amphibian · Fern · Fish · Coral · Trilobite · Jellyfish · Worm

Figure 2.1 **Some of the evolutionary events of the last 600 million years.**

Stephen Jay Gould, a professor of paleontology at Harvard University, recently wrote a lively summary of researchers' efforts to piece together the Cambrian-age fossils of the Burgess Shale of the Canadian Rockies. The Burgess Shale is a rock formation formed 530 million years ago by a mudslide that trapped thousands of organisms at one particular location. Although most of the animals that have been reconstructed measure just a few centimeters in length, Gould claims that they are the world's most important animal fossils.

Gould's claim is not hyperbole, because these fossils have led to a complete revision of evolutionary thinking. The most significant finding made by scien-

15

tists studying these fossils is that, in the past, there were many more basic body plans for life than exist today. Envisage a mudslide in your neighborhood. The chance of its trapping representatives of a moderate number of phyla is quite high, though to trap a representative from every phylum on Earth is most improbable. Yet in this mudslide of 530 million years ago, at least 15 and possibly 20 body plans sufficiently novel to be classified as distinct phyla have been described. Furthermore, many of these phyla have no modern representative.

The Cambrian Period was important in evolutionary history, for during this period of about 65 million years, almost all the modern phyla emerged in recognizable form. No new phyla have been added since the end of the Cambrian. Of the 15–20 phyla identified in the Burgess Shales, only 4 have modern representatives; the others went extinct before the end of the Cambrian Period. Among the 4 phyla in the Burgess Shales fossil record that survived the Cambrian were early members of chordates and insects.

Gould concludes that to yield 15 phyla with no modern counterpart in just one mudslide, the Cambrian must have been a time of much greater fundamental biological diversity than has been known since. Perhaps an explanatory analogy can be found in the early attempts at aviation. Those first pioneers of flight tried gigantic wings of silk or feathers, machines with flapping wings, umbrella-like structures that pumped up and down, bicycles with stubby wings, planes with eight wings stacked one above another, biplanes, triplanes, and balloons of various shapes and sizes filled with a variety of gases. For most uses, all but the jet engine–powered aircraft with one pair of wings have become curiosities. Although many types of jet now exist, the basic body design of flying machines today is less diverse than at the beginning of the twentieth century. Where this analogy fails is that the jet aircraft as a flying machine is superior to a bicycle with wings, but the phyla that emerged from the Cambrian were not necessarily the best—they might simply have been the luckiest.

One hypothesis to account for the surge in body types during the Cambrian Period is related to the first occurrence of eukaryotes. Eukaryotes could evolve to reproduce sexually and produce great variability in their offspring. In addition, the possibility of building bigger, more complex bodies offered the chance for the development of specialized groups of cells, such as mouthparts or tentacles. The oceans of the Cambrian may have been rich in resources, with little competition for them. In such a world, without the constraints of ferocious competition, many new body designs

could be initiated. Natural selection may not have been a significant factor until plant and animal populations expanded to the point that they started to compete with each other. Predatory behavior evolved during the Cambrian, and there was the first occurrence of a hierarchy of consumers, the large eating the small.

Why Did So Few Phyla Survive?

An ecologist's view of extinction is very simple: Any population that fails to reproduce goes extinct. One of the few certainties in biology is that every species will eventually go extinct; some just do it sooner than others. However, the extinction event that terminated the Cambrian Period was more profound in its biological implications than any since. The phyla lost at the end of the Cambrian Period were the last phyla to go extinct on this planet. Although individual species, genera, and even families have gone extinct, there has never been the loss of an entire phylum since that time.

The winnowing of the Cambrian diversity of body plans to the 20–30 modern phyla may have been caused by increasing competition that drove the least-adapted phyla extinct. An alternative, though not mutually exclusive possibility is that the Cambrian Period may have terminated with an environmental cataclysm. Paleontologist David Raup of the University of Chicago has suggested that the impact of a large meteorite changed Earth's climate, drying out many of the shallow oceans where these Cambrian life forms thrived. A meteorite could have caused a pattern of population decline where only the luckiest of species could have survived. A once-healthy population would have become increasingly weakened by sickness or starvation. Weakened animals often have low reproductive success, meaning that so few young survive that their numbers decline. If there was complete reproductive failure for all members of a species, extinction would follow. The forms that survived the Cambrian extinction event may have been the ones living in slightly deeper ocean environments or in a geographical area protected from the worst effects of the climate change. If this were the case, chance would have played a significant role in determining survival or extinction.

2.2   The Theory of Natural Selection

In 1859, Charles Darwin rocked Victorian society with his book *On the Origin of Species by Natural Selection*, which suggested a mechanism to drive evolution. Evolution had been championed more than 50 years earlier by the great French biologist Phillippe Lamarck and by Erasmus Darwin (grandfather of Charles), and it was broadly accepted within science. However, the apparently orderly march of one species replacing another was often interpreted as a procession orchestrated by God. Natural selection offered an explanation that accounted for the appearance and disappearance of species as part of a wholly natural process and that explained that humans, like any other species, were the product of natural change. The argument that species arose from natural, rather than supernatural, forces and that evolution was not directed was shocking. An enormous social inertia, the product of many centuries of theological teaching about the Creation, had to be overcome before the theory of natural selection could be accepted. As evidence of genetics, inheritance, and the ways that populations can be influenced by changing environments mounted, Darwin's theory became more familiar, less threatening, and easier for most people to accept.

The theory of **natural selection** is founded on three basic observations, the truth of which we all have witnessed. The first is that *all plants and animals produce more offspring than are needed simply to replace the parents.* For the continuity of a family lineage, only two offspring are needed (two young are needed to carry 100% of both parents' genes); why produce armies of young? The oak tree produces thousands of acorns each year for tens of years. Similarly, a female cat may have six or more litters of five or six kittens in a reproductive lifetime. If all the parents survived and all their young survived, we would soon be knee-deep in butterflies, robins, and kittens, and our oceans would be filled to capacity with seahorses, turtles, and jellyfish. That most young (almost all young of many species) die without reproducing is apparent in that we can walk and swim without constantly bumping into other life forms.

When thinking about this first observation of the theory of natural selection, we must isolate humans from the rest of nature—not because of any innate superiority, but because we have the power to think. Humans have developed (not evolved) the technology to control their reproductive rate and may choose to limit the number of their young. Our natural reproductive potential could exceed 10 children per couple, but in some countries the number of offspring averages fewer than two per couple. Humans are the exception, however, and maximal reproduction is the rule for all other species.

The second of Darwin's observations is that among sexually reproducing species *all the young are different from one another and some are better suited for survival than others.* The differences may be from the profound to the subtle—from gross abnormalities to only slight variations in size, color, behavior, or health. It is easy to understand that the runt of a litter, unable to force its way to its mother's teat, may stand less chance of survival than its stronger siblings (brothers and sisters). Similarly, the seedling that cannot grow tall enough to keep up with the other seedlings is overtopped by them and soon dies from lack of light. Less obvious is the fact that the biggest and boldest also may die young. The overly bold may stray too close to the edge of the nest and fall, or leave its mother's side and be picked off by a predator. Other behavioral or physical differences may determine success at the expense of siblings. The first to hatch among a brood of young birds may kill its nest mates or may be so much larger that it claims all the food brought to the nest. In either case, hatching first has the ultimate reward of an unshared food supply from the parents. Whatever the cause, a difference that results in death before reproduction, or a survivor that fails to reproduce, is a genetic dead end; the organism's genes will not be passed on.

Darwin's third observation was that *many of these behavioral or physiological differences, which we term* **traits**, *are inherited from the parents.* Through observation and breeding experiments, we know that such basic characteristics as size, eye color, wing pattern, seed weight, and disease resistance are heritable traits. It may not be immediately evident why a certain blood group or eye color should confer an advantage or disadvantage in terms of individual survival, but many other variations obviously affect survival. Sickle-cell anemia, Down's syndrome, and color blindness are heritable, and each would reduce an individual's chance of survival in a wild human population. Physiological differences are clearly passed from parent to young; so, too, are some behavioral (not learned) traits. Cognitive decisions are beyond plants and almost all animals; their behavior is effectively "programmed" to respond to circumstance. For example, as a shadow falls on a settled butterfly, it will fly away. This is an escape response, because the shadow may be that of a predator. The butterfly cannot have learned this behavior from a parent because there is no nurturing in the butterfly life cycle; flying away is a fixed, genetically driven response to possible danger. Butterflies that lack this response are quickly caught and killed by predators and have no chance to pass on their gene for nonresponse. The programming, both physiological and behavioral, that determines that response is the genetic material passed from parents to offspring.

Combining these three observations results in a hypothesis: Most offspring of massive reproduction will perish as a result of bad luck and poor biological programming, and the surviving young will be those with the appropriate program inherited from their parents.

Clearly the world is a dangerous place. Young of all species are struggling for individual survival, locked in a mortal rivalry with their siblings and with other, unrelated individuals of their species. They compete for such resources as food, light, water, and space, and the weakest (all but the optimally programmed) will die. Thus, although natural selection is commonly thought of as an active process that pushes the evolution of an organism toward an ideal form, it is not. Rather, it is a passive process of losing all the population except the survivors. Repeated testing of Darwin's initial hypothesis has produced so much evidence in its favor that it is now known as the Theory of Natural Selection.

Natural Selection and the Peppered Moth

The peppered moth is named for its coloration, black sprinkles on a white background. When resting on the lichen-covered trunks of trees, the pale moth blends into the background, and this superb camouflage protects it from predators such as birds. Occasionally, about 1 moth in 10,000 (0.01% of the population), is a rare genetic variant that is almost completely black (Figure 2.2). This black form remains rare because it has relatively poor camouflage and cannot hide from predators; its genes determine that it will be predated quickly, probably before finding a mate. Under such extreme selection the black form of the moth would remain rare compared with the white form. This example corresponds with our understanding of natural selection, but there could be other explanations: Perhaps the black moth is sterile and fails to reproduce, thus is always rare. The beauty of this example is that a vast unintentional experiment was run on the moth population, an experiment that took more than 150 years and that covered all of England.

The industrial revolution of the 1840s was driven by coal-powered engines. Smokestacks belched soot across the industrial heartland of England, fouling the air and blackening buildings and trees. In the polluted air, lichens died. The woodlands of pale birch trees and lichen-crusted oaks were transformed into whole forests of black-trunked trees. The polluted landscape was reflected in the peppered moth population as the white form became rare and the black form, first recorded in 1848, became abundant. With the change

Figure 2.2 Camouflage in the black and white forms of the peppered moth. On a tree growing in unpolluted air the white form blends into the background and is camouflaged from predators, whereas the black form stands out and quickly falls prey to birds. On a tree blackened by air pollution the black moth blends in and the white moth stands out.

in tree-trunk color, the white moths stood out and were quickly gobbled up by birds, whereas the black form now had the successful camouflage. By the 1890s, this reversal of predatory selection caused a form with a prior population frequency of 0.01% to increase to more than 99% of the population (Figure 2.3). Extra confidence in this interpretation of events is provided by further twists to the tale.

The completeness of the replacement of white moths with black moths was strongly correlated with the level of industrial activity and prevailing wind direction. In the central and northern areas of England, where industry was most important, all trees were blackened, and all moths were black. However, the southwestern region of England has almost no industry; its trees remained clean and the white moth dominated. This is the equivalent of having an experimental control because it shows that there was nothing inherently wrong with being white (apart from the changed surroundings). In East Anglia there is little industry, but the prevailing winds carried soot eastwards, blackened the trees, and allowed the black

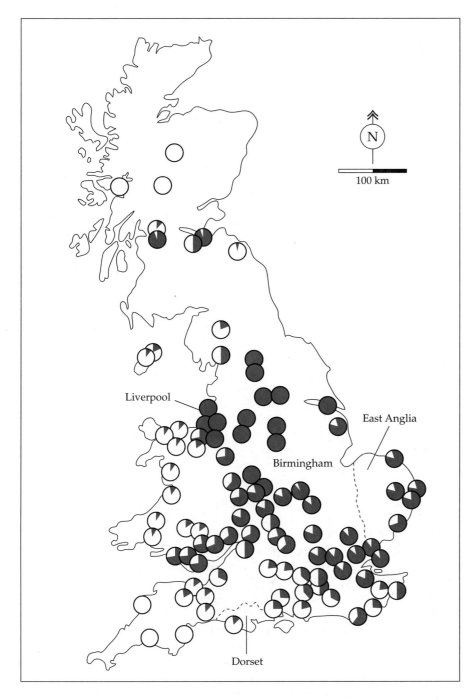

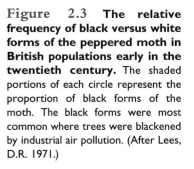

Figure 2.3 The relative frequency of black versus white forms of the peppered moth in British populations early in the twentieth century. The shaded portions of each circle represent the proportion of black forms of the moth. The black forms were most common where trees were blackened by industrial air pollution. (After Lees, D.R. 1971.)

moths to succeed. In this setting the land use did not change, and this provides a subtly different experimental setting in which the only factor to change was pollution. Consequently, it is evident that the transition from white to black moths was not due to noise or the changed use of the landscape, but was due to pollution.

Last, but not least, clean air legislation passed in Britain in 1956 put an end to the air pollution. The trees are once again pale trunked and covered in lichens,

and the white form of the moth is the optimal form to avoid predation. By 1994, in most areas the white form of the moth accounted for more than 80% of the population.

A scientist could scarcely have designed a better experiment to demonstrate the power of natural selection on a population. The experiment contained large "experimental" areas that were polluted and became industrialized and areas that were polluted but remained agricultural. The experiment also had a "control" area where there was no pollution. Having established that pollution brought about a change in moth populations, a confirmation of cause and effect would be to reverse the selection process to see if, as predicted, the white moth resumed dominance when pollution abated. The clean air acts and removal of polluting industry provided this last experimental phase.

The best experiments are replicated to make sure that the same thing happens every time to ensure that the first results were not a fluke. It is not possible to repeat history, but the next best thing is to see if the moth was similarly affected by industrialization elsewhere. In the United States, around the industrial center of Detroit, the pattern was repeated, and white peppered moths were replaced by black ones. In some areas, as much as 89% of the U.S. population of the peppered moth were black prior to the Clean Air Act of 1963. As the air around Detroit has gradually become cleaner, so the proportion of black moths has declined to about 16%. As in England, unindustrialized areas in the United States maintained populations dominated by white moths.

The outcome of this study is unmistakable: Natural selection can exert profound pressures on poorly adapted forms of a species and can be the vehicle for rapid change in the genetic balance of a population. Evolution is the process of genetic change within a lineage, and natural selection is the driving force underlying almost all evolutionary change.

Evolution and Genetic Drift

Another force that can induce evolutionary change, in addition to natural selection, is genetic drift. Genetic drift can, through a purely random process, gradually cause a change in the abundance of genes in a population. Compared with natural selection this is a weak force, but it becomes very important when dealing with small populations, such as the last remnants of a species, zoo populations, or species colonizing islands. It is significant because it may decrease the genetic variability in a population and may increase the abundance of individuals with undesirable traits, such as low sperm count or disease susceptibility. (For a fuller description see Chapter 21.)

2.3 Fitness and Genetic Immortality

A fundamental observation about individuals in natural populations is that they attempt to reproduce near to their maximum potential; every reproductive individual of every population is locked into a battle for genetic immortality. But **fecundity**, a measure of the number of eggs or sperm that are produced by each individual, does not tell the whole story of reproductive success, because many eggs fail to be fertilized and countless sperm never reach an egg. Perhaps we should instead consider **fertility**, which in this sense is defined as the number of eggs produced by a female that are fertilized. But think of the thousands of baby turtles that are gobbled up by gulls and fish when they are just a day old, or the chick that dies in the nest. These youngsters played no real part in their parents' quest for genetic immortality. A better measure of reproductive success is **fitness**. The fitness of a parent is measured by the number of its young that reach reproductive age (this number is divided by two if sexual reproduction is involved, because there are two parents). If a thousand young were produced, but all died at a prereproductive stage, the lineage would go extinct; the parental fitness would be zero. Conversely, if two offspring survived, no matter whether they were the only offspring or two among a thousand, the parents would have succeeded in passing on, on average, a full complement of their genes to the next generation.

With an ecological fitness of 1, each parent is represented by a reproducing offspring in the next generation, and the lineage has been maintained. If all members of a population were achieving a fitness of 1, the population would be stable through time, and there would be the same number of individuals in one generation as in the next. But such modest reproductive success is a sure way for a lineage to become extinct. In the real world, populations can be hit hard by catastrophes such as disease epidemics, habitat destruction caused by hurricanes or volcanic eruptions, or savage winters. Any such chance events can wipe out entire lineages. In most cases, the number of individuals of the affected species will rebound close to its former level after a few years, but with a different balance of individual lineages. Conversely, the species might expand its population for a few years because of a string of mild winters or because its chief predator is in a population slump. Populations are constantly rising and falling, varying around a mean. Overall, the fitness of the population may be 1, but the fitness of individual lineages may vary from 0 (they die out in that generation) upward.

If instead of having a fitness of 1, however, the parents produce three offspring that reproduce, the parents have a fitness of 1.5. If this rate of reproduction is maintained, it will in time overtake all lineages with lower reproductive rates. This race to produce babies explains the observation that all individuals are trying to maximize their reproductive potential; extinction of their lineage always looms, and if they do not succeed, another lineage will.

The observation that the majority of individuals in each generation die while still young implies that the environment in which they live prescribes a narrow range of behavioral and physiological variation. By definition, the parents were survivors and were successful in that environment. They inherited good programming to avoid predators, defy disease, survive climatic adversity, and reproduce successfully. In short, they had a genetic recipe for success. If that recipe is repeated exactly, a high probability exists that it will be successful again unless the environment changes. The corollary is that the further the young deviate from the model set by the parents, the less their chances of survival. This conservatism goes against the commonly held belief that natural selection promotes diversity and the attractive idea that a new, mutant, better variety may emerge with each new generation. In fact, natural selection acts to restrict the amount of variation between generations and guarantees that the surviving progeny will be very similar to the parents. It is as a result of natural selection that we do not have 10-legged lions or trees with white leaves.

2.4 Drifting Continents and Evolution

The continents are not fixed entities, moored in one spot on the surface of Earth. Through time, the great landmasses have roamed the oceans, riding from one hemisphere to the other, colliding and separating, their edges thrust up into mountains, continually reconfiguring Earth's surface. The drifting of the continents induces volcanic activity, isolation of populations, and climatic change, events that can prove critical in the survival, extinction, and evolution of species.

Francis Bacon was, perhaps, the first scientist to suggest that the continents had once been arranged differently. In 1620, he wrote of the apparent fit of the western coast of Africa with the eastern coast of South America, but the idea was so outlandish that it was largely ignored for the next 300 years. By the early 1900s, dinosaur and other fossil remains provided strong evidence that South America and Africa must have been connected in the past. Similarly, South America appeared to have had a connection with both Antarctica and Australia. Two rival hypotheses emerged. Many geologists favored the idea that great land bridges tens or hundreds of kilometers wide once existed, connecting these continents. The bridges provided a connection that allowed animals and plants to migrate and accounted for the apparent similarity of the organisms. They believed that geological upheaval eventually caused the land bridges to sink beneath the oceans, leaving the familiar arrangement of continents. This hypothesis has since been shown to be false. The opposing view was that over the last 300 million years the continents had drifted. The theory of **continental drift** describes how continents ride across the surface of Earth, propelled by powerful volcanic forces.

The first comprehensive theory seeking to explain the present arrangement of the continents through the process of continental drift was raised by German geophysicist and meteorologist Alfred Wegener in 1912. Wegener's basic idea was right, but he suggested that the power moving the continents was a gravitational pull toward the equator and a westward movement caused by lunar attraction. Geologists who were resistant to any theory stating that major landmasses could move rejected Wegener's theory on grounds that both gravity and lunar attraction were too puny to move continental masses. Rather than look for alternative explanations to account for the movement of the continents, geologists dropped the whole idea. As recently as the 1950s, the notion that continents could roam around the globe was still considered by most scientists to be a preposterously unscientific idea.

The key research supporting the movement of continental plates came from the study of ocean ridges. These ridges run like vast seams across the seabed and can be followed for thousands of kilometers. The geologic and magnetic record of the rocks on either side of these ocean ridges provided an unmistakable signal that the seafloor was moving steadily away from the ridges and being replaced by new molten material welling up from below (Figure 2.4). Unless Earth was getting larger—and it wasn't—somewhere a balancing portion of the seafloor had to be sliding back into the mantle of Earth where it would become molten. The regions where the seafloor disappeared into the mantle were termed subduction zones. The location of these areas is often marked by intense volcanic activity that produces great arcs of volcanoes, such as the Southeast Asian "ring of fire" that extends from Indonesia to Japan.

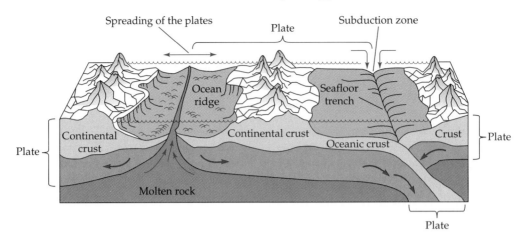

Figure 2.4   Seafloor spreading. Oceanic plates pulling apart allow an upwelling of molten material from vents along great rifts in the ocean bed. The molten material solidifies and is incorporated into the oceanic plate. The youngest rocks are closest to the vent, with progressively older rocks farther away. This process is called seafloor spreading. (After F.T. MacKenzie and MacKenzie, 1995.)

Instead of thinking of the surface of Earth as fixed, we now see that it is composed of a set of **tectonic plates** that form a crust. These plates are in constant motion, inching their way across the globe at the rate of a few centimeters per year. Two basic types of plate exist: **continental** and **oceanic**.

The continental plates form the great landmasses, but present-day continents do not necessarily correspond exactly to plates. For example, Europe and most of Asia are part of the same plate, and India, though part of the modern landmass called Asia, is a different plate. If two continental plates meet, one does not slide beneath the other; they grind together. The result may be that they buckle upward to form great mountain chains, as in the case of the Himalayas, where the Indian continental plate is ramming into the Eurasian plate. Alternatively, two plates may just slide past one another in spasmodic lurches. The San Andreas fault is prone to such movement, and the earthquakes experienced in California are the result of the gradual shifting of two continental plates.

Oceanic plates, which are denser than continental plates, will slide under a continental plate when the two meet. Consequently, it is the oceanic plates that are subducted. When two ocean plates meet, they may clash together to form underwater mountain chains that are volcanically active. These chains are the great ocean ridges. Or one of the two plates may be subducted, in which case a deep ocean trench is formed. The continued activity of volcanoes and the occurrence of earthquakes serve as reminders that continental drift is not a historical curiosity but rather a continuing force changing the surface of our planet.

The Lasting Effects of Continental Drift

From an ecological and evolutionary perspective, continental drift is significant because it explains some basic patterns of similarity and dissimilarity of flora (plant life) and fauna (animal life) around the world. Until about 200 million years ago, all the continents were clustered together as a single landmass (Figure 2.5a) centered on the tropical latitudes. Life was evolving on this land; even what is now Antarctica was at that time home to many species of plants and animals. This cluster of continents, known as Pangaea, began to break apart about 190 million years ago. Because the various plates that formed Pangaea broke off at different times, millions of years apart, each plate carried a different set of life forms and developed a unique evolutionary history. As each breakaway piece pulled free of the main landmass, the organisms living on the two parts were separated. From that moment on, the population on the mainland and that on the breakaway plate followed different evolutionary tracks. Thus the continents that separated first have had the longest evolutionary isolation and would be expected to have the most unusual collections of species. Those continents that separated comparatively recently, such as South America and Africa, have the most evolutionary history in common and should show considerable similarity in their flora and fauna.

The first separation of continents resulted in two supercontinents: Gondwana, which consisted of present-day Australia, Antarctica, South America, Africa, and India, and Laurasia, which contained

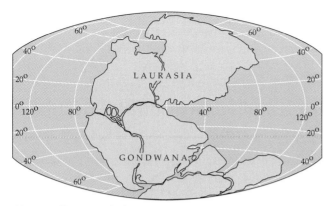

(a) 225 million years before present, Pangaea

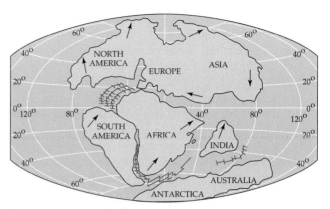

(b) 135 million years before present

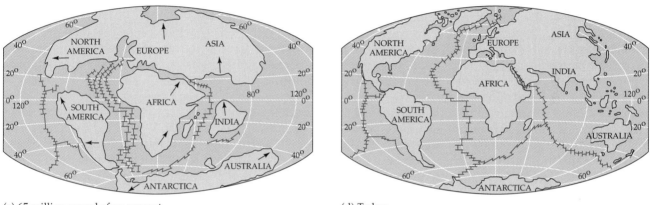

(c) 65 million years before present

(d) Today

Figure 2.5 **The drifting of continents.** Sketches map the relative locations of the continents at various times since about 225 million years ago. Note the position of the Indian plate as it crosses from one side of the equator to the other. (a) In Pangaea, all the landmasses are connected. (b) Gondwana separates from Laurasia about 190 million years ago; Australasia and Antarctica separate from South America, Africa, and India about 150 million years ago. By 135 million years ago, India has also separated from the other landmasses. Arrows indicate the direction of movement. (c) South America pivots away from Africa and separates from Antarctica. India drifts northward. (d) The modern arrangement of continents. Major oceanic plate boundaries are shown by hatched lines. (After McKnight, Tom, *Physical Geography: A Landscape Appreciation*. 5th ed., 1996. Adapted by permission of Prentice-Hall, Inc., Upper Saddle River, NJ.)

North America, Europe, and Asia (Figure 2.5b). The long isolation of Australia and Antarctica might be expected to have resulted in a highly distinctive flora and fauna, and indeed it did.

A dramatic example of one effect of continental drift came to light in 1994, when a grove of trees, distant cousins of pines, were discovered in seldom-visited forests about 200 km (125 mi) from Sydney, Australia. These trees were of a type previously known only from the fossil record of sediments deposited before the breakup of the continents 135 million years ago. More than 30 of these trees, the botanical equivalent of dinosaurs, were found growing in a secluded cove. Before the results of the initial study of these trees were published, a second botanical find was made in the

forests of Queensland, Australia. This was another "living fossil," a member of the family Proteaceae, one of the first flowering plant families to evolve. The nut of the plant is an almost exact match to fossils of 60 million years ago. In both these cases, the Australian population managed to survive while the counterpart population on the other plates went extinct because of competition from other species, climatic change, or disease.

It is not only the plants of Australia that hint at its long evolutionary isolation. The Australian plate separated from Africa and South America during the reign of the dinosaurs. The dinosaurs included the largest predators and herbivores, but a new group of small, largely nocturnal animals were becoming established; these were the mammals (animals that suckle their young by milk-producing mammary glands). The earliest forms of mammals in the evolutionary record are animals similar to the egg-laying monotremes, such as the duck-billed platypus, and to marsupials (mammals that raise their young in a pouch), such as the kangaroo. Placental mammals, which retain their young within the womb longer than either monotremes or marsupials, represent a later evolutionary origin. Humans, cats, dogs, horses, and whales are all examples of placental mammals. It is possible that Australia separated from Gondwana so early that mammal diversity was still very low, and only monotremes and marsupials had evolved. Alternatively, it is possible that placental mammals lived in Australia when it separated from Antarctica 45 million years ago, but that subsequent competition eliminated them. For tens of millions of years no terrestrial placental mammals existed in Australia; instead all were marsupials and monotremes. When Australia's northward drift caused it to plow into Southeast Asia about 16 million years ago, some rodents managed to cross to Australia and establish populations. Until the arrival of humans about 60,000 years ago, the only placental mammals in Australia were rats and bats. For over 50 million years, Australia was dominated by marsupials.

Other separations of plates, such as that of South America and Australia, came so late in evolutionary time, finishing only 70 million years ago, that many of the families of plants and animals in the two areas are shared. Evolution continued in both places, thus the flora and fauna are not identical, but they have a discernible common root. These data should give us some indication that evolution, although it continuously shapes species through natural selection, is essentially a very slow process.

Gould has stated that "history is contingent," by which he means that chance has played a large part in which species survived and which went extinct. Continental drift, climatic change, and the impact of meteorites are all evolutionary pressures over which an individual organism has no control. No amount of good genetic programming could avert the destructive power of a meteorite or avoid isolation as one continent broke away from another.

Gould has likened evolutionary history to a cassette tape. He suggests that small and large chance events in the past are so important that if we were to rewind the tape and play history again, we would probably end up with a different result. For instance, it would have taken only a minor twist of fate to have had an entirely different group of phyla emerging from the Cambrian. The consequence would, of course, have been an entirely different evolutionary history for life on our planet.

2.5 Biodiversity and the Bush of Life

Biodiversity is the term used to express the number of life forms in a given area. Often biodiversity is referred to in terms of the numbers of species, but it can also reflect the genetic variability within a single species. On a larger scale, we can think of biodiversity as the diversity of body plans for living things. It appears that, in terms of absolute diversity, the Cambrian Period was much more diverse than our modern world. This is an important realization because it reveals a fallacy that is commonly held about evolution.

One of the common images of evolution (Figure 2.6a) is the tree of life, which at its extreme form can show humans perched in the top branches. The top of the tree represents modern times and the base of the tree the origin of life. The implications of this image are that evolution has been a process of continuing expansion of diversity—the tree gains more branches and twigs as it grows—and that humans sit at the pinnacle. This is an anthropocentric view of the world and one we must abandon. Evolution is better depicted as a dense, twiggy bush (Figure 2.6b). Many of the lower branches do not reach up through time to the present; they represent past extinctions. If we were able to draw this bush in full, 99.9% of all the twigs would not reach the present, because only about one-tenth of one percent of all the life forms that ever existed are still found on the planet. Raup estimates that an average species has a "life span" of between 4 and 22 million years before it goes extinct. This extinction may be due to some catastrophic event, to random fluctuations of a population beyond a critical point, or to both. Populations are continually going through boom and bust cycles as their living conditions become more or less favorable. On a geologic scale of time, sooner or later the population dip for a species will be terminal. Some extra pressure or combination of pressures will drive the population to zero, at which point the species goes extinct (Figure 2.7).

This process of extinction is easier to understand when it is realized that most species have small geographical ranges and most are relatively rare. There-

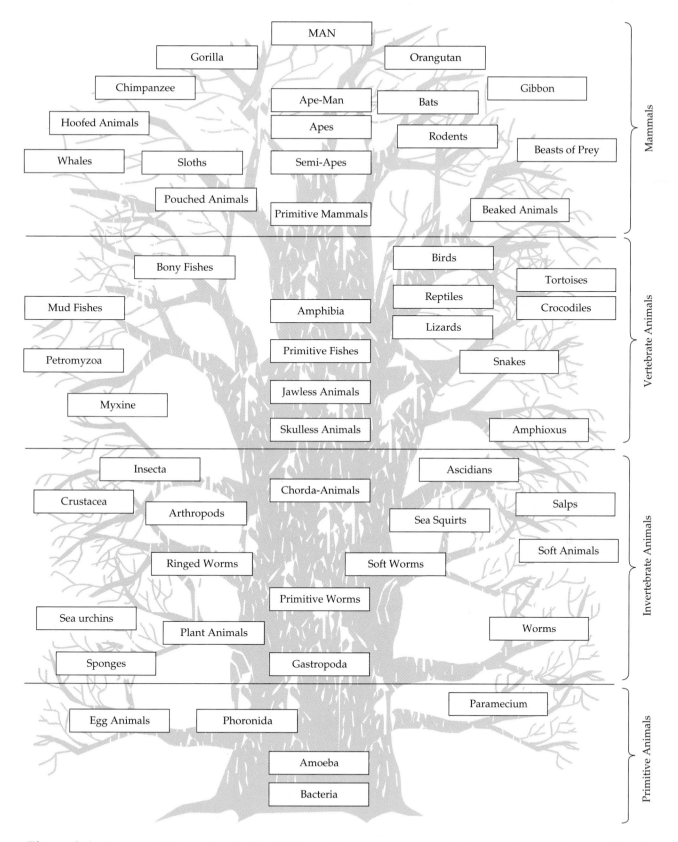

Figure 2.6 **Representations of evolution.** (a) The evolutionary tree as presented at the turn of nineteenth century. Note that Man is central and apical. There is no reason to place humans in the center of the evolutionary tree. Nor should we sit higher in the branches than any other living thing.

25

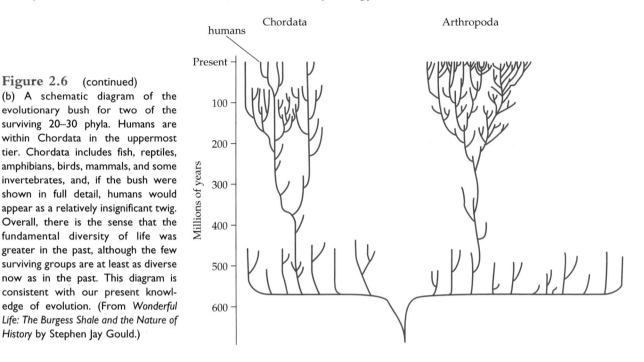

Figure 2.6 (continued) (b) A schematic diagram of the evolutionary bush for two of the surviving 20–30 phyla. Humans are within Chordata in the uppermost tier. Chordata includes fish, reptiles, amphibians, birds, mammals, and some invertebrates, and, if the bush were shown in full detail, humans would appear as a relatively insignificant twig. Overall, there is the sense that the fundamental diversity of life was greater in the past, although the few surviving groups are at least as diverse now as in the past. This diagram is consistent with our present knowledge of evolution. (From *Wonderful Life: The Burgess Shale and the Nature of History* by Stephen Jay Gould.)

fore, a relatively small event, such as a hurricane or a large flood, may coincide with a population low for a species and push it over the brink to extinction. Humans once again are an exception to the rule, as we are neither local nor rare (along with the pests, weeds, vermin, and cultivars that accompany us). In evolutionary terms, our place on the bush is at the upper level not because we are in any sense better than other organisms, but merely because the upper level has yet to go extinct. In the bush model of evolution, we are no longer centrally placed at the pinnacle of evolution, as we are in the tree model. *Homo sapiens* is merely the last surviving hominid (a member of the human ancestral line and its various dead-end branches), the last twig of a rather obscure side branch of the evolutionary bush.

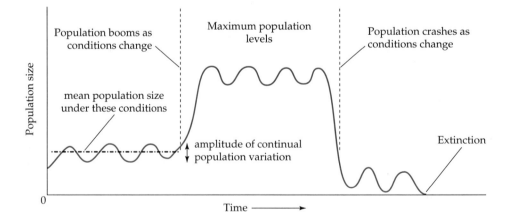

Figure 2.7 **The number of individuals within the natural population of a species is constantly fluctuating around a mean value.** The amplitude of natural change in the population is approximately constant. Sometimes, changed conditions cause the mean to shift upward as the species becomes consistently more abundant. Equally, a change in living conditions can cause the species to decline in numbers. If the species becomes sufficiently rare, the natural downswing of the population below its mean may drive the species to extinction.

2.6   What Causes Speciation?

Wave upon wave of species have inhabited the planet, each one playing a role in the intricate history of evolution. Many ancient organisms died out, leaving no modern representatives of their families or phyla. Some organisms, such as sharks, have graced the oceans for tens, even hundreds of millions of years with very little anatomical change. Others look quite different from their ancestors, but, nevertheless, an ancient lineage can be traced. For instance, the lineage of birds can be tracked all the way back to the dinosaurs. No modern bird species pecked at the seeds dropped by *Apatosaurus* or cleaned ticks from the fleshy folds of *Tyrannosaurus*. But an evolutionary sequence of species, albeit a sequence with gaps, can be followed back to the dinosaur ancestors of modern birds. The many species involved in this sequence underwent a cycle of events: They evolved, established breeding populations, survived generation upon generation, then died out.

At some stage in the rise and fall of the species, a subgroup of individuals established a new population, one that differed from the main lineage and formed a new species. This branch of the evolutionary family spawned new species before dying out, thus the process continued. For every lineage that reaches through into the present, there were many more that petered out in an evolutionary dead end. These long-extinct species often are known only from impressions on rocks forming part of the glorious geologic puzzle.

A **species** may be defined as a group of potentially or actually interbreeding populations that are reproductively isolated from all other populations. Note that a species is not being defined by how it looks, but by its capacity to breed successfully with other members of its type and its failure to reproduce with a member of another type. Under artificial circumstances, it is possible to get a horse and a donkey to mate to produce a mule, which is sterile. Other examples can be found of inducing mating between animals of different species, but the progeny are almost always sterile. Some exceptions can be found, particularly among species that are so similar that they belong to the same genus. Different species of oak (*Quercus* species) can hybridize and produce fertile acorns, and Galápagos finches of the genus *Geospiza* can mate with an individual from a different species within the genus and produce fertile young. Most species have a genetic code that determines them to be specialists in their lifestyle or niche (for a fuller definition of niche see Chapter 9). When the genetic codes of two similar species are merged, as is the case with a hybrid, the resulting young occupy a lifestyle partway between those of the parent species. The hybrids are in competition with both species, without being a specialist for either niche; even if they are fertile they will often have low fitness. Consequently, hybrid lineages are rarely sustained in nature. Most plants and animals can reproduce successfully only with one of their own kind, and this constraint provides our working definition of a species.

Within a sexually reproducing population, a genetic reshuffle takes place with each generation. The grossest variations from the ancestral theme are eliminated through natural selection. Behavioral or physical deviants are highly unlikely to succeed, and, under most conditions, only the young that conform to the parental type will survive. Sexual reproduction allows an exchange of genes throughout the population, but all the while the young are being constrained to a narrow range of behaviors and morphologies. This restriction and the mixing produced by sexual reproduction serve to homogenize the species. However, if something happens to split the population, the gene exchange between the two subgroups may be zero. Each subgroup will then reproduce and respond to local evolutionary pressures independently of the other. It is likely that, over the course of time, they will drift apart, morphologically or behaviorally. If this drift is profound enough, they may no longer be able to interbreed even if reunited: They have then **speciated**, and two species exist where originally there was one. A common example is found among animals that colonize distant islands.

The chance arrival of a pregnant female, or a pair of animals capable of starting a population, on an island may lead to a speciation event. If there are no others of their species on the island, these individuals are immediately reproductively isolated. Their offspring will be forced to breed among themselves. Interbreeding of the offspring may result in young with previously hidden deviations. Island life requires some adjustment, and, if they survive, it may be because a deviant used a different style of hunting or a different food plant. Evolution might then be relatively rapid, as forms different from the mother may prove progressively better hunters in the new environment.

For example, if a rodent manages to colonize an island, it may be faced with different predation, food availability, and physiological stresses than its mainland counterpart faces. Perhaps on the mainland the rodent is nocturnal in order to avoid being preyed upon by hawks. On the island, where there are no hawks, some of the rodents might start to hunt by day. Daytime foraging requires some different skills and some physiological change. The structure of the eye may reflect less emphasis on light gathering and more emphasis on a sharp image. Similarly, with better vision, there is less need to rely on smell and hearing. Natural selection will now eliminate some of the traits that served well on the mainland and allow others to succeed. The process of evolving to new conditions never stops, but it may need only the first stage of behavioral or physiological modification to prevent the daytime and nighttime populations from interbreeding. It is easy to see how these pressures would eventually cause speciation to occur.

Although islands offer a classic site for speciation, many other divisive events can befall a population. A mountain chain such as the Rockies or the Andes is thrust up, dividing populations to the east and west of this new barrier; a giant river starts to flow, carving its way through a once-unified forest; a road has been shown to seriously limit genetic exchange among a species of small beetles (though it has yet to be shown to demonstrate a full speciation event); even climate change can fragment populations, allowing speciation to take place. Speciation, like evolution, is a gradual but perpetual process, constantly emphasizing the differences between isolated populations. No plant or animal (including humans) has reached its "final" form.

Estimates for the number of species that currently inhabit the planet range from about 4 million to 100 million. The geological record suggests that our modern complement of species is a tiny fraction of all the life that has ever evolved. Paleontologist David Raup estimates that 99.9% of species that ever evolved are now extinct! If this statistic is even close to true, Earth has been inhabited by as many as 4 billion to 100 billion species. For each of these there was a speciation event, thus it would seem that generating a new species is an easy matter. Yet, since 99.9% of them have died out, it would seem an equally easy process to snuff a species out of existence.

2.7 Why Does a Species Go Extinct?

If there is one lesson that the geological record offers, it is that all species will ultimately go extinct; some just do it sooner than others. Extinction is really very simple: A population fails to reproduce. The events that lead to reproductive failure range from disease,

competition, and predation to climate change and habitat loss. If the average ecological fitness of a population (the number of young needed to replace adults) falls below 1, the population has started to drift toward extinction. A series of such declines over several generations of a species may take a very long time. If generation after generation of a species fail to replace themselves under the latest environmental conditions, eventually the species will go extinct. Such losses of species that are not correlated with a single catastrophic event that eliminates many species at the same time, forms the **background extinction rate**.

Some paleontologists maintain that the losses due to such background extinction are trivial when compared with losses that came about through catastrophic events. Raup takes an extreme stance and suggests that almost all extinction is the product of meteorite impacts. A more widely accepted view would be that meteorite impacts are an important component of our evolutionary history, but they are not the only cause of extinctions.

When a meteorite, or meteorite shower, smashes into Earth, a vast dust cloud is raised into the atmosphere. The dust blocks sunlight, significantly lowering the solar energy input and leading to the widespread death of photosynthetic organisms. Without the primary producers providing the conversion of sunlight into carbohydrate, food chains collapse, and mass extinction results. Most scientists agree that the evolutionary history of life on our planet has been strongly influenced by a number of these events. While they may be viewed as catastrophes for the species that go extinct, the recovering ecosystems are rich in vacant niches, which are filled by newly evolved species. Thus, in this view of history, evolution is a stop-and-go process, with long periods of relative stasis punctuated by profound change as new species burst onto the landscape. This model of evolution, forcefully espoused by evolutionary biologists Niles Eldredge and Stephen Jay Gould, has become known as the **punctuated equilibrium** model.

The best evidence for meteorite-induced extinction is the final demise of dinosaurs in a global catastrophe about 65 million years ago. In 1980, Luis Alvarez, of University of California at Berkeley, and his research team found that rock strata of that age in Italy, Denmark, and New Zealand contain a band of iridium. Iridium is an element that is extremely rare on the surface of Earth, but is a relatively common component of meteorites. Dinosaurs had been important ecological components for about 130 million years, during which time many forms of dinosaur had arisen and gone extinct. Mammals coexisted with dinosaurs for much of this time, but when a large meteorite, tens of kilometers in diameter, struck the Yucatán Peninsula of Mexico, all species of dinosaurs were wiped out. Mammals were less hard hit; even so, 43% of mammal genera disappeared along with the dinosaurs. A good deal of controversy exists as to why mammals should have been able to survive while the dinosaurs were not, and no convincing explanation has emerged to date.

Summary

- Evolution is the genetic change within a lineage, if the change is sufficiently profound a new species is formed.
- Natural selection is the loss of all members from a population other than those with optimal genotypes.
- Chance events have played a large part in shaping evolution.
- Continental drift is a continuing process that reshapes the configuration of our major landmasses.
- Past continental drift accounts for some of the most profound patterns of species distribution.
- The three basic tenets of Darwin's theory of natural selection are:

  (1) Plants and animals produce more offspring than are needed simply to replace the parents.

  (2) The young exhibit variability, and some are better suited for survival.

  (3) Traits are inherited from parents.

- Natural selection is not a procession leading to improved species.
- Natural selection can radically change gene frequency in a population in the space of a few years.
- A species is a group of potentially or actually interbreeding populations that are reproductively isolated from all other populations.
- All species go through a cycle of evolution, survival, and extinction.
- Speciation usually requires that the original population be split, so that one group is reproductively isolated from the other.
- Extinction events punctuate history. Some are caused by meteorites and climate change; the most recent are caused by humans.

Further Readings

Darwin, C. 1859. *On the Origin of Species by Natural Selection.* London: Murray.

Erwin, D., J. Valentine, and D. Jablonski. 1997. The origin of animal body plans. *American Scientist* 85:126–137.

Gould, S. J. 1977. *Ever Since Darwin.* London: Penguin Books.

Gould, S. J. 1989. *The Burgess Shales.* New York: W.W. Norton.

Grant, B., D. F. Owen, and C. A. Clarke. 1995. Decline of melanic moths. *Nature* 373:565.

Weiner, J. 1994. *The Beak of the Finch.* New York: Knopf and London: Cape.

Web Connections

On-line resources for this chapter are on the World Wide Web at:
http://www.prenhall.com/bush
(From this web-site, select Chapter 2.)

The Ecological Efficiency of Living Things

3

Amassing energy for successful reproduction is a unifying factor among all living things. Every organism must harvest energy needed to ensure that its genetic lineage will continue into the next generation. For some plants and animals, such as salmon or dandelions, reproductive effort is a single, massive outpouring of life force, followed by death. Among other organisms, such as oak trees or humans, a significant but proportionately lesser expenditure of energy allows reproduction to take place year after year. Without acquiring energy, an individual cannot survive, thus the availability of energy within an ecosystem is important to all the organisms that live there.

Solar energy is the primary source of energy that powers Earth. Light, biological processes, weather patterns, ocean currents, and the waxing and waning of glaciers over thousands of years can all be attributed to the flow of solar energy. The only other energy sources, but trivial by comparison, are volcanic activity, geothermal energy, cosmic radiation, and the occasional impacts of meteorites. For our present purposes, we can ignore all but solar energy.

Only recently have humans thought about tapping solar energy as an "alternative" energy source, but photosynthetic organisms have been doing it for 3.5 billion years. A tree, for example, is a giant solar-powered water pump. These impressive organisms are capable of raising hundreds of liters of water more than 120 m vertically; they function silently, they maintain themselves, and they can keep going for more than 3000 years. Building such a pump is a feat beyond modern engineering.

Not only are plants aesthetically pleasing, they are essential to life as we know it because they provide the link between an input of solar energy and the food needed by animals such as humans. Plants, many bacteria, and algae have the ability to derive their energy directly from their environment. The energy source is usually sunlight, but chemical and thermal energy, such as that provided by a deep-sea vent, occasionally provide the initial energy input to an ecosystem. Organisms that can make this energy conversion are termed **autotrophs**. A plant is an autotroph because it converts sunlight to carbohydrate through the process of photosynthesis. No animal can do this, so ultimately all animals are dependent upon autotrophs to supply them with energy. Animals, many microorganisms, fungi, and rarely some plants, are **heterotrophs**, organisms that cannot synthesize energy directly from the environment. The most common means for a heterotroph to gain energy is through the predation, parasitism, or decomposition of another organism.

3.1 Photosynthesis: Converting Sunlight to Carbohydrate

Our sun is beaming energy in all directions into the universe; only a tiny fraction reaches us here, 149 million kilometers from its source. Solar energy arrives

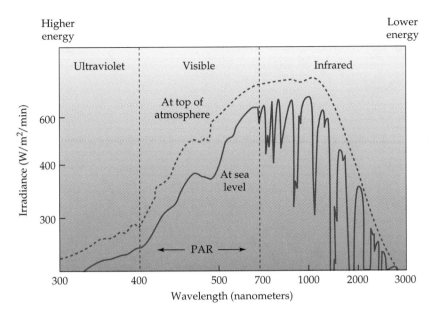

Figure 3.1 **Incoming solar energy arrives in a series of wavelengths.** The smooth upper line is the energy arriving at the upper edge of the atmosphere. The lower jagged line is the energy in each wavelength at sea level. The infrared range becomes jagged because of absorbance of specific wavelengths by water vapor and other atmospheric gases. The visible spectrum from violet to red contains consistently high intensities of energy. (After P. Colinvaux, *Ecology.* © 1992 John Wiley and Sons, Inc. Reprinted by permission of John Wiley and Sons, Inc.)

in a range of wavelengths (Figure 3.1). The maximum input of energy is contained in what we call the visible spectrum, that is, the familiar colors of the rainbow. Ultraviolet radiation at Earth's surface provides fewer watts of energy per square meter per minute than the visible wavelengths and less than infrared, which is also invisible to us. Although the infrared spectrum has a smooth curve of energy input when sunlight enters the upper atmosphere, by the time those wavelengths reach the surface of Earth some wavelengths carry relatively little energy. The jagged sea-level curve in Figure 3.1 is caused by absorbance of specific wavelengths by chemicals such as water vapor (H_2O), oxygen (O_2), and carbon dioxide (CO_2) as the light passes through our atmosphere. Our eyes have evolved to absorb the signal from radiation in the wavelengths where irradiant energy is most plentiful (between 0.4 and 0.7 micrometers). In just the same way, plants have evolved to absorb the wavelengths of visible light, termed *photosynthetically active radiation (PAR).* PAR is absorbed by a group of pigmented chemicals, the best known of which is chlorophyll. Chlorophyll is an excellent chemical for absorbing violet, blue, and red light (Figure 3.2), but it is relatively poor at absorbing green, yellow, or orange light. If the plant does not absorb this light, it is not making full use of the energy that falls on the leaf. Secondary pigments, such as xanthophyll, carotenoids, and phycobillins, absorb the yellow and orange light, thereby reducing what is known as the "green gap." Most plants are poor at absorbing green light and consequently reflect it. That is why plants look green to us. If they could absorb all light, they would look black; if

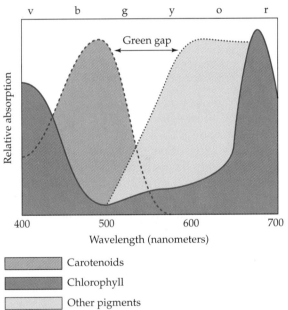

Figure 3.2 **Photosynthetic pigments absorb the same wavelengths of light that form the spectrum visible to humans.** Chlorophyll absorbs blue and red light effectively, but the spectrum from green to orange has to be absorbed by other pigments. The relative inability of most plants to absorb green light gives rise to the "green gap" and the characteristic green color of most plants.

we could see in ultraviolet, as a bee can, then plants would appear violet because they are also reflecting ultraviolet light.

As Figure 3.3 shows, plants use photosynthesis to convert absorbed solar energy into a form of storable chemical energy. In the most common form of photo-

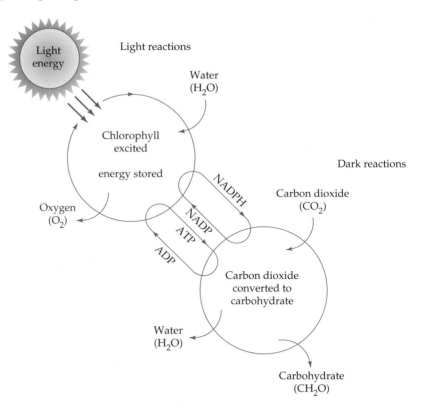

Figure 3.3   A schematic diagram showing the chemical processes of the light and dark reactions of C3 photosynthesis.

synthesis, the reactions take place in two stages, the first phase of which is called the *light reactions*, because they require sunlight.

In the first stage of photosynthesis, chlorophyll absorbs solar energy (hence light reactions) and produces an excited electron. Two chemical processes can temporarily store this energy. In one set of reactions, a water molecule (H_2O) is split, the O_2 molecule is released as a waste gas, and the hydrogen ion H^+ attaches to NADP (nicotinamide adenine dinucleotide phosphate) to form NADPH. Here the energy is stored in the charged hydrogen atom bound to the NADP molecule. Alternatively, energy is stored when ADP (adenosine diphosphate) forms ATP (adenosine triphosphate) by the addition of a phosphorus molecule. Here the energy is held in the phosphate bonds of ATP.

The energy stored in the light reactions is used in the dark reaction (so called because it does not need light to proceed). In this reaction, carbon dioxide (CO_2) is converted to carbohydrate (CH_2O) in the presence of ATP and NADP (Figure 3.3).

The source of the carbon dioxide for these reactions is generally the atmosphere. On the underside of leaves are stomata, openings that open or close in response to the internal water pressure of the leaf. When the stomata open there is gas exchange; moist air leaves the leaf, and fresh outside air enters. The in-

coming air contains the carbon dioxide for photosynthesis.

At its simplest, we can look on photosynthesis as a set of inputs and outputs. Water, carbon dioxide, and sunlight are the inputs; carbohydrate (the product) and oxygen (the waste) are the outputs. The chemical equation is just a formalization of this description

$$\text{energy from sunlight} + CO_2 + H_2O \rightarrow CH_2O + O_2$$

Variations on the Photosynthetic Theme

The photosynthetic process that has just been described is known as **C3 photosynthesis** because one of the critical compounds in the reaction is a three-carbon sugar. Most plants use C3 photosynthesis and are termed C3 plants, but an important variant is **C4 photosynthesis**. C4 plants have a more active chemical pump to absorb carbon dioxide from the atmosphere. As carbon dioxide enters the stoma, it reacts with an enzyme called PEP (phosphoenolpyruvate) carboxylase, which scavenges the carbon from the air and attaches it to an acid. The acid has a molecular structure with four-carbon molecules rather than the three-carbon molecules used in C3 photosynthesis, hence the process is referred to as C4 photosynthesis. The remainder of the photosynthetic process is very similar to that of the C3 plants.

It is legitimate to ask why we are interested in what appears to be a minor chemical variation on the theme of photosynthesis. The answer will become truly apparent when we consider global warming (Chapter 23). For now, however, we should note that plants that use the C4 method can gain enough carbon for their photosynthetic needs in less time than a C3 plant can. Plants absorb carbon dioxide when their stomata are open; therefore, a C4 plant needs to open its stomata for less time than a C3 plant. Every time the stomata are open, gas exchange is taking place, and the plant runs the risk of losing essential moisture. Thus, in hot, dry environments the shorter stomatal opening time of the C4 plants will be a distinct advantage. Sure enough, we find that C4 plants are most abundant in desert environments and areas where water availability is severely limited. A variation on the theme of C4 photosynthesis is crassulacean acid metabolism (CAM). CAM plants generally live in deserts and store water within thickened leaves that are so full of water that they snap if you bend them. These plants are called succulents, and many succulents have the ability to use CAM. The water storage is an adaptation to living in the desert and so, too, is CAM photosynthesis. The CAM plants open their stomata at night when the air is cool and use a C4 pathway to fix carbon. During the day the stomates close, and the carbon dioxide is released inside the plant and refixed using the C3 pathway. In a sense, CAM photosynthesis is a photosynthetic pathway intermediate between C3 photosynthesis and C4 photosynthesis.

If an organism has an ecological advantage, it should oust other species. Why then are there any C3 plants left? Why have C4 and CAM plants failed to take over Earth? The answer lies in the costs to the plant of carrying out the C4 stage of photosynthesis common to both C4 and CAM plants. A range of conditions must be met to gain the full benefits of C4 photosynthesis. The plant must grow in strong sunlight, because this process works poorly in subdued light. Consequently, we would not expect to find many C4 plants living in shady conditions. C4 photosynthesis works best at high temperatures, thus these plants will do best in the tropics and have a harder time surviving in colder locations. An important factor is that the C4 pathway also requires the plant to invest more energy in manufacturing enzymes and other chemicals than does the C3 pathway. These costs to plants offset the benefits that are incurred from C4 photosynthesis in many environments.

The predictions of climate change discussed in Chapter 23 are that Earth will become warmer and drier, conditions that may favor the expansion of C4 and CAM plants. The changes that this expansion might induce in the ecological balance of our forests, wetlands, and grasslands can only be guessed. Of economic importance is the fact that some of our crops such as wheat, oats, and potatoes, are C3 plants, while others, such as corn, are C4 plants.

3.2 The Fate of Carbohydrate

Each carbohydrate molecule has one of three fates: It can be used as an immediate energy source; it can be stored for later use; or it can be used to build new body parts for the plant. The conversion to produce either energy or body parts requires that the carbohydrate molecule is broken back down into its constituent parts. This reverse process is called **respiration**. In it, a molecule of oxygen is used to release energy as a carbohydrate molecule is broken apart. The chemical equation for respiration seems to be the near opposite of the photosynthetic reaction.

$$CH_2O + O_2 \rightarrow CO_2 + H_2O + \text{energy}$$

Note that respiration to a biologist is not breathing in and out, for breathing is merely the way to provide oxygen for the respiration reaction and to rid the body of waste carbon dioxide.

Photosynthesis and respiration are two halves of the cycle of energy production and usage. For every molecule of carbohydrate produced in photosynthesis, a molecule of oxygen is released to the atmosphere. During respiration, for each molecule of carbohydrate respired, a molecule of oxygen is taken back from the atmosphere. The gas inputs and outputs balance, and the only net input to the pair of reactions is sunlight.

Herein lies the refutation of the misconception that trees are our "green lung" and that without them we would soon become oxygen starved, turn blue, and asphyxiate. All three of the fates of a carbohydrate molecule—storage, immediate use, or incorporation into plant body parts—ultimately result in respiration and, therefore, in a consumption of oxygen. Immediate conversion to energy translates to immediate respiration; being stored just delays conversion to energy or conversion to body parts. The fate of all body parts is to become a corpse, which decays. Leaves fall and die, the whole tree falls and dies, or the parts are eaten by animals. Being eaten by an animal merely postpones the process of decomposition, for ultimately the animal (or its predator) will die and be decayed. The only escape from this fate is to become a fossil, a possibility to which we shall return. For now, the vast likelihood is that all the organic material produced by a plant will decompose. The process of decay is one of respiration in which the carbon-based bodies of plants and animals are broken back down to basic chemical compounds. For every atom of carbon tied up in the tissues of an organism, one molecule of oxygen was released into the atmosphere when the carbon atom (as carbon dioxide) was photosynthesized. As the organism decomposes, a molecule of oxygen is withdrawn from the atmosphere for every molecule of carbon decayed. Thus, even though plants release oxygen while they are alive, their subsequent decomposition uses up the same amount of oxygen that they released: They are neither net producers nor net consumers of oxygen.

Furthermore, if we calculate the amount of oxygen in a column of air from the ground to the upper atmosphere, we find that it contains about 64,000 moles of oxygen, while the amount "loaned" to the atmosphere by the storage of carbon in plants is about 8 moles. As ecologist Paul Colinvaux has observed, if we were to rid the planet of green plants, we would starve before we asphyxiate.

3.3   The Ecological Efficiency of Plants

Ecologists use the term **biomass** to describe the mass of organisms in a given area. Biomass can be expressed either in units of energy per unit area (joules per square meter) or as weight (tons per hectare). If we then add time as a factor, considering the amount of new organic matter accumulated per day or per year, we are dealing with the **productivity** of the area. Only photosynthetic organisms can carry out the initial transformation, turning carbon dioxide into carbohydrate. Consequently, this first stage of productivity determines how much energy will be available for animals. **Primary productivity** describes the rate at which plants and other photosynthetic organisms are accumulating biomass. Ecologists use two variations on primary productivity: **gross primary productivity (GPP)** and **net primary productivity (NPP)**. GPP is the total amount of energy used by the plant. It includes all the energy used to build the body of the plant and the energy the plant uses just to live, its respiration.

Subtracting the respiration component from the GPP leaves NPP. NPP is more widely used because it is easier to measure than GPP. Looking at the NPP of different types of vegetation, we find that there is considerable variability. One hypothesis that could account for this variability is that some plants are much better than others at converting the sun's energy into carbohydrate. This hypothesis can be tested by comparing how efficiently plants assimilate the sun's energy.

For 3.5 billion years, some form of photosynthesis has been taking place on Earth, and for about 2 billion years it has probably been a process similar to that observed in modern green plants. Given that natural selection has been laying waste all but the most efficient members of the plant kingdom, it seems reasonable to assume that plants are very efficient at photosynthesis by now. Just how good are they?

By **photosynthetic efficiency** we mean the percentage of received solar energy a plant uses. From the total weight of the plant we can determine the number of carbon atoms required to build the plant, and from there we can calculate the energy absorbed to form the body of the plant. Energy used in respiration is also calculated, and when combined with the body-building energy, a value for the GPP of the plant is obtained. By dividing GPP by the solar input of energy per unit area per unit time, we can derive the photosynthetic efficiency of a plant.

$$\text{photosynthetic energy} = \frac{\begin{array}{c}\text{energy equivalent}\\\text{of carbon compounds}\\\text{produced per unit area}\\\text{per unit time}\end{array}}{\begin{array}{c}\text{energy input per unit}\\\text{area per unit time}\end{array}} \times 100$$

It is very difficult to measure plant respiration outside the laboratory, so the first estimate of photosynthetic efficiency was calculated for relatively small plants. Corn was used in the initial study in the 1930s. After extensive experimentation, an estimate was obtained that only about 1.6% of incoming solar radiation was used by a corn plant. Respiration accounted for about 30% of the energy used, leaving only slightly more than 1% of the initial input of solar energy as the NPP available to grazers. Subsequent studies, looking at a range of different plants from different environments, have approximately replicated this result with GPP values close to 2%. Our hypothesis that different types of plants have different photosynthetic efficiencies is refuted, at least among land plants. Aquatic algae have been shown to have higher photosynthetic efficiency, using as much as 3%–4.5% of the available energy.

As we shall see, the amount of energy that is converted into plant material is of critical significance to the length and richness of food chains. The most useful productivity statistic is that of net primary productivity, because this represents the amount of food available to herbivores.

3.4   The Ecological Efficiency of Animals

Animals eat either plants or other animals to gain energy. An animal that feeds on plants is said to be living on the second trophic level (the first trophic level is the plant). A predator that feeds on this animal is at the third trophic level, and so on to the highest trophic level in the food chain, the top carnivore. Perhaps because animals can move around and select what they eat, they can be more energy-efficient than plants. Ecologists have devised various measures of animal efficiency, but we shall use the calculation that offers the best comparison to the efficiency of plants. **The Lindemann efficiency**, named after the scientist who first developed the calculation, estimates the amount of energy that is passed from one trophic level to the next as a percentage of the total amount available in the lower level. Thus the energy used in respiration of both the prey and the predator is part of the equation.

$$\text{Lindemann efficiency} = \frac{\text{respiration} + \text{increase in body weight of higher trophic level}}{\text{respiration} + \text{increase in body weight of lower trophic level}}$$

Lindemann was the first ecologist to try to estimate the ecological efficiency of animals in aquatic systems. His work established a figure of 10% efficiency, which is widely quoted in many texts. However, Lindemann's calculations involved an element of double counting, and this figure of 10% should be regarded as an absolute maximum for endotherms (animals that rely on internal sources of heat). Since Lindemann's work, many estimates have been made for animal efficiency, but no single statistic has emerged. Endotherms tend to waste energy on heating their bodies and rushing around, whereas sedentary ectotherms (animals that rely on external sources of heat), such as most reptiles and insects, may have relatively high ecological efficiencies. Primary consumers (animals feeding on plants) may have efficiencies as low as 1% (calculated for sheep) or as high as 7%, estimated for the desert-dwelling kangaroo rat. Both of these values may be flawed, but they remain the best estimates available. In the examples we consider, the 7% value

for an endothermic herbivore will be accepted, with the reservation that it may be an exaggeration of ecological efficiency. Endothermic predators, such as wolves and birds, appear to have ecological efficiencies close to just 1%; insects and other small ectothermic creatures may have efficiencies approaching 15%.

What Happens to the Other 99% of Wolf Energy?

To see what happens to the vast amounts of energy "wasted" by animals, let us look at the energy budget of a typical endotherm—the wolf. If just 1% of consumed energy was assimilated to become the wolf, what happened to the rest? Our immediate answers—that it was never absorbed or that it was used in panting, locomotion, or reproduction—are correct, but they are not a complete explanation of the general phenomenon taking place. In fact, in asking our basic question, we are encountering some of the most important laws that govern the natural world.

The **first law of thermodynamics** states that as energy is transferred from one state to another, none is lost; for, like matter, energy can be neither created nor destroyed, but it can be transformed. That statement leads us to the **second law of thermodynamics**, which states that as energy is converted from one state to another, the quality of the energy deteriorates as a result of increasing entropy.

Entropy is the natural progression of energy from a highly ordered form toward a less-organized, or more chaotic, form. For example, the energy added to a food chain in the form of visible light is energy of very high quality. Light in the visible spectrum has tightly defined, predictable wavelengths. The plant that absorbs the light for photosynthesis converts that light to stored chemical energy, still a high-quality energy, but of lower quality than light. If the plant is eaten, the chemical energy represented by its body is converted into various forms. Some of the energy will be stored as tissue or fat by the animal, but much of it will be used for kinetic energy (movement) or heat. Both of the latter forms of energy are of much lower quality than the chemical energy input. Thus, at every link of a food chain, most of the energy is degraded (entropy), leaving less and less high-quality energy for each higher trophic level.

In the case of the wolf, most of the absorbed chemical energy obtained by eating, say, a moose, is dissipated in the more chaotic energy states of mechanical energy and heat. It is critical to note that energy is not lost, for that implies it has been destroyed. At each stage in the food chain energy conversion has caused more and more of the initial input to be converted to an unusable state.

3.5 Energy Flow Through a Food Chain

Armed with information on the efficiency of plants, herbivores, and their predators, it is possible to construct a flow diagram that shows the reduction in the amount of energy being passed from one trophic level to the next (Figure 3.4). At the left of the diagram (our starting point) is the energy input to the primary producers, plants, and the passage of about 2% of the available energy to the herbivores. Herbivores constitute the second trophic level. An endothermic herbivore, such as a cow, will pass along to the third trophic level only about 7% of the energy entering the second trophic level. It takes a large, ferocious predator to kill a cow. In this example, the predator is endothermic, a lion. This predator is likely to respire a great deal of its energy either maintaining body temperature or pursuing quarry. About 99% of energy entering this trophic level will be degraded and leave

the food chain. Thus, after three trophic levels, only 0.0014% (1% of 7% of 2%) of the energy that entered the system is available for the next trophic level.

In the second example (Figure 3.5), an insect is the herbivore, and, because it is ectothermic, it may pass to the third trophic level as much as 15% of the energy it takes in. If its predator is another, larger insect, such as a praying mantis, perhaps about 10% of energy in the third trophic level is passed to the fourth. Here, a endothermic insectivore with an ecological efficiency of about 1%, such as a shrew, feeds on the praying mantis. In the fifth and final level of the food chain, a hawk (1% efficient) eats the shrew. Thus, after five trophic levels, 0.000003% (1% of 1% of 10% of 15% of 2%) of the initial energy entering the food chain remains. Note that the proportion of energy left within the system at the end of the food chain differs from one chain to another. The principal factors determining this value are the type of food chain, the initial productivity of the habitat, and the size of the last predator.

These food chains demonstrate that, at each step, the body size of the predator increases markedly (the reasons are discussed in Chapter 12). Because a large animal requires more energy to keep going than does a small one, the energy requirements of a predator at

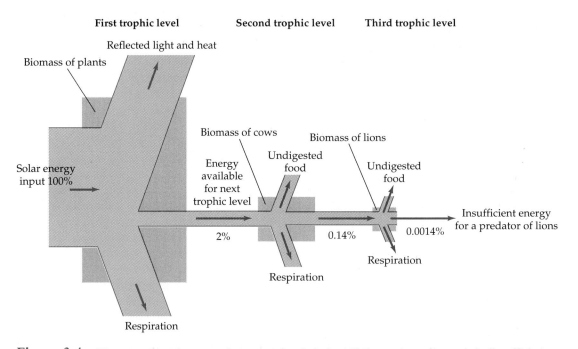

Figure 3.4   The transfer of energy through a food chain will depend on the metabolic efficiency of the organisms involved. A food chain with two large endotherms retains approximately 0.0014% of the initial energy within the system. In this example the cow is assumed to be about 7% efficient and the lion about 1%. The energy available for a third trophic level is insufficient for the immense ferocious predator that would be required to evolve as a predator of lions.

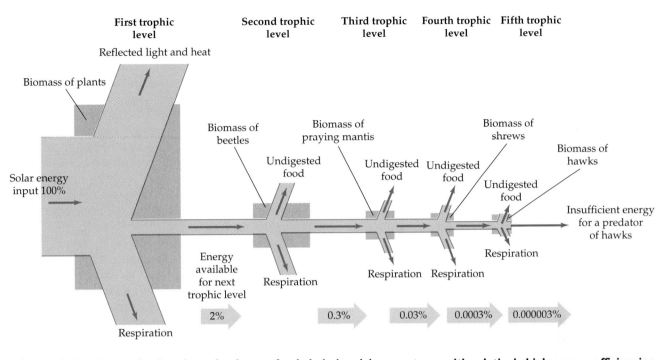

Figure 3.5   **Energy flowing through a longer food chain involving creatures with relatively high energy efficiencies.** The beetle is as much as 15% efficient, and the praying mantis is about 10% efficient. The two endothermic predators (shrew and hawk) are about 1% efficient. Four trophic levels is the maximum found in most terrestrial ecosystems.

one trophic level are higher than those of its prey. However, the entropy of energy results in a decreasing flow of energy in the food chain. The antagonistic pressures of less energy combined with the need for an ever bigger and more fearsome predator means that ultimately there is not enough energy in the system to support another trophic level. It is rare to find a food chain with more than four trophic levels, although a fifth level is possible as in the case of Figure 3.5. However, under exceptional circumstances a sixth trophic level may occur in aquatic systems because the initial energy efficiency of aquatic algae is sometimes considerably higher than that of land plants. For instance, killer whales and the great white shark, the "Jaws" of Hollywood, sit atop a six-tier food chain. The food chain starts with tiny, drifting photosynthetic organisms termed *phytoplankton*. The phytoplankton are food for *zooplankton*, microscopic animals that float and swim at the surface of the ocean. The zooplankton, in turn, fall prey to fish, which are eaten by larger fish, and so on.

phytoplankton → zooplankton → small fish → large fish → seal → white shark
primary   second   third   fourth   fifth   sixth
producer   trophic   trophic   trophic   trophic trophic
              level   level   level   level   level

The entropy of energy in natural systems also explains why the largest, most energy-expensive preda-

tors are scarce. There simply will not be enough energy to support large numbers of predators feeding high on the food chain: By definition, their way of living means that they must be rare. Such long food chains are truly exceptional and as rare in nature as the top carnivores they support.

Exploitation Efficiency

Another way to think about the energy flow through a food chain is to consider only the net primary productivity and then to look at the net productivity (available biomass) of each trophic level as a proportion of the one beneath it. This is termed the **exploitation efficiency**.

$$\frac{\text{exploitation}}{\text{efficiency}} = \frac{\text{energy ingested by a predator}}{\text{net production of its food species}}$$

This measurement is much easier for field biologists to use than is the Lindemann efficiency because they do not have to measure energy degraded in respiration. Also, if the organism they are studying eats only the seeds of a plant, they do not need to calculate the roots, leaves, and trunk mass (which are difficult to measure accurately). Exploitation efficiency provides the only good measure of the efficiency of an entire trophic level. This measure provides some important insights into the fate of plants and animals. For instance, on Isle Royale in Lake Superior, a population of moose supported a population of wolves. Observations of the moose herd revealed that the only likely fate for a moose was to be eaten by a wolf. Equally, the wolves were almost completely reliant on the moose for their food. This relationship allowed ecologists to calculate the Lindemann efficiency of the wolves and found it to be about 1%, while the exploitation efficiency was close to 100%; only the inedible portions were left. Mammalian carnivores generally are found to be about 50% to 100% efficient in exploitation. Mammals grazing the great African plains may have exploitation efficiencies as high as 28% to 60% of the grassland NPP. As a rule of thumb, if pasture is to be maintained in a healthy state for domestic cattle, the exploitation efficiency should not exceed 30% to 45%. Grazing pressure exceeding that range is likely to degrade pasture quality, and ultimately the pasture will be able to support fewer head of cattle. Such has been the case in many areas that have been overgrazed. Thus, exploitation efficiency could be used as a tool to monitor the sustainability of managed animal populations within an area.

3.6 The Costs of Control: Endothermy and Ectothermy

Because most chemical reactions are temperature sensitive and organisms rely on chemical reactions for energy conversion and movement, most multicellular organisms have evolved the means to regulate their temperature. Ectotherms gain their heat from outside sources such as basking in sunshine, lying on a warm rock, or living in a warm ocean. In contrast to this, endotherms generate heat internally as a result of respiration. Most life forms are ectothermic, and endotherms have only been on the planet for about 140 million years. Let us start by looking at the advantages and disadvantages of being endothermic.

An endotherm expends energy to maintain a steady internal temperature. This process is termed **thermoregulation**, and it allows chemical reactions in the body to run at a steady rate. One of the greatest advantages of being endothermic is that the creature is always ready for action (unless it is hibernating). On the other hand, an ectothermic animal relies on solar heat to warm up to an optimal operating temperature. On a cold summer's day, or early in the morning, some insects, such as butterflies, are barely able to fly. Conversely, the outside temperature, within limits, does

not greatly affect the running speed of a wolf, which is always ready to pursue its prey. A further advantage of being endothermic is that the muscles can provide more sustained power. A horse can gallop for miles, but a crocodile (an ectothermic reptile) can rush for only a few yards.

The greatest disadvantage of endothermy is that it is energy expensive. The process of maintaining constant internal conditions, such as a body temperature of 37.4°C or a specific pH, requires a considerable input of energy. A house that is unheated in winter and not cooled in summer will have lower fuel bills than one that is maintained at 27°C no matter what the outside conditions. The fuel bills to an endotherm come in the form of the need to eat more, and more often, than an ectotherm of the same size. The success of the hunt is never certain, so endotherms have adopted a strategy of high costs and an associated risk of failure for the potential of high benefits in terms of speed, endurance, and readiness.

Why Do Crocodiles Bask in the Sun?

The effect of raising the body temperature of an ectotherm is to accelerate the rate of metabolic activity, which in turn provides the raw energy for muscle contraction and hence movement. The increased energy availability per unit temperature change will vary from one physiology to the next, but a typical figure would be a 2.5 increase in metabolic rate for each 10°C increase in temperature. Thus, basking in the sun raises the surface temperature of an ectotherm. Blood flowing close to the skin absorbs warmth, then transports this heat energy to muscles. Once warmed, the crocodile is capable of moving very quickly, but its physiology determines that these episodes are sprints rather than enduring gallops.

Many organisms have not evolved to fight off the effects of chilling and simply cannot survive in regions where there are significant changes in temperature. For others, the only way to survive periods of cold is to become dormant. Endotherms such as bears and squirrels may hibernate, conserving what warmth they can and at the same time lowering their metabolic activity to the barest minimum for survival. Some endotherms take advantage of solar heating to reduce their energetic costs. The roadrunner, a desert bird of the American Southwest, allows its body temperature to fall from 38°–39°C to 33°–35°C during the night. By basking in the morning sun, roadrunners restore their normal body temperature. Some butterflies and other ectotherms may survive the winter by hibernation, during which they secrete an antifreeze into their bloodstream. Other survival strategies for insects are to overwinter in forms that are less temperature-sensitive, such as eggs or as grubs that burrow into the soil.

Although most fish are ectotherms, tuna and some sharks generate significant amounts of heat in large muscles through metabolic activity. Through muscle contractions, these creatures can maintain higher temperatures than the surrounding seas; they are endotherms. This physiology gives them the capacity to behave almost as a warm-blooded, pursuit-oriented predator. In another example of temperature regulation, social bees congregate in great swarms, and in their over-wintering hive they form a tight clump of bodies. Each bee shivers its wing muscles and thereby creates heat. The bees at the center of the clump are warmest, as they are heated by their tightly pressed neighbors. The bees at the outer edge of the swarm are losing some of their heat to the air and so are cooler. A steady migration of bees from the edge to the inside of the cluster ensures that mortality due to cold is kept to a minimum.

Rather than thinking of whether ectotherms are more advanced than endotherms, or vice versa, we should remember that evolution has resulted in all organisms being as good as they need to be at what they do. Different attributes, such as intelligence, size, speed, or blood temperature, do not make one creature superior and another inferior.

Summary

- The sun provides the energy input for almost all natural processes.
- Primary producers are autotrophs, consumers and decomposers are heterotrophs.
- The first law of thermodynamics: As energy is transferred from one state to another, none is lost; energy can be neither created nor destroyed, but it can be transformed.
- The second law of thermodynamics: As energy is converted from one state to another, the quality of the energy deteriorates as a result of increasing entropy.
- Photosynthesis converts solar energy into stored chemical energy, such as carbohydrate.
- Animals are ultimately dependent on autotrophic organisms for their energy.
- Energy and material are transferred from the first trophic level of a food chain (primary producers) to the highest trophic level (top carnivores).
- Relatively little energy (1%–15%) is passed from one step in the food chain to the next.
- Animals feeding high on the food chain have access to only a tiny fraction of the energy available at the base of the food chain.
- The maximum number of trophic levels in most food chains is four (terrestrial) or five (aquatic).
- Endotherms can thermoregulate, but require more energy to support than ectotherms.

Further Readings

Colinvaux, P. A. 1975. *Why Big Fierce Animals Are Rare.* Princeton: Princeton University Press.

Lindemann, R. L. 1942. The trophic dynamic aspect of ecology. *Ecology* 23:399–418.

Schmidt-Neilsen, K. 1972. *How Animals Work.* Cambridge: Cambridge University Press.

Slobodkin, L. B. 1962. Energy and animal ecology. *Advances in Ecology* 4:69–101.

Web Connections

On-line resources for this chapter are on the World Wide Web at:
http://www.prenhall.com/bush
(From this web-site, select Chapter 3.)

Climate

Weather and climate are different things. **Weather** is the short-term variability that describes a snowfall one day and warm sun the next or a deluge of rain breaking a drought. Weather is affected by changes in the distribution of wind patterns half a continent away, sometimes half the world away. Because so many factors are involved in producing weather (the jet stream, sea temperature, the formation of high- and low-pressure systems, solar activity), it is impossible to predict weather in North America more than five days in advance. In general, tropical areas have more predictable climates than do temperate regions, but even in the tropics storms and hurricanes can arrive with little warning. Indeed, weather seems almost random. Climate, on the other hand, is highly predictable. The **climate** of a region is defined by the typical conditions for a certain time of year. For example, the summer climate in Ohio might be described as being about 25°C, with rainfall of about 40 mm per month. This climate does not mean that every day will be 25°C with just over 1 mm of rain falling; this is the average condition for that time of year based on the data from many years of observations. Although we cannot predict weather very far in advance, we can characterize the climate of a region for the next 10 years.

The climate of an area is strongly influenced by the intensity and pattern of solar energy received. These factors are a function of latitude. Other factors that can play a strong local role are proximity to an ocean, elevation above sea level, and the presence of mountains.

4.1 The Solar Connection

Every 24 hours Earth rotates one full turn about its axis, providing us with night and day. The length of daylight at any given location is determined by the way the axis tilts toward or away from the sun (Figure 4.1). A much larger cycle, our 0.9 billion kilometer orbit around the sun, takes a year to complete and provides the familiar passage of the seasons. For any given point on Earth, the combination of these two cycles determines the solar energy input and hence the climate.

Intensity of Received Solar Energy

The sun is our ultimate source of energy. The output of the sun has often been treated as being constant, but it varies a little (by about 0.1%) in a cyclic fashion. Peaks and troughs in solar output occur about 11 years apart. Sunspots, appearing as dark blotches on the surface of the sun, and their abundance correlates with variation in solar energy output. Although sunspots themselves are cool areas on the surface of the sun, their appearance corresponds to increased solar flare events. The flares increase the brightness of the sun and more than compensate for the duller sunspots. Thus, when sunspots are abundant, the sun is giving off more energy. The regularity of the sunspot cycle can be seen in Figure 4.2. A peak of sunspot activity will be experienced about the year 2000.

Sunspot activity has been correlated to climate patterns and has been used to argue that global warming is not due to human actions, but due to sunspot activity (Chapter 23). However, the sunspot cycle brings only about a 0.1% variation in the solar output, which seems to be too small to induce any climatic change on Earth. It is entirely possible that the apparent relationship between sunspots and climate is coinciden-

tal. The alternative, that sunspots are a signal of a stronger mechanism—perhaps related to the sun's magnetic field, which can induce climate change—remains highly speculative.

Although sunlight falling on Earth as a whole is almost constant, the energy input to different locations on Earth's surface is variable. Latitude is the primary determinant of the amount of energy received at any given point. Because we inhabit a sphere (or nearly so, Earth is slightly flattened at the poles), the intensity of sunlight reaching the ground is not equal all over the

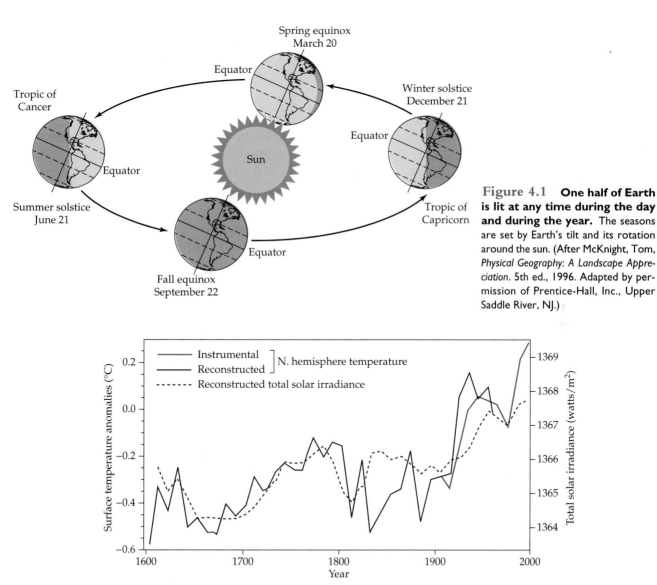

Figure 4.1   One half of Earth is lit at any time during the day and during the year. The seasons are set by Earth's tilt and its rotation around the sun. (After McKnight, Tom, *Physical Geography: A Landscape Appreciation,* 5th ed., 1996. Adapted by permission of Prentice-Hall, Inc., Upper Saddle River, NJ.)

Figure 4.2   The apparent linkage between sunspot cycles and recent climatic variability. Sunspot cycles go from maximum to minimum occurrence every 11 years. Over the last 300 years there appears to be a correlation between climate variability and sunspot cycles. No linking mechanism has been discovered, and this seeming correlation may be purely coincidental . . . or maybe it is not.

world. We are 149 million kilometers from our sun. This distance is so great relative to the sun's diameter (about 1.4 million kilometers) and that of Earth (about 12,800 km) that we can think of light as arriving in parallel beams. Imagine two beams of light of equal energy arriving at Earth, one at the pole and one at the equator (Figure 4.3). The beam of light arriving at the equator is perpendicular to the surface where it touches Earth, whereas the light reaching the pole hits Earth obliquely. Consequently, the "footprint" of the beam of energy at the equator is smaller than the footprint at the pole. Because the two beams of sunlight contain the same amount of energy, more energy falls per unit area at the equator than at the pole. Thus there is more light energy per square meter in the tropics than at the poles. A further, but lesser, factor is that solar inputs to the equator have a shorter passage through the atmosphere and thus suffer less scattering and absorption by water vapor and other gases.

The Effect of Daylength

Another cause of solar energy imbalance around the globe is the difference in daylength. Over the course of a year, the distribution of daylight hours from one place to the next is equal. An equatorial location has 12 hours of light every day, with dawn breaking at 6 A.M. and a short dusk beginning at 6 P.M. At the poles midsummer is marked by a day with 24 hours of daylight. Thereafter, the daylight hours gradually become shorter and shorter until at midwinter there are 24 hours of night. Because the pole is receiving a greatly reduced energy input in the dead of winter,

the temperature difference between the tropics and the pole will be greatest at that time of year.

4.2 Priming the Climate Engine

The differential energy inputs to various parts of Earth have immense climatic significance. It is this disparity that drives all our weather systems, our ocean currents, and patterns of life on the planet. That some areas receive more solar energy than others establishes a temperature imbalance between hot and cold areas. Heat energy flows from warm to cold, and the global temperature gradient establishes a pressure gradient that moves heat around the planet. Winds, air masses, and ocean currents are great heat transporters, redistributing heat around the globe.

Hadley Cells and the Coriolis Effect

The strong input of solar energy striking equatorial latitudes is stored temporarily as heat. Much of this heat is quickly radiated back into the atmosphere, heating the air close to the ground. Hot air rises, so over equatorial lands there are strong updrafts of heated air. However, because the axis of Earth is tilted relative to the sun, the strongest heating of Earth moves north and south through the course of a year. In the northern summer, it is the northern tropics that are perpendicular to incoming energy from the sun (Figure 4.4). Therefore, in midsummer these areas will receive the most heat energy. In December, it is the Southern Hemisphere tropics that receive the maximum solar input. The point on Earth's surface receiv-

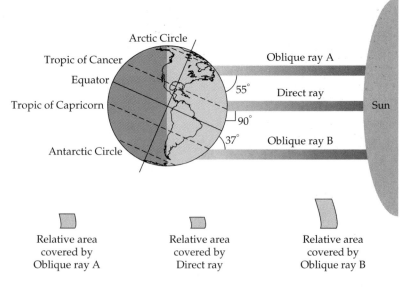

Figure 4.3 Variability in the intensity of sunlight at the surface of Earth is due to the incident angle of solar radiation. The parallel incoming beams of radiation hold the same amount of energy, but the beam hitting tropical latitudes has a smaller "footprint" on the surface of Earth than the beam hitting the polar region. Consequently, the intensity of energy input per square meter is higher at the tropics than at the poles. (After McKnight, Tom, *Physical Geography: A Landscape Appreciation.* 5th ed., 1996. Adapted by permission of Prentice-Hall, Inc., Upper Saddle River, NJ.)

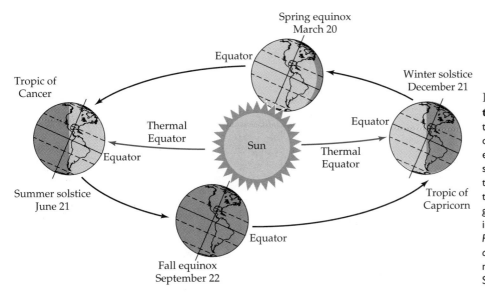

Spring equinox
March 20

Equator

Thermal
Equator

Tropic of
Cancer

Equator

Sun

Equator

Thermal
Equator

Winter solstice
December 21

Equator

Summer solstice
June 21

Tropic of
Capricorn

Equator

Fall equinox
September 22

Figure 4.4 Movement of the thermal equator. As Earth orbits the sun, the thermal equator, the area of Earth perpendicular to incoming energy from the sun, moves north or south of the geographical equator. In the Northern Hemispheric summer, the thermal equator is north of the geographical equator and moves south in the winter. (After McKnight, Tom, *Physical Geography: A Landscape Appreciation.* 5th ed., 1996. Adapted by permission of Prentice-Hall, Inc., Upper Saddle River, NJ.)

ing the most solar energy on any given day is the **thermal equator**. Because of the rotation of a tilted Earth, through the course of a year the thermal equator tracks north and south of the geographic equator. The movement of this thermal equator is a significant factor in determining the timing of rainy seasons in many, but not all, tropical regions.

To explain this relation, let us start by simplifying things. Assume that the thermal equator and the geographic equator are the same. Heated moist air rises over the equator, and as it rises it cools. The cooling air cannot hold as much moisture as heated air, so it rains. Eventually, high in the atmosphere, the packet of air is cool and dry. Air cannot continue to pile up over the equator, so it flows toward the poles. The "push" given to this air comes from its own momentum. A person, tree, or packet of air at the equator is revolving in space at about 38,400 km per day (the circumference of Earth). In contrast, someone standing at the North Pole is barely moving at all. (You can demonstrate this for yourself by making a mark on an orange at its equator and another at its pole and then spinning the orange about its polar axis. The dot on the equator obviously moves faster than the one at the pole.) As the packet of air high in the atmosphere moves poleward, it has a momentum that is greater than that of the air it encounters. The spinning of Earth gives the air an initial eastward momentum; although the air packet may start by traveling due north, as it encounters slower air its own momentum will cause it to deflect to the east. This process of the deflection of large masses (water or air) to the right in the Northern

Hemisphere or to the left in the Southern Hemisphere is the **Coriolis effect** (Figure 4.5). As the air moves about 20° to 30° north or south of the equator it slows and descends. The air is still dry, but as it descends it is warmed, and by the time it reaches Earth's surface it is hot and dry. The downdraft of this hot dry air causes the formation of the great desert regions of Earth such as the Sahara, Sonoran, Australian, Gobi, and Atacama deserts (Figure 4.6).

Because these deserts are receiving the downdraft of air, they have high atmospheric pressure. In contrast, the updraft of air at the equator forms an area of low atmospheric pressure. Air will always flow from an area of high pressure to one of low pressure, so there is a net flow of air from the deserts toward the equator: These are the trade winds (Figure 4.5). This air gains moisture as it passes across oceans and forests on its way to the equator. Thus the air arriving at the equator is hot and wet.

This circulation of air rising at the equator, spinning away poleward, sinking in the subtropical latitudes, and returning to the equator is termed a **Hadley cell**. Matching Hadley cells exist north and south of the equator and are important in defining the climate of tropical and subtropical regions (Figure 4.6).

At the start of this explanation of Hadley cells, we ignored the fact that the thermal equator and the geographic equator are not the same. Now let us take into account the fact that the thermal equator moves north and south of the geographic equator, according to the seasons. Consequently, the Hadley cells are not moored at the geographic equator, but they too will

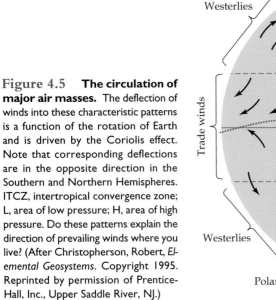

Figure 4.5   The circulation of major air masses. The deflection of winds into these characteristic patterns is a function of the rotation of Earth and is driven by the Coriolis effect. Note that corresponding deflections are in the opposite direction in the Southern and Northern Hemispheres. ITCZ, intertropical convergence zone; L, area of low pressure; H, area of high pressure. Do these patterns explain the direction of prevailing winds where you live? (After Christopherson, Robert, *Elemental Geosystems.* Copyright 1995. Reprinted by permission of Prentice-Hall, Inc., Upper Saddle River, NJ.)

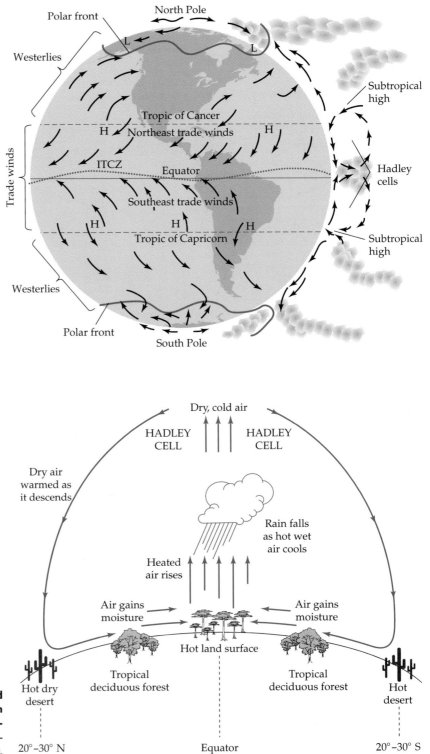

Figure 4.6   The updrafts and downdrafts of air that result in the wet tropics and desert regions of the world. These great circulations of air are termed Hadley cells.

track the thermal equator. Heating of the land, and particularly of the ocean, is slow, thus the areas at the northernmost and southernmost ranges of the thermal equator do not heat sufficiently to support Hadley cells. Consequently, as the thermal equator migrates away from the geographic equator, the Hadley cells lag behind. That the Hadley cells do not keep up with the full range of movement of the thermal equator is even more marked in the Southern Hemisphere than in the Northern Hemisphere because so much of the area is ocean. The updraft of the Hadley cell forms where the north and south trade winds converge and is termed the **intertropical convergence zone** (ITCZ). The ITCZ migrates a few degrees north and south of the equator, pulled by the movement of the thermal equator (Figure 4.7). As the ITCZ arrives it brings rain, and when it leaves a dry season can begin. The ITCZ is also important to weather conditions in eastern North America because hurricanes are spawned along the edge of the ITCZ at its most northerly extent. Consequently, the hurricane season for Florida and the East Coast is from June until October.

4.3 Frontal Systems

Two other pairs of circulation cells, similar to Hadley cells, have been suggested to exist: one in the midlatitudes and one near the poles. However, as our understanding of climatology improves, it becomes apparent that the climate of the temperate and polar latitudes is more affected by the movement of air masses and fronts than by circulation cells.

An air mass is a large body of air, at least 1600 km wide, that has characteristic properties of temperature and humidity. The basic properties of the air mass are gradually modified as it flows away from its source. Thus, a cold air mass that forms in the Arctic will gradually warm as it moves southward. Nevertheless, the arrival of that air mass in, say, Texas, can bring some of the coldest weather known to that region.

Five major types of air mass drive much of the weather in North America. Local weather conditions will often be determined by the presence or proximity of one of these major air masses. The **Arctic** and **continental polar** (cP) air masses form over northern Canada and the Arctic and can pour cold, dry air down the continent. Another source of cold air comes from **maritime polar** (mP) air masses (Figure 4.8). This cold, wet air can sweep across from the northern Pacific or down the East Coast from Labrador. Incursions of these air masses often bring heavy winter snowfalls. The position of these air masses is closely tied to the polar **jet stream** (Figure 4.9). High in the atmosphere, about 12 km above the land surface, a river of air rushes eastward around the planet. This body of air is the polar jet stream (there is a corresponding one in the southern hemisphere and weaker ones at the tropics). The average position of the jet stream is about 50°N, the latitude of Winnipeg, but it is continuously bending and flexing. Sometimes it loops deep into the southern half of North America. The continental polar air mass is bound at its southern edge by the polar jet stream. If the jet stream flexes south it brings cold air

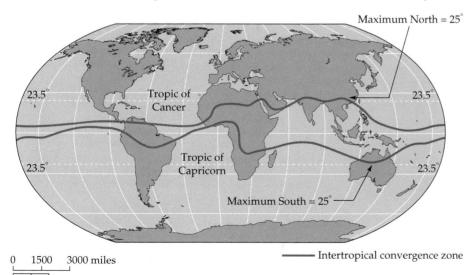

Figure 4.7 **Average positions of the intertropical convergence zone at its seasonal extremes.** (After McKnight, Tom, *Physical Geography: A Landscape Appreciation.* 5th ed., 1996. Adapted by permission of Prentice-Hall, Inc., Upper Saddle River, NJ.)

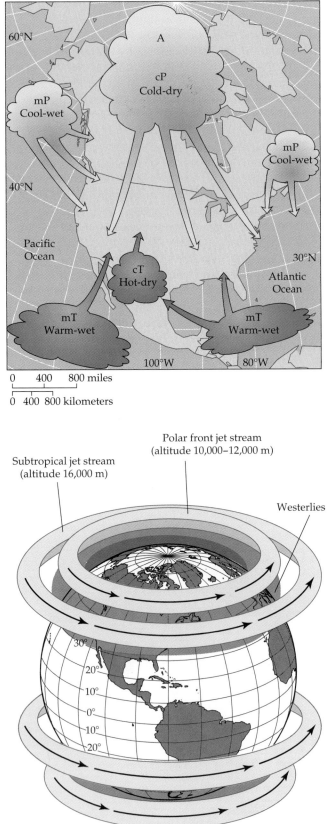

Figure 4.8   The source areas of major air masses that influence climate and weather patterns over North America. cP, continental polar; mP, maritime polar; mT, maritime tropical; cT, continental tropical. (After McKnight, Tom, *Physical Geography: A Landscape Appreciation.* 5th ed., 1996. Adapted by permission of Prentice-Hall, Inc., Upper Saddle River, NJ.)

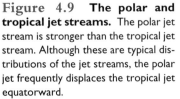

Figure 4.9   The polar and tropical jet streams. The polar jet stream is stronger than the tropical jet stream. Although these are typical distributions of the jet streams, the polar jet frequently displaces the tropical jet equatorward.

with it; if it pushes north, warm air will flow up from the tropics.

Maritime tropical (mT) air forms warm, wet air masses that can move northward from source regions in either the Pacific or the Gulf of Mexico—Caribbean region. Warm, dry **continental tropical** (cT) air can push northward from Mexico and the southwestern states.

Where air masses or even smaller bodies of unlike air meet, a **front** is formed. Commonly, one of these air masses is warm and the other is cold (or they differ in humidity or density); it is their failure to mix that forms the front. A front is a three-dimensional feature that is somewhat like a sloping wall. Just as a wall can be long but narrow, so too can a front. A front can extend for hundreds of kilometers, but may be only a few kilometers across (Figure 4.10). If a front is moving quickly, a person standing on the ground will notice a sharp change in temperature in a matter of hours. If the temperature change is from cold to warm, then it is a warm front that is passing through; conversely, a cold front brings cooler weather.

Because cold air is denser than warm air, when two air masses of different temperature meet, the cold air will force the warm air upward. As the warm air rises, it cannot hold as much moisture, and rain will fall.

Thus, it is common to find that rain falls along the front. Warm fronts tend to produce prolonged, relatively gentle rains, whereas cold fronts produce stormier conditions of brief high-intensity rains.

4.4 Oceanic Influences

That oceans heat and cool more slowly than land and that water flows from one place to another are important climatic factors. It is evident that land and water have different thermal properties. For example, a beach in summer may be too hot to walk on, but the ocean might be just pleasantly warm. Oceans heat less than the land, and they help to keep coastal districts cool relative to the inland areas. Hence the summers of San Diego are likely to be cooler than those of Phoenix. The oceans, therefore, are great climatic moderators, and the continents of the world can be divided into areas that have oceanic climates and areas that have more extreme continental climates.

Oceans moderate the temperatures of adjacent lands, causing them to vary less from one season to the next than areas far from the ocean. During winter, the land cools rapidly, but the ocean temperature falls slowly. Consequently, in the dead of winter snow may

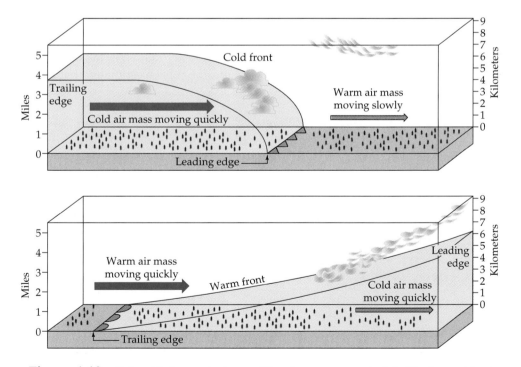

Figure 4.10 **A front is formed when unlike air masses meet.** Rain falls along and just behind the leading edge of the front. (After McKnight, Tom, *Physical Geography: A Landscape Appreciation.* 5th ed., 1996. Adapted by permission of Prentice-Hall, Inc., Upper Saddle River, NJ.)

cover the sand, and the water temperature might be well above freezing. The ocean has the effect of being an offshore radiator that heats the adjacent land. Land close to the ocean will therefore be warmer in winter than land located in the middle of a continental land-mass. For example, the winter temperatures of Minnesota or Kansas are considerably colder than those of New England.

The Ocean As a Heat Transporter

Oceans have far more than just a local influence on climate. The circulation of air masses has a parallel in the oceans, where long-distance currents act as vast energy transporters, bringing cold water equatorward and warm water poleward (Figure 4.11). The circulation of the oceans is driven by subtle temperature and salinity differences (Chapter 13). What cannot be seen from Figure 4.11 is that the circulation within the ocean is three-dimensional. Water flows through the oceans as if carried on a vast conveyor belt, gaining warmth as it loops upward and cooling as it plunges to the ocean bottom. This circulation pattern is known as the **ocean conveyor belt** system. We'll join the conveyor belt as it passes the coast of Greenland headed south (Figure 4.12). Salt water is denser than freshwater, thus has a tendency to sink to the bottom of the ocean. Similarly, cold water sinks below warm water. In the North Atlantic dense, cold, salty water sinks to the bed of the ocean and is pushed southward. This cold water flows south as the **North Atlantic Deep Water** until it merges with another major cold-water body, the **Cir-**

cumpolar Current, which flows around Antarctica. The cold current now swings east and north toward the equator. In the Pacific Ocean and in the Indian Ocean the water is warmed and loses its excess saltiness. The stream of water becomes less dense and floats toward the surface. The conveyor belt doubles back through the Indian Ocean and crosses the South Atlantic, before arriving in the Caribbean. As a surface current, this body of water is subject to evaporation, a process that removes water from the ocean surface but leaves salt behind. Consequently, the saltiness of the surface water in the ocean conveyor belt increases. All the while, this water is warming. When the flow leaves the Caribbean, it is known as the **Gulf Stream**. The Gulf Stream swings across the Atlantic, gradually cooling as it flows north past Iceland. By the time it reaches the Greenland coast, it is cold, salty water that will sink and become part of the North Atlantic Deep Water. Once more the water of the conveyor flows south. A complete loop of the ocean conveyor belt takes a molecule of water about 1000 years.

The flow of the Gulf Stream is 50 times greater than all Earth's rivers combined, and it carries a huge amount of energy from the tropics northward. The warming of Europe by the Gulf Stream allows arable farming and good living conditions north of an equivalent latitude in Canada. Britain is at almost the same latitude as the Hudson Bay, so one would expect the same desperately cold winters in the two places. But the heat provided by the Gulf Stream is enough to ensure that heavy snowfalls in London are relatively rare events. It has been estimated that in winter, when Ice-

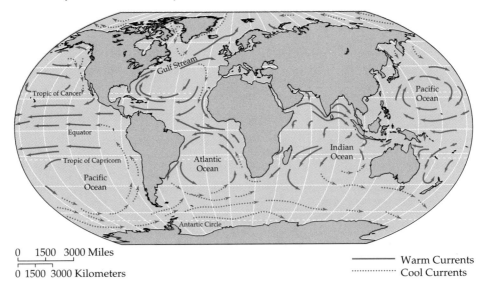

Figure 4.11   **The circulation of great surface ocean currents transporting heat around the globe.** Note how the currents are affected by the Coriolis effect (compare with Figure 4.5). (After F.T. MacKenzie and MacKenzie, 1995.)

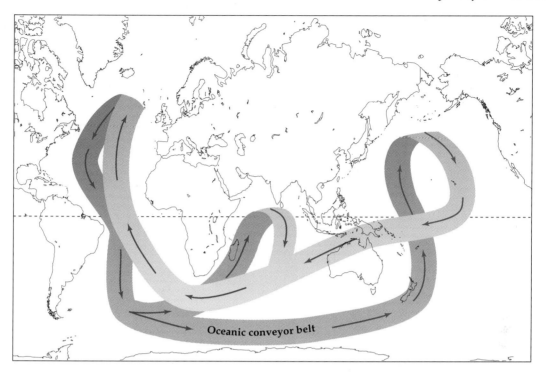

Figure 4.12   **The ocean conveyor belt circulation pattern.** Dense, cold, salty water sinking off the coast of Greenland sets in motion this immense flow of water through the oceans. After looping through the southern oceans and into the Pacific, the water is warmed, less salty, and rises to the surface of the ocean. The return flow on the surface across the Atlantic is the Gulf Stream that transports vast amounts of heat energy northward. (After Broecker 1989.)

land has only 2–3 hours of sunlight, the Gulf Stream may provide 50% of the heat input to the island.

4.5   Cycles of Climate Change

Climate is predictable, at least in the short term, but that does not mean that it is unchanging. Climate change is an integral part of ecology. The changes can vary on both spatial and temporal scales. Two examples that can change the climate of the entire Earth, though functioning on different timescales are glacial-interglacial oscillations and the El Niño Southern Oscillation (ENSO).

Glacial to Interglacial

For more than 100 years, geologists have known that, at times, great ice sheets developed, engulfing lands at high latitudes and carving new landscapes in Europe and North America. The Rockies, Antarctica, the Andes, and the Alps of Europe, bear the scars of ice grinding its way downhill. It is now understood that in the last 4 million years there have been at least 22 ice ages, or **glacial** periods, and that the warm periods between, the **interglacial** periods, were relatively brief.

The last ice age lasted for almost 100,000 years and finished about 10,000 years ago. In fact, about 90% of the last 2 million years has been glacial, during which time the temperature of Earth was between 4° and 9°C cooler than the present. A drop of 9°C may not sound all that extreme, but it is enough to trigger the expansion of ice sheets over all of Canada and the lowering of sea levels by about 125 m. During the last ice age, the ice sheets came south across New York and Wisconsin, extending as far south as Columbus, Ohio (Figure 4.13). In Europe ice sheets swept down out of the Alps into the lowlands below, or they came out of the north to cover much of northern Europe as far south as London, England. From an evolutionary point of view, for the last 10,000 years plants and animals have been living in an unusually warm environment. Present conditions are by no means favorable to many of these species, and they can only survive in the coolest settings available, such as mountaintops or at high latitudes. As ecologists, we should not assume that now is "good" and that the ice ages were "bad" for species; in fact, the opposite is more likely to be true.

During the ice ages sea levels fell because water evaporated from the oceans and was trapped in glaciers and ice-sheets where it was held as ice. A com-

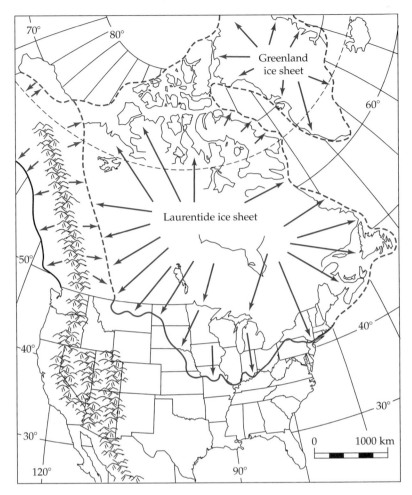

Figure 4.13 The extent of the North American ice sheet at 20,000 YR. B.P. (years before the present). Most of Canada and parts of the northern United States were covered by glaciers up to 3 km thick from 100,000 years ago until 10,000 years ago. The actual front of the ice sheet waxed and waned with warmer and colder cycles within the last ice age.

bination of a cooling ocean, which causes water to contract and occupy less volume, and the vast amount of water held in the ice, produced a lowering of sea levels by 125–150 m at the peak of the last ice age (21,000 years ago). At this time, you could have walked across a lightly wooded plain from England to France, or, standing on the modern shoreline near Jacksonville, Florida, you would have had another 100 km to walk to reach the Atlantic Ocean.

El Niño Southern Oscillation

Each year around Christmastime a warm ocean current appears along the coast of Peru. Usually this current fades away early in the new year, but if the warm water lasts until July it is named an **El Niño** event. The origin of the warm water is an atmospheric see-saw termed the Southern Oscillation that was first described in 1910 by a British climatologist, Sir Gilbert Walker.

Under normal conditions the air pressure over Tahiti is higher than that over Australia and Indone-

sia. Because air flows from high pressure to low pressure, steady westerly winds (the trade winds) result from this pressure difference (Figure 4.14). The air rushing from east to west across the ocean surface pushes water westward. Consequently, warm surface water is piled up around Indonesia. Periodically, the see-saw of air pressure flattens out or even tips in the other direction (hence it is called the Southern Oscillation). When air pressures off Australia are higher than those of Tahiti, the westerly winds weaken or even blow the other way. The warm water piled up off the coast of Indonesia sloshes back across the Pacific Ocean and results in unusually warm water temperatures from Tahiti all the way back to Peru. As the warm water hits the South American landmass it is forced north and south by the water pressing from behind. The effect of this reversal of surface water currents can be plotted as departures from normal temperature. For example, if the normal temperature is 29°C, and the recorded temperature is 30°C, a departure of +1°C is recorded. A false-color satellite image showing departures of surface water temperature in

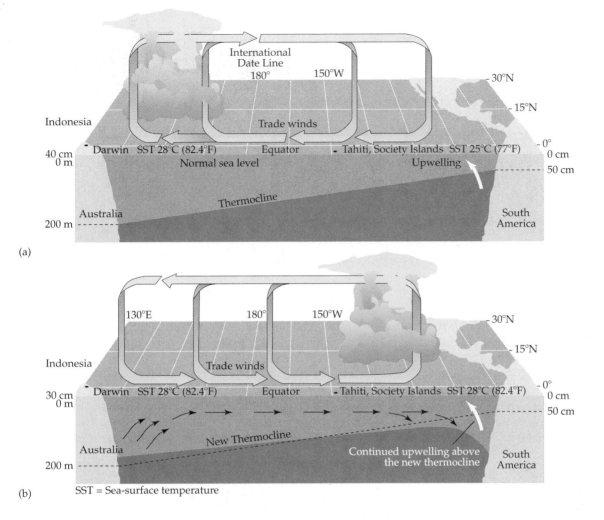

Figure 4.14   The El Niño Southern Oscillation (ENSO) forms as a result of reduced air pressure over the central Pacific Ocean. This leads to a reduction in wind speed and less warm surface water being piled up in the western Pacific. The ocean slops back eastward causing sea temperatures along the Pacific rim of North and South America to rise. This simple change in air pressure sets in motion a chain of events that affect global climate every three to seven years as they induce an El Niño event.

the Pacific reveals a characteristic plume of warm water from Tahiti east to the South American coast and two tongues of warm water poking north and south along that coastline (Color Plate 1). The satellite image taken during the peak of the 1997/1998 El Niño event shows temperature departures as high as 7°C during the strongest ENSO event yet recorded. Although the actual departure values shown in Color Plate 1 are high, this temperature pattern is typical of El Niño conditions. The combination of a strong El Niño and a strong Southern Oscillation give the phenomenon its scientific name the **El Niño Southern Oscillation**.

The Multivariate ENSO Index is a standardized method to measure the strength of an ENSO event.

The index is based on the following six variables:

1. sea-level atmospheric pressure difference between Australia and Tahiti
2. east-west length of the warm water tongue
3. north-south length of the warm water tongues
4. departure of sea-surface temperature from normal
5. departure of surface air temperature from normal
6. total amount of cloudiness

The scores calculated from the Multivariate ENSO Index allows a quantitative comparison of different ENSO events (Figure 4.15). The pattern of ENSO over the last 100 years suggests that the oscillation may be

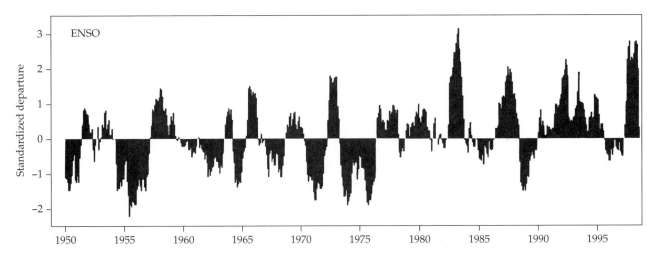

Figure 4.15 **The historical record of ENSO events 1950–1998.** A pattern of increasing El Niño intensity is evident since 1975. Data are expressed as a standardized departure from the mean value for the Multivariate ENSO Index for the period 1950–1993. (Data from NOAA-CIRES climate diagnostics center, University of Colorado at Boulder.)

intensifying. Indeed the strongest El Niño yet documented was the 1997/1998 event.

El Niño brings torrential rain to western South America, but the 1997/1998 event also brought widespread climate change to North America, Europe, and Asia (Figure 4.16). The principal way that ENSO has worldwide effects is through displacing the polar jet stream. During a strong ENSO, the area of warm surface water in the Pacific is unusually large. Heating of the overlying air mass drives vast quantities of warm, wet air upward through convective activity. This great warm air mass presses against the polar jet stream, causing it to change shape and move further north. With the jet stream displaced to the north, the influence of tropical air masses is strengthened, and warm, wet air flows into California, leading to strong rains.

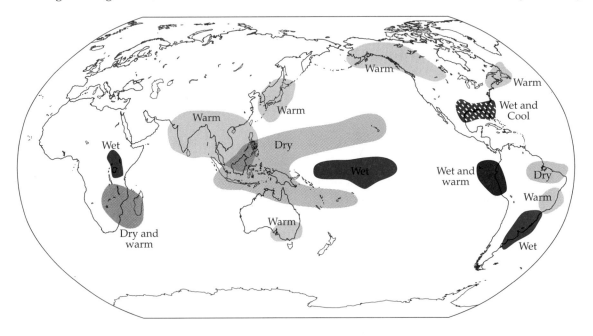

Figure 4.16 **Worldwide changes in precipitation and temperature induced by a "typical" El Niño event.** Each El Niño event produces a somewhat different climate pattern, but on average they would conform to those shown here.

Throughout the center of North America and other northern landbodies, the displacement of the jet stream results in unusually mild winters. In the Southern Hemisphere, the principal change is that the weakening of the westerlies results in decreased rainfalls, even droughts, across parts of Australia and Indonesia. The two strongest ENSO events in 1982/1983 and 1997/1998 have resulted in uncontrollable fires ravaging great areas of Indonesia. Following some ENSO events the see-saw tilts back to provide unusually strong trends in the opposite direction. These conditions are known as La Niña.

ENSO is now recognized to be the largest single cause of interannual climate variability over the last several centuries.

Summary

- The daily variability of weather makes it somewhat unpredictable.
- *Climate* describes the normal conditions for a given place at a given time.
- Differences in solar input to Earth's surface induce pressure gradients that drive climate.
- The greatest variable in tropical climate is precipitation caused by the migration of the intertropical convergence zone and Hadley cells.
- The climate of temperate regions is governed by frontal activity.
- In North America the polar jet determines the southward extent of polar air masses.
- During the last 4 million years, there have been more than 20 glacial cycles.
- For the last 10,000 years Earth has been in a warm interglacial period.
- The ocean conveyor belt system circulates heat around the planet. If it shuts down, the planet cools rapidly.
- ENSO activity is the largest single cause of interannual climatic variability.

Further Readings

Broecker, W. S., and G. H. Denton. 1989. The role of ocean-atmosphere reorganizations in glacial cycles. *Geochimica et Cosmochimica Acta* 53:2465–2501.

Morgan, M. D., and P. M. Pauley. 1996. *Meteorology: The Atmosphere and the Science of Weather*. 5th ed. Upper Saddle River, NJ: Prentice-Hall.

Kerr, R. A. 1996. A new dawn for sun-climate links? *Science* 271:1360–1361.

Kerr, R. A. 1997. A new driver for the Atlantic's moods and Europe's weather? *Science* 275:754–755.

Oppo, D. 1997. Millennial scale climate oscillations. *Science* 278:1244–1246.

Trenberth, K. 1991. *General Characteristics of El Niño–Southern Oscillation*. Cambridge: Cambridge University Press.

Web Connections

On-line resources for this chapter are on the World Wide Web at:
http://www.prenhall.com/bush
(From this web-site, select Chapter 3.)

Biomes: The Great Vegetation Types

The first maps of world vegetation, drawn in 1855 by Augustin de Candolle, a French plant physiologist, revealed that forest types appear to grow in belts that circle Earth. The temperate forests rich in broad-leaved trees (trees without needles) are found at about the same latitudes all the way around the world, both north and south of the equator. Poleward from these forests are great expanses of needle-leaved (coniferous) trees such as spruce and fir. The deserts lie in a belt at the subtropics, and the tropical forests form a distinct belt close to the equator. It appears that a global-scale phenomenon regulates the distribution of these vegetation types.

Climate is a major factor in determining growing conditions over large geographical areas. It should not be surprising to find that plants have evolved to maximize productivity under each climatic regime. Furthermore, we should expect a plant whose greatest problem is subzero winter temperatures and huge snow loads to look different from one that faces death from the beating tropical sun. General growth forms of vegetation—evergreen trees, deciduous trees, cacti—can characterize the vegetation of whole regions and can be classified into biomes. A **biome** is a large geographical region having a defining climate to which plants show a similar physiological adaptation. Rainfall and temperature are probably the largest factors controlling the biology of the regions (Figure 5.1). The

global distribution of the major biomes is shown in Figure 5.2, and their net primary productivity is given in Figure 5.3 and Table 5.1.

The following descriptions aim to characterize a biome, but it must be recognized that biomes have soft edges; the characteristics of, say, chaparral merge seemlessly into desert along a gradient of decreasing rainfall or into tundra as temperatures decline. Growth forms first observed in the chaparral as adaptations to dealing with drought will become more pronounced as we examine desert lifeforms. Biomes are very useful as basic building blocks to frame our ecological thinking, but they are underlain by evolutionary adaptations to overcome stress. It is important to see these areas, not as groups of plants and animals that have always lived together, but as individuals that have, at least temporarily, overcome the stringencies of a particular setting.

5.1 Tundra

Tundras are windswept expanses where nothing stands taller than grasses and sedges (Figure 5.4, p. 57). In these high-latitude wildernesses, the average temperature is 25°C, and water is held as ice, both on the surface and in the soil, for much of the year. Only during the brief summer does the surface of the soil thaw. Water held as ice is inaccessible to plants,

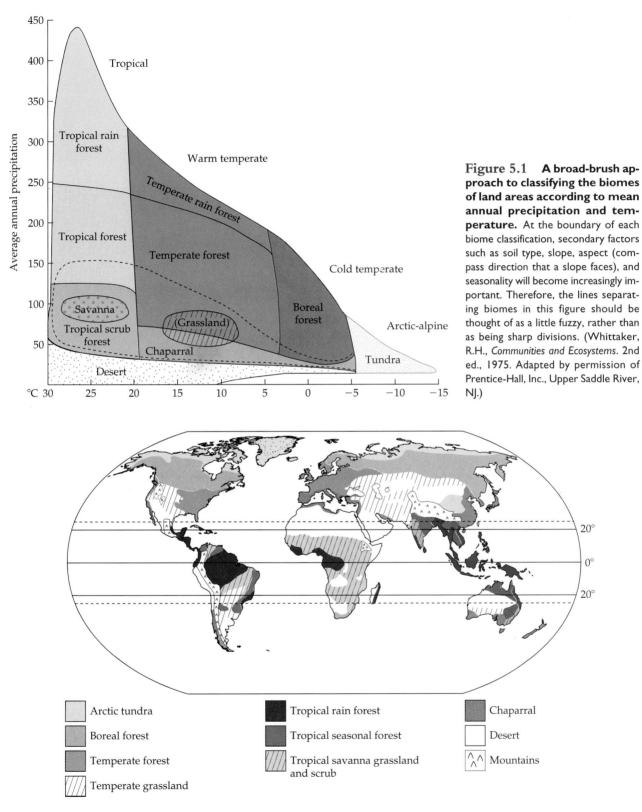

Figure 5.1   A broad-brush approach to classifying the biomes of land areas according to mean annual precipitation and temperature. At the boundary of each biome classification, secondary factors such as soil type, slope, aspect (compass direction that a slope faces), and seasonality will become increasingly important. Therefore, the lines separating biomes in this figure should be thought of as a little fuzzy, rather than as being sharp divisions. (Whittaker, R.H., *Communities and Ecosystems*. 2nd ed., 1975. Adapted by permission of Prentice-Hall, Inc., Upper Saddle River, NJ.)

Figure 5.2   A generalized map of the global distribution of biomes. (After Cox and Moore, 1993.)

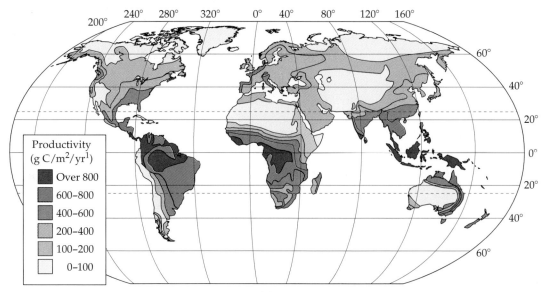

(a)

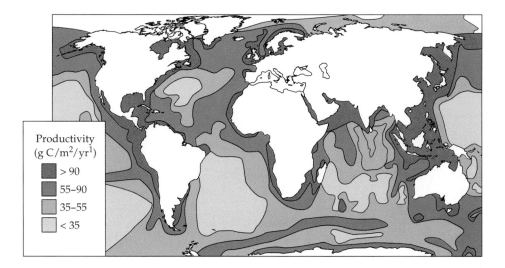

Figure 5.3 **Global net primary productivity.** (a) Net primary productivity of the land. (After Reichle 1970.) (b) Net primary productivity of the oceans. (After Koblentz-Mishke 1970.)

therefore plant growth is inhibited. A more inhospitable environment for plants is hard to envisage. Not only is liquid water in short supply, but the high winds intensify the cooling effect and produce such a wind chill that the growing tips of plants are killed. Under such conditions, the only chance for survival is to stay out of the wind. Trees that stand tall would be killed by the wind; hence they cannot colonize these latitudes. Close to the ground surface is a relatively still layer of air where grasses and herbs can survive. Shrub species of willows and alders can live in the tundra, but they must hug the ground to do so. Small depressions in the land surface may be filled by one of these prostrate

trees, but, because they cannot stick their twigs up into the wind, they can grow no taller than the general surface of the herbs, sedges, and grasses.

In tundra areas, the growing season (the time in which plants can photosynthesize and build leaves, stems, roots, and flowers) lasts only from May to August. Because the period is so short, the total amount of plant growth (net primary productivity, or NPP) is very low (Figure 5.3a; Table 5.1).

Although the NPP of the tundra is low, its soils are rich in organic matter. Soil organic matter is derived from dead organisms that have partially decayed. In the tundra, the decay process is so slow that the bod-

Table 5.1 The net primary productivity (NPP) of biomes (in grams carbon per square meter that is converted into plant matter per year).

Biome	NPP (g C/m²/yr)
Terrestrial systems	
Tropical rain forest	900
Tropical dry forest	675
Temperate evergreen forest	585
Temperate deciduous forest	540
Boreal forest	360
Tropical grasslands	315
Cultivated land (USA)	290
Chaparral	270
Prairie	225
Tundra	65
Desert	32
Extreme desert	1.5
Aquatic systems	
Swamp	1125
Algal beds and coral reef	900
Estuaries	810
Upwelling zones	225
Continental shelf	162
Open ocean	57

(*Source: Ecology: Theory and Application*, by P. Stiling, © 1996. Reprinted by permission of Prentice-Hall, Inc., Upper Saddle River, NJ.)

Figure 5.4 Tundra. The short growing season, intense cold, and strong winds greatly restrict the diversity of organisms that can inhabit this biome.

ies of the dead do not decompose as quickly as fresh corpses are added. As a tundra plant dies, it withers to the ground, and the fallen leaves and stems enter the deep freeze of the soil. Because the animals and bacteria that carry out decomposition cannot function in a block of ice, the process of decay and the recycling of essential nutrients in these environments are desperately slow. During the brief summer, the thaw extends only about 1 m down into the soil; below that, the water always remains as ice, and this continuously frozen soil is termed **permafrost**. Decomposition is limited to the surface few centimeters that thaw each May. After a few decades, the buildup of subsequent layers of organic material buries the partially decomposed body in the permafrost. The body of the plant or animal is then preserved in this state. This organic-rich soil, in which decomposition does not keep pace with organic input, is termed peat.

The short growing season of the tundra dictates that all growth and reproduction must be compressed into a short period of time. Tender young shoots appear as the plants build a new body; these shoots are rich in nutrients and make good grazing. Food is so abundant in the short growing season that flocks of wading birds, geese, swans, and ducks commonly migrate north from temperate or tropical areas. Herds of caribou and reindeer arrive and, in turn, support a host of biting flies, midges, and mosquitoes. All the insects must complete their life cycles and reproduce in the growing season. These insects survived the Arctic winter as eggs, larvae, or pupae. With the arrival of summer, the adult insects emerge, eggs hatch, and adults and larvae set about feeding on animals or tender young plants. Birds, chief predators of insects, arrive to nest and use the glut of insects to feed their young. Thus, the tundra has a brief but very active summer period that is important to the maintenance of organism populations that overwinter in less extreme climate zones.

Despite the arrival of wildfowl in the Arctic summer, the number of species of birds is still low. Similarly, although there are vast numbers of mosquitoes and blackflies, they are represented by just a few species. Few plant species have managed to evolve to live under the harsh climatic conditions and short growing season of the tundra. Ecologists use the term **species diversity** to describe the number of plant and animal species that live within an area and as a measure of that area's biological richness. The tundra has a low species diversity. Note that low species diversity does not equate with being biologically unimportant. The tundra is essential to the well-being of animals such as ducks, geese, seals, polar bears, and caribou.

Human activities in the tundra have been limited traditionally to harvesting pelts. Plans to expand oil-drilling operations in tundra regions threaten to be new sources of disturbance and pollution. One facet of the tundra biome that must be recognized is that, because of the extremely low growth rates of plants in this biome, biological recovery from disturbance is extremely slow. For example, the tracks of a vehicle crossing tundra are visible for several decades.

5.2 Boreal Forest

The **boreal** forests are the great fir forests of Canada and northern Eurasia (Figure 5.5). In North America, this vegetation type forms a biome that is defined very clearly by the seasonal expansion and contraction of the Arctic and continental polar air masses (Figure 5.6). In the winter months, the cold polar air extends as far south as the Great Lakes, and this defines the southern border of the boreal forest. In the summer, high-pressure systems over the eastern United States prevent the southward movement of the polar air. The cold air forms a front around the latitude of the southern edge of Hudson Bay. The biomes vary according to the exposure to the polar air. If the region has polar air all the year round, it is tundra; if it receives polar

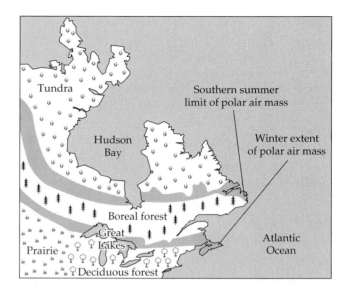

Figure 5.6   The movement of the polar air mass defines the boundaries of four biomes within North America. Land that is permanently covered by this cold air is the tundra region. If the polar air is replaced by southern pressure systems in the summer, the vegetation will be boreal forest. The southern winter limit of the polar air defines the northern boundary of both temperate forest and prairie biomes. (After P. Colinvaux, *Ecology*, © 1992 John Wiley and Sons, Inc. Reprinted by permission of John Wiley and Sons, Inc.)

Figure 5.5   Boreal forest. Fir and spruce forests dominate these landscapes and are the basis for intense forestry activities.

air in the winter and air masses from the south in summer, it is boreal forest.

Thus the boreal forests have very cold winters, but relatively warm summers. Trees can survive under such a regime, though they almost all have needles rather than broad leaves. Many trees of the boreal forest look like Christmas trees and not like oak trees because the overall shape of the tree, a cone, is good for shedding snow. Other tree designs, such as the upward sloping branches of an oak, would trap snow. The sheer weight of snow piled on such branches would cause them to break. Damaged trees are poor competitors; hence the only growth form that will work under these conditions is one that does not accumulate snow. The few broad-leaved trees that do grow in the boreal forests, such as birch, aspen, and poplar, do not form large spreading crowns like those of oak or beech trees.

Another factor in adaptive morphology is leaf shape. Why do most of the trees in a boreal forest have needles rather than leaves? A leaf is a solar panel, the collector of energy for conversion by photosynthesis. The evolution of needles reflects a balance between energy expenditure to make a new leaf and the need to withstand cold. To defy damaging frosts, the leaf must be strong and rich in lignin. The lignin stiffens the cell wall and helps the cell survive the replacement of tissue water with ice during winter freezes. Lignin is a carbon-rich molecule and therefore takes a great deal of energy to make. Because it makes the leaf durable, though, the leaf can be kept all year. Hence, the trees of the boreal forests grow stiff, almost woody leaves that can withstand the cold and photosynthesize year-round.

Another reason that the leaves are needle-shaped rather than flat has to do with the conservation of heat in winter. As the air temperature drops, the tree begins to lose heat. Big, flat leaves would have a large surface area through which heat would be lost. Because the needles are round in cross section and they hang in bunches, less heat is lost. The needles radiate heat outward in all directions, but because half of this heat is going toward neighboring needles, it warms the air trapped within the bunch of needles, and thus helps ameliorate some of the worst of the chilling.

Boreal forest soils are rich in peat and may be permafrosted in some locations. Species diversity is low, although it is higher than in the tundra. These forests were originally settled by trappers exploiting a wide diversity of fur-bearing mammals, such as sable, mink, lynx, and beaver. They are now major timber-producing regions.

5.3 Temperate Forest

The **temperate forests** can be divided into evergreen and deciduous forest types. Some of the largest trees in the world live in the temperate evergreen forests. These forests have a cool, moist climate throughout the year, receiving so much rain that they are described as temperate rain forests. Giant redwoods and Douglas fir in the old-growth forests of northwestern United States and Canada can grow over 120 m in height.

The high-quality wood and huge size of individual trees make these forests a valuable timber resource (Figure 5.7). Old-growth forests contain trees of all ages, but it is the oldest trees that are often the most valuable for wildlife. Hollow trunks provide nesting holes for rare bird species such as the Northern Spotted Owl, and the dimly lit forest floor is rich in ferns, mosses, and seedling trees. The ancient trees are also a valued timber resource, but their felling can lead to a degradation both of local wildlife and soils. In the Pacific Northwest, logging has greatly reduced the

Figure 5.7 **A stand of old-growth redwoods.**

area of old-growth timber. The supply of ancient trees on private lands has been virtually exhausted. Substantial holdings of old-growth forest timber still exist on federal lands. Considerable tensions have developed between conservationists and the timber industry over the appropriate use of the remaining old-growth stands. Rivers running through these forests are important spawning grounds for salmon and support salmon fisheries.

The deciduous temperate forest biome of Europe, the eastern United States, and some parts of Asia has hot summers and cool winters, though temperatures seldom drop below -12°C. Rainfall is fairly plentiful, 75–200 cm per year. The growing season is long, the forest soils are generally fertile, and species diversity is moderate. The typical trees of these forests are broad-leaved deciduous species, such as oaks, beech, and maple. Each year a deciduous tree produces big, new, healthy leaves for photosynthesis. This production entails a high investment of energy to make each fresh batch of leaves, but so many of the temperate forest trees adopt this strategy it must be energy efficient. The leaves are not kept through the winter, because it would take an even greater energy investment to make them frost resistant. Having water in the tissue of a plant turn to ice is a huge physiological hurdle to overcome. The expanding ice would fatally rupture the cell walls of most plants. With a trunk insulated by a corky bark, trees can withstand freezing. Or they can simply shut down, effectively draining the free water out of their vascular system and allowing those vessels to die. Early in the spring, a new set of vessels is formed from top to bottom of the plant. If you cut through a tree of the temperate or boreal forests, you can see the growth rings (Figure 5.8). These are the vessels built each spring and abandoned in the fall; only the outermost ones are in current use. Thus a deciduous tree must put on a prodigious growth spurt in the spring to build a new vascular system and new leaves. This investment of energy is repaid by having large, flat, efficient solar collectors.

The time when leaves are being built by the trees in a temperate forest is a time of high light and nutrient availability for ground-dwelling herbs. Spring ephemeral herbs take advantage of this abundance and produce a splash of early color as anemone, bluebell, trillium, and wild daffodil rush through their cycle of growth and flowering before being shaded by the emerging tree leaves. Even when the trees are fully leaved, a lot of light leaks through, allowing an understory of shade-tolerant holly, dogwood, and ivy to survive.

The soils of the temperate forests often make good farmland, and the timber of the trees provides hard-

Figure 5.8 Tree rings. The number of rings provides an age for the tree and the width of each ring provides evidence of the growth rate of the tree during that growing season.

wood ideal for building or furniture making. Few mature temperate forests survive in the United States or western Europe, because they have been converted to farmland.

5.4 Prairies and Grasslands

Prairies are climatically similar to temperate forests except that they are drier. A broad-leaved tree is an excellent solar collector, but the leaves would quickly overheat and die if it were not for a plentiful supply of water, which the tree uses as a coolant. Where the summers have too little rain to support broad-leaved trees, a prairie flourishes. Prairies are herb-rich grasslands, the height of which is determined by rainfall; the wetter the area, the taller the prairie. In the driest North American prairies close to the Rockies, with just 40 mm of rain a year, the prairie grasses rarely exceed 0.5 m in height, whereas in the tallgrass prairies of Indiana, which receive 80 cm of rain each year, the grasses and herbs can be more than 2 m high (Figure 5.9).

Figure 5.9 A tallgrass prairie. Fire is an important component of the prairie ecosystem, speeding the cycling of nutrients and inducing bursts of fresh regrowth.

The grasslands of Australia, the African veldt, the Russian steppes, and the prairies of North America are home to herd animals: kangaroos and wallabies in Australia; zebra, antelope, and giraffe in Africa; bison and pronghorn antelope in North America. Grazing is an important ecological factor in these areas.

Heat and aridity become a serious evolutionary factor when water is in short supply, and the following observations apply to dry grasslands and desert. Panting, sweating, or evaporation (including the loss of water through a leaf surface) are particular adaptations to help control temperature, but they all require loss of water. If there is plenty of water either in the soil or at water holes, this cooling system is efficient, but it is deadly in a desert setting. A common adaptation has been to conserve water by dissipating body heat through large radiators such as ears. If heat can be lost in this manner, the amount of panting can be reduced. The ears of the African elephant and the jack rabbit are well supplied with blood that transports heat from the body core to the radiator, whereupon it is lost into space.

The native grazers shaped the landscape, preventing any single plant species from outcompeting others and maintaining high species diversity. But as they are replaced by sheep, cattle, or other domesticated species, grazing patterns and intensities are changed, resulting in major reassortments in the floristic composition of the grasslands. The soils that underlie these grasslands are often extremely rich in well-decomposed organic matter and make very productive farmland. The breadbasket of the world, the midwestern states of the United States and the central Canadian provinces, supports domesticated grasses, such as wheat and corn, on what were once natural grasslands.

Fire can be a natural component of almost any terrestrial ecosystem, but it is especially important in drier biomes such as grasslands, chaparral, and dry woodlands. Lightning strikes are the most common natural causes of wildfires, and some areas will burn every three to five years. Many species, including most species of trees, cannot stand being burned this frequently and will be excluded from this landscape. Grasses, herbs, and quick-growing shrubs will be the main component of a landscape that burns this often. Consequently, fire is critically important to maintaining grasslands. As people move into grassland regions, we attempt to eliminate fires that would threaten homes and farms. In so doing, we may radically alter the fire regime, allow trees to invade the grassland, and establish the risk of infrequent, large uncontrollable fires (Chapter 6).

5.5 Chaparral

The scrublands of the Californian coastal area, Israel, parts of South Africa, and Australia are examples of the **chaparral** biome (Figure 5.10). The climate of these areas is mild and wet during the winter, but the summer is a time of drought. These climates are referred to as Mediterranean, and it is the summer drying that shapes the evolution of plants in this biome. Two fundamental strategies exist to deal with water shortage. The plants can limit the water they lose, or they can gain as much available water as possible. Most plants do both. To limit water loss, the plants of the chaparral have small hard leaves. The hardness is partly due to a waxy outer layer that helps prevent evaporation from the leaf surface. Most of these plants are evergreen shrubs, and this growth form brings several benefits. By being evergreen, the plants can photosynthesize and grow during the moist winter months. Also, by retaining their leaves, they lose no time in starting active growth when the rains start. To gain the maximum amount of water from the soil, many plants have deep roots that tap any remaining water deep in the soil. The vegetation of chaparral often appears to be in patches with an area of bare ground surrounding each

Figure 5.10 Chaparral. The leaves of chaparral plants are often leathery to resist desiccation.

plant. The bare ground is still being used by the plants. Because water and nutrients are in short supply, the plants grow root systems that extend to cover a greater area than the above-ground, leafy portion. Nutrients and water are withdrawn from the bare ground to support the plant. Consequently, other individuals cannot move in to colonize these bare surfaces.

Humans have a long history of settlement in Mediterranean areas and have changed these ecosystems profoundly. It is widely believed that the earliest agriculture was carried out in this climate region (in what is now Iraq). More than 10,000 years of human settlement has resulted in the eradication of top predatory animals such as wolves and lions from most chaparral areas. The native herbivores (plant eaters), such as ground squirrels, deer, and elk, have either developed unnaturally large populations in the absence of predation or have been replaced by domesticated grazers, particularly goats. Goats are indiscriminate browsers that will eat all the leaves from the low shrubs and eventually kill them. The loss of vegetation is a major degradation of local habitats. Overgrazing by goats has reduced many Mediterranean hillsides to bare rock with just the sparsest patches of vegetation.

In California, the cities of Los Angeles and San Francisco lie within the chaparral biome. The loss of chaparral habitats has accelerated as urbanization has sprawled along the Californian coast. The scrublands

did not claim the same place in public esteem as the giant redwoods of northern California, and for many years little attention was paid to this unique habitat. The protection of the chaparral biome really began to gather momentum in the 1980s and is centered around maintaining populations of endangered bird species such as the California Gnatcatcher.

5.6 Desert

Deserts are areas where dry air descends, often around latitudes 20–30°N and 20–30°S, and are too dry to support most life forms. Cloudless skies ensure a merciless noonday sun and cold nights. Few environments offer such a range of temperatures within a 24-hour period. Maximum temperatures can be more than 50°C, and in the same day nighttime temperatures can dip close to or just below freezing.

Desert plants and animals must be obsessive about water retention. Hot days and months, even years without rain, can bake life to a crisp. The only evolutionary paths are either to develop a strategy that minimizes water loss from the body or to develop a metabolic system that can store water and subsequently use it frugally. Animals can follow either route, but most survive by hiding from the full heat of the sun. They lie beneath rocks or in burrows, foraging in the cool of dusk and dawn, or they may be fully nocturnal. Usually, the top predators of deserts are reptiles. Snakes and lizards need less water to survive than their mammalian counterparts. Even so, some mammals can survive in the desert. The camel is the classic example: These "ships of the desert" tank up with water whenever they find it, but between stops they conserve as much as possible. An elaborate nasal structure prevents the loss of water in exhaled air, and the mangy-looking coat is an adaptation to reflecting heat.

Plants cannot move to avoid the full intensity of the sun, but evolution shapes them to minimize the surface area exposed to the strongest light. Thus the saguaro cactus of the southwestern United States stands tall and thin, exposing its flanks to the cool light of early morning and evening, but presenting just the top of the column to the sun at noon. The cactus is then growing in its own shade (Figure 5.11). In a superb example of **convergent evolution** (unrelated organisms in different places evolve similar strategies), plants from different evolutionary lines living in similar desert environments have all reached the same solution to water shortages and hot days (Figure 5.12).

Desert plants have also adapted to retain as much water as possible, so that their stems and leaves ap-

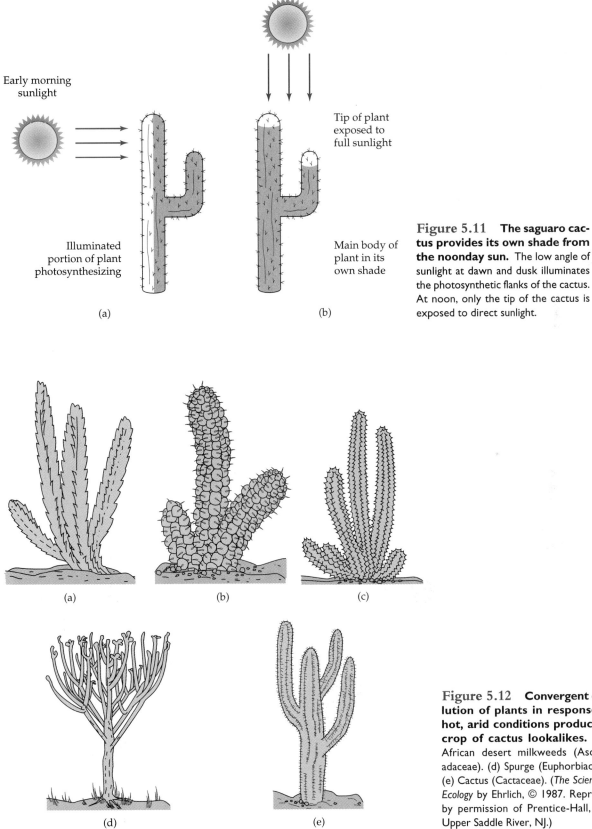

Figure 5.11   The saguaro cactus provides its own shade from the noonday sun. The low angle of sunlight at dawn and dusk illuminates the photosynthetic flanks of the cactus. At noon, only the tip of the cactus is exposed to direct sunlight.

Early morning sunlight

Illuminated portion of plant photosynthesizing

(a)

Tip of plant exposed to full sunlight

Main body of plant in its own shade

(b)

(a)

(b)

(c)

(d)

(e)

Figure 5.12   Convergent evolution of plants in response to hot, arid conditions produces a crop of cactus lookalikes. (a-c) African desert milkweeds (Asclepiadaceae). (d) Spurge (Euphorbiaceae). (e) Cactus (Cactaceae). (*The Science of Ecology* by Ehrlich, © 1987. Reprinted by permission of Prentice-Hall, Inc., Upper Saddle River, NJ.)

Figure 5.13   Evenly spaced desert shrubs. Extensive root networks compete for water and nutrients in the bare ground zones between shrubs. This landscape is "full" and new shrubs cannot colonize the bare ground due to root competition.

pear to be thick and fleshy, while the skin of the leaf is leathery. Water loss cannot be prevented completely, but if that moisture can be retained close to the surface of the leaf, it will act as a buffer against the dry ambient air. Spines on the leaves deter predators, but they also trap pockets of still air around the plant. These desert plants, like the needle-leaved trees of the boreal forest, are creating their own microclimate in which they live.

As found in chaparral, the plants of a desert tend to be widely but fairly evenly spread. A common characteristic of a desert shrub is the very extensive, shallow root network, so that any available moisture and nutrients can be gathered from an area many times the size of the plant canopy (Figure 5.13). The gaps between plants become a competitive zone in which water and nutrients are rapidly absorbed. This intense competition makes it difficult for new plants to colonize the bare ground. Leaves that are shed fall directly beneath the plant, decompose, and release nutrients back to the plant. Because each plant is gathering nutrients from a large area and its leaves will tend to drop and decompose beneath the plant, over time there will be a local accumulation of nutrients beneath the plant. Clearly such an adaptation is going to favor long-lived plants. Growth rates in this arid environment will be slow. The combination of long-lived individuals and slow growth means that recovery from disturbance will be very slow.

Some deserts form as a result of lying within the rainshadow of a mountain. A **rain shadow** effect can be found where the two sides of a mountain have radically different rainfall regimes. A common cause of this is moist air blowing in from the sea being blocked by a mountain chain. The air mass is forced up over the mountain, and as the air rises, it cools and cannot hold the same amount of water vapor; the water it cannot hold falls as rain (Figure 5.14). This seaward side of the mountain is likely to be cloudy and very wet. The vegetation growing there will be lush in compar-

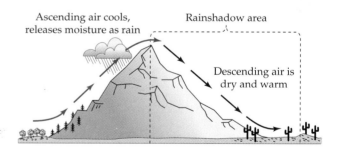

Figure 5.14   The contrast in rainfall and vegetation between the coastal and rainshadow side of a mountain. Air is forced upward by the mountain, cools, and can no longer hold as much water vapor. It rains. When the air crosses to the other side of the mountains, it is cold and dry, produces no rain, and results in a desert in the rainshadow area. (After McKnight, Tom, *Physical Geography: A Landscape Appreciation.* 5th ed., 1996. Adapted by permission of Prentice-Hall, Inc., Upper Saddle River, NJ.)

ison with the landward side of the mountain. Once the air has passed over the mountain it descends and is warmed by the land. Warm, dry air does not provide rain, so on the inland side of the mountain is a desert. The difference between the moist climate of San Francisco and the desert climate of Nevada is an example of a rainshadow effect.

5.7 Tropical Forests

The key feature of **tropical** climates is that, at a given place, the temperature of the air, the soil, or the water is remarkably constant over long periods of time. From one month to the next the average daily temperature varies by only ±2°C. In contrast, the change in temperature from night to day is large. For example, the Amazon lowlands might be as warm as 34°C in the afternoon and cool to 20°C at night, a range of ±7°C about a mean of 27°C. This observation of no seasonal temperature change, but a large daily range of temperature, led the great German explorer Alexander Von Humboldt (1769–1859) to say that "the night is the winter of the tropics." The annual rainfall may vary from 2000 mm to more than 15,000 mm. There is no winter or summer, but the rainfall can be concentrated into a wet season followed by a dry season.

The length and intensity of the dry season determine the character of the forest. In areas with a six- or eight-month dry season, the tropical dry forests, the vegetation is scrublands of deep-rooted deciduous trees and shrubs. The deciduous habit of these plants is not to escape the effects of cold as in temperate forests, but rather to prevent overheating when water is comparatively scarce. Semideciduous tropical forests (Figure 5.15) grow in areas with a dry season lasting three to five months. In these forests, some species retain their leaves; others are deciduous for the driest months. In these strongly seasonal forests, the peak flowering period tends to be at the start of the wet season. The female flowers, once fertilized, develop their fruit, and fruit production generally follows a few months later. One consequence of this timing is that there is a glut of fruit in the mid- and late wet season and a scarcity of fruit in the dry season (Figure 5.16).

Wet tropical forests, often called tropical rain forests (Figure 5.17), are evergreen communities. Water is abundant, with between 2300 mm and 20,000 mm of rain per year and a dry season of three months or less. Evergreen leaves function as solar panels all year, so rates of growth and net primary productivity are both higher than in temperate forests. The leaves are generally large compared with those found in temperate

Figure 5.15 **A semideciduous tropical forest.** Many, but not all, of the trees in this forest will drop their leaves during the long dry season.

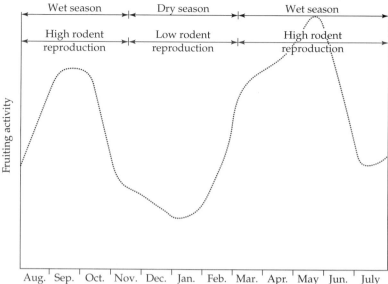

Figure 5.16 **Fruiting activity of trees on Barro Colorado Island, Panama.** Lack of fruiting activity in trees can bring food shortages to fruit eaters in the dry season. Reduced food supply leads to decreased reproductive activity among rodents. (Data are from E.G. Leigh Jr., A.S. Rand, and D. Windsor, 1982.)

forests and have a pointed, down-curved tip. The cause for the repeated evolution of this leaf shape in the tropics is still debated. One explanation is that the "driptip" helps water to drain from the leaf and consequently reduces the growth of light-blocking mosses and algae on the leaf surface.

Another characteristic of tropical rain forests is that they contain a huge species diversity. Perhaps 50% to 75% of all organisms living on Earth inhabit these forests. A temperate forest has about 20 to 30 tree species per hectare (1 hectare = 2.47 acres), whereas a tropical forest can support more than 350 to 450 tree species in the same area. Just to fit 350 tree species into a hectare means that each species is represented only

once or twice. As a result, almost every plant in the tropical forest is a rare plant. In contrast, there are few rare species in the forests of the temperate zone, where three or four species account for almost all the individuals.

The tropical rain forests are the most productive of all terrestrial ecosystems; that is, they have a higher NPP than any other system. High temperatures, moisture, and uninterrupted growing seasons all contribute to the teeming productivity of these habitats. **Decomposers**, the organisms that break down fallen leaves and recycle the nutrients essential to maintaining high rates of productivity, seem to work overtime in these forests. Leaves are often being decomposed by fungi

Figure 5.17 **A wet tropical forest.** Rapid nutrient cycling, high rainfall, and year-round warmth make this the most productive terrestrial ecosystem.

and bacteria while still attached to the tree. When a leaf falls, the dark, warm, moist conditions of the forest floor are ideal for rapid decay. The cycle of decomposition is so rapid that nutrients are recycled from the leaf before a year has passed.

The shallow roots of the trees of these forests ensure a rapid uptake of nutrients. Consequently, the nutrients are not retained in the soil but are reincorporated into the trees. It is a paradox that the soils of these immensely productive forests may make poor farmland (see Chapter 19). In some areas of the tropics, especially on wetlands and richer volcanic soils, the land has supported harvests for millennia, demonstrating that a sustainable extractive use of tropical forests is possible. Changes in the patterns of land use, poor management, and intensification of agriculture are frequently leading to erosion and loss of soil quality. The expansion of agriculture and urbanization into former forest areas is a major threat to biodiversity in tropical settings.

5.8 Tropical Mountains

It takes just five hours to drive from the Amazonian lowlands up to the ridge of the Andes. On the way, the steep road with hairpin curves passes through many different habitat types, passing some so fast that you might not notice them. The giant trees of the steamy **lowland** forests form a canopy that encloses the road. As the road starts to climb through the foothills, a change in the forest can be seen as the tall lowland forests with their three canopy layers give way to a smaller **premontane** forest with just two layers. This premontane forest extends from about 800 m elevation to about 1200 m and grades almost imperceptibly into the montane forests.

As we climb to 1800 m elevation, clouds swirl like fog outside the window of the vehicle, and the **montane** forest is obscured. Air rising from the plains below cools, and water condenses to form a ground-level cloud on the mountainside. To see anything here you have to hop out of the car and explore. The trees are perhaps half the size of those in the lowlands, and their branches are festooned with mosses, orchids, and bromeliads. Just brushing against the trailing skeins of moss that dangle from the branches is enough to saturate your shirt. Everything here is wringing wet. After half an hour of driving in whiteness, we reach about 2500 m elevation; here we emerge above the clouds. The air is colder still, and the dense forest has been replaced by shrubs and a few small trees. This is the **subpáramo** of the Andes. The road continues to

climb steeply, eventually breasting a pass that lies above the limit of where trees can grow. Here, at 4000 m elevation, the air is getting noticeably thinner, and the vegetation is a grassland with some tall, tough-looking herbs poking out of it. This is the Andean **páramo**, a terrain rounded by glaciers and dotted with the relics of the Incan empire. Going higher still, we will come upon the boulder-strewn forelands and ice caps that sit atop the highest Andean volcanoes.

This journey can be replicated (more or less) on any tropical mountain. The tropics have huge mountain systems and peaks—for example, the Andes, Kilimanjaro (Kenya), and Mount Kinabalu (Malaysia)—and their slopes gradually change from tropical lowlands to alpine meadows and even glaciers. The *latitudinal* gradient (gradual or incremental change) of tropical forest, temperate forest, boreal forest, and tundra shown in Figure 5.4, is reproduced *altitudinally* as one ascends the mountain (Figure 5.18).

5.9 Oceans

Open **oceans** have been described as blue deserts, but nutrient concentrations rather than water are the limiting resource. On the whole, it is true that life does not teem in the oceans; if it did, oceans would look green with chlorophyll. The oceans look blue because light penetrating deep into the water is scattered and then reflected back up to the surface, where it emerges as blue light. Blue light would be very effectively absorbed by chlorophyll, thus the fact that we see blue is a clear sign that photosynthetic organisms are relatively scarce in the sea (Figure 5.3b).

Salt water is toxic to many land plants, but this fact does not explain the lack of ocean productivity; after all, life evolved in seawater. The reason for the lack of plant life is the scarcity of essential fertilizers, such as nitrate and phosphorus, in marine systems. Terrestrial plants can root and take advantage of the recycling of the dead, drawing their nutrients from the soil. In the ocean, the corpses sink into the deeps, where there is no prospect of a plant's rooting to the oceanbed and still having access to sunlight for photosynthesis.

Productivity is not uniformly poor in the oceans. Indeed, some areas where nutrients are stirred from the bed of the ocean to the surface waters can be highly productive. These areas are termed upwellings and are essential to species ranging from phytoplankton to whales (Chapter 7). The very shallow margins of the sea, around reefs and in the in-

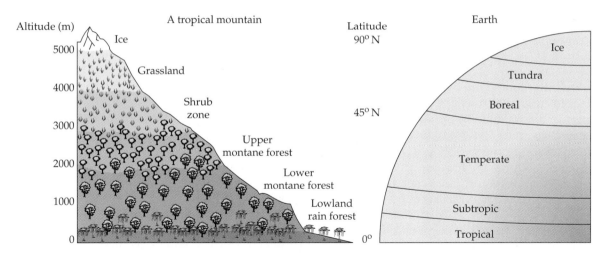

Figure 5.18   **The zonation of biomes on a tropical mountain reproduces the pattern of biomes between equator and pole.** Temperature decreases upslope, reproducing a similar environmental gradient to the one experienced by moving poleward. Thus, the lowland rain forest (tropical) is replaced by a cool upland (upper montane) forest equivalent to temperate forest. The "alpine" and high-elevation grasslands of tropical mountains resemble steppe or tundra grasslands near the Arctic region. (After Christopherson, Robert, *Elemental Geosystems.* © 1995. Reprinted by permission of Prentice-Hall, Inc., Upper Saddle River, NJ.)

tertidal zone (the area between the high-tide and low-tide lines) can also be highly productive. In the near-shore environments, algae are the primary producers, but here it is possible for them to root to rocks or sand, and they may grow to almost resemble trees. An example is the giant kelp beds of California, in which individuals of these attached algae can be 3 m or more long.

Those beautiful blue tropical seas are among the least productive of all oceans, and yet they support the luxuriant growth of coral reefs that are highly productive. This apparent paradox is resolved when it is realized that the reef does not share its nutrients with the sea. The reef organisms scavenge from the water column all the available nutrients, whether in the form of dissolved chemicals or phytoplankton. Once absorbed by the reef organisms, these nutrients stay within the reef. The reef organisms continually recycle the available nutrients, requiring very little input of fresh nutrients from the outside.

5.10   Estuaries

Estuaries are tidal water bodies located where a river enters the sea. The salinity of water in an estuary is intermediate between that of freshwater and seawater. The influx of organic material being washed down-stream by the river provides a rich supply of nutrients.

As the river water runs into a wall of seawater, it loses energy and deposits its muddy load. Estuaries are muddy, smelly places where decomposing, once-

terrestrial organisms become food for a mass of invertebrates. Primary productivity in these systems is high. Although grasses and trees can be important in estuaries (particularly in tropical mangroves), algae are frequently the major primary producers. In many estuaries, low tide reveals a wide expanse of mud that does not at first glance appear to support much life. However, the mud often has a coating of algae that are primary producers, and below the surface a vast population of decomposers recycles nutrients. These decomposers form the base of a rich food chain in which the most visible predators are the long-billed shorebirds patrolling the mudflats at low tide (Figure 5.19). The shorebirds probe the mud in search of mollusks and worms, while ducks, geese, and swans sift small organisms from the mud surface or feed on the shoots of grasses and sedges that poke up through the mud. Many of these waterbirds are the same individuals that migrate north to the tundra. Consequently, the estuaries of the midlatitudes are rich feeding grounds and biologically important, as either overwintering sites, migratory stop-offs, or nesting grounds.

In some estuaries, the rich mud also supports dense growths of saltmarsh vegetation that can survive inundation by seawater at high tide. These plants must be highly tolerant of salt, because often they have to endure concentrations of salt greater than that of seawater. As water evaporates from shallow pools and the mud surface, the salts are left behind. The remaining water around the plant roots becomes an in-

Figure 5.19 Mudflats exposed at low tide provide a rich feeding ground for shore-birds. The primary productivity of mudflats can rival that of any biome.

creasingly concentrated salt solution. The biggest threat to these plants is the danger that, when the outside salinity increases, either salt will flow into the plant or that water will be drawn out of their bodies. Therefore, the plants that can survive these conditions commonly show adaptations that conserve water—such as leathery leaves or a large stock of water in the leaf itself—or that rid the plants of salt. Some plants exude salt through their leaves; others can block the intake of salt through their roots.

In tropical and subtropical estuaries, the mudflats support salt-tolerant trees. These plant communities are called **mangroves**. Mangroves, like other estuarine habitats, are very important breeding grounds for wildlife, especially shrimp and fish.

Many fish that are near-shore (as opposed to deep-ocean) species, such as snapper, menhaden, spot, croaker, and flounder, need estuarine habitats for part of their reproductive cycle. These fish require the rich food resources of the estuary or the cover provided by plant stems either during spawning or as young fish. Similarly, the estuaries are essential for species such as salmon, sea trout, and shad, which migrate from open ocean to spawn in freshwater.

Estuaries are economically important as harbors and fisheries and, more recently, for the development of commercial shrimp farms. Estuarine habitats are also under pressure from boaters and aquatic sport enthusiasts.

Summary

- Biomes are defined by climatic conditions and characteristic vegetation.
- Biomes exist within environmental gradients that are reflected in the adaptations of the organisms.
- Plants and animals within a biome display adaptations to the prevailing climate.
- Rates of decomposition and nutrient cycling are highest under warm, wet conditions, so that trop-ical rainforests have the highest net primary productivity and cold deserts the lowest.
- Plant growth and productivity are highest under warm, wet conditions.
- Going up a mountain reproduces a poleward environmental gradient.
- Open oceans are relatively unproductive, while estuaries generally have high NPP.

Further Readings

Barbour, M. G, and W. D. Billings, eds. 1998. *North American Terrestrial Vegetation*. 2nd ed. Cambridge: Cambridge University Press.

Field, C. B., M. J. Behrenfield, J. T. Randerson, and P. Falkowski. 1998. Primary production of the biosphere: Integrating terrestrial and oceanic components. *Science* 281:237–240.

Frank, D. A., S. J. McNaughton, and B. F. Tracy. 1998. The ecology of the Earth's grazing ecosystems. *BioScience* 48:513–522.

Horn, H. S. 1971. *Adaptive Geometry of Trees*. Princeton, NJ: Princeton University Press.

Kricher, J. C. 1997. *A Neotropical Companion*. 2nd ed. Princeton, NJ: Princeton University Press.

Schmidt-Neilsen, K. 1979. *Animal Physiology: Adaptation and Environment*. New York: Cambridge University Press.

Web Connections

On-line resources for this chapter are on the World Wide Web at:
http://www.prenhall.com/bush
(From this web-site, select Chapter 5.)

Ecosystems, Nutrient Cycles, and Soil

All organisms, be they bacterium, robin, or spruce tree, are linked by ecological processes. Animals are dependent, either directly or indirectly, upon photosynthetic organisms to provide them with energy. Plants need sunlight, water, carbon dioxide (CO_2), and nutrients if they are to grow. The availability of nutrients in the soil can be affected by temperature, rainfall, floods, fire, and the kind of rock beneath the soil. Hence, a variation in rainfall can affect the availability of essential nutrient chemicals for plants, which in turn affects the animals that feed upon them. This is not a one-way street leading from soil to animals. Animals and microorganisms living in the soil actively contribute to the formation and maintenance of the soil. Living organisms interact with each other (**biotic** processes) and with nonbiological (**abiotic**) processes, such as temperature, rainfall, and soil. These interactions are critical to defining which organisms can live in a given area. The term **ecosystem** is used to include all relationships between organisms of a given area and their interactions with the physical (abiotic) environment.

This chapter will deal with the link between abiotic processes of ecosystems and the biological world. The interactions of the plants and animals that form the biological component of an ecosystem are dealt with in Chapters 9 through 13.

6.1 How Large Is an Ecosystem?

Earth is made up of ecosystems that range in size from the vast to the tiny. An immense ecosystem, for example, is formed by the Mississippi River and its tributaries, which drain most of the eastern United States and the Midwest. On a smaller scale, the area drained by a stream forms an ecosystem, and within that system we may find a woodlot that forms another identifiable ecosystem. Each tree could support another suite of ecosystems, with colonies of mosses and lichens forming miniature forests for microscopic organisms.

An ecosystem can be as small as the gut of a flea, which is home to all the microbes that live there. Some of the microbes are flea pathogens, some are bacteria that form the digestive gut flora of the flea; all are bathed in the digestive juices of that system. Ecosystems, therefore, can exist at many scales and can be nested one within the other (Figure 6.1).

Nested Ecosystems

An example of a suite of nested ecosystems would be the flea gut ecosystem fitting inside the woodlot ecosystem (because the flea is living on a squirrel that inhabits the woodlot). The woodlot fitting into the stream watershed ecosystem in turn forms part of the Mississippi ecosystem.

The Earth ecosystem, or biosphere

An island ecosystem

A watershed ecosystem

Lake

A tree as an ecosystem

Bird and nest

Leaves

A squirrel as an ecosystem

Decomposing leaves

Worms

Fur:
Fleas, lice, and mites

Stomach and digestive tract:
Bacteria (gut flora)
Parasites (intestinal worms)

Figure 6.1 **Ecosystems can be nested one inside another.** The microbial flora of the squirrel's gut forms an ecosystem lying within the larger ecosystems formed by the squirrel, the woodlot, the stream watershed, the island, and Earth.

If the Mississippi ecosystem as a whole is damaged, clearly the smaller ecosystems within it will be affected. Damage to small nested ecosystems does not necessarily lead to an ecological collapse of the larger one. However, damage to the smallest ecosystems does have a cumulative effect on the largest ones. If enough small ecosystems are damaged, the larger ecosystem will be destabilized. For instance, the Mississippi valley was once a nearly continuous sequence of wetland habitats. Because wetlands absorb and retain peak flows of water, the abundance of wetlands in the Mississippi drainage helped prevent flooding. The loss of a single wetland had no effect on the entire Mississippi drainage, but as more and more of the wetlands were drained, the larger system became unsta-

ble. As wetlands were lost from the ecosystem, severe floods became increasingly probable. Over the past century, 90% of the wetlands along the Mississippi were drained, and the capacity of that ecosystem to absorb heavy rainfalls has been lost. The consequence of the loss of wetlands combined with upwarping of the delta was evident in 1992 when prolonged rains resulted in extensive flooding along the Mississippi.

Sometimes all the ecosystems of Earth are referred to as the **biosphere** and are thought of as a single self-contained ecosystem. (Note that this concept does not imply that Earth is a "superorganism," for it is not alive, nor does it "attempt" to regulate atmospheric composition, temperature, or any other process.) The concept of the biosphere is valuable because it suggests that actions on any scale anywhere on the planet could have a cumulative effect upon the whole system.

Although space is not an ecosystem because there is no known life there, random and regular events in space periodically drive our ecology here on Earth. Meteorites smashing into Earth are random events that can change the course of evolution (Figure 6.2). Finding the telltale hole in the ground left by an ancient impact is often impossible, but it is strongly suspected that such impacts caused mass extinction events about 440 million years ago, 240 million years ago, and 65 million years ago. (The last was the impact that probably eliminated the dinosaurs.) It is not just irregular, catastrophic events in space that affect Earth's ecosystems. The regular, predictable variations in the orbit of Earth around the sun are believed to drive the rhythm of the ice ages (Chapter 13). On a shorter time scale, sunspot cycles appear to have an underlying pattern, and these patterns seem to be well correlated with changes in the energy output of the sun.

Finding the Edge of an Ecosystem

Dividing the natural world into ecosystems is very useful as a way to think about ecology, but it is nevertheless an artificial construct. We define the characteristics of each ecosystem and then try to locate them in space. However, we will almost always find that ecosystems do not have hard, identifiable edges. Standing on the sand of the beach you cannot say exactly where the sea starts. Tides, waves, and changes in the shape of the beach as sand is piled up or washed away make it impossible to draw a line in the sand and say "this is where the ocean ecosystem begins." Most ecosystems have similarly blurred edges, and so the division between one ecosystem and another is

Figure 6.2   The meteorite impact crater near Winslow, Arizona. A larger meteor that hit Chicxulub, Mexico 65 million years ago may have caused the extinction of the dinosaurs.

usually seen as an intermediate area or **ecotone**. An ecotone may be very narrow, just a few meters, or it might be many kilometers in extent, gradually becoming more and more similar to the nearest recognizable ecosystem.

6.2   Getting to the Root of Productivity

All ecosystems require an input of energy, but they also need chemical inputs of water, carbon, and nutrients. A critical difference exists between the input of energy and that of chemicals. Energy use is a one-way process. As energy is used or stored, it changes state and degrades in quality (the second law of thermodynamics; Chapter 3). Unlike energy, nutrients, carbon, and water can be used, recycled, and reused time and again. These chemicals can be stored, transformed into different chemical configurations, and then returned to their initial state. The recycling process, known as a **nutrient cycle**, can have a number of stages, and at its simplest would be:

Nutrient absorbed by plant and incorporated into leaf

Leaf dies and falls

Leaf decomposed by microorganisms

Nutrients released into soil

Nutrients absorbed by plant

Rapid nutrient cycling leads to increased nutrient availability and promotes high growth rates of photosynthetic organisms. The speed at which nutrient, carbon, or water cycles run is primarily determined by abiotic factors, such as temperature and humidity. A simple rule of thumb is that rates of chemical reactions double for each 10°C increase in temperature. Thus, decomposition would be expected to be fastest under warm conditions. Similarly, moisture increases rates of decomposition. Warm moist environments, consequently, experience the fastest nutrient cycling.

Biotic factors also play a role in decomposition. The death of an organism is not the end of its relationship to other organisms, for in its death it becomes the provider of nutrients. The body, be it oak leaf or squirrel, becomes the scene of intense activity as organisms vie for their share of the newly available nutrients and carbon to be recycled. In fact, the number of organisms living in, and feeding on, a dead tree is about the same as on a living one. Certainly, the number of organisms living in the soil decomposing organic material can easily outnumber the area's more obvious wildlife. A single acre (0.4 hectares) may contain more than a million earthworms, and a gram of soil is likely to contain more than 3 trillion microorganisms. Fungi, molds, beetles, worms, and bacteria play a vital role as decomposers in our ecosystems. Because the rate of nutrient cycling greatly affects the growth rates of plants, decomposers are a vital component of ecosystems; they are the unsung heroes that maintain the productivity of our land.

The Carbon Cycle

The piece of bread that you eat today could contain an atom of carbon that was once part of a dinosaur. Carbon is the basic molecule in all organic chemicals, such as amino acids, proteins, DNA, and cellulose, and it is present in the atmosphere as carbon dioxide (CO_2). This rare gas, making up just 0.03% of the atmosphere, is the origin of all the carbon used to build plant and animal bodies. Plants absorb carbon through their leaves in the form of carbon dioxide (a tiny amount of carbon may also be absorbed through the roots as part of complex molecules). The absorbed carbon dioxide is converted to carbohydrate through photosynthesis. Animals can neither convert atmospheric carbon dioxide to body tissue nor concentrate carbon in their bodies.

If a plant is dried and all the water removed, about 45% of its body weight is carbon. In other words, the plant has concentrated the atmospheric carbon (0.03%) by a factor of about 1500. After the initial concentration of atmospheric carbon by plants, however, there is no further carbon concentration in the rest of the food chain. Hence, animals are composed of approximately the same proportion of carbon as plants. The dry weight of carbon in a human is about 48% (a typical value for other animals, too), effectively the same concentration as in plants. The riches of carbon stored in the bodies of plants and animals make them nutritious in both life and death.

The carbon cycle (Figure 6.3) maintains atmospheric concentrations of carbon dioxide through the continuous uptake of carbon dioxide by plants and their respiration of stored carbohydrate. The decomposition cycle also is extremely important in the carbon cycle. Decomposers feed on dead organic matter and respire the carbon of the corpse. The chemical reaction of respiration calls for oxygen to be used as carbon dioxide is released. Thus the release of carbon dioxide from decomposers in the soil is called **soil respiration**. This carbon dioxide is released into the atmosphere, where it may once again be used in photosynthesis.

An alternate fate for a plant body part is to be eaten by an herbivore. The herbivore either releases carbon dioxide via respiration, dies, and is respired by decomposers, or becomes food for a predator. And so the cycle goes on. It is the ultimate fate of every carbon

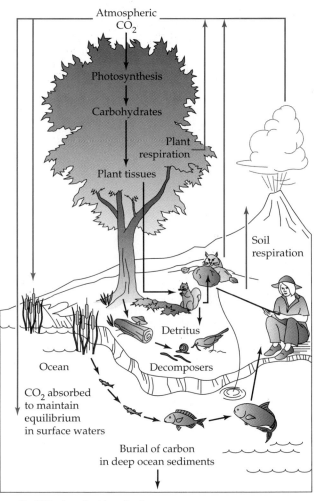

Figure 6.3 The carbon cycle. Carbon from the atmosphere is absorbed as carbon dioxide (CO_2) through the leaves of plants and converted to organic material through photosynthesis. Eventually all organic material is respired and returned to the atmosphere as CO_2. In a natural system, CO_2 is also added to the atmosphere by volcanic activity, but these additions are matched by absorption of CO_2 by the oceans. The proportion of CO_2 remains relatively stable in the atmosphere until the climate changes or humans start to pollute the system.

molecule in our bodies to be respired and returned to the atmosphere as carbon dioxide, unless a dead body is fossilized.

Each year there is a rhythmic pulse of slightly higher and lower atmospheric concentrations of carbon dioxide in phase with the seasons (Figure 6.4). In the winter of the Northern Hemisphere, because so many trees lose their leaves and stop photosynthesizing, there is a reduction in the biological demand for carbon dioxide. The growth of new leaves and twigs

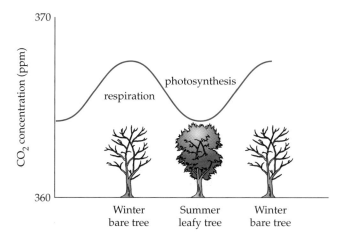

Figure 6.4 **Seasonal fluctuations in concentrations of atmospheric carbon dioxide in the Northern Hemisphere.** Carbon dioxide concentrations decrease because of photosynthesis in summer and increase as a result of soil respiration and decomposition in winter.

in spring prompts increased rates of photosynthesis. The trees increase their demand for carbon dioxide, causing a slight drop in the concentration of this gas during the summer months. Within our present climatic conditions, the natural carbon cycle appears to be a stable system that does not lead to a long-term change in carbon dioxide either up or down. However, on longer time scales, atmospheric carbon dioxide concentrations can be variable, and in the past century they have risen sharply as a result of the burning of fossil fuels and forests (Chapter 23).

A Generalized Nutrient Cycle

Sulfur and nitrogen come primarily from the atmosphere. The initial source of all other nutrients is weathered bedrock. **Weathering** is the decay of rock brought on by the concerted action of biological, physical, and chemical processes. As the rock decays, the nutrient chemicals dissolve and enter the soil. A soil nutrient is drawn up through the roots into the plant (Figure 6.5), where it is incorporated into the body structure. In general, the nutrient is released only when the plant, or body part, dies. Dead plant and animal material and animal wastes are called **detritus**.

A dead leaf or a dead animal can be thought of as a small package of organic material and nutrients that has fallen back to the forest floor. Decomposers, fungi and bacteria, feed on detritus. Many chemicals essential to plant growth, such as potassium, phosphorus, calcium, manganese, magnesium, and boron, are recycled through the process of decomposition. Before

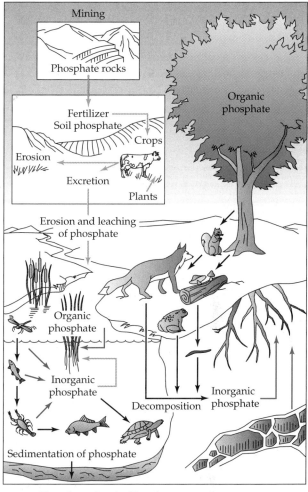

⟶ Phosphate absorbed by plants.

⟶ Movement of organic phosphate through food chains.

⟶ Release of inorganic phosphate into environment.

Figure 6.5 **A generalized nutrient cycle.** In this case the cycling of phosphate (PO_4) is documented. Cycles for many other nutrients would be similar. The source of the nutrient is rock, either naturally weathered or mined. Terrestrial plants absorb the nutrient through their roots. The nutrient is recycled from fallen parts or from food chains by decomposers. Some of the nutrient will be lost to the terrestrial system through erosion and will enter the aquatic system, where it fertilizes aquatic food chains.

decomposers can gain the nutritional rewards stored in the leaf, they must break open the cells of the plant and digest the cell walls; this is the process of decay. Some of the nutrients are absorbed by the decomposers and others leak into the soil. Here, they go into solution in the soil water and again become available for uptake by plants. Through the excrement and death of the decomposers, the remainder of the nutrients are eventually returned to the soil and dissolved in the soil water.

Many plants have **mycorrhizal fungi** that live in their roots. The plant and the fungus help each other, thus the relationship is said to be **mutualistic**. The fungi speed the uptake of nutrients by the plant, helping to provide it with a concentrated nutrient input that allows faster growth and increased photosynthesis. The fungus benefits by tapping into the vascular system of the plant, from which it harvests a fraction of the products of photosynthesis.

Phosphorus follows the pattern of this general nutrient cycle and provides an example of how a mineral element becomes an integral part of an organism. Phosphorus is a common element in many rocks, especially those that formed as sedimentary deposits in ancient oceans. In the millions of years since the phosphate was deposited, the rocks have been heaved upward to form part of the terrestrial landscape. Weathering releases phosphorus from the rock and, dissolved in the soil water, it becomes available to plants as phosphate (PO_4).

Within the plant, the phosphate becomes an essential component of many complex chemicals such as DNA (deoxyribonucleic acid), RNA (ribonucleic acid),

and ATP (adenosine triphosphate). When the plant is eaten by an herbivore, the plant chemicals are digested and reconfigured by the animal as it forms its own DNA and other phosphorus-requiring chemicals.

The Nitrogen Cycle

Nitrogen constitutes almost 79% of our atmosphere and is an essential nutrient for plant growth. However, nitrogen is an unreactive gas that cannot be used by plants until it has been converted, or **fixed**, to a different chemical state, namely nitrate (NO_3) or ammonium (NH_4). Thus, despite the rich store of nitrogen in the air, there is no direct atmospheric input of nitrogen to plants (Figure 6.6). Nitrogen can be fixed during storms, when lightning provides the energy to make the conversion. Consequently, rainwater contains some fixed nitrogen and so is a dilute fertilizer. Nitrogen fixation might also be accomplished by cyanobacteria that live in the soil or water or by a bacterium, *Rhizobium*, which forms nodules on the roots of plants (Figure 6.7). These bacteria are called **nitrogen-fixing bacteria**.

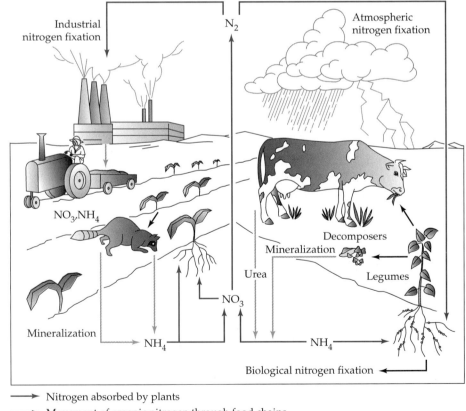

Figure 6.6 The nitrogen cycle. Nitrogen from the atmosphere (N_2) cannot be absorbed by plants and animals. It must first be fixed (converted to a different chemical state) by lightning or bacterial activity. The conversion of nitrogen to nitrate (NO_3) or ammonium (NH_4) provides an essential nutrient for plants.

→ Nitrogen absorbed by plants

→ Movement of organic nitrogen through food chains

→ Release of inorganic nitrogen into environment

Figure 6.7   Root nodules formed by *Rhizobium* a nitrogen-fixing bacterium.

Although nitrogen fixation is the real starting point of the nitrogen cycle, its importance in terms of nutrient flow is dwarfed by the amount of nitrate reabsorbed from dead organisms. When a plant or animal dies, the body is subject to **mineralization**. Mineralization is the process of breaking down protein and converting it to ammonia (NH_3) and then to ammonium (NH_4). Another source of ammonium for plants comes from the mineralization of the excrement and urine of animals. Urea can be mineralized as follows:

$$N_2H_4CO + H_2O \rightarrow 2NH_3 + CO_2$$
$$\text{urea}$$

and then

$$NH_3 + H_2O \rightarrow NH_4^+ + OH^-$$
$$\text{ammonia} \qquad \text{ammonium}$$

The ammonium ion can be absorbed by plants, providing them with nitrogen. Another reaction is carried out by the microorganisms *Nitrosomonas* and *Nitrobacter* that sequentially decay ammonium to produce nitrate (NO_3), which is also absorbed readily by plants. It is this recycling of proteins held in plant and animal tissues that accounts for most of the nitrogen moving through the nitrogen cycle.

It is important to bear in mind that the bacterial and fungal decomposers engaged in cycling nitrogen are not altruists. They have evolved to take maximum advantage of a resource so as to promote their own reproductive fitness. They are not intent on completing nutrient cycles.

How Long Does it Take a Leaf to Rot?

In general, net primary productivity (NPP) is closely related to the speed of nutrient cycling. Tracking the decay of a leaf and the cycling rate of nutrients provides an indicator of biome productivity (Table 6.1). Note that the difference in net primary productivity between the biomes is roughly correlated with nutrient cycling time. The biomes in Table 6.1 are listed from coldest to warmest, and the times represent the length of the decay cycle: the time it takes for a leaf to fall from a tree, decompose, and have its released nutrients taken up by a plant to form a new leaf. These values, therefore, are estimates of the average residence time of a nutrient in the soil. The values vary little from one nutrient to another, although potassium and calcium seem to cycle a little faster than the others. The strongest differences are from one biome to another. The average recycling time for organic material in a wet tropical system is just four or five months, whereas in a boreal forest it is about three and a half centuries. In other words, the leaves that fell in the boreal forest in the winter of 1650 are now just completing their decomposition. Cold winters and short summers slow decomposition in the boreal forests.

Although climate sets the general pattern for decomposition times, not all leaves decay at the same pace. If the leaf is relatively rich in nitrogen, it may be preferentially attacked by decomposers because it offers a relatively rich nutritional reward. The presence of large amounts of lignin, the leathery supporting tissue of leaves, slows decomposition because it is a relatively poor source of food for the decomposers. Pine, spruce, and fir needles are rich in lignin to help withstand the freezing conditions of the boreal forest. In contrast, leaves of beech and maple are relatively poor in lignin because they are dropped in the fall, and the plant does not need to invest as much energy in a strong leaf. Leaves with a low ratio of lignin to nitrogen (little lignin and lots of nitrogen), decompose faster than leaves with a high ratio (Figure 6.8). The uppermost point of the figure indicates that almost 120% of the leaf was decomposed in just one year. Clearly, more than 100% of something cannot be decayed, so this is a way of showing that the entire leaf had been recycled in less than a year. Another relationship that emerges from this figure is that the leaves in New Hampshire take longer to decompose (their weight loss per year is lower at every ratio) than those from North Carolina. New Hampshire has much longer and colder winters than North Carolina, rein-

Table 6.1 The mean residence time of organic matter and nutrients and the net primary productivity (NPP) of four biomes.

	Mean residence time (in years)*						NPP (g C/m²/yr)
Biome	Organic matter	Nitrogen	Phosphorus	Potassium	Calcium	Magnesium	
Boreal forest	353	230	324	94	149	455	360
Temperate forest	4	5.5	5.8	1.3	3.0	3.4	540
Chaparral	3.8	4.2	3.6	1.4	5.0	2.8	270
Tropical rain forest	0.4	2.0	1.6	0.7	1.5	1.1	900

*Mean residence time is the time for one cycle of decomposition.
(*Source:* W. H. Schlesinger, 1991.)

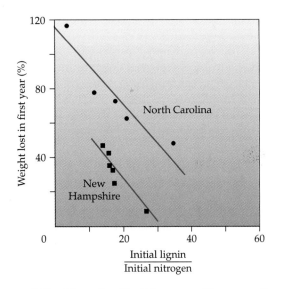

Figure 6.8 **The rate of leaf decomposition according to the lignin-nitrogen ratio and geographic locality.** High amounts of lignin in a leaf slow the decay rate, as does a prolonged winter (as is typical of New Hampshire, but not North Carolina). (From J.M. Melillo, J.D. Aber, and J.F. Muratore, 1982.)

forcing the idea that climate will play a critical role in determining the rates of nutrient cycling.

Nutrient Loss from Ecosystems

At any stage in a nutrient cycle, it is possible for nutrients to be lost from the system. For example, a leaf rots and releases nutrients into the soil water. The water carrying the nutrients drains into a river, which carries them out of the ecosystem. How large are the leaks from these systems under natural conditions, and how are they affected by changed land use? These questions are faced repeatedly as development encroaches on natural ecosystems. For example, a farmer may need to know whether topsoil rich in nutrients will be lost if the trees are removed. A developer may need to know whether deforestation will affect the local water table (the depth below ground to a water-saturated layer of soil or rock). Both farmer and developer may need to know what the consequences of their actions would be on local water supplies.

A research team led by Gene Likens of Cornell University conducted a long-term experiment at Hubbard Brook, New Hampshire, to answer some of these questions. Six experimental study areas were selected. Each site formed a catchment (an area in which all the surface water drained into a single stream), and the underlying rock was impermeable to water. Because of this impermeability, any nutrients leaking from the system could not be lost downward. The only loss of dissolved nutrients from the ecosystem would be in the stream exiting the catchment. Measurements were made of streamwater flow rates and concentrations of nutrients and suspended particles in the water. The study revealed that under natural conditions about 0.1% of the nitrogen contained in all living and dead organic material on the site leaked out into local streams each year. It is evident from this statistic that nitrogen is held very tightly within the nutrient-cycling system.

Several years after the experiment started, all the trees in one of the catchments were cut down by the experimenters. The trees were cut in January, and by the end of the summer it was evident that the water flow and nutrient budgets had been altered substantially (Figure 6.9). Trees are cooled by water; they steadily pull water up out of the ground, and it evaporates from their leaf surfaces. When trees are re-

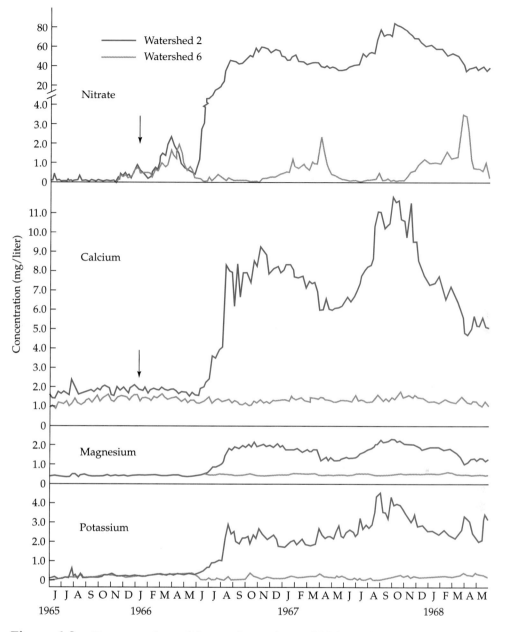

Figure 6.9   Concentrations of the nutrients nitrate (NO₃), calcium (Ca), magnesium (Mg), and potassium (K) in stream water from two watersheds in the Hubbard Brook study. Watershed 2 was deforested but watershed 6 was left undisturbed. The arrow indicates the completion of the cutting in watershed 2. (From G.E. Likens, F.H. Bormann, N.M. Johnson, D.W. Fisher, and R.S. Pierce, 1970.)

moved, more water remains in the soil and will eventually flow away in streams. At Hubbard Brook, stream flow increased by 40% after the trees were cut. The increase in soil water has an important consequence for nutrient budgets. Soluble nutrients, such as nitrate, are no longer absorbed by trees, but lie in the soil. These nutrients are dissolved in the soil water, and as the water drains from the land, the nutrients flow out with it. The streams leaving the cleared area had 60 times the concentration of nitrate, 14 times the level of potassium, and 7 times the level of calcium found in streams leaving undisturbed forest.

Small amounts of nutrient leakage are balanced by the input of nitrogen fixed from the atmosphere. However, such hemorrhaging of nutrients observed at Hubbard Brook cannot be remedied rapidly by natural systems.

The lessons learned from the Hubbard Brook study could be repeated in many settings. Soluble nutrients are lost rapidly after deforestation. Water tables rise and stream flow increases if the land is cleared. After a brief time lag, the uppermost soil on sloping ground will be washed away. All of these conditions present a farmer or a developer with problems, and they also affect people downstream. The extra water in streams can contribute to flooding. The suspended particulates reduce the quality of water and increase treatment costs to make it drinkable. This sediment may be deposited behind dams, causing damage to hydroelectric projects, or it may clog the water intakes of industrial plants. The biological consequences of nutrients washing into rivers can be profound. The influx of nutrients radically changes aquatic ecosystems; it leads to algal blooms, deoxygenation, and fish mortalities (Chapters 7 and 18).

Opening a Closed System

In the humid tropics, the speed of nutrient cycling promotes high productivity, even when soils are poor in nutrients. Nutrients are cycled so quickly in these forests that there is little opportunity for them to leak from the system. When there is virtually no loss of nutrients, a system has a **closed** cycle (the leakage of nutrients has been closed off). The opposite would be an **open system**, in which nutrients are being washed out rapidly. Highly erosive environments, where the soil is being washed or blown away, would lead to such large nutrient losses.

Many tropical forests have virtually closed nutrient cycles. Decomposition is completed so quickly that nutrients are reabsorbed into the vegetation before they can be washed away. A number of factors promote this rapid cycling. The combination of a warm climate, no winter to retard decomposition, the presence of armies of decomposers, and abundant mycorrhizal fungi on shallow roots ensures the rapid cycling of nutrients in a tropical forest. Anyone who has lived in the tropics is familiar with the array of fungi and molds that will turn leather green overnight, or the denizens of the crannies between your toes that do not wait for you to die before trying to help you rot! Here nutrient cycling is at its most rapid.

That nutrients do not wash into local streams and rivers can be demonstrated by analyzing the water for its electrical conductivity. The higher the conductiv-

ity, the more nutrient ions are dissolved in the water. A stream flowing through undisturbed forest in the Amazon is the color of Coca-Cola®. The red-brown hues are from organic chemicals, called fulvic and humic acids, released as vegetation rots. Although the water appears discolored, if an analysis is made for dissolved nutrients, the water is found to contain as few nutrients as rain.

There are many kinds of tropical soils; some are fertile and some are not. A common type of tropical soil, termed an *oxisol*, is an ancient soil that is poor in nutrients. Seemingly, this soil presents a paradox. On the one hand, it supports a tropical forest with the highest NPP of any terrestrial ecosystem. On the other hand, when that forest is cleared, the soils prove to be poor, supporting just three or four harvests. This is a riddle that can be solved through studying the rates of nutrient cycling. The nutrients available for plant growth exist either in the soil or in the **standing biomass** (all the plant matter of a given area). In a temperate forest system, recycling is relatively slow, taking more than a decade and, consequently, at any given time a high proportion of nutrients are lying in the soil. When that forest is felled and the trees removed, many of the nutrients remain because they are in the soil. Hence the cleared land is fertile and, with care, will support many years of agriculture. In the humid tropics, as little as 10% of the nutrients are in an oxisol at any one time. If this forest is felled, as the logging trucks leave they are carrying away the nutrients in the trees.

But the fertility problem is more complex than simply low nutrient concentrations in tropical soils. A second factor appears to be a change in soil acidity. Specialists in the nutrient cycles of tropical soils have demonstrated that, although nutrients are lost from the system as a result of timber removal and post-clearance erosion, the decomposition of leaves and branches left over from the felling quickly provides a balancing input of nutrients. Nevertheless, soil fertility still declines rapidly because of increasing soil acidity. As the land is cultivated, the soil becomes more acidic, and phosphorus is transformed to an insoluble form that plants cannot absorb. Phosphorus is a key nutrient, and without its addition as a fertilizer, crop yields may be reduced by half after three years.

6.3 Soil: Our Ultimate Resource

Soil has many bad connotations: We talk of things being soiled (meaning spoiled) or refer to the soil as dirt. Having a dirty mind and using dirty language are less-than-desirable social attributes. Coaxing growth from the soil in the cultured and tranquil art of gardening is demoted to "yard work": a thrash round the lawn with a mower and a spasm of leaf blowing in the fall. In modern American society, any job that involves becoming dirty is thought to be bad. However, this denigration of earth is unjustified because, for us land dwellers, soil literally is the stuff of life. Plants draw a range of essential nutrients from the earth, and without these nutrients neither plant nor animal growth would be possible.

Soils are a source of great variation in the landscape. Soil types can stretch for tens of kilometers without significant variation, or they may form a fine mosaic across the land, changing in character every few hundred meters. Indeed, where ecosystems come closest to having sharply defined boundaries, it is usually a soil property that has changed and is causing the discontinuity from one system to another. Each set of soil conditions could result in a markedly different set of plants that can flourish on either side of the divide. Indirectly, soil boundaries also determine animal distribution patterns: The animals are dependent upon specific plants, the distribution of which is determined by soil type. Because we are so dependent on plants for our food chains, we can regard soil as our ultimate resource.

Five key factors influence the formation of soil: parent material, time, climate, biotic processes, and topography.

Parent Material

The **parent material** is the source of the inorganic components of the soil. Often the bedrock beneath a soil profile forms the parent material (Figure 6.10). The nutrient richness of the soil will vary according to the concentration of inorganic elements, such as phosphorus, potassium, calcium, iron, and manganese, that are released as the bedrock decays. The amount of nutrients released to the soil per unit time will depend on two factors: the rate of decay of the parent material and the initial concentration of nutrient chemicals in the parent material. Consequently, the physical properties of the rock, such as hardness or solubility in water, will be important to soil fertility. Rocks that are only weakly cemented, or that are soluble in water, are likely to release nutrients relatively quickly and give rise to a fertile soil. Conversely, a very hard rock (such as basalt or granite) that breaks down very slowly and releases a meager supply of nutrient chemicals will yield an infertile soil.

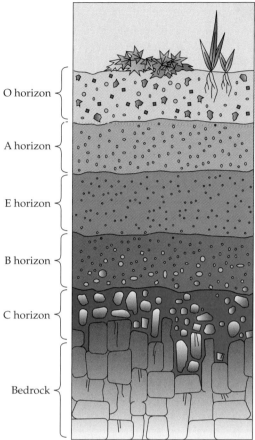

Figure 6.10 **A typical soil profile for a temperate forest.** Total profile depth, relative thickness of horizons, and presence of additional layers depend on the local history of the site.

The process of weathering gradually breaks up the bedrock until it becomes a series of fragments. Major factors in weathering are dissolution by water and freeze-thaw cycles, in which successive expansion and contraction of the rock cause it to crack and become fragmented. Even in areas where temperatures do not drop below freezing, but where there is a big difference between daytime and nighttime temperatures, exposed rocks can be fractured as a result of repeated expansion and contraction. The result is often a rock that looks like an onion with its layers peeling away. This degradation of the rock by weathering produces a layer of fragmenting rock termed the **C horizon** of the soil profile (Figure 6.10).

Above this lies the **B horizon**, which is the layer of soil receiving materials washed down from the overlying horizons. Sometimes minerals, especially iron, form crusty layers that may impede water drainage through the soil. The **E horizon** is a pale soil layer separating the B horizon from the A horizon. Clay particles and minerals are washed from the E to the B horizon, leaving high concentrations of sand or silt particles. The **A horizon** is often black or dark brown as a result of the presence of decomposing organic matter. The surface of the soil is the **O horizon**, the layer of dead plant and animal matter. Together, the A and O horizons are colloquially known as **topsoil**.

Exceptions to this soil-forming scheme occur when the soil is carried onto the site by wind, flood water, glacial activity, or as volcanic dust. In these instances the original source rock could be hundreds of kilometers away, and the products of many different rock types could be blended to form the new soil. Examples of such soils include the clay-rich soils of the American Midwest that were deposited by an ice sheet at the end of the last ice age; the valley soils of the Amazon, Nile, and Mississippi that have their parent materials far up in the headwater regions of those mighty rivers; and the great expanses of wind-blown (loess) soils that cover much of China.

As the process of weathering proceeds, it produces partially decomposed mineral matter such as clays. Clay minerals are small, flat particles that stack together, loosely bonded to each other by shared oxygen molecules. The clay particles have a relatively large surface area and a negative electrical charge. Dissolved substances with a positive charge can cling to the surface of the clay particle (Figure 6.11). Many nutrients such as potassium (K^+), calcium (Ca^{2+}), ammonium (NH_4^+), and magnesium (Mg^{2+}) are **cations**; that is, they are positively charged ions and will be attracted to the surface of the clay. The clays, therefore, play an important role in holding nutrients that have either been released from parent rock or are products of decompositional cycles. Nitrate (NO_3^2) has a negative electrical charge and will not be attracted to the clays; it is more readily washed out of a clay soil than are cations. Ammonium is a cation and therefore is held by the clay. Thus, in some soils, ammonium is more likely to be an effective fertilizer of plants than nitrate, and nitrate is more likely to wash from the land, polluting local streams and lakes.

In very ancient soils, even the clays have decomposed, and ultimately little remains other than quartz sand. The oldest, most infertile soils, such as the deserts of Australia and Africa, commonly are little more than quartz sand stained red by oxidized iron bound to the surface of the sand.

Time

Very young soils, such as volcanic ash settling to blanket the surface of the land, may be rich in nutrients,

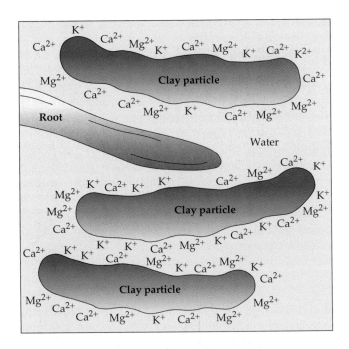

Figure 6.11 **Clay particles have a large surface area and carry a negative electrical charge.** Positively charged ions, cations, that are dissolved in water bond to the surface of the clay by electrical attraction. Many cations are essential nutrients for plants. (After McKnight, Tom, *Physical Geography: A Landscape Appreciation.* 5th ed., 1996. Adapted by permission of Prentice-Hall, Inc., Upper Saddle River, NJ.)

but are incomplete because they lack the texture and full range of organisms to help cycle the nutrients. Equally, very old soils may be nutrient poor because all their nutrients have gradually been washed away. Thus a soil is going to mature, reach a peak of productive potential, and then start to deteriorate. For every soil this process will differ, but in general soils become productive quickly—within just a few years of being deposited—and take hundreds of thousands, even millions, of years to lose that productivity. The great glaciers that covered much of northern North America during the last ice-age melted just 10,000 years ago (Figure 4.13). As they melted, they deposited a layer of clay that is rich in nutrients and readily broken down to form highly fertile soils. In the southern states, where there were no ice sheets, the soils are older and less fertile. In continents such as Africa and Australia, where the soils are millions of years old, few nutrients remain, and the soils are very infertile.

Climate

The greatest influences of climate are temperature and precipitation. We have already seen how climate influences nutrient cycling through the type of vegetation cover (Chapter 5), affecting rate of decomposition and the likelihood that nutrients would be washed out of a system. Another very significant way that climate affects soil formation is in the rate of weathering. Warm, wet climates will cause rock to decay faster than cold, dry ones, primarily because the process of carbonation will run more rapidly.

Carbonation is the dominant reaction that brings about the decay of rock through contact with carbonic acid (H_2CO_3). If there were no carbon dioxide being added to the soil from biological processes, the concentration of CO_2 in the soil would approximate that of the air (0.035%). However, plant roots and decomposing organic matter release CO_2 within the soil and lead to soil CO_2 concentrations 200 times that of air. Soil CO_2 concentrations were recorded in excess of 7% beneath wheat fields in Missouri. Carbon dioxide is highly soluble, and it combines with water in the soil to form carbonic acid. This acid can dissolve rocks, speeding the release of nutrients.

Biotic Processes

The biotic aspect of a soil includes the organic content of the soils and interactions between soil and organisms, especially that of decomposers. About half of the volume occupied by a soil is made up of mineral matter; the other half is mostly air and water, with a small component of organic material. The organic fraction, although small, is the most important component because it plays a large role in determining the texture and moisture characteristics of the soil. In most soils, less than 5% of its volume is organic matter. A wide variety of decomposers degrade the organic material until all that is left is **humus**, a black gelatinous material that is chemically stable. Humus is an important structural component in the soil, helping to bond soil particles into loose-fitting nodules. Between the nodules are air spaces that ensure the presence of oxygen around plant roots and provide drainage. Soils containing many air spaces have the appearance of bread crumbs, hence the term **crumb structure**. A good crumb structure ensures good drainage, as well as the maintenance of soil moisture during dry periods. Soil scientists refer to water that moves downward through the soil and bedrock as percolating water. Humus absorbs water percolating down through the soil, and thus helps it to hold water. Like a sponge, the humus can store water and later release it as the soil dries out. This property helps to even out the supply of water to plant roots between rainfalls.

The burrowing activity of organisms also helps to maintain an open soil structure. Their burrows and passageways allow air to penetrate the soil, thus helping to ensure that the roots of plants have access to oxygen. Creatures such as worms perform a dual function as both decomposers and aerators of the soil.

The biota also serve to fix nitrogen, an essential nutrient for plant growth, and one that is obtained from the air. Indeed, the recycling of nutrients by biota may be more important in terms of the amount of nutrients being made available than the supply of fresh nutrients. Thus, biotic processes are essential to maintaining nutrient availability.

Topography

The slope of the land surface, the configuration of hills and valleys, and the elevation above sea level, are all important attributes that can affect soil formation. Slopes may be too steep to allow soils to form any great depth, especially when rains can turn the soil into slippery muds that slide into the valley bottom where they accumulate. Consequently, flat areas such as plateaus, valley bottoms, and coastal plains tend to develop deeper soil profiles than sloped lands.

It is important to note that parent material, time, climate, biotic processes, and topography interact to form soils. For example consider the difference in forming a soil at high elevation compared with low elevation. High on a mountain the climate is colder than in the lowlands (climate), slowing biochemical reactions and nutrient cycles (biotic). The soils are probably thin and unstable because of the sloping mountainsides (topography), and consequently the soils never mature (time). Furthermore, the mountains are formed of hard rock (parent material) that resists weathering and only slowly releases its nutrients. Thus all five processes lead to the formation of a poor soil.

6.4   Soil Maps

The international classification of soil types used to describe soils around the world is shown in Table 6.2. A map of the soil types of Earth reveals their major distributions (Figure 6.12). Such maps are useful as guides to overall patterns. For example, ultisols, which are very old soils, are abundant in the southeastern United States, whereas the largest areas of oxisols are found in tropical latitudes. However, large-scale maps such as Figure 6.12 are only useful generalizations that cannot possibly show the complexity of the real world.

Table 6.2   Soil taxonomy and summary of soil characteristics.

Soil type	Characteristics	General location and climate
Alfisols	Moderate weathering, rich in nutrients	Midlatitudes and moderate environments
Andisols	Low weathering, derived from volcanic ash, rich in nutrients	Volcanic regions
Aridisols	Moderate to low weathering, dry, lacking organic matter, nutrients variable	Desert soils, hot dry areas
Entisols	Low or no weathering, least-developed soils	Widespread
Histosols	Low weathering, highly organic, rich in nutrients, waterlogged	Wetlands
Inceptisols	Low weathering, young soils, widely differing properties	Widespread
Mollisols	Moderate weathering, rich in nutrients and humus, basic	Temperate grassland
Oxisols	Maximum weathering, very low nutrients, very high iron content	Tropical, hot, humid forest
Spodosols	Moderate weathering, nutrients leached out, infertile	Northern temperate zones, boreal forest
Ultisols	Highly weathered, low nutrients, acidic	Subtropics and tropical forest
Vertisols	Low weathering, high clay content, swells when wet and cracks when dry, often rich in nutrients	Strongly seasonal tropics or subtropics

(*Source:* U.S. Dept. of Agriculture.)

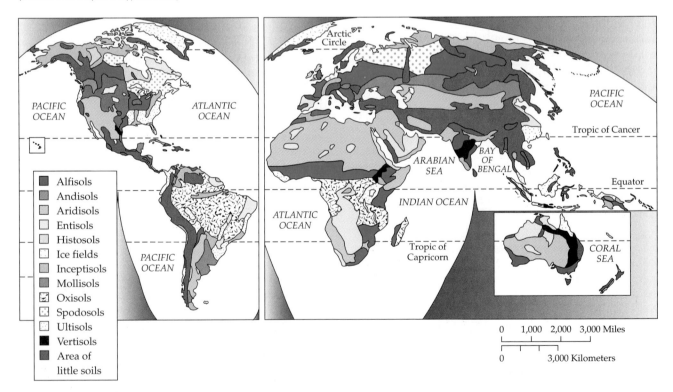

Figure 6.12   A generalized map of the distribution of soil types. (After McKnight, Tom, *Physical Geography: A Landscape Appreciation.* 5th ed., 1996. Adapted by permission of Prentice-Hall, Inc., Upper Saddle River, NJ.)

They cannot provide local details and should not be taken literally. At this scale, North Carolina appears to be uniformly covered by ultisols, with some inceptisols in the Appalachian portion of the state. But if you were to survey the soils of North Carolina, you would find different soils according to the slope of the land and whether you were close to the coast or in the center of the state. Wetter and more organic soils would lie in valley bottoms or around lakes and wetlands. This lesson is important because it applies to almost anything that is mapped. Maps are generalizations, and the scale of the map will determine how generalized, or simplified, the data are.

6.5   Soil Erosion

The erosion of the A and O horizons can greatly reduce the primary productivity of an ecosystem. Although wind can be an important erosive factor, most erosion results when water washes soil from the land. The rate at which water drains into a soil from the surface is termed an **infiltration rate**. Rapid infiltration means that surface water drains away quickly, whereas a slow infiltration rate may be found in soils prone to waterlogging. **Waterlogging** occurs when the air spaces in the soil become filled with water. An intense rain is most likely to cause soil erosion where soils become waterlogged.

In a storm, we instinctively head for cover, and by standing beneath a tree we may be spared a wetting as the canopy of branches overhead intercepts the rain. If the rain continues, the canopy will become saturated and unable to hold any more water. All the extra water hitting the canopy will then flow down the trunks of trees or drip from leaves to the forest floor below. Standing in the sodden forest, you might see that the soil is soaking up the rain, until it, too, becomes saturated. In the heaviest downpours, water may puddle on the forest floor. If the forest is on a slope, the water will run off downhill. Such water movement across the surface of a soil has the potential to cause erosive damage, with only the subterranean tangle of tree roots holding the soil in place. The forest and its soil act like a sponge absorbing the rainfall. Over the next few days, much of the water that fell in a single deluge will gradually drain out of the forest into streams. Stream levels will rise slightly, but the streams will probably stay within their banks. The forest has lessened the possible erosive and flood impact of the storm.

In another setting, one where the forest has been removed, nothing breaks the impact of the rain as it drives onto the exposed soil. Almost immediately, the soil becomes waterlogged, and water starts to flow downhill across the surface of the land. If the slopes are steep and without bonding roots, a heavy rain will wash the soil away. As was shown at Hubbard Brook, water moving across the surface of the land has the capacity to wash away both nutrients and topsoil. Topsoil erosion is a significant threat to many agricultural regions, and the danger of erosion is increased by soil compaction. Heavy machinery driven over the soil can cause the air spaces in the soil to collapse. Smaller and fewer air spaces decrease infiltration rates, increasing the chance of waterlogging and erosion.

The 40% increase in streamflow rate observed at Hubbard Brook was a factor in carrying away nutrients and topsoil. In that study, the streams were small and the downstream damage was probably slight. When large areas are deforested, the scale of the problems is increased. Deforestation in the Himalayas has been a prime cause of soil erosion, but the removal of the forest can also threaten lives downstream. In 1982, torrential rains swept the headwater regions of the Ganges, the great river that drains from the Himalayas to the Bay of Bengal. Without forests and forest soils to trap the water and even out the flow, water rushed straight into the mountain streams and from there into the mighty Ganges. The river rose, burst its banks, and flooded the densely settled farmland of the Ganges Delta with a swirling, muddy deluge. It is estimated that more than 100,000 people lost their lives as their lands were inundated.

Although deforestation and water erosion are the principal causes of global soil loss, overgrazing and general bad soil husbandry can result in the soil literally being blown off a field. On a summer day in eastern England, it is common to see a red-brown haze in the air over fields where farmers have removed hedgerows, opening up their fields to wind erosion. The haze is the fertile topsoil drying and then being blown off the field. The farmer has just lost part of his or her best resource, soil. The classic case of soil erosion was the dustbowls of the midwestern United States in the 1920s and 1930s. Decades of overexploitation of the land culminated in a thin soil with poor structure. A series of long, hot, dry, windy summers turned the air dark with soil being blown from the fields. The land became untenable, and many farms in Oklahoma and the surrounding states were abandoned.

One estimate is that about 1% of Earth's topsoil is eroded each year, soil that ultimately washes into the oceans. One percent is hardly a dramatic statistic, but remember that the uppermost soil holds the most nutrients and that this soil loss will be concentrated in the areas that are farmed: the areas where we need it most.

6.6 The Importance of Fire

Fire and Ecosystems

Fires are a natural part of most ecosystems, and the drier the environment, the more important fire is as a force driving the structure of an ecosystem. Wildfires need a source of ignition, usually lightning, and fuel. The fuel for the fire is the accumulation of organic material such as fallen leaves, twigs, and branches on the ground. The dryness of the fuel determines the probability that a fire will start, but once the fire generates enough heat, even damp material will ignite. It is the amount of fuel per hectare that determines the intensity of the fire. If there is little fuel, the fire just creeps across the land; it never develops the intensity to leap up into the crowns of trees. However, if there is plenty of fuel, the fire builds in intensity and can move quickly across a landscape, engulfing everything in its path. The intensity of the fire makes a huge difference to wildlife. A low-intensity fire that moves slowly may do no more than scorch plants, and it gives mobile animals the chance to flee. Because the heat from this low-intensity fire does not penetrate more than a few centimeters into the ground, some animals can survive by burrowing into the soil. Because many animal species can neither flee nor burrow, their numbers are seriously reduced during a fire. This relatively cool fire may not damage the root mats of trees, shrubs, long-lived herbs, and grasses. Trees may have their bark blackened and may lose all their leaves, but if the fire has not penetrated the bark, the trees can usually resprout from fresh buds lying beneath the charred surface. After a year or two, the regrowth is so complete that it is hard to tell that the area ever burned. By turning leaves and wood to ash, the fire releases mineral nutrients and actually speeds the cycling of nutrients. Consequently, ash is a fertile growth medium rich in potassium, phosphate, and many micronutrients. One nutrient that is lost as a result of fire is nitrogen. As much as 70% of the nitrogen in burned leaves and wood goes off in smoke. It is common to find that the first plants to flourish after a fire are adapted to living in nitrogen-poor soils and can fix their own nitrogen.

Ecosystems that burn every two to three years, such as the long-leaf pine barrens of the southeastern United States, are fire-adapted systems. Here, the seeds of many plants only germinate following fire. These seedlings have an excellent chance of lying in bare ground (because the previous occupant has been burned), and nutrients are plentiful due to the ashing of organic matter. The fires crawling through the understory of the pine forest also prevent invasion by hardwood species such as hickory and oaks. Each time the fire scorches the forest floor it burns the seedlings of the hardwood species, preventing them from growing up into the canopy. Without regular fire the pines would be outcompeted by the oaks, and the entire ecosystems would change. Some endangered species, such as the red-cockaded woodpecker, are absolutely reliant on mature pine woodlands. These systems require a delicate balance of fire in which the burns are sufficiently frequent to eliminate competing trees, but never so intense that they kill the pines.

Less benign fires are the large, intense wildfires that incinerate whole forests. These fires may have 10 to 100 times the energy per square meter of the low-intensity fires, and the heat can penetrate deep into the earth, killing roots and burrowing animals. The above ground portion of trees are killed, and the organic component of the topsoil may catch fire, leaving fertile ash but a poor soil structure that is very vulnerable to erosion. The ecosystem will recover from such fires, but the process takes longer, and it is more likely that isolated populations of rare animals will have been eliminated. A forest burned this completely may take more than 300 years to regain its full measure of habitat diversity (Chapter 14).

The dry Mediterranean climate of California creates forests and scrublands that burn readily. Wildfire is a natural part of that ecosystem, but as urban development invades the forests, fires are suppressed to protect the real estate. Ecologist Richard Minnich analyzed satellite imagery to compare the number and size of fires in southern California (with fire suppression) and in Baja California (no fire suppression; Figure 6.13). He found that without fire suppression there are more fires that cover less than 800 hectares. With fire suppression, there are fewer fires in total, but there are more large fires that burn 1000 to 10,000 hectares. These data indicate that, although fire suppression reduces the number of fires, the fires that do occur burn hotter and longer. The explanation lies in the process of fire suppression, which relies on the ability to extinguish fires quickly. An unwanted result of this control is that there is an accumulation of fuel on the forest floor. When a fire does start, it has a huge supply of fuel and generates a blaze that cannot be held in check. An uncontrollable inferno results; witness the

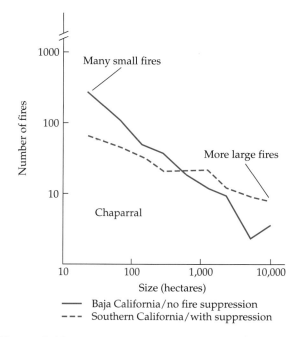

Figure 6.13   The frequency and size of wildfires in California under different regimes of fire management. When fire suppression is practiced, the burns are fewer, but when they do occur they are larger and more damaging than are fires in unmanaged areas. (From R.A. Minnich, 1983. Fire Mosaics in Southern California and Northern Baja California. *Science* 219:1287–1294. Copyright 1983 American Association for the Advancement of Science.)

fires of Yellowstone National Park in 1992, California in 1993, and Florida in 1998.

Wildfires may damage our property, but the ecosystem has not been damaged. The blaze is a part of a natural cycle of burn and resprout, burn and resprout. In the wake of the fire, old stems are replaced quickly by vigorous young ones, which provide improved grazing for the remaining herbivores. The exception may be under a regime of fire suppression, when a system is converted from high-frequency, low-intensity fires to one of low-frequency, high-intensity fires. The trees that would survive burns in the natural system are killed in the more intense fires associated with fire suppression gone awry. The heat of the blaze may also combust the organic fraction of the topsoil. Such fires alter the habitat significantly, and an ecosystem may take many years to recover.

For nature reserves set up in areas that burn naturally, such as a prairie or a fire-adapted forest, fires must be an essential part of the management program. In the tallgrass prairie regions, there is sufficient precipitation to support oak trees and other hardwoods, and it is only the frequent burning of these areas that maintains them as grasslands. Surrounding land use, such as irrigated farmland or where there is deliberate

fire suppression, may alter the fire regime within the reserve area. In these cases, fires may need to be set to mimic nature, so that the desired balance of plants and animals is maintained. Managers know this to be true on nature reserves, yet the lesson is forgotten when dealing with developed areas. There, the tendency to suppress fire has proved damaging. A new management strategy is emerging: to carry out controlled burns that reduce the fuel load and thus reduce the risk of catastrophic fires. Such an undertaking is one thing in a national forest or park, but quite another in a developed area, where any mishap could result in human or financial disaster.

6.7   Ecosystem Functions and Values

As development of the natural landscape proceeds, arguments over land use emerge. Farmers, foresters, urban planners, and industrialists may require land for development; the population as a whole might have a conflicting set of uses for that land. A change in land use, such as clearing trees to provide arable cropland, will bring a set of benefits. Those benefits can be conveniently measured as the profits made from the land. However, the change from forest to cropland will change the local ecosystem. Before an effective valuation of the financial gains and losses (Chapter 27) can be made, evaluators must understand the full ecological implications of the change in land use. Because all aspects of an ecosystem are interdependent, the changing of one facet brings repercussions throughout the ecosystem. For example, how will cutting a drainage ditch through a wetland affect users of the wetland? How will loss of a wetland affect distant ecosystems? (Remember the Mississippi story.) What are the characteristics of the wetland that we value, and how much can we change the system before we lose those attributes?

The interaction of abiotic and biotic processes define an ecosystem. If one of those processes were to change, other processes within the ecosystem would be affected. For example, if an ancient forest is felled, the local climate would change as temperatures and windspeeds increased. The soil surface would dry out and decomposition rates would decrease. Therefore, we can observe that the presence of trees on the landscape had an importance beyond simply being another organism. Ecosystems have a characteristic set of attributes, effectively the roles that they play in a natural system, that can be termed **ecosystem functions**. For example, a woodland has the ecosystem function of cycling nitrogen. This ecosystem function is a product of local temperature, wetness, flood duration, numbers of decomposers, presence of nitrogen-fixing bacteria, plant species composition (the array of species that grow there), and gross primary productivity (which serves as an index of the nitrogen input from dead vegetation). Note that the function takes place independently of humans.

Ecosystem values, on the other hand, are the uses that humans find for the ecosystem; these may be a direct use, such as hunting wildfowl within a marsh, or an indirect use. A farmer many miles downstream from the wetland may value it because the wetland helps to regulate floods that would damage crops.

The concepts of ecosystem functions and values help us to assess the importance and connectedness of each ecosystem within a landscape. Increasingly, managers of ecosystems plan their actions according to the effects that they will have on these functions and values.

Summary

- An ecosystem includes both the biotic and abiotic processes of a region.
- All ecosystems are affected by those that surround them.
- Natural cycles replenish supplies of nutrients, water, and carbon.
- Macronutrients are nitrogen, phosphorus, and potassium; other chemicals essential for plant growth such as iron, manganese, and boron are micronutrients.
- Soils provide plants with nutrients. Most nutrient uptake is from the upper 3 cm of the soil.
- The speed of nutrient cycling lies at the heart of agricultural and forest productivity.

- Parent material, time, climate, biotic processes, and topography interact to form soils.
- Soil erosion can lead to reduced crop yield, floods, and mudslides.
- Natural fires are essential to maintaining the structure of many ecosystems. The frequency and intensity of fires will have a strong effect on community structure.
- Ecosystem functions are the natural services that the ecosystem provides to neighboring ecosystems and to the organisms that live within it.
- Ecosystem values are the uses, or worth, that humans ascribe to an ecosystem.

Further Readings

Brown, J. H. 1995. *Macroecology*. Chicago: University of Chicago Press.

Kilham, K. 1995. *Soil Ecology*. Cambridge: Cambridge University Press.

Likens, G. E., F. H. Borman, N. M. Johnson, D. W. Fisher, and R. S. Pierce. 1970. Effects of cutting and herbicide treatment on nutrient budgets in the Hubbard Brook watershed ecosystem. *Ecological Monographs* 40:23–47.

Minnich, R. A. 1983. Fire Mosaics in Southern California and Northern Baja California. *Science* 219:1287–1294.

Pynne, S. J., P. L. Andrews, and R. D. Laven. 1996. *Introduction to Wildland Fire*. New York: John Wiley and Sons, Inc.

Web Connections

On-line resources for this chapter are on the World Wide Web at:
http://www.prenhall.com/bush
(From this web-site, select Chapter 6.)

Aquatic Ecosystems

7

Water moves continuously between the oceans and the atmosphere as it evaporates and precipitates. If rain falls onto land, some of it flows into streams and then to rivers, eventually returning to the sea. Alternatively, the rainwater might evaporate back into the atmosphere or drain downward through the soil to form groundwater. But even groundwater eventually makes its way back to the sea. Water held in the bodies of plants and animals will be released by the transpiration of plants and the respiration of animals or when the organism decomposes. All water is in the process of being cycled from one place to the next, until it ultimately returns to the sea to start the cycle all over again. This cycle is termed the **hydrological cycle** (Figure 7.1). Like other cycles on Earth the hydrological cycle is powered by solar energy.

The amount of water in circulation hardly varies, although it is evident that climate plays a role in determining where the water resides within the system. Under present conditions, the oceans hold about 97.4% of the water on the planet, leaving just 2.6% as freshwater. Of the freshwater, 80% is held in ice caps and glaciers, and almost 20% is held as groundwater. Lakes and rivers, our most apparent freshwater, form just 0.4% of the freshwater, or just 0.01% of all water. Although large quantities of water are evaporated and precipitated each year, at any one time the atmosphere holds only 0.0001% of water on the planet.

The oceans, atmosphere, groundwater, and surface freshwater can be thought of as reservoirs of water. The proportion of water held in each reservoir is more or less constant in the short-term. However, over the long-term, climatic change can induce subtle but important variations in the size of each reservoir.

For instance, if Earth were to heat up, glaciers would melt, reducing the size of the ice reservoir. Meltwater flowing into the sea would increase the amount of water held in the ocean reservoir. The atmospheric reservoir of moisture would also increase due to higher rates of evaporation from the land and sea surfaces. The net effect of this would be an increase in sea level (Chapter 23).

7.1 Marine Systems

Few beaches plunge directly into the abyssal depths of the ocean. Along continental margins there is a portion of the continental plate that lies submerged. This area, the **continental shelf**, varies in width from a few kilometers to tens of kilometers, forming a shallow sea with depths of 100 m to 200 m. The **continental slope** is the seabed that drops from the continental shelf to the full depth of the ocean floor. Classically, the marine environments can be divided into the oceanic, neritic, and intertidal environments (Figure 7.2). The **oceanic** environment is any portion of the ocean beyond the

Figure 7.1   The hydrological cycle. The natural cycle of water falling as rain, draining into the ground, or flowing on the land surface to the sea. Any standing body of water, ranging from a droplet on a leaf to the surface of an entire ocean, is subject to evaporation. Evaporation recharges the reservoir of water vapor in the atmosphere. Even water stored for thousands of years in ice caps, in the ocean deeps, or in the ground is connected to the land surface and the atmosphere by fresh inputs of water arriving as rain or snow. (From Nebel, Bernard J. and Wright, Richard T. *Environmental Science*. 5th ed., 1996. Reprinted by permission of Prentice-Hall, Inc., Upper Saddle River, NJ.)

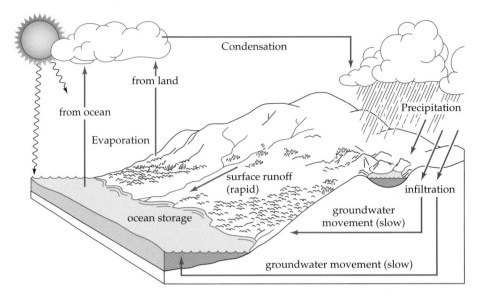

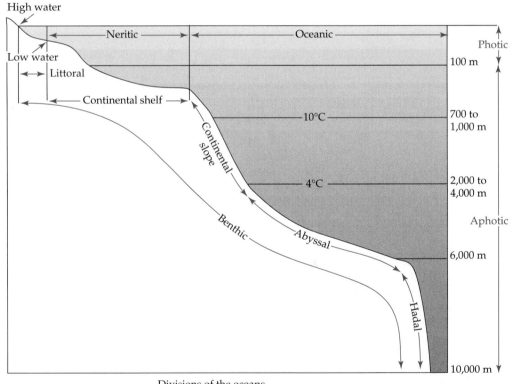

Divisions of the oceans

Figure 7.2   Divisions of the ocean (not to scale). The photic zone receives enough light to power photosynthesis and, hence, is the principal region of primary productivity. A small amount of primary productivity is also associated with ocean vents in the sea floor of the hadal region.

continental shelf. The **neritic** zone is the section of the ocean that lies above the continental shelf, and the **littoral** zone is the shoreline between the highest and lowest tide limits.

The Vertical Structure of the Ocean

In the open ocean, light, temperature, and nutrient availability are the three variables with the greatest affect on ecology. Different wavelengths of light can penetrate water to different depths. Red light is the first to be absorbed, and it only penetrates a few meters. Progressively, orange, yellow, and green light are absorbed with depth. As any scuba diver will tell you, by the time you are 25 m to 30 m underwater everything looks blue. At this depth the only light that can penetrate is blue light. At 100 m deep, only 1% of the light that fell on the surface of the ocean is available for photosynthesis. Clearly, the primary productivity of the oceans will take place in the uppermost waters. Although the ocean may be thousands of meters deep, the portion with almost all primary productivity is just the top 100 m; this is termed the **euphotic** zone (Figure 7.3a).

Temperature also has a strong vertical pattern in the ocean. Surface waters are lit and warmed by sunlight, while dark bottom waters remain cool. The transition between these two water segments is a short vertical distance in which temperature changes abruptly; this section is known as the **thermocline** (Figure 7.3b). Latitude and ocean currents influence where the thermocline forms, but it is generally found between 120 m and 240 m depth.

Nutrients have low concentrations in ocean surface waters, but often increase their concentrations beneath the euphotic zone. In the surface waters, active growth of phytoplankton scavenges nutrients from the water column, depleting the nutrients. Below the euphotic zone, nutrient concentrations increase due to the decomposition of organisms dropping out of the productive layer of the ocean (Figure 7.3c).

Chemical Composition and Productivity

Seawater has an average salinity of 35‰ (parts of salt per thousand parts of water), much of which is attributable to high concentrations of sodium and chloride ions. The ocean is more than just salt and water (Table 7.1), but most ocean waters are extremely poor in nutrients such as phosphate, nitrate, ammonium, and iron. Although the oceans cover 71% of the surface of Earth, they account for only 50% of global primary productivity. The surface waters of most oceans have low productivity compared with terrestrial systems (see Figure 5.3b and Table 5.1) producing 80 to 100 g $C/m^2/yr$ (about one-tenth of the productivity of some tropical forests). However, some areas of open ocean

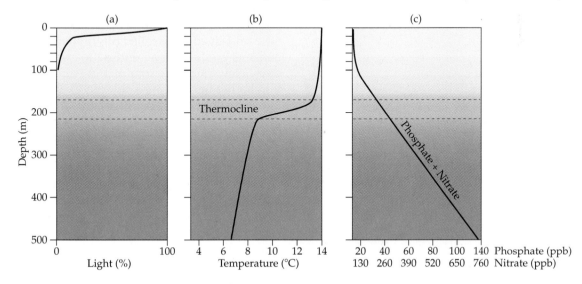

Figure 7.3   The temperature structure of an ocean. (a) Light is rapidly absorbed or refracted by water and light availability is greatly reduced within the upper few meters of the ocean. (b) The thermocline acts as a physical boundary, restricting the mixing of water above and below it. (c) Nutrient concentrations are rapidly absorbed in the photic zone by photosynthetic organisms. Concentrations of nutrients increase below the photic zone because of the decomposition of cells falling through the water column. Below about 500 m these concentrations decline once more.

Table 7.1 Some of the most abundant ions in seawater compared with their occurrence in freshwater.

Constituent	Concentration in seawater (mg/kg)	Concentration in river water (mg/kg)
Chloride	19,350	5.75
Sodium	10,760	5.15
Sulfate	2,712	8.25
Magnesium	1,294	3.35
Calcium	412	13.4
Potassium	399	1.3
Bicarbonate	145	52
Strontium	7.9	0.03

(*Source:* Data are from W. H. Schlesinger, 1991.)

are moderately productive and may have a net primary productivity between 100 and 200 g $C/m^2/yr$. Unlike terrestrial productivity, the primary production in oceans does not increase from pole to equator. Instead, productivity in the open ocean responds to the local presence of nutrients that may be provided by the surfacing of deep ocean currents or turbulence in the waters of the continental shelf.

Upwellings occur when offshore winds push surface waters seaward and when water flows upward from the ocean deeps to replace the surface water (Figure 7.4). The deep ocean waters are nutrient rich, and

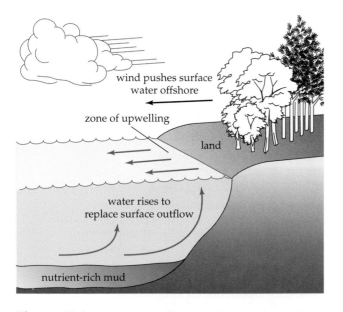

Figure 7.4 A diagrammatic representation of upwelling, which brings nutrients to the surface waters.

when they rise to the surface there is a local surge in primary productivity. Upwellings commonly occur on the western side of midlatitude landmasses (a function of the Coriolis effect; see Chapter 4) and in the polar regions. Where these upwellings occur, phytoplankton populations increase markedly and form the basis of food chains. In the polar waters, the productivity is concentrated in the summer months, when there are long days and the surface waters warm. The seasonal boom in plankton is short-lived but intense. Some species of whale, such as the blue whale, migrate from their equatorial breeding grounds to feed on the abundance of krill (tiny planktonic animals) in the areas of polar upwelling. An indication of the sheer productivity of these areas is that an adult blue whale can take in 7 tons of krill during a five-hour feeding period.

Continental shelf seas are shallow, seldom more than 100 m deep, and are moderately productive. Net primary productivity in these seas can be as high as 162 g $C/m^2/yr$ because the water is well supplied with nutrients washing from the land. In addition, wave action and turbulence stir these waters and prevent the nutrients from settling to the seabed. Because of their productivity and proximity to land, most ocean fisheries exploit continental shelf populations (Chapter 11).

Inverted Pyramids of Numbers

The primary producers of ocean systems are generally short-lived algae called **phytoplankton**. These organisms go through an entire life cycle in just a few days. New carbon is continually being incorporated into bodies, but, because older individuals are dying at an equal rate, the **standing crop** (the total number of living organisms) or standing biomass can remain stable. Where nutrients increase in concentration, there is not only an increase in productivity, but also an increase in biomass.

In Chapter 3, we discussed trophic levels, biomass, and energy flow through a food chain. Figures 3.4 and 3.5 illustrated that the biomass and available energy in each trophic level declined as you moved up the food chain. Those generalizations hold true for terrestrial systems, but in the oceans the biomass of primary producers alive at any one time might be small compared with that of their predators. Commonly, the pyramid of biomass is narrowest at the base in ocean systems (Figure 7.5). The key to this difference is that food chains are not so much reliant on available biomass as upon available energy. For example, phytoplankton

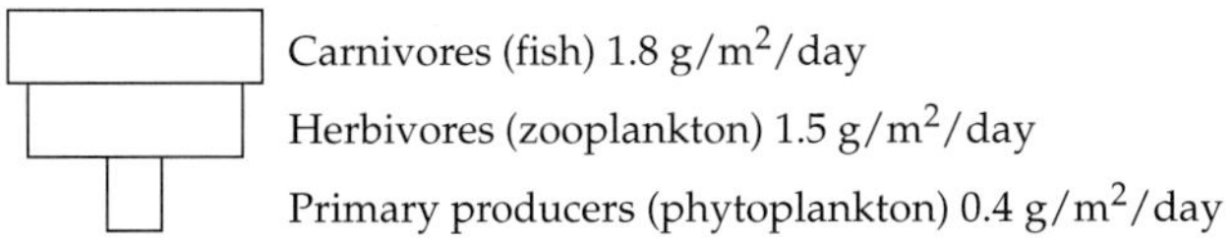

Carnivores (fish) 1.8 g/m²/day

Herbivores (zooplankton) 1.5 g/m²/day

Primary producers (phytoplankton) 0.4 g/m²/day

Figure 7.5   The pyramid of standing biomass for the English Channel, the sea separating England from France. Biomass is the mass (when dried) of organisms in a given trophic level and is expressed here as grams dry mass per square meter per day. Note that the standing biomass of primary producers is lower than that of predators. This inverted pyramid is possible because the phytoplankton are very short-lived and have a high rate of reproduction, compared with their predators. (After I. Valiela, 1995.)

grow and reproduce rapidly and die after just a few days. Their bodies are consumed or decompose almost immediately allowing a rapid turnover of the energy held in that tiny body. Although the phytoplankton can provide a lot of energy to the next level in the food chain, their population has a relatively low biomass at any one time. Conversely, their predator, a blue whale that lives to be 70 years old and has a generation time of 20 years, has a massive body that stores energy in the form of fat, muscle, and bone. Therefore, a relatively small planktonic biomass that is constantly replenished supports a larger biomass of predators. We do not see this inversion of biomass pyramids on land because trees and other plants do not turn over energy faster than the herbivores (primarily insects) feeding on them.

Harmful Algal Blooms

Concern has been growing that enrichment of coastal waters with fertilizers and sewage has caused a major change in the balance of these ecosystems. Coral reefs that thrive where nutrients are scarce are being overgrown by algae that flourish in nutrient-rich waters, and blooms of floating algae, known as **red tides**, have become more common. A red tide is formed by dense populations of microscopic phytoplankton, each of which is red or brown in color. Their bodies literally give the water a reddish appearance. In the 1960s, red tides were uncommon, localized events around the U.S. coastline. Their occurrence has increased dramatically over the last 40 years so that every coast is now affected by them. Red tides have been associated with the deaths of fish, shellfish, and marine mammals such as dolphins and manatees. As scientific research on red tides has progressed, it has become apparent that some red tides are relatively harmless, whereas others are potentially deadly. Equally, it has become clear that some of the fish and mammal deaths

were caused by algae, but there was no red tide present. Consequently, scientists prefer to use the more specific label of **harmful algal blooms**, rather than red tide.

Harmful algal blooms can bring about deaths through physically clogging the gill structures of fish and shellfish, through deoxygenation of the water column, and through toxicity.

The amount of oxygen required to support biological activity (including decomposition) is termed the **biological oxygen demand (BOD)**. When there is a massive bloom of short-lived algae, there is a corresponding boom in the number of dead cells to be decomposed. The increase in dead phytoplankton leads to an increase in decomposer activity and, consequently, an increased demand for oxygen. The absorption and movement of oxygen through water take a lot longer than the movement of oxygen through air. Therefore, when demand for oxygen in an aquatic system is high, it is possible to exhaust the local supply. If the BOD exceeds the available supply of oxygen, and the concentration of dissolved oxygen in the water drops, local populations of fish and shellfish die. In the larger harmful algal bloom events, millions of fish die in a few days. The decomposition of the bodies of the fish increases BOD. As BOD increases, dissolved oxygen concentrations in the water plummet, and more fish die.

Some algae that belong to a group called *dinoflagellates* may contain very potent toxins. Shellfish, such as oysters, clams, and mussels, feed by sucking water through filters that sift out organic solids. If toxic dinoflagellates are drawn into this filtering process, the shellfish digests the alga, but fails to digest the toxin. The toxin is not excreted, but remains in the body of the shellfish. Even when they have accumulated to relatively high concentrations, these toxins do not appear to harm the shellfish. However, a human, or other animal that ate the shellfish might become very sick or die.

These dinoflagellates may be part of a red tide, sometimes even forming the main population of the bloom, but they can also be deadly when there is no tell-tale discoloration of the water. One species that has achieved a degree of infamy in recent years is the "new killer," *Pfisteria* (Figure 7.6). This dinoflagellate has shown up repeatedly at sites of major fish kills. For much of the time *Pfisteria* exist as a resting cyst in the bottom muds of shallow seas. Upon some chemical cue, perhaps provided by an exceptionally dense shoal of fish, the dinoflagellate becomes active, leaves the bottom mud, and starts to release its toxins. The

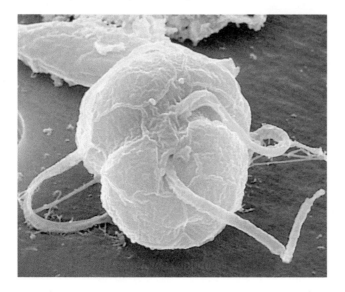

Figure 7.6 ***Pfisteria*, a fish-killing dinoflagellate.** *Pfisteria* have at least 24 morphological forms, one of which releases toxins that can cause skin lesions in fish and humans. Fish kills in the New River, North Carolina, are thought to be associated with this dinoflagellate.

toxins cause sores on the fish that the dinoflagellate enters. Embedded in the sore, the dinoflagellate absorbs particles of flesh from the fish. Humans are also at risk from these toxins through skin contact with contaminated water. Generally, considerable exposure is needed before there is acute damage to humans, but fishers working in areas where *Pfisteria* are releasing toxins have reported neurological damage. The scientific researchers are not immune either, they too have experienced numbing of limbs, confusion, and loss of memory as a result of exposure. One fatality has been attributed to exposure to *Pfisteria* toxins. *Pfisteria* has at least 24 different morphological stages that it can adopt. Sometimes it could be easily mistaken for an amoeba, at other times a unicellular green alga; much of its biology remains unknown, but, as you can tell, this is not a typical species of alga!

Where fish deaths are occurring, or where toxins are detected in the water column, fisheries and beaches must be closed. A single harmful algal bloom lasting a few weeks can result in millions of dollars in commercial losses to the shellfish, fishing, and tourist industries.

Algal blooms occur naturally, but the increased frequency and wider geographic range of these events in the past 40 years has not been explained. Some areas that have been particularly affected are the Atlantic and Gulf coasts of the United States, parts of the Asian Pacific, and the Mediterranean Sea. The precise cause of the harmful algal blooms is not known, though it is widely suspected that nutrient enrichment caused by human pollution may play a role. For example, an eightfold increase in harmful algal blooms in Hong Kong harbor since 1960 can be correlated with a sixfold increase in the human population living within the watershed and the discharge of untreated sewage into Hong Kong harbor. However, other factors, such as the introduction of exotic species (ones that are not naturally found in those waters) from the ballast of ships or the influence of changing climates, cannot be dismissed. Until the ecology of these phenomena is understood better, a causal role for human pollution in the growth of harmful algal blooms remains a likely but unsubstantiated hypothesis.

Carbon and the Oceans

Land plants obtain carbon from atmospheric carbon dioxide, but marine algae and phytoplankton must gain their carbon from seawater. Carbon dioxide is readily dissolved in water, where it becomes bicarbonate HCO_3^- and is the source of carbon for phytoplankton photosynthesis. At the surface of the ocean there is a continual exchange of carbon dioxide with the atmosphere. An overabundance of bicarbonate in the ocean creates a concentration gradient of carbon dioxide between water and the atmosphere, releasing more carbon dioxide into the air than is absorbed by the sea surface. If carbon dioxide becomes overabundant in the surface layer of water (so that its concentration is higher in the water than in the atmosphere), it breaks down and releases carbon dioxide back to the air. Thus the concentration of carbon dioxide in the surface of the ocean should be in equilibrium with concentrations in the atmosphere. However, the story becomes more complex. Organic chemicals absorb the bicarbonate and sink to the bottom of the ocean. Also, bicarbonate may combine with calcium to form calcium carbonate ($CaCO_3$). Calcium carbonate forms the skeletal structure or protective shells of zooplankton, corals, some algae, and mollusks. Some of these organisms live on the bed of the sea, and others fall to the floor of the ocean when they die. Either way, the result is that carbon is continuously being removed from the upper waters of the ocean and transferred to the deeps. Removal of the carbon disturbs the equilibrium of bicarbonate with atmospheric carbon dioxide, but the balance is restored as carbon dioxide is absorbed from the atmosphere. Thus the oceans are continually absorbing carbon dioxide, taking in an amount equivalent to the carbon that goes into deep storage in dead organisms and the ocean sediment.

Burning fossil fuels releases carbon dioxide, and its increased atmospheric concentration may result in global warming (Chapter 23). Can the oceans absorb this additional carbon dioxide? The surface waters are likely to maintain an equilibrium with the atmosphere (or nearly so). However, the oceans do not mix fast enough through their vertical depth for the concentration of carbon dioxide to become distributed uniformly throughout the entire ocean system from top to bottom. The surface of the ocean will become saturated with carbon dioxide relatively quickly—long before all the added carbon dioxide has been absorbed. Consequently, the vast majority of the ocean volume will not take up extra carbon. Therefore the hope that the oceans will cleanse the air of the added carbon dioxide is ill-founded.

Coral Reef

Where seawater temperatures seldom dip below 20°C, great colonies of corals form islands and reefs. A reef is much like a cake made of skeletons that is iced with living corals and algae. The great limestone mass is formed by previous generations of corals or sometimes by ancestral clones of the modern coral. Although corals undergo spectacular episodes of spawning when they release gametes into the water in a blizzard of sex, much of their reproduction is asexual. A single individual may be made up of thousands of interconnected polyps that form a giant coral head. The skeletons of the dead polyps that support the modern ones may also have belonged to that same individual. Corals open their tentacled arms at night. Each tiny limb is lined with stinging harpoons that can catch and kill plankton. The prey is brought to the mouth of the polyp, ingested, and the digested products are often shared with adjacent polyps. It should be remembered that corals are animals; therefore, they are not primary producers. However, the brilliant colors of a coral are due to photosynthetic organisms called **zooxanthellae** that live within their tissues. The zooxanthellae are microscopic algae that receive shelter from their coral hosts and supply the corals with liquid nourishment in the form of dissolved organic carbon.

Coral reef habitats are among the most productive on Earth, with as much as 900 g $C/m^2/yr$. This productivity is all the more remarkable because corals thrive only where nutrients are scarce, for they will quickly be outcompeted, smothered, and killed by algae if nutrients are abundant. In the blue desert waters of tropical oceans, the reefs teem with life and sustain huge species diversity, but none of this requires substantial inputs of nutrients. Nothing is wasted, everything is rapidly recycled by armies of crabs, shrimp, slugs, snails, and fish. A coral reef is an excellent example of a highly productive ecosystem with a closed nutrient cycle.

With the record temperatures of the late 1990s, the surface temperature of waters around many reefs has risen by as much as 2°C. When water temperature exceeds 30°C, there is an increased probability that the corals will eject their zooxanthellae. The loss of their photosynthetic partners means that the corals receive less nourishment, and no one can yet explain why the corals should expel the algae. The expulsion of the zooxanthellae, the organisms that give the coral color, leaves the coral plain white. Hence, this process is commonly known as **coral bleaching**. Record amounts of coral bleaching were observed during the hot summers of 1997 and 1998. It is not clear that this was truly indicative of an increase in coral bleaching, or just that it was being noticed and documented more thoroughly than ever before.

Rocky Intertidal Zonation

Where the ocean crashes or laps against a rocky coast there exist some of the most fascinating and diverse marine habitats. The rock forms a solid base for seaweed such as kelp or bladderwrack and many marine organisms such as limpets and barnacles. The growth of vegetation provides food and shelter for animals. Underwater crags and crevices shelter lobster and sleeping fish, while sea urchins and sea stars cruise the rocks grazing the nutritious scum of algae and bacteria that encrust them.

The rise and fall of the tide leads to a vertical zonation of habitats in which the greatest variable is exposure to air. At the lower limit of the tide range, the organisms are submerged almost all the time, only briefly facing the challenge of desiccation. In contrast, near the high tide limit the organisms are submerged for just a few minutes each tide cycle and must survive long periods of drying. A lesser factor, but one that is directly correlated to the amount of exposure to the air, is exposure to damaging ultraviolet radiation. These zonations are highly variable between locations, but at any given site the zonation is usually clearly visible and can be divided into three main classes: the sublittoral zone, the littoral zone, and the spray zone (supralittoral).

The sublittoral zone starts at the lowest low tide mark and continues upslope to the upper limit of large seaweeds. This zone is only briefly exposed to the air, so the organisms do not have to withstand extreme desiccation. The wave environment is important in determining which species can survive, because this can be an extremely harsh environment if large waves batter onto the rocks. The pounding and sucking of the surf can dislodge any organism that is not securely fastened to the rock. Algae growing in this zone have evolved root-like structures called holdfasts that actually penetrate the rock. Their floating portion is also adapted to the rough environment by being tough and leathery. The stipe and blade (an equivalent to a stem and leaf) of the algae are flexible, bending and swaying with each rush of water. These algal beds have extremely high primary productivity, often matching that of a terrestrial forest (Table 5.1). The animals of these zones range from sea otters and seals to a huge diversity of fish that move in at high tide. At low tide invertebrates dominate, with a large diversity of mollusks, crabs, anemones, sea squirts, and starfish.

Above the kelps lies the littoral zone. Here, exposure to the air increases with elevation up the intertidal slope. Red algae encrust the rocks, and small and medium-sized seaweeds, mussels, snails, anemones, and barnacles may be scattered. Barnacles are not the ecologically most important members of this community, but their shoreward limit is used to define the transition from the littoral zone to the spray zone.

Parts of the spray zone are occasionally submerged by an extreme high tide, but mostly it is the shoreline that is wetted by wave splash and spray. Organisms living here will experience a completely dry environment during low tide. This zone is usually characterized by the presence of periwinkles and some snails. The primary productivity of this zone is low and comes from black encrusting lichens.

In comparison with rocky shores, sandy beaches are hostile environments to most organisms. Without a fixed substrate onto which they can grip, plants and algae cannot become established on the shifting sands. Burrowing worms, some shellfish, and crabs abound, but the diversity of species is low compared with a rocky coastline.

7.2 Groundwater

Water that exists beneath the surface of the land is referred to as **groundwater**. Rainwater percolates downward through most soils until it enters a saturated layer. The upper level of this layer, referred to as the **water table**, is likely to move up or down according to cycles of rain or drought. During wet times, when the water table has risen to the land surface, pools of water will form and will not drain away until the water table falls again. If the rains stop, the same location may have a water table 20 or 30 cm below ground level just a few weeks later. Whereas the term *water table* refers to a depth of the saturated layer beneath the surface, the term *groundwater* refers to the water itself. Porous rock that holds underground water is termed an **aquifer**. The rock of an aquifer contains spaces, ranging from microscopic to huge, that can fill with water. In general, such porous rocks are bounded below and to the sides by rocks impervious to water; this boundary is termed an **aquiclude** (Figure 7.7). Water flows continuously, but slowly, through an aquifer as groundwater gradually makes its way back to the ocean. Rates of flow can vary considerably, but 1 m to 10 m of lateral flow per century would be normal. If the water supply to the aquifer is from the side, it is called **lateral recharge**; if it is from above, it is termed **percolation** or **infiltration**.

One way that knowledge of groundwater movement can be applied is in calculating the sustainable use of a well. So long as the rate of water supply (lateral recharge and infiltration) exceeds the rate of extraction, the well will be sustainable. If, however, the extraction exceeds the rates of supply, the well will, in time, run dry. Another application is in determining the likely spread of pollution through groundwater. A chemical spill, say from ruptured chemical

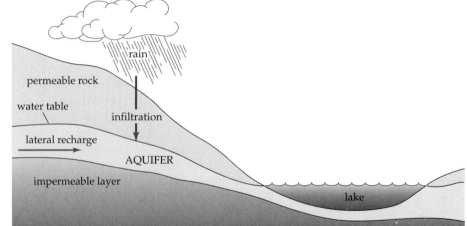

Figure 7.7 Movement and storage of water underground. Percolating water that enters the groundwater of an area may be stored in an aquifer. Water gradually drains downhill and generally goes out to sea, but in some geologies this lateral flow may be very slow, just a few centimeters each century. In other systems, underground rivers speed the movement of water.

drums, may seep downward into the groundwater, then spread outward as a plume of pollution. If the rate of lateral recharge is known, the distance and direction moved by the plume of pollution and its likely concentration can be calculated.

7.3   Surface Freshwater

A quick look at a map of North America reveals the abundance of lakes in some areas and their rarity in others; this disparity reflects different landscape histories and regional geologies. Even the deepest lakes are but dimples on the surface of Earth's crust and are generally the product of erosion. Most lakes are relatively young, at least on a geologic scale. The oldest lakes on Earth are only 5 to 20 million years old, and the vast majority of lakes are just a few thousand years old. We know that there were lakes in the time of the dinosaurs and before, yet not one of those ancient lakes survives to the present day. Thus, lakes are a transient feature of our landscapes, and whenever we see a natural lake, we can look for a recent (geologically speaking) or contemporary process that has led to its formation.

Lake Formation

Glacial activity is one of the great lake-forming processes in portions of America, Europe, the Andes, and northern Asia. In the last 4 million years there have been at least 22 glacial cycles when ice sheets advanced from the poles toward temperate latitudes. Each advance of the ice erodes the land surface, working like sandpaper to smooth even the hardest and craggiest geologies.

A moving glacier carries a huge amount of ground-up rock and debris, which is gradually pulverized into a rock paste or clay. The mixture of clay and partially broken rocks is eventually deposited by the glacier and is then termed **glacial drift**. Great smothering sheets of drift were deposited during the last ice age, reshaping Canada and much of the northern United States and northern Europe. The glacial drift is often pocked with depressions where ice once lay. Today, these depressions are filled with water to form **kettle lakes**. These kettle lakes form the modern lake districts of Minnesota and Wisconsin.

Probably the most common types of nonglacial lake originate from abandoned river channels or from low-lying areas that are seasonally flooded by adjacent rivers. Rivers are continually eroding their channels, and if the slope of the land is shallow, the river will start to form a series of tight loops called **meanders**. Rivers do not follow the same course year after year because the erosion on the outer bank of a river's meander loop is always stronger than that on the inside bank. The point of maximum erosion is also slightly downstream of the apex of the arch of the meander (Figure 7.8). Consequently, the meanders in a river channel are continuously working their way downstream. In general, the sediments carried by a river fill in the old channel as the new one is being carved. Oc-

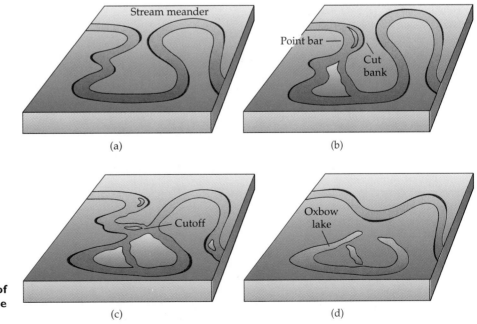

Figure 7.8   The pattern of bank erosion that leads to the formation of oxbow lakes.

casionally, as a particularly large meander loop forms, the river cuts a new path through the neck of the meander (Figure 7.9). The decapitated portion of the meander is left as an **oxbow lake**.

Lakes can also be formed where the local geology contains rocks such as limestone that are readily soluble in mildly acidic rainwater. Often, the limestone is buried beneath a thin veneer of rock or soil that allows water to seep down to the limestone. Progressively, the limestone is dissolved to form a series of caves or subterranean channels (Figure 7.10a). Through the millennia, such a cave enlarges until, if the crust of rock covering the limestone is thin enough, the chamber collapses. Although the process of cavern formation is slow, the collapse to reveal a basin can happen overnight (Figure 7.10b). If the water table is high, or

Figure 7.9   **Oxbow lakes forming beside a river.**

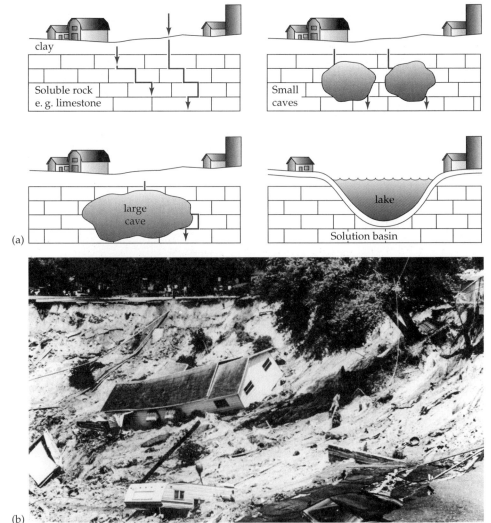

Figure 7.10   **Formation of a solution basin.** (a) If the local geology contains a soluble rock layer, the dissolution of the rock can form caverns. As a cavern enlarges, the strain on the roof increases, and ultimately the roof of the cavern will collapse. The depression in the ground surface becomes a lake. In the United States, this type of lake formation is especially common in Florida, where the underlying rock is limestone. These lakes can form suddenly, occasionally resulting in the sudden collapse of the land. (b) The sudden collapse of a subterranean cave swallowed up this house in central Florida. Caverns like this have been forming for many years out of sight. The collapse of the surface may take just a few minutes.

if there is sufficient rainfall, the basin will then fill with water and form a lake.

Volcanic cones and calderas, the craters left after huge volcanic explosions, provide steep-sided bowls that can hold lakes. If the volcano is still active, these provide unusual habitats for life. The lake water may be hot and soupy with volcanic mud. It will be extremely rich in sulfur, lacking in oxygen, and prone to being vaporized as steam during eruptive phases. These conditions may seem an extremely hostile environment for life; however, it is perhaps one of the most ancient habitats on the planet. The bacteria that dwell in these boiling sulfur mud pools are some of the closest living relatives of the sulfur-metabolizing bacteria that evolved 3.5 billion years ago (see Chapter 1).

7.4   What Happened to the Lakes Where Dinosaurs Wallowed?

If no lake remains from the time of the dinosaurs, something must have happened to them: What natural force would actively purge the landscape of lakes? Most lakes are fed by streams or rivers that drain a local catchment area. These streams carry sediment into the lake, and one school of thought is that it is the destiny of every lake to gradually fill with mud until there is no room for standing water. As the mud builds up, the area of open water gets smaller and smaller until there is just a small pool in the middle of the old

basin. Ultimately, even this pond is filled with mud. As the water recedes to the middle, a succession of plants colonizes the mud. First come the cattails and reeds, later some shrubs, and eventually trees grow right over the old lake. This sequential colonization by semi-aquatic plants and later by dry-land species is called a **hydrarch succession**. Not all ecologists agree on this scenario, and possibly a more likely view is that, although the edges of the lake may indeed fill with mud, the central portion will retain a water table at or near the surface, and there will always be a swamp (Figure 7.11). If this argument is true, there must be other processes at work to rid the landscape of lakes and swamps.

Many lakes have an outlet stream, the bed of which is continually being eaten away by erosion. The effect of this erosion is to lower the streambed relative to the lake bed in a process called downcutting (Figure 7.12). Eventually the stream may erode downward and reach the level of the lake bed, at which point the lake will drain, leaving just a river running through a wetland. Another way a lake may disappear is through the reshaping of landscapes as a result of earthquakes. Almost all parts of Earth periodically suffer from earthquakes (these tremors indicate that the crust of our planet is not stable). Severe earthquakes can cause Earth's crust to be warped locally, and this warping can cause the bed of a lake to tilt. The water could then spill out downslope. Yet another possibility is that past climatic change may have caused the lake simply to dry up. The dry basin could fill with wind-blown sed-

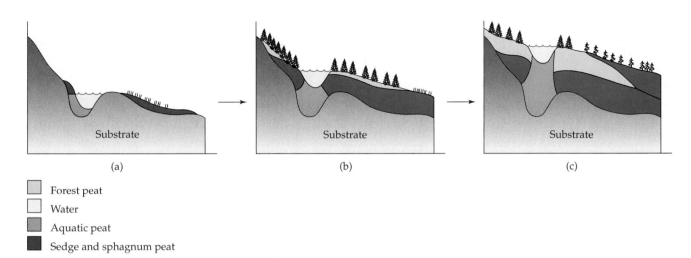

Forest peat
Water
Aquatic peat
Sedge and sphagnum peat

Figure 7.11   Hydrarch succession leading to a swamp. This model suggests that an area will always remain wet, even if it does not maintain an open body of water. Factors such as tectonic movement, climatic change, or a leaking natural dam are required to cause the wetland to dry out. (After M. L. Heinselman, 1963.)

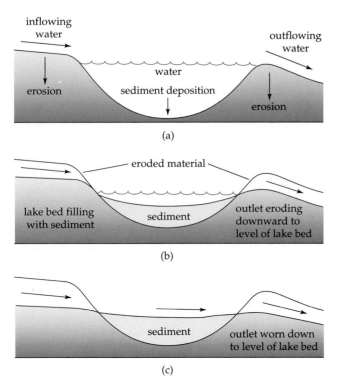

Figure 7.12 **A lake empties when its outlet stream downcuts to the level of the lake bed.**

iment, or it could erode so that it will not hold water again.

Any of these fates may have befallen those ancient lakes where dinosaurs drank and wallowed. Through geologic time, the lake sediments were buried and transformed by geothermal heat or the pressure of overlying sediment. The soft muds gradually were transformed into rock, coal, or oil and are now incorporated into the sedimentary geology beneath our land and oceans.

7.5 The Variability of Natural Lakes and Rivers

Freshwater bodies come in all shapes and sizes, and no two are quite the same. They range from being fertile green soups of organisms to pure blue pools with low productivity. Indeed, a single lake changes in character from one time of the year to the next. Freezing in winter, heating in summer, or wet-season deluges can all change the physical and chemical properties of a lake.

Nutrient availability and climate are the major determinants of the fertility of a lake or river. Just as the fertility of soil depends on the composition of the local rock (Chapter 5), the fertility of a lake depends on the geology of the watershed. The **watershed**, or catchment basin, may be defined as the area drained by a river's tributaries (Figure 7.13). The erosion of local rocks and soil by flowing water results in nutrients being carried in suspension or in solution into the lake or stream. High nutrient concentrations in local rock and soft, or soluble, geologies are likely to result in high nutrient concentrations in local streams.

Lakes that are rich in nutrients are **eutrophic**. Such fertile lakes are generally murky green or brown and have very poor underwater visibility. The rich nutrient supply in a eutrophic system supports dense blooms of algae whose photosynthetic pigments absorb most of the light entering the lake. Like photosynthetic plants on land, the algae are poor at absorbing green light, so green is the main color of light that reflects back to our eye.

Lakes that are poor in nutrients are **oligotrophic**. Because of the lack of nutrients, productivity is low. Phytoplankton populations are small, and these lakes appear blue. Light entering the lake is absorbed by the water, and only blue light penetrates. It is then reflected back to the surface. The light that leaves the

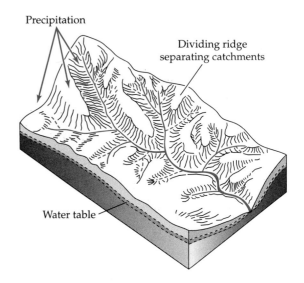

Figure 7.13 **A watershed, with streams draining into a river.** Note that, at their junction, the stream and the river form a V that points downstream. (After McKnight, Tom, *Physical Geography: A Landscape Appreciation.* 5th ed., 1996. Adapted by permission of Prentice-Hall, Inc., Upper Saddle River, NJ.)

lake is, therefore, mainly in the blue portion of the spectrum. Furthermore, because the lake supports few phytoplankton, the water is clear, and such lakes have good underwater visibility. Thus the picture-postcard lakes of mountain regions, the lakes that we find the most aesthetically pleasing, are nutrient-poor systems with low primary productivity.

Although some lakes are truly oligotrophic or eutrophic, most lakes are somewhere between these extremes (**mesotrophic**). Many lakes are affected by human actions and are becoming more eutrophic. Natural waters that receive sewage, fertilizers running off fields, or soil erosion have elevated nutrient concentrations and a changed ecology (Chapter 18).

Temperature Profiles

The same pattern of thermal layering apparent in oceans can be found in many lakes. By midsummer, lakes that were frozen in winter are often warm enough for swimming. But even in the height of summer, the swimmer has only to dive a meter or two beneath the surface to find colder water. The warm surface water, the **epilimnion,** is separated from the cold bottom water, the **hypolimnion,** by an abrupt thermocline (Figure 7.14). It is as if two lakes are occupying the same basin, one floating upon the other. To explain this separation, we need to start with some of the thermal properties of water.

One of the key factors in the evolution of life is that water reaches its maximum density at 4°C. As water cools below 4°C, it becomes less dense and floats on top of the denser (warmer) water. This property determines that water will always freeze from the top downward, and it allows ice to float. If water density increased as temperature decreased, then as air temperatures fell below freezing, ice would form at the bottom of a lake and rapidly freeze all the way to the surface. Fish and aquatic plants rooted to the bed of a lake could not survive, as they would rapidly become encased in ice. Had it not been for this property of water, life on land and in the oceans would have been forced to evolve in an entirely different manner.

When the surface of a lake cools and freezes, the ice forms a thin crust that gradually thickens. Although air temperatures may plummet to −40°C or −60°C, ice can be no colder than 0°C. As the ice thickness increases, it forms a layer that insulates the underlying water from the temperatures above. One consequence is that ice growth at the base of the ice layer slows down. Even under the coldest Arctic conditions, it is rare to find ice forming a layer more than 3 m thick (an ice sheet or glacier is built up of snow that accumulates and freezes into a solid mass, not of liquid water freezing). Water under the ice will range in temperature from freezing, just beneath the ice, to as warm as 4°C at the bed of the lake. Thus, the fish and plants have only to survive a fraction of the temperature change experienced on land, and, with water permanently available as a liquid, they can survive until the spring thaw.

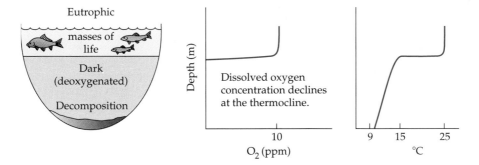

Figure 7.14   Thermal stratification and oxygen concentration in a temperate-latitude eutrophic lake during the summer. The epilimnion is a layer of warm water floating on the cold hypolimnion. The thermocline is the discontinuity between the two water segments. If you were to dive down through the thermocline you would feel a sharp cooling, because the temperature gradient is very steep. The nutrient-rich waters support a dense bloom of plankton at the surface. Light is blocked, and the hypolimnion is dark. The decomposition of dead organic matter leads to high BOD and the deoxygenation of the bottom waters, hence oxygen concentrations drop to zero just beneath the thermocline. The water of these lakes appears green and turbid.

7.6 Seasonal Changes in a Lake

Lakes undergo seasonal changes in their temperature, oxygen, and nutrient profiles. These events are partly driven by climate and partly by the organisms living and dying within the lake.

Temperature and Oxygen

Linsley Pond, Connecticut, freezes each winter. The surface of the lake is 0°C when it is frozen, and the dense water at the bottom is about 4°C (Figure 7.15). During this phase, any wildlife will either be hibernating in the bottom mud or living in the water beneath the ice. The ice seals the lake from the atmosphere, and the water is without a fresh supply of oxygen until the spring thaw. Decomposition rates in such a cold lake are relatively low, and the adult populations of many creatures will be at a minimum. Nevertheless, BOD in the sediment can cause the oxygen concentration of bottom waters to fall. With the thawing of the ice, the surface water starts to warm, although the bottom water stays close to 4°C. The lake in early spring has a cold portion of water (0°C to 3°C) lying above a warmer, denser portion of water (4°C). The upper lake water is absorbing oxygen from the atmosphere and thus becomes oxygen rich relative to the bottom water. As the surface layer warms and approaches 4°C, the lake has a uniform temperature profile from top to bottom. At this time, the two layers in the lake are so similar in temperature that the layering is unstable. A spring storm sends the waters of the deep swirling to the surface in what is called the spring

overturn. Immediately after this event, temperature and oxygen are evenly distributed throughout the water column; but this state does not last long. The summer sun warms the surface waters more than the deeps, and the warm water is less dense than the cold water. Consequently, the warm water floats above the cold water. Once this pattern is established, the disparity between the warm and cold layers increases. At this point, the epilimnion (the upper warm water), hypolimnion (the deep, cold water), and the thermocline (the zone of rapid temperature change) have been formed, and the dense bottom water scarcely mixes with the lighter water. Once again two separate lakes occupy the basin, but this time the coldest water is at the bottom (Fig. 7.15).

With the onset of fall, the surface waters start to cool; eventually they will reach approximately the same temperature as the bottom water. The layering of the lake is now unstable, and a strong wind can be enough to cause the autumn overturn, in which the whole lake mixes and the thermal layering is lost. The lake cools further with the onset of winter. Contact with cold air cools the surface waters rapidly while the bottom water maintains a temperature of 4°C. Because at this temperature water attains its maximum density, the colder water floats above the dense bottom waters and the winter layering is established.

The Seasonal Availability of Nutrients

Eutrophic systems commonly experience a roller-coaster pattern of high and low nutrient availability

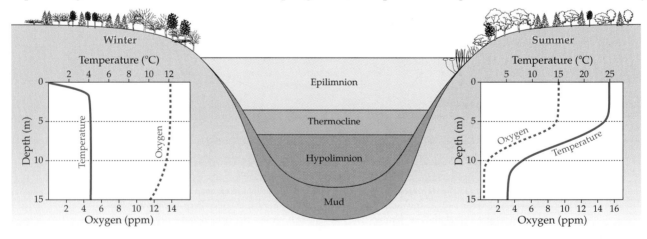

Figure 7.15 Seasonal changes in the thermal stratification and oxygen concentrations in Linsley Pond, Connecticut. The summer conditions are shown on the right and in the cross section of the lake. The thermocline between 5 and 10 meters depth separates the warm, oxygen-rich epilimnion from the cold, oxygen-poor hypolimnion. Winter conditions are shown on the left. In winter the pond surface freezes, and temperature and oxygen are evenly distributed throughout the water column. (After E. Mose in E. Deevey, Oct. 1951, *Scientific American*, pp. 70–71.)

within a growing season. As a temperate, eutrophic lake warms in the spring, plankton and algae build large populations and absorb nutrients from the water column. By early summer, these populations may run short of nutrients and their populations decline. The dead bodies are added to the layer of decomposing organic matter on the bed of the lake. Progressively, the organic material is broken down to its constituent nutrients. As nutrients become available once again there is a second bloom of algal populations in the early autumn. This bloom is cut short by falling water temperatures and shorter day-lengths. During the fall overturn (and again in the spring overturn), the slurry of decomposing organisms that forms the mud–water interface is mixed back into the water column. This event redistributes nutrients throughout the lake (Figure 7.16). This pattern is common to most nutrients such as, nitrogen, phosphate, iron, and silica in lake systems. In oligotrophic lakes nutrients are consistently scarce, and there are not enough of them to support these cyclic explosions of productivity.

Consequences of the Stratification Cycle for Wildlife

The layering of the lake has profound effects on oxygen and light availability, both of which are strong environmental gradients for aquatic organisms. However, the strength of these gradients *in the summer* is largely determined by the fertility of the lake. Thus, we need to consider oligotrophic lakes and eutrophic lakes separately, because they develop characteristic environmental gradients for oxygen, light, and nutrients.

In the early summer the thermocline forms, and the bottom waters, the hypolimnion, are sealed off from the atmosphere by the warm waters of the epilimnion. In an oligotrophic lake, there are too few nutrients in the water to support substantial growths of phytoplankton, but the mud on the lake bottom will often contain enough nutrients to support rooted aquatic plants (Figure 7.17). So long as the lake is not overly deep, light can penetrate to the bed, allowing rooted aquatic plants to photosynthesize. Water will absorb all colors of light apart from blue light, but chlorophyll can absorb this wavelength of light effectively; thus photosynthesis is possible. Because the byproduct of photosynthesis is oxygen, these green plants oxygenate the bottom water of the lake. Once the thermocline is established, the photosynthetically derived oxygen is the only source of oxygen for the hypolimnion. Meanwhile, the surface waters of the epilimnion are oxygenated either by diffusion from the atmosphere or, more significantly, as a result of little bubbles of oxygen being dissolved into the water whenever there are ripples or waves on the lake surface. Thus the lake is oxygenated both above and below the thermocline (Figure 7.17). The lack of nutrients in the epilimnion ensures that almost all the

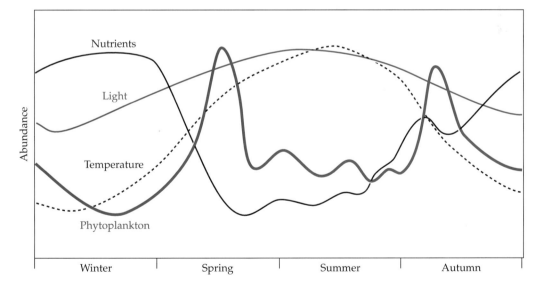

Figure 7.16   Schematic diagram relating lake phytoplankton populations to abiotic factors. Spring warmth, longer days, and nutrient availability promote a spring phytoplankton bloom. However, the booming population depletes nutrients; populations decline and remain low during the summer. As decomposers release nutrients from dead phytoplankton, there is an autumn surge in phytoplankton population. Cool autumn temperatures and decreasing day length cause the population to contract once more. (After E. Odum, 1971.)

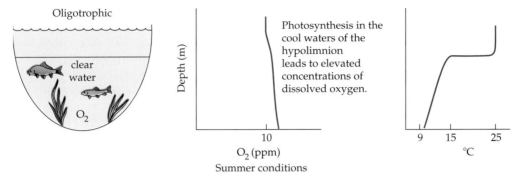

Figure 7.17   Thermal stratification and oxygen concentrations in an oligotrophic lake.
The lake is nutrient poor, thus populations of plankton are small (they are nutrient limited), but there are
enough to allow a slow buildup of organic sediment on the lake bottom. The decomposition of the organic matter releases nutrients that can be absorbed by rooted aquatic plants. The plants photosynthesize, so the hypolimnion is provided with O_2. The water of these lakes appears clear and is oxygenated from top to bottom.

biological activity is in the hypolimnion. The system is
nutrient limited, so populations are small; therefore
there is not much demand for oxygen from zooplankton or fish. Because the populations are sparse, the
amount of dead organic matter on the bottom of the
lake is small. Therefore the BOD from decomposers is
modest even in the height of summer.

Early in the summer, the eutrophic lake will also
develop a thermocline. In this lake, the fertile waters
provide an abundant supply of nutrients throughout
the water column. Dense growths of short-lived phytoplankton grow at the surface, and zooplankton
abound as they prey on the phytoplankton; the lake
is a teeming mass of life (see Figure 7.14). The surface
of the lake becomes a dense mat of algae; these plants
absorb and block the penetration of light. Consequently, the epilimnion is lit, but the hypolimnion is
inky dark. Beneath the thermocline there is no supply
of atmospheric oxygen, and, although there are abundant nutrients, no plants grow in the hypolimnion.
The deeps are simply too dark to permit photosynthesis; all the light has been absorbed by phytoplankton and plants in the epilimnion. Thus the upper lake,
the epilimnion, has both an atmospheric and photosynthetic oxygen supply, but the hypolimnion has
none. Furthermore, the BOD of the hypolimnion is extremely high. The BOD is caused by the decomposition of bodies that tumble down through the water
from above, a steady pulse of the freshly dead. As the
corpses of algae, zooplankton, and fish fall to the lake
mud, their bodies are invaded and respired by decomposers, a process that requires oxygen. With no
fresh supply of oxygen and this high demand for the
dissolved oxygen left over from the spring mixing of
the lake, the hypolimnion becomes oxygen depleted.

The bottom of a eutrophic lake is often completely deoxygenated before the end of the summer. The autumn overturn of an oligotrophic lake is not a particularly spectacular affair, but for a eutrophic lake it can
be an extremely smelly one. As the layers attain similar temperatures and are mixed by an autumn storm,
oxygen is mixed throughout the lake, but so too are
the products of decay that have been accumulating on
the lakebed. The semidecomposed remains rise to the
surface, and for a few days they can provide a rich and
distinctive aroma. The dead settle back to the bed of
the lake, but nutrients have now been mixed throughout the water column, ensuring another fertile growing season next summer.

It is not uncommon to find the shoreline of a eutrophic lake plastered with dead fish. Late in the summer, after algal populations have boomed, the bottom
of the lake is a mass of decomposing organic material.
The water temperature is nearing its maximum, and
this means that its ability to hold dissolved oxygen is
at a minimum. During the night, there is no photosynthesis taking place in the surface waters, and therefore no oxygen production. But throughout the lake,
especially in the bottom waters, respiration is removing oxygen from the water column. Under these conditions, the BOD is very high, and dissolved oxygen
concentrations can sag dangerously low. The result is
a "fish kill" in which most fish in the lake die for lack
of oxygen. Although this chain of events can happen
under natural conditions, it is made much more likely
by human activities that lead to the enrichment of a
lake with nutrients (Chapter 18). Clearly, this sequence
of events is similar to those when marine algal blooms
lead to fish mortalities. That the freshwater fish kills
are exacerbated by increased nutrient concentrations

in the water provides strong circumstantial evidence that marine harmful algal blooms are also the product of eutrophication.

River Systems

Rivers can be variously eutrophic or oligotrophic, but their movement makes them a very different habitat from that of a lake. The kinetic and potential energy of moving water has long been harnessed by people to float logs downriver, to turn millstones that grind corn, and more recently to turn turbines to provide hydroelectric power.

The power of a river is a function of the slope of the riverbed and the volume of water that is flowing. The slope will determine the speed of the river, and this in turn will determine the size of the rocks that a river can move. A fast-flowing mountain torrent can move large boulders and is likely to have a steep-sided streambed with a rocky bottom. At the other extreme, a slow-moving river meandering across a floodplain is likely to have a broad, flat riverbed that is covered in mud. However, because the wide, muddy river carries so much more water than a gushing stream, it is likely to have much greater energy.

The environment of a stream or river is not constant throughout the year. Spring snowmelt or winter rains often increase the flow rate, increasing the energy of the stream and, hence, its ability to carry suspended loads of sediment. Conversely, in summer, some smaller rivers may dry out completely, leaving just a few isolated pools of water. Organisms inhabiting these rivers must be adapted to living under such changeable conditions.

To avoid being washed out to sea, most organisms in a stream or river are capable of swimming or are attached to the bottom or sides of the channel. The extremely dynamic habitat of a high-energy stream results in the rocks' being rolled around the streambed, reducing opportunities for plants and animals to establish colonies. Given the turbulence of the water, these rivers are well mixed and usually have plenty of dissolved oxygen.

In the slower-moving rivers, the muddy bottom and sides of backwaters and creeks, and even the main channel, provide plenty of opportunity for algae and larger plants to grow. If the muds carried by the river are rich in nutrients, these environments can be extremely productive. Most rivers are well oxygenated, but it is possible that large, sluggish, muddy rivers can become relatively oxygen poor as a result of high productivity leading to high BOD during the summer.

Rivers that are shaded by overhanging trees are cooler than those where the trees have been removed. In the Hubbard Brook experiments (see Chapter 6), after sections of woodland flanking a stream had been removed, the temperature in the shaded and unshaded portions of the stream differed markedly. At midday on a sunny summer's day, the shaded river had a temperature of about 15°C, whereas the unshaded sections had temperatures of about 20°C. Such a temperature difference would profoundly affect the ecology of a stream. Some of the changes that might be anticipated would be a decrease in the dissolved oxygen in the water (warm water cannot hold as much oxygen as cool water) and increased productivity because of increased light availability. These changes may be manifested by a dense growth of filamentous green algae covering the rocks, riverbed, and stems of aquatic plants. The species composition of the phytoplankton may change, leading to changes in other tiers of the food chain. Fish such as salmon and trout are intolerant of both the increased sediment input that generally accompanies deforestation and the warmer water temperatures. Such sensitive species would quickly be lost from the river.

Summary

- All water, whether in the ground, in the oceans, or on the land surface, is in the process of moving through the hydrological cycle.

- The main reservoirs of water are formed by oceans, groundwater, freshwater, ice, and atmosphere (water vapor).

- Oceans cover 71% of the surface of Earth and provide about 50% of all NPP.

- Open-ocean productivity is generally less than 100g $C/m^2/yr$, but upwellings and continental shelf areas may be two to three times as productive.

- Coral reefs are closed nutrient systems that maintain high rates of productivity under oligotrophic conditions.
- Water bodies can be vertically thermally stratified into epilimnion, thermocline, and hypolimnion.
- Increased nutrient supply can lead to algal blooms and fish kills.
- Old lakes are lost through geological factors and hydrarch succession, and new lakes are formed

by glaciers, rivers, chemical dissolution, and volcanic activity.

- Lakes vary in their natural productivity according to the local geology and climate.
- In ascending order of fertility water bodies are oligotrophic, mesotrophic, or eutrophic.
- Different communities of aquatic life will be found according to the nutrient levels and temperature and oxygen profiles of lakes.

Further Readings

Abramovitz, J. N. 1996. Imperiled waters, impoverished future: The decline of freshwater ecosystems. *Worldwatch Paper 128*. Worldwatch Institute, Washington, DC.

Barker, R. 1997. *And the Waters Turned to Blood: The Ultimate Biological Threat*. Cupertino, CA: Simon & Schuster.

Baskin, Y. 1995. Can iron supplementation make the equatorial Pacific bloom? *BioScience* 45:314–317.

Larson, D. 1993. The recovery of Spirit Lake. *American Scientist* 81:166–177.

Peterson, C. H. 1991. Intertidal zonation of marine invertebrates in sand and mud. *American Scientist* 79:236–249.

Pielou, E. C. 1998. *Fresh Water*. Chicago: Chicago University Press.

Sparks, R. E. 1995. Need for ecosystem management of large rivers and their floodplains. *BioScience* 45:168–182.

Wetzel, R. G. 1975. *Limnology*. Philadelphia: Saunders Publishing.

Web Connections

On-line resources for this chapter are on the World Wide Web at:
http://www.prenhall.com/bush
(From this web-site, select Chapter 7.)

Why Wetlands Aren't Worthless

8

Bog, mangrove, marsh, mudflat, slough, peatland, pothole, and swamp have two things in common: They are all types of wetland and seldom form part of a fashionable address. Despite being ecologically and evolutionarily fascinating, wetlands are some of the least appreciated of all habitat types. Swamps conjure images of mosquitoes, leeches, smelly mud oozing between your toes, and realtors desperate to sell you a "wonderful piece of rural Florida." Until relatively recently, wetlands were considered worthless by both the public and the government. Indeed, the Swamp Lands Acts of 1849, 1850, and 1860 expressly encouraged the drainage and reclamation of wetlands. Subsidies and policies promoting the destruction of wetlands persisted until the mid-1970s. Of the 42 million hectares of wetland once present in the lower 48 states, 53% have been destroyed since the arrival of Europeans (Figure 8.1). In some areas, such as the Mississippi valley, 95% of wetlands have been destroyed. Only in the last 30 years has there been a realization that these habitats are important components of our landscapes.

Recent appreciation of the value and diversity of wetlands is based on the realization that they often provide a wide range of ecosystem functions that affect not only the wetland itself, but also the adjacent ecosystems. Wetlands prevent floods, improve water quality, and provide breeding and feeding grounds for many plants and animals. Wetlands are vital to the survival of more than one-third of the species listed as endangered or threatened in the United States (Chapter 28) and are essential to the recruitment of commercially important fish, shellfish, and birds.

8.1 What Is a Wetland?

The gradual blending between the edges of one habitat and another makes it almost impossible to define, ecologically, a precise boundary. The transitional area between the two habitats has features of both and is termed an *ecotone* (Chapter 6). Wetlands often form the ecotones between open water and dry ground.

The soils of wetlands are saturated long enough to induce chemical changes. The chemically changed soil makes ordinary plant growth difficult, favoring specially adapted wetland plants. Already we can see that the plants and soils of a wetland will be unusual and will play a part in defining it. The third factor, the physical presence of water, is highly variable. Long droughts can dry up lakes, let alone wetlands, so we cannot always expect wetlands to be permanently soggy. For the moment, we will assume that a wetland is an area that has water at, or close to, the soil surface for part of the year, but does not normally have standing water more than 2 m deep. In other words, the open water of a deep lake is not a wetland, but the

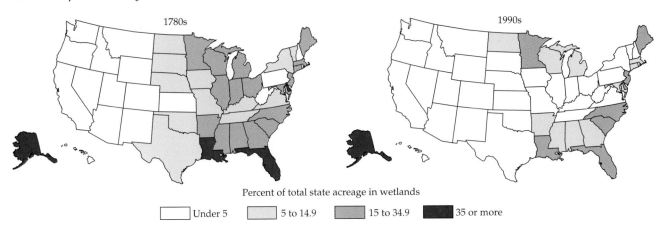

Figure 8.1 **The past and present distribution of wetlands in the United States.** (After R. Doyle, June 1998, *Scientific American*, 178(6), p. 25.)

shallow water at the edge of the lake and everything that forms the transition to dry ground is. Wetlands can be any size, from a tiny puddle to the immense expanse of the Canadian peatlands (111 million hectares).

A **marine wetland** forms in the intertidal zone along shorelines away from a river mouth. Here, the principal factors that determine the type of habitat will be the intensity of wave activity on the beach, the tidal range (the difference between the highest high tide and the lowest low tide), and whether the intertidal area is composed of sand or mud. A common marine wetland along the eastern U.S. seaboard is a grassy marsh with large areas of exposed mudflat. Although the mudflat may look bare, it often supports dense colonies of algae, burrowing worms and shellfish, and shorebirds. These areas are important to the survival of larval and juvenile fish and as feeding grounds for waterfowl and wading birds. Mangroves are important marine wetlands, though they need a source of some freshwater to survive. The wetlands protect the coastline from beach erosion and are commercially important for harvesting shellfish such as clams, cockles, and shrimp.

An **estuarine wetland** lies within the mouth of a river, in the area affected by saltwater. Freshwater from the river mixes with salt water pushed upstream by tides and wind. At the ocean mouth, the concentration of salt is about 30 parts per thousand, and it steadily diminishes in an upstream direction. The upper limit of the estuarine wetland is defined as being where the tidal effect is so reduced that the river is "fresh" and has less than 0.5 parts per thousand of salt. Examples of estuarine wetlands include salt marshes, mudflats, or in warmer climates, mangroves. The wetland vegetation and sediments buffer coastal settlements from flooding and erosion caused by storms. Estuaries have immense primary productivity and hence support rich food chains. These wetlands are essential feeding stations for waterfowl and wading birds during long-distance migrations. Estuarine wetlands are important areas for the recruitment of fish and shrimp.

Freshwater wetlands make up 91% of all wetlands in the lower 48 states of the United States. These wetlands range from being lush, highly productive sites teeming with life, to bare-bottomed peatland pools that are very acid and support little life other than bacteria and midge larvae. Every region of the country has its own range of wetland habitats, and they will vary according to local climate, soils, bedrock, and ecology. Although these habitats look remarkably different, they all have ecological functions that are shaped by wetness.

How Wet Is a Wetland?

Most wetlands will have some standing water covering the soil surface for at least part of the year. However, if you visit the site during a dry time you may be able to walk around without getting your feet wet. The duration and the depth of flooding are the most important ecological variables that determine differences between freshwater wetlands. The upper limit of the flooded level within the soil is described as the water table. Water tables are fed by lateral movement of water through the soil or through an aquifer and by infiltration (Chapter 7). If the pore spaces of the soil fill with water as the water table rises, it is said to be **waterlogged**. It is a characteristic of wetlands that for much of the year the water table will be at, or close to, the soil surface. But the extent and duration of flooding and the rhythm of wet and dry cycles vary greatly between sites. The seasonal variation in water level in a wetland is described by its **hydroperiod**. Some examples of different hydroperiods reflect major seasonal patterns, such as the wet and dry season of an Amazonian forest (Figure 8.2a). Here, a four-month dry season from August to December is followed by a rainy season that floods the landscape to a depth of 8 m. In the American Midwest (Figure 8.2b), the bottomland forests are dry nine months of the year, but the snowmelt and spring rains cause the forest to flood from March until June. Note that a water level below

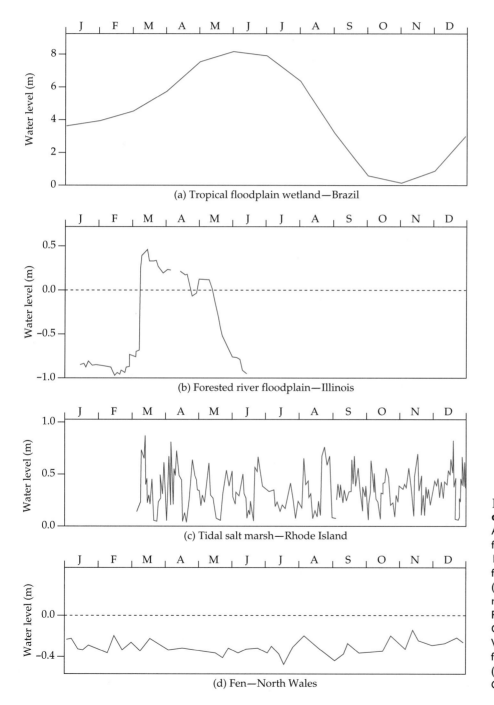

(a) Tropical floodplain wetland—Brazil

(b) Forested river floodplain—Illinois

(c) Tidal salt marsh—Rhode Island

(d) Fen—North Wales

Figure 8.2 Contrasting hydroperiods for four wetlands: (a) Amazonian forest with deep flooding for much of the year. (Data from Junk, 1982.) (b) Spring flooding in the forested flood plain of a river in Illinois. (Data from Mitsch et al., 1979.) (c) The rhythmic inundation of a tidal marsh in Rhode Island. (Data from Nixon and Oviatt, 1973.) (d) A fen peatland in Wales with soil saturation, but no flooding. (Data from Gilman, 1982.) (Figure modified from Mitsch and Gosselink, 1993.)

the ground surface is indicated by minus values on Figure 8.2b. The hydroperiod of the coastal marshes of Rhode Island is determined on a daily basis by high and low tide and on a monthly basis by the waxing and waning of the moon that causes tides to be higher or lower than average (Figure 8.2c). These rhythms provide a regular cycle of flooding. However, the irregularity of the curves is the result of random events, such as storms and high winds, that push water into or out of the marshes, being superimposed on the basic cycles. Some wetlands are wet all the year round, but may never have standing water at the surface (Figure 8.2d). An underground spring is permanently soaking the soil of the Welsh fen (a fen is an alkaline wetland), and even though it never has surface water, it is still a wetland. Because hydroperiod controls the chemical reactions that take place in a wetland, it is the single most important variable in wetland types.

8.2   Water and Wetland Chemistry

Wetlands are harsh environments for plants. Most plants thriving in these settings have undergone evolutionary change that makes them wetland specialists, slaves to a twilight environment between land and water. These species have sacrificed the ability to live on dry land, but have largely overcome a huge evolutionary hurdle: living with low oxygen availability.

The single most important statistic affecting the chemistry of a wetland environment is that oxygen diffuses about 10,000 times slower through water than it does through air. Underwater, or in the saturated sediments of a wetland, biological and chemical oxygen demand can exhaust the supply of oxygen, leaving these systems entirely **anaerobic** (oxygen free).

A dry soil has oxygen in its airspaces, and chemicals such as manganese, iron, and nitrate exist in an oxi-

dized form. As the soil is saturated with water, say due to a flood, the free oxygen in the airspaces is lost. Because oxygen diffuses so slowly through water, only the layer of the sediment in contact with the air remains oxidized. At depths greater than a few centimeters, the remaining oxygen is quickly depleted by microbial decomposers.

The biological and chemical demands for oxygen in these organic-rich soils are high. Under conditions of high oxygen demand and low oxygen supply, the chemicals are progressively reduced. Chemical reduction means that an atom, molecule, or ion gains an electron. One of the most common ways that this happens is when a molecule gives up an oxygen atom. Some molecules surrender their oxygen atoms more easily than others and can be thought of as oxygen donors.

$$O_2 \rightarrow NO_3^- \rightarrow Fe(OH)_3 \rightarrow MnO_2 \rightarrow SO_4^{2-} \rightarrow CO_2$$

first to
surrender
oxygen

last to
surrender
oxygen

As a soil floods, the oxygen donors should disappear in sequence, starting with free oxygen (O_2) molecules. In experiments it was found that within one day the initial supply of free oxygen is exhausted in flooded soil (Figure 8.3). With the loss of free oxygen, nitrate (NO_3^-) became the oxygen donor, and after two to three days all the nitrate had been converted to nitrite (NO_2^-). After about a week of saturation, iron was reduced from its ferric state (Fe^{3+}) to its ferrous (Fe^{2+}) state. This reduction provides one of the most visible changes in a soil. The oxidized iron is rust-red, but when it is reduced to the ferrous form it turns blue gray and provides the dominant color of many wetland clay soils. In the experiment, manganese was reduced at almost the same time as the iron. With further reduction, sulfate (SO_4) is converted to a range of chemicals including iron sulfide

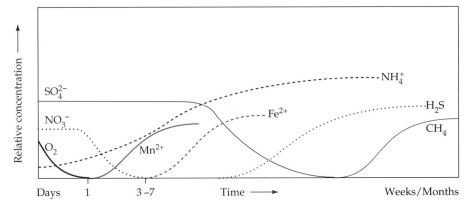

Figure 8.3   The sequential change in soil chemistry following flooding. Oxygen depletion is followed by the reduction of nitrate, ferric iron, manganese, sulfate, and carbon dioxide. As waterlogging continues, soil concentrations of ammonium, ferrous ions, hydrogen sulfide, and methane increase. This graph can also be interpreted as relative concentrations, with depth in a wetland soil.

(FeS) and hydrogen sulfide (H_2S), which has the smell of rotting eggs. As the reduction process nears completion, carbon dioxide (CO_2) is reduced to methane (CH_4). The entire reduction sequence can take several months of inundation, although temperature and rates of biological activity influence the speed of these processes.

The oxygen content of the soil can be replenished during times of lowered water table. As water level drops below the ground surface, air enters the pore spaces in the soil, and the soil becomes oxygenated. The reducing reactions are reversed, leading to **oxidation** of the soil as minerals lose an electron or gain an oxygen atom. Reducing conditions are re-established as the soil floods upon the return of wet conditions.

Another way to use Figure 8.3 is to think of the x axis as representing increasing depth in the sediment. The chemical conditions at the top of the soil profile approximate those at the left-hand end of the x axis. Deeper in the soil, water saturation lasts longer, and the reduction of chemicals progress further. Hence, conditions lower in the sediment profile are represented farther to the right on the x axis. From a biological viewpoint, the loss of O_2 is the most critical factor because plant roots, decomposers, and burrowing animals all require O_2 for respiration. Some bacteria can use other oxygen donors, thus they can function in anaerobic conditions. The depth to which free oxygen is available in the soil determines the abundance and activity of decomposers and burrowing animals, such

as earthworms and beetle larvae, and the densest growth of plant roots (Figure 8.4).

Decomposition and Nutrient Storage

Organic material accumulates quickly on the soil surface of many wetlands. Each layer of falling organic debris is covered by the next before decomposition is complete. Once the organic material is buried deep enough to enter the anaerobic zone, the lack of oxygen donors greatly reduces rates of organic decomposition. The deeper the organic material is buried in the soil, the more reducing the environment and the slower the rate of decomposition. The sediments that accumulate in a wetland have a high organic content, containing at least 20% dry weight of organic matter. Some wetland soils are almost pure organic material; these are known as **peats**.

A dramatic example of the power of preservation in a peat bog came to light in 1953. A human body, the victim of death by strangulation, was found in a peat bog near the village of Tollund in Denmark (Figure 8.5). Tollund was a sleepy place, and finding a body with a noose around his neck in the local bog caused a stir. The police were called in to investigate what appeared to be a particularly grisly murder. Tollund Man may well have been murdered, but the felons were not around to be punished. On closer examination and with the aid of radiocarbon dating, it was found that the body had lain in the bog for 2000 years prior to its discovery.

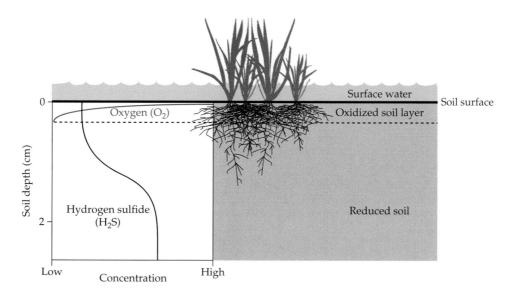

Figure 8.4   **Schematic diagram of plant rooting relative to oxygen and sulfide concentrations in a flooded wetland soil.**

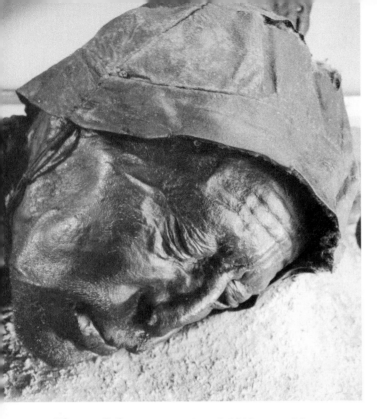

Figure 8.5 Tollund Man: A 2000-year-old corpse recovered from a Danish peat bog. Note the preservation of such fine features as eyebrows, eyelashes, beard stubble, and the noose that killed him.

Tollund Man was not the first bog body discovered, but he was one of the best preserved. The preservation makes these bog bodies an archaeologist's treasure trove. The gut contents provided evidence of the last meal and his parasite load. The state of the teeth indicated dietary quality, and the pattern of scratches on the tooth enamel indicated the meat-to-grain ratio in his diet. Basic dentistry and medicine, such as bone setting, can also be seen in some bodies.

To a wetland scientist, apart from its ghoulish interest, Tollund Man demonstrates some of the most important principles of wetland chemistry. The local hydrology was similar to that in Figure 8.2d, in which there was little fluctuation of the water table, leaving the soil anaerobic. His preservation is due to the lack of oxygen that inhibits the process of decomposition. In the absence of decomposition, the carbon and nutrients of his body (apart from the calcium that dissolved from some of his bones) are still held in his corpse. Chemically, Tollund Man's body is essentially the same as that of a cattail or a dragonfly, and we could expect them to be similarly well preserved. Therefore, the lack of oxygen in a wetland leads to the long-term storage of nutrients and carbon.

The property of nutrient storage by wetlands provides another clue to their ecological value in a landscape. The ecosystems of many rivers and lakes are adversely affected by excessive nutrient loads. Soils eroding from hillsides and nutrients washed from decomposing organic matter in terrestrial ecosystems are often absorbed and held by wetlands before reaching the rivers and lakes.

Wetlands are usually, broad, flat landscape features that allow water to spread out and slow to a crawl. Dense vegetation, such as a reed bed, also slows the flow. If the water is carrying dissolved nutrients such as phosphate or nitrate, these chemicals are rapidly scavenged from the water column by plants, algae, and bacteria. The nutrients are temporarily stored in the bodies of the organisms. Unlike a terrestrial system where death is followed by decomposition and nutrient release, in a wetland the lack of decomposition results in the nutrients being permanently trapped in the sediment. When a wetland releases fewer nutrients than it receives, it is functioning as a **nutrient sink**. A wetland that releases more nutrients than it receives is a **nutrient source**. The most common way for a wetland to shift from being a sink to a source is by drying up. Drying reverses the chemical processes that lead to nutrient storage, and previously stored nutrients are released. So long as the wetland is functioning as a sink, concentrations of nutrients in water leaving the wetland may have been reduced by 50%–90%. Because wetlands act as biological filters, removing chemicals that would otherwise damage the wildlife of downstream ecosystems, they have been called the kidneys of the landscape.

8.3 Wetlands As Hydrologic Regulators

Wetlands, like forests, are sponges that absorb runoff. Water is soaked up by peat soils; indeed, 80% of the soil volume and 90% of its weight may be water. Although flow entering a wetland may vary greatly with recent rainfall, the amount of water leaving a wetland is much more constant; wetlands temporarily store water. The wetland has a much greater capacity to absorb excess water than does a river channel. If more water is added to the wetland, it causes the water level in the wetland to rise. This is a flood, but it is a comparatively gentle and predictable event. Wetlands absorb as much as 80% of the peak discharge that would otherwise gush downstream as a damaging storm surge. Long after the rains have ceased, the flood water that has been stored will continue to flow from the wetland. This gradual release supplements minimum flow periods and helps alleviate the effect of drought on downstream aquatic systems. Thus, wetlands moderate erratic flow regimes and increase the diversity of organisms that can live in the downstream environments.

In a valley where the wetlands have been removed, there is no buffering of rainfall events. In a downpour, the upland soils quickly become saturated with water. Puddles form, and water starts to sheet off the soil surface running downhill to the nearest stream. Leaves, topsoil, and twigs are carried along as the water washes the soil surface. Streams swollen with runoff drain every corner of the catchment and surge into the main river channel. Without the wetlands to absorb this flow, the river rushes seaward gaining energy as more streams add to its volume. Racing water has a huge destructive potential. River channels can be reshaped in a single night, and human constructions such as levees, dams, and bridges can fail under the raging torrent.

The problems caused by flood events are compounded by the ways humans use wetlands. People like to live near rivers and build homes on riverbanks, often within the floodplain of the river. By building within the floodplain, natural flow regimes are altered as water flows much more rapidly from roof shingles, asphalt, and concrete than from forest or wetland. Development, therefore, leads to higher peak discharges during storm events. To make matters worse, the wetlands that would have reduced peak flows are often filled or drained in the course of development. Consequently, our development pattern enhances peak flow events and places homes in the path of flood water. The inevitable happens, and the homes are flooded. Rather than seeing this as a clear message that a floodplain is not a good place to build a home, the homeowners are subsidized by federal funds to restore or rebuild flood-damaged houses. Federal flood insurance programs have encouraged people to take the risk of building in what are essentially unsuitable locations. In the United States, about $5 billion of tax money is spent each year to repair flood damage. Much of this expense could be avoided if restrictions were placed on building in floodplains, and the money was used to assist people to relocate rather than rebuild.

8.4 Adaptations to Living in a Swamp

The first organisms to make the transition from life in the ocean to life on land did not come ashore in a single bound. Many generations would have been required to evolve from aquatic to terrestrial life-styles. Transitional stages would have exploited previously untapped feeding opportunities in wetland environments. Little by little, organisms evolved to live entirely on dry land. Thus, wetlands played a pivotal role in our evolution. Many of the animals that live in a wetland, such as fish and frogs, provide insights into the evolutionary sequence that occurred as life moved onto land. Gills were replaced by lungs, fins replaced by legs and feet, and scales were exchanged for a slimy skin that allowed improved gas exchange. Given that the ancestors of terrestrial animals were able to make this transition, we tend to see wetland creatures as an evolutionary intermediary between fully aquatic and terrestrial organisms. For example, toads have gills and a finlike tail as a tadpole and metamorphose into an adult with lungs and feet. Toads are wonderfully adapted to a wetland environment, specializing at different life stages in both its aquatic and terrestrial aspects. Similarly, the life history of a dragonfly combines both submerged and aerial phases. In temporary wetlands, the larva of a dragonfly may be the top predator in the food chain. These larvae are voracious underwater hunters that can catch and kill small fish, tadpoles, and other insects. Through a series of stages, the dragonfly emerges to rule the skies above the wetland, where it feasts on mosquitoes and other small insects. Female dragonflies often return to the water by crawling down the stem of a reed to lay their eggs. Both toads and dragonflies can survive as adults away from the wetland, but are absolutely reliant on it for their reproduction. Indeed, animals that have evolved to live fully on dry land still have a "wet" phase in their reproduction. Mammalian young spend a period of gestation in the amniotic fluids of their mother's womb, while birds, insects, and reptiles develop in a fluid-filled egg. Dependency on a moist beginning is a vivid reflection of our aquatic origins.

Flowering plants evolved from ancestors that had already made the transition to life on dry land, but numerous evolutionary lines of flowering plant have adapted to live in wetland settings. The challenge posed by anaerobic soils has led to the evolution of new ways to supply the roots with oxygen. A common response of trees to saturated soils is to develop swollen trunk bases. Because oxygen can be adsorbed through the trunk and swelling increases the trunk's surface area, there is an increased surface for gas exchange near the root zone. Other means to supply roots with oxygen are to produce breather roots or **pneumatophores**. These are root extensions that grow upward from below ground level into the air and are the botanical equivalent of a snorkel tube. Even when flood waters cover the soil surface, cypress knees and the pneumatophores of black mangroves ensure a flow of oxygen to the roots (Figure 8.6).

A common adaptation of wetland plants is to have some elongated, swollen cells that are filled with air. These specialized cells, termed **aerenchyma**, form a continuous chain from the leaves down to the roots. Oxygen can flow freely through these air chambers, and they allow plants to maintain active growth despite a lack of soil oxygen. Some plants even maintain a slight pressure in the aerenchyma that forces air out of the root. This exudation of oxygen causes the sediment and the molecules immediately surrounding the root to remain oxidized. Within this layer, molecules that are potentially toxic, such as manganese and sulfides, are oxidized into harmless forms.

In many wetlands, the organic soils may be rich in nitrogen, but because of the anaerobic conditions it is present in the reduced form of nitrite that plants cannot absorb. In wetlands that are only periodically saturated, during the dry times nitrogen is likely to be lost to the atmosphere as a result of denitrification (Chapter 6). Consequently, nitrogen deficiencies may severely limit plant growth in many different wetland settings. Wetland plants have evolved numerous ways to overcome nitrogen shortage, but two ways have evolved independently in many different lineages. One means to overcome nitrogen deficiency is to exude oxygen from roots, causing adjacent nitrite molecules to be oxidized to ammonium or nitrate, which can be absorbed through the root. A second way to gain nitrogen is to become a predator.

Insects are packets of nitrogen with wings, and some plants have evolved to attract and then ambush insect visitors. Pitcher plants, sundews, and Venus flytraps are all wetland plants whose sweet scent lures insects to a sticky end. These plants are unrelated and so each represents an independent evolutionary path in which the problem of nitrogen shortage was overcome by carnivory. All of these plants have evolved digestive enzymes to break down the insect and solubilize its nutrients. However, the separate evolutionary paths have led to different ways for carnivorous plants to capture their prey: Pitcher plants form a pitfall trap from a modified leaf tip, sundews have a sticky surface, and Venus flytraps have a modified leaf

Figure 8.6 Cypress knees. The knobby structures that surround a swamp cypress tree are modified roots that increase the uptake of oxygen. Cypress knees are an adaptation to living in saturated soil with low oxygen availability for at least part of the year.

Figure 8.7 **The Venus flytrap.** This carnivorous plant has leaves that snap shut as a fly lands on them. The insect is then digested by enzymes secreted by the plant. Carnivory among wetland plants is an evolutionary response to nutrient shortage, particularly to low nitrogen availability in many wetland soils.

that is touch sensitive and snaps shut when an insect lands on it (Figure 8.7).

8.5 Wetlands and Wildlife

A human can live about three weeks without food, but only two to three days without water. A waterhole in a dry region may be the single most important com-

ponent of the landscape because it is the only place where animals can drink. Although many organisms are highly adapted to wetland environments, these wet areas support an even larger number that use the wetlands on an occasional basis as a watering hole, for reproduction, or for hunting. It is estimated that 80% of all species of birds breeding in the United States and 80% of fish living in coastal waters spend part of their life cycle in a wetland. Immature forms of many species that live elsewhere as adults rely on wetlands as their nursery ground. Vulnerable young can shelter from predators among the maze of stems, venturing out to feed when the coast is clear. The high productivity of these systems provides a substantial crop of phytoplankton and zooplankton, which in turn assures rapid growth of young insects, fish, and amphibians.

Although many species of duck spend the majority of their time at sea or on large inland water bodies, all nest within wetlands. It is estimated that 50%–80% of all ducks in North America breed in the prairie potholes of the U.S. Midwest and southern Canada. These potholes are depressions in the landscape that are dry for part of the year, but fill with water each spring, providing there is enough rainfall. Between 1955 and 1985 the estimated duck population ranged between 35 million and 16 million birds, with no consistent trend up or down. However, when the number of ducks was compared with the number of potholes that are flooded each year (Figure 8.8) a strong relationship is evident: The more flooded potholes at the beginning of the breeding season, the larger the number of ducks. The flooding of potholes is dependent upon the level of local water tables. Water tables can be lowered as a result of reduced precipitation or by

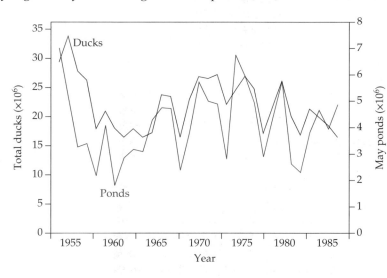

Figure 8.8 **Populations of ducks relative to water level in the prairie potholes of the Midwest and southern Canada, 1955–1985.** (From Batt et al., 1989; copyright Iowa State University Press, reprinted with permission.)

extraction of water from aquifers for agriculture. This data set clearly demonstrates the relationship between hydrological change and the reproductive success of wetland-dependent species.

It is important to realize that animals as well as plants shape an ecosystem. Beavers and alligators are examples of animals that create new habitats within their ecosystem. Beavers are forest animals, and their presence can radically change a local landscape. The dams that they build block streams and create a pond. Generally, this pond would be considered to be a wetland, and it may cover tens of hectares. In mountainous regions, these beaver pools may represent the great majority of still-water habitats. Trees whose root bases have been inundated by the beaver pond are likely to die and become important nest habitats for woodpeckers and other hole-nesting creatures, such as owls and Wood Ducks.

Alligators shape the wetlands in which they live by creating a wallow. As the dry season approaches and water levels drop in the shallow marshes of the southeastern United States, alligators excavate a hole where they can stay moist and cool even during the driest times. Falling water levels concentrate fish and amphibians into the last remaining pools, many of which are these "gator holes." The alligator lying in the hole is ensured a food supply, and the pool, though occupied by a predator, is the best chance of survival for the fish. Without the gator hole, many more fish would die in the annual dry season. Another feature of a gator hole is that the sediment excavated from the hole forms a tiny island that is a little drier than the surrounding wetland. Trees can take root on the island and the trees in turn can support a wide array of insects and birds. Thus the presence of the predator adds to the biodiversity of the system.

8.6 Indicators of Ecosystem Quality

The quality of a wetland can be judged by the functions that it performs for surrounding ecosystems. High-quality systems are those with high biodiversity and with the ability to clean water and store nutrients. The local benchmark of quality is the level of function offered by a natural wetland that has not been heavily disturbed by human activities. Wetlands that may have been degraded by urban or agricultural development, or those that have been created recently, can be compared with high-quality wetlands to determine the completeness of their ecological functions. Certainly, a wetland can perform the functions of nutrient storage and flow regulation and yet offer poor biodiversity. Some wetland species are much more demanding of their habitat than others. For example, cattails and pickerel weed can grow under almost any nutrient regime, whereas sawgrass can only flourish under low nutrient conditions. Adding nutrients to a system causes cattails to outcompete sawgrass and all the other oligotrophic-habitat specialists. In terms of ecosystem function, exchanging one suite of plants for another does not unduly affect hydroperiod or nutrient storage (actually storage would increase, because of increased productivity). However, the eutrophication of the wetland would be seen as a form of pollution and a loss of habitat quality. Ecologists, therefore, look beyond the strictly biochemical or hydrological aspects of the wetland and also evaluate aspects of biodiversity in deciding the quality of a wetland.

Biodiversity measures vary considerably, but they should be more than a simple estimate of species number. Incautious estimates of species diversity may include upland or invasive species that swell apparent biodiversity, but are indicative of problems within the wetland. Other estimates of community and population parameters are preferred, such as completeness of food webs, reproductive success, and population sizes.

Food-web studies can compare the breadth of species on each trophic level in a high-quality wetland with the wetland under scrutiny. A strong food web has many alternate species at each trophic level. If the population of one species crashes due to disease, it does not cause the whole web to collapse. Predators simply switch to other prey items until the fallen population recovers. Consequently, a food web with only a few species at its base, or the absence of top-level predators, would reveal a biodiversity problem in the system.

The reproductive success of all species in the system is important. Variation in reproductive success and population fluctuations are part of nature. A given stimulus can induce a widely differing response among species. For example, a fire will cause one species to lose members while another flourishes. When whole suites of species start to decline, it is likely that the system is undergoing major ecological upheaval. An example of such a decline is found among the world's amphibians.

Frogs: The Wetland Canary?

Canaries die at carbon monoxide levels slightly lower than those fatal to humans. By carrying a canary into a coal mine, the bird served as an early warning system for the presence of toxic mine gas. Similarly, by monitoring frogs in wetlands, we may receive warning of chemical problems in our environment.

In the 1990s a global decline in amphibian populations has been observed. The decline is not restricted to disturbed areas, but is also evident in areas of remote rain forest and high mountains where human activities are minimal. It is possible that the amphibians are being adversely affected by global climate change and that the downturn in their populations is natural. However, most scientists suspect that a combination of factors are at work. Likely contributors to the decline in amphibians include the following:

1. Human introduction of new predators, such as trout and bass that decimate tadpole and froglet populations

2. Habitat destruction, such as draining wetlands

3. Acid deposition, because low pH makes immature stages much more susceptible to toxic effects of other chemicals

4. Global climate change and increased frequency of drought

5. Pesticides and other chemicals entering wetlands from human activities

In 1995, a class outing by students from New Country School, La Seuer, Minnesota to a local wetland set in motion a wave of publicity that has heightened interest in and awareness of the plight of amphibians. The schoolchildren noticed that many of the frogs in the pond were deformed. A few deformed individuals is natural in any population, but to find a high percentage with major deformities is a clear indication that something is wrong. Since then a mixture of good and alarmist reporting has swept through the media and the internet.

The root fear is that the frog deformities are new, therefore a new factor, possibly toxic chemicals in the local water, are causing the birth defects. Because the earliest developmental stages of both tadpole and the human fetus are chemically and sequentially very similar, a chemical that induces deformities in frogs may do the same in humans. This could result in increased frequencies of birth defects among babies born to women drinking water contaminated with these chemicals. Any time there is the prospect of human birth defects, the potential problem must be taken very seriously. But concern does not constitute proof, and rigorous scientific investigation is needed to discover the cause of the frog abnormalities.

One possibility is that there are no abnormal levels of frog deformities. Malformed frogs were recorded as early as 1740 in North America and across Europe. Furthermore, the intensified interest in wetlands and declining frog populations may mean that there are unprecedented numbers of observers to document birth defects.

A wider census of frogs showed that birth defects appeared to be concentrated in the agricultural region of the upper Midwest and the adjacent provinces of southern Canada. In this region, many frogs had developed an extra set of partially formed, but paralyzed, legs (Figure 8.9). Others were missing legs or eyes, or had contorted spines. In Quebec, a survey of four species of frogs and toads found that among 854 individuals, 12% had hindlimb malformations. That the birth defects seem to be concentrated in the upper-midwestern states and southern Canada suggests that at least in these areas there is a real problem to be investigated.

Another possibility is that parasitic trematodes (flukes) may be causing the deformities. Trematodes have a complex life cycle that involves at least two hosts. In the case of trematodes that parasitize frogs, snails are a secondary host in their life cycle. The cysts of trematodes have been found in unusually high numbers in malformed limb buds from which a leg should have grown. But why should trematodes be more abundant in agricultural areas? The snails needed by the trematodes as intermediate hosts are generally favored by eutrophic conditions. Agricultural runoff rich in fertilizer or animal waste would boost productivity in the ponds, increase the snail population, and hence increase the likelihood of building large trematode populations. This hypothesis is still

Figure 8.9 A deformed frog from a midwestern pond.
The extra pair of legs is partially formed, but paralyzed. The cause of apparently increasing numbers of such birth defects in frogs is not known.

being investigated, but it does not easily explain the growth of additional limbs.

The third possibility is that a pesticide, or complex of pesticides, is causing the deformities. Residual DDT and the myriad of chemicals that are currently entering the environment could cause, or interact to cause, birth defects. One particular chemical group used in mosquito control is being subjected to intense scrutiny. The pesticides, one of which is methoprene, mimic developmental hormones in the larval mosquito. Shortly after the egg has been fertilized, the cluster of cells that forms the embryo starts to partition into body sections. The sequence of development determines the orientation of head (anterior), tail (posterior), front (ventral), and back (dorsal) surfaces. Next comes the partitioning of sections of the body that will later form brain, spinal cord, and legs. It is important to note that this does not mean that these features all become formed at this early stage, but within the first few cell divisions those cells "know" what they will later become. The instruction comes from messenger hormones, one of which is retinoic acid. Experimentation on a variety of organisms shows that a lack of retinoic acid leads to gross deformities, among which is a failure of leg development. If too much retinoic acid is supplied to the embryo, additional legs form.

The pesticide, methoprene, is not retinoic acid, but it has such a similar chemical structure that mosquito larvae absorb it, become deformed, and fail to develop. That is the intent of the mosquito-control operation. This type of pesticide was introduced to be less dangerous than insecticides such as dieldrin and DDT (Chapter 18), which are much less selective in what they kill and are known to cause cancer in humans. Unfortunately, it is possible that in addition to retarding the development of mosquitoes, methoprene also affects frog development. Methoprene is not the first mimic of retinoic acid suspected of causing birth defects in vertebrates. Accutane, a retinoid-based acne treatment, was shown to cause birth defects in humans. The concern is real that the pesticides applied for mosquito control could affect human fetal development, but the causal link has yet to be established.

The current consensus among scientists working on this issue is that there is a genuinely high rate of birth defects among some populations of amphibians. However, they also state that it is unlikely that any single factor is responsible and that a complex interaction of biological and chemical changes may have induced the deformities.

8.7 Altering Wetland Functions and Values

Natural wetlands have distinct ecosystem functions to which humans can assign a value (Table 8.1).

The alteration of any physical function (such as hydroperiod or water quality) has repercussions on other functions, and this chain reaction affects the values attributed to the wetland. It is possible to construct a model that shows how one change in function affects all the wetland values. The assessment of value for such uses as recreation or hunting may best be described by environmental economics. Monetary values can be attributed to an organism if it performs a clear service or if it appeals to the general public (Chapter 27). But many species do not fall into these categories, and, ultimately, assessments of the value of endangered species and biodiversity may be best done through ethical considerations (Chapter 22).

The values of a natural ecosystem are derived from its functions. Land-use changes that minimize disruption of those functions are likely to be ecologically the most benign. Many values resulting from wetland functions are felt outside the wetland, such as downstream flood control and water quality. Consequently, the link between the value and the wetland is not ob-

Table 8.1   Some common functions and uses of wetland ecosystems.

Ecosystem function	Ecosystem value
Nutrient storage	Removal of fertilizing nutrients helping farmers attain compliance with water quality targets
Accumulation of organic material for fuel or agriculture	Storage of carbon, source of peat
Filtering solids from waters	Wastewater and sewage treatment, trapping sediment
Animal habitats	Fishing, wildfowl hunting and, fish and shrimp hatcheries
Plant habitats	Forestry, agriculture (e.g., harvesting cranberries)
Regulating water outflow	Flood and erosion prevention
All functions	Recreation, research and education

(*Source:* C. Richardson, 1994.)

vious. For the last 30 years, environmentalists and ecologists have publicized the value of wetlands, but marshy areas remain cheap tracts of land that appear ripe for agriculture or development.

The ecosystem function that is most frequently targeted for change is water storage, and this is achieved through drainage. By cutting a ditch through the wetland below the level of the water table, water will flow out of the wetland soils. The flow of water into the ditch lowers the water table in the wetland and brings about a number of chemical changes in the soil that increase its usefulness as a cropland. The foremost of these changes occurs as water drains from the soil and is replaced by air. This is a transition from anaerobic to **aerobic** (oxygen rich) conditions. Wetland plants are specialists at living under low-oxygen, or no-oxygen conditions, for at least some of the year. Even wetland plants need oxygen for their metabolism and have either sacrificed some metabolic efficiency to accommodate the low-oxygen conditions or they have adapted structurally to deal with the low-oxygen levels. Such specialization, while allowing them to live under the extreme condition of a wetland, reduces their ability to compete with dry-land plants in aerobic conditions. As the land dries out it will be invaded by plants adapted to dry conditions that outcompete

the wetland plants. Wetland specialists such as ducks, wading birds, and dragonflies find that their wetland is transforming into a woodland that does not offer them suitable living conditions. Thus, the drainage of the marsh affects the habitat functions for both plants and animals.

Drainage of a wetland reverses the process that led to the storage of nutrients and carbon. The drainage allows oxygen to enter the soil, and under aerobic conditions rates of decomposition increase and nutrients are cycled more rapidly. Because organic matter is now decaying and not being preserved, the wetland is no longer functioning to store nutrients. Far from working as a kidney, the drained wetland will now release the chemicals previously cleansed from the river. Thus, draining a wetland has a twofold effect on downstream water quality. The loss of the wetland eliminates the filtering process so that all the upstream contamination is now experienced downstream of the wetland. Second, the oxidation of the peat releases previously stored sediment and nutrients that will cloud the water and cause eutrophication.

Drainage can have several physical effects on peat, the greatest of which may be shrinkage. Wetland peat has the structure of a fine sponge with water-filled pores. Water makes up 80% of the volume of many

Figure 8.10 Peat shrinkage and oxidation following drainage in the Florida Everglades. This house was built directly on top of the soil surface in 1930. The structure is supported by columns reaching down to the bedrock, thus the house has not moved even though the underlying peat has shrunk 2 m vertically.

peats and provides structural support to the peat bed. If the water is drained from the peat, the structure collapses and the peat shrinks. A combination of shrinkage, decomposition, and oxidation reduces the volume of peat in a wetland when it is drained. In the Florida Everglades, where drainage has resulted in a lowering of the water table, as much as 2 m of peat have been lost from the upper portion of the soil profile in the past 50 years (Figure 8.10, p. 119).

8.8 The Restoration of the Florida Everglades

The Florida Everglades has been severely damaged by the alteration of its natural hydrology. In this century, the populations of more than a dozen species of ibis, stork, herons, and egrets has fallen by about 90%. Invasive species, such as cattail, that are indicators of poor water quality are invading the great expanses of sawgrass, and the whole ecology of the system is teetering on the brink of change. The Everglades is affected by upstream changes in water quality, and, in turn, it affects Florida Bay, which receives the outflowing water (Figure 8.11). In the past 20 years, the ecology of Florida Bay has changed as water quality has deteriorated. It appears that too little freshwater is flowing out of the Everglades, and the water that does leave the park is rich in nitrogen. Increased nutrient levels have fostered algal blooms, and surges in salinity are thought to have precipitated a die-off of 27,500 hectares of sea-grass. Recognition that changes must be made if the Everglades and Florida Bay are to be enjoyed by future generations has led to the most ambitious restoration project yet undertaken. The cost of this restoration process is estimated to be $1.5 billion to be spent between 1997 and 2003.

In the Everglades the goal of restoration is to restore the natural functions of the wetland. "Natural" in this context is taken to be the state of the ecosystem before the first major drainage canals were installed in the 1880s. Two basic factors must be restored: natural hydrology and natural water quality.

The natural hydrology of the Everglades has been substantially altered as a result of canal construction and flood-control structures installed over the past century. Under natural conditions the water flowed from the Kissimmee River into Lake Okeechobee and then spilled across the entire southern end of the lake basin as a sheet flow. The entire Everglades area from Okeechobee south to the coast is a shallow river, seldom more than 2 m deep and about 50 km broad. In the 1940s, a system of canals and dikes was installed

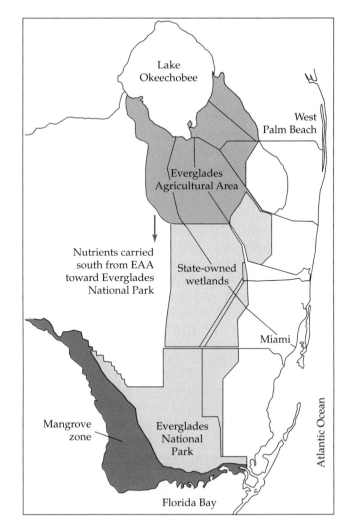

Figure 8.11   The south Florida drainage. Note how the natural north-to-south drainage has been altered by canals that divert water to the east.

that controlled flooding south of Lake Okeechobee. Drainage of almost 5 million hectares south of the lake formed the Everglades Agricultural Area (EAA). The EAA is a nationally important agricultural producer and during the winter provides more than 40% of farm produce supplied to the eastern United States. Cane sugar is one of the most important crops. The crops and jobs of this agricultural region make it a multibillion dollar industry.

The Kissimmee, which pours water into Lake Okeechobee and from there into the Everglades, was once a gently winding river that meandered through a marshy plain. In the 1960s, the Kissimmee River was straightened, and its floodplain turned over to cattle pasture. The straightening, loss of wetlands, and spread of cow pasture disrupted flow and increased the input of nutrients to Lake Okeechobee. The lost water storage function of the Kissimmee wetlands resulted in a net loss of water to the Everglades. Instead of draining gradually out of a wetland, water rushed into Lake Okeechobee in storm surges. The storm input forced so much water into the canals of the EAA that it was simply allowed to flow directly into the Atlantic. The hydroperiod of the national park was completely changed due to a reduced volume of inflowing water, which was channeled rather than in a sheet flow, and the timing of its flood events.

The proposed restoration project includes buying 34 km of the Kissimmee floodplain and restoring those marshes. Also, the U.S. Army Corps of Engineers will undo the canalization of the river and restore it to its natural, winding channel. To the south of the Everglades Agricultural Area the plan calls for almost 20,000 hectares to be purchased from the sugar growers and established as a biological filter to clean water leaving the EAA. Nine canals are to be redirected, and seven new pumping stations are to be built in an attempt to restore a near-natural hydroperiod to Everglades National Park.

Almost as important as the amount of water and its hydroperiod is the quality of the water. The ecological balance of the wetlands of the Florida Everglades is threatened by elevated concentrations of nutrients arriving in water draining from the EAA. The natural ecosystem of these areas is limited by phosphate availability, meaning that a slight increase in the availability of this chemical is likely to bring about profound changes in the productivity and species composition. The threat of phosphate pollution comes from excess fertilizer washing from the fields of the EAA. The water leaves the EAA through a complex of canals; some is guided to the metropolitan area of Miami and some goes south to the Everglades.

The central problem is that a single source of water, Lake Okeechobee, supplies water to users with different needs. The principal users are: South Florida Water Management District (SFWMD), The National Park Service, Dade County, and farmers. Each of these groups has a goal and each employs ecologists to represent their interests.

SFWMD has the obligation to divide water resources between farmers, the 5 million people who live in southern Florida, and the Everglades National Park. They are also concerned that the water quality should be of an appropriate standard for each of their users. Dade County has a similar set of goals, since they represent the urban population of the greater Miami area and the Everglades National Park. However, Dade County has no direct interest in the farmers of the EAA because they lie in a different county.

The National Park Service has the goal of protecting the wetlands of the Everglades National Park. The park is under severe pressure from a variety of ecological changes that have resulted in a 90% reduction in breeding bird populations over the last 50 years, and seven of the native species are recognized as being close to extinction. Excess fertilizer washing out of the EAA has already changed the ecology of areas in a belt at least eight kilometers wide south of the EAA. The concern is that this fertilization will continue to spread southward into the wetland sanctuary of the Everglades National Park. To protect this habitat, the National Park Service is seeking the maximum possible water flow with phosphate concentrations of about 8 ppb (parts per billion) entering the Everglades.

The crops (sugarcane, rice, and citrus) and jobs of the EAA add billions of dollars to the local economy, which gives the farmers a lot of political power. The farmers need to control water on their land to ensure a good harvest, and they also need to use fertilizer. Water leaving the farmers fields has concentrations of phosphate between 100 ppb and 200 ppb, compared with the natural concentration of 8 ppb. Removing this excess phosphate would be expensive. Consequently, the farmers want the allowable phosphate concentration to be as high as possible.

The dilemma is clear; if the farmers are allowed freedom of action, the ecology of Everglades National Park and other wild lands in Florida will be at risk. If, however, the farmers are required to emit no phosphate pollution, the profitability of farming, local employment, and food prices will all be adversely affected.

Forthcoming legislative changes require that a standard be set for phosphate concentration in the natural waters of South Florida. Existing law dictates that this new standard must ensure that water quality (including the amounts of pollutants) would not result in an "imbalance in the natural flora and fauna." Ecologists have been given until the year 2001 to recommend an appropriate phosphate concentration for the new water-quality standard.

The scientists working for the farmers recognize that phosphate is a problem and that some limit must be set. However, they emphasize that the environmental problems of the Everglades are not wholly caused by phosphate. In addition to researching the phosphorus issue, their work demonstrates the ecological imbalances that result from varying the water level in the Everglades, the invasion of non-native species, and of the effects of nitrate or mercury contamination.

The ecologists working for the National Parks suggest that because phosphorus is the limiting nutrient in this system, any change could lead to a cascade of ecological change. They recognize that other factors are involved but state that unless the phosphate pollution is held in check, there is no chance of maintaining the Everglades as a natural system.

Given the high priority that conservation receives in Florida, it is likely that the legislated value for maximum phosphate concentration will be closer to 10 ppb than the 50 ppb hoped for by the farmers. But this is a debate that is bound to intensify as the deadline of 2001 approaches.

Perhaps the most important message to emerge from the whole Everglades water-quality issue is that ecology, like other sciences, is not truly objective. Although the data set that emerges from such a study appears as a set of numbers on a page and, therefore, seems free of bias, this is not necessarily the case. All the scientists working on an issue will be inclined (consciously or subconsciously) to interpret the data according to previously held beliefs. This is not a question of intellectual honesty or dishonesty, it simply happens because scientists are not computers, but human. The scientific method may be objective, but human beings are not.

8.9 Wetlands and the Law

There are no laws that specifically protect habitats. Some wetland protection is granted by a series of laws designed to do other things. For example, a wetland that contains an endangered species receives some (though not certain) protection from the Endangered Species Act (1973). In essence, this act will prevent harm being done to a particular species and its home (Chapter 28). However, most wetlands are not graced by an endangered species, and a more common form of protection is found under Section 404 of the Clean Water Act of 1972. This act requires that, if possible, wetlands are preserved. If a developer or government agency cannot avoid damaging the wetland, the damage must be minimized and the wetland restored as much as is practicable. If the wetland is to be damaged or destroyed, a permit is required from the Army Corps of Engineers. This section of the law is designed to protect wetlands from hydrological changes caused by filling or dredging and applies equally to government agencies such as the Department of Transport and the general public (Chapter 28). However, wetland destruction for the purpose of forestry or agriculture is not covered by this law. Relatively few permits are denied, though the mitigation (see Ecology in Action: Wetland Restoration) that is required on projects that affect more than 0.5 acres of wetland can be expensive.

The feasibility of a development project, both in terms of permits and finances, may hinge on the quantity and quality of wetlands that would be destroyed. Developers will need to minimize the area that is classified as a wetland, while conservationists and regulators may want a more inclusive definition of a wetland. When large sums of money are involved, lawyers and court battles can't be far away. An ecologist's rather vague statement that a wetland will be an ecotonal area between upland and aquatic habitats, or an area that is at least intermittently flooded, cannot be used in court to draw a definite line delineating what is, or is not a wetland.

In 1989, President Bush laid out a government policy that henceforth there should be "no net loss of wetlands." Though sounding tough on developers, the policy had a considerable amount of fine print in which exceptions would be made if the loss of the wetlands was "unavoidable" or "insignificant." These qualifiers provided plenty of flexibility in the interpretation of the no net loss policy. "No net loss" is not the same as "no loss." At the heart of this policy lies the belief that creating or restoring degraded wetlands can compensate for the destruction of a natural wetland. There then remains the thorny issue of determining what is the potential net loss caused when a wetland is destroyed. Should this be seen strictly on a hectare for hectare basis, so that if a hectare of wetland is destroyed, a hectare is created elsewhere? Or should it be seen in terms of habitat function? A hectare of high-quality natural wetland may perform the same functions as 10 hectares of created wetland. In which case, if less than 10 hectares were created there would have been a net loss of function.

The compensation made by a developer for damaging a wetland is termed **mitigation**. Mitigation can take a number of forms, any of which will result in roughly equal cost per hectare to the developer. The quality of the wetland being destroyed and the quality of the mitigation sites cause substantial variability in the amount of mitigation required (Table 8.3). Another variable is the size and degree to which the natural wetland is damaged. Mitigation is not needed if less than 0.5 acres of wetland is damaged, but can become very expensive if a large area of high quality wetland is affected.

Different types of wetland mitigation require different amounts of compensation. The range of mitigation to destruction is the ratio between wetland that must be created, restored, or purchased for every acre destroyed or damaged. In the case of mitigation banking, a cash payment may be made to offset damage to a wetland. A typical ratio is shown in parentheses.

Ecology in Action

Wetland Delineation

Millions of dollars may ride on where the boundary is drawn between an upland and a wetland. Pressure from adversaries battling over defining what is, or is not, a wetland has led to a new field of applied ecology: wetland delineation. The courts needed a procedure that could result in a firm line being drawn separating wetlands from uplands. A precise formula needed to be invented to take the fuzzy-margined ecotone concept advocated by ecology and turn it into a legally defensible boundary. Ideally, any such formula is reproducible; that is, any skilled observer conducting the delineation would identify the same boundary. However, because of the ecotonal nature of many wetlands, it simply is not possible to have such a hard and fast definition. In practice, when a wetland boundary is contested, experts representing both sides will conduct a delineation and then attempt to reach a compromise.

Legally, wetlands are defined as

areas that are inundated or saturated by surface or groundwater at a frequency or duration sufficient to support, and that under normal circumstances do support a prevalence of vegetation typically adapted for life in saturated soil conditions. (Clean Water Act of 1972; Section 404)

This definition identifies three components that form the basis of wetland delineation: hydrology, wetland-adapted vegetation, and soils. In 1987 the U.S. Army Corps of Engineers produced a manual of wetland delineation that sets out criteria to define a jurisdictional wetland. Because of regional variation, each state may add criteria and define its own list of wetland plants.

Wetland Vegetation

Each state has an official list of plant species that can be used as **wetland vegetation indicators**. The definitions in Table 8.2 indicate the reliance of a species upon wetland habitats. Thus an obligate species is essentially one that can only grow in wetlands. The facultative categories reflect that some species can do well in both wetland and upland settings. The subdivision of facultative types reflects whether a species is more or less likely to be found in a wetland setting.

Exotic species (Chapter 21) are, by definition, not occurring in their natural habitat and are placed in the category FAC, indicating that they neither indicate the site to be a wetland or an upland.

If an entire area is covered by OBL plants there is little doubt that this will prove to be a wetland. However,

Table 8.2. Categories of vegetation used in the delineation of a jurisdictional wetland.

Wetland vegetation indicators	Percentage of individuals that occur in wetlands
OBL (Obligate wetland)	99%
FACW (Facultative wetland)	66–99%
FAC (Facultative)	33–66%
FACU (Facultative upland)	1–33%
UPL (Upland)	<1%

toward the margin of the wetland, a mixture of all five classes of plant will be found, and this is where the delineation becomes critical. The test is based on the canopy coverage of plants. This is taking a bird's-eye view of the land and deciding the relative spatial abundance of each vegetation type. The upper canopy is generally the one used, which would be trees in a woodland, but could be the herbs of a meadow wetland. The basic test used is

OBL + FACW + FAC > 50% of total canopy cover

If the combined value of these three categories is more than 50% of the cover, the area meets the vegetation index requirement.

Hydric Soils

A **hydric soil** is one that is saturated for long enough during the year that the upper layers develop anaerobic conditions. The following list are indicators of a hydric soil:

1. A dark surface

2. At least 24% organic matter by weight

3. Gleying (the blue-gray coloration from reduced iron) immediately below the A horizon

4. Oxidized layer around plant roots

5. Sulfidic materials (such as H_2S)

6. Iron and manganese concretions

7. Shiny or mottled appearance to soil below the A horizon

Generally, two of these features is enough to identify the soil as being a hydric soil.

continued

Hydrologic Indicators

Hydrologic indicators are used to determine if the water table is at, or close to the surface for at least a week during the growing season. The simplest way to verify this is by direct observation of seven (not necessarily continuous) days of flooding during the spring and summer. If this flooding is seen or if a test pit dug in the soil immediately fills to within 10 cm of the soil surface, the hydrologic requirement will have been met. However, because of the variability in hydroperiod between wetlands, it is possible that the site visit is done during a dry phase, in which case other indicators need to be studied. These include:

Water marks on trees (reflect highest level of recent inundation)

Weed wrack (vegetation and litter trapped in branches during periods of flooding)

Drift lines (leaves and twigs washed to form a line along the shoreline during flooding)

Morphological plant adaptations (swollen tree bases or roots sprouting from the trunk above ground level)

The area must test positively for wetland vegetation, hydric soils, and hydrological indicators for the area to be declared a wetland. Even with this manual-based approach, experience is needed to determine whether the area truly is a wetland. For this reason, in addition to plants, soils, and hydrologic indicators, a fourth aspect of delineation is Reasonable Scientific Judgment. The scientist is required to take into account all the available data and add to that a knowledge of the site history, recent climate conditions, and any human alterations to the site. Because this judgment requires ecological expertise, the growth of wetland delineation has provided a new employment market for ecologists.

Table 8.3   Wetland mitigation requirements for the State of Florida.

Type of wetland mitigation	Mitigation: destruction ratio or $ range
On-site creation	1.5:1–4:1 (2:1)
Restoration/enhancement	3:1–10.1 (4:1)
Purchase of wetland	10:1–100:1 (12:1)
Mitigation banking per acre	$25,000–$50,000 ($28,000)

On-site creation is when a developer creates a wetland on the same piece of property as the one that is being destroyed. The benefit of this is that the created wetland will still serve the adjacent ecosystems and minimize the loss of local habitat functions. Restoration and enhancement of wetlands is usually required to be carried out within the same watershed from which the developed wetland will be lost. Similarly wetland purchase is generally required to be within the same watershed and to be of the same type of wetland as the one being lost. A developer can purchase wetland and donate it to the state as a form of compensation. Alternatively, the developer can pay what is effectively a tax, the proceeds of which are used to buy and protect natural wetlands or to restore degraded wetlands; this is termed **mitigation banking**.

Despite a no net loss policy for more than a decade and attempts to restore and enhance existing wetlands, there is still a decline in wetland quantity and quality in the lower 48 states. In February 1998, President Clinton advanced a clean water action plan that calls for an increase in wetland area of 100,000 acres each year, beginning in 2005. If this is achieved, it would be the first expansion of wetland habitats since Europeans arrived in North America.

8.10   Creating Wetlands

When wetlands are altered and mitigation is required, one form of mitigation is to construct a new wetland. Creating a habitat from scratch is a lot like trying to build a car from a pile of sheet metal, nuts, and bolts: It is possible to build some kind of a car, but whether it is the Rolls Royce that you wanted is another matter. There have been some successful wetland creation projects with high biodiversity as their goal, but many more have been failures. If maximum biodiversity and a habitat as near to a high-quality natural wetland as possible is the goal, it is best to devote resources to restoring or enhancing a natural wetland, rather than trying to create a new one. In wetland restoration the land in question was a wetland before a change in land use. Many times a wetland can be restored simply by blocking drainage ditches that have artificially lowered the water level.

In contrast, to create a wetland the entire ecology and hydrology of an upland area must be altered. Leveling the land so that the slope is less than 1%, scooping out earth to create wet hollows, or building a dam to retain water on the site, are expensive operations. Even if these operations are achieved, the soils of the site will be mineral rich, rather than organic rich, offering an unsuitable growing medium for most wetland plant species. A common solution is to salvage the organic topsoil from a wetland being destroyed and spread it across the created wetland. This soil transfer is a partial solution. But the depth of soil is often too thin to provide all the functions of a true wetland soil.

A comparison conducted by the U.S. Geological Survey noted that a marsh creation project took six months of site preparation, whereas a similar area of marsh restoration took only two weeks, and the cost of the creation project was 15 times higher than for the marsh restoration. However, there are circumstances that make wetland creation a cost-effective and desirable strategy.

Created wetlands have been successful as part of a sewage treatment program or to remove acids and toxic metals from mine discharge. Another success for created wetlands is in regulating storm runoff. It is common to see large depressions in the ground around shopping malls and apartment complexes; these hollows or ponds are water retention areas. Water sheeting off parking lots and other paved areas are stored in the retention area before being gradually released over the next few days. Water retention areas perform a single function, and if they happen to provide wildlife habitat, that is a bonus. Thus wetlands can be created that have the desired functions, even though they may not be high-quality wetlands in terms of their biological attributes. Most successful created wetlands that have been studied were de-

signed to treat municipal wastewater. Municipal wastewater is anything that flows through a drain from residential dwellings or light industrial facilities. Industry is required to pre-treat water they discharge into municipal drains to remove toxins and heavy metals. The wastewater entering a municipal treatment facility will include fecal matter, sanitary products, detergent, food waste, and other odd things that we flush away, but should not include high concentrations of toxic metals or acids.

An alternative to the expense of building a traditional water treatment facility is to use a constructed wetland to cleanse the water (Figure 8.12). Often the water is guided through a series of wetlands, being retained for 7 to 14 days in each. During this time, particulates settle out of the water column and nutrients are absorbed by bacteria, algae, and plants growing within the wetland. Removal of nutrients is very variable according to local conditions, but at loads of less than 100 grams of nitrogen per square meter per year, 40%–50% of the nitrogen will be removed. If higher loads of nitrogen than this are applied the nitrogen simply washes through the system, and it is not cleaned from the water.

Lush vegetation will grow in the constructed marsh, and there will be a substantial buildup of organic sludge that would fill the wetland if it were not removed. Periodically, these marshes must be drained,

Figure 8.12   A White Ibis feeding in a constructed wetland. The water enters the wetland with high loads of nutrients and particulate matter. After about three to six weeks, the particulates have settled to the bed of the wetland, and plants, algae, and bacteria have absorbed most of the nutrients. The water leaving the wetland is much cleaner than that entering it. Such constructed wetlands can also provide valuable habitats for a wide variety of wetland organisms.

cleaned out, and replanted. In this process much of the stored organics and nutrients are removed from the system. The sludge is dried and can be used as a soil conditioner for local agriculture (unless toxic metals are present).

By having three marshes operating in series, the water is purified progressively, so that when draining from the system it appears clean. The water is suitable for crop or lawn irrigation, but is not pure enough to drink and may not be clean enough to discharge into a natural river system. The created wetland has functioned to remove many of the impurities from the water, and it is also providing other natural services. Often these areas provide habitat for wildlife, and may even be sufficiently scenic to attract recreational users, ranging from hunters and fishers to hikers and birders.

Summary

- Wetlands perform a variety of ecological functions, including flood regulation, nutrient and carbon storage, provision of plant and animal habitats.

- Wetlands are essential recruitment grounds for many species that spend their adult lives in other habitats.

- Salt concentrations separate marine from freshwater wetlands.

- Hydroperiod is the key determinant of freshwater wetland types.

- Slow oxygen diffusion through water (10,000 times slower than through air) dominates wetland chemistry.

- Plants and animals that live in wetlands exhibit evolutionary adaptation to habitat conditions, such as low soil-oxygen availability.

- Mammals, birds, some insects, and flowering plants that live in wetlands have evolved from upland-dwelling ancestors.

- Created wetlands are expensive and seldom replace the biodiversity functions of a natural wetland, but can provide water treatment functions.

- Wetland delineation is based on proportion of wetland vegetation cover, presence of hydric soils, hydrologic indicators, and reasonable scientific judgment.

- Wetlands continue to be destroyed despite a policy of "no net loss."

Further Readings

Cowardin, L. M., V. Carter, F. C. Golet, and E. T. LaRoe. 1979. *Classification of Wetlands and Deepwater Habitats of the United States.* U.S. Department of the Interior, Fish and Wildlife Service. (http://www.wes.army.mil/el/wetlands/wlpubs.html)

Pechmann, J. H. K., D. E. Scott, R. D. Semlitsch, J. P. Caldwell, L. J. Vitt, and J. W. Gibbons. 1991. Declining amphibian populations: The problem of separating human impacts from natural fluctuations. *Science* 253:892–895.

Gilbert, O.. and P. Anderson. 1998. *Habitat Creation and Repair.* Oxford: Oxford University Press.

Mitsch, W. M. and J. G. Gosselink. 1993. *Wetlands.* 2nd ed. New York: van Nostrand Reinhold.

Richardson, C. 1994. Ecological functions and human values in wetlands: A framework for assessing forestry impacts. *Wetlands* 14:1–9.

Tiner, R. W. 1998. *In Search of Swampland: A Wetland Sourcebook and Field Guide.* New York: Rutgers.

Web Connections

On-line resources for this chapter are on the World Wide Web at:
http://www.prenhall.com/bush
(From this web-site, select Chapter 8.)

Population and Community Ecology

Populations and Resources: A Balancing Act

9.1 *Assessing trends in populations*
9.2 *The drive to compete*
9.3 *Populations and natural processes*

9.4 *Ecological niche; or, how to be your favorite organism*

Ecologists use the term **population** to describe the number of individuals of a species occupying a defined area at a given time. Populations may be described in terms of **density**, which is expressed as the number of individuals of a species per unit area. These densities may be of great importance, as some species require a certain minimum density in order to trigger their reproductive effort. If the population density should fall below this "critical mass," reproduction fails and extinction follows.

The extinction of the passenger pigeon is an example of what can happen to a species when conservation efforts are implemented only after a population falls below its critical density. Until the mid-1800s, flights of passenger pigeons darkened the skies of the midwestern United States for days at a time. One flock was estimated to contain at least 2 billion birds. Hunting drastically reduced the population of the pigeon in the 1870s, and by 1885 only one breeding colony was left. Hunting of the pigeon was eventually prohibited, but the various groups of pigeons that formed separate populations repeatedly failed to reproduce. In just 50 years, the passenger pigeon went from being North America's commonest bird to being extinct. In hindsight, it is evident that the presence of thousands of birds flying, flocking, and roosting together stimulated breeding. Once flock sizes were reduced below this critical mass, breeding failed. The last passenger pigeon died in the Cincinnati Zoo in 1914.

When the effort to save the passenger pigeon started, there were still many thousands in the total population. However, because each local population was small, none had a large enough critical mass to trigger breeding. This observation highlights the difference between the total population of a species and local populations, or the number of individuals within a breeding group. Often, conservation efforts must prioritize the well-being of local populations.

9.1 Assessing Trends in Populations

All populations change in abundance through time. Most changes are just random fluctuations, increases or decreases varying about an average abundance. Depending on the scale of these variations, particularly over short time periods relative to the life span of the organism, populations may appear more or less stable. However, if the population size is observed closely, it will be seen to vary. If the trend is consistently upward, the species may become a pest. If a downward trend is maintained, the population will go extinct.

A lot can be learned by monitoring the population of even a common species to verify that it is maintaining a strong population. Given the power of humans to change entire landscapes, species that are common now may easily be rare in 50 or 100 years.

Population Demographies and Life Tables

One way to determine the well-being of a population is to study the age of its members. Other estimates might include the proportion of males to females, birth rates, death rates, and ecological fitness (the replacement of the parents in the next generation). Taken together, these various measures constitute **population demography**.

An investigator can follow an entire generation of an insect such as a grasshopper, which lives for a year, from birth to death. A similarly aged group of individuals in such a study is termed a **cohort**. The life cycle of the European common field grasshopper starts in the fall when eggs are laid in the soil. In the spring, the eggs hatch into nymphs. The nymphs go through a series of metamorphic changes in which they shed their outer skin to reveal a new individual beneath. Each nymphal stage (referred to as an *instar*) brings it a little closer to the adult form. By midsummer, the fourth instar metamorphoses into a winged adult grasshopper. The adults feed, reproduce, lay eggs, and die. By mid-November, all the adults are dead, but the next generation of eggs lies in the soil, and the life cycle is complete.

A grasshopper life table

The perilous journey from egg to adult can be tracked in a **life table** (Table 9.1). These tables, subdivided by age or developmental class, provide statistics such as the life expectancy, the proportion living, and the reproductive output.

The grasshopper life table is simplified by the fact that only the adults can reproduce. Life tables commonly draw upon the following statistics:

Time	x = a time interval appropriate for the organism studied.
Survival	l_x = The proportion of individuals that are alive at the start of interval x.
Mortality	d_x = The proportion of individuals that die during interval x.
Mortality rate	$q_x = (d_x/l_x)$.
Life expectancy	$e_x = (T_x/l_x)$. Where T_x is the sum of L_xs, where $L_x = (l_x + l_{x+1})/2$ for each of the time classes, starting at the bottom of the table working upward until the l_x for the age interval under consideration is reached.

The average ecological fitness of adults is an important statistic that can be calculated from the life table. In this case, the 1300 adults (stage 5) laid 22,617 eggs. For an average fitness of 1, it would be necessary to start each generation with the same number of eggs. It is clear that the 22,617 eggs produced by this cohort is less than the 44,000 eggs laid by the previous generation. Dividing 22,617 by 44,000 yields an average ecological fitness of 0.51.

Table 9.1   A life table for a cohort of grasshoppers.

$Stage_x$	n_x	l_x	d_x	q_x	F_x
	Number observed at start of each stage	Proportion surviving to start of each stage	Proportion dying in each stage	Mortality rate in each stage	Number of eggs produced
Egg (0)	44,000	1.000	0.920	0.92	—
Instar I (1)	3513	0.080	0.022	0.28	—
Instar II (2)	2529	0.058	0.014	0.24	—
Instar III (3)	1922	0.044	0.011	0.25	—
Instar IV (4)	1461	0.033	0.003	0.11	—
Adult (5)	1300	0.030			22,617

(*Source:* Data are from O.W. Richards and N. Waloff, 1954.)

The average fitness is a measure of the rate of change in a population (R_0). If R_0 is less than 1, then the population will contract through time, and if this trend continues, the population will go extinct. If R_0 is 1, the population is stable, and if R_0 is greater than 1, the population is increasing in size. In the case of the grasshoppers, R_0 can be expressed mathematically as:

$$R_0 = \frac{\sum F_x}{a_0} = \frac{22,617}{44,000} = 0.51$$

where F_x equals the total of fertilized eggs produced in each generation, and a_0 equals the number of original individuals. With an R_0 of 0.51, the grasshopper population is in a steep decline for this reproductive generation. It should not be assumed, on the basis of one year of data, however, that the grasshopper is lurching to extinction. Populations can show considerable variation in their fitness from year to year, going through cycles of boom and bust. This example emphasizes the need for long-term studies, in which many years of data are compiled, before a pronouncement is made about the overall trends of population increase or decrease.

A more usual way to calculate R_0 from a life table is to tabulate the average number of eggs, seeds, or young produced at each stage (m_x). Among grasshoppers, only adults can produce eggs, so only the final stage has a value for m_x ($m_x = 22,617/1300 = 17$). The value for m_x at each stage in the life table is multiplied by the appropriate value for l_x. However, in the case of the grasshoppers, only the adult value of l_x has a corresponding value of m_x.

$$R_0 = \sum l_x m_x = 0.03 \times 17 = 0.51$$

Note that the value for R_0 is the same in the two equations. The latter formula for calculating m_x is especially useful when comparing the reproductive output of different ages of adult. In such cases, the l_x categories may become year groups rather than stages in a life cycle.

Dealing with Overlapping Generations

The grasshopper study was simplified by having only one reproductive age alive at one time. In populations of other organisms, such as humans, many ages of potentially reproductive individuals coexist. In these cases, it is often impossible to tease apart the fitness of an individual cohort, so the overall population **age structure** (the proportions within a population of individuals of various ages) is used. If enough young mature to replace the previous generation, it may be

assumed that the species is reproducing successfully. If, however, the majority of individuals are postreproductive and there are relatively few young, the species is declining in number. The best documented demographies are for humans (see Chapter 16), and the repeated census of our population allows the comparison of birth rates and other demographic statistics for more than a century.

Ideally, demographies should be compiled for populations across time intervals. A comparison of demographies taken at different times will provide the strongest data of the trends in population growth or shrinkage. The ideal time interval between demographic studies will vary according to the organism. If the study were to be of the future of oak trees in our forests, taking demographic data a year or so apart would be virtually meaningless given the life span of each oak tree. However, if the subject of the study were a bacterium that reproduces every few minutes, studies conducted a few hours apart would provide meaningful information.

For nonhuman populations with overlapping reproductive generations, most demographic data come from a survey at one point in time. This approach produces a **static life table**. In these data sets the observer infers past populations from the age structure of the present population or evidence gathered from dead members of the population. If the age of an individual at the time of its death can be calculated, the probability of death at a given age can be estimated.

The classic study of this type was conducted on Dall sheep (Figure 9.1) on Mount McKinley, Alaska. More than 600 skulls were collected from the slopes of the mountain over several years. The number of growth rings on the horns indicates the age of the sheep when it died. Each spring, the horn of the sheep puts on a growth spurt, and during the rest of the year there is a slower, denser growth of the horn. The horns appear banded, with each pair of bands representing one year. When a life table was constructed for the sheep (Table 9.2), it became clear that their average life expectancy was about 7 to 8 years, but some individuals lived to be 13 years old.

An odd statistic in this data set is that newborn lambs have lower mortality (death rate) than do those that are six months to one year old. Most populations exhibit a very high mortality among the newly born, and the same would be expected among these sheep. The explanation for the unexpected data may lie in the way they were gathered. Because the life-table data were based on skulls of dead sheep, if a skull had been eaten completely by a predator there would have been nothing to find. Newborn lambs have small skulls

Figure 9.1 **Dall mountain sheep.**

with no horns, and it is possible that many were killed and consumed completely. The consequence would be to underestimate mortality rates of newborn lambs and to artificially increase the statistic of average life expectancy for those in the 0–0.5 year age class. Once the lambs are six months old, the skulls probably provide a more accurate basis for the life table. As the data stand, the average life expectancy of a newborn Dall sheep is 7.1 years. However, the sheep face different perils throughout life, and the young are especially vulnerable. Mortality rates of the young sheep are high as the weaker members of the flock die. If the sheep survive the first year, their life expectancy actually increases, so that a one-year-old sheep is likely to survive for another 7.7 years, giving a total age at death of 8.7 years. Similarly, a sheep that is six years old still has a life expectancy of 3.4 years and would be 9.4 years old when it dies.

As indicated by doubts over the accuracy of the statistics for the newborn sheep, numerous sources of error can creep into static life tables. Such tables are founded on real data, but the full complexity of the table is the product of educated guesses and assumptions, and the product should always be treated with

Table 9.2 A life table for a population of Dall sheep on Mount McKinley, Alaska.

X Age (years)	l_x Proportion surviving to start of each stage	d_x Proportion dying in each stage out of 1000 born	q_x Mortality rate of individuals within each stage	e_x Expectation of life; or mean lifetime (years) remaining to those attaining age interval
0–0.5	1.000	0.054	0.054	7.1
0.5–1	0.946	0.145	0.153	—
1–2	0.801	0.012	0.015	7.7
2–3	0.789	0.013	0.017	6.8
3–4	0.776	0.012	0.016	5.9
4–5	0.764	0.030	0.039	5.0
5–6	0.734	0.046	0.063	4.2
6–7	0.688	0.048	0.070	3.4
7–8	0.640	0.069	0.108	2.6
8–9	0.571	0.132	0.231	1.9
9–10	0.439	0.187	0.426	1.3
10–11	0.252	0.156	0.619	0.9
11–12	0.096	0.090	0.937	0.6
12–13	0.006	0.003	0.5	1.2
13–14	0.003	0.003	1.0	0.5

Note: Data were derived from 608 sheep skulls. All of the sheep were killed by wolves.
(*Source:* Data are from E. S. Deevey, Jr., 1947.)

some caution. However, in marine ecosystems, where it is often impossible to monitor whole populations on a year-round basis, and for many long-lived terrestrial species ranging from elephants to oak trees, static life tables are the only ones available.

Population Stability

If the goal of a population study is to track whether populations are rising or falling, an additional pool of information is needed beyond a measure of fitness. Populations are mobile; even trees and other plants can move substantial distances as seeds. Hence, for a more complete view of a population, it is necessary to calculate arrivals to (immigration) and departures from (emigration) the population. Thus, population numbers are boosted by births and the immigration of individuals from other populations. Similarly, population numbers are drawn down by deaths and emigrations. If these variables are known, the likely future trajectory of a population can be calculated.

$$N_{future} = N_{present} + (B - D) + (I - E)$$

where N_{future} is the predicted population size at a given time in the future and $N_{present}$ is the present population size. The population will change according to the difference between number of births (B) minus number of deaths (D) and number of individuals immigrating (I) minus the number emigrating (E). The success of an immigrant to an area will depend on its finding enough of the appropriate resources to allow it to reproduce and join the breeding pool.

9.2 The Drive to Compete

The race to exploit the environment is often decried as a human trait and something that is "unnatural." Far from being unnatural, such exploitation is a vestige of our truly natural behavior. Holding back and not stripping resources would distinguish us from other organisms. All organisms change the conditions for life in their surroundings as they use up available resources. Ecologists refer to anything that is essential for the continuance of life and that is finite, as a **resource**. In natural systems, the growth of an individual or a population is limited by the resource that is the hardest to obtain. For example, water, light, nitrogen, and phosphorus are all potentially limiting resources for a seedling, but even if these were all abundant, another resource such as potassium might be too scarce to allow growth. In that case, potassium would be the **limiting resource**.

Rivalry for resources by members of a species intensifies as the population grows and becomes crowded. Competition for resources among individuals of the same species is termed **intraspecific** (within species) **competition**.

Intraspecific competition is difficult to measure in the wild because of interactions with other species. More than 60 years ago, biologists started conducting competition experiments on simplified ecological systems in their laboratories. In the 1930s, the Russian microbiologist G.F. Gause conducted a classic experiment on the response of a population to resource depletion. Realizing that the ideal laboratory organism should be small, easy to rear, and have fast reproductive rates, he chose *Paramecia* for his study. The *Paramecium* is a microscopic unicellular animal that produces a generation in a matter of hours. Gause added a known number of *Paramecia* to a tube of oatmeal broth. The oatmeal was food for yeast and bacteria, which in turn were food for *Paramecia*: The whole formed a very simple food chain.

Gause observed an exponential growth in the *Paramecium* population for the first 10 days (Figure 9.2). After this, although their food supply was replenished regularly, the numbers of *Paramecia* began to stabilize, reaching an asymptote after 14 days. The *Paramecium* population had run into an invisible ceiling: One of their resources was now sufficiently depleted to limit population growth. One possibility was that food availability had become limiting. At the beginning of the experiment, food would have been overwhelm-

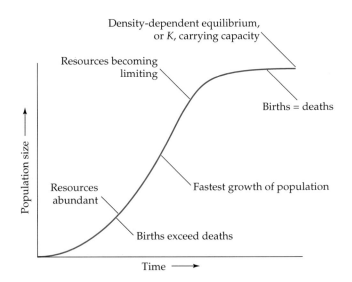

Figure 9.2 A schematic representation of population growth. Density-dependent equilibrium is reached as the population of (*N*) rises to the carrying capacity (*K*), where it is limited by resource availability.

ingly plentiful, but as the numbers of *Paramecia* grew, more and more food was being consumed. Progressively, each individual searched a little longer for each meal. If more and more energy was being spent on movement relative to food intake, the amount of energy left over for reproduction was reduced; hence population growth would have slowed. Other factors, such as space or a change in the chemistry of the nutrient broth due to a buildup of dead *Paramecia* or fecal matter, may also have been factors that limited population growth.

Another way to think of this experiment is in terms of birth and death rates. At the beginning of the experiment, birth rates greatly exceeded death rates, and the number of births relative to deaths (per unit time) defines the rate of population increase. As the population grew, the balance between births and deaths began to shift as deaths became proportionately more common. At the point where the maximum population was reached, the birth rate exactly equaled the death rate (Figure 9.2). When births equal deaths, a population is in **equilibrium**.

The Logistic Growth Model

The ascent and plateau of population growth can be summarized in a simple mathematical formula or **logistic equation** that, on a graph, produces a symmetrical S-shaped curve with an upper asymptote. The change in population size (N) between two sampling sessions is dN. This is divided by the change in time between sampling points (dt). Thus dN divided by dt equals the rate of change in the population size. The value r is the maximum per capita rate of population increase in the absence of limited factors: a constant value. The maximum population size (k) is also constant. In addition to r, the only variable is the size of the actual population (N) at a given time (t). Thus:

$$\frac{dN}{dt} = rN\left(1 - \frac{N}{K}\right)$$

As the population nears its upper limit (K), N/K will get closer and closer to 1. When $N = K$, the value for $(1 - N/K)$ is $(1 - 1)$, or zero; consequently, the whole equation equals zero. When the rate of population increase dN/dt is zero, equilibrium has been reached.

One important point to appreciate is that K represents the equilibrium population. This asymptote is also the **carrying capacity** (perhaps remember it as "Karrying Kapacity"), which is the maximum population that can be supported by the available resources. Another finding of this experiment is that, as the pop-

ulation growth nears the carrying capacity, the rate of population increase (dN/dt) becomes smaller as N approaches K. Population growth rates slow down as the crowding of individuals increases. Indeed, it seems that population growth is restricted *because* of the increased density of organisms. As the density of individuals increases, the intraspecific competition intensifies. Such an equilibrium can be described as being **density dependent**. Characteristic of a density-dependent equilibrium is that at least one resource has become limiting.

The Lotka–Volterra Competition Model

An extension of the logistic growth model was devised independently by two mathematicians, Antonin Lotka and Victor Volterra, in the early 1930s to describe simple interactions of competition and predation. In the Lotka-Volterra model for competition, equations describe the fates of species 1 (represented by subscript 1) and its competitor, species 2 (represented by subscript 2). The competition is for a single resource. Although the basis of this model is intraspecific competition, it also describes **interspecific** (between species) **competition**. The effects of intraspecific and interspecific competition are made comparable by introducing a competition coefficient into the equation. The competition coefficient (α_{12}) is literally the intraspecific and interspecific competitive effect of species 2 on species 1. This effect is expressed by multiplying the number of species 2 in the population by the coefficient. The resulting value translates the competitive effect of species 2 into species 1 "equivalents." For example, if the competition coefficient is 3 and there are 50 individuals of species 2, their use of the resource is equivalent to the use of 150 individuals of species 1. The competition model for the effect of species 2 on species 1 is

$$\frac{dN_1}{dt} = \frac{r_1 N_1 (K_1 - N_1 - \alpha_{12} N_2)}{K_1}$$

Equally, a competition effect of population N_1 on N_2 can be calculated

$$\frac{dN_2}{dt} = \frac{r_2 N_2 (K_2 - N_2 - \alpha_{21} N_1)}{K_2}$$

Essentially, there are four possible outcomes from this model. One outcome would be that species 1 (if it were a weak competitor) goes extinct, leaving species 2 to reach its carrying capacity. A second would be that species 2 is the weak competitor and that it goes extinct, leaving species 1 as the winner. A third outcome would be that the two species are evenly matched and neither goes extinct. A fourth outcome would be that the two species are evenly matched, but one wins because it started with a much larger population.

In an experiment to test these predictions, Gause grew two species of *Paramecium* in separate test tubes (Figure 9.3a) and in the same test tube (Figure 9.3b). Both *Paramecium* species increased in abundance when there were abundant resources and very little competition (Figure 9.3a). However, when in the same test tube, these two species soon locked in vigorous competition, and as the population densities grew, one of the species was outcompeted. *Paramecium aurelia* was the stronger competitor under these conditions and was more successful at converting the available food into offspring. The fitness of *P. caudatum* started to decline 8 days into the experiment, and the species went extinct in the test tube after 25 days (Figure 9.3b). The population of *P. aurelia* had meanwhile increased to a density-dependent equilibrium.

In another experiment, *P. caudatum* was grown with *P. bursaria*. After 20 days, it was evident that both species of *Paramecium* had established stable populations (Figure 9.4), although the densities of both species were lower than when either was grown alone. Why were both species surviving? Why hadn't the more abundant *P. caudatum* driven *P. bursaria* to extinction? Although the two species were competing for the same pool of resources, they were spatially sep-

arated within the tube. *Paramecium bursaria* fed on yeast cells in the sludge in the bottom of the tube, while *P. caudatum* fed on bacteria within the main portion of the tube. This experiment demonstrates that two species can coexist, so long as they are not in head-to-head competition for the same set of resources.

The Concept of Limiting Resources

As organisms constantly deplete available resources, the pressure mounts to harvest the scarcest resources maximally. Intraspecific and interspecific competition

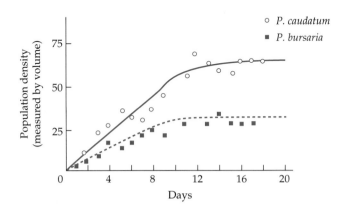

Figure 9.4   Populations of *Paramecium caudatum* and *P. bursaria* grown together in culture. The two survive by avoiding direct competition for resources. (After G.F. Gause, 1934, reprinted by Hafner, 1964.)

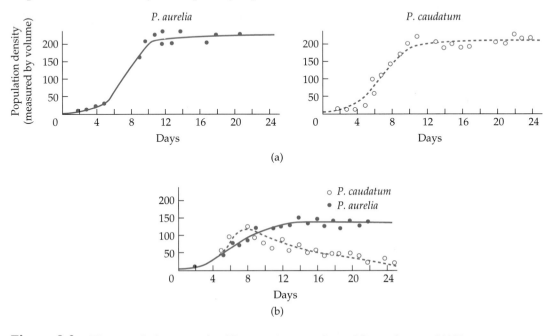

Figure 9.3   The population growth of *Paramecium aurelia* and *P. caudatum*. (a) When grown in separate cultures the population sizes level off where resources became limiting. (b) The fate of populations of *Paramecium caudatum* and *P. aurelia* when grown in the same test tube. As resources start to become limiting, *P. aurelia* outcompetes *P. caudatum*, which declines to extinction after 25 days. (After G.F. Gause, 1934, reprinted by Hafner, 1964.)

for the rarest resource will be intense. Species that are poor competitors for that resource may well go extinct at that location. However, at another site, a different resource may be the limiting factor, and here the fortunes of the competing species may be reversed. Population ecologist David Tilman has suggested that limiting resources are the prime determinants of population sizes and distributions among plants. He hypothesizes that each species of plant is a specialist for a single limiting resource, say, nitrogen. Some species are able to thrive despite low concentrations of nitrogen in the soil conditions that are unfavorable for other species. However, such nitrogen specialists have had to develop a particular physiology to cope with the low availability of this nutrient. This specialization costs the organism dearly, to such an extent that it will not be a good competitor for any other resource. In much the same way, a 200-kg sumo wrestler is a specialist and a truly formidable adversary in his domain, but he is unlikely to excel at pole vaulting, 110-m hurdles, or ballroom dancing.

Such specialization narrows the range of conditions in which a plant species can compete. Each subtle variation in resource availability will have a set of plants optimally adapted to grow under those conditions. For example, dry, sandy soils would be expected to support an array of plant species different from those growing on wetter sands or on soils with a slightly higher nutrient content. That plants are specialists in resource requirements and use helps to explain why there are so many different species of plants.

9.3 Populations and Natural Processes

Wild populations seldom demonstrate the same growth and equilibrium that Gause found in his experiments. The experimental world for the *Paramecium* was constant. Food inputs and temperature were stable, and the whole system could attain equilibrium. The real world is neither stable nor equilibrial, and few organisms attain their density-dependent carrying capacity. Consequently, demonstrating the existence of density-dependent equilibria in nature is virtually impossible.

Little doubt exists that plants are specialists for nutrients and that some populations are indeed limited by resource availability. However, chance events generally control the numbers of individuals in populations. Hurricanes, fires, severe winters, summer droughts, and lightning are all examples of events that can disturb a natural population. Each of these factors could reduce population size. Many populations are continually building up, only to be scythed down by the next natural catastrophe.

The size of natural populations is subject to continual change. Rather than being the smooth exponential growth and equilibrium described by Gause, the record of a natural population looks jagged (Figure 9.5). Each time the population crashes, it starts on a new episode of exponential growth, only to be cut down once more. Many causes of the fluctuations of populations—fire, drought, climate change, and so on—are independent of the population density. Such causes are termed **density-independent** factors.

The population of the Grey Heron (closely related to the American Great Blue Heron) in England and Wales, was repeatedly reduced by especially severe winters (Figure 9.6). Herons feed on amphibians and fish in the shallows of lakes and swamps. During severe winters when ponds freeze, the fish are inaccessible beneath the ice, and amphibians are overwintering in the bottom mud or in sheltered spots away from the lake. The herons cannot find enough to eat, and some starve. Because winter weather is not determined by the population size of herons, this is an example of a density-independent factor affecting a population.

The factor that causes one population to crash, such as a flood, might provide ideal conditions for the population growth of a species that needs a lot of moisture. Consequently, different species of plants and animals will have different patterns of population crash and recovery. Evidence that populations are differentially affected by series of unpredictable factors is a major contributor to the growing awareness that natural systems are disequilibrial. To understand more about why particular species are affected by specific environmental changes it is necessary to consider the niche of a species.

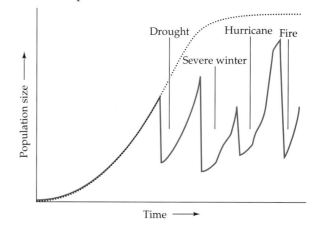

Figure 9.5 **Density-independent factors cause a population to fluctuate through time.** Most populations are continually undergoing such crashes and recoveries.

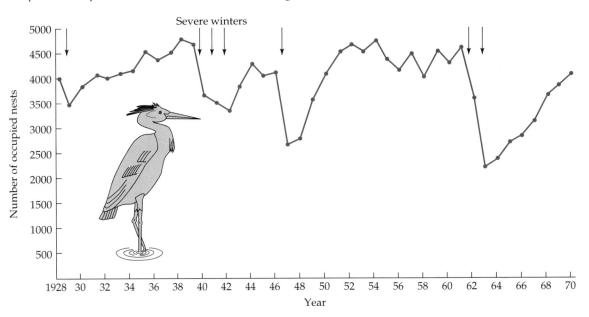

Figure 9.6   Cold winters cause slumps in the population of the Grey Heron in England and Wales. This is an example of density-independent factors causing populations to fluctuate. (After J. Stafford, 1971.)

9.4 Ecological Niche; or, How to Be Your Favorite Organism

Anyone who has owned a pet has at one time or another watched that cat/dog/tortoise (let's assume it's a cat) as it performed some bizarre feat and thought, "What on Earth are you doing?" The answer is simple; it is being a cat. All species must be good at what they do, otherwise they would be extinct. When we see kittens playing, it is likely that they are developing hunting skills or defining their group social structure; very little that an animal does is truly aimless. If there was ever a kitten that took no part in acquiring these skills, it would have been a poor adult hunter, and it would have failed to maintain fitness.

Charles Elton, an ecologist and keen observer of animal behavior, noted that we can liken the life-style of a wild animal to a human's job. Just as a butcher or a banker carries out a predictable set of actions and plays a role within the life of the community, so too does a fox or a squirrel. Elton termed this role the **niche** of the animal. Thinking in this anthropomorphic way comes naturally to us, but we need to always remember that the "decisions" made by almost all animals come from programming in their brain, not from learning (some vertebrates being the exception). To return to the analogy of the job, animals are born into their "job," rather than training for it. In the case of a cat, the kittens are honing their skills to fulfill an instinct. Their programming determines their instinct to be hunters of small animals, yet they learn to be better hunters within their

niche. Any deviation from conformity to their niche, such as abandoning hunting mice and birds in favor of hunting venomous snakes, spells almost certain death. Each organism has a well-defined niche, so well defined that no other species can occupy it. A cat is a hunter of mice, insects, and birds, which it eats after the search, the stalk, the pounce, and the kill. This description of behavior is the definition of niche in terms of the "job" of the animal.

A More Detailed View of Niche

The Eltonian view of niche is a good starting point, but if we are to use the concept of niche it must be defined more rigorously. G. Evelyn Hutchinson of Yale University described the concept of niche in terms of the many needs of each organism. He observed that many environmental factors, such as acidity, temperature, and rainfall, contribute to the living conditions of a species. Furthermore, he noted that each of these factors has a wide range of values in nature. Temperature varies from extreme cold at the poles to extreme heat in tropical deserts, and this range is continuous between the two extremes. The full range of values for one of these factors is an **environmental gradient**. Take the example of temperature: Most organisms flourish at a specific temperature or through a narrow range of temperatures. At other temperatures, both warmer and cooler, there is a range in which they survive, but their growth and reproductive rates may be adversely affected. Beyond this range of temperatures that the organism can barely tolerate, it will die. An environmental gradient can be defined for the organism: from fatal temperatures, toward optimal temperatures, and then back to fatal temperatures. The species will be most abundant at its optimal temperature and will get scarcer further away (Figure 9.7). The range of temperature that can be survived by the organism defines the temperature component of its niche and can be thought of as the first dimension of its living space.

A range of pH values of local soil or water could form another environmental gradient. For example, under extremely acidic conditions the organism dies, at neutrality it flourishes, and with increasing alkalinity its numbers start to decline. Plotting temperature against acidity defines a two-dimensional area within which the organism can survive (Figure 9.8a). If a third gradient is added, such as rainfall, the living space can be plotted in three dimensions as a volume (Figure 9.8b). There are no more dimensions that can be drawn on a page, but there are many more gradients that will determine where a species can live: the number of frost-free days, the number of hours of sunshine per day, soil depth, or the number of days of flooding. Each of these environmental gradients is a dimension that determines where the organism will, or will not, be found. Because there are so many of these gradients Hutchinson described his view of niche as an *n*-**dimensional hypervolume** (where *n* is the number of relevant gradients).

The tolerable range of each of these gradients is important in defining the areas where that organism can live. Hutchinson included behavioral requirements as some of the *n* dimensions of a niche. Thus, the niche of the cat can be defined using environmental gradients,

then additionally by such factors as prey availability, territory size, access to water, or lack of predators. If all these *n* conditions are met, it can get on with the job of "catting." Natural selection will then ensure that the cat will be as good as it needs to be to use these resources.

Niche Separation and Competition

Gause's experiments on *Paramecia* and the work of Hutchinson suggested that two species occupying the same niche would result in the extinction of one of them: This is the principle of **competitive exclusion**. For a while it appeared that an exception had been found to the rule of "one species, one niche."

Five species of migratory warbler were seen to flock together. In spring, they arrived in the forests of the northern United States, where they nested. During the nesting season, they preyed upon caterpillars, particularly the spruce budworm. All five species could be seen foraging for the same prey in the same tree and apparently occupying the same niche. If this were true, the principle of competitive exclusion predicted that the warblers should have been locked in such intense competition that they could not coexist. Robert MacArthur of Princeton University spent many days documenting the precise feeding behavior of the different warbler species and found that, although they were foraging in the same tree and hunting for the same prey, their hunting styles differed (Figure 9.9). MacArthur subdivided each tree into 15 zones: 5 vertical ones and 3 zones from outer branch tip in toward the trunk of the tree. He found that the warblers seldom overlapped in their search pattern as they foraged for food. Where spatial overlap was evident,

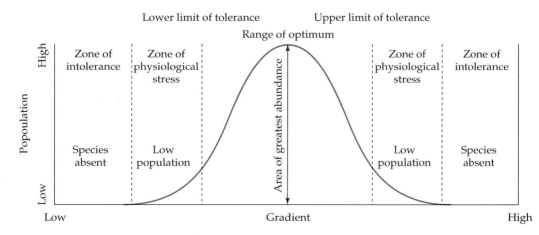

Figure 9.7   An environmental gradient, such as temperature, can determine the survival of an organism across a range of conditions. The organism will achieve its maximum population where it experiences an optimal environmental gradient. (Modified from C.B. Cox, I.N. Healey, and P.B. Moore, 1976.)

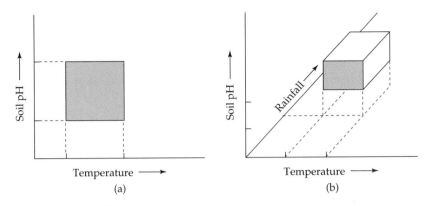

Figure 9.8   Environmental gradients as dimensions of a niche. (a) Two environmental gradients, such as temperature and soil pH, defining the conditions a species can occupy, can be represented in two-dimensional space. (b) As a third environmental parameter, such as annual rainfall, is added, the space that the organism can occupy must be plotted in three dimensions. Further gradients will affect the distribution of the species but cannot be diagrammed. Hence Hutchinson said that a niche should be thought of as an *n*-dimensional hypervolume, with each dimension representing a resource needed by the species. (After M. Begon, J.L. Harper, and C.R. Townsend, 1986.)

there were further behavioral differences. Some species searched methodically, picking off hidden caterpillars, whereas others would flit from branch to branch picking off the most obvious prey items. The warblers' different approaches to hunting served to minimize competition between them and affirmed that these birds were functioning in different niches. The concept of the species-specific niche was upheld.

A more complex example of niche separation is provided by the herbivorous megafauna of Africa. The great grasslands of the Serengeti support many dif-ferent grazing animals that seemingly occupy the same niche. Some behavioral overlap is evident. It is partly compensated for by their migratory behavior, but most of these species have evolved to coexist with a minimum amount of direct competition. The time when resources are likely to be limited, therefore the time when maximum population size is determined, is the dry season, when the grazing becomes sparse. Great migrations take place at this time, with more than a million animals in a vast procession constantly moving to less-grazed areas.

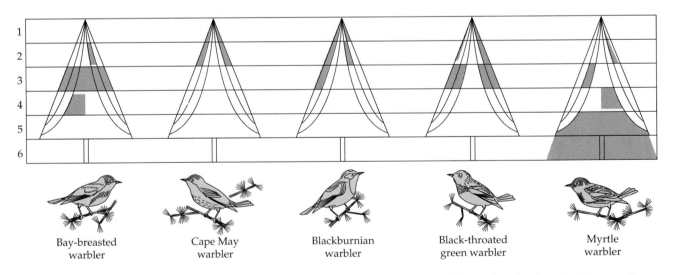

Figure 9.9   Five species of migratory warblers appear to occupy the same niche as they feed on similar prey items and can be found foraging together on the same tree. Robert MacArthur demonstrated that the warbler species feed in different portions of the tree, and, where their hunting overlaps spatially, their hunting practice differs behaviorally. Hence, the warblers are occupying separate niches. (After R.H. MacArthur, 1958.)

Zebra move through an area quickly, grazing the upper stems of the grass. Members of the horse family have a very active digestive process that runs twice as fast, but is not as efficient, as that of a wildebeest. Consequently, zebra need the most nutritious forage, which is why they graze the growing tips of the grass. After a herd of zebra pass through an area, the bases of grass tussocks remain. The great herds of wildebeest generally arrive later in the season than the zebra, and these mowing machines eat the bases and leaves of the grass. Unlike zebra, wildebeest are ruminants; that is, they have an extra stomach for fermenting partially digested vegetable matter. The process of regurgitation and second mastication is known as "chewing the cud." This way of digesting cellulose, the principal component of grass, is highly efficient at extracting the greatest possible nutritional value from each mouthful. Consequently, wildebeest can survive on poorer fodder than can zebra. The grazing of these large herbivores thins out the vegetation and provides the third species of the mass migrations, Thompson's gazelle, with access to low-growing, protein-dense herbs. The gazelle, another ruminant, has a much smaller muzzle than either zebra or wildebeest and thus can graze much more selectively. By being able to chew the cud, selecting the best of the fodder that is left, and having a small body size, the gazelle can gain enough nutrition to survive. Topi and antelope are nonmigratory, and, as the great herds arrive, they avoid competition by moving into the woodlands, where they browse on the leaves and seeds of trees. Giraffe may mix with the herds, but these are not in competition with gazelle, antelope, zebra, or wildebeest because they graze the leaves of trees that are beyond the reach of the other herbivores. Once again, the niches of these animals are subtly different. Evolution appears to have driven each species into a specific pattern of behavior that minimizes the stress of competition.

When a group of species are harvesting a similar resource and co-occur within an ecosystem they are referred to as a **guild**. MacArthur's warblers form a guild that preys on spruce budworm, and the various antelopes, zebras, and topi form a guild of grazers. Plants, too, can form guilds. The species of tree that can live within a mangrove are a recognizable guild, as would be the grasses and herbs of a prairie or the cactuses of a desert.

It is evident from the study of guild members that subtle differences in behavior or adaptation help to limit interspecific competition. Yet clearly some competition does exist. This raises the interesting question of how much competition can be tolerated by coexist-

ing species within a guild. Perhaps a 15% difference between species in their morphology (such as beak size or leg length) or behavior would be enough to allow coexistence. More likely, the critical factor will be how the species respond to the worst environmental conditions. In other words, when resources are plentiful a more significant overlap can be tolerated than when times are tough and competition intensifies. Peter and Rosemary Grant of Princeton University have monitored finch populations on the Galápagos Islands since 1973. They observed how similar species appear to occupy identical niches during times of plenty, but during droughts the birds become specialists, with small-beaked finch species feeding on small seeds and large-beaked finch species feeding on larger, tougher seeds. Here the niche overlap disappears. Increasingly, ecologists are led to the belief that it is the variability of the natural world that plays a key role in maintaining species diversity.

Competition and the Occupied Niche

If all n dimensions required by a species are known, these conditions will describe the entire potential range of that organism (anywhere outside these limits would be fatal). This maximum area of occupation is the **fundamental niche**. In practice, organisms seldom fill this entire potential range. The range that the species succeeds in occupying will be smaller, reduced by competition with other species.

Where the fundamental niches of two species overlap, only one species can survive (Figure 9.10). In a two-dimensional representation of the niche (not as a geographical area), it is evident that, as competing species are introduced, they erode the area of niche where species A has an absolute advantage. Even though species A still has the same fundamental niche, it is now constrained to occupy a much smaller range on the environmental gradients. The area that it actually occupies in the presence of competition is its **realized niche**. The realized niche is thus a more realistic definition of where a species will actually be found than is the fundamental niche.

Niche and Evolution

Another aspect of niche and evolution is that different species can evolve to fill the same niche so long as the species are geographically separated. Thus, we can now modify the ecological rule of one species, one niche to: "one species, one niche, one place." Sometimes the two species will be closely related. For example, the American robin (*Turdus migratorius*) pulls worms from lawns and rummages among fallen

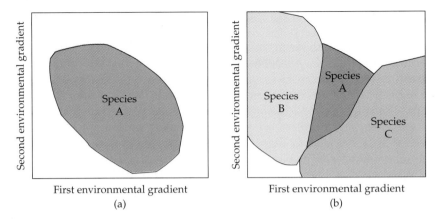

Figure 9.10 **The fundamental versus the realized niche.** (a) For the sake of simplicity, the niche of species A is viewed in two dimensions. The maximum area on those two environmental gradients that species A can occupy is its fundamental niche. (b) As competing species B and C are introduced, their optimum on the environmental gradient may be different. They outcompete species A over part of its potential range, and in those portions of its potential range, species A will go extinct. The portion of the environmental gradients in which species A is the best competitor will be where it is actually able to exist. This is the realized niche of species A.

leaves for insects. No other American bird has this combination of hunting strategies. However, in Europe its close evolutionary relative, the European blackbird (*Turdus Merula*) occupies exactly the same niche.

Particularly on islands (see Chapter 20 for an explanation), evolution leads to the island niches being filled by species that are unrelated to the occupants of the equivalent mainland niche. An example comes from the mainland niche filled by woodpeckers. These birds have strong bills to dig grubs and larvae from rotting wood, but woodpeckers are not known in Hawaii or Australia. On these and other islands, unrelated birds whose ancestors were seed-eaters with stubby beaks have evolved longer bills and have acquired the grub-hunting behavior of woodpeckers (Figure 9.11). Even stranger is the evolution of the aye-aye, a primate mammal that lives on the island of Madagascar. Madagascar has no woodpeckers, and the aye-aye has evolved to fill the equivalent niche. The middle digit of the aye-aye has evolved to become long and slender, considerably longer than its other fingers. The aye-aye taps on rotting wood with its finger to sense hollow passages made by an insect grub (experiments have shown that it does not necessarily listen for the writhing of a grub). The aye-aye then bites into the trunk and uses its long finger to pry the grub from the log. The niche is similar to that of the woodpeckers, although, unlike woodpeckers, the aye-aye is active at night. Aye-ayes and the different types of woodpecker-like birds have completely separate ancestral lines, yet they have converged on a similar way of hunting for grubs and provide an example of

convergent evolution. The key here is that, even though they are *without a recent common ancestor*, the organisms have evolved a similar form or function.

Another example of convergent evolution comes from the marsupial fauna of Australasia (Australia,

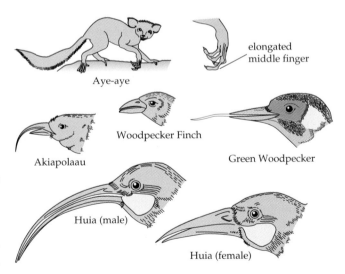

Figure 9.11 **Convergent evolution of birds and a mammal to fulfill the niche of "woodpeckering."** Unrelated organisms have arrived at a similar anatomy to prey on grubs in rotting wood. The European Green Woodpecker is the only true woodpecker in this figure. The others live on islands where there is no true woodpecker, and hence there is a vacant niche. This has been filled by unrelated birds: the Akiapolaau in Hawaii, the Woodpecker Finch of the Galápagos, and the Huia of New Zealand (extinct). The aye-aye of Madagascar is a primate that occupies a woodpecker-like niche. The long probing bird bill is replaced by an elongated middle finger and strong teeth. (After D. Lack, 1947.)

New Zealand, and Papua New Guinea). Because it was long ago separated from the great landmasses due to continental drift (Chapter 2), Australasia was geographically isolated from the great expansion of mammals that took place about 70 million years ago. Until the time of separation the only mammals in Australia were ancestors of the echidna and duck-billed platypus, both very peculiar and ancient mammal groups. The best known mammals of Australia, the marsupials, are not known in the fossil record until about 15 million years ago. Either they invaded Australasia after it had separated from Gondwana, or their early fossils have yet to be found. Modern Australian marsupials are fairly closely related in evolutionary terms because they have a common ancestor that was a rat-like marsupial. However, the form of modern marsupials varies very widely, and some even appear very similar to completely unrelated placental mammals (Figure 9.12). Placental mammals, be they humans, cats, or squirrels, are more closely related to one another than to any marsupial. Thus, the fact that the placental mammals have a lookalike among the marsupials of Australia is a result of evolution to fill the same niche rather than a result of relatedness. For example, because there are ants in both places, a niche exists to eat ants. Eating ants requires powerful digging claws for opening up ant nests, a pointed snout for poking into the hole, and a long sticky tongue for scooping up ants. It is not surprising to find that two creatures that do the same thing evolved to share a similar anatomy.

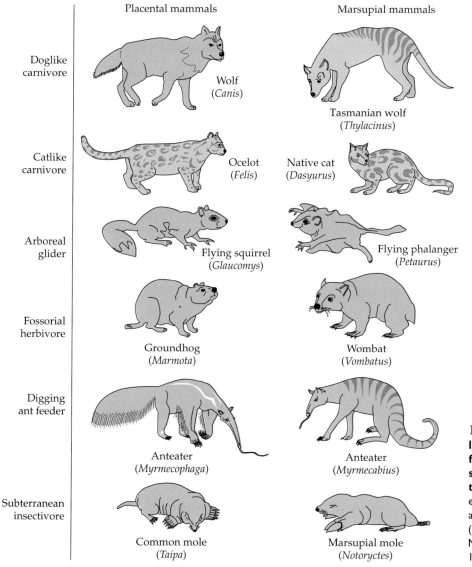

Figure 9.12   Convergent evolution of placental mammals from South American and marsupial mammals from Australia to fulfill a range of niches. The evolutionary lineage is strongest within a column, and the ecological equivalents (rows) are completely unrelated. (After M. Begon, J.L. Harper, C.R. Townsend, 1986.)

Summary

- Individuals are part of a total and a local population.
- A population at equilibrium with available resources has reached its carrying capacity.
- Natural populations seldom reach carrying capacity and are continually cut back by density-independent factors.
- Organisms will evolve to minimize competition with their neighbors.

- Most deviations from a species' "normal programming" are detrimental and cannot survive.
- Organisms evolve to occupy a tightly defined niche.
- Only one species occupies a niche in one place at one time.
- Niche will determine the morphology of the organism.

Further Readings

De Angelis, D. L. 1991. *Dynamics of Nutrient Cycling and Food Webs*. New York: Chapman and Hall.

Gause, G. F. 1932. Experimental studies on the struggle for existence. 1. Mixed population of two species of yeast. *Journal of Experimental Biology* 9:389–402.

Hutchinson, G. E. 1959. Homage to Santa Rosalia, or why are there so many kinds of animals? *American Naturalist* 93:145–159.

Tilman, D. 1987. The importance of mechanisms of interspecific competition. *American Naturalist* 129:769–774.

Web Connections

On-line resources for this chapter are on the World Wide Web at:
http://www.prenhall.com/bush
(From this web-site, select Chapter 9.)

Who Needs Sex Anyway?

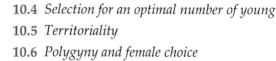

The reproductive biology of a species is of vital interest as we try to save it, eradicate it, or control it. Applied ecologists are often given the task of managing an environment to favor an endangered species or to eliminate an invasive one. Understanding the reproductive biology of these organisms is a prerequisite to successful management.

Reproduction is a selfish act. The goal of reproduction, for any organism, is to ensure the survival of its genetic lineage. Such genetic immortality can be attained only by having as many members of the next generation as possible carrying your genes. Genes can be passed on through **asexual reproduction**, in which all the offspring are exact genetic copies of a single parent, or through **sexual reproduction**, in which chromosomes of two parents are segregated and recombined so that no two offspring are identical to each other or to either parent. An organism that reproduces asexually would need only one youngster to maintain an ecological fitness of 1 (because it has 100% of its parent's genes). Organisms that reproduce sexually must raise two offspring to achieve an ecological fitness of 1. Each of the young inherits half of its genes from either parent. From the parents' point of view, the youngster represents a 50% replacement of their genes in the next generation. Therefore, to achieve an average of 100% representation of its genes, each parent must raise two youngsters.

If twice as many young must be raised by sexually reproducing organisms to maintain their fitness, it seems that asexual reproduction places less of a burden on a parent. An asexual world would seem to be a more efficient way to ensure maximum ecological fitness. In this asexual population, every individual is "female" in the sense that all can bear young. Furthermore, because 100% of the asexual "mother's" genetic material is in each of her "daughters," they will be exact copies, genetic clones, of their mother. Even in asexual species mutation may provide some variability, but natural selection will weed out deviant forms, leaving those most like the mother.

10.1 Ways to Produce Clones

The simplest of all reproductive methods is cell division. Single-celled organisms, such as amoeba or yeast, employ this means of asexual reproduction in which the mother cell divides to produce two identical daughter cells. So that each daughter cell has a full complement of DNA, the encoded programming for life, the chromosomes housing the DNA must be replicated before the division. This is done through the process of mitosis. In the last stage of the mitotic division, two complete sets of chromosomes lie within the nucleus. The chromosome pairs are polarized, so that at each pole of the nucleus there is a complete set of genetic material. The nucleus then divides so that each half has a full complement of chromosomes. The cell then divides to produce two daughter cells, each with a nucleus. The daughter cells are identical to each other and are clones of the original mother cell. This kind of reproduction can be extremely rapid, with several generations being produced each hour under optimal conditions.

Some larger organisms, such as *Hydra*, can produce clones through the process of budding, in which a miniature version of the parent organism begins to form as a side branch from the main trunk of the body. When fully formed, the base of the bud constricts so that the bud falls away from the parent and is then a free-living clone. Again, all the cell division undertaken in this process is mitotic.

A third means of asexual reproduction for plants is through vegetative spread; shoots radiate from the body of the plant, and at the end of each shoot a new plant forms. The whole is still a single plant and is genetically uniform. Although the new plant may remain attached to the mother plant, if the connecting shoot is severed it will continue to grow and is then a clone of the mother plant. This reproductive method is very common among grasses, and gardeners will recognize this as the way that crabgrass (*Digitaria* spp.) spreads through a lawn. Some larger plants may be clones spread vegetatively, such as elm trees that grow in European hedgerows or willows and alders that can sprout from a fallen twig that embeds in mud. Whole stands of these trees may be genetically identical.

Parthenogenesis is a means by which more complex animals can reproduce asexually. In a parthenogenetic organism an egg is produced that can develop into an embryo without fertilization. Some lizards, fish, and insects use parthenogenesis to produce clonal young. Parthenogenesis should not be assumed to be a primitive type of reproduction. Animals so closely related that they are in a single genus, such as *Cnemidophorus* lizards, can contain both parthenogenetic and sexual species. Furthermore, among minnows, parthenogenetic species appear capable of reproducing faster and outcompeting closely related sexual minnows.

In summary, asexual reproduction guarantees young to be just like their mother, reproductive rates can be very rapid, and all members of the population can bear young. Asexual reproduction seems to be an almost ideal system, yet 97% of organisms have sacrificed it for sexual reproduction. As believers in natural selection, we have to search for an evolutionary benefit underlying the adoption of sexual behavior, because the benefit is not intuitively obvious.

10.2 The Ecological Costs of Sex

For sex to be adopted, three basic costs of sexual reproduction must be overcome: the costs of meiosis, of recombination, and of mating. The **cost of meiosis** can also be thought of as the cost of producing males. There are actually two arguments here, both relating to the expectation that half of the progeny will be male. Males do not give birth, thus a female must produce twice as many young as her parthenogenetic counterpart to maintain the necessary number of daughters in the next generation. Furthermore, all of her offspring will carry only half of her genes, whereas if she were to reproduce parthenogenetically she could raise the same number of young, each with a full complement of her genes, without extra effort.

The **cost of recombination** recognizes that the female has to accept a 50% genetic input into her young from a male. The splitting of chromosomes in meiosis is far more likely to produce variation than is the chromosome duplication of mitosis. Variation in behavior, anatomy, or physiology may pervert one of the many characters that determine the ecological success of an individual. Consequently, almost all serious variations from the parents are likely to be immediately fatal to the offspring. The genetic code that determines success frequently results from complex pairings or sequences of genetic information. Chromosomes from both the egg and the sperm are needed to complete these sequences. The female is surrendering control over 50% of the genetic information in each of her youngsters (compared with a parthenogen, who could pass on 100% to her daughter). A substantial risk exists that the male's chromosomes will not complete the optimal genetic pairings or sequences. The resulting offspring may be weaker than a comparable youngster born to a parthenogen.

Examples of the cost of recombination can be found in humans who have certain medical conditions. For example, the two genes that confer resistance to malaria and sickle-cell anemia are genetic switches that can be tripped by inherited DNA codes. The switches are operated by specific DNA combinations, or **alleles**, one inherited from either parent. The Off position is often maintained by having unmatched (heterozygous) alleles. In the case of the sickle-cell gene, a heterozygous state is the best, for this individual does not develop sickle-cell anemia *and* is resistant to malaria. But the switch can be turned On if the alleles from both parents match, resulting in a homozygous child. A child who carries two sickle-cell anemia alleles will probably die in infancy. The third possible combination is that the child has two non–sickle-cell alleles, in which case it will be susceptible to malaria. Thus, a heterozygous female is taking a considerable risk in accepting the DNA from a male, because some of her young are almost bound to be susceptible to disease. If she were able to reproduce without a male, all her clones would be free from malaria and from sickle-cell anemia.

The **costs of mating** are a significant drain on the female. These can be listed as follows:

1. The cost of sexual mechanisms. Chemical attractants, sexual organs, flowers, and so on all require a substantial diversion of energy from the basic processes of building a bigger body or making more young.

2. The cost of mating behavior. Courtship and the rituals of mating can be time consuming and energy expensive. A parthenogen can spend this time productively feeding or sleeping. Courtship may involve display flights, calling contests, or other showy demonstrations. Often these involve the males' displaying to the females, but such displays are also likely to attract predators and parasites. Any time the female is exposed to increased danger her chances of successful reproduction have been diminished.

3. Injury inflicted by the male. Mating itself can be dangerous. Males of some species (e.g., sea lions) are much larger than the females and can unintentionally injure them. Deliberate injury to the female is understandably rare, because this would reduce her chance of reproducing and consequently would fail to pass any of the male's genes into the next generation. Don't forget, the male is also questing genetic immortality through the perpetuation of his lineage.

4. Disease transmission. Any time two creatures come close enough to mate they increase the likelihood of transmitting a disease, for example, a sexually (through intercourse) transmitted disease or a skin-, insect-borne or pneumonic infection. Disease may kill the female, render her sterile, prevent successful delivery, or debilitate her so that she cannot raise the young to maturity.

5. The cost of escape from unwanted sexual attentions. The last thing that most females need is to be remated once they have been fertilized, but males approach life differently: If the female still has sexual cues indicating that she can be approached for mating (or even if not, in some cases), it is always worth attempting to inseminate her. The female may find herself pestered by male attention to the extent that she cannot feed, at a time when it is important that she maintain or increase body weight. Worse still, she may be injured by would-be mates.

What Good Is a Male?

It is clear from the preceding list that in evolutionary terms males start with at least two strikes against

them. Still there must be something good about them . . . in fact there is. The stock answer may be that sexual reproduction allows a greater variation in the offspring. Clones are likely to respond poorly to changing environmental conditions, whereas among a brood of genetically different youngsters, one of them may be well adapted to the new conditions. Variability does confer an advantage under some circumstances, but for most generations the benefit is dubious. For example, a tropical butterfly that lives as an adult for three weeks and has a generation time (the time between the birth of a parent and the birth of its offspring) of about three to four months has a good chance that its offspring will face the same environment that it faced. So why reproduce sexually? One answer is that among the many offspring from a sexually reproducing adult there will be a small, but significant, proportion of the "superfit" (Figure 10.1). By comparison, a parthenogen will produce more moderately fit individuals. The superfit are few, but those individuals can outcompete all others, including the parthenogens.

This argument can be extended into the circumstance of environmental change, when a new area is colonized. This element of uncertainty favors the sexual reproducers because they have a broader range of genotypes (genetic constitutions), any one of which may prove to be the superfit strain under the new con-

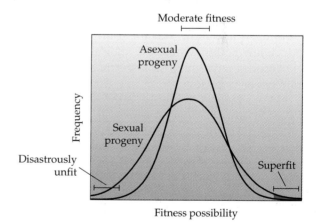

Figure 10.1   Comparison of two large populations, one parthenogenetic and the other reproducing sexually. The parthenogenetic female will have young that are clones. They will all have a very similar chance of survival and thus can be depicted as almost all having a similar fitness. The young of a sexually reproducing population will contain a proportion of runts with very low fitness, and most young will be moderately fit. But a small proportion of the young will be superfit. This superfit group will ultimately outcompete all other lineages (including the parthenogens, if they are in direct competition). (After P. Colinvaux, *Ecology*. © 1992. John Wiley and Sons, Inc. Reprinted by permission of John Wiley and Sons, Inc.)

ditions. Ecologists compare the likelihood of having superfit offspring to the purchase of lottery tickets. If a lottery ticket represents each potential offspring, a sexually reproducing individual could have thousands of tickets. Each acorn produced by an oak tree, or each fertilized salmon egg, is genetically distinct and would have a different ticket number. Among all the offspring are thousands of chances for hitting the jackpot: superfitness. On the other hand, the thousands of offspring produced by a parthenogen could be likened to a single ticket that has been photocopied again and again; all stand the same chance of being superfit.

Sex seems to have a real benefit for those organisms that produce vast numbers of young, but what is the advantage to species that produce relatively few young, such as mammals and birds? A possibility exists that these species have come so far down an evolutionary track of sexual reproduction that they can't "go back" to asexual reproduction. An alternative view is that the presence of a male may not be wholly detrimental. Males can gather food for the female or the young, and they may protect the young or train the young. If through any of these activities a male doubles the number of offspring that survive compared with a female's raising young on her own, then he has repaid the costs of sex. The female is then placing as many genes into the next generation as her parthenogenetic counterpart, thus she is not at a competitive disadvantage. For stickleback fish, pair-bonding birds, and many mammals, paternal help in raising young is apparent. But for some creatures that produce few young, where the male's only contribution to reproductive success is sperm, a further explanation needs to be sought.

Perhaps the most insightful contribution to explain the presence of sex in organisms with low fecundity comes from American population ecologist Leigh Van Valen, who posits that plants and animals are driven to constant evolution by the insistent pressure of pathogens. This idea has become known as the **Red Queen hypothesis** after the character in Lewis Carroll's *Through the Looking Glass*. Alice and the Red Queen raced and raced, ending precisely where they began, drawing the cryptic comment from the queen, "Now here, you see, it takes all the running you can do to keep in the same place."

Pathogens are disease-inducing organisms, such as viruses and bacteria, that can infect and kill. Most organisms have an inherited mechanism for resisting pathogens (our immune system), which limits the seriousness of an infection, effectively locking the invader out. However, some pathogens such as bacteria freely swap portions of their DNA with other bacterial cells, giving rise to many novel forms. Viruses can change their genetic structure by incorporating host cell DNA. The continuous mutation of these pathogens may produce a strain that is superfit. A new, superfit strain would be able to evade or "pick the lock" of the host's immune system. In defense, the host must continually change its defenses (effectively changing the lock). This deadly race against disease may be the single biggest factor that drove most organisms from asexual to sexual reproduction.

Why Stop at Two Sexes?

If two sexes are enough to provide some variability to defend against disease, would it not be better to have three or four sexes and confound our pathogens further? This question brings us neatly to the explanation of why we have two, rather than more, sexes. A male, through diligent "husbanding" may be able to offset the costs of sex by doubling the number of young raised by a female. If the definition of the female is the sex that bears the progeny, the addition of sexes will be to add extra "male" sexes. As extra non-childbearing sexes are added, we have to find a way to compensate for increasing costs of meiosis (50% cost with one male sex; 66% cost of meiosis with two male sexes; 75% with three male sexes, etc.). With more sexes, more energy would also be lost in extra mating, rituals, squabbles, and so on. Each additional sex must increase the number of young raised by the female to compensate for this extra cost. Whereas a male must double the number of young raised compared to those that could be raised by the female alone, the combined efforts of the male and the third sex would have to treble the number of young raised. The chance that an additional sex can do this declines so rapidly that no example of a three-way sexual reproductive strategy has ever been found.

The Best of Both Worlds?

Consider those organisms that have evolved to reproduce sometimes sexually and sometimes asexually. Such heterogonic life-styles are common among creatures that inhabit very unstable habitats such as puddles. The amoeba used as an example of asexual cell division can also reproduce sexually. Sexual reproduction is usually presaged by a downturn in living conditions, as when the amoeba's puddle dries up. Two amoebae then fuse to form a zygospore, which has a double load of DNA. Within the zygospore, meiosis takes place, reinstating the original number of chromosomes in each of the new cells. During this process, some DNA exchange occurs so that none of the daughter cells are alike. If the zygospore settles on a new puddle, it bursts open, and the genetically diverse offspring enter the lottery of survival. The sexual stage provides the diversity of young needed to colonize a new area, where the original maternal recipe might not be the best suited. Once established, the "winning" form will reproduce clonally, because this is the quickest way to produce a big population. Thus, for these transient habitats, a heterogonic lifestyle of clone when you can, and have sex if you must, appears ideal.

10.3   Many Babies or Big Bodies: An Energetic Trade-off

Growth, reproduction, and daily metabolism all require an organism to expend energy. The expenditure of energy is essentially a process of budgeting, just as finances are budgeted. If you spend all your money on clothes, then you may have none left to buy food or go to the movies. Similarly, a plant or animal cannot squander all its energy on growing a big body if none is left over for reproduction, for this is the surest way to extinction. All organisms, therefore, allocate energy to growth, reproduction, maintenance, and storage. No choice is involved; this allocation comes as part of the genetic package from the parents. Maintenance for a given body design of organism is relatively constant, therefore we need not consider it. Storage is important, but ultimately that energy will be used for either maintenance, reproduction, or growth. Therefore the principle differences in energy allocation are likely to be between growth and reproduction.

Two extreme sets of energy allocation are immediately obvious. Almost all the energy could be diverted to reproduction, with very little allocated to building the body. At the other extreme, almost all resources could be invested in building a huge body, with a bare minimum allocated to reproduction. Both of these extremes exist in nature, and they bring with them some predictable lifestyle consequences.

The plant or animal that allocates most of its energy to reproduction will be small, because resources are not being devoted to body size. The defenses of the organism are likely to be weak, because building a stern array of teeth, claws, or distasteful chemicals takes a considerable investment of energy. Conse-

quently, this organism will be vulnerable to attack. Under such circumstances, the best means of defense would be to avoid predators or pathogens by being highly mobile. This can be achieved, even by plants. The solution is to grow fast to a minimum size, invest every iota of energy, including the energy reserved for maintaining life, in a single reproductive outpouring, then die. Of course, with no parent around, the young are completely on their own and defenseless. If this is going to be the reproductive strategy, there had better be hundreds of offspring produced, because the chance of survival of any individual is inordinately low. This provides a good description of the reproductive strategy of a dandelion or a common house fly. Such organisms are said to be highly fecund because they produce large numbers of young. A distinction now needs to be drawn between fitness and fecundity. Fitness is the number of the offspring that live to breed (divided by two), whereas fecundity is the total number of young born to the female.

Some further observations can be made about being a dandelion. Dandelion clocks are the seedheads, raised just high enough above the ground to catch the wind. The plant is no bigger than it needs to be, its stems are hollow, and all the rigidity comes from its water content. Thus, a minimum investment has been made in the body that becomes a platform for seed dispersal. These plants are very short-lived, surviving for just one reproductive effort, and, because they do not live beyond one year, they are termed annuals. Their profligate reproduction means that they provide a constant rain of seed in the neighborhood of parent plants. A new plant will spring up wherever a seed falls on a suitable soil surface, but because they do not build big bodies they cannot compete with other plants for space, water, or sunlight. These plants are termed **opportunists** because they rely on their seeds' falling into disturbed settings where competing plants have been removed by natural processes, such as along an eroding riverbank, on landslips, or where a tree falls and creates a gap in the forest canopy.

Opportunists must constantly invade new areas to compensate for being displaced by more competitive species. Human landscapes of lawns, fields, or flowerbeds provide disturbed settings with bare soil and a lack of competitors that are perfect habitats for colonization by opportunists. Hence, many of the strongly opportunistic plants are the common "weeds" of fields and gardens.

Because each individual is short-lived, the population of an opportunist species is likely to be buffeted by drought, bad winters, or floods. If their population is tracked through time, it will be seen to be particularly unstable, soaring and plummeting in irregular cycles. Because chance climatic events have nothing to do with the density of individuals in the population, the populations of opportunists are said to vary in a density-independent manner (see Chapter 9).

The opposite of an opportunist is a **competitor**. These organisms tend to have big bodies, are long-lived, and spend relatively little effort each year on reproduction. An oak tree is a good example of a competitor. A massive oak claims its ground for 200 years or more, outcompeting all other would-be canopy trees by casting a dense shade and drawing up any free water in the soil. The leaves of an oak tree taste foul because they are rich in tannins, a chemical that renders them distasteful or indigestible to many organisms. The tannins are part of the defense mechanism that is essential to longevity. Although oaks produce thousands of acorns, the investment in a crop of acorns is small compared with the energy spent on building leaves, trunk, and roots. If the tree lives for two centuries it will have almost that number of attempts (times the number of acorns produced each year) to maintain its fitness. Once an oak tree becomes established, it is likely to survive minor cycles of drought and even fire. A population of oaks is likely to be relatively stable through time, and its survival is likely to depend more on its ability to withstand the pressures of competition or predation than on its ability to take advantage of chance events. As competi-

tion and predation are often related to the concentration of individuals, the population of a competitor may be influenced by density-dependent factors (Chapter 9).

An animal that is a good example of a competitor is a bear. The defensive capabilities of a bear are not in doubt, and, like humans, they have antibodies that provide their defense against disease. During the 20 years of a bear's lifetime, a female may produce a pair of cubs 10 times, providing 20 opportunities to maintain her ecological fitness. Some cubs will be sickly, some will be unlucky, and on average two will survive to reproduce.

I have used the terms opportunist and competitor to describe these lifestyles, but you will also see them termed r-strategists (opportunists) and K-strategists (competitors). The attributes of opportunists and competitors are summarized in Table 10.1.

If opportunists are outcompeted by competitors, how do they survive? If habitats were constant, and there were no disturbance events, it is hard to see how opportunists could survive. But habitats are continually perturbed by disturbance events. These events may be small, such as a tree falling, or they may be huge, such as a forest fire. In either case, the disturbance provides a fleeting opportunity for an opportunist to occupy the disturbed area and produce a new generation. Where a patch of woodland has burned, in the time it takes a tree to resprout, the dandelions and fireweed have colonized, produced a generation of young, and died. The disturbance factors are the den-

Table 10.1 Pianka's list of attributes that can be assigned to opportunists and competitors.

	Opportunist	Competitor
Climate needed	Variable and/or unpredictable	Fairly constant and/or predictable
Mortality	Often catastrophic	Seldom catastrophic
Population size	Variable through time, seldom reaching carrying capacity, disequilibrial	Fairly constant, approaches carrying capacity and equilibrium
Ability to compete	Low	High
Selection favors	Rapid development, early reproduction, small body size, single reproductive effort	Slow development, postponed reproduction, larger body size, repeated reproduction
Length of life	Short, usually 1 year	Long, usually 1 year

(*Source:* E. R. Pianka, 1970.)

sity-independent factors, and they serve to prevent equilibrium from being established in an ecosystem. Therefore, the more the system is disturbed by density-independent factors, the further it is from equilibrium and the larger will be the populations of opportunists. In the absence of disturbance, the longer-lived competitors have the advantage.

It is important to note that the pure opportunist or competitor are the extremes of a continuum. Most organisms exhibit a blend of some opportunistic and some competitive aspects to their ecology.

Many Small versus a Few Large Young

A female cod may lay as many as 5 million eggs in a single breeding season. Many of these will never be fertilized, many more will prove infertile, or produce hopelessly weak hatchlings. The lives of those hatchlings that emerge are extremely dangerous, and yet a few survive. Certainly, it is during the first few weeks of life as a salmon that the greatest number of a given generation will die. Compare the fate of human young: Our children and middle-aged adults have a relatively high chance of survival, so humans are most likely to die when they are old. From this simple observation of the natural world a set of hypothetical curves can be drawn to represent the proportion of a population that survives in each age class (Figure 10.2). The type 1 curve is for a population that has highest mortality among its oldest members. Such a curve typifies humans and large mammals. The type II curve is for a population in which a constant proportion of the members of the population die each year. Examples of this linear rate of mortality with age are some species of birds, invertebrates such as *Hydra*, and seeds of large rainforest trees buried in the soil. The type III curve is for a population where the greatest losses are among the very young. Many insects, fish, mollusks, and small-seeded plants such as orchids and dandelions would be among this group.

Salmon produce thousands of eggs that never even hatch. Thousands more of their young die or are eaten within days of birth. Although the individual investment of energy in any given salmon egg is minute, any young that fail to survive represent a wasted investment. If one egg in 10,000 survives to become a mature fish, of the total salmon energy invested in eggs, 0.01% is represented in the next generation of fish. That so many young die suggests that this is an extremely inefficient use of energy. But, as the majority

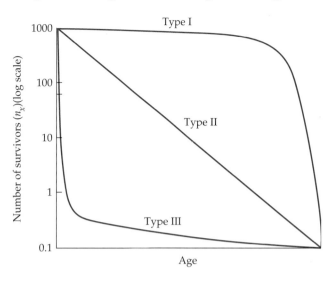

Figure 10.2   Hypothetical survivorship curves for three different reproductive strategies. Type I: large-young strategists that nurture their young; Type II: populations that have a steady rate of mortality; Type III: small-young strategists with low, or no, investment in nurturing. (After R. Pearl, 1928.)

of organisms reproduce this way, and, by definition, over the medium-term they all have an ecological fitness of 1 (otherwise they would go extinct or smother the planet surface), it seems to work.

Most large animals and trees are **iteroparous**; that is, they live to reproduce repeatedly. But many species of insects, plants, and microorganisms are **semelparous**; after a single reproductive effort one or, often, both of the parents die. Examples of this reproductive behavior are not limited to small creatures. Salmon and a large rainforest tree of the pea family, *Tachigalia*, are both examples of large organisms that are semelparous. The reproductive efforts of a semelparous species represent the ultimate expenditure of energy in order to produce as many eggs as possible.

The strategy of producing a few large young has a trivial investment in egg or sperm, but a massive investment in nurturing. That the onus generally falls on the female to raise the young is particularly exaggerated in mammals, such as ourselves; a fairly long period of gestation inside the female's body precedes a lengthy period of absolute dependency. The large-young reproductive strategy is relatively rare and is found most commonly in mammals, birds, and reptiles. The young are often nurtured for a year or more (arguably in humans for up to 20 years) and woe betide the predator that attempts to separate mother bear from cub. Each cub represents a very high investment of energy and is, therefore, savagely protected. The likelihood of any individual young reaching maturity is much greater than among the small-egg strategists, but overall there is the same net fitness of 1. Even though the young are protected by a parent, "faulty programming," disease, famine, or plain bad luck will pare the numbers of surviving offspring until, on average, each parent couple produces just two young.

Humans have broken free of the restriction of a fitness of 1, as evidenced by our soaring global population. We have achieved this feat not because of any different reproductive strategy, but because we have evolved intelligence. We can house, heat, feed, and nurse our infants more effectively than any other creature. We are the first organisms capable of beating the lottery, but the cost of this has been to deplete global resources.

10.4   Selection for an Optimal Number of Young

Among large-young strategists a division can be made between those organisms, such as a whale, a cow, or an elephant, that have one offspring at a time and those that have a clutch of young at a time. What determines how many eggs the bird will lay or the number of cubs born to a lioness? Natural selection should result in an optimal number of young. If you raise one fewer youngster than your neighbors, they have a higher fitness than you and will ultimately drive your lineage to extinction through competition. Surely then, it is better to raise one young more than your neighbor and ensure genetic immortality.

Variable Clutch Size

The number of eggs laid by some species of bird is the result of an environmental cue. For example, the European Robin (*Erithacus rubecula*), unrelated to the American Robin (*Turdus migratorius*), is a visual hunter that specializes in catching flying insects. Because it relies on sight to locate its prey, its hunting activity is limited by the number of hours of daylight. In the breeding season (May–June), the near-Arctic will have 18 to 20 hours of daylight compared with perhaps 15 hours of daylight in North Africa. The longer days allow the northern robins to gather more food than those in the south. The northern robins are also taking advantage of a boom in insect populations that takes place in the northern forests each spring. The likelihood that both greater hunting time and greater prey abundance enable the robins to raise more young

at high latitudes is seen in the average number of eggs found in robin nests (Figure 10.3). In southern Europe and North Africa there is an average of four eggs in each nest, whereas in the northern latitudes the average is 6.3. Northern robins probably suffer greater mortality as chicks and prereproductive adults because of the harsh climate. Robins maintain broadly constant numbers in both regions, thus the overall fitness of both populations approximates 1. It should be noted that to think of this disparity in egg numbers as the northern robins attempting to compensate for a higher mortality rate is absolutely wrong. Each population is programmed to lay as many eggs as can be raised: The later fate of the brood is not a consideration.

Too many eggs in a nest can be as bad as too few. A study of the European swift revealed the danger of having one egg too many. The swift typically has a brood of two or three chicks each year. If enough nests are studied, a typical survival rate can be calculated for the young that reach the age when they start to fly. This statistic provides an indication of average reproductive success that can be related to the number of eggs laid. In an experiment, when a fourth nestling was added to each nest, instead of increasing the num-

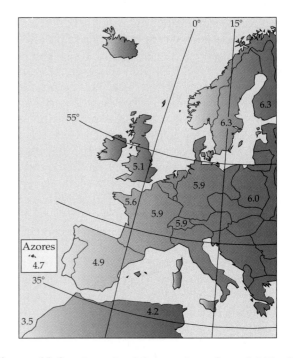

Figure 10.3   A study of the number of eggs laid by the European Robin. The more-northern populations had more eggs in each nest than the southern populations. (From D. Lack, *British Birds*, 39:98–100, 130–135. Reprinted by permission of British Birds Ltd.)

ber of chicks that fledged, the number declined to fewer than "normal." The swifts were trying to feed all four young, but failed to collect enough food. Not only did the extra chick starve, but so too did some of the chicks that would otherwise have survived. In an extreme case, by adding one extra chick the whole brood would be lost.

A detailed study of the Great Tit (Figure 10.4), a relative of the chickadee, demonstrated how clutch size can vary. A population of Great Tits in Wytham Wood near Oxford, England, was monitored for three years to determine the number of eggs laid by birds of known age. In the first year of the study, all the birds were tagged with individualized leg bands so that they could be recognized. As new birds entered the population, they too were tagged. Great Tits feed on caterpillars and grubs, and their reproductive effort is closely tied to food abundance. Prey availability may be related to natural cycles within prey populations, to

the severity of winters, or to the presence or absence of disease. Consequently, the birds lay a different number of eggs each year (Table 10.2).

Another consistent trend is that birds that are two years old lay the most eggs. It may be that two-year-old birds claim better territories than the less-experienced birds and that they are physically stronger and more able to gather food than older birds. Of the 128 birds recorded at the start of the study in 1961, only 29 were still alive in 1963. From these data it is evident that the majority of Great Tits live to be only two or three years old.

Gathering the Great Tit data involved a time-consuming study that was complicated by the fact that several generations of individuals were participating in the breeding population. Consider how much more complex such a study would become if there were many differently aged individuals, such as an oak forest in which some trees may be 300 years old. In such cases, following a cohort throughout its life is impractical.

Preprogrammed Clutch Size

A bird that is genetically programmed to always lay the same number of eggs is the Galápagos Penguin (Figure 10.5). As a graduate student of Ohio State University, Dee Boersma researched the reproductive success of the penguins. She documented the reliance of the birds on an upwelling of cold nutrient-rich water off the coast of the Galápagos to trigger their breeding cycle. The upwelling brings dense shoals of fish within range of the penguins' breeding grounds, and the birds take advantage of this food bonanza to breed. The penguins always produce two eggs, one laid a few days before the other. Consequently, the chicks differ in size, the older being considerably the largest. The upwelling does not last long, and as soon as it ceases the fish move away, and the penguins are once again hard-pressed for food. If the upwelling lasts throughout the breeding season, both chicks can be raised; but that rarely happens. In general, the food runs out (swims away) before the chicks are old enough to fend for themselves. As soon as food becomes scarce, the younger chick is ignored by the parents and left to starve. All their food-gathering effort is then fixed on the older chick. Often, the parents cannot catch enough food for the survival of even this chick, and it, too, is abandoned. Thus, in many of their attempts to breed, the penguins will abandon both chicks. An important insight is gained from this example: It is better to quit now than to struggle against

Figure 10.4 The European Great Tit.

Table 10.2   Variations in Great Tit average clutch size at Wytham Woods according to age of nesting birds.

Age years	1961		1962		1963	
	Number of birds	Average clutch size	Number of birds	Average clutch size	Number of birds	Average clutch size
1	128	7.7	54	8.5	54	9.4
2	18	8.5	43	9.0	33	10.0
3	14	8.3	12	8.8	29	9.7
4			5	8.2	9	9.7
5			1	8.0	2	9.5
6					1	9.0

(Data are from C.M. Perrins, 1965.)

Figure 10.5   The Galápagos penguin.

a food shortage when raising young. The Disneyesque view of animals battling nobly against all adversity to feed their starving brood is romantic nonsense. The reason is simple: An adult can try to breed again next year, and it is never worth sacrificing an adult for a chick. Each pair of penguins may have 20 years in which to strike the jackpot of a sustained upwelling; self-sacrifice is evolutionary suicide.

One Young at a Time

The single-young option can be viewed as the extreme form of predetermined clutch size. Some organisms are not equipped to care for more than one offspring at a time; they would almost certainly suffer complete reproductive failure if they had a second youngster. Their reproductive fitness would be zero, and their lineage bound for extinction. Consequently, the only successful lineages of these species have a single youngster at a time. Among whales and some large mammals, this youngster is so dependent that its mother will not reproduce for several years. Animals with such low reproductive output will prove to be exceptionally vulnerable to extinction from hunting.

10.5 Territoriality

The dawn chorus that wakes us in springtime is a shouting match that determines reproductive success among many bird species. The first song contests of the year are often between all-male groups. The song is a demonstration by the male proclaiming possession of a **territory** and warning other males to stay away. A territory in this sense is not so much a piece of real estate as a set of resources essential for successful breeding. A good nest site, an abundant food supply, and an apparent absence of predators or disturbing influences would all contribute to a good territory. Red-winged Blackbirds (Figure 10.6) and

Figure 10.6 **A male Red-winged Blackbird proclaiming his territory.**

Table 10.3 Territorial migratory birds shot in a Maine woodland in a two-week period in two consecutive years.

Year 1:	148 males singing from perches
	302 males shot
Year 2:	154 males singing from perches
	352 males shot

(*Source:* R. E. Stewart and J. W. Aldrich, 1951.)

buntings, such as the European Yellow-hammer and the American White-throated Sparrow, are migratory species. The entire population shifts south to avoid the cold of winter, and then floods back north in the spring. The males are the first to arrive on their breeding grounds, and, in general, the first males claim the best territories. The male claims his territory by puffing out his chest and yelling at the top of his lungs from song perches sited at the corners of the defended area. As the females arrive, males holding territories will sing not only to proclaim their grasp on the land but also to attract a female.

Other territorial animals, such as rhinos, lions, and dogs, establish their breeding domain by defecating or scenting along the boundary of their territory. This is also why a cat rubs its face against your legs; you are being scented by glands at the corner of its mouth and eye, and hence declared part of its domain.

The two principal advantages to territoriality that result in increased fitness are that (1) it safeguards the supply of food, and (2) the female can be protected by the male from the unwanted attentions of other males.

A group of researchers called the Maine Gunners conducted a somewhat gory study on the effect of predatory birds (warblers) on caterpillars. The study produced an interesting byproduct: It revealed that many male warblers are forced to sit out the breeding cycle each year. In this study, the number of bird territories in a 140-acre woodland was estimated by counting the number of males singing from perches (Table 10.3). The gunners then went through the woodland for the next two weeks shooting all the males they encountered. At the final count, they found

that they had shot twice as many males as there were territories.

Males that did not hold territories lived in the adjacent hedgerows and trees. As territory holders were shot, the vacant territory was immediately claimed by one of these "outside" males. By moving into a territory, these males entered the breeding pool. The birds without territories were most unlikely to breed that year. These warblers are migratory, so the next spring new migrant males arrived to the woodland and established their territories. The shooting experiment was repeated a second year, and two remarkable patterns emerged. The number of territories in the second year was approximately the same as in the first, even though all the inhabitants establishing territories were new to the area. Once again, the number of males appeared to be about double the number of territories. These findings lead to the identification of two potential disadvantages of territoriality: (1) energy must be expended in the defense of the territory; and (2) a substantial portion of the male population does not get the chance to breed.

The cost of defending a territory is borne primarily by the male rather than the female. For most organisms, it is the female that devotes the largest share of energy expenditure to raising young—through gestation, birthing, and nurturing. Energy expended by the female on nonreproductive activities at this time could endanger the brood. Energy expenditure by the male is less likely to be directly deleterious to breeding success. Therefore, any energy he expends defending a territory that benefits the female is likely to have an indirect reproductive benefit. Thus, because it is the male that bears the brunt of the energy cost of territoriality, the energy expenditure is of less significance than if this cost had been incurred by the female.

The cost of having a nonreproductive portion of the population appears to contradict our principle of maximum reproductive effort. Why should some males simply accept a nonbreeding status? It would be a mistake to think that the nonbreeding males accept their status passively, as considerable competition takes place for territorial space. Among birds, the competition between males rarely ends in bloodshed; much posturing, fluffing of feathers, and singing generally settles these disputes. Usually a demonstrably weaker bird will cede; this often is an immature or first-year bird, and, usually, the territory owner will win. The resident generally wins the dispute because he has more to lose than the interloper, who can retreat and try to find a less well-defended territory. If all else fails, the losing male can wait for next year, when he has a better chance of gaining a territory. This provides a fine example of "he who lives to run away gets to breed another day."

In some cases male competition is partly territorial, but more important than "resource space" is a social hierarchy that determines breeding priority. An example is seen in the behavior of the northern elephant seal. In late December, male seals arrive on the breeding beaches of California and proceed to spar for supremacy. The battles are generally vocal and postural, but they can become bloody. When the pregnant females arrive in January to give birth to their pups, the ferocity of male contests increases. By late January, the females have birthed and are ready to be mated. Male competitiveness reaches a crescendo because the top-ranked males will dominate that year's breeding. In one study of 115 males on the beach, the 5 most dominant males performed 123 of the 144 witnessed copulations. The male mating stakes are high. The battles and mating are utterly draining; after a single breeding season, successful males may be so injured and depleted in fat reserves that they quickly succumb to disease. But they have bred! The evolutionary die is cast: For a male to stand any chance of reproductive success in this system he must become a bigger, fiercer, more-determined competitor than surrounding males. The chance of any individual male pup's surviving to reproduce may be less than 1 in 100, but compete he must.

10.6 Polygyny and Female Choice

When one male has more than one mate, the relationship is said to be **polygynous**. An important understanding underlying polygyny is that, although it is in the male's best interest to mate with as many females as possible (this will increase his ecological

fitness), similar promiscuity is not advantageous to females. A female is likely to be as reproductively successful with a single mate as with multiple mates.

Males of territorial species such as deer, wild cattle, lions, and red-winged blackbirds commonly have more than one mate. Under these circumstances, females choose the male on the basis of the quality of his territory, his apparent strength, and the number of females already in his harem. This process is referred to as sexual selection. From studies of reproductive success among migratory birds, it is apparent that the first female to arrive in a breeding area has the pick of males and breeding sites. This advantage determines that females arriving first generally raise the most offspring. Females arriving late are forced to choose between polygyny on the best territories, where the optimal nest location has already been secured, or monogamy on the worst sites. In general, females entering as secondary females to a harem have breeding success equal to that of females that are monogamous on middle-quality territories, and they do better than those on the worst territories (Figure 10.7).

Among humans, animals we have domesticated, and those that scare us, such as wolves and lions, the norm is that males are larger than females. The more extreme the sexual dimorphism—that is, the larger the size differential between male and female—the larger the typical harem of that species (Figure 10.8).

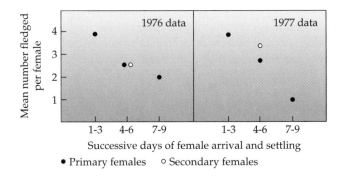

Figure 10.7   The first females to arrive in an area enter the best territory. Their reproductive success is measured in the number of young that fledged. Females arriving later choose between being the primary female in a lesser territory or one of the secondary females in the best territory. Being a secondary female was not disadvantageous, so long as the territory occupied was a good one. Birds arriving still later could occupy only marginal territories, where their reproductive success was poor. (After P.J. Garson, W.K. Pleszcynska, and C.H. Holm, 1981.)

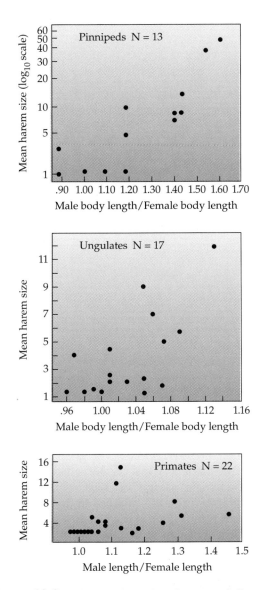

Figure 10.8   Harem size related to sexual dimorphism (difference in size between males and females) for pinnipeds (seals, sea lions, walruses, etc.), ungulates (deer, goats, and sheep), and primates. The strongest relationship between harem size and sexual dimorphism is among pinnipeds. (Note that pinniped harem size is plotted on a log scale.) (After M. Daly and M. Wilson, 1983.)

The most extreme relationship is seen among the pinnipeds (seals, sea lions, and walruses; Figure 10.9). The biggest males may have a harem of 50 or more females, and the size of the harem is directly related to the fighting weight of the male. The dimorphism of ungulates (deer, cattle, sheep, antelope, goats) is again a strong predictor of harem control. The pattern among primates is not so distinct, perhaps because they have

smaller social groups and a greater tolerance of secondary males having some reproductive role.

The advantage to the dominant male of having a harem is that he can inseminate as many females as possible, thereby greatly increasing his chance of fitness compared with a monogamous male. The benefit to the female is less obvious and is still debated, especially when the male plays little or no part in nurturing the young. By entering the harem of the dominant male she has opted for what appears to be the winning genotype among available males. This mate is perhaps her best bet in terms of reducing the chance of acquiring "weak" genes. Also, the strongest male can keep other males away, which may reduce predation on her young and interruptions to her feeding them.

Regarding a harem's effect at a population level, one male holds the best breeding ground. Often, all the other males are excluded from this territory and are forced to live solitary lives apart from the herd. In such circumstances it could be argued that the breeding pool has achieved almost total femaleness. The best resources are for the exclusive use of females, apart from the single stud male. Thus the prime resources wasted on males are kept to a minimum.

A practical human application that comes from these observations is that a proportion of males can be harvested in many species without reducing the overall population level. A standard way to express the proportion of each sex in a population is a sex ratio. This ratio considers the balance of males and females among 100 randomly selected individuals. Given that the sex ratio commonly approximates 50:50 and that a substantial number of males are outside the breeding pool, killing a male for food may have less impact on a population than killing a female. Among pheasants, the ratio of male to female birds needed to maintain full female fertilization could drop to 12:88 (male: female). In other words, approximately three-quarters of male birds could be shot and there would be little effect on the next year's population under ideal conditions. Deer and other polygynous mammals commonly form healthy herds with a sex ratio of 30:70, allowing us to shoot some bucks without reducing the population. However, before we merrily blast all males from the skies and fields, some cautions should be noted. Many species exist in a monogamous pair bond for their entire lives, and shooting males of these species would have serious population implications. Among polygynous

Figure 10.9 Sexual dimorphism of elephant seals. The larger animal is the male of the harem. Male elephant seals are shown battling for supremacy. The victor will gain control and earn the right to fertilize the female seals. The females are the smaller seals shown in the background.

species, the dominant male, the "king of the herd," has the most impressive antlers, tusks, or teeth and becomes the preferred target of hunters. If the king is shot, a lesser, perhaps weaker male will take his place, potentially harming the genetic quality of the herd. Hunting does not necessarily weaken populations, but regulations such as bag limits (the number that may be killed), closed seasons (when hunting is prohibited), areas of complete protection, and a careful watch on the impact of hunting on populations, are necessary.

Finally, it should be noted that mammals are unusual in having a sexual dimorphism in which the male is larger than the female. The great majority of species on the planet are insects, and among these organisms the female is larger because of the reproductive onus placed on her. It takes a bigger body to develop several hundred eggs than it does to produce a similar quantity of sperm.

Summary

- Asexual reproduction facilitates rapid replication of clones.
- 97% of organisms reproduce sexually.
- There are three costs of sexual reproduction: cost of meiosis, cost of recombination, cost of mating.
- Sexually produced offspring have great genetic diversity.
- Males can offset reproductive costs by increasing the survivorship of the young (through feeding, protecting, or nurturing them).
- Opportunists tend to be short-lived, have small bodies, and have many young.
- Competitors tend to be long-lived, have large bodies, and have few young.
- An individual's ecological fitness is measured by the number of its young that live to breed.
- An adult is worth more than a juvenile: if necessary, sacrifice the young to survive.
- Animals will have the optimal number of young to achieve an ecological fitness of 1.
- Territoriality and polygyny may offset some of the costs of having males.

Further Readings

Dugatkin, L. A., and J. G. J. Godin. 1998. How females choose their mates. *Scientific American* 278 (4):56–61.

Daly, M., and M. Wilson. 1983. *Sex, Evolution and Behavior.* 2nd ed. Belmont, CA: Wadsworth.

Pianka, E. R. 1970. On r- and K- selection. *American Naturalist* 104:592–597.

Thornhill, R., and S. W. Gangestad. 1996. The evolution of human sexuality. *Trends in Ecology and Evolution* 11:99–102.

van Valen, L. 1973. A new evolutionary law. *Evolutionary Theory* 1:1–30.

Web Connections

On-line resources for this chapter are on the World Wide Web at:
http://www.prenhall.com/bush
(From this web-site, select Chapter 10.)

Making Connections: Fisheries

11

11.1 *Fishing isn't what it used to be*

11.2 *Fish, fisheries, and productivity*

11.3 *A simple model of fisheries*

11.4 *Further ecological thoughts on fisheries*

11.5 *Prey switching and fishing*

11.6 *Local solutions to fishery problems*

11.7 *Are fish farms the answer?*

11.8 *National and International protection*

11.1 Fishing Isn't What it Used to Be

Ask any old-timer about the past and you hear that the streets were safer, the politicians were all honest, and the fishing was better. Perhaps the first two of these observations are largely the product of nostalgia, but is it true that the fishing was better? Throughout this century the total tonnage of fish caught has increased steadily (Figure 11.1). This trend might lead us to think that fish are an abundant resource and that catches can continue to rise in the future. An alternative possibility is that we are approaching a maximum harvest and that catches will stabilize at this point. This hypothesis might explain the slight flattening of global fishing yield in the 1990s. A third hypothesis is that we are already overharvesting this resource and that fisheries are collapsing, but that we are too good at catching the remaining fish for this to be evident yet.

The demand for fish is likely to increase as human populations continue to grow. In affluent societies, the demand for fish has paralleled interest in a healthy diet. Many species of fish, particularly those with white flesh, have less fat per gram than red meat and have become a popular health food. The poor of many countries rely on fish as their only affordable source of animal protein. Globally, fish provide about 16% of animal protein consumption by

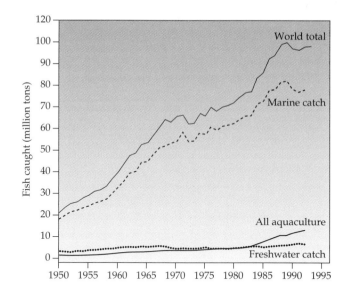

Figure 11.1   The global fish catch from 1950–1993. (Data from FAO.)

humans or about 5.6% of total protein from all sources. With a burgeoning human population, it is critically important to understand whether this resource is presently under- or overexploited. To understand the trends in fisheries and, more importantly, what needs to be done to make their use sustainable, it is necessary to understand both the economics and ecology that affect commercial exploitation of fish.

11.2 Fish, Fisheries, and Productivity

A strong relationship exists between primary production and fishery yield (compare Figure 11.2 with Figure 5.3b). The highest productivity of marine systems is in estuarine and continental shelf environments. Here, nutrients washed from the land fertilize the waters and provide a rich basis for food chains. Turbulence in these shallow seas keeps the nutrients suspended near the surface waters where they promote phytoplankton growth and thus raise primary productivity. Because inshore fisheries are the most productive and the easiest to access, they have supported the greatest fishing intensity. Fish caught landward of the continental shelf form the **inshore fishery** and represent more than 90% of the world's fish yield.

The blue water of the open oceans reflects a low nutrient status, low gross primary productivity, and, therefore, low prey abundance. Predators of the open oceans are relatively scattered with huge hunting ranges compared with predators of more productive environments. Hence we would not expect the open ocean beyond the continental shelf to provide high concentrations (number of animals per cubic volume of ocean) of predatory fish. Because most of the economically important fish species are predators of other

smaller fish and feed on the third and fourth trophic levels, open oceans are generally poor fisheries. An exception to this rule is found around upwellings (Chapter 7). Upwellings provide nutrient-rich waters that promote the growth of dense blooms of phytoplankton. Gross primary productivity in an upwelling may be 10 times that of other areas of open ocean. Consequently, all members of the food chain are likely to occur at higher concentrations in the upwelling rather than in an area of low productivity. Upwellings form the center of major fisheries, such as the anchovy fishery of Peru, the sardine fishery of West Africa, and the whaling grounds of Antarctica. Fisheries that exploit the open ocean beyond the continental shelf are **deep-sea fisheries**.

Fisheries can also be divided into small-scale and large-scale fishing operations. No precise definition exists to separate these two ends of a continuum. A small-scale fishery could be typified by an individual's or community's effort to harvest fish using small boats or pulling nets by hand. At the other extreme, a large-scale fishery could be a fully mechanized trawler fleet supported by a mother ship. The infrastructure, such as boats, processing plants, and transport, needed to support small-scale fisheries are simpler than those needed for large-scale fisheries. Compare the investment needed for a group of villagers to cooperate in pulling a seine net (Figures 11.3) with the resources

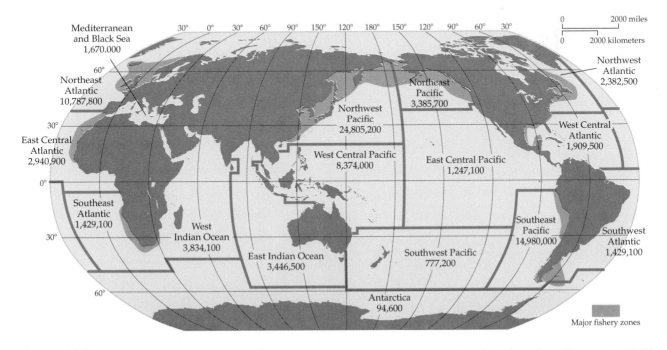

Figure 11.2 **Principal fishing grounds of the world.** Boundaries are defined by the Food and Agriculture Organization (FAO) of the United Nations. Data are for catches in millions of tons made in 1993.

Figure 11.3 **Villagers in Sri Lanka collaborate to pull ashore a seine net.**

needed to support a large-scale trawling operation (Figure 11.4). The villagers need muscle power, a canoe, a net, and some baskets to carry their fish home. The trawlers are built in a shipyard, fueled by a petrochemical industry, and crewed by skilled sailors. The villager's catch is consumed immediately or preserved by drying in the sun. The deep-sea catch is processed by canning or freezing and then distributed by land, sea, or air to market, all of which carries an energetic cost. Consequently, inshore small-scale fisheries are most important for less-developed countries (LDCs). The more-developed countries (MDCs) have the resources to build larger-scale fishing fleets capable of accessing both inshore and deep-sea fisheries.

Since the 1960s there has been a huge investment in improving deep-sea fishing efficiency. Ways to achieve this include reducing the amount of time that a fishing boat spends traveling between the fishing grounds and the port, reducing spoilage of the catch, and increasing the ability to find and catch fish. In the past, a major constraint on a fishing boat was dealing with the catch before it spoiled or filled the ship's hold. The solution was to return to port and unload the catch for processing. Because the fishing grounds may be thousands of kilometers from the port, this return to land was expensive in fuel and lost fishing time. Modern fishing fleets no longer have to return to port every few days. They are tended by a huge mother ship and can stay at sea for months at a time. The mother ship stores and processes the catch, leaving the fishing boats free to continue fishing.

Are these large-scale fishing fleets more efficient than small-scale fishing efforts? Efficiency can mean

Figure 11.4 **A modern trawler with her mother ship**

many different things, so let us compare an economist's, agriculturalist's, and ecologist's view of fishing efficiency. An economist's definition of efficiency would be that the benefits derived from an action must always be greater than the costs incurred. At a personal level, the costs and benefits of catching fish might reflect the number of fish caught and the amount of money earned. Data from the Food and Agricultural Organization of the United Nations (FAO) indicate that each fisher in a large-scale operation may catch 50 to 100 tons of fish per year and earn about $15,000. In comparison with this, a fisher in a small-scale operation catches about 1 to 2 tons of fish per year and has an annual income of $500 to $1500. On this basis, the large-scale fishery appears to be a better use of human effort.

An agriculturalist's view of efficiency might be to study the ratio of food energy produced to energy spent in food production. According to the laws of thermodynamics, all transfers of energy lead to some level of entropy and wasted energy, so energetic inputs would always exceed outputs if all inputs are considered. In the following analysis, photosynthetic inputs are not considered and these ratios only take into account human effort, materials, and fossil fuel as energy inputs. A ratio of 1:1 indicates that energy contained in the food is equal to the energy expended to obtain the food. Low ratios indicate that more energy was invested in catching or raising the food than can be obtained from eating it. A comparison of large-scale and small-scale fisheries with other food harvesting strategies (Figure 11.5) reveals that small-scale inshore fishing is much more efficient than large-scale deep-sea fishing.

An ecologist could view fishery efficiency in terms of the energy captured versus the energy acquired by the hunter. When the seine net is brought ashore by the villagers, all the fish, crabs, and other edible animals become food. In this sense, the efficiency of the hunting effort is close to 100%. In most commercial fisheries, there is a "target" group of species that is the principal object of the fishing effort. The fisheries are specialized and cannot deal economically with species other than the target group. Other species that are caught during the netting operation are regarded as "trash," and although they are killed in the fishing process, they are either ground up for use as fertilizer or thrown back. The amount of "trash" fish caught is referred to as the **bycatch** of a fishery. The shrimp industry has the highest bycatch to target ratio. When trawling for shrimp, a typical bycatch would be 8 to 10 kilograms of fish per kilogram of shrimp landed. The bycatch of one fishery may have been the essential recruits to another. For example, red snapper support a fishery in the waters of the southern states, but this fishery has declined to 14% of its former size. It is estimated that in the Gulf of Mexico, where there is an active shrimp fishery, as much as 80% of each cohort of red snapper are lost as shrimp bycatch. The worldwide estimate for shrimp bycatch is about 15 million tons per year. Other fisheries have lower bycatch ratios, but even so, more than 30% of all fish caught and killed are classified as bycatch. On this basis, the more highly commercialized fisheries have the lowest efficiencies.

Clearly, there is more than one way to think about fishing, and the different philosophic approaches provide different insights into how fisheries should be managed.

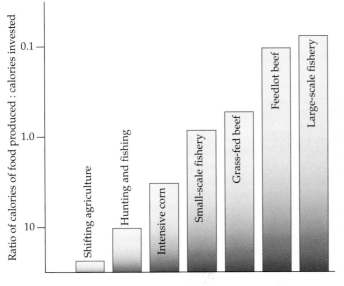

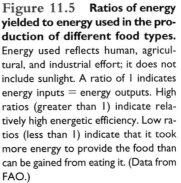

Figure 11.5   Ratios of energy yielded to energy used in the production of different food types. Energy used reflects human, agricultural, and industrial effort; it does not include sunlight. A ratio of 1 indicates energy inputs = energy outputs. High ratios (greater than 1) indicate relatively high energetic efficiency. Low ratios (less than 1) indicate that it took more energy to provide the food than can be gained from eating it. (Data from FAO.)

11.3   A Simple Model of Fisheries

To make predictions about future populations, scientists use models. A model is a simplified view of the world that allows the cause and effect of different events, such as fishing intensity, to be played out. Once a basic pattern is observed, additional complexity can be added to the model to make it more realistic. If the initial observed patterns persist, it is likely that the initial model is robust and credible.

The critical question for a fisheries manager is what is the maximum number of fish that can be removed from the population without causing the population to decline? The starting point for a model to answer this question is the logistic growth equation. Our initial model has three starting assumptions.

1. Fish population growth follows a logistic growth pattern (Chapter 9).

2. Fish populations are governed by density-dependent factors (Chapter 9).

3. If the rate of catch is less than the rate of growth in a population the fishery is sustainable (it would never run out of fish).

The first step in this model is to determine the logistic growth of the population through time (Fig-

ure 11.6a). However, a fishery manager wants to know at what population size is the population growing the fastest? To find this information we can determine the number of extra fish being added to the population between any two population intervals. These data provide a rate of population growth at any given size of population. By looking at the growth rate at population increments of 100, we would have a rate that could be graphed for population sizes of 100, 200, 300, and so on. This is simple, but time-consuming. A mathematical shortcut that gives precisely the same answer is to take the first derivative of the logistic growth curve for the species (Figure 11.6b). When the total population is plotted on the x axis, and the growth rate at each population size is plotted on the y axis, a domed curve results (Figure 11.6b). When the population is small, it can only grow slowly. As the population increases in size, the rate of growth accelerates until reaching a maximum rate at point G. Further increase in population size results in increased competition and lower individual fitness as resources start to become limiting. At this point, population growth rates decline. When the population reaches its carrying capacity (K) the growth rate is zero.

Assumption 3 listed above states that if the rate of catch is less than the rate of growth in a population, the fish extraction could go on indefinitely. Think of this as a bathtub half full of water with the faucets full on

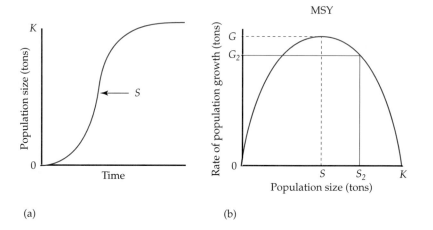

(a)   (b)

Figure 11.6   The logistic growth of a population through time and an economic model of fishery populations. (a) S = the fastest population growth. K = the carrying capacity of the population. (b) Population growth rate is plotted against the size of the population. The fastest growth rate occurs when the population size is S. At this population size, G tons of fish could be removed from the population without causing the population to decline. Because this is the largest number that can be harvested without damaging population levels, it is the maximum sustainable yield (MSY). The catch of S_2 fish at population G_2 is also sustainable. Catches that are equal to or less than the rate of growth of a population will be sustainable. Catches that exceed population growth will be sustainable so long as they never exceed the MSY and populations are greater than at MSY (to the right of MSY).

and the plug pulled out. If water flows into the bath faster than it drains out, the bath continues to fill, albeit slower than with the plug in place. If more water is flowing out of the plughole than is running into the bath, eventually the bath drains. Thus, if the drain on the population is less than recruitment, the fishery is sustainable.

The maximum rate of population growth identifies a turning point. Where populations are large (to the right of S), the number of fish harvested is replaced by recruitment, and the population continues to grow. The removal of fish reduces competitive pressure and actually promotes a faster rate of population growth. To the left of S, removing members from the population causes reduced population growth rates, and consistent harvesting would result in a declining population.

How many fish can be removed without reducing the population level? For any population, so long as the annual catch does not exceed the annual growth of the population, the harvest is sustainable. As an example, the sustainable catch size at population S_2 is G_2. The maximum catch size G that results in a sustainable fishery is seen to be at population size S. Hence, S represents the **maximum sustainable yield (MSY)** for the population.

Setting a harvest to exactly equal the maximum sustainable yield does not allow for natural fluctuations in populations that may also serve to reduce fish numbers. Therefore, a harvest lower than the MSY would be recommended. However, as we shall see, the MSY is almost always exceeded. To understand why this happens we need to convert our population diagrams to show the earnings and costs incurred in fishing. Three further assumptions are introduced to help us to make this conversion.

1. Fish prices are constant.

2. The additional cost of catching an extra fish is constant.

3. The amount of fish landed per unit of fishing effort is directly proportional to fish population size.

The effect of assumption 1 is that if every fish caught is worth $1, then rate at which fish are removed from the population will equal our rate of earning. The rate of earning is termed the marginal benefit. In Figure 11.6b, the unfished population (K) would represent no earnings, whereas if we were able to catch all the fish this would represent the most money that could be earned. However, the rate of earnings would be greatest where the rate of fish growth was greatest. Thus the peak rate of earning would be at the MSY. In other

words, a graph of marginal benefits has the same shape as the graph of population growth rate, but runs from low to high in the opposite direction.

Fishing effort includes the number of ships, the amount of time spent fishing, and the distance traveled to fishing grounds. When there are many fish, it takes little effort to catch one; but as fish become scarce, catching a fish takes a lot of effort. When fishing effort is low, fish are not being caught, and thus populations will be high. Conversely, high fishing effort will result in low fish populations. Figure 11.7 is a graph of the effect of increased fishing effort on the marginal benefit. It can be seen that the marginal benefits of fishing are highest at the MSY. Because the population of fish on the x axis is reversed relative to Figure 11.6b, to be safe, the fishery needs to operate to the left of the MSY in Figure 11.7.

Three economic factors lead to unsustainable use of the fishery: the lack of ownership of either the fish or the fishing grounds, new technologies, and subsidies.

Fisheries As a Commons

Resources that no one owns are said to be **common property** or a **commons**. The origin of the term *commons* comes from the English medieval system of having an area of shared land, called a common, beside each village. The common belonged to no one, but could be used by all. The land provided essential resources for the villagers, such as grazing for cattle, firewood, and medicinal herbs.

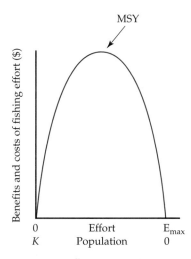

Figure 11.7   The relationship between marginal benefits, fishing effort, and population size. The marginal benefits will peak where population growth is fastest (MSY). Note that when fishing effort is high, fish populations are reduced. When little effort is expended, fish populations increase. Also note that the direction of the population axis is reversed relative to Figure 11.6b.

The modern usage of the term commons, still addresses resources that are shared by everyone and owned by no one, but it is also used at a larger scale. Modern commons include such resources as air, rain, and fish in the ocean. Ecologist and environmental writer Garret Hardin describes the state of these resources in a classic paper entitled the "Tragedy of the Commons." Hardin develops the idea that because no one owns the commons, no one takes responsibility for a sustainable use of the resource. When there is clear ownership of a resource, the owner is likely to make well-considered use of it and will not be wasteful in making decisions about harvesting. For example, a catfish farmer could raise a crop of catfish, clean out the pond, and send them all to market. But he or she must then purchase more fish and start over. Rather than do this our farmer would estimate the rate of reproduction of the catfish and calculate a maximum sustainable yield. The optimum number of fish is harvested, leaving a viable population to breed. In other words, the farmer is adhering to the static equilibrium model, because sustainable income is preferable to short-term gain.

Compare this situation with finding a shoal of catfish in a local creek that anyone is allowed to fish. These catfish are "free," and to receive any amount of money for them is a profit. Is it best to catch all the fish immediately, or to catch a few each month and maintain a sustainable population? This is where ownership becomes a critical issue. Because anyone can catch these fish they are a commons. The fish are available on a first-come, first-served basis, and by leaving fish in the creek there is no guarantee that someone else will not come along and catch them. Thus, the logical tactic is to catch as many fish as quickly as possible, because if I do not, someone else will. What is the optimum number of fish to catch under these circumstances? It is worth fishing until the cost exceeds the benefit of catching an extra fish. Although no money is paid to catch the fish, it takes effort to catch them, and this is an economic cost. At point E (Figure 11.8), the added effort required to catch an extra fish (marginal cost) equals the benefit received from selling it (marginal benefit). Thus the catch removed from a common property fishery is at point E. If E lies to the right of the MSY (as it does in Figure 11.8), the population of fish has been reduced below the point at which it can maximize recruitment. Theoretically, an equilibrium fishery could be maintained at this point, but in practice this is extremely unlikely. Overfishing will continue and fish stocks will decline to the point where it is no longer economically viable to exploit them. The fishery is collapsing. Two further factors

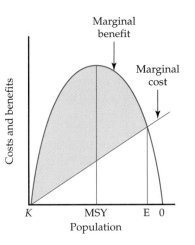

Figure 11.8 **The economically efficient point at which to harvest a commons.** Marginal benefit is the added income derived from catching an extra fish. Marginal cost is the added cost (or effort) incurred to catch an extra fish. The shaded area represents profit. Beyond point E, costs outweigh benefits, thus E is the maximum harvest that is economically efficient. The harvest at E exceeds the maximum sustainable yield (MSY): The fishery collapses.

that hasten the decline of fisheries are subsidies and new technologies.

Subsidies Corrupt Economics

The economic model assumes that marginal benefits and marginal costs regulate the number of fish caught and, by extension, the number of fishing vessels operating. Who would invest millions of dollars in building a new trawler when there is not a profit to be made? Unfortunately, many governments build economically (and ecologically) unsustainable fishing fleets through subsidies. Globally, about $54 billion are spent each year to subsidize fishing fleets. The subsidies affecting a fishery take many forms, ranging from subsidized ship building, which leads to overly large fishing fleets, to cheaper fuel oil. A common feature of subsidies is that they hide the true costs of the fishing operation. Because the economic models predict that the size of the fish catch is a function of cost, subsidies serve to increase catch size. An example of this is found in the fishery of the Bering Sea where fishery models predict that the efficient number of trawler groups is 9. However, because of subsidized fleets the actual number of trawler groups in the 1980s was 140.

In economic terms, the effect of these subsidies is to reduce the marginal costs of the fishers. The effect of subsidies in Figure 11.9 is to cause the harvest to increase from E_1 to E_2. In this example, E_1 is to the left of the MSY, indicating that the harvest is sustainable. The subsidy lowers costs, flattening the cost curve, and es-

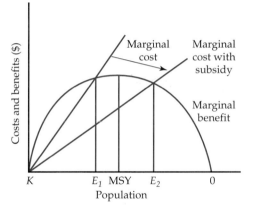

Figure 11.9 Subsidies reduce marginal costs (MC). In this example a sustainable fishery (E_1) is made unsustainable by a subsidy that allowed catch sizes to increase to (E_2).

tablishing the new harvest of E_2. Because E_2 lies to the right of the MSY, fish removal exceeds replacement, and the fishery will collapse.

Why would a government allow its economy to be distorted by subsidies? For the major fishing nations, subsidies are needed to keep their fleets competitive with other subsidized fleets. Allowing the industry to collapse would cost thousands of jobs and lead to higher food prices. For many democracies the short-term prosperity that accompanies sustaining jobs in fisheries and lower fish prices is consistent with political popularity and re-election. The longer-term view of fishery sustainability is too distant to be a political concern. Therefore, politicians are likely to enact subsidies because that is what the public wants. Another major source of subsidy is through international aid. In response to famine, or in attempting to improve the food supply to a less-developed country (LDC), more-developed countries (MDCs) might provide money to modernize their fishing fleet. Such a subsidy costs the recipient country nothing and is unlikely to be refused.

New technology used to increase fishing success is often the result of subsidies. Locating a good fishing hole, a wreck, or a sea mount where fish are abundant once took great skill and navigational expertise. The development of cheap and accurate sonar to monitor the exact contours of the ocean bottom has enabled seabed features to be relocated accurately. Sonar was developed by the military to locate submarines. The civilian versions are cheap imitations of the naval equipment that reflect none of the research and development costs that went into inventing the system. Military spending is subsidizing fishers.

A revolution in navigation was achieved with the development and cheap availability of the global positioning system (GPS). GPS units are small comput-

ers that receive and interpret signals from satellites. By locking on to the signals of three or more satellites, an exact position accurate to within a few meters (or millimeters with a more sophisticated system) can be determined within seconds. Nowadays, when a good fishing ground is located, the skipper records a GPS reading. At any time the fishing boat can be guided back to that precise location using the GPS system. GPS technology greatly intensifies fishing pressure on key habitat areas and allows species with very local abundances to be harvested. We should also note that GPS is a subsidized service. The cost of inventing, building, and deploying satellites that provide GPS information is met from tax dollars raised by the U.S. government. Even though the cost of placing satellites in orbit is enormous, using the satellites to receive GPS information costs nothing. The precise navigation afforded by the GPS units reduces time and fuel used to relocate sites and improves hunting efficiency, thereby reducing costs. The free GPS service therefore subsidizes fisheries.

11.4 Further Ecological Thoughts on Fisheries

The commons model of fishery decline is compelling, but it does not go far enough to describe the ecological state of fisheries. Key factors that must be addressed are that fish populations are variable; many fish populations are not resource limited, but recruitment limited; and some fish are disproportionately important to the welfare of the population.

Population Fluctuations among Fish

Like all other populations, fish fluctuate in abundance through both space and time. Fish populations can be driven down by disease, predation, and density-independent factors (Chapter 9) such as circulatory changes in the ocean, the destruction of nursery grounds such as wetlands, or pollution. Thus we can expect that fish populations will not follow the logistic growth curve, but rather be subject to the peaks and valleys caused by density-independent factors. Consider then, a fishery in which there is an intrinsic growth rate of 1.3. If the population was allowed to grow at this rate it would have an exponential growth curve that can be described by the logistic equation (Chapter 9). The upper line on Figure 11.10 represents this growth curve. The maximum annual increase in population is 173 fish and occurs between years 3 and 4. Therefore, the MSY for this population is 173 fish per year. Let us now introduce fishing to the model.

Figure 11.10 A fisheries model demonstrating the effect of natural perturbations that cause periodic population decline and fishing. The model is based on the logistic growth curve $R = rN(1 - K/N)$. r (the intrinsic rate of increase) is 1.3 (a 30% population growth rate as found among flounder). N (the number in the population) starts at 800. Without fishing or perturbation the MSY of the population is 173. With a removal of 60 fish each year (7.5% of starting population) the MSY is 113 fish. According to the static equilibrium fishery model this is a sustainable fishery. Repeated stochastic population declines of 25% and 40% reduce population size, but do not endanger the population in the absence of fishing. With fishing, so long as the natural declines are well spaced the fishery remains sustainable (years 0–40). However, if natural declines bunch together (years 40–60), the fishery collapses.

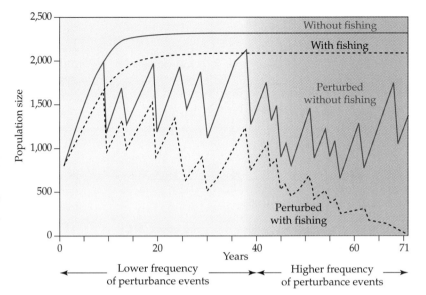
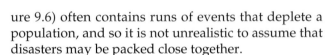

Every year 60 fish, 7.5% of the starting population, will be caught and removed. This curve is represented by the lower of the smooth curves on Figure 11.10. The flattening of this line indicates that the population is not growing as fast, therefore the MSY is lower (113 fish). Nevertheless, despite the fishing, the population is large, and the catch is sustainable as predicted by the static equilibrium model.

To make this model more realistic we can introduce natural perturbations to the system. Periodic reductions in the population by either 25% or 40% are included to simulate disease, poor recruitment years, or other density-independent disasters that may befall a population. The resulting curves with and without a fishery are strikingly different. Natural population downturns, even when clustered together, provide a temporary low from which the population recovers. However, with the addition of fishing pressure the chance of a population extinction is greatly increased. Because the population is driven up and down by density-independent factors and unrelenting fishing pressure, the actual ability of the population to recover in numbers is reduced. The extraction overwhelms replacement, and the population goes into a steady decline.

Note that the spacing of perturbation events is important to the survival of the fish. When there are only occasional downturns, such as between years 0 and 40, the fishery can recover. With such low-frequency population downturns, this level of fishing could be supported indefinitely. However, closely spaced perturbations such as between years 41 and 60 cause the population to slide downward. A real data set (Figure 9.6) often contains runs of events that deplete a population, and so it is not unrealistic to assume that disasters may be packed close together.

This model has some important messages for us. Setting catch quotas, even as low as half of the maximum sustainable yield, does not guarantee fishery stability. External factors beyond the control of the fishery may play a larger role in the population dynamics of the fishery than the actual catch, but it is the additional mortality resulting from fishing that makes the difference between survival and extinction.

Recruitment Limitation

With the exception of habitat limitation among some reef fish and a few special cases, fish populations are not governed by density-dependent factors. Far more important to most species are the vagaries of density-independent factors that affect reproductive success. In other words, the size of the population is not determined by the amount of resources available, but by the rate of recruitment.

Fish occupy a full range of niches from opportunists to competitors. An anchovy is an opportunist, exhibiting small body size, fast growth, a short life cycle, and profligate egg production. At the other extreme are competitor species, such as a great white shark or the less-threatening orange roughy. These fish are slow growing, long lived, and in the case of some sharks, give birth to fully formed fish. Given this breadth of lifestyle, it is not possible to generalize about rates of growth or maturation among fish. We can, however, make some observations about how these various reproductive strategies work under hunting pressure.

Fisheries that exploit opportunist species can harvest larger catches than those exploiting competitors. Opportunistic adaptations accommodate rapid population fluctuations (Chapter 9). The short generation time, quick maturation, and large numbers of young allow the opportunist population to rebuild rapidly after a catastrophe. The competitor relies on long-term population stability and is slow to re-establish a population. Clearly, if a large part of the population is being removed each year, competitors are more likely to go extinct than opportunists. Many of the fish species that we like to eat, such as marlin, shark, grouper, orange roughy, sturgeon, and halibut, are competitors and cannot withstand consistent hunting pressure. All of these populations are in decline as a result of fishing.

Given that opportunist species should be able to withstand moderate fishing pressure, it is both surprising and alarming to find that their populations are also declining. In the last two decades, catches of anchovy, sardines, salmon, flounder, and cod have shrunk. The Atlantic cod provides a good example of the effect of fishing on an opportunistic species. A large female cod can produce 4,000,000 eggs per year, and with such high fecundity, they are an ideal fishery species. Historically, cod were one of the most abundant fish in the cooler oceans and formed the mainstay of North Atlantic fisheries for more than 100 years. Intensified fishing pressure in the 1970s caused stocks to decline (Figure 11.11). As fish became scarcer, more effort was expended to catch them. The scarcity of cod caused its price to rise, which, in turn, induced larger trawler fleets to spend more time at sea. The unrestricted harvesting of the fish caused population numbers to spiral downward. The George's Bank cod fishery in Newfoundland was so depleted that in 1994 it was closed entirely in an attempt to promote a population recovery. Such closures are rare, and globally only about 1% of potential fisheries are closed.

Some Fish Are More Equal than Others

The economic model does not differentiate between fish of varying sizes and ages, but just as among the inhabitants of George Orwell's *Animal Farm*, we find that some are more equal than others. Young fish invest a huge amount of their energetic intake into growth, some energy is spent on metabolism, but almost none is invested in reproduction. These small fish are members of their population, but are not contributing to its sustainability. Contrast this with a large mature fish. She has almost ceased growing, her metabolism is proportionately lower than for a small fish, and most of her energy is spent on reproduction. Through this reproductive investment, the matriarchs of the shoal ensure the future of the population. The individual fitness of these large fish is massive compared with

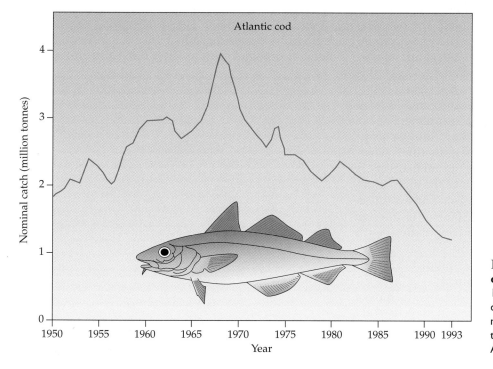

Figure 11.11   The rise and fall of cod catches in the Atlantic 1950–1993. Despite a huge reproductive capacity, cod cannot withstand modern fishing pressure. (Data from the National Ocean and Atmosphere Administration.)

average members of the shoal. Fishers selectively target the larger fish and consequently remove the reproductive stars, thereby greatly reducing the ability of the remaining fish to rebuild their dented population.

Not only are fish becoming scarcer, they are also becoming smaller. Part of this effect is purely because the large members have been removed from the population, and the only ones remaining to be caught are smaller. However, this is not a complete explanation of the observed data. An interesting consequence of intensive fishing activity is that it becomes such a strong selective agent that it drives fish evolution.

If the larger fish are being removed from a population, it follows that the burden of replacing the population is falling on the smaller fish. This leads to two evolutionary predictions. If there is a strong selective pressure against growing large (you get caught and killed), only fish that are small survive to reproduce. Therefore, genes that lead to large growth will be selected against, and genes that confer smallness will increase in abundance in the population. Overall, the maximum size to which fish grow is genetically determined to be smaller. Remember, fish are not choosing to be small, it is simply that those that grow large enough to be caught, fail to maintain a fitness of 1.

The second evolutionary prediction is that if staying small is favored, genes that allow reproduction at a smaller size will also be favored. The fish whose genes determine that she becomes large before forming eggs will be selected against. She never gets to reproduce, and her lineage is terminated. The fish who can produce eggs at a smaller size will become an increasingly important component of the population because they now have a selective advantage. Under this intense size selection, through time the sexual maturation of fish will shift to smaller and smaller size classes (though these fish are not necessarily younger).

One way to test our hypothesis that fishing is shaping evolution is to assess the size of individuals when they first become sexually mature. Sexual maturity is easily determined in fish by the size of the internal reproductive organs (testes and ovaries). If no evolution is taking place, there will be no difference in the size of the fish at first sexual maturity in populations with and without fishing pressure. If, on the other hand, evolution is taking place, we would predict that fishing will result in earlier sexual maturity.

The black-chinned tilapia provides a good subject for this experiment. This species is native to West African mangrove habitats and in Ghana is the subject of an intense fishery. A population of these fish, accidentally introduced from West Africa to the Indian River Lagoon in Florida, flourishes without fishing pressure. Both populations started from the same genetic stock, and the U.S. fish have been isolated only for a few decades, not long enough for a major change in growth habit without a strong selective pressure. The mangroves of Florida provide a very similar habitat to the West African mangroves, and thus environmental factors should not cause rapid evolutionary change. In the Florida mangroves, the tilapia grow to a maximum length of 25 cm and become sexually mature when they are 15 cm long. It is a safe assumption that this is their "normal" growth character. Jon Shenker and Junda Lin of the Florida Institute of Technology reported that the fishers in Ghana had responded to falling catches by using finer and finer net sizes. The maximum length of tilapia in these waters is now 10 cm, and sexual maturity occurs at 3 cm long (Figure 11.12). These data provide strong evidence that the intense selection that accompanies overfishing can drive the evolution of fish.

Such declines are not limited to LDCs. Data for five of the most sought after sport fish in the waters around the southeastern United States indicate that over a 15-year span, the size of the average fish fell by almost 75% (Figure 11.13).

Fishery managers are interested in the tonnage of fish landed. Declines in the size of fish being caught suggests that overfishing is working to change both the size spectrum of the remaining population, as well as their genetic quality. The MSY is based on growth rate, which can be thought of as tons of fish. If fish are maturing at a smaller size, their energy is diverted from growth into reproduction. Each individual fish not only does not grow as large, but it also grows more

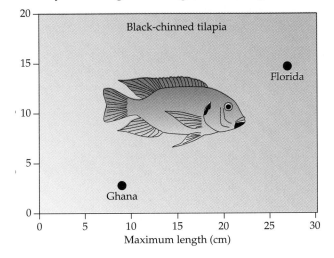

Figure 11.12   Maximum growth size and size at sexual maturity among black-chinned tilapia in Florida (unfished) and Ghana (overfished).

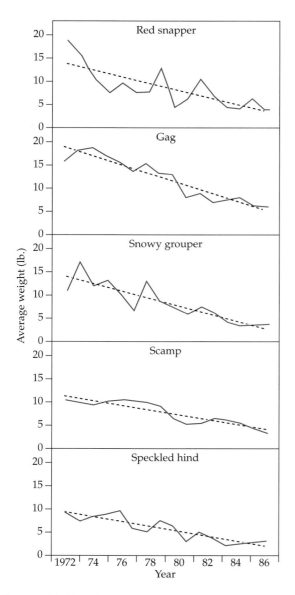

Figure 11.13   Declines in the average size of the five most important fish species landed by sport-fishing boats in the southeastern United States from 1972 to 1988. (After Huntsman and Willis, Plan Development Team.)

slowly. Therefore, the growth in tonnage of *catchable* fish declines, and this translates into a lower MSY.

11.5   Prey Switching and Fishing

Humans' behavior conforms with the expectations of optimal foraging theory. As our prey items become scarce, we switch to an alternate. Perhaps the best example of this comes from the history of Antarctic whaling. Although whales are mammals and not fish, the story of Antarctic whaling sends a powerful message about the overexploitation and abuse of a com-

mons. The largest whales were exploited first. The largest of all creatures living on the planet, the blue whale lives for about 70 years, reaches sexual maturity when it is about 20 years old, and has a gestation period that lasts a year with at least one or two years between pregnancies. With such slow reproductive rates, the whale populations could not rebound from the mortality inflicted by the hunters. As blue whales started to decline in numbers during the 1940s, humpbacked and fin whales were targeted next (Figure 11.14). Fin and humpbacked whale populations proved equally vulnerable, and their populations were greatly reduced by the late 1950s. The 1960s was the decade of the sei whale hunt, but their numbers also collapsed under hunting pressure. By the early 1970s, none of the large whales had commercially viable populations. Attention was then turned on the minke whales, a relatively small whale that had been largely ignored until that time. In 1987, in response to global declines in all populations of large whales, commercial whaling was banned, although some whaling continues under the guise of science by the Japanese.

The fishing industry closely parallels the whaling industry, with fish being overexploited until a replacement species is sought. An example of prey switching among fish species comes from the fishery based in New England. Data from the National Marine Fisheries Service document the landings of cod, haddock, and six species of flounder between 1950 and 1995 (Figure 11.15). By the late 1950s, haddock catches were declining, but the tonnage was made up by increasing the fishing pressure on flounder. The 1960s saw a precipitous decline in the haddock fishery and the 1960s and 1970s a steady downward trend in flounder landings. An intensification of effort in the late 1970s and early 1980s resulted in increased landings of all species. However, the catch rate had exceeded the MSY and precipitated a rapid decline in the catch rate for all species.

Look back at the model of the effects of natural population variability on a fishery (Figure 11.10) and compare that model with the catches of cod (Figure 11.11) and the New England fishery data. The bumpy decline in fish stocks are remarkably similar in appearance in all three figures, suggesting that natural fluctuations in fish density must be taken into account in fishery management. The fishery model of Figure 11.11 was based on a 30% population-growth rate and an initial catch of 7.5% of the total population. The flatfish of New England have a 30% population-growth rate, but the harvesting rate is about 60% of the local population each year. With such intensive fishing, natural fluctuations or not, this population

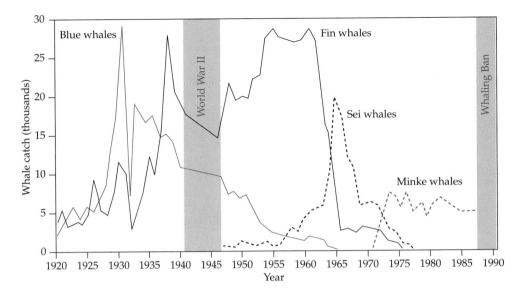

Figure 11.14   An example of prey switching by commercial whalers. The sequential harvesting of different whale species in Atlantic and Antarctic waters between 1920 and 1987. (Data from FAO.)

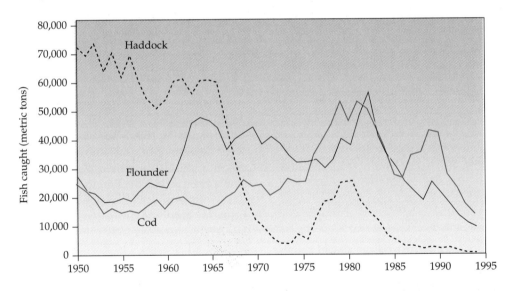

Figure 11.15   An example of prey switching and progressive overexploitation of fish species in New England. The apparent increase in catches in the 1970s reflects increased fishing effort. (Data from the National Marine Fishery Service.)

will become commercially unviable in a very short time. Despite switching between species, the New England flounder catch has declined by 85% from its peak in 1983.

The orange roughy is another dramatic example of how a fishery switches onto a new prey item and then quickly exhausts it. In 1986, large populations of orange roughy were found in the deep oceans around New Zealand. The fish aggregated around rock pinnacles that rose high from the seabed. The fish, there-

fore, were easy to find, and a profitable fishery developed quickly. Within 10 years, the initial population, estimated to have been about 200,000 tons, was reduced to about 30,000 tons. With a little ecological hindsight the reason for this is obvious. An orange roughy takes about 30 years to reach sexual maturity and may have a life span of almost 100 years. Such slow-growing fish will have very low population-growth rates, and once a substantial harvest began, there was virtually no replacement taking place.

Eating Our Way Down a Food Chain

Generally, the fish that we prefer to eat are predators, feeding on the third or fourth trophic level of a food chain. The energetics of food chains determine these fish to be rare relative to their prey items (Chapter 9). One solution, therefore, to diminishing catches of these predatory fish is to try to catch fish lower on the food chain. By shifting the fishing effort to these smaller fish, a previously untapped resource is harvested by the fishing boats. To make this decision on ecological grounds would be one thing, but to be forced to do it because the higher trophic levels have been exhausted is an indication of a very serious management problem of fishery resources. In a study of more than 220 species of fish landed by global fisheries it was found that the average trophic level of a fish being caught in 1955 was 3.4, and by 1992 it had declined to 3.1 (Figure 11.16). The odd dip in the data set from 1960 to 1972 was due to a massive increase in the Peruvian anchovy catch. In the late 1950s the government of Peru invested heavily in a trawler fleet. The seemingly endless supply of anchovies would provide protein and a source of export revenue. The anchovy harvest was huge, and during the 1960s it represented as much as 30% of the global marine landings. Anchovies feed on the second trophic level, so when these catches peaked, it had the effect of reducing the average trophic level of fish caught. In 1972, the great anchovy

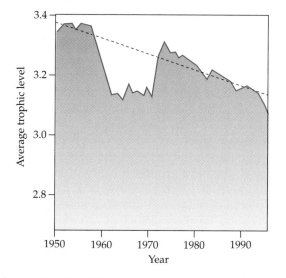

Figure 11.16 **The average trophic level of fish landed 1950 to 1992.** The rate of decline is about 0.1 trophic level unit per decade, a sign that we are emptying the seas of fish. The sharp dip in trophic levels from 1960 to 1972 was due to intensive exploitation of Peruvian anchovies that feed on the second trophic level. This fishery collapsed due to overexploitation in 1972 and has not recovered since.

shoals were exhausted, and the fishery collapsed. Since that time a steady downward trend is evident in average trophic level, declining at a rate of about 0.1 unit per decade.

If the fishery is removing predators, one might expect that their prey items would increase in abundance. However, other species that are of no commercial interest, such as jellyfish, are filling the vacant predatory niche. The jellyfish benefit in two ways from the fishing activities. First, jellyfish predators such as turtles and rays are being killed, and second, the fish that are in competition with jellyfish, such as sardines, are being removed from the oceans. Jellyfish, released both from competition and predation, are thriving! That we are fishing down the trophic structure of the oceans may be the most powerful indication yet received that global fisheries are seriously depleting fish stocks. The steady gain in fish catch achieved during the period from 1950 to 1997 was achieved at the cost of reaching down into the lower trophic levels of the oceanic food web.

The impact on the oceanic food web is not limited to the species that we deliberately target. Populations of fish-eating birds, turtles, and mammals such as seals and dolphins are declining. Although some of this decline is due to their being accidentally caught in nets and hooked on long lines, they are also being outcompeted by fishing boats. The removal of 100 million tons of fish from a previously balanced ecosystem is thought to be causing a food shortage among animals on the highest trophic levels. For example, Stellar's sea lions that live in the North Pacific feed on pollack (a fish similar to a cod). The global overfishing of cod increased the demand for pollack. The pollack fishery intensified its activity, effectively denying the fish to the sea lions. Stellar's sea lion has declined in abundance from an estimated population of 300,000 animals in 1960 to 66,000 in 1990.

11.6 Local Solutions to Fishery Problems

An ecologist would argue that the surest way to enable a fishery to recover is to close it. However, closing fisheries puts voters out of work, causes hardship, and may prove to be political suicide for the decision maker. A common compromise between unrestricted access and closure has been to limit the number of days that fish can be captured, to limit the number of fish that can be caught, or to impose size restrictions on fish that can be killed. Size limits generally forbid the killing of fish below a minimum size. Clearly, such limits do not explicitly protect the larger fish that are so important to recruitment, but the intent is to save at least some that are reproductive.

Zonation of a coastline into areas with different uses has been suggested for Australia's Great Barrier Reef and the Florida Keys. The zones would divide the coast into areas where all fishing is allowed, sport-fishing only is allowed, and fishing is prohibited. If recruitment is strong in the totally protected area, it is hoped that fish will spread and repopulate even the areas that are heavily fished. Also, it has been recognized that tourists and commercial fisheries do not mix well. Tourists may spend a lot of money to dive and snorkel undamaged reef or to swim with large fish. Indeed, recreational divers can provide the basis of a flourishing ecotourism industry for countries with high-quality reef environments. The Great Barrier Reef of Australia supports a $1 billion dollar tourist industry with 1.5 million visitor days per year. By comparison, the local fishing industry generates $250 million per annum. A strong case can be made for ecotourism on the Great Barrier Reef, but in areas where hunger and poverty are pressing, it may not be feasible (Chapter 20).

Fishery Leases

The economic argument that a fishery can only be used sustainably when it ceases to be used as a commons leads to the establishment of fishing leases. A lease is granted for a particular area to a particular fisher. It is then in the interest of the fisher to manage the lease sustainably. Such rights to an area have been successful in maintaining the quality of trout fisheries in small streams, but applying the same principle to the oceans is challenging. The mobility of fish means that they do not stay within a lease area (given that most leases would have to be relatively small because of the number of fishers). However, management on a lease basis for lobsters, clams, and mussels have been successful because shellfish are less mobile than fish. These shellfisheries have proven successful at maintaining populations and substantiate the idea that we must move away from regarding ocean resources as a commons.

Not all the woes of fisheries can be blamed on the fishers. As many as 80% of the most valuable species to inshore fisheries are absolutely reliant on mangroves or wetlands for some stages of their life cycle. The shelter provided by the dense vegetation of these shoreline habitats helps to protect young fish from predators, and in the nutrient-rich waters, the young

fish can grow rapidly. Coastal development, deforestation, and the conversion of mangrove to shrimp-farming ponds has removed about 50% of these habitats. With their nursery grounds damaged, the recruitment of many fish species is precarious, making them yet more vulnerable to overfishing.

Inshore water quality is also threatened by eutrophication caused by sewage effluent and runoff from agricultural and residential areas that are enriched with fertilizers and pesticides. Regulations that protect wetland environments, enactment of laws that regulate waste and sewage disposal in coastal regions, education programs to lessen the use of chemicals on lawns and gardens, and fining polluters are all steps that would help salvage fisheries.

11.7 Are Fish Farms the Answer?

Raising fish and shellfish for commercial use is known as aquaculture. Fish and shrimp have been raised in ponds for thousands of years, but in recent years aquaculture has become big business. Aquaculture production doubled between 1984 and 1993, and by 1997 it accounted for 14% (by value) of fish sold.

Aquaculture can be as simple as having a pond into which small fish or shrimp are added, grown until marketable, and then sold. Carp and shrimp thrive in highly eutrophic systems, and their growth can be enhanced by fertilizing the lake with the waste from chicken farms. Another low-technology form of agriculture is to partition an area of a lake or estuary with nets and grow fish within the enclosures. A wide range of fish have been successfully raised in this way, including salmon and sea trout. However, the high densities of fish in these confined spaces can lead to increased disease problems and local eutrophication. For example, in the summer of 1998, the large aquaculture industry in the bays around the city of Hong Kong suffered enormous losses. Fish were killed by a red tide, a bloom of algae and dinoflagellates caused by eutrophication. It is very likely that although the dense shoals of fish being raised in floating nets added to the nutrient load, the real culprit was untreated sewage being released from the nearby city. When fish, especially shellfish, are grown in water tainted with fecal matter from urban or agricultural sources, the possibility of the fish harboring pathogens and parasites that can affect humans is greatly increased. Fatal cases of botulism and other forms of food poisoning are particularly common when shellfish, such as oysters, are eaten raw.

To increase fish production and decrease the health risks, a more controlled approach to aquaculture has been developed. The farmed organisms are raised in artificial ponds or concrete tanks through which there is a continuous flow of water. The outflowing water carries away waste material and is filtered and treated before being recirculated through the tanks. The fish, or shellfish, are fed a controlled diet that is tailored to their species and their size. Diseases are carefully monitored and fish can be treated as soon as they show signs of distress. Harvesting is done as soon as their growth starts to slow, ensuring a maximum ratio between food provided and body mass.

Brood fish can be selected on the basis of ideal characteristics, such as rapid growth, large size, and disease resistance and raised to provide eggs and milt (sperm). This is the precise equivalent of selecting the best cow and bull to breed. A female fish ripe with eggs can be manually stripped and all the eggs squirted into a waiting bucket. This does not harm the fish and greatly increases the protection that can be afforded the eggs. Similarly, sperm can be stripped from a male and used to fertilize the eggs.

The costs incurred in aquaculture are higher than in regular fishing. Consequently, the greatest successes in aquaculture have come from raising highly valued species such as salmon and shrimp. Although the fisheries of these species are overexploited, the shortfall in catches is being made up by aquaculture production. By the 1990s, a quarter of salmon and shrimp sold had been farm raised.

Another role that aquaculture can play is in providing tropical reef fish to aquarium enthusiasts. Collecting wild fish is extremely damaging to the reef and wasteful, because 90% or more of fish collected die in transit. Each fish raised in captivity can be sold for tens of dollars, making this a lucrative market. A further conservation role that aquaculturists can play is in raising and releasing to the wild endangered fish populations. Red drum, a marine sport fish, had all but disappeared from the southeastern fishery in the 1980s. In the 1990s, a combination of increased protection and stock replenishment with farm-raised fish has greatly boosted populations and improved the local fishery.

Aquaculture has its critics, and one of the environmental problems associated with shrimp farming has been the destruction of mangrove ecosystems. It has become a common practice in Asia, Central America, and South America to cut down mangrove and establish shrimp farms in place of the wetlands. In the short term, the shrimp farms are highly profitable, because the mudflats previously occupied by the mangroves make an ideal environment for culturing shrimp. In

the longer term, the removal of mangrove, an important nursery ground for fish and shellfish, leads to declining wild populations. Local fisherman find their livelihood is draining away as the mangroves are destroyed by the shrimp farmers.

Another cost associated with mangrove removal is increased flood damage. The dense mangrove forest forms a protective barrier that insulates the uplands from storm-induced flooding. Southeast Asia has typhoons that are more powerful than our worst hurricanes, and without the mangrove fringe, cities and villages in the lowlands are susceptible to severe flood damage. The cost of erecting a seawall in Malaysia that provides the same degree of protection as a mangrove fringe is estimated to be $300,000 per kilometer.

11.8   National and International Protection

In the United States, the first major legislation to conserve national fish stocks was the Magnuson Fishery Conservation Act of 1976. This act divided the nation into eight fishery regions that would determine regional quotas for landings of key fish species. The act also established a 200-mile exclusive fishery zone that could exclude foreign fishers. The 200-mile limit had previously been proposed by the Third United Nations Conference on the Law of the Sea in 1973. A convention was proposed that would expand the territorial water that a nation controls from 12 to 200 nautical miles beyond their coastline. Within this limit, a nation has sole authority over the use of marine resources, and national laws apply regarding net sizes or the definition of exclusion zones where fishing is prohibited. The area within the 200-mile limit is known as the **exclusive economic zone** (EEZ). In 1993, the law was finally enacted, and it became international law as of November 16, 1994. By restricting access to the EEZ, it was hoped that national fisheries would take more responsibility for the long-term sustainability of the resource and stop treating the fish as a commons. Remember also that the great majority of the fish catch is landed from the continental shelf regions and that these would generally lie within an EEZ. Thus, it was hoped that if the most productive regions of the ocean cease to be used as a global commons, fish populations would stabilize. However, the United States, Canada, and Iceland had begun to enforce a 200-mile EEZ since the mid-1970s (even before it was international law), and it is clear that this has not saved the fisheries. Most nations have so many fishers that even though other nationalities may be excluded, the fishery is still treated as a commons. The National Marine Fishery Service estimates that more than 40% of assessed species are overfished in U.S. coastal waters. Claude Martin, director-general of the World Wide Fund for Nature, considers 60% of the 200 most important commercial species to be overfished or fished to the limit. It does not seem likely that the expanded EEZs will solve this problem.

Beyond the 200-mile limit, control of the fisheries can only be achieved through international treaties. Each treaty will only be as good as the number of signatories and the willingness to enforce it. Once again, the picture is bleak because there is no international policing of these open-ocean fisheries.

Specific Solutions for Particular Problems

Problems that claim public attention can generate sufficient political pressure so that a quick cure is found. With the discovery that stocks of dolphins and turtles were being threatened by fishing as a consequence of bycatch, a groundswell of public concern in Europe and North America led to changes in the fishing industry.

Dolphins were used by fishers as a guide to the presence of tuna shoals. The dolphins are tuna predators and often swim directly above the main concentration of tuna. By encircling the dolphins with nets, fishers also captured the shoal of tuna. This technique was highly successful at catching tuna, but the dolphins were generally drowned as the nets were hauled in. About 400,000 dolphins were killed each year in tuna nets in the 1970s and 1980s. When this fishing strategy became public knowledge, there was a grass-

roots movement to outlaw the technique. Intense public pressure led to legislation in the United States that banned tuna importation unless it was "dolphin safe." Dolphin safe does not mean that no dolphins were killed; it indicates that after the dolphins were encircled by the nets, one edge of the net was dropped and the dolphins could escape. Nations that do not practice dolphin-safe fishing are seeking to overrule the import restriction claiming that this U.S. law breaches international free-trade agreements.

Similar public outcries came in response to the number of turtles drowned in shrimp trawl nets; the largest cause of human-induced ocean turtle mortality. The consequence was the requirement for all shrimp trawlers operating close to turtle nesting beaches to install turtle excluder devices (TEDs) on trawl nets (Figure 11.17). A TED is effectively a trap door in the roof of the net that allows a turtle scooped into a trawl net to swim to safety. Turtles drown in trawl nets, but by having the TED the turtles natural inclination to swim for the surface leads it to the trap door and freedom. The TEDs and dolphin-safe fishing are good examples of how the North American fishing industry has modified its practices to minimize environmental damage.

Solutions for General Problems

Fishery management needs to be changed if its goal is to maintain a sustainable harvest and a balanced ecosystem. The economic model clearly demonstrates the need to reduce the size of fishery fleets and to remove subsidies that lead to overfishing. Wherever possible, the fishers must be given control over local fisheries so that these areas may be "owned" rather

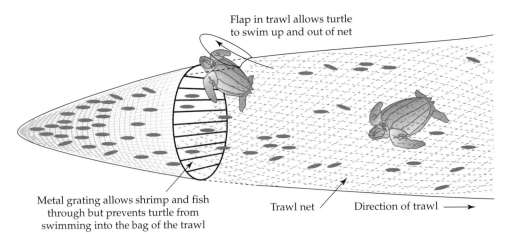

Flap in trawl allows turtle to swim up and out of net

Metal grating allows shrimp and fish through but prevents turtle from swimming into the bag of the trawl

Trawl net   Direction of trawl ⟶

Figure 11.17   A turtle excluder device (TED) fitted to a shrimp trawl net. The turtles, although swept into the net, may find escape through the trap door of the turtle excluder device.

than treated as a commons. The ecological data can be used as a guide to determine which species can withstand harvesting and at what level they can be harvested. International agreements need to be reached that identify and enforce appropriate harvesting quotas for open-ocean fisheries. Lastly, the political will needs to be found to curtail fishing activities if populations are threatened. By monitoring fish populations and landings, warning signals of a future population crash can be seen. Action should be taken at an early stage to reduce fishing pressure, not as a last resort when the population has already been drastically reduced and recovery is very slow.

Summary

- The static equilibrium model of fisheries suggests that fisheries would never be overexploited if they were "owned."
- Oceanic fish are a commons and will be exploited to the point where marginal costs exceed benefits, regardless of sustainability.
- The large fish are the most fecund and are disproportionately important in allowing fish populations to recover, but these larger individuals are the prime targets of fisheries.
- Density-independent variables cause fish populations to naturally rise and fall. Adding fishing pressure can prevent a recovery from a natural downswing in populations and lead to a more serious population decline.

- More than 40% of U.S. fish stocks assessed are in decline, and 60% of the 200 most globally important species are overfished or fished to the limit.
- Switching from one target species to another extends the life of a fishery, but cannot postpone a fishery collapse indefinitely.
- Fishing is such a strong selective agent that it causes fish to evolve to be smaller and to reproduce at a smaller size.
- International and national laws protecting fisheries from overexploitation are weak and poorly enforced.

Further Readings

Casey, J. M., and R. A. Myers. 1998. Near extinction of a large, widely distributed fish. *Science* 281:690–691.

Pauly, D., Christensen, V., Dalsgaard, J., Froese, R., and F. Torres, Jr. 1998. Fishing down marine food webs. *Science* 279:860–863.

Rogers, R. A. 1995. *The Oceans Are Emptying*. New York: Black Rose Books.

Teitenberg, T. 1992. *Environmental and Natural Resource Economics*. New York: HarperCollins Publishers, Inc.

Weber, P. 1994. Net loss: Fish, jobs and the marine environment. *Worldwatch Paper 120*. Worldwatch Institute, Washington, DC.

Web Connections

On-line resources for this chapter are on the World Wide Web at:
http://www.prenhall.com/bush
(From this web-site, select Chapter 11.)

Predators, Parasites, and More

12

The natural world is composed of a web of interactions as plants and animals go about their specialized lives. No species lives independently of all others, so its actions will affect other species. The strength of these interactions can range from lethality to a benign coexistence. The lion clawing down an antelope is a predatory interaction; the oak tree shading the ground and preventing the growth of a dandelion is a competitive interaction (Chapters 9 and 10). Both of these relationships have a clear winner and a clear loser. Some interactions can be mutually beneficial; some can be essential for one of the species yet virtually unnoticed by the other; and some can benefit one while harming the other. Let us start by considering predation.

12.1 The Evolutionary Success of Cowards

That predators are brave is a common fallacy; noble they may look, but foolhardy they are not. To be evolutionarily successful, a predator must be a bully. Heroics of any kind are deservedly rare in nature. Hollywood may enjoy pitting two almost equally matched adversaries, such as *Tyrannosaurus rex* and King Kong, in a life and death struggle, but animals do not fight like this as one hunts the other. The process of killing and feeding should be as uneventful (for the predator)

and mundane as possible. Natural selection determines that a predator will kill with a minimum of fuss; it will kill only animals that it can overpower with ease.

12.2 Pyramids of Power

Charles Elton observed that food chains could usefully be thought of as food pyramids. He noted that in a given area there were large numbers of herbivores, many fewer predators feeding on the herbivores, and very few predators of the predators (Figure 12.1). These observations are easily explained in terms of the energetics and the dissipation, or entropy, of energy from one trophic level to the next (see Chapter 3). As only a fraction of the energy is passed from one trophic level to the next, so the number of individuals supported declines abruptly. Thus, the sides of the pyramid, which represent the number of organisms in each trophic level, are abruptly stepped, not smoothly sloping. Another insight gained from thinking about the pyramid of numbers is that there is not only a large difference in numbers of predators and their prey, but that there is also a large difference in body size from one trophic level to the next. A robin is much bigger than a worm, a cat is much bigger than a robin, and so on. Predators must be large enough to overwhelm their prey with ease; a struggle cannot be endured.

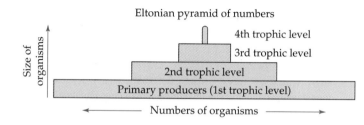

Figure 12.1 The Eltonian pyramid of numbers. The pyramid represents sharp discontinuities in both the number and size of organisms at each trophic level.

Take the case of a falcon. It must kill every day in order to sustain its metabolism: At least 365 times a year, for the 10–20 years of its life span, it strikes, overwhelms, and feeds. The peregrine falcon eats birds, rabbits, mice, squirrels, maybe even large insects. Why doesn't it kill something larger and avoid having to kill so often? Why doesn't it try to kill an adult sheep? By folding back its wings and swooping onto the unsuspecting prey from a great height, a peregrine falcon might, with one powerful strike of the talons (arriving at a speed of about 120 mph), break the neck of the sheep. Although this would result in a large food supply, it is also a high-risk predatory strategy. The impact with such a solid object may hurt the falcon. Most of its prey killed in a full-speed swoop are flying birds. The impact kills the prey but, because the birds are in the air, for the falcon it is a soft landing. If the falcon survived the impact with the sheep, but the sheep was not killed, then the falcon would have to fight. It may well be kicked, butted, bitten, or rolled on. How many times could the falcon kill cleanly before being wounded in such a struggle? This question is of prime importance, because a wounded falcon is a very poor predator and likely to starve to death before healing. Any time the odds of injury are increased, fitness is decreased.

Let us assume that the falcon is successful, kills the sheep, and starts to feed on the carcass. It would take days for the falcon to eat its prey, but it may be only minutes before wild dogs, cougars, coyotes, or vultures arrive to share in the feast. The falcon will be driven from its kill. Thus the predator is unlikely to derive the full caloric benefit from the death of a large prey item. It is apparent that there is, after all, a benefit to killing something small, eating it quickly, and moving on to the next prey item.

It is evident that a falcon is too puny to feed regularly on sheep. Perhaps more surprising is evidence that a wolf will be slow to attack a healthy sheep. The rugged slopes of Mount McKinley, Alaska, are home to both wolves and Dall sheep. The wolves prey on the sheep, but the large curled horns of the sheep provide some measure of defense. In the 1930s, Adolpho Murie, a local scientist, collected the skulls of 608 Dall sheep on Mount McKinley (see Table 9.2). All the skulls showed signs of wolf predation, and it seems that wolves were indeed the primary cause of mortality among sheep. However, a closer analysis of the data by Edward Deevey of Yale University revealed that the wolves were seldom killing healthy middle-aged sheep (Figure 12.2). This predation pattern is reflected in the relatively high rates of survivorship, once

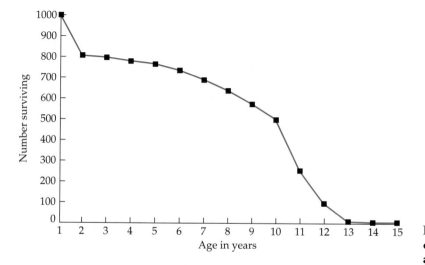

Figure 12.2 The survivorship of Dall mountain sheep plotted against time.

the sheep attained maturity and before they grew old and frail.

An important insight from this data set is that wolves kill the very young and the very old but seldom kill a sheep in its prime; hence the life expectancies of a middle-aged sheep are high (Table 9.2). The predators simply avoid confronting a strong, fit sheep.

We have all seen film of lions running down a wildebeest. Doesn't this refute the argument that hunters kill only the small or the weak? Lions and wolves, often operating in packs, tackle large prey items, but if closely observed they are seen to select the very old, the very young, or the sick. It is rare for these predators, even when working as a group, to kill a large animal in its prime. We then have to modify our assessment of hunters to say that they kill the small, the weak, or the hopelessly outnumbered.

Predators may be summed up as scavengers without the patience to wait for the prey to die. When predators do kill large animals, the risk that they will lose their meal to scavengers is very real. Having to protect a big kill entails potential danger even to a lioness or a leopard. Hyenas can be true predators, but they have also become expert at driving other predators, even a pride of lionesses, from their kill. The hyenas, operating as a pack, will outnumber the lionesses. They will snap at them, circling, trying to maneuver and isolate each lion, before rushing in to attack. Occasionally, the fights are fatal, either to the lion or the hyena, but more often the lionesses slink away into the night, leaving the hyenas to fight for the rights of the carcass.

12.3 Optimal Foraging Theory

All predators face a point of compensation between maximum prey size that can be tackled *safely* and optimizing the energy gained from killing the largest possible prey item. **Optimal foraging theory** states that predators will evolve to hunt in the manner that is most efficient to their physiology and that will lead to the greatest fitness. Hunting is then seen as an exercise in cost control. The costs are energy expended in searching for prey and in handling the prey when it is caught. The handling covers the period from the pounce to leaving the kill site. The time it takes for the predator to subdue a prey item, chew it, and swallow is time lost from hunting the next prey item. Therefore, it may be better to hunt more, smaller animals, although each offers a lower energy reward to the predator. An example comes from an insect-eating bird, the Pied Wagtail, that lives in short grasslands in Europe. The wagtail walks across the grass, driving up small insects as it goes, and it appears to be highly selective about the size of its prey (Figure 12.3). In an experiment, the wagtail was confronted with prey of different size classes, with the 8-mm class being the most abundant. Although the wagtail caught some of these insects, it showed a strong preference for rarer insects in the 7-mm-size class. Even this small difference in sizes allowed the wagtail to handle the smaller prey item more quickly and return to the hunt. Advocates of the optimal foraging theory interpret this behavior as an indication that the wagtail gains more energy from being a specialist hunter of the 7-mm-size

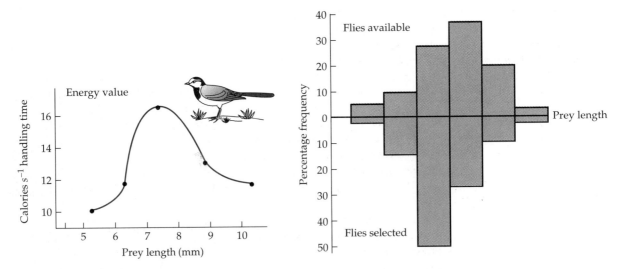

Figure 12.3   Under experimental conditions, the Pied Wagtail, which eats insects, chooses prey items in the 7-mm-size category, even though prey in the 8-mm-size class are more abundant. More, small prey give a higher energy yield than fewer large ones. This is an example of optimizing energy yield per kill. (After N.D. Davies, 1977.)

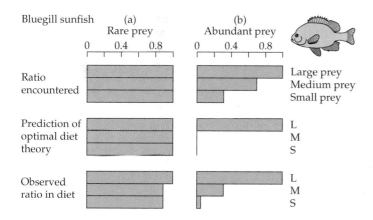

Figure 12.4   Prey selection by the bluegill sunfish when prey are rare and abundant. (a) The bluegills are not discriminating when the food supply is reduced. (b) The bluegills are choosy feeders, picking only the largest water fleas when food is plentiful. This is an example of prey switching as food abundance changes. (After E.E. Werner and D.J. Hall, 1974.)

class than from hunting the more abundant 8-mm prey. Evolution has programmed the wagtail to pick the prey item that will provide the most energy that can be channeled into reproductive effort and that will result in the highest fitness.

Another prediction of optimal foraging theory is that most predators will switch prey items, if they can, as one type of prey becomes scarce. This may mean selecting a different size of prey, or it may mean switching to an alternative prey species. In extreme cases, a specialist hunter will be forced to generalize and take all prey that comes within its grasp. When prey were abundant, bluegill sunfish (*Lepomis macrochirus*) were seen to forage almost exclusively on the largest water fleas (*Daphnia* spp.). Smaller *Daphnia* were ignored as the bluegills concentrated their search on the large ones (Figure 12.4). As prey became scarce, the bluegills were forced to capture all the potential prey items they encountered. Humans exhibit the same pattern. If we are well fed, we insist on fresh, well-prepared food, but if we are threatened with starvation, cockroaches, worms, or caterpillars may sud-

denly become part of our diet. We become much more generalist feeders.

That organisms switch between prey items according to food availability suggests that a **food web** (Figure 12.5) is a more realistic model of an ecosystem than is a series of simple food chains. By viewing an ecosystem as a food web, one can easily understand that changing the population size of any species in the web will have some repercussions (some good, some bad) for all the other species in the web. Studies of ecological communities, or all the species that are bonded by a food web, reveal some interesting patterns.

Some communities—for example, the communities at the edge of a polar desert or a newly formed volcanic island—are very simple; they contain relatively few species. Others—such as a lowland tropical rain forest or a coral reef—are hugely complex, with thousands of species. Despite the massive species diversity of some of these systems, studies reveal that the number of potential links in a food chain does not increase. Food chains are limited by the energy flow retained within the system, and they rarely exceed five

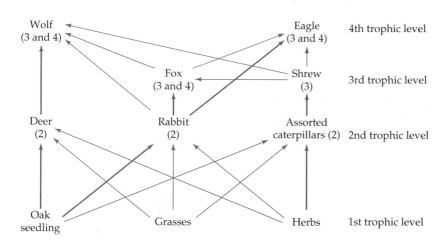

Figure 12.5   A vastly simplified food web for a temperate forest community. Thickened arrows indicate that the prey item is the primary source of energy for the predator. The number in parentheses is the trophic level at which the predator is feeding. For example, the fox is feeding on the third (grass → rabbit → fox) and fourth (herb → caterpillar → shrew → fox) trophic levels.

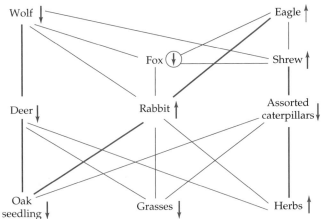

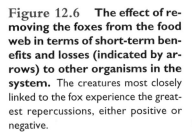

Figure 12.6   The effect of removing the foxes from the food web in terms of short-term benefits and losses (indicated by arrows) to other organisms in the system. The creatures most closely linked to the fox experience the greatest repercussions, either positive or negative.

trophic levels, including the plants at the base of the food chain. Another observation is that if a particular species is lost from the food web, the whole does not collapse. In a hypothetical woodland, let us assume that foxes are exterminated and then use the food web to predict short-term changes in the new system (Figure 12.6). The strongest effect would be an increase in the number of rabbits, which formed the foxes' principal prey. The extra rabbits would increase the grazing pressure on plants of the forest floor and particularly on oak seedlings. Deer populations may do poorly because of increased competition for seedlings from the rabbits. A reduced deer population could result in a decline in wolf numbers. The population of shrews might also increase because of the loss of the foxes, which prey on them. The shrews eat insects, so the population of insects would decrease, and, in turn, the herb population would increase because of the decrease in the number of insects. Eagles may take advantage of the glut of rabbits and the increase in shrews, so their population would increase. Another likely effect is that, with deer numbers reduced, wolves might switch to preying on rabbits and shrews more extensively. In time, the rabbit and shrew populations would be reduced, and the balance of the system would be restored, although the proportions of each species would be somewhat altered.

The food web analogy allows us to twang a thread in the web and watch the reverberations pass through the system. Removing the fox has an effect on the system, but it does not change the community radically. If a second species is removed, particularly if it is one occupying a similar niche (e.g., the eagle), the effects experienced by the community will be more profound. If a whole trophic level is removed from the community, then the changes may be so profound as to change the community beyond recognition. As one ascends the pyramid of numbers, it takes fewer and

fewer deaths, or species extinctions, to remove an entire trophic level.

12.4   Do Hunters Control Prey Populations?

Whether a predator controls a prey population or prey population size determines the number of predators has long been debated by ecologists. Both cases can be true. Let us consider the example of a predator introduced to a habitat in which there are abundant potential prey items. The predators will thrive and increase in number (Figure 12.7). After a while, the prey numbers will start to decline. The predator is, therefore, restricting the size of the prey population. We can say that this system is subject to **top-down control**, because it is the number of predators that will determine the number of prey. Alternatively, we can think about a world in which populations are driven by limiting resources and the entropy of energy along food chains. Here, we start with a finite pool of resources that are used by the primary producers. The population size of herbivores in this situation may be controlled by the quality of grazing. Predators in this system are seen as limited by the amount of energy in the herbivore trophic level. As herbivore numbers swell, more energy is available to the predators, so their numbers grow. But when the herbivores do badly because of lack of food, disease, or some other factor, the number of predators will fall because of lack of food. This system is being driven by factors at the base of the food chain, with repercussions into the upper trophic levels; thus it is termed a system of **bottom-up control**.

No book on ecology would be complete without the data of the Hudson's Bay Trading Company between 1850 and 1940 showing the number of pelts

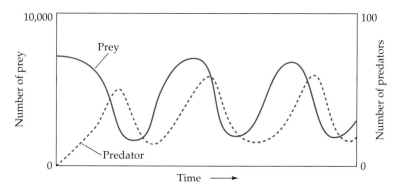

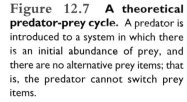

Figure 12.7   A theoretical predator-prey cycle. A predator is introduced to a system in which there is an initial abundance of prey, and there are no alternative prey items; that is, the predator cannot switch prey items.

brought to trading posts each year over almost 100 years of trapping. It is assumed that the amount of trapping time devoted to each animal was relatively constant, so the number of pelts reflects the number of animals each year. These data are generally used to demonstrate that number of predators controls number of prey (Figure 12.8). A quick look at the figure reveals that the numbers of snowshoe hare pelts (the prey) and those of the predatory lynx vary on about a 10-year cycle, with the lynx populations usually peaking just after a peak in the hare population. One explanation is that, as prey numbers build up, the lynx do well; they breed and ultimately overexploit the hares, causing hare populations to crash. A food shortage then causes predator populations to decline. With the loss of predators, the hare population starts to grow again, and the cycle repeats. This data set is often used as a classic study of top-down control. However, the quality of the data and whether they truly reflect a predator-driven system should be considered further. At two points in the cycle, in the early 1880s and

late 1890s, it may be seen that the lynx population apparently increased before there was a rise in the snowshoe hare population. According to the predator-controls-prey model this cannot happen, because there would be nothing for the extra lynx to eat. However, there are many potential sources of error within the data. One might be a bias among trappers: When populations of both hare and lynx were low, they could get more money for a lynx pelt and so devoted more energy to trapping lynx. When hare populations boomed, the trappers went with the flow and returned to trapping hares. This interpretation does not invalidate the top-down forcing model; it just makes us aware that we are not looking at the actual data for population, but at a surrogate that is one step removed.

A more serious criticism of the interpretation of these data comes from the study of snowshoe hare populations on an island where there are no lynx. In this system, we might expect the hare populations to behave like Gause's *Paramecium* and reach an

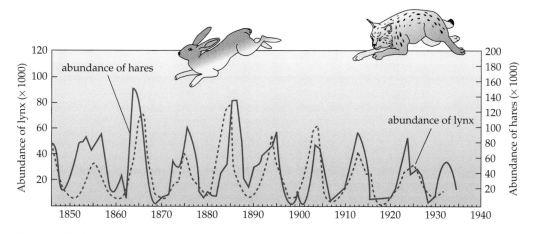

Figure 12.8   Oscillations in the predator (Canada lynx, *Lynx canadensis*) and prey (snowshoe hare, *Lepus americanus*) populations, based on the number of pelts bought from trappers by the Hudson's Bay Trading Company between 1850 and 1940. A general 10-year cycle is evident, but note that, according to these data, lynx populations rose ahead of snowshoe hare populations in the early 1880s and late 1890s. (After D.A. MacLulich, 1937.)

equilibrial number. But they don't. The island hare populations go through boom and bust cycles just as the Hudson's Bay hares did, even without a predator. This finding has led ecologists to suggest that the snowshoe hare population is being governed by the availability of young shoots of willow, their prime food source. If the willows are abundant, the hares boom; when the willows are scarce, they bust. Furthermore, if willows are subject to intense grazing pressure, they respond by developing higher concentrations of alkaloids in their young shoots. The alkaloids make the leaves unpalatable to hares. Thus, as hare populations grow to a maximum, they are faced with increasing competition for a decreasingly palatable diet. Reproduction fails, and hare populations crash. The willows have had to divert scarce resources to make the additional chemical defense, so shortly after the hare population crashes, when grazing pressure has relaxed, that energy is redirected into growth and reproduction. Lynx populations in this explanation would just be piggybacked and be totally dependent on the hares. They would play no part in driving hare populations up or down. Thus, this situation would represent a system of bottom-up control. Bottom-up control is prevalent in systems where plant productivity is markedly limited by nutrients, water, or other resources. Bottom-up control is also the norm in food chains that just involve plants, a herbivore, and a carnivore (although there are exceptions). Because most terrestrial food chains fit one or the other of these descriptions, bottom-up control is more common than top-down control.

However, genuine cases of top-down control can be found in numerous systems, such as where grazing sheep and rabbits control plant growth in a pasture, where wolves control deer populations, and in lake ecosystems where large predatory fish may be holding other populations in check. If those large fish are removed, a frequent result is that the populations of smaller fish increase rapidly because they are now free of predation.

12.5   Predators and Prey Behavior

The lurking peril of the thing under the bed or the possibility of encountering a shark when swimming in the ocean may affect our behavior at certain ages or under certain circumstances. Arguably, our most ferocious predator is the motor car; hence we look before

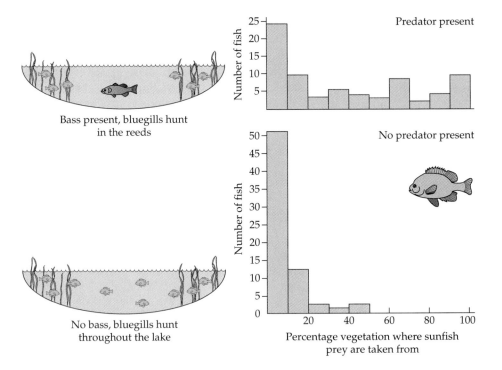

Figure 12.9   The spatial distribution of prey (bluegill sunfish) in response to predation by bass. Without predation, the bluegills will be found throughout the lake. When bass are present, the bluegills forage in the weedy margins, where they can hide from the larger fish. (After E.E. Werner, J.F. Gilliam, D.J. Hall, and G.G. Mittelbach, 1983.)

we cross the street. Other organisms live in a world where predation is a constant threat and respond accordingly. Alarm calls, feigned displays of broken wings to lead predators away from a nest site, traveling in herds, and camouflage are all life-saving strategies to avoid becoming the victim of a predator. Some behavior is revealed only after the predator has been removed. One example is the flying fox (*Pteropus vampirus*), a fruit bat with a wing span of almost a meter. This bat, best known as the vampire bat of Dracula movies, does not live in caves, nor does it suck blood. Its relationship to the undead has yet to be established, but it is known to be fond of bananas. These fruit bats spend their days in communal treetop roosts, flying out at dusk to forage in distant forest (or banana plantations), but their nocturnal way of life may be purely a strategy of predator avoidance. The principal predators are eagles and hawks, which are diurnal (daytime) hunters. If these predators are absent, then the bats forage more during daylight hours.

Another example of predator avoidance is spatial rather than temporal. Bluegill sunfish are the prey of largemouth bass. When bass are present, the hunting range of bluegills searching for *Daphnia* is restricted to the weedy areas of the lake, where they can hide from the bass. If there are no bass in the lake, the bluegills will forage throughout the lake (Figure 12.9). It is important to note that although the presence of the predator may cause changes in the behavior and abundance of the prey species, the prey still survive.

12.6 Predators Can Increase Species Diversity

Predation leads to the whittling of a prey population, and one might intuitively think that this process threatens the very survival of the prey species. But the converse may be true. It may be that predation is essential for the survival and coexistence of many species. Predation and the continued existence of healthy predator populations may be an integral part of the biodiversity equation.

Predation ensures that the weak and sickly are removed from the prey population. They seldom get the chance to breed, so the breeding stock remains strong. But the predators may have an even more profound effect as they limit the intensity of competition among their prey. To minimize energy expenditure, predators will hunt the most abundant, preferred prey item. Then, as that becomes scarce, they will switch to an alternative. As that prey population declines, the predator will be forced to search for a third type of prey. A diversity of prey populations may be held in check by a single predator.

In a classic study of the communities of rock pools on the craggy coastline of the state of Washington, Robert Paine investigated the relationship between a predator and a diversity of prey. The predator was the starfish *Pisaster ochraceus*, a formidable hunter of mollusks and barnacles that are attached to rocks in the intertidal zone. The experiment took the form of an exclusion study. Paine wanted to know the effect of the predator, so he excluded the starfish from some areas by handpicking them off the rocks. For six years, a section of rock 16 m^2 was kept free of the starfish. Neighboring rocks served as a control: that is, an area where the starfish were allowed to hunt normally. After six years, the control area had unchanged populations of the 15 species of mollusks and barnacles that form the prey of *Pisaster*. However, the experimental area had just 8 species of potential prey items left. The explanation is that without predation, 1 species of barnacle flourished and displaced some of the mollusks through intensified competition. In turn, this barnacle was later replaced by 2 species of bivalve. Competition had been fierce for the limited resources on the rock, and without the predator holding the various prey populations in check, some species had been ousted through competitive exclusion (see Chapter 9).

Harvard ecologist Edward Wilson noted of this study that it demonstrates that at times predation can be less dangerous to the prey than competition. An example from the plant kingdom is the chalk grasslands of England. These shortgrass meadows are rich in herb species. In fact, they form the most diverse flora per square meter that can be found in Britain. However, if grazing animals, such as sheep and rabbits, are excluded from these meadows, in just a few years the diverse community of wildflowers is replaced by just two or three species of tall grasses, and ultimately shrubs and trees will colonize the area. The woody plants shade out the grasses and herbs as a woodland develops. Thus, the herbivores that graze the wildflowers and are, in fact, their predators ensure their survival by also grazing the tall grasses, seedling trees, and shrubs. So long as the vegetation is cropped close to the ground, it is the wildflowers that prosper.

Another important insight that arose from Paine's study of the rocky intertidal zone was that one species may play such a pivotal role that if it were to go extinct the entire community would disappear. Because such a species is disproportionately important to the ecosystem, it was likened to the keystone in an arch. When building a stone arch, the structure was completed by the placement of a wedge-shaped stone in the apex of the arch. This, the keystone, held all the other blocks in place; if it were removed, the arch would collapse. In Paine's study he suggested the starfish is so important that it be termed a **keystone species**. A keystone species is now recognized to be broader than just a top predator. A keystone species can occupy any trophic level, but its influence on the local community is disproportionately large compared with its biomass. Removing a keystone species results in a cascade effect that causes all the trophic levels to change. Another realization is that keystone species are essentially local phenomena. In one habitat the starfish is a keystone species, in another the same species of starfish is just an ordinary member of the local community. The starfish, gray wolf, and alligator are examples of predatory keystone species.

Even a primary producer can be a keystone species. In most tropical forests, fruit availability is somewhat seasonal (Figure 5.16); a pattern that results in a glut of fruit production at some times of the year and a dearth at others. Animals such as monkeys, chimpanzees, toucans, and hornbills, rely on year-round fruit availability, and when fruit are scarce, their populations are vulnerable to starvation. Figs (*Ficus* species) are fairly unusual rainforest plants in that they produce nutritious fruit throughout the year. During the hungry times for the fruit-eating animals, the difference between survival and extinction is the availability of ripe figs. In these settings, fig trees are considered to be a keystone species.

12.7 The Defensive Weapons of Plants

The willows that deterred the grazing of snowshoe hares were defending themselves with chemicals. Plants can form deadly toxins such as cyanide or cardiac glycosides such as digitalin that can induce heart failure. Small amounts of these chemicals may prove deadly to an herbivore. Because it is the inherent quality of these chemicals to be highly toxic, they are known as **qualitative defenses**.

Plants can also produce an array of chemicals that upset digestion or are bitter or distasteful. The more of these chemicals that are eaten, the worse will be the stomach upset. Therefore, the more of these chemicals that a plant has in its leaves, the less an herbivore can eat. Because the defense of the chemicals depends on the amount ingested, these are termed **quantitative**

defenses. Tannins, a group of chemicals that bond to plant proteins, thereby making them indigestible, are an example of this group of chemicals. An animal that eats a tannin-laden leaf is unlikely to be able to extract adequate nutrition from it, and natural selection will favor those individuals that avoid leaves that have chemical defenses. A plant that invests in a chemical defense is likely to have reduced predation. Consequently, although the manufacture of these chemicals may be a significant energy cost to the plant, the reduced leaf damage is likely to be a net benefit. Plants with chemical defenses will therefore have a competitive advantage over those without them.

The defenses of a plant are not uniform throughout its body. The leaves of plants in the passion flower genus, *Passiflora*, contain cyanic glycosides, which form cyanide in the body of an herbivore. Cyanide is deadly to humans, yet we can eat passion fruit. The plant has evolved to defend its leaves with cyanide but to produce a fruit that is edible. Birds, rodents, and monkeys are essential seed dispersers for the passion flower. The flesh of the fruit is useful to the animals as a food source, but they do not digest the seeds. The seeds pass through the animal's gut and will be ready to germinate if they fall on a suitable patch of soil. The spreading of the seeds enhances the fitness of the plant, therefore it has evolved not to poison the animals that eat its fruit.

The property of plants to produce different chemical compounds to protect their stems, their leaves, their roots, and their fruits is important in the human quest for natural medicinal products. Many of our drugs are derived from the defensive chemicals of plants and animals. Consequently, the various body parts of a single species of tree or vine may each provide a potentially useful chemical.

12.8 Other Species Interactions

Species that live together may have evolved adaptations over a long period of time that can benefit one or both species and that, usually, one (or both) species depends upon. For example, each of the hundreds of species of fig has a dedicated species of fig wasp that pollinates it. The entire reproductive cycle of the fig wasp takes place inside a fig. Both species are totally reliant on the other. A long evolutionary history such as this, which binds the two species closer and closer together (even if the relationship is not mutually beneficial), is termed **coevolution**.

Finding two species with interwoven life-styles does not necessarily mean that they have a shared evolutionary history. Some species are **preadapted** as a result of evolving adaptations in a community that no longer exists. For example, the behavior and anatomy of a parasitic wasp that lays its eggs on butterfly caterpillars may evolve in an interaction with one particular prey species. But because of environmental changes, the once common caterpillars now become extinct. The wasp may also go extinct, or it may be able to switch to parasitizing an entirely different species of caterpillar. The caterpillar has encountered other parasites in its evolutionary history and already has some defensive adaptations, such as dropping off the leaf if touched or having irritating bristles. Both species have previously held adaptations that arose from former interactions, not through the present relationship. If you did not know their histories, these two species look to be co-evolved; but their encounter is only in ecological not evolutionary time.

Symbiosis is the close ecological association of two species, and although it is generally taken to be beneficial to both species, it does not have to be. Parasitism, commensalism, and mutualism can all be symbiotic relationships.

Parasitism

Parasites are organisms that derive benefit from a relationship that harms a host organism (an individual of the species upon which the parasite lives). Small-bodied parasites, such as many viruses and bacteria, are **microparasites**, while larger organisms, such as fleas, ticks, tapeworms, and fungi, are **macroparasites**. Most organisms carry parasites that steal products of metabolism, but at relatively low population levels the parasites do not cause great damage. However, should those populations increase, the energy drain may become fatal. Many diseases are caused by parasitic infections and are socially, economically, and ecologically important. It is not only human diseases that are caused by parasites. Millions of dollars are lost each year due to fungal infections of crops and parasitism among livestock.

Fleas, lice, intestinal worms, rusts, and smuts are **biotrophic** parasites: That is, they only thrive so long as the host is alive. If the host dies, they have to leave it and find a new one. The blowfly is an example of a **necrotrophic** parasite. The blowfly lays its eggs on a live sheep. The maggots burrow into the sheep and, if there are a large number of them, they may kill it. Even after the sheep has died, the larvae continue to feed and mature inside the corpse.

Parasites that always kill their host are termed **parasitoids**. Most parasitoids are flies and wasps, many of which lay their eggs on the larvae of other insects. For example, a parasitoid wasp injects its egg beneath the skin of a larval butterfly. The eggs hatch, and the parasitoid larvae eat the host from the inside. The host, though ailing, survives and continues to grow, providing nourishment for the parasitoids. When the parasitoid larvae are mature, they emerge through the skin of the host and pupate. It is this final emergence of the parasitoid larvae that kills the host.

Commensalism

The thought of parasites might make your skin crawl, but commensalists are actually crawling on your skin. Your hair follicles and sweat ducts are home to microscopic mites. These tiny creatures cause no harm or discomfort as they glean a living from the oils on your skin. If you want to see some of these creatures, hold the skin on your forehead taut and gently scrape the skin with a suitably dull blade. Carefully rinse the blade with a drop of water and transfer the drop to a microscope slide. Place a coverslip on the drop and view it at 400✕; your worst nightmare will be realized (Figure 12.10).

Our relationship with these mites is termed commensalist because it is essential for the mites' well-being and is of no consequence to us. In a **commensalist** relationship, one organism derives a benefit from the relationship while the other is unaffected.

Mutualism

Some of the most interesting of all biological relationships are mutualistic; that is, they benefit both species. We mentioned one such relationship in Chapter 6; plants and their mycorrhizal fungi benefit from each other. A common form of mutualism occurs when animals disperse the seeds of plants. A ripe cherry makes

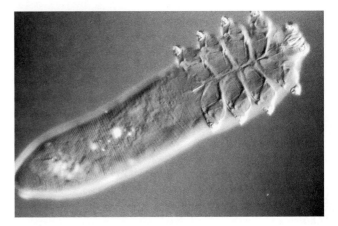

Figure 12.10 **A follicle mite is a commensalist that lives in human hair follicles.**

a tasty, nutritious mouthful for birds such as jays, waxwings, and thrushes. Apples, berries, and figs are eaten readily by wild pigs, bats, and monkeys. The fleshy fruit is an important part of the animal's diet because it may be rich in vitamins, carbohydrate, and water. The seed in the middle of the fruit, although swallowed, is not digested and passes right through the animal's gut. In the meantime, the animal has flown off or is roaming the land, carrying the seed farther and farther from the parent plant. Eventually, maybe 20 minutes to a day later, the seed is defecated. If conditions are good, it will start to grow. The animal was fed and the plant was spread; both parties benefited from this mutualism.

A more complex interaction involves the leafcutting ant of the genus *Atta* and a basidiomycete fungus. The leafcutting ants are a familiar sight to anyone who has visited a South or Central American rain forest. Long ribbons of wobbling green squares thread their way across the forest floor. Each square is a leaf fragment, many times the size of the ant, being carried back to the nest (Figure 12.11). Sometimes the load will be a flower petal instead of a leaf, and the column is a swaying chain of yellow or pink. Thousands of ants, all of one species and from a single nest, work together to harvest a particular tree in the forest. The column, which may be more than a kilometer long, leads back to an underground nest that can cover tens of square meters. Why do the ants carry such gigantic loads? Why don't they simply eat the leaves or petals where they find them?

The ants do not eat the leaves because they cannot digest cellulose. Instead, they carry the leaves back to the ant nest, where they farm a basidiomycete fungus. The ants feed the leaves to the fungus and maintain a great decomposing mat of leaves as a fungal food supply. They will tolerate only one species of fungus in their nest and pick any foreign fungus away from the basidiomycetes. The fungus gains nutrition from the leaves and grows. The ants then harvest the fungus, eating a portion of it, but always maintaining a healthy fungal mass. The fungus cannot live outside of the ant nest, and the ants can eat only the fungus. But to do so, the ants must devote all their energies to feeding it. Indeed, the fungus is so important to the ants that when the queen ant changes nest locations, she carries a piece of the fungus in her mouth to establish a new fungal colony. The mutual benefit of this relationship is clear.

Another classic example of mutualism comes from pollination. Bees, flies, butterflies, moths, birds, and bats are attracted to flowers by color, shape, or scent. The flower is a feeding station with a reward of a natural high-octane fuel, nectar. Nectar contains sugars that provide energy for the visitor. As the animal burrows into the flower to capture the nectar, it brushes against the pollen-bearing stamens. The pollen sticks to the body of the animal, who may then carry it to another flower of the same species. When the animal enters the second flower, the pollen is rubbed off onto the female portion of the flower. The process of pollination, the transfer of the male gene stock to meet the female gene stock, has been completed, and fertilization ensues. This kind of pollination accomplishes the spread of pollen between different plants of the same species. The plants benefit by achieving sexual reproduction, and the pollinator gets a meal (usually).

Plants are not above trickery to accomplish pollination. The plant that produces the largest flower in the world (about 1 m in diameter), *Rafflesia*, smells like

Figure 12.11 **Leafcutting ants carrying their loads back to the nest.**

a dead animal. Flies are tricked into settling on the flower as they search for a meal of carrion. The flies wander around the flower looking for decaying meat, and, finding none, they leave. In the meantime, the flies have brushed against stamens and are coated with pollen. The flies do not learn, and they will fly to the next *Rafflesia* flower in search of a carcass, only to complete the pollination.

Some orchids are the masters of deception. Species of bee orchid have evolved a flower that looks (to a male bee) like the enticing hindquarters of a female bee sticking out of a flower. The male bees attempt to mate with the supposed female, only to become coated in pollen. The truly remarkable part of this deception is that the flower is more sexually alluring than the real female. In experiments in which males could choose between a flower and a female bee, most chose the flower! Not surprisingly then, the misguided male bee is quite likely to try to mate with the next bee orchid that it finds, once more completing the pollination process. Clearly, when the deception is so complete that the bee gains nothing, the plant is no longer a mutualist, but is really a parasite.

Accidental poisoning of pollinators with insecticides has greatly reduced the number of bees and other pollinating insects. Without them there is no pollination and no fruit formation. A lack of pollinators is one of the most serious ecological problems facing fruit growers.

12.9   Mimicry

Mimicry is a form of defense against predation that either reinforces a lesson that this prey is not palatable or fools the predator into avoiding potential prey. In nature bright colors, such as yellow and red, and bold juxtapositions of these colors with black, such as the stripes on a yellow jacket, are warning colorations. The warning may be of a sting, a venomous bite, or that the creature is poisonous to eat. Predators would learn to recognize the toxic species, based on their vivid coloration. If more than one species has a similar appearance and all are toxic, the lesson of toxicity will be reinforced every time a predator tries to attack one of them. The species evolve to mimic each other. The more similar they look, the stronger the message being sent to the predators that this body form is bad to eat. Thus, natural selection could result in unrelated species looking very similar and for both sexes to look similar. The yellow and black banding of stinging wasps and bees is an example of this form of mimicry.

The message of the warning coloration is reinforced with each sting. The fitness of individuals in all the mimetic species increases as a result of reduced predation. This kind of mimicry is termed **Müllerian mimicry**.

Cheats that Prosper

If coloration can warn a predator away from poisonous prey items without the predator ever taking a bite, there is the opportunity for the evolution of cheats. Building and storing enough toxic chemicals so that every bite is distasteful is energy expensive and, therefore, carries a penalty in fecundity; energy goes to defensive compounds rather than reproduction. However, copying the appearance of another species is a relatively light energetic load. **Batesian mimicry** occurs when a harmless species mimics one that is toxic. This kind of mimicry has a standard pattern involving model, mimic, and dupe. The **model** is the species that will be copied, the **mimic** is the species doing the copying, and the **dupe** is the predator that is fooled by the mimicry.

Our example is taken from a group of butterflies, members of which have evolved to copy the pipevine swallowtail. Pipevine swallowtails are found from Ontario and New England south to Mexico and Florida, living in wetlands and moist areas. The caterpillar of the pipevine swallowtail feeds on pipevines (*Aristolochia* species), which are rich in toxins. The caterpillars store the poison, and it is still present even after metamorphosis to pupa and adult. Consequently, even though the adult butterfly only sips nectar, it contains enough toxins to deter predators such as Blue Jays. Pipevine swallowtails have iridescent blue on the inside of their wings and red blotches on their underside that serve as warning colors (Color Plate 2). When they are resting, the red spots on the undersides of the wings are clearly visible, and when flying, the red and blue flashes of color warn predators to stay away. The pipevine swallowtail is the toxic model that other harmless species mimic.

The pipevine swallowtail is mimicked by female butterflies of six species: the Diana, the red-spotted purple, the black swallowtail, the spicebush swallowtail, and dark forms of the eastern and western tiger swallowtails (Color Plate 2). All of these are harmless, palatable butterflies that would make a good snack for a predator; however, they shelter behind a masquerade of toxicity. Natural selection has favored individuals that are most confusing to the predators, hence those forms most similar to the pipevine swallowtail have the best chance of survival (remember that they cannot choose to look like the model). This kind of mimicry only works well when the mimic is rare, relative to the model. Otherwise, the predator does not learn to avoid the brightly colored prey. As long as the great majority of bright, obvious prey are toxic, the predator will learn to ignore them and, in so doing, will also ignore the mimics. Although there are potentially six species that mimic the pipevine swallowtail, they do not all occur in the same habitats and geographic ranges. Consequently, the abundance of mimics does not overwhelm the abundance of the model. Indeed, the two tiger swallowtail species are normally bright yellow, and their mimetic dark forms only occur if there is a local abundance of pipevine swallowtails. Another important aspect of Batesian mimicry is that generally only females are mimics and the males look totally different (Color Plate 2). This is simply explained by observing that there is selective advantage to preserving a female longer than a male. The female needs time to seek out flowers and oviposit, whereas the male only has to inseminate a female (and one male can impregnate many females). Because mimicry is less effective the greater the dilution of models with mimics, the benefits of mimetic protection are not squandered on males. Other examples of mimicry, both Batesian and Müllerian, can be found among other insects, fish, and reptiles.

12.10 Predation and Management

Most nature reserves and remnant patches of natural vegetation are too small to support the predators that feed highest on the food chains. This imbalance may lead to an overabundance of herbivores, leading to overgrazing and environmental degradation. As plans are made to preserve natural areas from development, it is necessary to predict future changes in population densities. For example, it may be necessary to decide if the area would benefit from limited deer hunting to prevent overgrazing of tree seedlings. If hunting is desirable, what is the optimum number, age, and sex of deer to be removed? Studies of natural predation patterns may lead us to these answers, and they may not accord with the answers preferred by hunters. In terms of reducing deer numbers, it is more important to shoot does than bucks, and shooting a prime male with a fine rack of antlers makes no ecological sense. Perhaps the most important message to conservationists and land managers is that it is possible to change many trophic systems from either the top, the bottom, or for that matter the middle. Before establishing a nature reserve or providing planning permission for a change in land use, it is essential to understand the local systems, their food chains, and predator-prey links.

Summary

- Natural selection determines that predators are cowards; the brave go extinct.
- Predators forage optimally by expending the least amount of energy necessary to achieve fitness and will switch prey items if necessary.
- Food webs can be used to predict the changes that will cascade through a community if the population size of one species is altered.
- Both top-down and bottom-up control can influence community composition.
- Predator avoidance is the best defense against predation.
- Predation can be an important factor in the maintenance of biodiversity because it may prevent competitive exclusion.

- A single species may be so important within a community that it is termed a keystone species. Loss of the keystone species cascades change throughout the community.
- Plants use chemicals in their interactions with animals: sugars to attract pollinators, toxins to deter herbivores.
- Several types of close symbiotic relationships are common among organisms, including parasitism, commensalism, and mutualism.
- Müllerian mimicry is the evolution of toxic or dangerous species to appear similar.
- Batesian mimicry is a form of defense that provides a selective advantage without the cost of producing toxins.

Further Readings

Buchmann, S. L., and G. P. Nabhan. 1996. *The Forgotten Pollinators*. Washington, DC: Island Press.

Jolivet, P. 1998. *Interrelation Between Insects and Plants*. Boca Raton, FL: St. Lucie Press.

May, R. M. 1983. Parasitic infections as regulators of animal populations. *American Scientist* 71:36–45.

Paine, R. T. 1966. Food web complexity and species diversity. *American Naturalist* 100:65–75.

Web Connections

On-line resources for this chapter are on the World Wide Web at:
http://www.prenhall.com/bush
(From this web-site, select Chapter 12.)

Communities Through Time: Changing Populations and Landscapes

13

Henry Ford, founder of Ford Motor Company and inventor of the mass production of cars, once commented that "history is bunk." Alternatively, a psychologist might tell us that a sense of our past is perhaps one of the prime characteristics of human intelligence. A knowledge of the past is often necessary to understand the present. For instance, unless one is familiar with the colonial history of Canada, it would be impossible to understand why both French and English are spoken in that country. Without knowing that half of Europe was recently communist, it would be impossible to comprehend the present political upheavals in Russia and the former soviet satellites. However, for prehistoric periods there is no written history, and interpretation of the past is based on the fossil record. Charles Lyell, the founder of modern geology, wrote that "the present is the key to the past." He believed that the basic physical processes of erosion, sedimentation, volcanism, and glacial activity were the same today as in the past. Therefore, by observing modern processes, he could understand the forces that had laid down the sediments of the ancient oceans and caused the uplift of mountain chains.

Ecologists recognize the need to study modern systems in order to understand ecological processes such as evolution and competition. However, to observe only what is happening in the present is not sufficient to understand a modern ecosystem, let alone one in the past. The populations, the physiological and morphological attributes of individuals, even the behavior of organisms, could have been determined by an event in the past that we have not witnessed. The coming and going of ice ages, the ravages of hurricanes, renewal through wildfire, even the millennia of human activities, cause changes in the distribution and abundance of species. Observing an ecosystem at one point in time cannot reveal whether the balance of populations is stable or if one species is plummeting in number while another is soaring. It is necessary to know the history of the system so that the trajectory of populations can be determined. Ecologists would argue that the past is the key to the present, and perhaps, more important, to the future.

13.1 The Coming and Going of Ice Ages

Chronologies of glacial and interglacial oscillations can be obtained from cores of mud raised from the seabed or from the bottom of ancient lakes. But the most detailed histories come from cores of ice collected from ice caps. Although these sources of information are all different, each relies on the steady accumulation of mud or ice through time. Every year supplies another input of mud to lake or ocean, or another snowfall that eventually freezes to become ice. Layer upon layer, the mud or ice accumulate. The oldest lie at the bottom, and successively younger deposits build toward the modern oceanbed, lake bed, or ice surface. Each layer is a time capsule of information that can tell us of past climate, vegetation, or animal communities at that site. By examining these deposits, we can track changes in temperature, rainfall, fire frequency, even the concentration of atmospheric gases through time. These time capsules also provide our tool for determining the trajectory of modern populations and long-term ecosystem change.

Isotopes and Ocean Histories

Atoms of an element have a fixed number of protons, but some elements can have a variable number of neutrons in their nucleus. The term **isotope** is used to describe an atom in terms of its neutron number. For example, ^{16}O is an oxygen atom with 8 protons and 8 neutrons, whereas the much rarer ^{18}O has 8 protons but 10 neutrons. The extra neutrons in ^{18}O make it a heavier atom than ^{16}O. The lighter ^{16}O form of oxygen will be evaporated more readily from seawater than will ^{18}O. Consequently, rain and water bodies such as ice sheets, glaciers, and freshwater are poor in ^{18}O relative to ^{16}O. The modern concentration of ^{18}O in the oceans reflects the relative abundance of ice and freshwater versus seawater. If the proportion of freshwater to seawater changes, so too will the concentration of ^{18}O relative to ^{16}O in the oceans. The changing ratio of these two isotopes in seawater documents changes in the size of the polar ice caps.

Microscopic oceanic organisms whose corpses were incorporated into sediments can yield a sensitive indicator of ^{16}O and ^{18}O concentrations. Because the con-

centrations of ^{18}O in seawater reflect the amount of water held in glacial ice at the poles, it provides an indicator of the timing and extent of past climate change. As climate cools, the ice sheets expand, and the rain and snow that form the ice are rich in the lighter isotope ^{16}O. The ^{16}O has come primarily from evaporated ocean water, leaving behind ^{18}O. High concentrations of ^{18}O in ocean water therefore represent cold, glacial periods, and low concentrations of ^{18}O represent warm interglacial periods. Our present interglacial conditions result in comparatively little water being held in ice caps, so the concentration of ^{18}O in the oceans is low (Figure 13.1).

Events taking place within the last 40,000 years can be dated accurately using radiocarbon dating. Older materials can be dated using uranium, potassium-argon, or argon-argon dating. The sedimentary record from the deep oceans can, therefore, be dated, and it provides a signal of the expansion and contraction of glaciers. From this and other fossil evidence it is possible to reconstruct the temperature of the oceans over the last 700,000 years.

It is evident from these data that Earth is now in one of the warmest episodes of its history in the last 700,000 years. In fact, at no time during the last 4 million years has Earth been more than 2°C warmer than present, and most of the time it has been as much as 9°C cooler.

Orbital Wobbles and Ice Ages

Over the last 2.5 million years, an ice-age Earth has been the norm, separated by relatively brief warm periods, such as now, which are the interglacials. Considerable debate centers around what causes ice ages to turn on and off. But, on the basis of the apparent rhythm of glacials (lasting about 100,000 years) and interglacials (lasting about 10,000 years), it seems likely that some long-term cycles are at work. One such group of rhythms is related to variations in the orbit of Earth around the sun. Three distinct rhythms emerge, and the consequence of each is a change in the amount

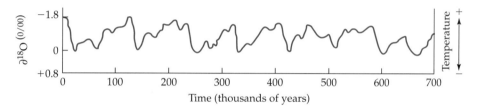

Figure 13.1   **The variation in the concentration of ^{18}O in ocean water over the last 700,000 years.** Positive values indicate more ^{18}O and therefore cooler water, while negative values indicate less ^{18}O and warmer water. (After E. Emiliani and N. Shackleton, 1974.)

of solar radiation received at a given point on Earth's surface (Figure 13.2a).

The **eccentricity** of Earth's orbit will affect the amount of sunlight received by the Northern or the Southern Hemisphere. It takes one year for us to orbit the sun, and if that orbit is nearly circular, then the two hemispheres will receive approximately the same amount of light energy. As the orbit becomes more elliptical, or eccentric, whichever hemisphere has its summer as Earth passes closest to the sun will receive the most energy (and grow warmer), while the other hemisphere will grow cooler. The eccentricity of our orbit changes very slowly, going through a full cycle once in 100,000 years.

The second rhythm, termed the **obliquity**, reflects changes in the tilt of Earth's axis toward or away from the sun. Think of Earth as being impaled through the poles onto a skewer; the skewer is the axis of the planet. Earth is rotating on that axis every 24 hours,

steadily moving a point on the surface between daylight as it faces the sun and night as it faces away. The axis is tilted so that the North Pole is angled a little toward the sun during the Northern Hemispheric summer; it will receive more sunlight at that time of year than at any other (Figure 13.2b). However, because of the way Earth rotates, the North Pole will be tilted away from the sun in winter, so it will receive less sunlight at that time of year. Thus, the more the axis tilts, the more the temperature difference will increase between summer and winter for both hemispheres. This rhythm from a stronger to a weaker tilt goes through a complete cycle every 40,000 years.

The last rhythm is called the **precession of the equinoxes**, and this is literally the amount of wobble that Earth has about its axis. This wobbling causes a gradual shift in the day that is Midsummer Day (Figure 13.2c). At present, Earth is closest to the sun on the Southern Hemisphere's Midsummer Day

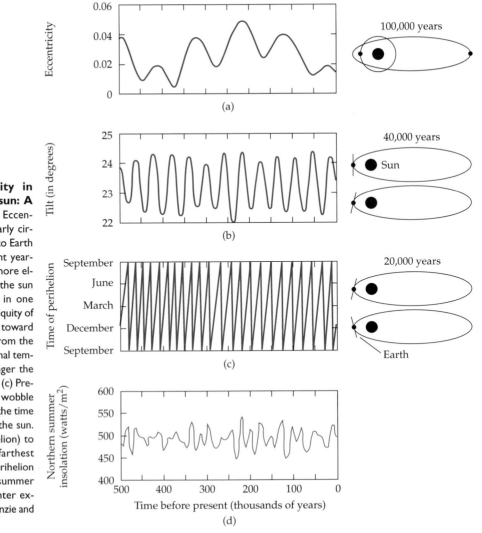

Figure 13.2 Variability in Earth's orbit around the sun: A set of three rhythms. (a) Eccentricity of the orbit. With a nearly circular orbit, the solar radiation to Earth as a whole is relatively constant year-round. As the orbit becomes more elliptical, the energy input from the sun becomes more concentrated in one season than in another. (b) Obliquity of Earth's axis. The tilt of the pole toward the sun in summer and away from the sun in winter creates the seasonal temperature difference. The stronger the tilt, the stronger the seasonality. (c) Precession of the equinoxes. The wobble of Earth about its axis changes the time of year when Earth is nearest the sun. Today, Earth is closest (perihelion) to the sun on January 4 and farthest (aphelion) on July 14. When perihelion coincides with summertime, summer will be extremely hot and winter extremely cold. (After F.T. MacKenzie and MacKenzie, 1995.)

(December 21) and farthest from the sun on the Northern Hemisphere's Midsummer Day. The consequence is that the north is experiencing an all-time low in seasonality (difference between summer and winter temperatures) as a result of this cycle, whereas the south has maximum seasonality. Every 10,000 years this situation is reversed, because this whole cycle takes 20,000 years to run its course.

These three separate rhythms are named the **Milankovitch cycles**. These cycles all affect global seasonality. Because they are independent of each other, their effects can combine to more or less cancel each other out, or they can combine to produce conditions of "superseasonality." Recent analyses of ocean and terrestrial climate records reveal that when these cycles combine to reduce the solar input to Earth, an ice age is the likely result.

The Milankovitch cycles suggest that the high solar inputs that characterize an interglacial period are maintained for about 10,000 to 15,000 years. The present interglacial period has lasted for about 10,000 years, and the next few millennia should see a return to glacial conditions. It has been suggested from the Greenland ice-core data that when the switch from interglacial to glacial conditions takes place, it may take as little as 3 to 10 years.

This switching does not mean that ice sheets zoom into place all across Canada and northern Europe in such a short time. The switch refers to a physical mechanism that turns on or off this rapidly, and that can set in motion events that will cause the expansion of ice sheets and a drop in global temperature. Although Milankovitch cycles are critical to determining the amount of solar energy falling on Earth, their change is too slow to account for the sudden onsets of glacial episodes: There must be an intermediary mechanism. This mechanism would need to be global in scale, it would need to be very finely balanced between "on" and "off," and it must be capable of inducing a chain reaction of events leading to cooling or warming.

A Mechanism of Rapid Climate Change

Recent research by Wallace Broecker of Columbia University and George Denton of the University of Maine suggests that the ocean conveyor system (Chapter 4) could be a switching mechanism that precipitates rapid climate change. The formation of the North Atlantic Deep Water is achieved because it is so salty. Saltwater is denser than freshwater, just as cold water is denser than warm water. Where the surface ocean water is cooled by very low air temperatures and is

very salty, the water can sink from the surface all the way to the bed of the ocean. This happens only in two places on Earth, around Antarctica and in the North Atlantic. It is critically important to realize that in the North Atlantic the water sinks because it is salty. The Atlantic Ocean is about 1 ppm saltier than the Pacific Ocean. This higher salinity of the Atlantic is enough to cause water at a temperature of +2°C to sink to the bed of the ocean. In the Pacific, the water would have to cool below freezing to achieve sufficient density to sink. Once ice forms, the water does not cool any more and hence does not sink. If anything happens to cause a reduction of salinity in the North Atlantic, the North Atlantic Deep Water would not form (because water would not be sinking deep enough), and the entire ocean conveyor system would shut down. Thus, our modern climates hinge on a salinity difference of 1 ppm between the Pacific Ocean and the Atlantic Ocean.

Data from analyses conducted in 1993 on ice cores from Greenland and Antarctica and ocean sediments from the North Atlantic gave rise to a new understanding of the pattern of glacial and interglacial episodes (Figure 13.3). Until that time it had been accepted that the climate of the last 2.5 million years had been a fairly simple sequence of glacial and interglacial events. The new data indicate that during the last ice age there were warm stages almost as warm as at present. These started and stopped abruptly, and these violent fluctuations in global climate are believed to be linked to the ocean conveyor belt system being turned on and off (Chapter 4). Our present interglacial climate appears to be unusually constant. Note how much more pronounced the climate oscillations were in the preceding interglacial period (140,000–115,000 years ago). In the last 10,000 years there have been none of the sudden intense fluctuations seen in the ice records of earlier times.

The ocean conveyor is always teeter-tottering on the point of shutting down and seems to have a cycle in which it strengthens and weakens every 1000 years or so. Events such as melting or partially melting the polar ice caps would add a lot of freshwater to the ocean and could transform a minor cycle of weakening into a more profound shutdown of the conveyor.

The last time the conveyor shut down was at the end of the last ice age, about 13,000 years ago, and that induced a final cold burst that lasted 500 years. Paradoxically, it appears that it is warm events that can trigger a shutdown of the conveyor system. This last shutdown is referred to as the Younger Dryas event, and it was caused by the melting of the Laurentide ice sheet. The ice sheet had blocked the St. Lawrence River, and all the meltwater from the decaying ice

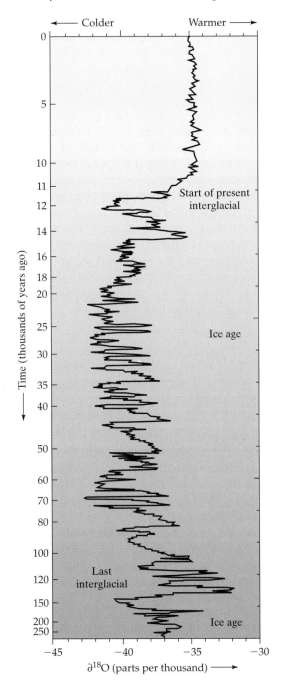

Figure 13.3   Climate records from the Antarctic and Greenland ice caps showing rapid fluctuations in local climate during the last interglacial and glacial periods. Note the relative climatic constancy of our present interglacial period. (After Greenland Ice-Core Project Members. Reprinted with permission from *Nature* © 1993 MacMillan Magazines Limited.)

flowed south into the Mississippi. With further melting, the ice sheet retreated past the St. Lawrence River and sent meltwater down the St. Lawrence valley into the North Atlantic. Glacial meltwater is freshwater, and the intrusion of this water reduced the salinity of the North Atlantic. With reduced salinity, the water of the North Atlantic was not dense enough to sink; the ocean conveyor shut down.

The history of the North Atlantic suggests that if large ice caps melt, or if a huge raft of icebergs drifts to a lower latitude and melts, it may cause a shutdown of the ocean conveyor. If circumstances are right, the shutdown sets in motion a positive feedback mechanism (a series of events that induce progressively greater changes) leading to cooling. As the conveyor stops, the Gulf Stream shuts down, and the seas around Iceland and northwestern Europe cool rapidly. Cold water that was being transported away from the poles now stays at high latitudes, and both the Antarctic and the Arctic cool. Colder conditions induce an expansion of polar ice sheets, and as the shiny whiteness of the ice expands, the amount of sunlight reflected back into space increases. Consequently, the amount of light absorbed by Earth is reduced, causing further cooling. The effect may snowball until global temperature has dropped about 6°C, at which point glaciers will form on mountain peaks such as the Rockies, the Alps, and even on tropical peaks of the Andes.

Shutting down the ocean conveyor does not always induce an ice age, and scientists are still working to discover the other factors that make the difference between a weakening of the ocean conveyor and the chain of events that triggers a full ice age. Part of the puzzle is undoubtedly the Milankovitch cycles, but it seems that there are still some missing pieces to be found.

Why should we care about whether this system is running or not? If, as is likely, human activities warm the planet by a further 1°C to 3°C over the next century (Chapter 23), this warming could see a significant reduction in the volume of the Greenland ice cap. The injection of meltwater from icebergs calving from that ice mass could be leading us toward another shutdown cycle. A possible outcome of such an event would be a new ice age. A more likely outcome would be a few hundred years of much colder weather, in which global temperatures would drop 5°C to 9°C, rendering much of our most productive farmland too cold for crops. Both outcomes would be socially and economically dire. A further consideration is that the weakening and shutdown cycles that occurred in the past took about 3–10 years to go from "ocean conveyor on" to "ocean conveyor off."

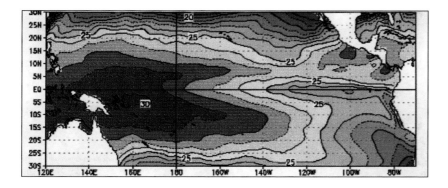

(a)

Plate 1

A comparison of sea-surface temperatures under normal and El Niño (c, d) conditions. False color satellite images document:

(a) The sea surface temperature of the Pacific Ocean in February 1997. Numbers are temperature in °C.

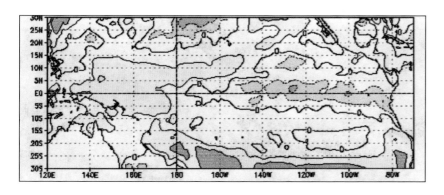

(b)

(b) Sea surface temperature anomalies (variations from average conditions for that time of year) in °C in February 1997. Note that anomalies are close to zero. Areas warmer than usual are shown in orange and red, areas cooler than usual are in blue.

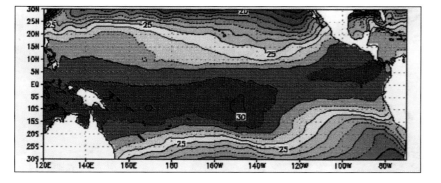

(c)

(c) The temperature of the earth surface in February 1998 during the strongest El Niño event yet recorded. Note how warm water extends all the way across the equatorial Pacific.

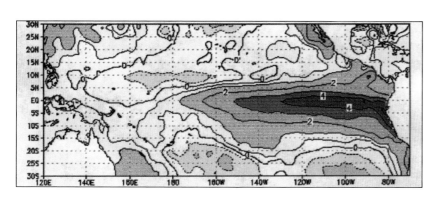

(d)

(d) Sea surface temperature anomalies in February 1998. Note how El Niño is marked by a warming of the eastern Pacific, and how this warmth extends north and south along the coastline of the Americas.

(Data from the National Oceanographic and Atmospheric Administration.)

(a)

(b)

(c)

(d)

(e)

(f)

Plate 2

Batesian mimicry among North American swallowtail butterflies. The poisonous model is the pipevine swallowtail (a, b); the black swallowtail (c) and the spicebush swallowtail (d) are harmless mimics of the pipevine swallowtail. The eastern tiger swallowtail butterfly is normally yellow (e), but when pipevine swallowtails are common, a dark mimetic female form of the eastern tiger swallowtail occurs (f).

13.2   Are Communities Stable Through Time?

In evolutionary terms, climates that frequently flip-flopped from cold to warm and back again induce physiological stress and force migrations, as populations suffer unaccustomed warmth or cold. Species that have survived the assault of climatic change (and this appears to have been most of them) have had to adapt, migrate, or survive in reduced numbers. The parental genetic recipe, or genotype, may no longer be the best one for the new conditions posed by warming, drying, cooling, flooding, or changes in atmospheric concentrations of carbon dioxide. If new conditions select for different behavioral or physical traits within a species, the result may prove to be the evolution of a new species. Even if these changes do not result in full speciation, they may lead to a change in the genotypic composition of the species through time.

During this century there has been an ongoing debate among ecologists as to whether communities are relatively stable entities or whether they are in a continual process of reorganization. The implications of this debate are rather profound. If communities are stable through time, they are likely to contain a high proportion of species that are coevolved (Chapter 12). If, on the other hand, communities are just chance assemblages of species with no shared evolutionary history, then more species would be preadapted than coevolved (Chapter 12).

To find if communities have been stable through time requires an analysis of the fate of a community through a major environmental upheaval. The last (Wisconsinin) ice age provides a disturbance profound enough to test community stability and recent enough that we do not have to worry about major evolutionary change. If I had a time machine we could just hop in and skip back 18,000 years to the height of the last ice age; but I don't. Instead we enter a branch of science called **paleoecology**, the reconstruction of past environments and ecosystems using the fossil record.

13.3   The Pollen History of Northeastern North American Forests

Each spring, trees and herbs produce massive quantities of pollen that is blown around on the wind. The pollen becomes thoroughly mixed before falling to the ground. Each species of plant produces a characteristic

pollen type, with a distinctive pattern of bumps, holes, or grooves that can be seen under a microscope (Figure 13.4). Fossilized pollen can be found in oxygen-free mud, such as lake sediment. The first goal is to find a lake that is more than 18,000 years old, for its sediments will contain the history of the region since its formation. Because lake mud builds up year upon year in a vertical profile, the oldest material will be at the bottom and the youngest will form the modern bed of the lake. To capture this column of mud and its enclosed history, investigators use a drilling rig to raise a vertical core of mud from the bottom of the lake. Back in the laboratory, almost everything can be dissolved away but the fossil pollen. Carefully identifying and counting the different pollen types found at different layers in the core documents a history of vegetation change. The more layers that are analyzed, the more precise is the image of vegetation change around that lake through time.

Silver Lake, Ohio, lies in an area that was covered by ice during the peak of the last glaciation. About 12,600 radiocarbon years ago, the ice melted, a lake formed, and the pollen record began (Figure 13.5). Pollen diagrams are drawn so that the oldest samples are at the bottom, and the youngest are at the top (just as they were found in the core of mud). A diagram is always read from the bottom upward, because this direction reproduces the order of events in the history

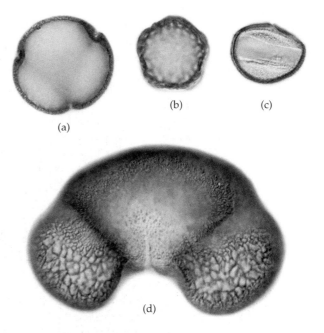

(a)

(b)   (c)

(d)

Figure 13.4   Fossil pollen grains. Pollen types are frequently identifiable as to their genus based on their architecture and surface patterns. Shown here are (a) beech, (b) elm, (c) oak, and (d) spruce.

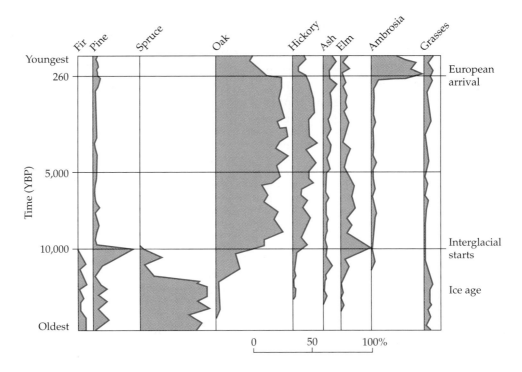

Figure 13.5   A 12,000-year record of vegetation change as indicated by fossil pollen from Silver Lake, Ohio. The vegetation progresses from a fir-pine parkland to a closed temperate forest at the start of the interglacial. Note how the forest has continued to change throughout the last 10,000 years. The increased abundance of *Ambrosia* is a signal that cultivation by European settlers has started. (After J.G. Ogden, III, 1966.)

of the region. At the bottom of the Silver Lake diagram, the pollen of fir, pine, spruce, and grass are common. This pollen record indicates that a mixed coniferous woodland or parkland existed in central Ohio at the end of the last ice age. The combination of these trees suggests cooler temperatures than at present, and the grasses suggest that it was relatively dry: conditions typical of the end of the ice age. About 12,500 years ago, oak pollen starts to increase in abundance, a sign that the climate was becoming warmer and moister. Spruce and fir could not compete with the oak and pine under the changed climatic conditions, and they died out at this site. The peak of pine pollen at 9800 years ago probably marks the last cold snap of the ice age. Afterward, the climate warmed quickly, and elms and hickory increased in abundance as the temperate forests formed.

Near the top of the diagram, at 260 years before present, the land was cleared for agriculture. The forest taxa all decreased in abundance, and *Ambrosia*, a weed, suddenly increased in abundance. The rise of *Ambrosia* is found again and again in pollen diagrams from eastern and midwestern states, as it tracks the invasion of European settlers and the first tilling of the

land. Thus, our pollen record provides a minimum date for the first agriculture in this part of Ohio of about A.D. 1700.

Silver Lake is one record among more than a thousand pollen diagrams from the eastern United States. This immense information resource illuminates the movement of tree species from 18,000 years ago until the present. What should we look for? Let us examine three hypotheses.

Hypothesis 1: Climatic change does not affect the forests. If there is no change in the distribution of vegetation through the last 18,000 years it is evident that ice ages have not affected the forests.

Hypothesis 2: Forest communities migrate to withstand climatic change. If all the species in a forest migrate together, this finding would support the idea that forest communities have evolved together and that each species is dependent on the others. Therefore they respond as one to major climate change.

Hypothesis 3: Species respond independently to climate change. This scenario would require species to have followed different migrational paths and for the forests of the past to be unlike those of the present.

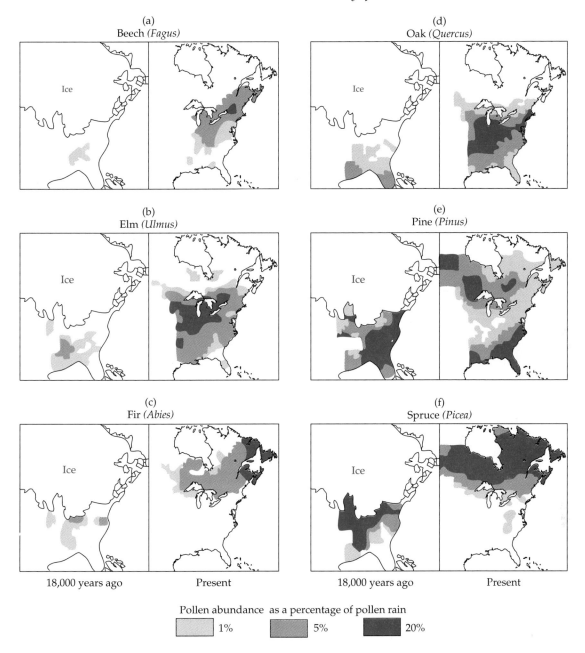

Figure 13.6   Changing distributions of tree species within the last 18,000 years. (a) The modern distribution of six forest tree genera that all occur in the northeastern United States. (b) The distribution of these six forest genera at the height of the ice age of 18,000 years ago, determined by the analysis of hundreds of pollen records. (After T. Webb III, 1988.)

Separating species within a genus is not always possible using fossil pollen, and so the following analysis considers the occurrence of genera. Figure 13.6 illustrates the modern and fossil abundance of pollen shed by six genera of trees. It is clear from these distributions that these forest genera occupied different ranges 18,000 years ago than they do today. The ice-age cooling forced the forest genera in the Northern Hemisphere to expand their ranges southward, and forests in the Southern Hemisphere were forced northward. In other words, the poleward populations were killed by cold, and most high-latitude species extended their ranges toward the equator. Hypothesis 1 can be rejected, because it is evident that the forests were affected by climate change. These six genera are important components of the modern forests of the northeastern United States, yet it is clear that they had separate strongholds 18,000 years ago.

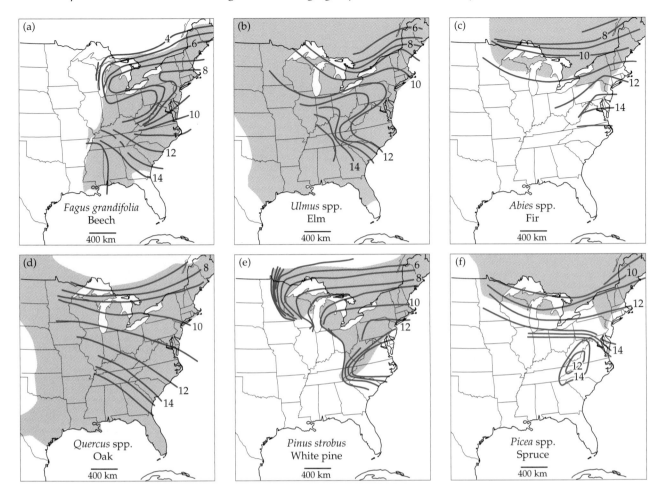

Figure 13.7   The migration routes taken by trees between 15,000 years ago and the present. The "contour" lines indicate the expanding frontier of this tree type at 1000-year intervals. The numbers refer to the age of the frontier in thousands of years before present. The shaded area represents the modern range of the tree type. Data were derived from the analysis of more than 500 pollen records. Note the different route taken by each type. (After M. Davis, 1983.)

Working independently, paleoecologists Margaret Davis of the University of Minnesota and Thompson Webb III of Brown University were the leaders in unraveling the next stage of this intellectual knot. Their analyses of a huge fossil pollen database that contained pollen sequences from hundreds of sites revealed the migratory patterns of the eastern deciduous forest genera. As the climate warmed, each genus followed a different migratory route to its present location. In Figure 13.7 the heavy lines indicate the northerly limit of the species at 1000-year intervals from 15,000 years ago until present. White pine started from the coastal lowlands and spread northwest. Oak populations had spent the glacial period in the southern states and spread rapidly northeastward early in the interglacial. Beech had also been restricted to the south, but its northward spread was much slower than that of oak. Spruce had a glacial-age distribution southeast of the Great Lakes in the

Appalachians. With the onset of warming, an upslope migration is evident in the Appalachians between 14,000 and 12,000 years ago, hence the smaller geographic range of the species as the climate warmed. With the additional warming between 12,000 and 10,000 years ago, spruce became extinct in the Appalachians. Because of the different rates and routes of migration, the past forests of the eastern United States would have had a different species composition than those of the present. Today's forests are a unique combination of species, different from those of any time in the past, and probably unlike those of the future. Hypothesis 2 is, therefore, rejected and hypothesis 3 accepted.

Modern temperate forest communities of the eastern United States, when viewed on a scale of hundreds of years, are fleeting, chance associations of species. Does this pattern hold true with other organisms and at other locations?

13.4 Plant Migrations in the Southwestern United States

In the western United States the presence of mountains prevents a simple analysis of south-north migrations. Not only did mountains affect local climate and hindered the migration of lowland species, but they also offered high habitat variability. Sheltered coves and valleys provided mild microclimates that allowed species to survive even though the regional climate was too harsh for them. Similarly, a windswept ridge or a cold-water spring that chilled a valley provided an unusually cool microclimate so that species normally associated with cold environments could survive a warm period. Although there was some migration north and south as conditions changed, most species were able to find suitable habitats by migrating up or down mountainsides. Because mountains become progressively cooler with increasing elevation, the change in habitats from low to high elevation was similar to those experienced going from lower to higher latitudes (Chapter 5). A migration upslope was the equivalent of migrating poleward, a migration downslope the equivalent of migrating equatorward.

In the dry conditions of the U.S. Southwest, few lakes provide long sedimentary records for pollen analysis. Instead, fossil pollen, leaf fragments, and seeds can be found in the burrows and caves inhabited by packrats (Figure 13.8). Packrats collect flowers, seeds, and leaves from the most abundant plants around their dens. Inside the den, discarded food items and feces accumulate. Often mounds of debris are glued together by the urine of the packrats, which dries into a hard, dark, shiny, solid. The debris mounds, termed middens, are a rich source of fossil pollen, seeds, and leaf fragments. Analysis of hundreds of packrat middens has provided an insight into the past ecosystems of the southwestern United States.

The middens of packrats living on the eastern side of the Grand Canyon provide a superb example of changes in local vegetation over the last 21,000 years. In Figure 13.9 the records from a number of sites are shown together. On the horizontal axis three time periods are represented: full glacial, early Holocene, and modern. Each of these panels is subdivided to represent a gradient of moisture. The driest sites are located to the left and the wettest sites to the right within each time period. The vertical axis represents increasing elevation, and the dots on the figure indicate the relative location of packrat middens. During the height of the last ice age, from 21,000 to 15,000

years ago, the base of the mountains (at a height of 1000 m) was a mixture of juniper, shadscale, and sagebrush growing as a scatter of low shrubs in a relatively open plain. Higher up the mountain, Douglas fir and pines grew, with spruce occurring at the highest elevations. As climates warmed, singleleaf ash invaded the lowlands, and many of the former inhabitants died as conditions became too warm for them. However, seeds of that species sprouted further upslope where cool conditions prevailed. Consequently, it appears that the species was marching up the mountain. Not all species could accommodate the warmth of the Holocene, and spruce that had been growing on the mountaintops went locally extinct due to the warming. The early Holocene period (11,000 to 8,500 years ago) was wet, and the vegetation was lush relative to both the ice age and modern conditions. As the Holocene progressed, the climate became drier and desert encroached. In drier locations, chaparral vegetation invaded the middle elevations, and woodlands were restricted to the highest elevations. In moister settings the woodlands were found lower on the mountain, but note that they are not the same in modern times as during the early Holocene. Notice that the species present have changed through time and so, too, has their relative abundance. Once more, the modern communities are different than those of the past.

13.5 A Mammal Community of the Past

Have mammal communities undergone parallel changes to those experienced in plants? The large mammal fauna of North America was devastated in the period between 20,000 years ago and 8,000 years ago. Therefore, we know that the large mammal communities of the past would have been different from those of today. A fairer test of mammalian community stability would be to look at small mammals that were not the subject of that extinction event. A good example of a small mammal community is provided by the fossilized remains of mammals that occupied a single cave in Pennsylvania 11,000 years ago. Four species of mammal were found to have shared the cave: Ungava lemming, smoky shrew, thirteen-lined ground squirrel, and heather vole. These species are never found living together in modern communities because they have widely differing ecological ranges. The Ungava lemming lives in tundra, the heather vole lives in boreal forest, the smoky shrew lives in eastern

Figure 13.8 **A packrat.** Packrats gather leaves, seeds, and flowers and carry them back to their den. Discarded plant parts pile up in the den as a midden heap. Scientists investigating the paleoecology of the southwestern United States have successfully recreated past environments based on the fossil record held in these middens.

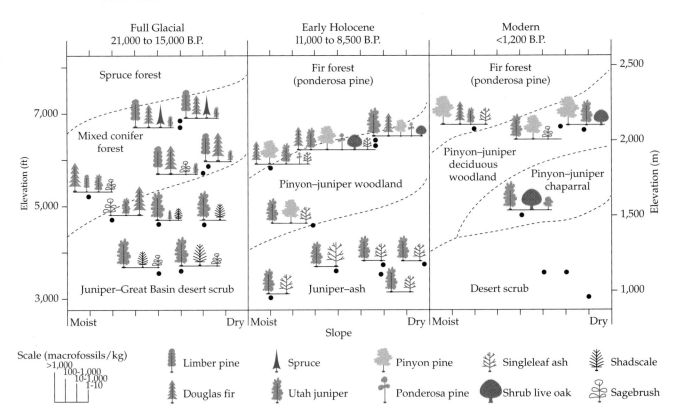

Figure 13.9 **Changes in plant communities of the eastern Grand Canyon during the full glacial, early Holocene, and modern times.** Data are based on pollen, seeds, and plant parts recovered from packrat middens. Note that species have migrated upslope with the warming of the Holocene and that the communities have been changing continually. Sizes of symbols (shown on the scale) are varied to illustrate the abundance of fossils in local packrat middens. (After K.L. Cole, 1990.)

deciduous forest, and the thirteen-lined ground squirrel lives on prairie (Figure 13.10). One inference based on their past coexistence is that these species could live in a habitat that mixed elements of these four biomes. The fossil pollen of this area indicates that the habitat around the cave 11,000 years ago was such a mixture, with herb-rich grassland dotted with clumps of birch, spruce, and pine trees. No modern vegetation type is an exact match, and apparently this novel habitat offered niches suitable for all four of these small mammals. As the climate warmed and the vegetation of eastern United States changed, three of these mammal species went locally extinct. Survivors either migrated to their present locations, or the modern distributions represent the remnants of much larger former ranges.

13.6 Instability in the Tropics

One of the explanations of the very high species diversity in a tropical forest is that they had experienced almost no climatic change during the last few million years. In the absence of such change, there was no climatically driven extinctions, and as a new species evolved it added richness to the flora and fauna. Without extinctions the biodiversity grew and grew. This argument is often repeated by the environmental move-

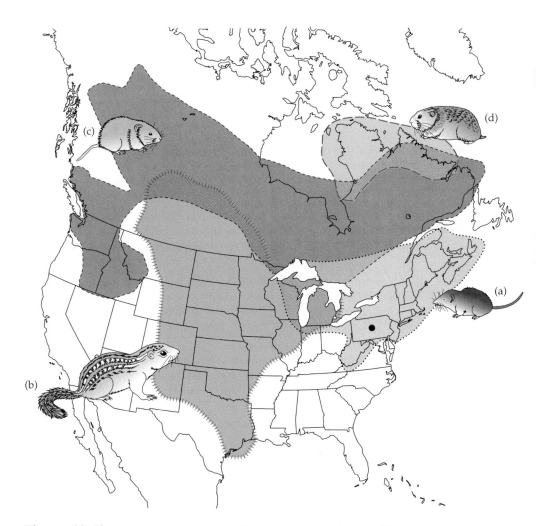

Figure 13.10 **Fossils reveal unique Pleistocene faunal assemblages.** In the Pleistocene, four small mammal species with widely differing modern ranges lived together. The figure shows modern distributions of (a) the smoky shrew, (b) the thirteen-lined ground squirrel, (c) the heather vole, and (d) the Ungava lemming. ● is the site in Pennsylvania where 11,000-year-old fossils of all four mammals were found (After E.C. Pielou, 1991. *After the Ice Age.* Chicago University Press.)

ment in referring to tropical forests as being "timeless," implying that their antiquity is a reason to save them.

Numerous researchers have demonstrated that the tropics have undergone huge climatic changes over the past million years. Just as the high latitudes were affected by ice ages, so were the tropics. Ice caps formed on high tropical mountains, and even the lowlands were subjected to considerable cooling (as much as 5°C to 9°C cooling in Africa and Amazonia) and, in some areas, drying. Fossil pollen analyses of lake sediments from Asia, Africa, and South America have shown that ice-age cooling was global and strongly affected the lowland tropical forests. These forests could not migrate to warmer latitudes because they were already at the equator, so individuals would have been greatly stressed by the changes in temperature and rainfall. Some species probably thrived, while others foundered, and it would certainly have been a time of great change in the population balance of tree species, their herbivores, and predators. The tropical records show mixtures of species that currently do not live together. Plants that are now associated with mountain habitats grew alongside species of the modern lowlands. The history of the reshuffling of these plants and animals into unfamiliar mixtures is a close parallel to the changing communities of North American forests over the last 20,000 years. Tropical environments have not been stable and appear to have experienced profound change similar to that of temperate regions. Consequently, the suggestion that tropical diversity depends on stability is firmly refuted.

13.7 So, Are Communities Stable Through Time?

We have seen four examples where communities have been subjected to profound change in their species combinations and abundance. In these settings communities are not stable. Certainly, we should not dismiss the possibility that *some* communities have a long history of coexistence and coevolution. For example, corals appear to form long-standing communities, but the majority of our most familiar communities are subject to change. Equally, we should recognize that even within communities subject to change, some species may always move together; for instance a butterfly and its foodplant may be evolutionarily, and hence ecologically, linked.

If plant and animal communities vary with each climatic change on a large scale, then it is reasonable to suppose that there are more subtle changes in community structure reflecting finer shifts in precipitation and temperature. Although climate drives these community changes, it takes time for plants and animals to respond fully. It may take hundreds of years for a stand of trees to die out and have others grow to maturity in a new location, as a result of subtle climate change. Thus the living world lags behind the climatic changes and is always catching up with the latest conditions. Many ecologists now see ecosystems as being in a permanent state of nonequilibrium (Chapter 14). Note that nonequilibrium in populations dove-tails with the view that populations are limited by density-independent rather than density-dependent factors (Chapter 9). Whether a nonequilibrium status is true of all ecosystems is a subject of active debate and research.

13.8 Another Note on Extinction: The Blitzkrieg Hypothesis

From the fossil record it appears likely that almost all North American tree species survived the climatic oscillations of the last 30,000 years. Despite being forced to migrate south from the ice advance and then being faced with the warmth of our present interglacial period, they survived. The same cannot be said of large mammals around the world. Huge bears, giant sloth, tigers, camel, and horse relatives all maintained ice-age populations, only to go extinct between 30,000 and 9,000 years ago. About half the genera of North American mammals that weigh between 5 and 1000 kg went extinct in this period.

A debate continues as to whether it was a series of rapid and unfavorable climate events that hit these creatures so hard or whether it was hunting pressure from humans. As a species, humans have an inglorious proven track record of decimating species that we encounter. Strong fossil evidence exists to suggest that the earliest colonists of the Americas caused the sudden demise of at least some of the largest mammals and birds (megafauna) in North America.

The "noble savage" as a term may fall short in terms of political correctness, but many of us would like to believe in the underlying romantic idyll: that humans can exist as innocents in harmony with nature. However, the more cynical view of life as "nasty, brutish, and short," particularly when applied to game animals that are not wise enough to run away from humans, is perhaps more accurate.

A wave of megafaunal extinctions appears to coincide with the first entry of humans into North America about 14,000 years ago. Anthropologist Paul Martin of the University of Arizona has summarized these events in his Blitzkrieg (lightning offense) hypothesis. Humans evolved in the Old World and were contained there until the end of the last ice age, when the first pioneers crossed from what is now Siberia. For a few thousand years, as glaciers retreated and sea level had not yet risen to its present level, it was possible to walk through a tundra environment from Siberia to Alaska. Martin suggests that the first humans to cross into the Americas about 14,000 years ago found a continent rich in large mammals, such as giant ground sloth, cave lion, horse, and mammoth (Figure 13.11).

These mammals, unfamiliar with predatory humans, apparently fell easy prey to the tool-using invaders. Human populations blossomed and fanned out, wiping out the megafauna in the space of a few millennia as the tide of hunters swept south and east. We can be fairly certain that it was these earliest human invaders that decimated the megafauna because the tools and spear points belonging to the Clovis culture of 11,000 years ago are highly distinctive and are unlikely to be confused with tool kits of later stone age peoples. Archaeological evidence links Clovis spear points to skeletal remains of some of the megafauna, indicating that, indeed, they overlapped with human arrival and were prey items. Another line of evidence is from the dung of the shasta ground sloth, a giant slow-moving herbivore that is known to have been hunted by humans. The ground sloth was a cave-dwelling creature, and its dung is preserved in the dusts that build up in caves. Because the dung is rich in the carbon of leaves eaten by the sloth, it can be dated accurately using radiocarbon ^{14}C. The radio-

carbon dates providing the most recent 35 ages for the ground-sloth dung illustrate that the population appears to have gone extinct at about 8000 years before present (Figure 13.12). This data set certainly seems to support the Blitzkrieg hypothesis in the case of the giant shasta ground sloth.

Another school of thought is that the climatic change from the ice age to the interglacial period was a time of tremendous stress for these populations and that they would have died out anyway. American anthropologist Donald Grayson has suggested that some of the megafauna had gone extinct thousands of years before humans entered the Americas. This explanation would appear to be most plausible for the camelids and woolly rhinoceros that roamed the unglaciated portion of the United States, because no well-dated remains are seen to overlap with the human presence. The argument over the extinction of megafauna became factioned into the "humans-only" and "climate-only" causal schools of thought.

An intermediate position might be that large-bodied, slow-reproducing mammals are vulnerable to sudden dips in their population number. A major climate change, and the vegetation changes that go with it, is likely to disrupt the breeding success of a number of generations. Hence the terminal millennia of the last ice age, which saw some very significant climatic wobbles, is likely to have had a severe effect on the populations of these large mammals. Some, such as the camelids and woolly rhinos, went extinct; others would have recovered slowly but for the intercession of human hunters. The hunters were the final straw breaking the mammoth's back (the camelid's had already snapped). Indeed, it seems that mammoth may

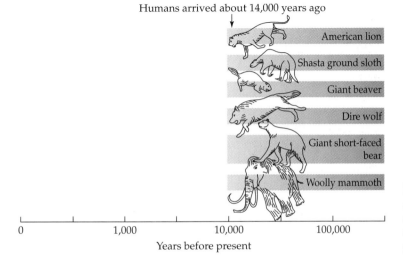

Humans arrived about 14,000 years ago

American lion

Shasta ground sloth

Giant beaver

Dire wolf

Giant short-faced bear

Woolly mammoth

0   1,000   10,000   100,000

Years before present

Figure 13.11 Some of the American megafauna that died out at the end of the last ice age or early in the post–ice-age period.

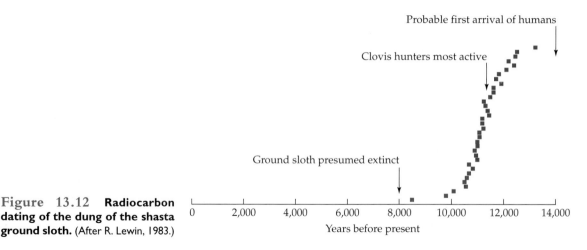

Figure 13.12 Radiocarbon dating of the dung of the shasta ground sloth. (After R. Lewin, 1983.)

have survived on islands in the Bering Straits until just 6000 years ago. These data suggest that these populations could endure the warmth of the interglacial but not the added pressure of human hunting. In other areas, human arrival and the disappearance of whole species complexes has been documented. For example, the killing of all the Moas, the large flightless birds of New Zealand, can be linked categorically to the arrival of humans.

The loss of the megafaunas because of climate change or hunting are spectacular losses to global biodiversity, but they are relatively unusual events. Habitat degradation and destruction by humans are the leading causes of extinction in the world today. The draining of wetlands, the felling of tropical rain forests, and the pollution of rivers and oceans are generating an unprecedented wave of extinction. These effects are compounded by the introduction of new predators and competitors that oust the local wildlife. The domestic cat is a major threat to the bird populations of the Galápagos, Hawaii, and New Zealand. An accidentally introduced snake has eliminated virtually all the native bird species of the island of Guam, and the Nile perch (a predatory fish that can grow in excess of 100 kg) is making huge inroads on the native cichlid fish species of African lakes. It would be comfortable to blame this latest wave of extinctions on extraterrestrial bodies or climate change, but we cannot. Humans alone are to blame.

13.9 Continuing Changes in Our Forests

Basic modern biogeographic patterns, such as the diversity of tree species across North America (Figure 13.13), reflect current climatic patterns. A gradient of decreasing species diversity from the southeastern United States toward the prairies of the Midwest is evident. This gradient does not reflect temperature as closely as it does the abundance of water in the soil. It seems that the greater the soil moisture, the higher the species number. As other life forms are so reliant on trees, it is reasonable to expect this pattern to be reflected in general biodiversity. Thus a change in the soil moisture balance could radically alter ecosystems.

Future climate change ensures that forest compositions will continue to change in both temperate and tropical settings. Through our manipulation of natural resources over the past hundreds of years, future tree species' migrations may be unlike those of the past. As an example, the loss of the Passenger Pigeon could have lasting consequences for the oak forests of the eastern United States. In the mid-1800s scientific observers recorded single flocks of Passenger Pigeons that were thousands of birds wide. The flocks blackened the sky above and took three days to pass by (Figure 13.14). These birds literally flocked in millions until the late 1800s. A flock would descend on an oak forest, strip it of acorns, and move on. The pigeons were probably the largest single transporter of oak seed from one part of the eastern forests to another. Hunters reported that in autumn the birds' crops were so full of acorns that they would explode on impact with the ground. Hunting pressure on the flocks was intense, and the last wild population of the pigeons was hunted to extinction in the late 1800s. Whether the genetic exchange facilitated by Passenger Pigeons carrying seed hither and thither was important to modern oak populations is uncertain. But given future climate change, the oak is now without its primary means of being moved long distances. How different would the post–ice-age migration of oaks (Figure 13.7) have been without the Passenger Pigeon?

Our forests are still changing. Pollution will selectively kill some species (such as birch and firs) while leaving others (such as plane trees, introduced for their tolerance to city pollution, and some maples) relatively untouched. Climatic change will lead to a reassortment of species best adapted to the new conditions. And manipulation of populations of predators (such as wolves, cougar, and coyote) will affect the populations of their prey, such as deer. Because deer are primarily grazing animals, increasing their populations will have a profound negative impact on the success of young seedlings. Deer will preferentially graze the tender leaves and stems of seedlings. This predation will greatly reduce the chance of successful tree regeneration. Thus, extinction of seed dispersers or population control events near the top of food chains can have unpredicted effects on the future composition of forests. Sometimes the effects are seen almost immediately. A strong drought may kill one species preferentially in a forest; a pollutant may do the same. Other changes are more subtle, because they affect the reproductive success of the trees. Such changes in seedling number, vitality, or survivorship may take several tree generations, each lasting 100 to 200 years,

to take full effect. Thus, some of the changes already set in motion in our forests have yet to be revealed.

Disease As a Factor in Forest Changes

For the last 30 years, the composition of European and American deciduous forests has been dramatically changed by disease. Two diseases—Dutch elm disease and chestnut blight—have virtually eliminated large elms and American chestnut trees from the North American landscape. The diseases are caused by fungi that infect and kill mature trees or young trees that are just old enough to reproduce. Very few trees survive infestation, so disease-tolerant strains among the native species of elms or chestnuts have yet to be identified. Consequently, there has been a precipitous decline in the populations of these trees in the forests of the eastern United States. These species were once major components of the forest canopy. As they have died, their replacement by other species has led to continued shifts in forest composition. Elm trees were particularly susceptible to infection because they often reproduce asexually. Runners sent out from the roots sprout and develop into an apparently free-standing

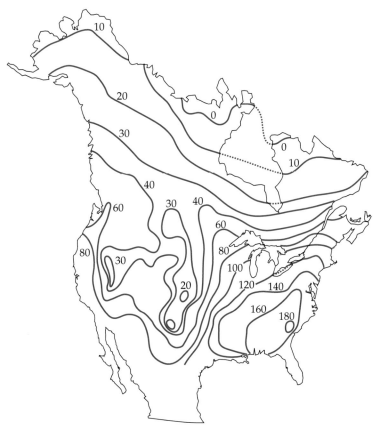

Figure 13.13   Number of tree species (defined as any woody plant over 3 m tall) found in different parts of North America. The contours indicate where particular numbers of tree species were documented within large-scale sampling areas (mean of 70,000 square kilometers). (After D.J. Currie and V. Paquin, 1987.)

SHOOTING WILD PIGEONS IN IOWA.

Figure 13.14   **Shooting Passenger Pigeons in Iowa.** Note the dark mass of pigeons yet to pass overhead and how the trees are filled with pigeons. (From *Leslie's Illustrated Newspaper*, 1867.)

tree that actually is attached to others through the root system. Thus, once one tree in the forest was infected, the fungus could easily infect other trees. The elm disease arose quite suddenly and swept through Europe, Canada, and the United States. It is likely that the disease will die down as the host trees are lost from the landscape (Chapter 14). This was not the first time elms have faced a precipitous population decline.

Fossil pollen evidence documents a sudden slump in elm populations about 5000 years ago in Europe. The cause of that elm decline is still discussed, but given the present evidence of the elm's vulnerability to disease, it is likely that a similar disease epidemic was responsible for their demise. Thus, long cycles of disease can be added to our list of factors that lead to the continual disruption of forests.

Summary

- The past is the key to the present and the future; geology is governed by basic physical processes.

- For the last 2.5 million years, ice ages have been the norm, and species are adapted to conditions 5°C–9°C cooler than present.

- The present interglacial period is an abnormally warm one that has driven species upslope or equatorward.

- The cycles of glacial and interglacial periods may be driven by changes in Earth's orbit relative to the sun.

- Major climate changes cause plant and animal communities to disaggregate as species respond independently to the new conditions.

- Plant and animal communities continue to change in response to changes in climate, predators, disease, and seed dispersers.

- Community composition is ephemeral, subject to change, and is the product of a mixture of predictable and chance events.

- Changes in climate and the arrival of humans in North America induced an extinction event among North American megafauna.

- Communities are nonequilibrial, responding to, but lagging behind, the latest climatic or ecological change.

Further Readings

Alley, R. B., and M. L. Bender. 1998. Greenland ice cores: Frozen in time. *Scientific American* 278 (2):80–85.

Davis, M. B. 1983. Quaternary history of deciduous forests of eastern North America. *Annals of the Missouri Botanical Gardens* 70:55–563.

Goodman, B. 1998. Where ice isn't nice. *BioScience* 48:586–590.

Pielou, E. C. 1991. *After the Ice Age*. Chicago: Chicago University Press.

Prentice, I. C., P. J. Bartlein, and T. Webb, III. 1991. Vegetation and climate change in eastern North America since the last glacial maximum. *Ecology* 72:2038–2056.

Pringle, H. 1998. New women of the ice age. *Discover* 19 (4):62–69.

Stahle, D. W., M. K. Cleaveland, D. B. Blanten, M. D. Therell, and D. A. Gray. 1998. The lost colony and Jamestown droughts. *Science* 280:564–567.

Web Connections

On-line resources for this chapter are on the World Wide Web at:
http://www.prenhall.com/bush
(From this web-site, select Chapter 13.)

Ecological Succession: Rebuilding Ecosystems

14.1 *Clements and the superorganism*

14.2 *Ashes to forest*

14.3 *Succession and ecosystem functions*

14.4 *From field to forest*

14.5 *Succession and coral reefs*

14.6 *Disturbance that maintains diversity*

14.7 *Succession and habitat management*

14.8 *The old-growth controversy*

14.9 *Equilibrium or nonequilibrium in our modern ecosystems*

If left alone, the fresh scar of bare earth left by a landslide will gather a mantle of green as grasses and herbs sprout. As the years pass, a procession of plants will file through the space, until after a few decades, only a practiced eye could discern that the vegetation had been disturbed. After a few centuries, all trace of the disturbance will have disappeared. This recovery process that leads to the revegetation of an area is an example of ecological **succession**, but it is much more than the re-establishment of herbs and trees. When an ecosystem is disturbed, whether by fire, wind, oil spill, or chainsaw, the balance between the biotic and abiotic processes is changed. The visible effect of disturbance may be dead trees, bare soil, or oiled birds, so it is natural to think first of the recovery of living things. However, the healing process has to include the restoration of the abiotic processes. The re-establishment of ecosystem functions, such as nutrient cycles, carbon storage, microclimate, and the hydrological cycle, may be essential precursors to the reinvasion of the flora and fauna.

14.1 Clements and the Superorganism

In 1916, Frederick Clements proposed the concept of succession leading to a climax vegetation, and proposed that the successional stages and the eventual climax were

a superorganism capable of shaping its surroundings. We now know this view to be mistaken, but his ideas have played such an important role in ecology and environmentalism that they are worth investigating.

After more than two decades of researching the regrowth of vegetation following disturbance, Clements proposed that each climate type would have a single **climax community**; an end point to the successional process. Because the climate determined the climax community, all successions within that climate zone would produce the same climax community. In other words, microclimate, soils, depth of water table, and other habitat variables would, in the end, be shaped by the vegetation to produce the climax system. For example, on the sand dunes around Lake Michigan, Clements believed (mistakenly) that successional sequences starting from a lake, or a sand dune, or an abandoned field would all lead inexorably to a beech-maple climax forest (Figure 14.1). Clements went a step further to liken the various stages of succession to the development of an organism and suggested that the climax community was an "organic entity."

When Clements first proposed his view of succession it was widely accepted, but a vocal critic of the notion of communities as superorganisms was H. Gleason of the New York Botanical Garden. In 1926, he prophetically stated that "... every species of plant is a law unto itself." Gleason argued that succession was not a

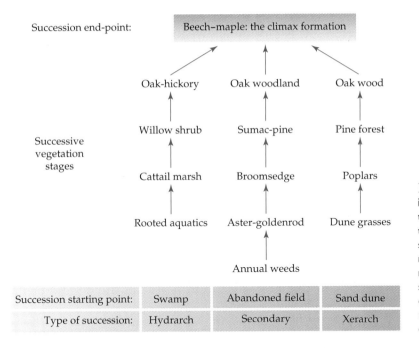

Succession end-point: Beech–maple: the climax formation

	Oak-hickory	Oak woodland	Oak wood
Successive vegetation stages	Willow shrub	Sumac-pine	Pine forest
	Cattail marsh	Broomsedge	Poplars
	Rooted aquatics	Aster-goldenrod	Dune grasses
		Annual weeds	

| Succession starting point: | Swamp | Abandoned field | Sand dune |
| Type of succession: | Hydrarch | Secondary | Xerarch |

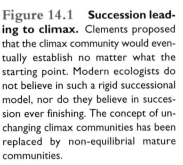

Figure 14.1   Succession leading to climax. Clements proposed that the climax community would eventually establish no matter what the starting point. Modern ecologists do not believe in such a rigid successional model, nor do they believe in succession ever finishing. The concept of unchanging climax communities has been replaced by non-equilibrial mature communities.

predetermined process of development, but something much more haphazard; it was dependent upon local growing conditions, the chance scatter of seed, and the vagaries of migration. The truth of this statement has been born out by later paleoecological studies (Chapter 13).

Modern ecologists dislike the term *climax* because it carries an underlying notion of stasis: What is going to happen to a climax community? Nothing, because it has reached its climax. Climax is a convenient, but wholly "now-centric" view of the world (our self-centered belief that *now* is normal). But our present is just another stage in the great procession of species; it just happens to be the stage that we have documented most fully. Future changes in climate, predation, or pathogen pressures will cause the composition of our plant and animal communities to alter again and again. Each time we undergo change, the changes in species composition will be roughly proportional to the degree of stress applied, as small climatic or pathogen pressures will be met by small responses in community composition and large stresses will cause entire rearrangements of community structure. Because change in community composition is normal and stasis is not, the more accurate term of **mature community** is preferred.

Despite the scientific rejection of the idea of ecosystems as a superorganism, this idea has proved attractive to some environmentalists; they refer to ecocide, or killing the environment. The environment is not a living entity, it cannot be killed. An extreme form of

this pseudoscience is presented in the Gaia hypothesis, which states that the whole Earth is a superorganism and that all living things interact to regulate and maintain an equilibrial environment. According to the Gaia hypothesis, plants emit oxygen and draw in CO_2 to maintain optimal growth conditions for themselves, and they regulate climate locally and globally. Plainly, this is nonsense. Cycles of climate change are sudden, severe, and erratic, and the atmospheric concentrations of gases vary greatly on both millennial and longer time scales (Chapters 13 and 23). Earth is not a superorganism, climate and atmosphere are not constant, natural cycles are not regulated by organisms, and all organisms survive or perish in the conditions that surround them; they do not *try* to alter them.

The modern view of succession recognizes two discrete processes, primary and secondary succession. **Primary succession** describes the colonization and ecological succession taking place on a new surface, such as a lava flow, an explosion crater, a sandbar in a river, the clay left by a retreating glacier, or a sunken ship. **Secondary succession** describes colonization and ecological succession following a disturbance. Secondary succession could take the form of the recovery of vegetation on a land surface where previously existing plants were disturbed, such as an old field, a landslide scar, or a burned forest. Or it could be the rebuilding of a coral reef after it has been hit by a ship. The essential difference is that primary succession occurs on an area that has never before supported life, whereas secondary succession is the reclamation of an area by organisms after a disturbance. The islands of Krakatau (formerly Krakatoa), Indonesia, provide the best example of primary succession resulting in a complex ecosystem.

14.2 Ashes to Forest

On August 27, 1883, the largest natural explosion ever recorded was heard from Australia to islands off the coast of South Africa. The island volcano of Krakatau, Indonesia had erupted. A tsunami (tidal wave) generated by the explosion damaged rain forests along the adjacent coasts of Sumatra and Java, claiming the lives of 36,000 villagers (Figures 14.2 and 14.3). Scientists visiting the Krakatau island group within a year of the eruption found that more than 20 km^3 kilometers of rock had been blown away and that the old land surface had been buried by hot ash as much as 100 m deep. A new volcanic landscape had been created. No plant or animal, seed or root, had survived the eruption; the islands had been sterilized. Scientifically, this is very important because the islands offer a definitive time for the start of the succession. Also, that the islands were

Figure 14.2 An imaginative depiction of the seas off Java after the eruption of Krakatau in 1883.

virtually free from human disturbance before and after the explosion allowed the regeneration process to proceed naturally.

Scientists visited Krakatau regularly between 1883 and 1934, observing the gradual return of a tropical flora and fauna. Within 13 months of the eruption, grasses and blue-green algae were present, and, by the early 1900s, a grassland with scattered trees and shrubs was gradually invaded by a light woodland. Late in the 1920s, the woodland became dense enough to form a closed canopy and exclude the grasslands on Sertung, Panjang, and Rakata (Figure 14.4). A new factor then entered the successional history of Krakatau as the islands experienced the first volcanic activity in more than 50 years. Between 1928 and 1932, Anak Krakatau (literally translated, this means "child of Krakatau"), a new active volcano, emerged from the sea in the middle of the volcanic caldera.

Since the 1930s, Anak's frequent volcanic activity has sent dust and poisonous gases onto the islands of Sertung and Panjang. The soil profiles of these islands are sometimes over 2 m deep, most of which is volcanic ash that has fallen onto the islands since 1930. The forest of Sertung and Panjang has been repeatedly damaged, and what started as a primary succession is now so disturbed as to be a secondary succession on these islands. Only the island of Rakata, where no ash from Anak is found in the soil profiles, has been unaffected by the volcanic activity, and a thin soil just 12 cm deep has formed. The soil is black with organic carbon, though still gritty from the parent ash material.

The centenary of the 1883 eruption kindled researchers' interest in the island group. Since 1979, numerous research expeditions have visited the islands to document the returning plants and animals. The tallest trees on Krakatau are now more than 40 m tall, and some of them have a trunk diameter of more than 1 m (Figure 14.5). The forest looks dense and lush, but the succession on Krakatau is far from finished. After more than 100 years, fewer than 300 species of plants have returned to the islands (Figure 14.6, p. 213), whereas a similar area of the mainland might support between 1200 and 1400 species of plant. Because the forest canopy is now closed, it is harder for a new species to compete for resources and grow. Consequently, the rate of forest change will slow down. From the work on these islands, it is estimated that it may take more than 1000 years for Krakatau to become a truly mature forest.

An important aspect of succession is that small differences in habitat type will lead to variations in the

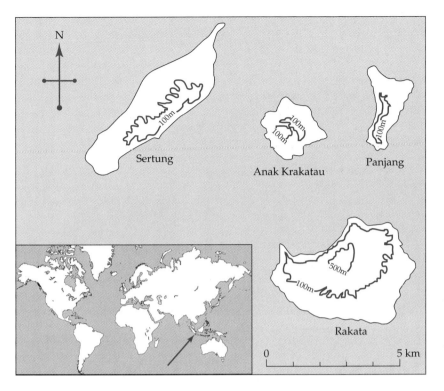

Figure 14.3 **A sketch map of the modern Krakatau island group showing the outline of the islands before the 1883 eruption.** Anak Krakatau is the new active volcano that emerged from the seas in 1930.

species composition of all stages of the succession. A coastal succession will have a progression of salt-tolerant plants, because salts are carried in the fine sea spray and may affect vegetation some hundreds of meters inland. A montane succession will have plants more tolerant of steep slopes, cooler temperatures, and increased ultraviolet radiation. A slight variation in soil moisture from one location to the next will enable one species to outcompete another. Thus, succession has a predictable sequence that depends on its stage of development and the growing conditions. However, there is also a random element. For example, if three species are all equally well suited to become the next major player in the successional game, which one actually moves in may be purely a matter of chance, especially in tropical settings, where there are so many more species to become involved in a successional sequence. Although the type of plant in the next stage can be predicted, often the actual species cannot.

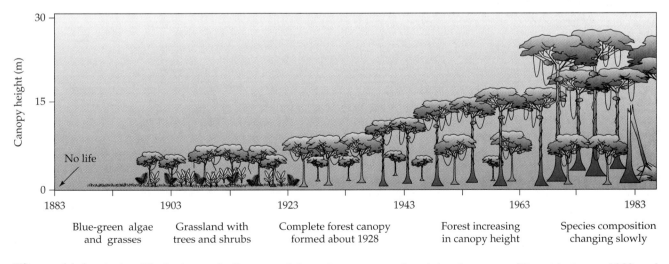

Figure 14.4 **A simplified schematic diagram of the primary successional development of forest between 1883 and 1995 on the Krakatau islands, Indonesia.**

Figure 14.5 (a) The shoreline of Rakata in 1989. Note the complete forest cover and the ash cliffs where the sea has eroded the shoreline. (b) The interior forest of Rakata contains large trees such as this fig (*Ficus pubinervis*). The scientist who serves as a scale object is over 6 ft tall. He is beating the vegetation to cause insects to drop from the leaves into the white tray. This is one way to document which insects have colonized the islands.

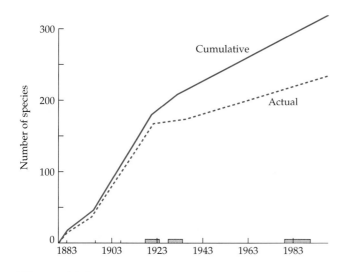

Figure 14.6   The colonization of the Krakatau islands by plant species between 1883 and 1989. (After M.B. Bush and R.J. Whittaker, 1991.)

14.3   Succession and Ecosystem Functions

We will concentrate on secondary succession here because it occurs more frequently than primary succession. The scale and nature of disturbance profoundly affect the degree to which ecosystem functions are damaged. In general, the larger the disturbance the more profound will be the dislocation of the processes. Examples of disturbances that would affect ecosystem function and, ultimately, succession are flooding, erosion, and climatic changes. Each would play a strong role in determining ecosystem function and, in turn, which species could live at each stage in the succession. Thus, we can think of disturbance as providing a shock to ecosystem functions. The recovery of ecosystem functions, such as the cycling of nutrients, the storage of carbon, or the provision of breeding grounds for organisms, affects and is affected by succession in a continual cycle of positive feedback. A positive feedback loop is a self-reinforcing series of events. As the loop is repeated, the consequences become more marked.

An example of how succession might affect an important ecosystem function is the effect of vegetation on water retention. In most intact forests, surface-water runoff is relatively rare, because the inflow of rainwater is slowed by the vegetation cover and absorbed by the soil. So long as plants are photosynthetically active, their roots take water up out of the soil and into the upper portion of the tree, where it is

transpired from the leaf surface to the atmosphere. When there are no leaves on the trees, there is no photosynthesis and no transpiration; thus it is during the growing season that plants lower the water table. In an experiment at Hubbard Brook, New Hampshire, in which trees were removed, the surface-water runoff increased 30%, and dissolved nutrients were washed from the land (see Chapter 5). Hubbard Brook was a **clear-cut**, that is an area from which all the trees have been removed, but increased runoff can also result when the natural vegetation is replaced by plantation trees.

In Malaysia, the replacement of rain forest with palm plantations resulted in a doubling of the peak discharge of water (the flow most likely to cause erosion and floods). In these plantations, the trees are planted in rows and spaced so that the tree canopies will not touch. The ground is weeded and kept bare to ensure rapid growth of the palms. The effect is that the sparse vegetation cover provided by the palms is insufficient either to absorb all the precipitation or to help prevent rapid runoff after the rain hits the ground.

Another factor that will be affected by succession is a change in the temperature and quality of the soil. Plants and animals have optimal living conditions that form part of their niche and so, too, do decomposers such as fungi and bacteria. These organisms have evolved to live within the normal temperature range experienced in a forest. The problem of changing the soil temperature with habitat disturbance is most acute in moist tropical ecosystems, where the land is shaded by giant trees. Within an intact forest canopy, the soil temperature scarcely varies between night and day and is an almost constant 24°–25°C (Figure 14.7). Where a small gap in the tree canopy is opened, a more pronounced midafternoon heating is apparent, with temperatures approaching 30°C. Such variation is natural within these systems, because forest canopies are continually developing small holes when individual trees collapse. However, the temperatures recorded in a clear-cut, effectively an unnaturally large gap, are almost twice that of the closed forest. The extreme habitat presented by the soils of the clear-cut will be too hot for the survival of many of the decomposers, and without them the nutrient cycling is slowed.

Nutrients, the chemicals essential for plant growth, often exist within the soil in soluble and insoluble forms. If an area has been deforested, it has lost the trees that wick moisture out of the soil. One

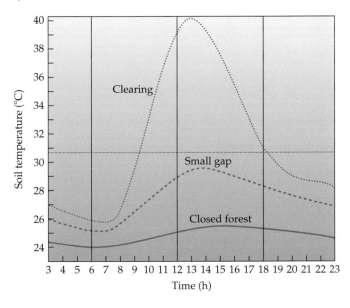

Figure 14.7 Changes in soil temperature measured over a 24-hour period in a tropical setting. Small canopy gaps are natural features of these systems, and soil organisms would have evolved to accommodate changes caused by an occasional gap. The greatly elevated temperatures of the clear-cut would be beyond the tolerance range of most soil organisms. (After J.P. Schulz, 1960.)

consequence is that the water table can remain at, or near, the surface all year. In such a saturated soil, the air spaces are filled with water, the biological oxygen demand (BOD) is still present from microbes, and it is common to find that oxygen concentrations fall to zero. The soil is then said to be **anaerobic**. The conversion of an oxygen-rich soil to an oxygen-poor soil may cause chemical changes (reduction) to nutrient molecules in the soil. For example, phosphates are released into the soil water as their bonds to iron molecules are broken, while nitrates are converted to gaseous nitrogen or other nitrogen compounds that cannot be absorbed by plants. If the soil water drains out of the soil or is washed out in a slurry as the surface soil erodes in a mudslide, the nutrients are lost from the system. The movement of soil nutrients either downward through the soil column or laterally out of the system is termed **leaching**.

On sites that have steep slopes and where the loss of forest does not result in saturated soils, there can still be a substantial loss of nutrients and soil erosion. The loss of topsoil may be slower than the loss of dissolved nutrients. One explanation is that there is a delay between the cutting of the trees and the washing away of the soil. Initially, the roots would hold the soil in place and so prevent erosion. After several years, the root mat of the dead trees breaks up; as the roots rot, the soil is held less securely and starts to wash downslope. The steeper the land, the more rapid is the loss of soils. Erosion can be sudden and severe, especially when forests on steep slopes are cut, the tree roots pulled out, and the land plowed for agriculture.

Erosion rates can also be affected by local wind speeds. Early successional and deforested landscapes of bare ground often have high wind speeds close to the ground. A moving stream of air will increase evaporation rates and thus speed the drying of the soil. Once the soil is dry, it can just blow away as dust.

Wind speed close to the ground will often decrease as succession progresses, having an effect on both the soils and the structure of the forest. Wind is also important in the heat budgets of high-elevation or high-latitude ecosystems. Here an increase in wind speed may prevent bud development and hence hinder the re-establishment of plants. A further influence that wind can exert on a system is through blowdowns. A forest draws much of its physical strength from having a relatively complete canopy. The protective crown and closely spaced trunks create a layer of still air within the forest. The trees growing in a dense forest effectively hold each other up. The tightly packed individuals prevent neighboring trees from swaying too violently and snapping their trunks. However, individual trees do not have trunks and root systems that can withstand strong winds. If an opening is made in the canopy, wind vortices can be created in the gap. These high winds knock down trees at the edge of the clearing, turning a small canopy gap into a large one.

It is evident that a disturbance to an ecosystem can result in severe soil damage, changes in soil-moisture conditions, and nutrient availability. The repair of these abiotic systems comes about through the biotic processes of succession. Because the functions of the ecosystem have been altered, the niches available to the species that will colonize the landscape are different from the ones that existed just before the disturbance. Consequently, the species that return will not be the ones that were lost, at least not at first.

14.4 From Field to Forest

One of the classic studies of succession was done on the old-field systems of North Carolina by Dwight Billings in the 1930s. By looking at different areas of land that had been abandoned at various times in the preceding 150 years, he was able to piece together a detailed successional sequence for these fields in the Carolinian Piedmont. The land surface had carried vegetation before, so this was an example of secondary succession.

The soils of this area are sandy clays with very little organic content. The abandoned fields had been used for growing tobacco and cotton, crops that drain nutrients from the soil. Consequently, the starting condition for the succession was bare ground composed of nutrient-poor dry soils that baked hard in the hot, dry summers. Plants colonizing these areas to start the succession would have to prosper in poor soils and be resistant to drought. The sequence described by Billings is typical, so we can generalize from it.

The first plants to occupy the bare ground (called **pioneer** species) are characterized by short life spans, rapid reproduction, and copious seed production. The first pioneer species to invade and build substantial populations on ground like that in Billings's area live for just one year and hence are termed **annuals**. These annuals, for example crabgrass, are followed swiftly by ragweed, horseweed, aster, and broomsedge (Figure 14.8). By late summer, the combination of these species is beginning to cover the ground surface. The pioneer annuals (crabgrass, ragweed, and horseweed) collapse and die with the first frost, and their bodies, leaves, stems, and roots, contribute the first major input of organic matter to the soil. Aster and broomsedge are **biennials**; they live for two years, reproducing in the second summer. The seeds produced by the annuals sprout in the second spring of the succession, but they are now competing with the larger biennial plants. By the end of the second summer, aster and broomsedge are beginning to displace the annual crabgrass, but once the biennials have reproduced, they also die.

Progressively, the land is colonized by herbs and shrubs that are **perennial**: plants that can reproduce multiple times and live many years. With each year, the amount of organic matter falling to the soil surface increases. Worms and other subterranean animals drag the rotting material down into the soil. The soil gains an organic component, and the depth of the organic horizon increases because of biological activity (Figure 14.9). These curves represent the changes taking place at a single site. The bumpiness of the curves is a result of chance variations that befell this particular site. If a large number of sites were studied, the average trend would result in smooth curves for soil depth and species number.

As the succession continues and the landscape fills with plants, competition will be increasingly expressed in bigger and deeper root systems to draw on

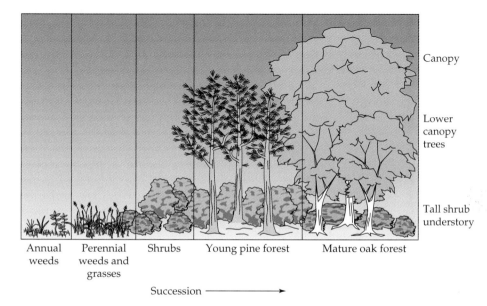

Annual weeds | Perennial weeds and grasses | Shrubs | Young pine forest | Mature oak forest

Succession ⟶

Canopy

Lower canopy trees

Tall shrub understory

Figure 14.8 Secondary succession on an abandoned field in the Carolinian Piedmont region. The pioneers are replaced by perennials, shrubs, softwood trees, and then hardwood trees in a series of predictable stages that lasts about 120 years. Even then the succession has not ended. There will continue to be changes in the composition of hardwood species for at least the next 200 years, probably longer. (After W.D. Billings, 1938.)

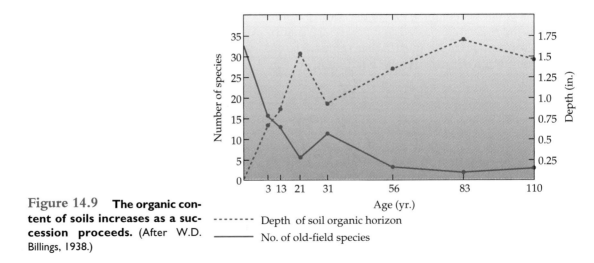

Figure 14.9 **The organic content of soils increases as a succession proceeds.** (After W.D. Billings, 1938.)

-------- Depth of soil organic horizon

————— No. of old-field species

more nutrients and water, larger stems, and more densely packed leaves that catch more sunlight and cast more shade. A big body thus has a twofold benefit. Increased leaf surface area allows the plant to produce more photosynthetic product, and, by casting shade and draining the soil of water and nutrients, a big plant can prevent would-be competitors from growing close to it.

Tree seedlings with tough leathery leaves or needles can survive in these shadeless conditions. After 5 to 15 years, loblolly pine, shortleaf pine, and sweetgum trees invade the land. The pines and rapidly growing broad-leaved trees are termed softwoods. The hardness of wood is largely determined by its density or the proportion of carbon to water and air inside the trunk. These softwood trees build big bodies but trade strength for fast growth. Instead of building densely packed cells with a high proportion of strengthening compounds, such as lignin, these trees rely more heavily on the incompressibility of water to give them strength. Consequently, the timber of softwood trees has less carbon and more water and air than slower-growing hardwoods.

The softwood trees rapidly form a complete woodland cover, shading out most of the herb (nonwoody) species. Beneath the branches of these trees, there is a new **microclimate** (as used here, defined as the climate beneath or within vegetation). The forest cover casts shade, reduces temperatures and wind speed, restores much of the hydrological budget, and increases humidity beneath the canopy. These conditions favor the growth of seedlings of late-successional hardwood trees that require relatively rich soils and some shade, particularly when very small. These seedlings need the moist, shaded understory of a forest to protect them from the burning sun. Seedling

oaks and beech, common hardwood trees of mature temperate forests, have large floppy leaves relative to their roots. Having large solar collectors (leaves) is an ideal adaptation for life on the shaded forest floor, where light, rather than moisture, may be the limiting factor. However, if exposed to full sunlight, these seedlings cannot regulate their temperature and will develop a water deficit. They lose water through their leaves and cannot absorb enough through their roots to replace it. The consequence of the water deficit is that they wilt and die. Thus the seedlings cannot become established until the earlier successional stages have occurred.

The fast-growing pines and sweetgum form dense stands of tall, thin, straight-trunked trees. Short-lived, these trees start to senesce (to grow old) after about 50 years. As these trees age, the hardwoods that have been growing in the understory burst through into the gaps created by falling softwoods. About 50 years into the succession, the hardwoods become important canopy components as the pines and sweetgums die out.

A combination of hardwoods that represents the most mature forest on the modern Carolinian Piedmont is generally reached after about 150 years. More-subtle changes then take place within the forest. As the community proceeds toward maturity, the proportions of the various tree species within the woodland community change. Past timber extraction denies us the opportunity to study an undisturbed old-growth forest in Eastern forests, but Dwight Billings estimated that the succession from old field to fully mature forest would take at least 350 years.

It is evident that succession is much more than merely changing plant species, it is also a process of soil improvement and microclimate modification.

Shortcuts in succession are unusual; a succession must run its full, predictable course.

It is important to understand that almost all of the landscapes that surround us are successional. The pines that flank the highways of the eastern United States, the oak forests of the Midwest, even the chaparral vegetation of California, are on a successional trajectory of change. For example, the oak forests that stretch from Wisconsin to New England are the product of 100 to 300 years of succession, the time that has elapsed since the earliest settlers logged the land. From historical records, it is apparent that oaks were a minor component of those pre-European forests. In fact, oaks are generally transitional species that bridge the gap from softwoods to late-successional hardwoods. Not surprisingly, therefore, the forests of New England are changing. The great red oaks that have dominated the forest for the last 100 years are failing to reproduce, and the next wave of hardwoods is growing underneath. The species that will replace the oaks are sugar maple, red maple, hackberry, cherry, or sweet birch, according to the part of the country.

Secondary Succession and Seed Banks

The nature, extent, and duration of a disturbance will play a large part in determining the recovery time and pathway of the successional sequence. Recovery from small-scale events will take relatively little time. For instance, a section of forest that was flattened where a tornado touched down will recover quickly. Root bases of snapped trees may resprout almost immediately, and patches of bare ground will quickly green up as seeds dormant in the soil germinate. These seeds, produced during earlier fruiting seasons and buried in the soil, form the **seed bank**. The species composition of the seed bank is likely to include all the plants that grew at the site immediately before the disturbance plus others that have been carried onto the site from farther away. For instance, under a closed forest canopy there may be no dandelions, fireweed, or grasses growing because they cannot tolerate the shade. Although the adult plants cannot grow in the forest, their seeds are blown on the wind and will disperse over a large area. Alternatively, if the forest site is already recovering from a disturbance, the seeds of pioneer plants from years ago may be lying dormant in the seed bank. These seeds can remain dormant for a number of years and will sprout only when the forest is disturbed again. Therefore, although the parent plants cannot grow at the site, the seed bank may well contain their seed. Following a disturbance that

breaches the forest canopy, these sun-demanding species will germinate and grow until they are shaded out. Such pioneer species are often termed opportunists (Chapter 10) because they have sudden population booms that are dependent on the chance formation of small disturbance sites.

The Survival of Opportunists

The progression from pioneer to competitor is commonly one from shade-intolerant to shade-tolerant species. The herbs are shaded out by birch and pine, trees that grow quickly but cannot tolerate shade. Characteristically, their energy investment in reproduction will be high, although the investment in each individual seed will be small. Each year they will produce a huge crop of small seeds. In contrast, the slower-growing shade-tolerant species of the mature forest, such as beech or chestnut, produce relatively few, larger seeds. The seedlings of shade-adapted trees have evolved large seeds that are energy-expensive to produce, but provide a nutrient packet for the emergent seedling. The relationship between seed size and shade tolerance was demonstrated in a study of seedling survival (Figure 14.10). The seedlings of small-seeded tree species cannot compete for light or water, hence they will do badly in a shaded environment.

Although these small-seeded trees are generally preceded by annual and perennial herbs, they still share many of the same living strategies. For example, in comparison with a large-seeded tree, such as

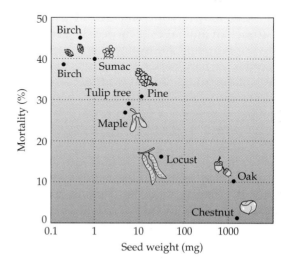

Figure 14.10   Results of a three-month study of the survival success of temperate-forest tree seedlings under shaded conditions. The smallest seeds had the highest rates of mortality. (After J.P. Grime and D.W. Jeffrey, 1965.)

an oak or a beech, the small-seeded trees are relatively short-lived and have a high energy expenditure on reproduction. A continuum of life strategies is evident from the pioneer herbs, through the early-successional (small-seeded) trees to the late-successional (large-seeded) species.

If a whole forest is considered, the persistence of opportunists is a significant contribution to the overall species diversity. All natural systems have some degree of disturbance. Trees die and fall; rivers erode their channels; storms cause blowdowns. Small populations of opportunist species continue to exist within the forest matrix of competitors. The opportunist populations are always low, always on the move as they colonize a new treefall gap, erosion site, or landslide, but they are readily outcompeted. Not all opportunist species will survive in all forest patches. In any one given area, just a few of the many opportunist species will be able to colonize and reproduce before being shaded out. However, if all the gaps, streambanks, and disturbance areas in a forest are considered, an almost full complement of opportunist species will be found.

Another difference between the early and late successional species is in the way that the trees evolve to spread their seeds. The tiny seed of early-successional trees often has a diaphanous wing, as in the case of pines or the flat blades of a birch seed. These forms are adaptations to seed dispersal by wind. The construction of a tiny flimsy wing is a relatively cheap energy investment in each seed, but it allows the seed to be blown clear of the parent. Late-successional trees have fewer seeds, but they invest more energy in each one. The large seed may be supplemented by a tempting bait to attract an animal that carries the seed away from the parent plant. An example is a cherry. The cherry seed is the hard part in the middle; the sugary, juicy flesh of the fruit is the bait to attract a disperser. It is important to recognize the difference between the flesh of a berry, which is a sugary reward for the animal, and the seed(s) contained in the fruit. The animal may collect the fruit, gnaw off the flesh, and spit out the seeds. Alternatively, it may eat the berry whole, digest the flesh, and excrete the seed. The seeds of some plants can germinate only after they have passed through an animal's digestive tract. It may seem a strange tactic for a species to evolve to rely on being eaten. Some animals digest and destroy the seeds, in which case they do not help the plant at all. Other animals—such as squirrels—may be primarily seed-eaters, but, because they drop or deliberately bury so many seeds, they can also be regarded as seed dispersers.

Nuts are particularly nutritious seeds that are collected by squirrels, monkeys, and birds. In each crop of nuts, many will be eaten, but some will have been carried to a new location by the animal and will survive to germinate. Blue Jays and squirrels both stash acorns and beech seeds by burying them. Blue Jays do not appear to return to these hoards, and squirrels rarely seem to remember where they have hidden theirs. The evolutionary reason for these creatures' stash-and-forget behavior is unknown, but two hypotheses have been raised. First, this behavior could be vestigial; that is, at one time they stashed and did remember to go back to the hoard. But the additional food supply must not have been a significant survival factor. If it had been, there would have been strong selection to remember where the nuts were buried. This hypothesis concludes that because the behavior pattern is unimportant, it is being lost by degrees. An alternative hypothesis is that although these organisms receive no reward from burying the acorns, they increase the seedling success rate of oak trees. The future descendants of these squirrels may reap the harvest of the acorns so diligently sown. Such arguments for altruism or foresight among animals are hard to substantiate because there would be just as much benefit to the lineage that did not waste time burying acorns, but kept on eating. This "selfish" line could rely on altruistic neighbors to bury the acorns and provide future harvests, while they worked to increase their individual fitness. Ultimately, the selfish line would "win" and would be the only surviving lineage.

Succession and Animals

Animals play an important part in successional systems as herbivores, seedeaters, and seed dispersers. However, animals are not generally classified according to a successional stage, although most species may be associated with particular successional stages. Insects such as butterflies that are dependent upon a particular food plant may display a strong association with successional stages. Different bird species are also characteristic of the various stages in a succession, although this association is often as much related to the physical structure of the forest (the size and spacing of trees) as it is to the presence of a particular species of plant. An example of the changes in bird species that accompanies the succession of fields to forest in the eastern United States is shown in Figure 14.11.

An example of a bird that can only survive in old growth forest is the Northern Spotted Owl of the

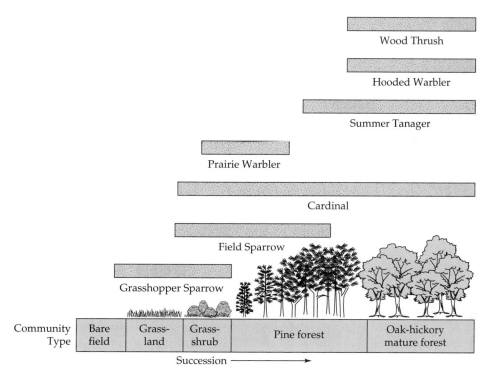

Figure 14.11 **The association of birds with successional vegetation types in the Piedmont region of Georgia.** (After D.W. Johnston and E.P. Odum, 1956.)

Pacific Northwest. This owl needs a particular structure of forest rather than a specific food item. Its niche defines it as a hunter working within a habitat of large trees, some of which are dying. It is a hole-nesting species and must have access to large standing dead or dying trees to provide the nest cavities. It cannot feed and reproduce successfully in other habitats, even though its prey may be abundant.

The Florida Scrub Jay has explicit habitat needs and relies on fire to modify succession. Fire can repeatedly disturb an area and prevent a succession from progressing, periodically kicking the succession back to an earlier stage. The scrub jay needs this frequent disturbance to maintain open areas that would otherwise become a dense tangle of palms and shrubs. Scrub jays are ground-feeding birds that need about 30% of the ground area to be sparsely vegetated. Yet they nest in slow-growing scrub oaks, trees that reach about 8 m in height. Hence, the jays need elements of both an early successional stage (the bare ground) and a late successional stage (the presence of hardwood species). Because the jays are strongly territorial and need to be able to forage close to the nest site, the apparently paradoxical needs of both early and late successional elements must be met at one site. Frequent low-intensity fires that burn the understory, but do not kill the trees, create the ideal habitat for the jays. In Florida, this is a vegetation type that once covered large portions of the state, but has been lost due to fire suppression. Less frequent fires have allowed the succession to proceed unchecked. The bare ground has been invaded by shrubs denying the birds their foraging areas. When fires do occur they are hotter and more destructive, killing oaks as well as other species (Chapter 6). Consequently, the scrub jays are extinct throughout much of their former range and are presently listed as a threatened species.

14.5 Succession and Coral Reefs

A succession that is dependent upon a sequence of animals rather than plants is the process of coral reef formation. Corals are small soft-bodied animals that secrete a hard shell of calcium carbonate. Vast colonies of these creatures, each with its own protective coat, form a coral block. The alga *Halimeda* is a green plant that also builds a body of calcium carbonate, which remains intact when the plant dies. Some reefs have a core of *Halimeda* bodies and an outer coat of coral.

The succession that leads to the recovery of a coral reef is more haphazard than in terrestrial systems because the larvae of the corals are carried by ocean currents. It becomes a matter of chance which species of coral establishes itself first. For coral reef to develop, the water must be clean, without suspended sediment, and must be above 17°C, with a suitable attachment surface. Once the corals are established, other species such as sea anemones, crabs, and sponges will find niches within the structure of the coral mass. Animals that prey on coral, such as parrot fish and the crown-of-thorns starfish, will also colonize the reef. The colonization and development of reefs is relatively slow, so they cannot withstand sustained damage from wave action or human use.

Coral reefs are popular with fishermen and divers. Boats anchoring on coral cause immense damage to the living reef, and some fishing techniques, such as the use of explosives and poisons to stun fish for aquarium collection, have led to a degradation of reefs. Scuba divers and visitors who treasure the reef may also damage these systems, either by collecting coral, touching coral, or disturbing the wildlife of the reef. Where tourism exerts a sustained pressure on the reef, some initiatives have been made to create artificial reefs. Metal frames or deliberately sunken ships provide a platform for coral attachment. Such attempts to create a reef must be well planned. The engine oil and fuel must be thoroughly cleaned from the hulk and the wreck must be positioned so that it is not a hazard to shipping and will not adversely affect local ecosystems. The wreck provides shelter for a wide array of fish species, and within a few years some coral growth will be seen. Divers and fishermen are then encouraged to frequent these newly created reef habitats, thus reducing the pressure of human disturbance on the older naturally formed reef.

14.6   Disturbance That Maintains Diversity

The disturbance inflicted on a reef by divers and anchors is damaging because it breaks large, slow growing structures. The frequency and scale of disturbance can vary both in magnitude and in frequency. Frequent massive disturbances such as volcanic explosions will lead to low local species diversity; however, very rare trivial disturbances do not lead to high diversity. Consider the experiment run by Paine on the

effects of starfish predation (Chapter 12). When the predatory starfish (a disturbing force) were removed, the diversity of prey species fell. When released from predation, the most competitive of the prey species excluded other species. In exactly the same way, a plant community that sustained no disturbance would go through a successional process in which a few dominant species prevail. In 1978, Joseph Connell of the University of California at Santa Barbara proposed the *intermediate disturbance hypothesis*. This hypothesis states that an ecosystem maintains its highest species diversity under conditions of moderate disturbance (Figure 14.12). Connell argues that all ecosystems are continually perturbed by disturbance events, such as fire, disease, volcanic eruptions, trampling by herds of animals, or climatic change. Even the death of an old tree is a form of local disturbance. Individual trees collapsing and dying provide an opening, or a gap, that can be filled by pioneer species. The gaps provide a continuous supply of sites to be invaded by pioneers. If it were not for continuing disturbance, poor competitors, such as pioneer species, would go locally extinct. The continuous change in the forest and the diversity of ages of patches within the forest, reduce the effects of interspecific competition and maintain the niche of early successional species. If interspecific competition does not develop fully, a superior competitor will not have the opportunity to outcompete and eliminate a weaker competitor. As the plants in the gap mature, they pass through a miniature successional sequence. This view of the forest emphasizes that the forest is like a patchwork quilt in which each patch represents a different stage of forest regrowth.

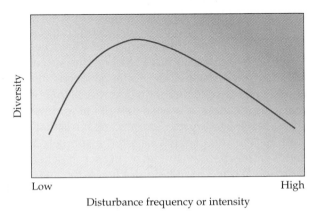

Figure 14.12   The intermediate disturbance hypothesis. The highest species diversity is found under conditions of moderate disturbance. With too much disturbance the habitat is hostile to colonization, with too little disturbance competitive exclusion reduces diversity. (After J.H. Connell, 1978.)

The intermediate disturbance hypothesis provides a good explanation of why mature ecosystems, whether a forest or a coral reef, maintain high diversity. This model also explains why increasing disturbance beyond a natural level is likely to reduce species diversity (see Chapter 20 for a fuller discussion).

14.7   Succession and Habitat Management

Changes in net primary productivity (NPP) during the successional sequence are important to foresters and conservationists. Bare ground has zero NPP, but productivity increases rapidly as a complete vegetation cover develops. When herbs colonize the land, their leaves form a single layer covering the ground, and this photosynthetic surface area determines what the NPP will be. A higher NPP can be attained if multiple layers of leaves, all of which are photosynthesizing, can be stacked above a given spot on the ground. By having vertically spaced branches, trees can have such stacks of leaves, and because photosynthesis can operate maximally even in less than full sunlight, the soft shade cast by the upper leaves does not prevent the lower ones from being productive. The highest NPP is found in relatively young forests, where every individual is growing vigorously. In older forests, the trees are often spaced farther apart and have bigger gaps in their crowns. Consequently, on average, there are fewer leaves stacked over a given spot in a more mature forest. Also, because some of the trees in an old-growth forest are senescent and dying, the NPP for the area will be lower (Figure 14.13).

Foresters maximize their profits by harvesting timber at the end of the fast-growth period, the peak of NPP in the young forest. In a forest left to mature beyond that point, the rate of increase in log size is decreasing. In other words, it is more profitable for foresters to cut old trees and replant with young vigorous stock than to let a forest reach maturity. Consequently, forestry stands are filled with young trees. But from a wildlife perspective, an older forest is more valuable than a young forest. In a mature forest, there are all ages of trees, from seedlings to the dying giants of the forest canopy, and, importantly, there are also dead trees. Large trees, especially dead ones, offer nesting holes and a rich supply of insects and grubs. Dead trees offer as many niches as do living trees, so a mature forest, with its more diverse age structure, will carry more species than a uniformly young forest. Consequently, conservationists place a priority on mature forests. Thus, foresters and conservationists are likely to remain at odds over the ideal use of a forest.

14.8   The Old-Growth Controversy

The old-growth forests of northern California, Oregon, and Washington have been a political football for more than a decade. These are the last large tracts of uncut forest in the United States, and they represent the fully mature forests of the northwestern ecosystem. Although the debate is often phrased as "owls versus jobs," the issues involved run much deeper. The Californian Northern Spotted Owl has become a symbol of the conservation effort. The owl is a cute and endearing logo, easier to sell to the public than the real argument that encompasses the broader issues of biodiversity, erosion, wilderness, deforestation (both domestic and international), and

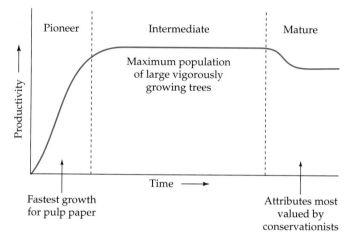

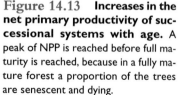

Figure 14.13 Increases in the net primary productivity of successional systems with age. A peak of NPP is reached before full maturity is reached, because in a fully mature forest a proportion of the trees are senescent and dying.

fisheries management. One of the largest issues at stake is the role that the United States can play in international conservation efforts. Until we protect the last of our own virgin forests, what possible right do we have to censure any tropical country for logging its rain forest?

The cutting of the old-growth forests is altering the ecology of the region. Clear-cutting leaves the land surface exposed, and the resulting erosion can wash away the topsoil from clear-cut areas (Figure 14.14). The soil washes into local streams and rivers, causing eutrophication and clouding the water. Rivers that have supported salmon and a salmon-fishing industry are now so altered that the salmon will not return. The loss of topsoil also slows the process of natural regeneration. The Douglas fir will take more than 200 years to become full-sized, and the cycle of succession from clear-cut to mature forest is likely to take at least 400

Figure 14.14   The landscape after a clear-cut. The exposed soil is in danger of eroding.

years. Once the forest has been clear-cut, further human land use ensures that there will be very little chance that the forests ever regain maturity. Species of the deep forests will be lost from these systems, as only those tolerant of regrowth forests will remain. The felling of the last of these forests would be an act of finality, for they would not, could not, regrow as before. Aesthetic questions are raised: Will we cut the last of our untouched forests? How much is it worth to us to decide that at least some forest in the United States should not be shaped by commerce?

The logging industry argument that conservation is costing jobs certainly has a foundation of truth, but to reduce the issue to owls versus jobs is to hide behind a smoke screen. Until the 1990s, about 30,000 hectares of old-growth forest, representing 4.6 billion board-feet of timber, were being felled each year. This rate of cutting virtually exhausted old-growth forest as a resource on private lands (Figure 14.15). By the mid-1980s, 90% of old-growth forests between Washington and California had been cut. Past rates of clear-cutting could not be sustained, even if the cutting were to include all the ancient forests on federal lands. If, as some logging industry advocates maintain, it is only the cutting of old-growth forest that is economically rewarding, the industry and all its jobs are doomed anyway within the next 15–30 years. Mismanagement and short-sighted profit-taking both contributed to an industry fighting for survival.

Simply put, timber was extracted without being replanted. Extractive logging with no replanting is not a recent phenomenon. In fact, it was the hallmark of the expansion of Europeans that wherever they went they cleared the forest. Between 1600 and 1920, 98% of the forests of the eastern United States were logged. Uncut forests are virtually absent in the eastern United States, but now that the same philosophy of extraction is to be applied to the last large stands in the Pacific Northwest, there is a public outcry. Massive reforestation projects for commercial timber have been undertaken in the southeastern United States and also in the Northwest. However, it is still more profitable to fell the old growth than to wait for the planted forests to become old enough to cut.

The cutting of old-growth forest was for many years tacitly encouraged, with federal subsidies being granted through the U.S. Forest Service. Tax money is spent to build logging roads so that trees may be felled from public lands (national or state forests) by private corporations. Tax money is then spent to reforest the land so that it may once again be cut by the private companies. The private companies keep the earnings

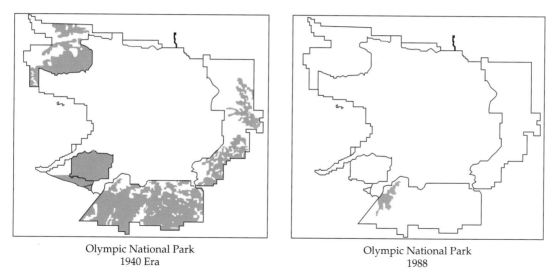

Olympic National Park
1940 Era

Olympic National Park
1988

Figure 14.15   **Deforestation of ancient forest patches on private land surrounding Olympic National Park, Washington, between 1940 and 1988.** Only forest patches greater than 10,000 acres are shown. The logging industry has used up the supply of ancient forests that they owned. Consequently, they are now petitioning to cut the ancient forests on publicly owned lands.

from the cut, apart from a stumpage fee. A stumpage fee is effectively a payment made by the logging company for the privilege of cutting the trees. The investment of tax dollars in logging roads and reforestation is not offset by stumpage fees. For every dollar of taxpayers' money spent, as little as 1 cent is repaid through stumpage fees. In the past decade, subsidies extended to the logging industry have amounted to about $2.3 billion.

A further criticism of U.S. forestry policy is that it does not encourage a domestic "finished-products" industry. One of the characteristics of a developing nation is that it ships cheap raw materials for processing to an industrialized nation. The finished product, an expensive item, is then sold back to the country that furnished the raw materials. The profit to the industrialized nation is the foundation of most economic empires. For years, federal subsidies existed to encourage the export of raw logs to Japan, China, and other trading partners. The goal was to generate export revenue, but instead the subsidies encouraged the U.S. timber industry to behave as an economic colony, selling cheap raw materials. The results of this trade relationship have been long-term damage to the logging industry, over-exploitation of what should have been a sustainable resource, and the failure to establish a finished-products industry in the Northwest.

Caught in the middle, through no fault of their own, are the loggers themselves. These people have

the greatest emotional and personal investment in the forests, and they would adamantly defend their position as environmentalists. For several generations, they have lived, worked, hunted, and fished beneath the great trees. The foresters see their role as forest stewards—integral to this landscape and equally deserving of consideration as the Northern Spotted Owl. The trees are seen as a resource that will go to waste, taken by fire or eaten by bugs, if they are not harvested. This is an accurate assessment of the fate of a tree in an old-growth forest in terms of timber production. The forester views the regrowth that follows clearance as a vibrant resurgence of young vigorous trees, promising future harvests—again an accurate assessment. Threatened with unemployment, the foresters have become vocal opponents of conservation measures.

The Clinton administration came to office on a platform of strong environmental measures, but quickly recognized that decisions regarding the northwestern forests have to be politically tenable. The new policy provides for the continued logging of 1.2 billion board-feet each year from old-growth forests on federal lands, and it includes $1.2 billion in spending to revitalize the region's sustainable industries. The plan also establishes a series of old-growth reserves to protect forest species and salmon populations. Like any compromise, it satisfies neither of the adversarial lobbies completely, but it may represent a settlement that is politically workable.

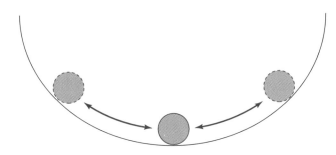

Figure 14.16 **The ball represents a species at equilibrium; if the system is perturbed the ball may move, but it will return to its equilibrial position.** Although this representation allows for some change around the equilibrial position, the assumption is that the equilibrium is the normal state. This representation may be true for short-term changes, but it does not allow for longer-term changes in community composition.

14.9 Equilibrium or Nonequilibrium in Our Modern Ecosystems

Ecological equilibrium implies that all the species likely to grow in a given area have established populations that are fluctuating within normal limits. Thus, even at equilibrium, some changes in species abundance can be expected. This state of a stable equilibrium has been likened to a ball lying in a hollow (Figure 14.16). Some change may occur in the population (the ball moves), but the population soon will return to its starting point (the ball rolls back). This view of ecosystems is reasonable over very short time scales of a few hundred years or less. However, the forests of New England, and even our most mature hardwood forests in the eastern United States (a few remote areas in the Appalachians and isolated stands in the Midwest), do not fit this model. These forests are not equilibrial, and if disturbed they do not return to their starting point. Forests tend not to be equilibrial because they are still responding to the last perturbation and are still successional.

An alternative model of an ecosystem (Figure 14.17) is represented by a ball rolling slowly between low hills. This movement could represent directional climatic change, such as the warming at the end of an ice age or a successional sequence following a volcanic eruption. As succession proceeds, a chance event—perhaps a disease—affects one of the dominant species, deflects the successional pathway, and results in a different forest type. In the model, this deflection is represented as the nudging of the ball into a different groove. At any time during the succession, a change in the environment caused by factors such as fire, drought, or pollution could set the process back. The setback is represented by the ball moving backward. As succession starts to repair the damage, the ball will move forward again, but perhaps along a different course. This process continues indefinitely, so that the ball is continually moving but never necessarily returning to any previous position. As nature is continually shifting the balance of equilibrium, many ecologists believe that, in the long term, ecosystems are essentially permanently nonequilibrial, lurching from one state of near-chaos to the next, albeit in a fairly orderly successional fashion.

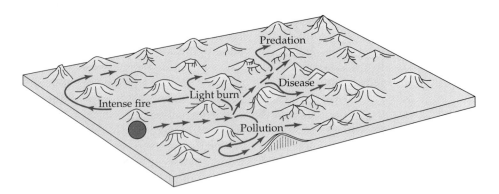

Figure 14.17 **A view of a world that is constantly changing may be represented as a ball that rolls along a groove between hills.** The forward movement of the ball represents successional change in response to a changing environment or recovery following disturbance. Any further disturbance (e.g., the arrival of a new predator or disease) may result in a new successional trajectory and push the ball into another groove. Profound disturbance can send the community back to an earlier successional state (drive the ball backward).

Summary

- Ecosystems are not superorganisms: They cannot be "killed."
- Ecological succession is the colonization of an area by organisms and the maturation of habitats through time.
- Primary succession is the colonization of a new surface.
- Secondary succession is the recolonization or recovery of habitats after disturbance.
- Disturbance events can alter ecosystem functions.
- The recovery of ecosystem functions and the recolonization of vegetation proceed as a positive feedback mechanism.
- Vegetation succession has predictable stages in which certain plant strategists thrive.
- Animals are dependent on plants; therefore they follow the succession, but they do not play as active a role in ecosystem function recovery as plants do.
- The time taken for a system to recover will depend on the type and scale of disturbance.
- Tropical forests may take more than 1000 years to recover from profound disturbance.
- Temperate forests recover from profound disturbance in about 350 years.
- The most valuable forests for wildlife are the most mature systems.
- Relatively immature forests are best for commercial timber production.
- Because habitats are continually disturbed, succession has no absolute end point, and most ecosystems are nonequilibrial.

Further Readings

Billings, W. D. 1938. The structure and development of old field shortleaf pine stands and certain associated physical properties of the soil. *Ecological Monographs* 8:437–499.

Reice, S. R. 1994. Nonequilibrium determinants of biological community structure. *American Scientist* 82:424–435.

Stott, P. 1998. Biogeography and ecology in crisis: The urgent need for a new metalanguage. *Journal of Biogeography* 25:1–2.

Turner, M. G., Dale, V. H. and E. H. Everham, III. 1997. Fires, hurricanes and volcanoes: Comparing large disturbances. *BioScience* 47:758–768.

Whittaker, R. J., Bush, M. B. and K. Richards. 1989. Plant recolonization and vegetation succession on the Krakatau Islands, Indonesia. *Ecological Monographs* 59:59–123.

Web Connections

On-line resources for this chapter are on the World Wide Web at:
http://www.prenhall.com/bush
(From this web-site, select Chapter 14.)

The How and Why of Tropical Biodiversity

15

In a mature rain forest, massive columns rise from the forest floor, some fluted, some ornate with vines; high overhead they merge in the somber light, supporting a vaulted ceiling of branches that shuts out the tropical sun. This depiction of a tropical rain forest as a leafy cathedral-like structure may conjure up some of the majesty of these places, but it does little to capture the almost palpable smell of rotting vegetation, the insane whine of cicadas, the chirruping of insects and tree frogs, the howling mosquito intent on flying deep into your ear, or the thorny vines clutching at your ankles. Nor does it convey the prudent trepidation that precedes sitting down. For here, there are more kinds of bugs and beasties that will chomp, sting, chew, or skewer you than in any other habitat. But, once you have visited a tropical rain forest, like malaria, it is hard to get it out of your blood.

Conservationists obsess about the fate of tropical habitats, and politicians of all persuasions vow their determination to "save the rain forests," areas that until a few years ago would have been written off as "jungle." What makes these forests so special? Perhaps part of their appeal lies in the romantic ideals of adventure; where would Indiana Jones and Tarzan be without jungle? These exotic forests also fascinate because of their potential for discovery. For some, it is the lure of gold in the muddy rivers, for others the complexity of ritual and life-style among indigenous peoples or perhaps the chance of discovering a wonder drug to cure AIDS.

Biologists are not immune to all of these images of the forest, but the overriding attraction of tropical regions is their sheer diversity of habitats and species.

15.1 Where Are the Tropics?

The tropics are defined as the latitudes where, at some time of the year, the sun is directly overhead. Because Earth's axis is tilted, the alignment of Earth relative to the sun changes during the course of a year. The North Pole is closest to the sun in June; the South Pole is closest in December (Chapter 4). The northern limit of the tropics is marked at 23.5° north by the tropic of Cancer, and the southern limit is marked at 23.5° south by the tropic of Capricorn. Within these parallels lie portions of Africa, Central and South America, Asia, and Australia.

Tropical ecosystems range from snow-capped peaks to coral reefs. Most international attention has centered on tropical rain forests because they are the largest collection of ecosystems and are of vast biological importance. The global distribution of tropical rain forests is shown in Table 15.1.

Table 15.1   The global distribution of rain forests prior to deforestation.

Region	Area (km^2)
Brazil	2,800,000
Other Neotropical countries*	1,860,000
Africa	1,600,000
Asia	900,000
Total	7,160,000

*Neotropical includes tropical South and Central America.

One country, Brazil, owns more than a third of all the tropical rain forests, so the policies implemented by this one country are disproportionately important to forest conservation. In the many types of tropical forest there is an unparalleled array of plants and animals. To an ecologist, the immediate question that comes to mind is, Why are tropical forests so diverse?

15.2   How Many Species Live in the Tropics?

About 1.4 million species of plants and animals have been described by science (Figure 15.1), but this is just the tip of the biological iceberg. No one knows how many species there are on Earth, but some scientists estimate there could be as many as 100 million. Good estimates are available for the number of species of birds and mammals on the planet. There is even a fair estimate of the likely number of plants; but when it comes to insects and other invertebrates, educated guesses vary wildly.

Because trees are immobile, measuring the diversity of trees in an area is relatively simple. In the most diverse tropical forests, 430 species of tree may be found per hectare, and there is a total of about 50,000 tree species in the tropics. Mammals and birds are relatively easy to observe or collect, and hence estimates for their numbers can be made fairly accurately, too. All mammals on Earth total only about 4000 species; birds, fish, lizards, and all other vertebrates add a further 38,000 species. In all these cases, a great asset is that the organisms are relatively large, and their species diversity is measured in tens or hundreds per hectare. Insects present an entirely different problem.

Insects can be almost microscopically small, and many thousands of different species could potentially exist in a hectare of forest. Many of these species have never been described by science, and no one scientist has the knowledge to identify all types with accuracy.

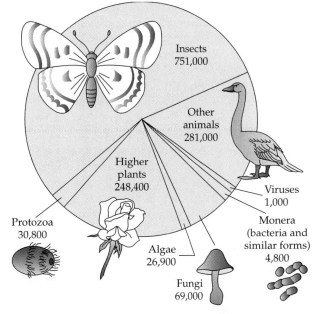

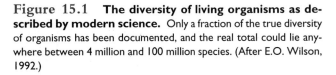

All described organisms: Total species: 1,413,000

Figure 15.1   The diversity of living organisms as described by modern science. Only a fraction of the true diversity of organisms has been documented, and the real total could lie anywhere between 4 million and 100 million species. (After E.O. Wilson, 1992.)

Estimations of total insect diversity are really just educated guesses, and because this taxon represents the largest portion of global biodiversity, it makes estimates of total species diversity highly speculative. For many years, the total number of species on the planet was estimated to be between 4 and 10 million, most of these being insects.

In 1982, Terry Erwin of the National Museum of Natural History set out to reevaluate the number of insects in a tropical forest system. Erwin developed a system for "fogging" the canopy of a tree with insecticide. The first step was to select a suitable tree and surround it with 1 m diameter funnels, which caught insects falling from the canopy. When the fogging started, the poison gas killed all the insects in the canopy of the tree, and dying insects cascaded downward in thousands. Insects falling into the funnels were preserved in alcohol and later identified.

After fogging the tree *Luehea seemanii*, Erwin recorded 163 species of beetle that were found only in this species of tree. From this observation, he built an argument that the total number of insect species is likely to be closer to 30 million than to 4 million. His argument was as follows.

A single species of tree contained 163 species of beetles that were exclusive to that tree species. There are about 50,000 species of tropical tree; therefore, there are 8,150,000 (163,350,000) species of beetles living in tropical trees. Beetles make up about 40% of all types of known insects. Therefore, the total number of insects living in trees would be about 20 million. The last step in the calculation is based on the observation that about 66% of insects live in trees versus 33% that live on the ground or in the soil. The total for all insects would then be 30 million species.

Critics of this calculation can argue that it is full of assumptions and that the tree species that Erwin selected may have been exceptionally rich in beetles. Other estimates in Asian and South American forests using similar methods have produced estimates of about 5–10 million species. More recently still, there has grown the suspicion that many insect species do not fall when they are poisoned. Instead, their last act is to grip tenaciously to their twig. These insects would never be caught, and once this number is quantified it could push the diversity estimate higher once more.

Microbes are another set of life forms that should be included in an estimate of biodiversity, yet their ecology is even less known than that of insects. Apart from a few species of medical, agricultural, or industrial interest, the vast majority of microbes have received little attention. Whether factoring in microbes to global biodiversity estimates would add another 5 million or 50 million species remains a matter of debate.

No matter whether the planetary diversity is 4 million or 100 million species, it is likely that the majority of Earth's current life forms, perhaps as many as 50%–80%, live in the tropical rain forests. Some examples of the sheer species diversity of the tropics come from data for butterflies and ants. From a *single tree* in the forest of Panama, Edward Wilson documented 43 species of ants, a number roughly equal to the ant diversity of the entire British Isles. In a 12-hour observational period in a single Brazilian forest plot, an entomologist identified 429 species of butterflies; compare this number with the 440 species in the whole eastern United States. The United States is relatively species rich for a temperate region, but we have a measly 2400 species of beetle, compared with Erwin's estimate of 8,150,000 beetle species for the tropics.

15.3 Why Are There So Many Species in the Tropics?

Two of the greatest natural historians, Charles Darwin and Alfred Wallace (a contemporary of Darwin and co-formulator of the theory of natural selection), struggled

to explain why there were so many species of tropical plants and animals compared with those of temperate or polar regions. A century and a half later, we still cannot give a simple explanation for this, the grandest of species distribution patterns. Few questions draw on so many strands of ecological thought, but if the threads are woven together we may come close to an answer.

Let us start from the basic observation that there is a steady swelling of species numbers from the poles to the equator. From the temperate zone to the humid tropics there is a tenfold increase in the number of species per unit area of land. Take, for example, the increase in diversity of bird species from North America southward to the tropics of Central America (Figure 15.2). This increase in diversity is reflected both in the number of families and, in most cases, the number of species within a genus. Exceptions to this rule of increasing diversity equatorward include penguins, which reach their maximum diversity in the Antarctic, and voles, coniferous trees, and salamanders, which reach their maximum diversity in the temperate regions. These exceptions aside, however, the gradient of species diversity from pole to equator is a strong one. At the poles, the short growing season, long nights, high winds, and extremely low temperatures prove insurmountable obstacles to most life forms. The steady increase in species toward the equator correlates with increases in temperature and precipitation. However, few ecologists find the correlation of temperature and rainfall with species diversity a satisfying explanation of the superabundance of species in the tropics. The link to climate is strong, but this may not be a direct cause-and-effect relationship because there are some tropical forests in Africa that do not show the same high species diversity found in similar climates elsewhere.

The explanations that account for the global gradient of increasing diversity can be divided into those that suggest historical causes, increased habitat diversity, increased niche diversity, decreased probabilities of extinction, or increased opportunities for speciation.

The Rise of Angiosperms and Global Cooling

About 150 million years ago, **angiosperms**, the flowering plants, started to take over the world, replacing the previous dominant vegetation of ferns and gymnosperms (ancestors of spruce and pine). Angiosperms developed in a time of high atmospheric CO_2 concentrations and warm, moist conditions. At this time, the continents were still largely bunched together and

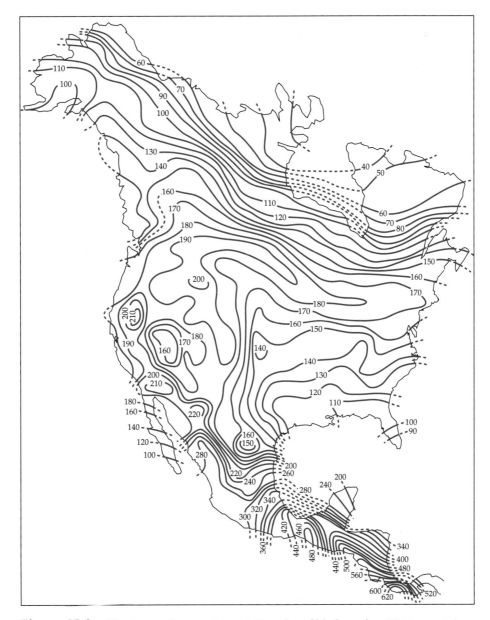

Figure 15.2   The increasing southward diversity of bird species. This is a typical example of species diversity increasing from the pole toward the equator. (After R.E. Cook, 1969. *Systematic Zoology* 18:63–84. Published by the society for systematic zoology, Washington D.C.)

Earth was much warmer than at present (Chapter 2). Even Antarctica supported a temperate forest, and virtually all the world was tropical or subtropical (Figure 15.3). Over the next tens of millions of years the continents drifted apart, but the incredible surge of evolution of new plant types continued for the next 100 million years. The angiosperms offered many new niches for animals. Flowers provided nectar for bees and butterflies; leaves provided a nutritious diet for dinosaurs and the larvae of many insects; fruit and seeds sustained reptile, bird, and mammal popula-

tions. Sure enough, the animals dependent on flowering plants show up in the fossil record shortly after the evolution of angiosperms. Almost all of this evolution took place under warm conditions. By about 50 million years ago most modern families of plants were represented on Earth.

About 35 million years ago the drifting of the continents produced a configuration of landmasses that allowed the circumpolar current that encircles Antarctica to form. This current changed ocean circulation patterns and led to an overall cooling of the planet.

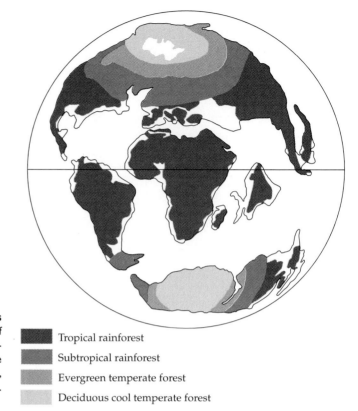

Figure 15.3   Tropical regions of the past. The configuration of continents and the distributions of climate types 90 million years ago. Note that much of North America, Europe, and Asia were tropical. (After J.H. Tallis, 1992.)

- Tropical rainforest
- Subtropical rainforest
- Evergreen temperate forest
- Deciduous cool temperate forest

Antarctica became a frozen mass, and snow and ice penetrated deep into the midlatitudes each winter. This cooling provided a severe challenge to most species. A critical evolutionary threshold is the ability to withstand freezing temperatures. Even a single frost is enough to kill most tropical species. A hard freeze, in which water within the cells of leaves or the stem turns to ice, is fatal to most plants. The plants that have evolved to live in climates that experience such temperatures either abandon their soft tissues each year, as in the case of most herbs and deciduous trees, or have a physiological response to avoid damage. This response may take the form of withdrawing all the water from cell spaces and allowing their conducting vessels to collapse, as in the case of the grape vine, or of building massive cell walls to withstand the pressures of freezing, as in the case of spruce trees. For animals, they must be able to survive long periods of reduced food availability in addition to withstanding cell-freezing temperatures. Clearly, withstanding freezing is a challenging evolutionary hurdle, one that relatively few plants and animals have negotiated successfully. Hence, the full diversity of many plant and animal families has never spread beyond the tropics. Evidence for this comes from fossil deposits of tropical plants and animals in northern Europe, Australia,

South Africa, and North America. They all perished as conditions cooled. The surviving descendants were but a poor representation of the former wealth of species. Only in the tropical lowlands did a full array of species survive and maintain the diversity of their individual evolutionary bushes. This argument is powerful, but the cooling of Earth began 30 million years ago. Surely there has been time for a new explosion of cold-adapted species. We have found part of the answer to high species diversity in the tropics, but there must be more.

15.4   The Diversity of Tropical Habitats

There is no such thing as *the* tropical forest. There are tens, perhaps hundreds of different types of tropical forest, and they vary in appearance from scrubland to majestic forests with trees more than 80 m tall. The type of forest that grows at a given location will depend primarily on the local rainfall (Chapter 5). The tropics support every form of vegetation from sparse grasses and cactus of deserts to the lush steamy rainforests. The length of the dry season is critically important in determining forest types and so is the degree and duration of flooding. Low-lying areas may or may not be subject to flooding (from 0 days to more than 200 days per year), and the depth of flooding provides a further example of environmental gradients that will determine the composition of the forest. Where the forest floods during the wet season, the trees may have to survive for months each year in more than 5–10 m of water. In the flood forests of the Amazon (Figure 15.4), the young trees must survive total immersion for several months each year if they are to become adult. Even the adult trees, although having their leaves and branches clear of the floodwaters, must be able to photosynthesize while their roots are under water and deprived of oxygen (Chapter 8). It takes a true specialist to survive these conditions. One form of this specialization is found among tree species that fruit while the land is flooded. Some of these species have evolved to have their fruit dispersed by fish. Many of these fruit-eating fish are in the same family that includes the infamous flesh-eating *Piranha*.

Soils and Forest Diversity

The wet-season flooding of the Amazonian forests often determines soil fertility. Rivers with headwaters

Figure 15.4 A tropical flooded forest. During the 6–8 month wet season, the water depth can be as much as 10 m in this forest, although 2–5 m would be more common. In the remainder of the year, there are still rains, but not enough to maintain the flooding. Annual rainfall is about 3500 mm.

that originate in the Andes carry a sediment load of fertile clay, the product of erosion in the mountains, that gives the river the color of milky coffee; hence they are called whitewater systems. The clays are deposited in the forests during flood episodes and provide an important input of nutrients, making these lands unusually fertile. The rivers themselves are rich in fish and have a long history of use by human hunter-gatherer-fishers. The sediments washing down from the mountains can also be rich in minerals such as gold. Rivers that originate within the Amazon forest itself have the appearance of Coca-Cola and are termed blackwater rivers. Their shores can have beautiful white sand beaches, but the sediment and the water in these rivers are almost devoid of nutrients; hence, they do not fertilize the land when they flood, are surrounded by forests with poor sandy soils, and do not support large fish stocks. The variations in periodicity, depth, duration, and nutrient load of sediments carried by floodwaters are factors that lead to the existence of many different kinds of flood forest.

A vast number of dry-land forest types exist at elevations above the level of the floodwaters, and soil types vary. The clays that commonly form tropical soils are millions of years old, and most of the nutrients have been weathered from them. In contrast, other tropical soils are young, derived from volcanic ash, and nutrient rich. The differences in nutrient availability offer different habitats to plants that would grow on them. The slope-angle of the ground also plays a role in determining the soil quality and depth. The steeper the slopes, the more likely that a soil will be washed away or shaken loose during an earthquake. Where only shallow soils develop, they offer little material to hold roots secure, and, consequently, trees are generally small and topple as they get large. Differences in soils such as nutrient availability, texture, soil depth, and degree of waterlogging lead to differences in the forest communities that they support and add complexity to the mosaic of forest types.

15.5 Structure and Niche Diversity in a Tropical Rain Forest

Ecologists use *Alpha* and *Beta* to describe different kinds of species diversity of a region. If the species number *within* a single habitat type is considered, this is **Alpha**-diversity. But most landscapes are not made up of a single habitat type. If all the neighboring habitats are very similar, perhaps forests that are of slightly different ages, many species could live equally well in any habitat. Therefore, the overall species diversity is not greatly affected by the habitat diversity in this landscape. Compare this with a forest, a wetland, and a lake that are all neighboring habitats. The number of shared species will be low, and, therefore, the species diversity in this landscape that is attributable to habitat diversity is high. A measure is needed to describe the ecological difference between neighboring habitats. Ecologists use **Beta-diversity** to describe the diversity of species *between* habitats. The *Beta*-diversity of the tropics is not greater than that of an equivalent temperate area, but it is important to realize that considerable *Beta*-diversity exists. To pursue the question of why there are so many species in the tropics, we must address the *Alpha*-diversity.

It has been suggested that one factor contributing to the rich diversity of tropical forests is that their structure offers a greater abundance of niches than does the structure of a temperate forest. In this context, niche is being used in the sense of a set of resources that can be used by a species. Therefore, a greater diversity of resource combinations provides a greater diversity of niches. One source of diversity lies in the varied architecture of tropical forests. In dry forests there may be only one canopy layer, whereas in some rain forests there can be three tiers in the canopy. Naturally, subtle differences in climate, flooding, soils, and age of the forest result in a

complete range of vegetation structures from grasslands to three-canopied forest. The shape and configuration of the tree canopies, trunks, and buttresses provide a wide range of highly local microclimates. Generally a microclimate is taken as being the climate within the vegetation layer. However, the dark moist conditions found among root buttresses provide a different microclimate from the hotter, drier conditions of the upper canopy. These highly local microclimates are reflected in microhabitats that support a characteristic flora and fauna specialized to those conditions.

Emergents and Epiphytes

Tropical rain forests differ greatly in their species composition and their structure. As a general rule, tropical forests are more structurally complex than temperate forests. Vines of differing sizes snake from the floor to the canopy, roots of strangler figs trail from high branches down to the ground, and, in some forests, three distinct canopy layers are formed by trees. From my experience of tramping through these forests, the actual number of canopy layers is often indistinct, but the structural complexity is always apparent. Ecologically, it is this complexity that is the truly important aspect of the forests, because it offers many unique niches. For ease of description, a forest with three canopy layers is considered here. The uppermost layer is a discontinuous series of trees

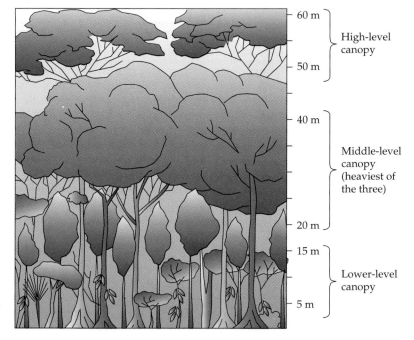

Figure 15.5   A cross section through a mature lowland rain forest reveals the classic three-tiered canopy structure. In this example, drawn from a real forest in Southeast Asia, the emergent layer is particularly full. Note the different canopy shape of trees in the lower canopies as they reach up for light compared with the spreading canopies of trees in the emergent layer.

60 m
50 m } High-level canopy

40 m
20 m } Middle-level canopy (heaviest of the three)

15 m
5 m } Lower-level canopy

that can grow to 50–80 m tall (Figure 15.5). These trees are termed **emergents** because of the way they poke up through the middle canopy. Many of the trees that have evolved to this gigantism come from unexpected sources to someone familiar with temperate forests. These huge trees are not related to oaks, sequoias, or maples, but they may be close relatives of violets, figs, and peas—families of plants that achieve their full diversity in the tropical forests and have just a few diminutive relatives that straggle into temperate latitudes.

The upper branches of the emergent trees are exposed to both wind and sun, but the lower branches are shaded and sheltered by those above. Consequently, the humidity of the air in the lower portion of the canopy is much higher than that in the upper portion. It is in the lower portion that the greatest density of epiphytes grow (Figure 15.6). **Epiphytes** are plants that use another plant as a platform for their growth. Instead of being rooted to the soil, epiphytes root onto the leaves, branches, and trunks of trees. The forest floor is too shaded for many species of orchids, bromeliads (pineapple relatives), and lilies. Consequently, they have evolved to live in the more lightly shaded portion of the canopy. High above the forest floor they have access to light, but the trade-off is that there is no soil from which to derive water and nutrients. Numerous solutions have evolved to overcome this problem.

Some epiphytic species live purely on atmospheric moisture and water that they can absorb from a tiny root system that secures them to the branch. Water use must be frugal, and these species generally have adaptations that help to restrict water loss or to store water when it becomes available. Obtaining nutrients also requires unusual strategies. Often, the leaf arrangement of an epiphyte forms a bowl that traps falling leaves, which decompose in the bowl. The nutrients released by the decomposition process are absorbed by the epiphyte. Some epiphytes are parasitic and drain nutrients and photosynthetic products from the host tree. Other species, such as strangler figs, may start as epiphytes and later send roots downward to reach the ground. This root development represents a huge investment of energy in the root system. But the reward is access to moisture and nutrients from the soil.

In temperate forest systems, relatively few species have evolved to become epiphytes. Frost damage to exposed roots and the lack of year-round humidity in the canopy are important factors in limiting the growth of epiphytes.

Fruiting and the Middle Canopy

The middle canopy layer is where much of the leaf and fruit foraging by the large mammals of the rain forest takes place. This is a continuous canopy, with the tree crowns often touching or overlapping, making it easy for herbivores such as monkeys, apes, sloths, and raccoons to move around high above the ground. Many mammals have a fairly broad diet and will eat fruit, leaves, or even insects. For example, the white-handed gibbon of Asian forests eats about 270 species of plants. Thus, the burden of feeding the primates does not befall a single species, and the primates do not exert a profound grazing pressure on any particular plant. Vines may account for as much as 40% of the leaf area of this canopy. The leaves, flowers, and fruit provided by vines can be enormously important to maintaining mammal populations.

One key aspect of tropical forest ecology is that the trees of almost all species fruit synchronously (at the same time). Furthermore, some trees may miss several years between fruiting episodes. One consequence of this timing is that fruit availability can be very erratic. One month there is a superabundance of many different types of fruit, a bounty that may not be matched for another four or five years. These periods of increased fruiting activity are termed **masting** events. Frugivorous (fruit-eating) species are faced with lean periods when few tree species are fruiting. At these times, trees and vines that do not mast but have a continuous fruit production become essential to the survival of the animals. Figs, for example, have continuous fruiting. If it were not for the presence of

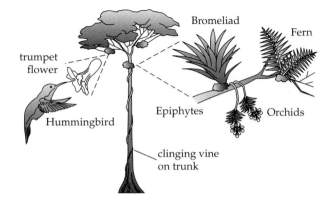

Figure 15.6 **Tropical forests have more niches than temperate forests.** Examples are niches occupied by epiphytes, vines, nectar-feeding birds and bats, and the microclimates provided by buttress roots.

the figs, monkeys, wild pigs, rodents, and birds would starve. A species of plant that has this disproportionate importance in the local ecosystem is a keystone plant species (Chapter 12).

That fruit and flowers are continuously available throughout the year allows the evolution of animals that specialize in these food items. For instance, most bats in tropical forests are fruit eaters; whereas in the temperate forests where suitable fruits are available for only two to three months, bats cannot succeed as fruit-eating specialists. Consequently, almost all temperate bats are insect eaters, taking advantage of a food source that is available for three seasons.

The Lower Canopy and Vines

The lowest canopy is a continuous layer composed of individuals about the height of the largest trees of the temperate forests. Some of these are understory specialists and others are the saplings of larger trees waiting for the opportunity to bolt into a canopy opening. Light is a limiting factor restricting the growth of herbs and saplings on the forest floor. Apart from sun-filled gaps green with herbs and ferns, the forest floor is brown with decomposing leaves. The most noticeable plants are vines looping down from the canopy and the scattered saplings of plants that would be canopy trees.

Vines are abundant and diverse in most tropical forests, yet relatively uncommon in temperate ones. Vines have evolved to leaf, fruit, and flower in the canopy, without building a large trunk. Instead, they rely on support from other plants. The stem of a vine twines, grips, hooks, or clings to the trunk of a tree as it grows upward into the upper canopy. It has a slender stem relative to the amount of leaves that it produces because it is "cheating," putting a large proportion of growth into leaf production and very little into stem production. Vine stems must stay light and slender so that they do not come crashing down out of the tree. Temperate trees avoid the consequences of frost damage by allowing last summer's vessels to die and growing a new set of conductive tissues each spring (hence they grow tree rings). Vines cannot afford to do this because they would quickly become too big and heavy to stay in the canopy. Their vessels must be reused year after year if their stems are to remain slender. The thin stems offer little insulation from freezing, and if ice crystals form in their vessels the plant will die. Consequently, vines are vulnerable to frost damage and very few vine species have evolved to overcome the effects

of freezing (grapes, honeysuckle, and ivy being notable exceptions).

The Ground Layer and Root Buttresses

The principal source of light energy that can be tapped on the forest floor are sunflecks, the shards of bright light that flare through a chink in the canopy. As the sun moves across the sky, the sunfleck moves in the opposite direction across the forest floor. The intensity and duration of illumination as a sunfleck lingers on the leaves of a seedling appear to be determinants of the probability of survival. Sailing has been said to be nine-tenths boredom and one-tenth terror. For a seedling on the forest floor, life is perhaps nine-tenths waiting and one-tenth shooting for the sky. The seedling of many tree species will grow to be about a meter tall, and then wait, gleaning only just enough light to maintain its metabolism. It has entered a period of suspended growth that can last for more than 30 years. Most of these plants will succumb to predation, but a few will bolt upward as an overhanging tree falls.

Rising from the forest floors are the trunks of trees. A common feature of the larger tropical trees is that they have buttressed root systems to support them (Figure 15.7). The buttresses are often about the size and thickness of the lid of a grand piano, flaring out from the trunk of the tree. Buttresses can also take the form of thin stilts that support the tree like a "flying buttress." Buttresses have evolved many times in different families and appear to serve a variety of purposes, the importance of which differs between habitats. In drier areas with steep slopes, the most important attribute of a buttress may be physical support. The buttresses prop the base of the tree, increasing its stability and reducing the likelihood of its being uprooted in storms. This property of buttresses and flying buttresses of reducing the strain on a wall has been known to architects since medieval times, but nature has been using the principle for millions of years. In wetland areas where soils are permanently or seasonally anaerobic, the buttresses provide a large surface area for the absorption of oxygen needed for cell respiration. In these habitats the role of the root as a prop may be secondary to its importance as an oxygen membrane.

Given the apparent advantages of buttresses, why are they not found in temperate or boreal forests? The answer may be that the thin, platelike buttresses have a large surface area-to-volume ratio and therefore cannot conserve heat during winter. In any forest that

Figure 15.7 **Tropical trees have evolved planklike buttresses to serve as load-bearing structures.** A variety of buttressed root systems have evolved over millions of years.

freezes, the vessels carrying sap through the buttresses would freeze and be ruptured by ice crystals. The same argument of frost sensitivity that explains why few species of temperate vines have evolved may also explain the lack of buttresses in temperate forests. If a new layer of growth were built each year as a response to replacing frost-damaged vessels, the buttresses could not maintain their thin flat shape.

The clefts between the buttresses have a moist dark microclimate and microhabitat that adds another dimension to niche availability in tropical forests. Even on dry land, crabs can be found living in these buttresses. These dark, dank growing conditions are ideal for molds, fungi, mosses, and ferns, all of which add to the diversity of the forest.

15.6 Niche Richness and Diversity

Tropical forests and coral reefs are immensely productive habitats, producing more food energy per unit area than any other habitat. Given this abundance, some models proposed that the effect of increasing productivity is to increase the number of species in the forest rather than the population of individual species.

235

However, proponents of these arguments have the uncomfortable problem of accounting for the well-documented decline in species diversity in grasslands and lakes that accompanies increasing the productivity of those systems through fertilization. In these cases, although primary production increases—and food availability also increases—the diversity of both plant and animal species declines. Food availability, therefore, does not seem to drive species diversity.

Given the abundance of energy in the tropical system, it has also been suggested that species can extract sufficient resources for survival from much smaller, more closely defined niches. Each species becomes even more specialized than in the temperate setting, and hence there is reduced competition. An analogy is found in slicing a cake. If the cake is small, a relatively wide slice of it will be needed to provide a satisfying portion. On the other hand, if the cake is much larger, the same amount of cake can be served from a thin (relative to the size of the cake) slice, and more people can be fed. The productivity of tropical systems allows niches to be very thin wedges and allows a greater number of species to coexist.

Some evidence exists to support the argument that very fine habitat distinctions, for example, a few days more or less flooding, can allow closely related species to exist side by side. The inference is that they are not competing unduly; otherwise, one would go extinct. However, the actual energy available within a niche, even what the boundaries of these fine-grained niches are, is very difficult to establish. The fine-grained niche-separation hypothesis seems a plausible explanation for some of the diversity in tropical systems, but it does not tell the whole story.

The abundance and quality of niches that exist only in the tropics contribute to the understanding of tropical diversity. However, they do not explain why there are so many species of trees in a hectare of forest in what appears to be an undifferentiated niche.

15.7 Are Extinction Rates Lower in the Tropics?

For many years the prevailing hypotheses used to explain tropical biodiversity centered on the favorable and stable climates of these latitudes. In other words, the forests accumulated species through time, and the environment was so conducive to survival that few species went extinct. Recent research has shown that the tropical forests were subjected to severe cooling (about 5°–9°C cooler than present) and, in places,

drying during the last ice age. These data indicate that the tropics have had a tumultuous climatic past, as have the higher latitudes; but it may be critically important that the lowest lowlands of the tropical regions would never have been exposed to freezing temperatures, even with a 9°C cooling. Fossil pollen evidence hints at higher species diversity among tropical plants before the onset of ice ages 4 million years ago. The cooling caused extinctions, demonstrating that tropical diversity is not the product of low extinction rates in unchanging forests.

15.8 Pest Pressure

Given that most tropical forests support a high diversity of tree species per unit area, it follows that each species is likely to be relatively rare. It has been suggested that this rarity is the result of predation. The prime predators of plants are not deer, cattle, or monkeys, but insects. The millions of caterpillars, grubs, aphids, and bugs all chewing and sucking a livelihood from plants do far more damage than the odd mouthful of leaves seized by a vertebrate. In a climate system that has winters, insect numbers are restricted by the cold, but in the tropics, the insect populations can reproduce continuously. Brood upon brood are raised in succession until their food resource is eliminated.

The plants defend themselves with an array of toxic and distasteful chemicals to discourage browsing and predation. Herbivorous insects are specialist predators that have found the combination to the defensive chemical lock of a particular plant. Most of these herbivores have diets restricted to a single plant species or to a complex of species that have similar defensive chemicals. The specialist herbivore has developed the ability to pick the particular chemical lock of one group of plants, but it cannot use plants with other defensive combinations.

David Tilman noted that a plant species must specialize to use a particular resource as that resource becomes limiting; because of that specialization, the plant is less well adapted to compete for a different limiting resource (see Chapter 9). The relationship between herbivores and their food plant is similar to that between a plant and its limiting resource. The predator population is held in check by a lack of its prey. If a particular species of tree becomes common, its predator population will increase, eventually killing the trees. If the species remains rare, the predator is less likely to find the trees. Thus, in this deadly game of hide-and-seek, the trees will live in isolated groups or as solitary individuals in a sea of other species, escaping would-be predators through rarity. Note, however, that the tree is reproducing maximally; it is not "trying" to remain rare. The scatter of trees results from the inability to grow as a dense stand, not from a choice to be rare.

In an extension of this argument Daniel Janzen, a leading tropical ecologist, describes how seedlings are unlikely to survive beneath their parent tree. Although the parent tree may lose the odd leaf to a caterpillar and survive, a seedling that has only two or three leaves would die as a result of such an attack. The large parent tree would attract predators, which then would feed on the seeds and seedlings. The result is that only seedlings dispersed a considerable distance from the parent stand a chance of maturing undiscovered. Here is yet another factor that will tend to decrease the chance that a single supercompetitor will outcompete other species.

These various arguments for reduced extinction rates are all likely to contribute to maintaining forest diversity, yet extinction is known to take place in tropical floras and faunas. A satisfying explanation of tropical diversity would also contain an element of increased speciation rates.

15.9 Speciation Mechanisms in the Tropics

Speciation is generally associated with the isolation of a reproductive population. When a population is split and the two subpopulations become new species, this is termed **vicariant speciation**. Various mechanisms have been proposed that could have provided such speciation in the tropics. Populations may have been divided by the uplift of a mountain chain, the presence of a mighty river, or as a result of populations' becoming fragmented during times of climatic upheaval.

Evolutionarily, humans are more closely related to chimps than to buffalo or birds. Scientists determine this relationship by comparing physiology, DNA, and behavior of the various organisms. A comparison of humans with chimps, orangutans, and gorillas would still find that we are more closely related to chimps. However, if the comparison were extended to include the ancient hominid *Homo erectus*, even the little that is known about that creature indicates that our closest relative of all those mentioned would be *H. erectus* (Chapter 16). Indeed, it was on the basis of such comparisons that this extinct species was classified as a member of the human

genus (*Homo*) in the first place. In the same way, a comparison of the relatedness of groups of birds, groups of primates, or groups of insects determines the sequence in which they evolved. From such studies, it is clear that much of the speciation of Amazonian populations is related to the rise of the Andes that started about 25 million years ago.

Physical Barriers That Separate Populations

The uplift of the Andes formed a barrier between populations to the east and west of the mountain chain, causing one to be isolated from another. Over many generations, the populations diverged and speciated. The building of the mountain chain did more than just separate eastern from western populations: It caused the Amazon River to reverse its flow. Before the Andes formed, the Amazon drained westward into the Pacific Ocean, but with the uplift of the mountains and the tilting of the Amazon basin, the river changed its course to flow east. This was an immense change to the ecosystem; it meant that all the tributary rivers shifted their courses. The new river courses split populations and led to further speciation events.

Along the Amazon and along many of its tributaries, the far bank is barely visible. In the foreground, the rivers race past, carrying trees toward the sea. The dangerous currents of these rivers bar passage to all but the luckiest swimmer. If the rivers are daunting to humans, they are impossibly huge to a cricket or a mouse. Populations are effectively isolated by the rivers, and speciation is likely to follow. DNA tests indicate that even for some species of small forest birds that could fly across the river, there is almost no gene exchange between populations on opposite banks. In other words, even though they could fly across the river, they don't. It is suggested that the sheer number of large tropical rivers would promote such speciation and lead to high diversity.

Climate Change and Speciation

Another theory raised to account for high speciation rates in the tropics was the refugial hypothesis. In 1969, a petroleum geologist and avid birdwatcher, Jurgen Haffer, introduced an elegant theory of ice-age speciation events that could explain the pattern of modern bird distributions in Amazonia and, by extension, the patterns of distribution of other species, as well. The assumption of this model was that temperatures in the lowland tropics were the same as those of the present during the ice ages, but rainfall was much lower (by about 70%). Consequently, the land became too dry to support forests, and savanna grasslands expanded. The model predicted that isolated pockets of forest would survive in the wettest locations. These forest remnants would form a refuge for all the forest wildlife, hence they were termed **refugia** (Figure 15.8).

Within these refugia, Haffer suggested, populations of birds, insects, and plants were isolated. Cut off from the populations of other forest organisms in other refugia, the fragmented populations underwent speciation. The model was the first comprehensive attempt to provide a mechanism that would lead to unusually high rates of speciation in some tropical areas. The elegance of the model attracted many other workers to develop similar refugial maps for butterflies, lizards, and plants. The refugial hypothesis is included as the definitive history of Amazonian forests in many texts; however, recent research shows that it is false.

From studies of fossil pollen and plant remains in South and Central America, Africa, and Asia, it has become clear that the dominant climatic feature of the ice ages in the tropics was not drying, but cooling. The rainfall was reduced not by 70%, but by only about 20%. The fossil record reveals that plants that are presently restricted to high mountain habitats grew in the lowlands of glacial times. Temperature is the largest environmental factor that controls the altitudinal distribution of plants. For the plants of cold mountaintops to flourish in the lowlands, there must

have been cooling. During the ice ages, the climate of the tropical forests cooled by as much as 8°C. This cooling did not lead to the fragmentation of the Amazon forest as predicted by the refugial hypothesis. The majority of the Amazon basin remained forested, although the cooling would have led to a change in the relative abundance of species.

Unlike the species of temperate forests, tropical species could not migrate to lower latitudes to escape the cold. The equatorial species were already as far south (or north if migrating from the Southern Hemisphere) as they could go. They were forced to withstand the cooling. Some species would have thrived under the cool conditions, while others would have struggled to survive. Thus, species that are common today would not necessarily have been common during the glacial times, and vice versa. The forests of the past in the tropics were novel assemblages of species without modern counterpart.

The ice ages were subject to rapid climate changes of drying and wetting, cooling and warming. Each change would have reconfigured the populations of plants and animals. During this reshuffling of species, it is possible that populations became isolated (although the overall landscape remained forested) and were subject to speciation. A map of the ice-age forests of the Amazon as suggested from the fossil pollen evidence indicates that the western Amazon would have maintained relatively intact forest (Figure 15.9). In the drier region of eastern Amazonia, there may have been a spread of grasslands in response to an estimated 20% reduction in rainfall.

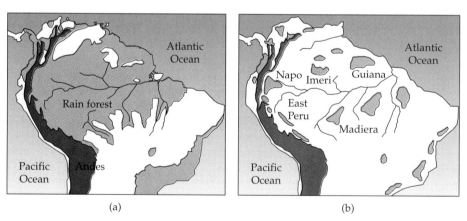

(a)　　　　　　　　　　　　(b)

Figure 15.8　The myth of ice-age Amazonian aridity (a) The distribution of modern Amazonian rain forests. (b) The ice-age fragmentation of Amazonian forests as suggested by Haffer. According to his refugial hypothesis, the Amazon basin dried out during the ice age and only the wettest locations could still support forest. Although widely quoted for the last 30 years, this hypothesis is almost certainly wrong. (Reprinted with permission from J. Haffer, 1969. Speciation in Amazonian Forest Birds. *Science* 165:131–137. Copyright 1969, American Association for the Advancement of Science.)

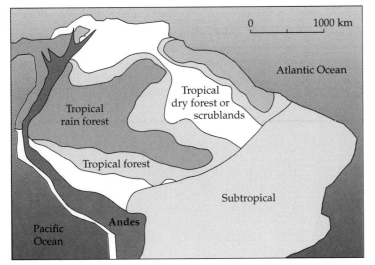

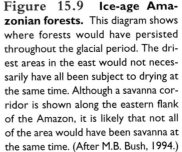

Figure 15.9 Ice-age Amazonian forests. This diagram shows where forests would have persisted throughout the glacial period. The driest areas in the east would not necessarily have all been subject to drying at the same time. Although a savanna corridor is shown along the eastern flank of the Amazon, it is likely that not all of the area would have been savanna at the same time. (After M.B. Bush, 1994.)

The ice age may seem impossibly remote and irrelevant to modern conservation, but the plans that prioritize areas for conservation in Amazonia have been shaped by the refugial hypothesis. A significant factor in assigning protected status to an area in the Amazon is whether it was an ice-age refugium. As refugia did not exist, it is unlikely that this means of selecting protected areas will provide a sound basis for conserving the forests. The more recent view of a forested ice-age Amazon has yet to be assimilated into conservation thinking. When it is, it will necessarily lead to a revision of thinking about the design of conservation areas in the Amazon basin.

No single explanation for high tropical diversity is likely to be correct. Rather, the answer will be found to lie in a combination of factors. It is evident that regional and local biotic and abiotic factors promote niche diversity, reduce competition, and retard extinction. In addition, historical factors such as the turbulent climatic and geologic history of the tropics over the last 25 million years may provide further critical sources of speciation and lasting diversification.

Summary

- There are many kinds of tropical forest.
- Tropical rain forests are the most species-rich ecosystems on the planet.
- It is estimated that as much as 50%–80% of organisms live in the tropical rain forests. Most of the diversity is made up of insect species.
- The main grazing area of a tropical rain forest is in the middle canopy.
- Many plant and animal families evolved in the tropics. Leaving the tropics for cooler biomes has presented huge evolutionary hurdles.
- The reason for high tropical diversity is unknown, but the answer is likely to include continental drift, climate and landscape change, lack of extinction, pest pressure, and fine-grained niche separation.
- Because so much biodiversity lies in the tropics, conserving biodiversity is closely tied to conserving tropical ecosystems.

Further Readings

Bush, M. B. 1994. Amazonian speciation: A necessarily complex model. *Journal of Biogeography* 21:5–18.

Caulfield, C. 1986. *In the Rainforest: Report From a Strange, Beautiful, Imperiled World.* Chicago: University of Chicago Press.

Erwin, T. L. 1991. How many species are there? Revisited. *Conservation Biology* 5:330–333.

Kricher, J. C., and M. Plotkin. 1997. *A Neotropical Companion: An Introduction to the Animals, Plants and Ecosystems of the New World Tropics.* Princeton: Princeton University Press.

Kellman, M., and R. Tackaberry. 1997. *Tropical Environments: The Functioning and Management of Tropical Ecosystems.* New York: Belhaven Press.

Wilson, E. O. 1992. *The Diversity of Life.* New York: W. W. Norton.

Web Connections

On-line resources for this chapter are on the World Wide Web at:
http://www.prenhall.com/bush
(From this web-site, select Chapter 15.)

Peopling Earth

Homo sapiens is a young, successful, species with a population that is growing vigorously. We have expanded beyond our ancestral lands in Africa to invade every continent, and as our population has burgeoned, there has been a corresponding increase in our demand for resources. Already our species uses 25% of the gross primary productivity of the land. More food, more living space, and more water will be minimum requirements of future human populations. Furthermore, it is our hope that all societies will develop to become secure from famine and disease. A side effect of that development must be that most humans require an increased standard of living. The drive to produce more food, more energy, and more consumer goods is likely to intensify environmental problems such as the loss of biodiversity and pollution. Even if human populations were to stabilize now, it is likely that our demand for resources would still increase. The historical growth of the human population reveals trends that lead to predictions of future growth and enable us to determine the carrying capacity of the human population.

16.1 Humans: A Late Arrival

Since the first suggestion that we have evolved from apes and that our closest living primate relative is a chimpanzee, the popular conception has been that somewhere in the evolutionary record there is a "missing link." Half-human, half-chimp, this creature would be the stepping stone from ape to human. But evolution does not work that way; it is much more complex, challenging, and interesting than that. As the intricate pattern of evolution is revealed, some of our basic assumptions are undone. Stephen Gould has emphasized some of the problems of our view of human evolution as depicted in the classic images of the "Ascent of Man" (Figure 16.1). The depiction of an ape turning into a crouching, beetle-browed, half-human, and then into a modern human, is both misleading and revealing. Certainly we have evolved into bipeds, and we have gone through an evolutionary phase of prominent brow ridges; to that extent the depiction reveals what is true. However, the image is misleading because it implies that evolution is a steady process of continual improvement; it is not.

Evolution from ape ancestor to modern humans is a history filled with false starts and hiccups. The first evolutionary step from ape ancestor toward modern apes and modern humans may have been taken many times; almost all resulting in immediate failure. Let us not forget that it is not just humans that have been evolving for the last 4 million years. The ancestor that we share with chimps and gorillas was neither human, chimp, nor gorilla, but rather an *ancestor* of these modern forms. We could trace an evolutionary bush for

Figure 16.1   **"The Ascent of Man" has been a popular way to represent hominid evolution, but it reinforces the fiction of evolution resulting in improvement.**

any of them back to that ancestor, but we shall concentrate here on our human lineage. At least one of these apelike lineages, through a mixture of a good genetic combination and good environmental conditions, survived. We are fairly certain that the apelike ancestor of humans was a forest dweller; recent evidence seems to confirm that *Ardipithecus ramidus*, our oldest known relative, was probably a woodland creature living about 4.4 million years ago. This creature was fairly large, standing about 1.2 m (4 ft) tall, and weighing about 29 kg (65 lb). It had a fairly small brain and was closer to an ape than to a member of the genus *Homo*. Whether *A. ramidus* was the only evolutionary line to split away from the apes and develop a larger brain is not known. Certainly there were several later evolutionary lineages of australopithecines (Figure 16.2), and they appear to have evolved from *A. ramidus*. In fact, throughout much of the period of transition from ape to modern human, two or three distinct types of Hominid (an ape that walks upright most of the time) may have coexisted. *Australopithecus anamensis* and the similar, but more fully documented species *Australopithecus afarensis* evolved about 4 million years ago. *A. afarensis* could be described as a bipedal chimpanzee, and like the other Australopithecines lived largely within forest environments. Thus the evolution toward intelligence and away from the ancestral ape does not necessarily hinge on the new forms living in grasslands. The change from forest to grassland probably was not choice, but was forced upon them. Between about 3 million and 2.3 million

years ago, ice ages gripped Earth. The plains of Africa became cooler and drier, causing a contraction of forests and an expansion of grasslands. The grasslands were new environments, with new evolutionary pressures and vacant niches to be filled.

There are various interpretations of the lineage that follows *Australopithecus afarensis*, and a fairly conservative, simple version is presented. About 3 million years ago, *A. afarensis* (Figure 16.2) gave rise to *Australopithecus africanus*, a light-boned creature with a brain about one-third the size of a modern human brain. *A. africanus* may have been the immediate ancestor to both the genus *Homo* and the small-brained and heavy-boned genus *Paranthropus*. About 2.5 million years ago, the genus *Homo* appears for the first time. *Homo rudolfensis* is distinguished from the other hominids by having hip and leg structure that indicates a striding gait. *Homo* were also tool users. For many years, anthropologists considered tool use as a basic behavioral distinction between the genus *Homo* and other hominid lineages; however, recent evidence suggests that Australopithecines and *Paranthropus boisei* were also tool users. *Homo erectus* evolved about 2 million years ago and almost immediately dispersed around the world, becoming the first hominid to invade a wide geographic area. Fossil remains of *H. erectus* dated to 1.8 million years have been found as far away as Indonesia and China.

Homo erectus was tremendously successful, surviving for about 1.7 million years and maintaining a geographical range from Africa to Asia. In Europe, a large-

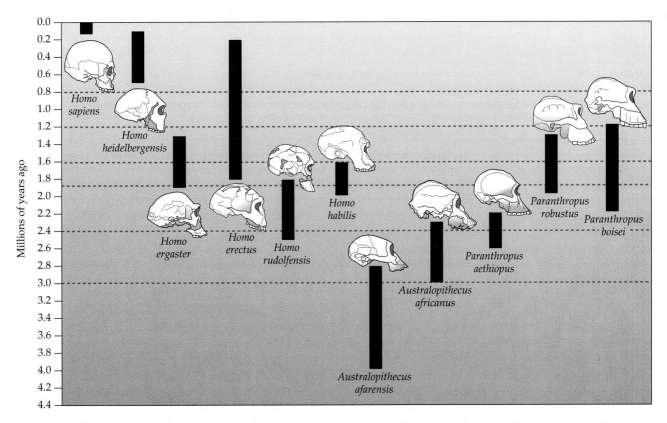

Figure 16.2   An evolutionary bush showing the origins of the genus *Homo*.   Considerable debate remains over the exact relationship and modes of branching of this bush. Details of the bush chosen for this figure are likely to be revised in the future, particularly with regard to the relatedness of hominids where the genus *Homo* emerges. (After R. Larick and R.L. Ciochon, 1996.)

brained hominid appeared about 600,000 years ago. This species, *Homo Heidelbergensis*, probably gave rise to Neanderthal man, whereas modern humans (*Homo sapiens*) evolved directly from *H. erectus*. Consequently, most anthropologists would now place Neanderthal man as our close evolutionary cousin, but not our direct ancestor. Some debate surrounds where and when the evolution of *H. sapiens* from *H. erectus* took place. It seems most likely that it took place in Africa about 100,000 years ago. For a while, it is likely that *H. erectus*, *H. neanderthalensis*, and *H. sapiens* all shared the planet. The last surviving nonhuman hominid, Neanderthal, went extinct just 35,000 years ago. Arguably, for the first time in 4 million years, the planet supported just one species of Hominid. Our branch of the evolutionary bush has thinned down to just one twig.

16.2   From Hunter-Gatherer to Urban Dweller

Until 10,000 years ago our ancestors lived in caves or temporary shelters where they were protected from the elements and from predators. Although humans appear puny against the might of the lion or the elephant, in fact, we are one of the largest creatures on Earth; two or three humans together are a match for almost any other terrestrial animal. Despite our lack of speed, large teeth, and claws, our intelligence and social behavior make humans truly terrifying predators. Hunting with sticks, spears, and fire, a group of humans could routinely kill potentially dangerous animals such as wildebeest, wild pigs, even mammoth.

Although stone-age humans killed animals to provide protein in their diet, plant matter was also essential to provide carbohydrates, vitamins, and minerals. Before agriculture was invented, humans were hunter-gatherers. The "gathering" refers to scavenging edible plant parts, such as fruits, seeds, leaves, and roots from the local area. Considerable knowledge of the local plants would have been needed to provide for a family on a year-round basis in this way. Inevitably, the diet of the hunter-gatherer varied with the season as animals migrated within or beyond hunting range and as plants fruited or died back in winter. Some periods of the year were times of plenty, whereas other times were lean. During the phases when pickings were lean, grubs and adult insects, rodents, and shellfish became important dietary components. Food shortages, even starvation, were permanent threats to all hunter-gatherers.

Even allowing for our tolerance to a range of diets, a band of humans living as hunter-gatherers and operating from a fixed base would rapidly deplete their prey populations. Consequently, for the hunter-gatherer life-style to be successful, human populations would have either moved with the prey herds, be they zebra or caribou, or carried out seasonal migrations between winter and summer hunting grounds. The life of the hunter-gatherer was one of near-constant mobility, and only a few possessions could have been carried between camps. Similarly, an adult female could not carry, nurse, and tend a mass of squalling infants when living this existence. In ecological terms, the clutch size of hunter-gatherers was small, possibly no more than two children that were similarly aged. Ecologist, Paul Colinvaux, observes that induced miscarriage and the abandonment of infants have been common practices among many cultures, including hunter-gatherers. Population regulation in these societies was driven by their life-style, and under such conditions, human population increase was gradual. Prior to the invention of agriculture, the human population is estimated to have taken 90,000 years to reach 5 million individuals.

16.3 Agriculture: The Springboard of Population Growth

About 11,000 years ago, agriculture sprang up at several locations in areas such as modern Iran, New Guinea, and Peru. The timing of these events, coincident with the end of the last ice age, suggests that the invention of agriculture may have been facilitated by climatic or atmospheric change. It is likely that the first agriculturalists did not abandon their hunter-gatherer life-style completely, but incorporated the cultivation of crops into their nomadic pattern. Initially, the harvest was a food bonus that added to dietary carbohydrate and protein intakes. However, as crops were improved and as farming methods became more reliable, the increased yields would have become more important. When the harvest was either too large to carry around or provided more food than could be routinely found from hunting and gathering, there would have

been a strong incentive to settle in one place. A dependable harvest was a prerequisite for the establishment of permanent village life. Over the next few thousand years, crops gradually improved, the hazard of a failed harvest was somewhat reduced, and an increase in village life is apparent. In Europe, neolithic (late stone-age) farmers cleared great tracts of forest about 5000 years ago, and in the New World farmers were cultivating maize in areas as diverse as the hills of Mexico and the lowlands of the Amazon basin. The human population was poised to break into the exponential growth that is possible once food supplies are abundant and when birth rate does not have to be artificially curtailed.

Growth in human populations was slow at first, but it gathered momentum around 3000 years ago. The technological progression in tools from stone to metal (first bronze and then about 2000 years ago, iron) allowed the construction of more efficient tools and weapons. Although still using stone or bone tools, the New World populations of 2000 years ago were acquiring more refined tool kits and ceramics as their cultures became more sophisticated and agriculture flourished. Population growth had its setbacks. The Black Death, caused by a bacterium carried in the saliva of fleas, swept through Europe and parts of Asia between 1346 and 1350. The symptoms of the disease, a raging fever and dark swollen lymph nodes (buboes) in the groin and armpit, gave the disease its other common name of bubonic plague. Death from this disease usually took less than a week, and from the writings of the day, it appears that one-quarter of the European population succumbed.

16.4 An Exponentially Growing Population

Despite setbacks, human population growth accelerated. It took about 90,000 years to accumulate the first 5 million people and another 10,000 years to reach the first billion (Figure 16.3). It took just 200 years to add the next 5 billion. We are now in a steep upward section of the population growth curve that is often referred to as a J-shaped growth curve. At the growth rate of 1.2% experienced between 1995 and 2000, we are adding a billion mouths to feed, clothe, and provide for every 10 years, and our population will double to 12 billion around the year 2050.

As long ago as 1798, when the population stood at nearly 1 billion, the English economist Thomas Malthus

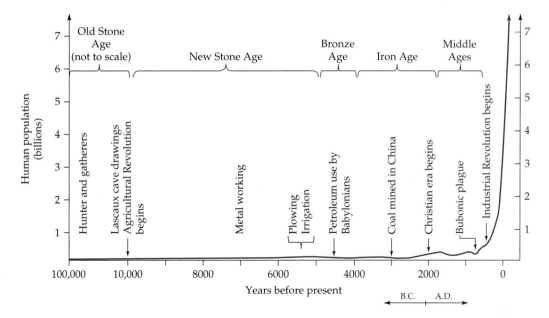

Figure 16.3   Human population growth. The J-shaped growth curve suggests that at the most recent rate of increase, human populations are doubling about every 50 years.

observed that the number of humans was growing exponentially while harvests were increasing only arithmetically. Predicting a shortfall in food and widespread famine, Malthus declared that the world was nearing its carrying capacity for humans. Two hundred years later the world's population stands at about 6 billion people. An important question is, How many more people can we fit on this planet?

Human Carrying Capacity

The Food and Agriculture Organization (FAO) of the United Nations estimates the biological carrying capacity of the planet to be about 50 billion people, approximately eight times our present population. But supporting this number would require that every resource be devoted to maintaining human life. No energy would be available for any luxuries, such as cars, air conditioners, flower gardens, forests, or pets. All nonessentials, the things that give us greatest pleasure, would be abandoned in the quest to feed an extra human mouth.

In addition, as populations become more crowded, there has to be greater regulation of the actions of individuals. Take traffic laws as an example. When there were very few cars on the road, speed limits did not need to be enforced in many areas, stoplights were unnecessary, and parking was unrestricted. However, as traffic built up, minimum and maximum speed had

to be regulated, wearing seatbelts became mandatory, parking was tightly controlled, one-way street systems were installed, and drivers were required to be licensed and insured and to have their vehicles inspected. All these regulations are absolutely necessary given the congestion of the roads, but all constrain our actions. As the planet becomes more congested with people, a corresponding increase in legislation restricting personal freedoms can be expected.

A difference is immediately apparent between the number of humans that we could support on the planet, the biological carrying capacity, and the number that we want. Clearly, we are striving not for our biological carrying capacity but for a modified version at which a maximum human population can be maintained at an acceptable standard of living. This compromise is the **cultural carrying capacity**. Although a single global statistic would emerge, there will be a strong geographical variation in this capacity because each nation has different societal expectations and values. The cultural carrying capacity will also change through time as technology improves agricultural returns, mitigates the effects of pollution, develops new energy sources, and as societal expectations are revised.

The basic possibilities for human population growth (Figure 16.4) are:

• Population will continue to increase exponentially until a biological carrying capacity is reached. This

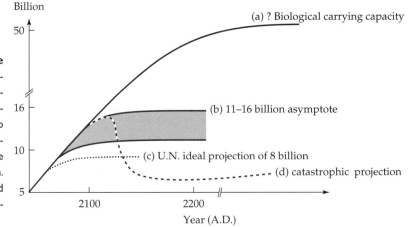

Figure 16.4 Possible future human population trends. (a) Exponential growth continues until the biological carrying capacity is reached. (b) A cultural carrying capacity limits population to at most 16 billion people. (c) The population stabilizes at 8 billion—more a hope of the United Nations than an expectation. (d) Populations rise to an asymptote and then crash as a result of overuse of natural resources, famine, disease, and war.

is unlikely to happen because we will work to limit population before then.

- Population will grow to an asymptote that is lower than the biological carrying capacity, and it will stabilize there. Predicted sizes for such a population, based on resource availability, vary from 11 to 16 billion people.

- In 1996, the United Nations suggested a goal of stabilizing the human population at 8 billion by the year 2050. For zero population growth (when the birth rate and death rate are equal) to be achieved in such a short time would require some changes in the social and economic order of most countries.

- Population will continue to grow until it reaches a threshold at which famine, disease, and warfare cause a decline. One influential prediction advanced by a think tank known as the Club of Rome suggested a slump of population in the twenty-second century that would prune our numbers back to about 6 billion.

16.5 Population Demographics

The demography of a population describes the rate of growth, age structure (proportions of a population in different age groups), gender ratios, birth rates, death rates, and fertility of any organism. A study of human demographies is the best way to make predictions concerning future populations. The age structure of a population is a strong clue as to whether the population is about to grow, to stabilize, or to shrink.

In a country with a zero population growth rate, the number of individuals yet to enter the breeding pool should be equal to those already in it. Thus, the

age distribution diagram (Figure 16.5). would have approximately parallel sides, tapering only near the top. This upper taper is due to death of the elderly. In a population that is stable, the higher the taper begins, the healthier the population is, or the better its medical services. Because these are percentage data, a change in the value for one age group affects all the others. The huge modern population of the very young in Chile, compared with the population of the very young 80 years ago, will result in a steep taper at the top of this data set regardless of whether any of the elderly have died.

Where there are fewer young than those in older age groups, as in Denmark, the population is likely to shrink in the future. For a population to stabilize, the number of children surviving to reproductive age should equal the number of parents in the preceding generation. Because of infant and prereproductive mortality and the slightly higher proportion of male than female babies, slightly more than two babies per couple need to be born to maintain this replacement level. In the United States, the replacement level is achieved with an average of 2.1 births per couple.

Chile has an age structure that forms a pyramid, indicating that a great army of the very young who have yet to breed is amassing (Figure 16.5). Each adult couple is having many more than two children, with the consequence that the next breeding group will be huge relative to the present one. A time lag exists between the social change to having fewer children and the stablizing of a population. For example, even if Chile were to persuade each couple to have just two children, effective immediately, their population would still continue to grow for the next 30 years. In a pyramid-shaped age distribution, such as that of Chile, in

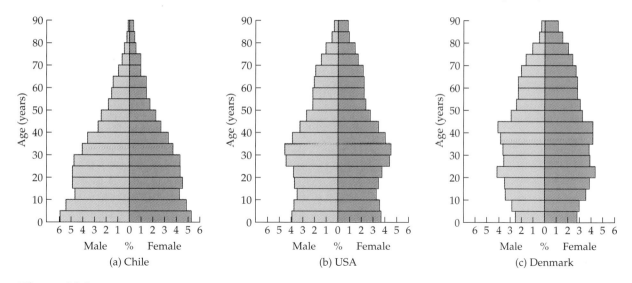

Figure 16.5   **Age distribution for a country with a rapidly growing population (Chile, 1988), a near-stable population (United States, 1988), and a shrinking population (Denmark, 1988).** (Rubenstein, T.M., *The Cultural Landscape.* 5th ed., 1996. Adapted by permission of Prentice-Hall, Inc., Upper Saddle River, NJ.)

which 42% of the population are under 14 years of age, most of the population has yet to reproduce. The babies born into the base of that pyramid will not reach reproductive age until their twenties, and some will not reproduce until their thirties. Even though they are going to have only two babies per couple and the new age distribution will become a straight-sided column, the actual population size will be much greater than it is today.

The age-structure data for Chile are typical of many African, Asian, and South and Central American nations, and they indicate that global populations will continue to grow in the next decades. In general, it is the LDCs that have the highest population growth rates, so they become the natural targets for projects aiming to slow population growth rates. At the present rate, we will reach a population of 8 billion people around the year 2020, but will we be able to stop population growth there?

Enacting a Population Policy

Given that population growth continues for 20–30 years after birth rates have been reduced to replacement level, it is imperative that programs designed to limit population growth be initiated very quickly. Enormous controversy surrounds the enactment of such policy. Fear that personal and religious freedom, social traditions, and economic power structures will be upset if population control measures are introduced hinders reform. Issues such as the morality of abortion, the use of contraception, enforced sterilization,

and the killing of daughters (in societies where male children are especially prized) overwhelm the debate. Furthermore, as the MDCs have near-stable populations, it is the LDCs that are being encouraged to control their populations. Some LDCs perceive population control programs to be another power play by the developed world, with MDCs attempting to exert authority over the most private and intimate aspects of life in LDCs.

16.6   The Emergence of the MDCs

Two hundred years ago, the MDCs were just beginning to enter a phase of industrialization. It was a time of huge population expansion throughout Europe and, a little later, in the United States. The industrial base of these nations grew, and the population became increasingly centered on cities. For the first time, women entered the paid workforce, and prosperity rose. With this newfound wealth, birth rates started to decline. Increased standards of living are reflected in the expectations for a child. A prosperous family expects to be able to feed, clothe, and educate their children and diminish the responsibility of the child as a wage-earner. Effectively, there is a disproportionate increase in the cost of having a child; consequently, parents choose to have fewer of them. This inclination is reinforced if the mother is working and decides to delay having children.

The data for birth rates in three MDCs over the past century provide an opportunity to see the effects of what has been termed the *demographic transition* (Figure 16.6). As in all real data sets, the big trend, an overall decline in birth rates, is overlain by "noise" (variations from the larger pattern) that could be confusing. However, the history of the twentieth century explains most of the noise. Throughout the first decades of the 1900s, birth rates in these countries fell steadily; this was the passage through the demographic transition (explained below). Women were starting to take their place in the workforce, becoming more educated, and demanding the right to vote. As the 1920s gave way to the 1930s, the Great Depression took hold of the country, and birth rates dropped even lower as couples were unsure of their future. Birth rates boomed with the return of troops at the end of World War II, manifested by a jump in birth rates in 1946. The trend of increased birth rates continued into the postwar era. The 1950s were a time of prosperity, optimism, and economic growth. The United States became an isolationist power and lived out a version of the American Dream in which parental roles were clearly defined and women left the workforce, typically reporting that they wanted to be homemakers with four children. This was the period when the "baby-boomers" were born.

The declining birth rates of the 1960s reflect the realization that unchecked population growth was detrimental to the country's welfare and social changes in the perception of the role of women. The proportion of women in the workforce again rose markedly, with the almost immediate consequence of lowered birth rates. Family-planning education and supplies of contraceptives became more freely available. Along with these changes there was a change in the number of children that most couples wanted. Consequently, in the 1970s, an average family had two children. This trend toward smaller family size has continued into the 1990s, and it is now quite acceptable for couples to have no children or a single child; this is a very real shift from the values of the 1950s.

Despite the falling birth rate, the U.S. population is still increasing at about 1.1%, or 2.5 million people, each year. About one-third of these are immigrants, and the remaining population increase is because the generation born during the baby boom is still in the reproductive age class. Once this cohort, particularly the women, passes through into the post-reproductive population, and if birth rates are maintained, the population in the United States should stabilize. Of course, immigration, which could change unpredictably with changing legislation, is still a significant factor.

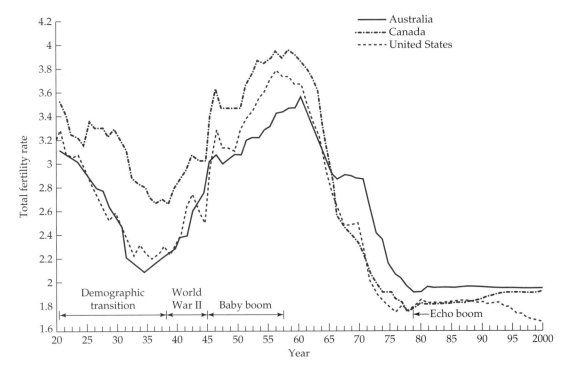

Figure 16.6 **Twentieth-century birth rates in the United States, Canada, and Australia (births per 1000 adults in the population).**

16.7 The Demographic Transition

The difference between birth rates and death rates always underlies the dynamics of a population, and they are the basis of the **demographic transition model**, which can be used to predict the development (not implicitly an improvement) of a culture. Four stages are identified that follow the progress of the country from a preindustrial to a postindustrial state (Figure 16.7).

- The preindustrial phase: This is a nation that is on the brink of becoming industrialized. Birth rates are relatively high, but so, too, is infant mortality, and adults tend to die relatively young. Therefore, death rates are relatively high and more or less balance the births, with the result that low population sizes are maintained.

- The early-industrial stage: Strong population growth occurs as a result of predictable agricultural returns, increased medical care, and improved housing. Death rates fall as life expectancy increases and infant mortality decreases. Meanwhile, the birth rate stays high, and there is enough food to feed the extra young. The youngsters can help to support larger areas of farmland, work in industry, and also provide for their parents as they age. Culturally, economically, and energetically, increasing populations are to be expected. Most LDCs are either entering, or have entered, the industrial phase. The pattern observed among countries that have passed through this phase indicates that it may last at least 50–100 years.

- Late-industrial phase: Birth rates drop markedly in response to increasing education and decreased in-

fant mortality. Nevertheless, birth rates exceed death rates and the population continues to grow.

- Post industrial phase: Birth rates decline to equal or be slightly lower than the death rate. If birth rates equal death rates, the population has achieved **zero population growth**. Any time the birth rate drops below the death rate, the population is getting smaller. Some European nations (such as Sweden, Italy, and Holland) have reached this point. One effect of this pattern is that because fewer children are being born than in previous generations, the average age of the population is increasing.

Achieving the Demographic Transition

Many economists believe that if the economic well-being of a country can be raised sufficiently, the demographic transition will slow population growth. By observing economic indicators, one can monitor the progress of countries toward the demographic transition. One of the leading economic indicators is the nation's per capita gross national product (GNP). This statistic documents the average earnings per person per year and is part of the definition of the economic development of a nation. As a rough guide, an annual per capita GNP in excess of $7000 is one definition of an MDC, whereas an LDC commonly has an annual per capita income of less than $1000. The comparison of GNP, population growth rate, total fertility rate (the average number of children born to a woman), and under-five mortality rate (deaths of children aged under five, per 1000 live births) are strikingly correlated (Table 16.1).

Less-developed countries (the top half of Table 16.1) have the fastest population growth and the highest fertility and infant death rates. All were predictions of

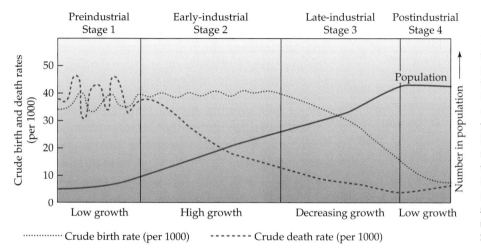

Figure 16.7 **The demographic transition undertaken as a nation proceeds from an early-industrial to a late-industrial culture.** Death rates decline during the early industrial stages in this development while birth rates start to decline in the late industrial stage. Population growth starts and ends close to zero but goes through a phase of rapid growth during the early and late industrial phases. (Rubenstein, T.M., *The Cultural Landscape.* 5th ed., 1996. Adapted by permission of Prentice-Hall Inc., Upper Saddle River, NJ.)

Table 16.1 Gross national product (GNP) and population data (per thousand) for 1995–2000.

Country	Per capita GNP (U.S. $)	Number of births	Number of children per women	Under-5 mortality	Couples using contraception (%)	Population increase (%)
Ethiopia	100	48.2	7.0	195	4	3.0
Rwanda	180	42.8	6.0	139	21	2.6
India	340	25.2	3.1	115	41	1.6
China	620	16.2	1.8	47	83	0.9
Russian Federation	2240	9.6	1.4	30	N/A	(0.2)
Mexico	3320	24.6	2.8	32	53	1.6
Brazil	3640	19.6	2.2	60	66	1.2
Chile	4160	19.9	2.4	15	43	1.4
United Kingdom	18,700	11.9	1.7	7	82	0.2
Canada	19,380	11.9	1.6	8	73	0.9
United States	26,890	13.8	2.0	10	71	0.8
Denmark	29,890	13.0	1.8	7	78	0.1
Japan	39,640	10.3	1.5	6	59	0.2

(Data are from World Resources Institute, UNDP, UNEP; *World Resources 1998–1999.* New York: Oxford University Press, 1998. Reprinted by permission of the World Resources Institute.)

the demographic transition model. The LDCs are in the industrial phase of the model and may take some years to break through into the postindustrial phase. Until that time, birth rates will exceed death rates, and populations will grow. The momentum given to a population by the prereproductive group ensures that even as the postindustrial phase is reached, the population will continue to grow for about another 30 years.

Before accepting this model uncritically, it should be noted that no country has ever been documented to pass through this demographic transition. The model is based on the hindsight interpretation of what happened in MDCs. On the available evidence, Asia and South and Central America are on course to pass through the demographic transition early in the twenty-first century. The demographic changes in Africa will probably follow suit, but many social factors could influence their path to the demographic transition. These uncertainties make it impossible to predict the time of population stabilization in Africa. The United Nations is urging nations to adopt policies that could stabilize populations at 8 billion people in the year 2050. For zero population growth and a stable population, it is estimated that a total fertility rate of 2.1 is required. The slight excess above the value of 2 reflects infant mortality and prereproductive deaths in a population that has attained MDC levels of sanitation, healthcare, and hygiene. On the present evidence, it seems that most of the world could attain

this goal, but it is less likely to happen on schedule in the poorer African nations.

16.8 Limiting the Expansion of the Human Population

If human populations are to rise until reaching a cultural carrying capacity, but to stop well short of the biological carrying capacity, there will have to be some mechanism to limit population growth. Catastrophic events such as warfare and famine are possible factors limiting growth for some subpopulations, but one hopes these will not provide a long-term limit to human population growth. The possibility of human populations being ravaged by disease has been raised, but disease seems to be an unlikely long-term controlling influence on population size. More likely is a demographic shift that brings with it the desire to limit populations. That shift is likely to result from economic, educational, and social change. However, relying solely on the demographic transition to curb population growth may be too slow for the common good. If reaching that transition can be speeded up through social reform, the human population might peak at a lower level.

Population Growth: Asset or Bane?

It should be noted that the description of the demographic transition was based on the attainment of that state by Western European and North American cultures. The subsequent rise in economic strength of Asian nations suggests that they are passing through a similar pattern of lowered childbirth, but the *theory* of the demographic transition should be treated as such. It is a theory with the inherent cultural bias of the Western nations and is likely to be revised as a greater diversity of nations join the ranks of the MDCs.

A knee-jerk reflex exists among many environmentalists that human population growth in developing countries is necessarily evil. Although, from the perspective of an MDC, the press of human population appears to be responsible for many ills, the populace of an LDC may believe, at least at a family level, that large families are desirable. Population growth is viewed as an asset by governments of some developing nations, peasant farmers, some religions (notably Catholicism), and some economists.

An economist might argue, and certainly some leaders of developing nations would agree, that a phase of rapid population growth was crucial to the success of the industrial revolution in MDCs, because it provided a young, vigorous labor pool. There was enough labor to supply agriculture, industry, and an empire-building military. Each person had to pay for food, housing, and consumer goods. An increasing number of consumers represented a growing marketplace of people ready to buy the goods being manufactured. In a growing economy, population growth can act as a spur to increase domestic sales. Thus, for some small LDCs that are expanding their industries, increasing population size could be a boon. But note, this benefit will hold true only where there is relatively full employment, the country is potentially wealthy, and where a small workforce has been the limiting factor in advancing economic development.

Government may favor a large population for a variety of reasons. If the country is expansionist, the growing population will provide young workers and members for the military. The spread of a population into remote areas helps to maintain claims to that land and prevents separatist movements or incursions by neighboring countries. If a tax system works in that country, the increased number of workers will contribute more taxes to fund improvement programs.

A peasant farmer in an LDC will benefit from having many children, because they can help with work on the land. If there is no mechanization, all planting, weeding, irrigation, and harvesting must be done by hand. A child of six or seven becomes a productive member of the household, caring for livestock, scaring birds from the fields, helping with the harvest, or doing domestic chores. Each child of working age in-creases the area that can be farmed and the potential earnings from the land. The eldest children may leave home to work in cities and send money back to their family. These earnings can be used to educate the younger children or raise the family's standard of living. One of the most important factors promoting a large family size is that it creates a social net to support the parents as they become old and frail. In countries that have no social security or welfare system, where pension plans are unheard of, the children will house and provide for their aging parents. A large family ensures that this burden can be spread among the young and guarantees the parents that, if some of the children die as young adults, there will still be someone to look after them. Religious leaders may advocate large family size because of their personal convictions that such is the correct interpretation of their doctrine.

The Goals of Family Planning

Installing a condom dispenser in a high school is not a family-planning program; it is simply providing access to birth control. Similarly, simply making condoms or contraceptive pills available in LDCs is not a family-planning program. The goal of family planning is not to prevent parents having children but to achieve a healthy, happy family.

It is virtually impossible to persuade people to reduce family size until infant mortality becomes rare. Ironically, the best population control measures are associated with increasing infant survival, ensuring a population boom in the future as those babies reach reproductive age.

A successful family-planning program is built on pre- and postnatal infant care, the education of women in motherhood and basic hygiene, and improvement of local sanitation. All of these aspects are necessary, because the children that are born must be safeguarded. Once the survival rate of infants improves, mothers may choose, or can be persuaded, to have fewer babies. At this stage, the provision and explanation of contraceptives such as condoms, IUDs, contraceptive pills, and implants can reduce birth rates. Male and female sterilization are often available (sometimes compulsory), and abortion is also included in many family-planning programs.

We will not deal with the various methods of contraception; rather, we will look at the first age of reproduction, because the timing of first reproduction is a major component in population growth. A woman who starts having children in her teens is likely to have more than two children. In addition, recent studies among teenage mothers in the United States indicate that their mothers were also likely to have become pregnant in their teens. To demonstrate the significance of that observation, let us consider two hypothetical families from a purely female demographic point of view. In this example, we will assume that everyone lives to be 75 years old. Both families have a mother who has one daughter, but in one family the generation gap is 15 years (so that a 15-year-old becomes a mother), while in the other family there is a 30-year generation gap. We'll now assume that these generation gaps are maintained in future generations and that, with each generation, each woman has one daughter. After 70 years, the family that breeds at 15 years old has five women alive (Figure 16.8), whereas the other family has just three. However, a more realistic model might be that teen mothers will have more than two children because they will be fertile for many years after having their first child. Let us look at what happens when two daughters are put into each generation. In the 15-year reproductive cycle, 31 women are alive after 70 years, whereas the 30-year cycle has just 7. Clearly, in population growth terms, it is not just how many children you have, but also the spacing of the generations that are important.

A prime goal of family-planning schemes, therefore, is to delay first reproduction as long as possible. Unless enforced by taboo or government policy, reproductive age, like all family-planning policies,

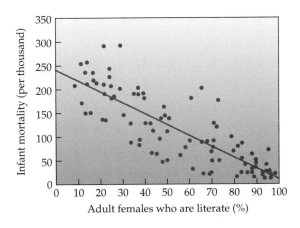

Figure 16.9 **Mortality rate for children under 5 worldwide, plotted against the proportion of the female population that is literate in 1990–1991.** (Data are from World Resources Institute, UNDP, UNEP; *World Resources 1994–95.* New York: Oxford University Press, 1994.)

hinges upon the education of women. Perhaps the most direct route to delaying reproduction is through the emancipation of women in LDCs. Until women are recognized as more than breeding machines, the best approximation to emancipation is to encourage education and family planning. The effects of educating women, both on infant survival and on the total fertility rate, are clearly shown in Figures 16.9 and 16.10. World Bank statistics indicate that for every additional year of schooling a woman's fertility rate is reduced by 10%.

Obstacles to Family-Planning Programs

Family-planning programs will be worthless unless couples are persuaded that they should have fewer

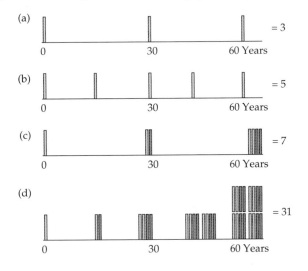

Figure 16.8 **The number of females in a lineage.** If there is one daughter per generation. (a) With a generation gap of 30 years, after 70 years there are just 3 women alive in that lineage. (b) With a generation gap of 15 years, after 70 years there are 5 women alive in the lineage, if there are two daughters in each generation, which may be typical of a woman who first reproduces at age 15. (c) With a 30-year generation gap, after 70 years there are 7 women alive in the lineage. (d) With a 15-year generation gap, after 70 years there are 31 women alive in the lineage.

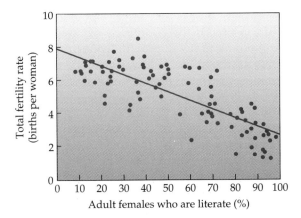

Figure 16.10 **Total fertility rate worldwide plotted against female literacy in 1990.** (Data are from World Resources Institute, UNDP, UNEP; *World Resources 1994–95.* New York: Oxford University Press, 1994.)

children. It is a basic human trait that we enjoy children, and in many societies the social status of a couple relies on their having a healthy brood. These values lie at the very core of concepts of self-worth, so they can be extremely hard to change. Furthermore, such values are not based on an irrational like of children, but have arisen as a result of the need for large families (as discussed previously). Thus, the first obstacle to a family-planning program is to persuade the target population that they will be better off with fewer children.

A further social factor that influences the success or failure of family-planning programs comes from the direction of religious leaders. Where the church and its missions are active participants in these family-planning programs they can offer a strong infrastructure for the dissemination of information, general education, medical care, and contraceptive supplies. Indeed, in some remote areas, the missions may be the only existing infrastructure that could achieve local support for family-planning programs.

Religious conviction that all forms of contraception break a moral code defines the position of some churches. The Vatican has maintained a consistent opposition to population control in any form. In a speech given in Brazil in 1990, Pope John Paul II urged Brazilians to have more children so that there would be a greater supply of males entering the priesthood. This sentiment seems oddly out of phase with the general strategies for economic and social development put forward by financial and humanitarian aid organizations.

Family-planning programs in developing nations face some obvious problems, such as societal or religious resistance, lack of funding, and the lack of a supporting infrastructure of clinics. Some less obvious problems are that the contact between the local population and the medical or social workers can be interrupted as wet season rains wash out bridges and make roads impassable for months at a time. These same rains can overwhelm sewer systems and cause contamination of drinking water supplies, undoing the work to provide hygienic living conditions. Contraceptive supplies have to be accessible throughout the year, even when villages or homesteads become isolated. The problem is exacerbated by the reduced shelf life of latex condoms in tropical heat. Although in many LDCs travel is becoming easier for much of the population, travel beyond the nearest market town is a rare event. Languages can change over short geographical distances, and historical rivalries can serve to separate populations. For all these reasons, the most successful family-planning programs have been those that are coordinated locally rather than from central government. Local instructors and nurses are likely to be the most effective communicators, and they are likely to engender the most trust.

Local family-planning groups have to be coordinated at a national level and funded at an international level, and here they become enmeshed in politics. As environmental problems, famine, and warfare are driven by soaring human populations, it might seem obvious that international efforts should be made to instigate family-planning programs. However, domestic politics of MDCs can have far-reaching consequences on these programs in LDCs. Abortion has become an ethical and political football in the United States, with the consequence that the Reagan administration felt unable to support any family-planning program that included abortion; the policy was continued under the Bush administration. For 11 years, the most powerful government in the world played almost no role in attempting to control global population growth. While the views of both sides in the abortion debate in the United States are sincere and well intentioned, it can be argued that the cost of this struggle has been borne in the LDCs. The withdrawal of U.S. backing from family-planning programs and reproductive education of women in LDCs may have prevented some abortions. But it also exacted a human cost in pre- and postnatal deaths of infants that were wanted, while doing nothing to reduce unwanted pregnancies.

Lessons from China

In the 1980s, China experimented with the most rigid population control program yet enacted anywhere. Families were coerced to have just one child. To understand the reason for the government's intervening in the lives of its people in such a draconian way, we must look to the Great Leap Forward of the 1950s. The Great Leap Forward was designed to bring China from an agrarian to an industrial economy. The goal was to increase agricultural yield through an expansion of cropland and modernized farming methods and, at the same time, build an industrial base. The resulting chaos was compounded by terrible flooding. In the winter of 1959–1960, 30 million Chinese died in the world's worst famine.

In the 1960s, the communist leadership recognized that their rapidly growing population and uncertain food supply were deadly liabilities. A program to reduce population growth was built around a rural health system funded by agricultural collectives. Contraception and advice were freely available. By the mid-1970s the average fertility rate was reduced from

5.7 to 2.8 children per woman. However, the Chinese population was still growing rapidly, with an increase from 600 million to 1 billion taking place between 1950 and 1975. The leadership decided that their goal would be a stable population by the year 2000.

Because China was (and is) a dictatorship, the government could impose its will upon the people. During the 1980s, the Chinese government implemented a program to limit each couple to one child. Only married couples could have children. Men were encouraged to delay marriage until they were 28 and women until they were 25. The forms of "encouragement" were:

- extra food and improved housing, pensions, medical care, and wages for couples pledging to have just one child;
- free access to contraceptives, abortion, and sterilization;
- abortion of babies carried by unwed mothers;
- social ostracism for couples having a second child;
- enforced abortion of the third child conceived.

These policies were always unpopular, and less rigidly enforced, among the rural poor. Farmers relied on sons to help with heavy labor on the land, and abortion of female fetuses became commonplace; unusually high death rates of female infants were also reported. If the human rights issues are set aside, the program was at least partially successful. The population surge slowed, the average fertility rate was reduced to 2.0, and the overall population growth rate was lowered from almost 3% in the 1960s to 0.9% in 1995–2000.

In the 1980s, the Chinese leadership had hoped for zero population growth in the year 2000, with a population of 1.2 billion. The United Nations predicts that the Chinese population may stabilize about the year 2100 at a population of about 1.7 billion. Unknown factors in this calculation are the effects of democratization and increasing development. These two aspects of cultural change could exert opposing influences on population growth. If democracy or a government less dictatorial than the present one is established, the constraint to have one child might be lifted, and the population might grow rapidly. In contrast, if China's economic development is maintained, the country may experience reduced birth rates associated with passage through the demographic transition. The Chinese authorities appear to be gambling that the influence of increased wealth will override the desire for a return to large families. In the late 1990s, China has changed its emphasis from enforcing the one child per family rule to promoting the status and role of women in the workforce.

16.9 Reforming the Role of Women

The greatest single step that could be taken toward controlling the exponential growth of human populations in LDCs is to educate and empower women. In many LDCs, the majority of women are economically invisible. For centuries, their roles have been confined to agricultural decision maker and laborer, water and wood gatherer, healthcare provider, mother, and cook—roles that are seldom reflected in the GNP of developing nations. In these LDCs, women will perform more hours of work than men, but will receive little credit in terms of financial return or social recognition. Women seldom share the institutionalized power of local, regional, or national government; often they are excluded by a vicious circle of discrimination and illiteracy.

One of the most damaging social values encountered when trying to implement a family-planning program, is that the worth of a woman is judged by the number of male children she produces. This value not only relegates a woman to the status of a brood mare but also ensures the cycle in which boys continue to take the lion's share of all resources. In many societies, available resources such as the best food, educational opportunities, and medical care are prioritized for boys. Girls are withdrawn from schools at an earlier age and are expected to perform household chores until reaching marital age. Consequently, illiteracy in some African nations is almost twice as prevalent among women as men. As shown earlier, female illiteracy is closely correlated with a high total fertility rate and high rates of infant mortality. It is also correlated with high rates of maternal mortality during childbirth.

Recently, it has been recognized that directing economic development aid toward women has more far-reaching environmental effects than does instigating a family-planning program alone. As their educational level increases, women become more employable and more likely to make financial decisions that affect the family. In many societies, because the men and women maintain separate finances within a household, the men retain spending rights. This pattern results from the division of labor in which men raise cash crops (produce that is sold) and women raise the family's food crops. The only cash to enter the family coffers comes from the man's crops: The money is "his" and he spends it. A study in the West African Cote d'Ivoire indicated that in households where the income was controlled by women, as income doubled, spending on alcohol fell by 26% and on tobacco by 14%. In a study of child nutrition in Guatemala, it took 15 times more spending when the earnings went to the father rather than to the mother to achieve a set increase in dietary quality for the children. Women tend to prioritize spending on nutrition and subsistence needs, whereas men are more likely to squander money on entertainment, alcohol, and tobacco. Employed women are also likely to want fewer children and are therefore likely to take advantage of family-planning programs. Thus, aiming development projects directly at women, particularly at their education, and then supporting this effort with innovative agricultural techniques to allow more earning capacity is a highly efficient use of funds. The provision of family-planning education and supplies at a local level is an integral part of this social development.

The wide recognition of the need to address the emancipation of women in LDCs is a positive sign, but before we get too gung-ho and pressure other cultures to change their ways, we should consider political undercurrents. Changing societal values is an uphill struggle, as witnessed at the 1994 United Nations conference on population in Cairo. Leaders, particularly of Muslim nations, were attempting to tread a tightrope between supporting the effort to improve the lot of women and avoiding a fundamentalist backlash. The lesson of the Iranian revolution in 1978 is not quickly forgotten. In the 1970s, oil revenues promoted prosperity in Iran, and cultural and economic changes indicated increasing Westernization of that culture. Women started to dress in the fashions of Paris rather than in the all-concealing clothing of traditional Muslims, and Iran became a playground for the rich. However, the rule of the Shah of Iran was oppressive, and fundamentalist leaders were able to convince the underclass that the dilution of Islamic values must stop. In 1978, the Shah of Iran was deposed by fundamentalist Muslims led by Ayatollah Khomeini. The revolution in Iran achieved its purpose: Islamic law prevailed, and the drift toward Westernization was reversed. The resurgence of fundamentalist religions is a sign that a vocal part of society is demanding intense conservatism to avoid rapid, unsettling change, particularly change threatening the status of male supremacy. Those who advocate societal change must temper their impatience

with the fear of an Iran-style backlash that undoes their efforts.

16.10 Human Population Growth and Consumerism

Environmental problems are generally adverse changes to ecosystems attributable to the actions of humans. The human population is, therefore, the root cause of most environmental problems, and the sheer number of humans plays a very significant role in degrading the environment. However, a second factor that must be considered is the amount of resources demanded by each person. Although it is often the countries with the lowest standards of living that have the highest population growth rate, these people cannot afford electric light or sewage, much less a car, boat, or air conditioner. The wealthy nations may have low birth rates, but they have a much higher per capita demand for resources. Globally, the United States and Canada have the greatest per capita resource demands of any nation. If that demand is converted into standard units, we find that the average North American uses approximately twice as many resources as a Briton, a German, or a Japanese and approximately 400 times the resources of a citizen of nations such as Rwanda or Ethiopia. The birth of an extra child in the United States places a proportionately greater load on the environment than an extra child elsewhere in the world. If all people on the planet used resources at the North American rate, there would be serious shortages of oil, gas, water, timber, and food in just a few years. Indeed, even though the United States is self-sufficient in some of those resources, it still subsidizes its present consumption levels by importing energy, timber, and foodstuffs, and mining its aquifers. Thus, although the effects of deforestation, desertification, and environmental degradation may be most apparent in the developing world, the cause of much of the damage is the appetite for consumption of the developed nations. The path toward sustainability can be followed only if the developed nations are willing to reduce their consumption, and that means a change in lifestyles for many of us. To date, the changes that would reduce the demand for energy and materials in MDCs, such as energy conservation in industry, agriculture, and the home; reduced use of cars; water conservation; and less shipping, processing, and freezing of foodstuffs have not received widespread public support.

Summary

- The development of agriculture allowed human populations to settle and boom.

- Human population growth is exponential, adding about 1 billion people every 10 years.

- The biological carrying capacity of the planet for humans may be about 50 billion. An acceptable cultural carrying capacity may be about 8 billion.

- Postindustrial economies are likely to show lowered birth rates and population stabilization, having passed through the demographic transition.

- The economic solution of the demographic transition is too slow to prevent the next doubling of population.

- The best single action to limit population is to educate women. The number of children born to a woman and the age of first reproduction determine the rate of population growth.

- Improved health care, sanitation, and family planning are effective tools to lower birth rates.

- Attempts to limit population growth may be resisted for social, economic, political, or religious reasons.

- The increased use of resources by humans precipitates environmental problems.

Further Readings

Cohen, J. E. 1995. *How Many People Can the Earth Support?* New York: W.W. Norton & Co.

Ehrlich, P. R. and A. H. Ehrlich, 1990. The population explosion. *Amicus* (4):23–29.

Freed, S. A. and R. S. Freed, 1985. One son is no sons. *Natural History* 10:12–13.

Hardin, G. 1986. Cultural carrying capacity: a biological approach to human problems. *BioScience* 36:599–606.

Jacobson, J. L. 1991. India's misconceived family plan. *Worldwatch* (6):18–25.

Larick, R. and R. L. Ciochon. 1996. The African emergence and early Asian dispersals of the genus *Homo. American Scientist* 84:538–551.

Leakey, M. and A. Walker. 1997. Early Hominid fossils from Africa. *Scientific American* 276 (6):74–79.

National Geographic, 1998. Special Issue on Population. October 1998.

World Resources Institute. 1998. *World Resources* 1998–99. Oxford: Oxford University Press.

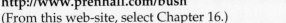

Web Connections

On-line resources for this chapter are on the World Wide Web at:
http://www.prenhall.com/bush
(From this web-site, select Chapter 16.)

P A R T

THREE

Ecological Impacts of Changing Land Use

Feeding the World

17

To a man with an empty stomach food is God.

—*Mahatma Gandhi (1869–1948)*

There should be no more people in a country than could enjoy daily a glass of wine and a piece of beef for dinner.

—*Thomas Malthus (1798)*

To a Hindu such as Mahatma Gandhi, the concept of eating beef would be repugnant. Nevertheless, the standard of living indicated by the meal that Thomas Malthus described would, for many peoples in less-developed countries, represent true luxury. The primary concern of people who are not on the brink of starvation is the quality of life rather than pure survival. Pleasure may be found in personal rewards, such as food, cars, houses, and clothes, but the quality of life for many people is also measured in less tangible assets, such as personal and religious freedoms, wildlife, clean air, and natural landscapes. If the number of humans on the planet continues to expand until it reaches the biological carrying capacity, huge sacrifices will have to be made in both the tangible and intangible aspects of the quality of life.

The farms of the world presently produce enough food to feed the population of 5.7 billion, but the food is not distributed equitably. Butter mountains and wine lakes result from overproduction, yet the world has an estimated 750 million hungry people, 20 million of whom die as a direct result of that hunger each year. The glut of food is in the MDCs of North America and Europe, while the majority of starving people are in the LDCs of Africa and Asia. To assert that all the food should be redistributed to feed the starving is simplistic; economic, logistic, and political barriers prevent it. Although hunger is present in every nation, it is most marked where natural calamities or warfare drive human suffering. In this chapter, some of the factors that have contributed to improved food production over the last decades and some of the factors that might constrain further improvement, are investigated.

17.1 Human Nutritional Requirements

The Food and Agriculture Organization (FAO) of the United Nations recommends an average nutritional intake to be about 2500 Calories per day. However, this figure is an average for the whole adult population, and the activity level and size of an individual will determine that person's actual food requirements. Caloric intake is an indication of the raw energy provided to the body. The unit of measurement, the calorie, refers to the amount of energy required to raise the temperature of 1 g of water 1°C. One thousand calories is a kilocalorie, which is often written as a Calorie (note the capital C). Food and energy expenditure are usually discussed in Calories. A person engaged in hard physical labor may require in excess of 4000 Calories a day, whereas a small sedentary individual may require only about 2000 Calories a day. Failure to maintain an adequate diet can lead to weight loss, listlessness, increased susceptibility to disease, reduced learning capabilities, and, ultimately, starvation.

To be complete, a human diet also needs—in addition to the Calories obtained mostly from carbohydrates—protein, vitamins, and minerals. Protein is made up of amino acids that are released through the process of digestion. Humans need 20 different amino acids, 8 of which cannot be manufactured within our bodies and can be acquired only by eating them directly. A deficiency of protein can result in the disease kwashiorkor or, if coupled with a general caloric deficiency, marasmus. Children are especially susceptible to these diseases, and those who suffer from them are characteristically undersized, pot-bellied, and skinny-limbed. They appear listless, with almost expressionless features, but stare disconcertingly with the eyes of the old. This gaunt face of malnutrition became all too familiar on the news and in magazines documenting the African famines of the 1980s and 1990s.

Vitamin deficiencies may result in disease and eventually in death. For example, a lack of vitamin A causes blindness; vitamin B deficiency causes the potentially fatal disease beri-beri; too little vitamin C leads to bleeding gums, loss of teeth and hair, and potential death from scurvy; a vitamin D deficiency causes weak bones and the bow-leggedness of rickets. Most of these deficiencies can be rectified by taking vitamin supplements, and the effects of the earlier stages of vitamin deficiency can be reversed if treated promptly.

Minerals are also an essential component of a balanced diet. Sodium and potassium are needed to maintain the correct balance of body fluids. Calcium is needed for the growth of healthy teeth and bones, and iron is required to avoid the debilitating condition of anemia.

Children are especially vulnerable to all kinds of dietary deficiencies because the process of growth requires a huge input of all dietary components. Similarly, pregnant and nursing mothers need enhanced nutrition because the physiological strain of feeding both themselves and a child is enormous.

A person is considered to be undernourished if his caloric intake is 90% of the recommended level for his size and occupation. If the caloric intake is below 80% of the normal level, then the individual is said to be severely undernourished. These are people with well-balanced diets but an inadequate energy input per meal. Sometimes the caloric intake is adequate, but the diet is missing some vital component, such as a particular mineral or vitamin; a person with this type of diet is said to be malnourished and suffering from **malnutrition**.

One of the principal obstacles to the development of LDCs is that a substantial portion of their population is malnourished. A poor diet generates a vicious circle (Figure 17.1) that reduces the chances of the sufferer's breaking free from the poverty trap either through learning or through hard work. Another consequence of this cycle is that if the diet is poor, infant

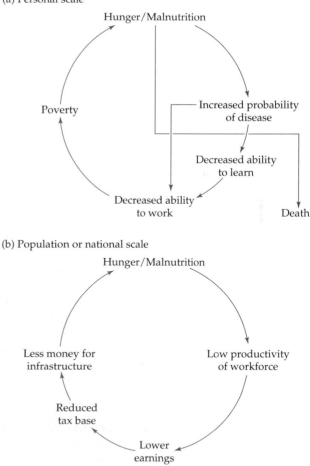

Figure 17.1   Social effects of hunger. (a) The cycle of poverty and malnourishment that can lock an individual into a downward spiral, reducing the ability to work, remain healthy, or even live. (b) Hunger and malnourishment among a workforce can have serious consequences for the economic development of a country.

mortality will be high and, as discussed in Chapter 16, birth and population growth rates will also be high.

17.2   Agriculture Versus Population Growth: A Deadly Race

Since 1950, global food production has more than doubled; in fact, it has increased by a factor of 2.6 (Figure 17.2). Food, at least in developed nations, now consumes less of our disposable income than at any time in history. The average North American family now spends about 9% of its income on food, whereas in LDCs families commonly spend 40%–60% of the family income on food. Even in MDCs, however, the agricultural news is not wholly good. For the past decade or more, per capita food production from U.S. and European farms has remained nearly constant. Indeed, in 1993, harvests in the United States were almost 5% by weight lower than the harvests of the previous year. This was due not to technological failure but to a wet spring and the flooding of the Mississippi.

Poorer than usual harvests are not a disastrous blow to food supplies in the United States, where there are grain stockpiles, but they do force up the world's price of grain. In countries with less robust economies and lower income levels, an increase in grain price can make staple dietary items, such as flour or bread, unaffordable. In general, the greater the world grain surplus, the more resistant the market is to short-term fluctuations. As population presses upward and food supply fails to increase at the same rate, the world markets for all food commodities will become more volatile. To understand the trends of modern and future food supply, we must look at the processes that drove agricultural improvement in the 1950s and 1960s.

The First Green Revolution

New agricultural practices, new crops, and mechanization all contributed to a jump in food production in MDCs in the 1950s and 1960s; these improvements together were called the **Green Revolution**. The resurgent economies of MDCs in the post–World War II era were built on cheap oil. Agriculture in the MDCs benefited in many ways from this abundance of energy; mechanization, the change from horse and mule to tractor and combine harvester, enabled farmers to work larger fields. Cheap energy could also be used to power irrigation schemes that allowed previously marginal land to be brought under the plow. Since 1950, the agricultural use of fuel has risen 400%, and agriculture now uses one-twelfth of all oil produced. In the 1950s, a huge effort was made to breed and introduce improved crops that provided higher yields. Previous attempts to increase the yield of corn, rice, and other cereal crops had not always been successful. A recurrent problem was that, as the seed production increased, the plant would collapse under its own weight, making it impossible to harvest. The weight of the head would literally break the stem of the plant. In

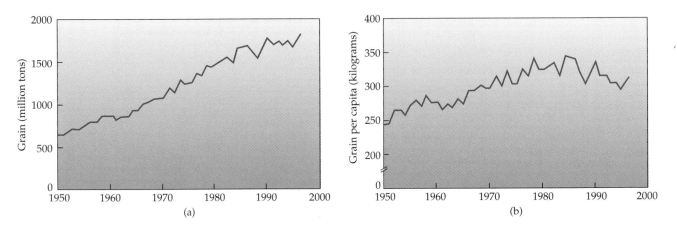

Figure 17.2   Trends in global grain production. (a) Increased global grain production between 1950 and 1996. The improvement is largely due to the adoption of Green Revolution farming technologies in both MDCs and LDCs. A recent decline in production has been largely due to the faltering economies of eastern European countries. (b) World grain production per person between 1950 and 1996. The peak abundance of food per person was in the early 1980s. Since then, world population has grown, changes in U.S. farm policy have reduced U.S. production, and reduced harvests in eastern Europe have contributed to falling per capita grain availability. (After L. R. Brown, C. Flavin, and H. French, 1997.)

the 1950s, crop breeders crossed (bred) crop varieties that produced huge quantities of seed but had floppy stems with varieties that had short, strong stalks and small seedheads. The resultant stubby plants with large seedheads provided vastly improved harvests. The gains in production were so significant that the newly bred crops became known as **miracle strains**.

The high-yield varieties were more demanding than the wild-type strains and needed more fertilizer. A huge fertilizer industry developed that synthesized **inorganic fertilizers** from byproducts of the oil industry. Previously, fertilizer had been based on dung, composted plant remains, and other organic (carbon-rich) materials. The new fertilizers were inorganic; they contained just the nutrients (such as nitrogen, phosphorus, and potassium) needed by plants, without the carbon. The inorganic fertilizers delivered a concentrated input of nutrients to the crop and were immensely successful. Since 1950, the amount of fertilizer used on U.S. farms has risen by 1700% (Figure 17.3).

Growing better crops is half the agricultural battle. The other half is protecting them from pests. The Green Revolution made a major effort to tackle the pests that destroy about 30% of all crops grown (Figure 17.4). Specific **pesticides** were used to protect stored grain from fungus before it was sown, and the crops in the field were doused with chemicals to fight off infestations of herbivorous insects (Figure 17.5), weeds that would compete for nutrients, and pathogens, such as molds and rusts, that could decimate a crop. The harvest, once gathered, was protected from rodents, fungi, and bacteria. Livestock were also protected from disease by being given preventive doses of antibiotics, and their growth was enhanced through hormone injections. The petrochemical in-

Figure 17.4   A swarm of locusts can strip a field in a matter of minutes.

dustry manufactured many of these pesticides and drugs, inputs of which have risen 3200% in the years since 1950 (Figure 17.6).

The introduction of miracle strains of wheat and corn and the pesticides to ensure that the harvest was not lost, encouraged the establishment of **monocultures**: fields planted with a single crop type year after year. The new farm machinery, crops, and chemicals were very expensive, so farms became more and more specialized, with the unwanted consequence that they were made vulnerable to changing market conditions for the crop they grew. If crop prices fell, the farmers

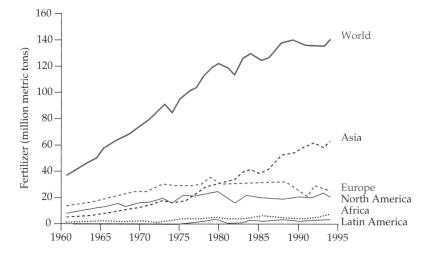

Figure 17.3   World use of fertilizer 1961–1994. (Redrawn from *World Resources 1998–99.* Data from Food and Agriculture Organization of the United Nations [FAOSTAT], 1997.)

Figure 17.5   A crop-dusting plane sprays pesticides.

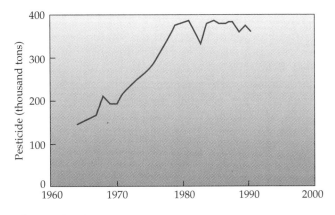

Figure 17.6   U.S. use of pesticides (active ingredients only) 1964–1991. (After L.R. Brown, H. Kane, and D.M. Roodman, 1994.)

from 1950 to 1980, farm yields grew rapidly; science and government policy had been applied to agriculture, and the public benefited as food prices fell.

The transition to this intensive agriculture, with its high energy inputs, machinery, improved grain, and chemicals, was not without social upheaval. Small farms proved to be uneconomic and were gradually sold and amalgamated into larger enterprises. These large, profitable farms were often bought by pension and insurance companies. Gradually, decisions about planting and harvesting were being made not by people on the land, but by people in the boardroom. Since 1950, about half the farms in the United States have gone out of business and the farmers have moved off the land and into the city. Large farms dominate the agricultural market: About 56% of food presently produced in the United States comes from 7% of the farms. This trend toward larger, more economically powerful farms is predicted by the U.S. Department of Agriculture to continue so that by the year 2000 50% of the food produced will come from just 1% of the farms. Such corporate farming, dubbed **agribusiness**, is the single largest industry in the United States. Although agribusiness leads to increased crop returns, it has some disturbing consequences: for instance, the connection between seed suppliers and petrochemical companies. In the past few years, petrochemical giants have bought more than 400 small seed-producing companies. The seed producers used to raise many genetic hybrids that gave distinctive flavors or high yields on narrow ranges of soil types; farmers could buy seed "tailor-made" for their soil. Recently, the emphasis has shifted to altering the land to fit the crop, rather than choosing the crop strain to fit the land. Soil properties can be altered by the addition of minerals or fertilizer. Fewer strains of crop are then needed, but those that remain have the unifying character that they require high fertilizer and high pesticide inputs. It is not a coincidence that the petrochemical companies have become involved in seed growing. By promoting the use of a narrow range of seed types, the companies can guarantee their sales of pesticides and fertilizers. The seeds they sell will produce high-yielding crops, but whether these are necessarily the best varieties for the future of agriculture is open to debate.

A further concern is whether the growth in harvests experienced on farms in the 1950s and 1960s can be sustained. The meteoric increase in agricultural production on U.S. farms tapered off in the 1970s and 1980s, and in the 1990s crop yields (in per capita terms) have fallen slightly. The lowered yields are largely due

could go bankrupt. To maximize farm yield and to stabilize market prices, the U.S. government implemented federal farm subsidies and guaranteed prices. This policy further encouraged the adoption of monocultures, because the subsidies were available only when the fields were being used for certain crops. Fallow fields and fields laid down to alfalfa, clover, or other crops that help to revitalize the soil, but are not marketable, were not subsidized. During the period to changes in federal subsidies that have resulted in less acreage being farmed than in previous decades. Without subsidies, some farmland has proved uneconomic and thus has been allowed to go fallow. Furthermore, the Farm Bill passed in 1985 provides subsidies to farmers who allow their fields to lie fallow. The decline in crop production, therefore, represents an economic rather than a biological adjustment, and it could be reversed as demand for grain increases. But the yield increase experienced early in the Green Revolution may not be matched again.

Improved harvests were based on high-yield crops and the use of fertilizers, but adding fertilizer has only finite benefits. It may seem surprising that continued watering and fertilizing does not necessarily increase crop yield. Irrigation water or the nutrients in the fertilizer are supplements for natural resources. If one of those resources is in short supply, it may limit the growth of the plant. Adding water or a nutrient supplement to the soil or leaves of the plant overcomes that limitation and the plant will increase in size until it is limited by the absence of another resource.

If large amounts of fertilizer are added, the plant no longer finds any of those chemicals limiting and some other resource becomes the limiting factor. A trace mineral that is not included in the fertilizer may become limiting, or it may be the carbon dioxide concentration in the atmosphere (the plant's only source of body-building carbon), or a climatic factor, such as nighttime temperature, that sets the limit to growth. Consequently, plant response to fertilizer is often good at first, producing increased yields, but ultimately, a point is reached at which adding more fertilizer brings no further gain in harvest. Thus, when farming intensified at the start of the Green Revolution, the benefits of increased fertilization were immediate. Now those inputs are a normal part of farming, and a further surge in farm output will take more than fertilizer.

The Mining of Aquifers

For 5000 years, irrigation of marginal land has been an important component of increased crop production. Wind- and gasoline-driven pumps have greatly increased the ability of farmers to draw water from the aquifers beneath their fields so that it can be piped to their crops. The ability to guarantee water supply, despite erratic rains, has played a huge part in increasing the probability of successful harvests and keeping food costs low.

Aquifers are supplied with water from the side (lateral recharge) and from above (percolation; see Chap-

ter 7). If the extraction of water is less per unit time than the inflow of water, the aquifer is being used sustainably. However, if the rate of extraction exceeds the rate of resupply, the aquifer is being pumped dry. Under these conditions, the aquifer is not being used sustainably, and its use can be likened to the **mining** of a nonrenewable resource such as coal or oil.

An example of an aquifer that has been mined is the Ogallala aquifer (Figure 17.7) of the central plains of the United States. The dry prairies from Texas to

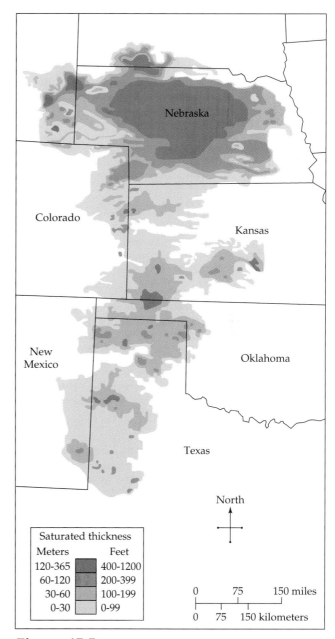

Figure 17.7 **Water withdrawal from the Ogallala aquifer to support agriculture exceeds rates of recharge.**

South Dakota have been transformed into fertile farmland by the pumping of water from the Ogallala aquifer. These lands provide 15% of the corn and wheat grown in the United States, 25% of its cotton, and 40% of its beef. The crops grown, especially cotton and sugar beets, have a high demand for water, but are highly profitable. However, the rate of water extraction is eight times the rate of recharge; by 1980 the depth of the aquifer had been reduced by between 3 m and 30 m. An aquifer is not uniformly rich in water; some farms may overlie areas that have deep reserves of water, whereas others, closer to the edge of the aquifer, may have only a shallow deposit. Because the lateral movement of water through the rock of the aquifer is so slow (about 1 m of lateral flow per century), it is possible for individual wells to be pumped dry while other wells continue to produce water. The first farms to be affected will be those overlying the shallowest part of the aquifer.

The aquifer as a whole will not be pumped dry; before that happens, there will be changes in both crops and management styles. These may come about through an increased awareness of the value of water as a finite commodity. Or, failing that, the increased cost of pumping water from deeper and deeper reserves will make it too expensive to continue to irrigate the land at the same rate. Farmers will be forced to adopt new crops that do not require irrigation or to abandon their land.

Any law requiring a rationed or sustainable use of water would be difficult to enact and enforce. Educating farmers to treat water as a scarce resource seems the most likely path to ensuring that irrigated lands can stay in cultivation.

17.3 Exporting the Green Revolution

There has been a long tradition of trying to export temperate-zone agricultural technology to the tropics. It has long been failing. The application of Green Revolution techniques during the 1960s and 1970s in LDCs (most of which are in or contain tropical regions) became known as the Second Green Revolution. Initially, the program met many obstacles. Monocultures became pest-ridden, crops developed for temperate climates withered or drowned in the tropics, and the monsoon downpours washed away expensive fertilizers and pesticides. When success came, it was based on crops raised specifically to thrive under tropical climate regimes. These crops featured the same combi-

nation of short stalks, large seedheads, and fertilizer dependency that they had in the temperate system, but they were better adapted to the growing season of a tropical climate. Multiple cropping, the raising of more than one crop in a year, is possible in the tropics because temperature changes little throughout the year. Raising multiple crops was facilitated by the invention of fast-maturing varieties of the miracle crops. As in the first Green Revolution, fossil fuel inputs were required for machinery, irrigation, and pesticides, and the increased demands on the soil were offset by the addition of fertilizer. Gains in tropical farm yield were as spectacular as temperate farm gains (Figure 17.8), with harvests increasing between 200% and 500% from previous levels.

Despite social problems, the agricultural changes brought about huge changes in crop yield and have enabled many countries to maintain per capita food production, even during times of rapid population growth. Asia and South America have experienced a doubling of food production, much of which is due to improved agricultural practices, but some of which is also due to expanding their agricultural lands. Although the increase in food availability is welcome, it should be recognized that it has come at a cost to tropical forests. The new farmland is land gained from clearing the species-rich tropical forests, and the price is lost biodiversity.

In the former USSR, the collapse of communism and the faltering path toward capitalism undermined their agricultural system. Food production plummeted, and, although populations are relatively stable in these countries, the per capita food availability declined markedly. The most troubling decline in food availability is in Africa. Rapidly increasing populations of African countries play a part, because many nations are still doubling in size every 20–25 years. A decade-long drought in the sub-Saharan region (Sahel) brought famine and lost food production. But the greatest ally of famine was warfare. The 1980s and 1990s have been times of almost continuous civil war in the poorest African nations. Eritrea, Sudan, Chad, Somalia, Ethiopia, Angola, Liberia, and Rwanda have all been immersed in bloody conflicts. Tides of refugees, abandoned farms, and social breakdown are all indicators of a land where harvests are lost. The warfare creates a power vacuum that is often filled by military rulers whose spending is more likely to favor tanks than tractors and soldiers than seed. The armies of such rulers often become the tyrants of the people they are there to "protect." Ethnic, cultural, tribal, or religious divisions fester and erupt into further fight-

(a)

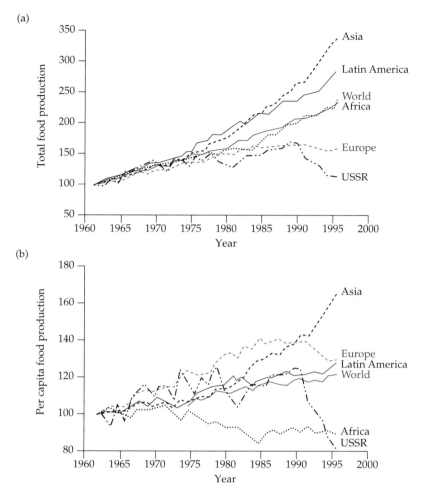

(b)

Figure 17.8   Global grain production by region. (a) Production for Asia, Africa, Oceania, South America, and the former USSR (1961–1996). Grain production increased in all regions during this time, although social change in the USSR has resulted in reduced harvests since 1990. (b) Grain production per capita 1961–1996 reveals that increased food production has kept pace with population growth in India and China (Asia), but has failed to keep pace with population growth in Africa and the former USSR. In the former USSR, this is a recent and probably reversible trend. In Africa, there has been a continuous reduction of food availability, suggesting a far more serious long-term outlook. Values are indexed against 1961 production levels. (Redrawn from *World Resources 1998–99*. Data from Food and Agriculture Organization of the United Nations [FAOSTAT], 1997.)

ing. Such warfare has characterized the recent history of many nations and can be argued to be the greatest threat to agricultural and economic development.

Where warfare threatens a population, there may be an international peace-keeping initiative, as in Somalia in 1992. But, as with most famine relief, such interventions are the equivalent of bandaging the mouth of someone with pneumonia. The remedy may (temporarily) prevent the symptom of coughing, but it does nothing to address the cause and is an inappropriate treatment. Providing food to starving people is a temporary solution, although it may allay our guilt when we see the newscasts of starving children. However, if the fundamental reason for the starvation lies in warfare, poor farming practices, overpopulation, or climatic change, deliveries of food will not correct the problem. Unless the root cause of the famine is addressed, further famine will follow. If a minimum acceptable diet is to be provided for every person, long-term changes must be sought in political, economic, and agricultural development.

17.4   Social Problems and the Second Green Revolution

A number of social problems have dogged the export of the Green Revolution. The adoption of intensive agricultural techniques requires training and an input of cash to pay for improved seed, fertilizer, pesticides, and irrigation equipment. Consequently, it was often the wealthiest farmers who were able to adopt the new technology. They reaped the rewards of increased yields. As more produce became available, prices fell and food became more affordable. This change benefited the urban population, who had cheap food, and the wealthy farmers, who could grow a lot of food. It was the poor farmers who could not afford the initial investment to intensify agriculture on their farms who were the losers. They suffered a loss of income as the crops they produced were worth less. In some countries, subsidies were made available to redress the balance, but often this money did not reach those who needed it most. Farmers who were literate or had political connections were most likely to obtain financial aid; once again, the poorest farmers saw scant benefit.

Another problem area lay in the targeting of education and aid. At first, the aid advisers and the recipients were almost all male, yet in many nations it is the women who raise the crops for household consumption. The gender bias had the effect of promoting the men's cash crops at the expense of the women's subsistence crops. It is now recognized that this approach was an error and that standards of living would have been raised more effectively if women had been given equal access to the new technologies and funds.

Large-Scale Economic Problems

A further cause of local food shortages is the result of policies to reduce national debt. Many nations use agricultural production to generate export revenue in an attempt to offset the cost of imported goods. In LDCs, the land is often given over to **cash crops**, such as tea, bananas, coffee, spices, or chocolate, to be exported in the effort to fight their debt burden. That land could be used for domestic grain production, but the politics of debt will not allow it. An individual farmer can make more money raising a cash crop than growing corn.

It can be argued that the ultimate cash crop of the Andean foothills is *Erythroxylum coca*, the source of the basic ingredient of cocaine. In this instance, it is not the government that is promoting the crop but the drug lords. The coca bushes are very easy to grow and harvest; the young leaves are simply plucked from the tips of the branches. A farmer can increase his annual income from about $800 a year growing corn to about $9000 a year for the same acreage of coca. The leaves are then sold and processed by the operations of drug lords into cocaine for illegal sale in the MDCs.

From before the time of the Incas, farm families in the Andes or the lowlands of western Amazonia have grown a coca bush or two, not to sell the produce, but to have a personal supply of painkillers. Chewing the leaves will cure minor aches, will delay feelings of physical tiredness, and will alleviate the gnawing pains of hunger. In efforts to control cocaine production, government agents have destroyed these bushes providing homeopathic remedies as well as the major coca fields. A farmer I met living in a remote area of Peruvian Amazonia had grown just one coca bush, but it had been removed by soldiers. Puzzled, he asked: "What am I meant to do next time my daughter gets a toothache?"

The potential for increased crop production in LDCs exists. On the grand scale, social and political systems have to be established to provide regional stability so that investments can be made in farms rather than in armies. On a more local scale, the agricultural practices of many farmers could be improved. Inferior crop strains and poor farming techniques that lead to a rapid degradation of the land, such as slash-and-burn agriculture in Amazonia, are still the norm in many areas.

Conservation goals, particularly the maintenance of biodiversity, can be met through farming improvement. If better land-management practices are taught and the appropriate farmers receive the aid to allow them to farm more sustainably, the need to clear mature rain forest can be reduced. Farm yields will rise, and countries will get closer to the demographic transition.

17.5   Desertification

The reduction of land to an infertile state, **desertification**, is usually the consequence of progressively losing the organic component of the soil. The resulting earth is sandy, does not retain water, and is liable to blow or wash away. This global problem (Figure 17.9) has two basic causes: overuse of the soil and changing climates.

The Human Factor

A human population that is near the biological carrying capacity of the land and that is doubling every 20–25 years, strains the capacity of farmers to provide enough food. Every hectare of land must produce a good harvest or there will be food shortages. A drought reduces the number of fields that can support a crop and reduces the yield on the remaining fields. Soils that are only just fit for agriculture are often the most sensitive to changes in land use or climate. Areas that barely provide an agricultural return worthy of the labor input are termed **marginal** farmlands. Such marginal land can be brought into production as grazing land, but there is always the danger of overgrazing. Many regions, from the tropics to the Mediterranean, have been reduced from marginal agricultural land to "goatscapes." Goats, through their complete dietary impartiality, have overgrazed these landscapes, and now nothing but this hardy beast can eke out a living. In other areas, cattle may prevent the natural succession and recovery of the land. The grazing and trampling by overstocked cattle damage soil struc-

Figure 17.9   **Areas threatened with desertification.** Climate change and changing land use are causing deserts to expand and invade marginal agriculture land. (United States Congress on Desertififcation, Nairobi 1977. After B.J. Nebel and R.T. Wright 1993. *Environmental Science* 4th ed. Prentice Hall, Upper Saddle River, NJ.)

ture, remove the protective plant cover, and are a leading cause of soil erosion in many tropical areas.

Compounding the problem of overgrazing is the collection of firewood. Wood is the main energy source for cooking and heating in the rural areas of many developing countries. The removal of wood can be a significant force in deforestation, especially in semiarid areas or where there are dense concentrations of humans. The influx of Rwandan refugees into Uganda in 1994 resulted in disease epidemics, the consequences of which will be temporary. However, the deforestation caused by a million people searching for firewood may pose a longer-lasting problem for the affected area. The deforestation has the immediate consequence of increased erosion, which will wash away the best soils. A future consequence of the removal of firewood is the loss of nutrients from the system; instead of decomposing and returning its nutrient hoard to the land, the wood has been carried away.

In many of the marginal lands of northern Africa, the only way to scrape a living from the land is to be a nomadic herder. The Masai of Africa are an example of a people culturally adapted to living as herders in marginal lands. To avoid overgrazing a particular spot, they and their herds move constantly. For this way of life, they require large expanses of land to sup-port relatively small herds of cattle, goats, sheep, or camels. As populations grow and political boundaries limit nomadic movement, the land available to the herders is reduced; they are forced to confine their livestock, and the grazing pressure intensifies. These lands are the least able to support grazing or the gathering of firewood. The result is that even the scrub forest of these lands is lost, erosion increases, grazing becomes even more impoverished, and the pressure on the remaining plants intensifies. The loss of plant cover reduces the supply of moisture to the air from evapotranspiration; that reduction reduces cloud cover and the chance of rain. The loss of shade and cooling rains results in increased soil temperature, which kills the microorganisms essential for nutrient cycling, and so the cycle goes on until the area is a desert. Such pressures on land in the Sahel have been too great. Throughout large areas, the combination of drought and overuse has resulted in severe desertification. In the Sahel, drought-induced crop failures provoked a regional famine that cost the lives of at least 1 million people; the misery has been and still is exacerbated by civil wars.

Even actions by international agencies trying to help the nomads have proven disastrous. After World War II, the Agency for International Development

(AID) tried to soften the hardships of nomadic living by drilling wells to provide clean drinking water. Other aid services such as hospitals and schools were sometimes sited with the wells. Villages sprang up around the wells as nomads with an ancestry of wandering the great semidesert of North Africa settled for the first time. Temporarily, living conditions improved as clean water and permanent housing lowered infant mortality rates. The provision of health care and a settled life-style led to women raising more children than before, and populations started to increase rapidly. Of more immediate concern, the flocks of animals were now concentrated around the wells, overgrazed the area, and led to desertification. There were now large aggregations of people with water but little food, and famine a short step away.

Climate Change

Not all desertification can be laid at the door of humans. Natural changes in Earth's climate are going to affect these marginally cultivable lands more than others. A slight shift to a drier or a warmer climate can lead to the desertification of even well-maintained land. The most harrowing example of desertification has come as a result of a deadly combination of climatic change and inappropriate land use. The northeast African countries of Angola, Ethiopia, Somalia, Sudan, and Chad have been savaged by a drought that has lasted since the late 1960s (Figure 17.10).

The spread of desert in Saharan Africa is not new. During the last ice age and the early post–ice-age period, the Sahara supported woodlands and grasslands. Rivers flowed northward through this land to the Mediterranean Sea, and the region was home to early agriculturalists. Progressive climatic change over the last 6000 years has steadily reduced rainfall in the re-

gion, and the desert has spread to cover the cities and once-fertile lands. The climatic cycles of the last several thousand years have minor wetter and drier phases superimposed on a general drying trend. Thus, the spread of the Saharan desert is basically a natural phenomenon. The most recent drought in northern Africa has caused a further expansion of the desert into the marginal semidesert areas at its edge, and actions of humans trying to survive in these areas have hastened the degradation of the land.

Human intervention to halt completely the process of desertification is unrealistic given our present technology, because desert boundaries move according to long-term climatic change. Far from reducing the effects of climatic change, human activities may increase the trends that lead to desertification. Pollution leading to a doubling of carbon dioxide concentrations in the atmosphere (Chapter 23) is predicted to increase desert areas by as much as 17%. These climatic changes are operating over relatively long periods of time, a hundred years or more, while the next famine sequence is likely to be much more immediate. Control over industrial emissions is needed, but shorter-term resolutions should also be sought. Changes in land use and even in the cultural values of the people affected will be needed to reduce risks of famine.

Is There a Simple Solution to Famine in the Sahel?

At the root of the immediate problem is the size of the human population. In marginal settings such as the desert fringe, the carrying capacity of the land for humans and their livestock fluctuates according to short-term climatic trends. A single wet-season failure is enough to induce famine. Farming a semidesert environment will always be a precarious endeavor, as crop failure through a chance lack of rain is inherently more likely and more catastrophic than in other areas. However, agricultural practices could be modified to make the best use of available water by planting drought-resistant crop varieties and introducing irrigation wherever possible.

Reforming the social values that lead to high birth rates and overstocking the land are necessary, but their implementation can strike to the core of local cultures. For example, among nomadic herders personal wealth is often measured in the number of livestock owned. Increasing herd size on land that is being progressively overgrazed accelerates degradation of the land, but persuading the herder to forego this enhancement of social status requires altering the long-held traditions of an indigenous society.

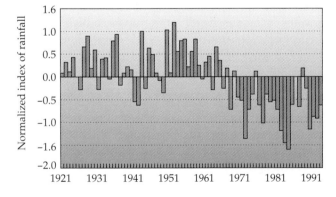

Figure 17.10   Rainfall statistics for the Sahel region, showing the onset of the drought that provoked regional famine in the 1980s. The vertical axis is a "normalized" scale.

Although applying best-management practices and Second Green Revolution technologies may solve part of the problems in the Sahel, nothing can be achieved without regional peace followed by substantial financial investment. Historically, the public in developed nations are more responsive to images of a crisis in progress than to appeals to prevent a crisis. Consequently, donations flow in to save starving children, but there is no equivalent outpouring for agricultural assistance that would prevent the problem in the first place. For a long-term resolution to food problems in LDCs, rather than throwing money into emergency relief flights carrying U.S. grain to starving people, we should be investing in their agriculture and infrastructure when times are relatively good. The salutary lesson to be learned from the tragedy of the Sahel is that any human population reliant on crops growing at the limits of their ecological tolerances is going to be highly vulnerable to climatically driven famine. Famines like those of the 1980s will be repeated.

17.6 Sustainable Agriculture

If all countries farmed as intensively as the United States, it is estimated that all known oil reserves would be depleted in about 12 years. While the intensification of agriculture has brought huge benefits, the improved harvests are built on the availability of fossil fuel. The reliance on nonrenewable commodities, such as oil to power machinery and for manufacturing fertilizer and pesticides, means that this farming style cannot be practiced indefinitely. A growing dissatisfaction with rising costs of chemicals, increased resistance of pests to pesticides, and concerns over creating a chemical soup instead of a healthy soil, have led many people to question modern intensive agricultural practices.

An alternative set of farming procedures has been developed, some of which hold commercial promise. Any farming strategy that leads to falling crop productivity, be it intensive overuse of the land or an overly "natural" approach, is morally dubious given our rising global population. The best of the **sustainable farming** practices can be likened to an individual's becoming health conscious. We know that if we eat well and exercise regularly, we will be fitter and healthier, will require less medical attention and fewer dietary supplements, and overall will be stronger and happier. Farming is little different. If the land can be returned to a naturally "healthy" state, the necessity for input of inorganic pesticide or fertilizer will be greatly reduced. The crux of this argument rests on the ability of these new farming techniques to maintain optimal farm yields, and that has yet to be proved.

No commercially viable technique for farming is completely sustainable; sooner or later fossil fuel will be used to power the harvester, the truck that takes produce to market, or the pump that irrigates the land. Complete sustainability of society would require greater changes than can be achieved by farmers.

Sustainable, in a farming sense, really applies to reducing reliance on high inputs of inorganic chemicals to the land, which will result in less pollution and reduced reliance on fossil fuels to make fertilizer. Organic farming is a means of sustainable farming, relying solely on natural inputs, such as dung and compost as organic fertilizers and plant extracts as organic pesticides. This method represents a desirable extreme in the sense that it minimizes pollution and chemical use, but it may not produce the yields that we need. Sustainable agriculture is not limited to organic farming. Indeed, sustainable agriculture is not philosophically opposed, if needs arise, to using inorganic compounds in order to increase farm yield.

Good farming centers around taking care of the soil. The soil must have a good structure, be safe from erosion, and have a lively set of decomposers and adequate nutrients. Whereas intensive farms rely on heavy inputs of inorganic fertilizers to maintain fertility, other paths can be followed with equal success.

Fundamental to maintaining soil fertility is avoiding soil erosion. **Contour plowing**, in which the field is plowed across rather than up and down the slope, reduces erosion during rainstorms. Soil erosion can also be minimized by not burning the stubble in the fall because burning reduces many of the plant nutrients to ash that washes away. The burning also makes the soil so friable (brittle) that wind erosion can be a problem. Planting a winter cover crop of vetch or some other plant that actively fixes nitrogen helps soil fertility while protecting the soil from erosion by winter storms. Come spring, the cover crop is plowed into the soil as green manure that will decay to provide more organic matter in the soil and improve the soil texture.

The Green Revolution relied on the planting of the same monoculture year after year. One consequence of this method of farming is that the nutrients most needed by that crop are exhausted and have to be supplemented by fertilizer. An alternative planting strategy, **crop rotation**, provides the means of improving the balance of nutrients in the soil and can be used in both tropical and temperate settings. This method is not new; it was the mainstay of medieval European agriculture. However, the combination and varieties of crops grown in modern rotations are different from those of yore. In a crop rotation, crops with different nutrient demands or replenishing characteristics are grown in succession. A modern rotation (Figure 17.11) in the temperate region would typically last four years. Crops such as wheat, which requires high levels of nitrate, are followed by millet, which has a lower demand for nitrate. A crop of beans is grown in the third year to replace nitrate, and alfalfa, another nitrate restorer, is grown in the fourth year.

A further advantage of crop rotation over monocultures is that it can prevent a buildup of pests. Monocultures present a steady food source for the crop's predators, organisms we consider pests. Controlling the damage to the crop requires the application of pesticides. Sustainable agriculture concentrates on pest avoidance, and one way to achieve this is through being rare. Rarity can be thought of in time or space. Crop rotation is rarity through time, because the host crop changes with each planting. If a different crop is sown at each planting season, the pest population is only beginning to build as the host plant is harvested and a substitute is planted. The second crop is one that the pests cannot use, so the pests of the first crop decline in number. Sure enough, the pests of the second crop start to appear just as it is harvested, but their growth is thwarted by the planting of yet another crop.

Rarity in space can be achieved by mixing the sowing so that, for example, wheat, peppers, and beans are planted together and, as each is spatially rare, it will not attract predators. This strategy, termed **polyculture**, closely replicates the modes of agriculture described by the Conquistadors at the time of European penetration into the Americas. Polyculture presents problems to machine-oriented agriculture; the use of seed drills and harvesters is awkward when a field contains more than one crop type. Polyculture is most likely to be adopted in LDCs, where labor is plentiful and mechanization is not cost-effective.

Another variation in crop management that has been used for many centuries is **interplanting**. When the land is cleared, some of the economically valuable forest trees, such as cashew, brazil nut, and mango, are left, and crops are grown between them. The trees provide some shade, prevent soil erosion, and provide a supplemental income because their fruits can be marketed. Another way to reduce reliance on chemicals is integrated pest management (Chapter 18). This system places more reliance on using natural predators and carefully timed pesticide applications than on a regular spraying with large doses of pesticide.

Most farms in the United States still practice intensive agriculture, but it is estimated that between 50,000 and 100,000 farmers use sustainable agriculture. Recent estimates of profitability for sustainable farms show them to be comparable with intensive farms. Yields are marginally lower than are those of intensively farmed land, but the costs of chemical inputs are greatly reduced and prices for organic produce are high. Cotton, one of the crops that conventionally requires intensive pesticide applications, fertilizer, and irrigation, is now successfully being grown organically. The "organic" cotton is being snapped up by trendy fashion stores as the latest in hypoallergenic fashion. Consequently, the organic cotton sells for considerably higher prices than intensively grown cotton. It is likely that if all cotton growers changed to organic methods, the market would become saturated with "organic cotton," and the price advantage would

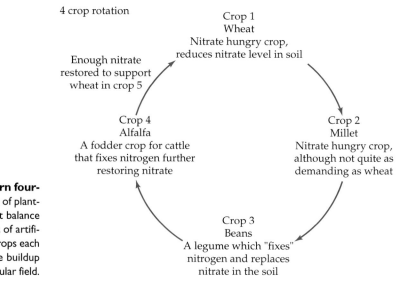

4 crop rotation

Crop 1
Wheat
Nitrate hungry crop,
reduces nitrate level in soil

Enough nitrate restored to support wheat in crop 5

Crop 4
Alfalfa
A fodder crop for cattle that fixes nitrogen further restoring nitrate

Crop 2
Millet
Nitrate hungry crop, although not quite as demanding as wheat

Crop 3
Beans
A legume which "fixes" nitrogen and replaces nitrate in the soil

Figure 17.11   A modern four-crop rotation. A sequence of planting that maintains the nutrient balance of the soil without large input of artificial fertilizers. The change in crops each year also helps to prevent the buildup of pest populations on a particular field.

evaporate. This price reduction, combined with the lower per hectare production of organically grown cotton, might well prove unprofitable. Hence, advocates of the status quo would argue that organically grown produce is economically viable only so long as it is rare enough to warrant premium prices.

The next few years will be an interesting time as consumers seem to demand more produce that is grown organically, or sustainably, while the petrochemical industry is pulling toward seed types that are increasingly dependent on pesticides and fertilizers. This apparent mismatch between what the consumer requests and what industry offers may prove to be just a lag effect. The public interest in sustainable agriculture is relatively new, and industry may be waiting to see if it is a passing fad or a lasting shift in consumer taste. When these changes occur (for example the recent increased customer demand for fat-free food items), industry profits from the development of a new market. If industry analysts judge that consumers will, given the choice, opt for free-range eggs, meat raised in a "sustainable" way, or cereals grown organically, we may expect that industry will regear to take advantage of the new market. However, until the transition from industrial to sustainable agriculture is perceived by agribusiness to represent financial opportunity rather than a threat, it is not surprising to find that progress is slow.

Summary

- In some areas of the world, population growth outstrips growth in agricultural harvests.
- Globally, hunger is not a problem of food production but one of distribution and social order.
- Food production has more than doubled in MDCs and in many LDCs since the 1950s.
- Per capita food production continues to increase for most of the world's population, but distribution of agricultural success is uneven.
- The Green Revolution was based on new crop strains that needed high inputs of fertilizer and pesticide that, in turn, required high inputs.

- High-yield intensive agriculture as practiced in the United States is not necessarily the appropriate technology to be adopted by other countries.
- Not all farmers can afford to use the new technology, nor do they trust unproven "improvements."
- Marginal agricultural areas will be susceptible to crop failure, and famine could follow.
- Desertification resulting from climate change and overexploitation threatens marginal farmlands in many areas.
- Sustainable farming aims to reduce reliance on fossil fuel inputs and maintain soil quality while maintaining harvests.

Further Readings

Brown, L. E. 1997. Facing the prospect of food scarcity. In *State of the World 1997*. Edited by L. E. Brown, C. Flavin, and H. French. Washington, DC: Worldwatch Institute.

Brown, L. E. 1997. The agricultural link: How environmental deterioration could disrupt economic progress. *Worldwatch Paper* 136. Washington, DC: Worldwatch Institute.

Cleveland, D. A., D. Soleri, and S. E. Smith. 1994. Do folk crop varieties have a role in sustainable agriculture? *BioScience* 44:740–751.

Ervin, D. E., C. F. Runge, E. A. Graffy, W. E. Anthony, S. S. Batie, P. Faeth, T. Penny, and T. Warman. 1998. A new Strategic Vision. *Environment* 40 (6):8–15; 35–40.

Lockertz, W. 1978. The lessons of the dust bowl. *American Scientist* 66:560–569.

Smil, V. 1997. Global population and the nitrogen cycle. *Scientific American* 277 (1):76–81.

Web Connections

On-line resources for this chapter are on the World Wide Web at:
http://www.prenhall.com/bush
(From this web-site, select Chapter 17.)

Pollution: The Other Face of Fertilizers and Pesticides

18

The river burned. Firefighters sprayed the blaze with water taken from the river; this water caught fire, too, and the fire spread. These are not the ravings from a fevered imagination, but a brief description of events on the Cuyahoga River, Ohio, in 1959. Flowing through the industrial zones of Akron and Cleveland, the Cuyahoga had been used as an unofficial chemical dump for decades. Local industry poured 155 tons of toxic chemicals and sewage into the river each day. Under this deluge of waste, the ecosystem collapsed, and the river became a lifeless chemical sewer coated in oil. The fire burned for eight days the first time. In 1968, the Cuyahoga River achieved national recognition by catching fire a second time.

Not all pollution is as obvious or as destructive as a river coated in oil. In this chapter and Chapters 22 to 24, the consequences of various types of pollution will be examined. The effects of pollutants range from promoting plant and animal growth to endangering the ecosystems in which we live. A fine distinction may exist between a chemical that we introduce to the environment to help us to grow a crop or to protect us from disease and a substance that threatens to damage natural ecosystems. Indeed, often it is the same chemical in different settings.

18.1 What Is Pollution?

Defining pollution is a little difficult. Flamingos can live in such dense numbers on African lakes that their fecal droppings can completely change the ecology of the system. Their dung is rich in nutrients, and as it fertilizes the lake, algal populations change; some fish species are killed by the new chemistry. Is this pollution? Flamingos are a natural part of that ecosystem. Although they contribute to chemical change in those lakes, the change is "natural," so most ecologists would not consider this pollution.

Although the fertilization of a lake by wild flamingos is not considered pollution, the herding of captive flamingos could result in pollution. If a flamingo farm were established in California, and they were introduced to a local lake, the chemical change that resulted from the arrival of the flamingos would constitute pollution because they were introduced into that ecosystem by people. Thus, a working definition of **pollution** is the unwanted alteration of natural chemical processes in an ecosystem as a direct or indirect result of human activity.

Pollution may originate from a clearly defined source, such as a heap of mine waste bleeding heavy metals into the local soil or an effluent outfall from an industrial plant. If such a precise origin can be attributed to the pollution it is termed a **point source pollutant**. If, on the other hand, there is no clear source, it is termed a **nonpoint source pollutant**. An example of this latter category would be inorganic fertilizers being found in groundwater; the pollution is evident, but no individual source can be established because the polluting chemicals probably came from a number of fields, golf courses, or gardens.

Pollutants can also be subdivided into two broad categories: biodegradable pollutants and nonbiodegradable pollutants. To be **biodegradable** the pollutant chemical must be able to be broken down by biological activity or through interaction with naturally occurring chemicals. The products of this decomposition are chemicals that will be simple nontoxic compounds.

A common way for biodegradation to occur is through metabolism by bacteria, fungi, plants, or animals. Examples are the absorption of carbon dioxide for photosynthesis; the nitrification of nitrous oxides; and the bacterial metabolism of carbon and nitrogen in sewage. However, it should be noted that simply being biodegradable does not make a substance environmentally "safe." For example, human sewage is biodegradable, but the decomposed sewage is rich in nutrients; if these were to wash into a lake or stream, the resulting fertilization could be severely harmful to the natural wildlife. At low concentrations, however, many ecosystems can absorb and neutralize these chemicals.

Nonbiodegradable pollutants are chemicals that cannot be metabolized and will not break down into harmless forms. These chemicals will simply accumulate through time, becoming more and more of a toxic threat as their concentration increases. Examples of these would be PCBs (polychlorinated biphenyls), DDT, dioxin, and heavy metals such as mercury, arsenic, and lead. Both biodegradable and nonbiodegradable pollutants can inflict serious damage on ecosystems. However, remember that pollution does not necessarily result in the death of a system. Indeed, some kinds of pollutant can greatly increase growth rates and primary productivity, although this will change the balance of the ecosystem.

18.2   Pollution That Increases Growth

The act of fertilizing a field to encourage the growth of a crop is not counted as pollution because it is a deliberate act. In applying a fertilizer, the farmer is pur-

posefully altering the chemical balance of that system to favor a particular species of plant. Frequently, the fertilizers are rich in nitrogen, phosphorus, and potassium, chemicals that will increase the growth of most plants, not just crops. Although the intent is to fertilize croplands, forests, or lawns, it is all too common that a portion of the fertilizer washes off the land and makes its way into local watercourses. At that point, the fertilizer becomes a pollutant. These fertilizers are nonpoint source (they come from a large area, from many fields or gardens), biodegradable (because they are absorbed into natural chemical cycles) pollutants. The effect of fertilizing a body of water on the growth rates and chemical cycles is called **eutrophication** and is generally the result of increasing the availability of nitrogen and phosphorus in the system. (We discussed eutrophic lakes in Chapter 7.) Potassium, the third of the "big three" chemical additives in terrestrial fertilizers, is of relatively little importance as a eutrophying agent in aquatic systems. Lakes naturally undergo a process of eutrophication as they age. More and more organic-rich sediment is deposited on the bed of the lake, and the surface portion of this sediment supplies nutrients to the water column or provides a fertile sediment in which aquatic plants can root. However, this slow process of lake eutrophication may take thousands of years, whereas the chemical changes induced by human pollution can occur literally overnight. The pollution can result when water carries dissolved fertilizer running off the land, or it can be caused by soil erosion, human sewage, or industrial activities (Figure 18.1).

Erosion that accompanies deforestation or the building of residential areas washes topsoil off the land and into rivers and lakes. The topsoil is rich in nutrients, which dissolve and fertilize the lake. Untreated sewage from human settlements or farms is rich in nitrogen and phosphates and is again a very effective fertilizer. As the fertility of the lake increases, its biology changes.

Changes reflecting increased nutrient availability in lakes are especially evident in the summer. As the lake heats, a strong temperature gradient divides the lake into an upper region of warm water floating above the cold bottom waters of the lake (Chapter 7). Under these conditions the surface waters of a fertilized lake teem with life as green algae form huge populations. The most obvious effect of fertilizing a lake is to see the change from a clear, blue lake to one that is green and soupy. During the summer, the dissolved oxygen concentration in a eutrophic lake could fall to zero in the hypolimnion (the cold bottom waters;

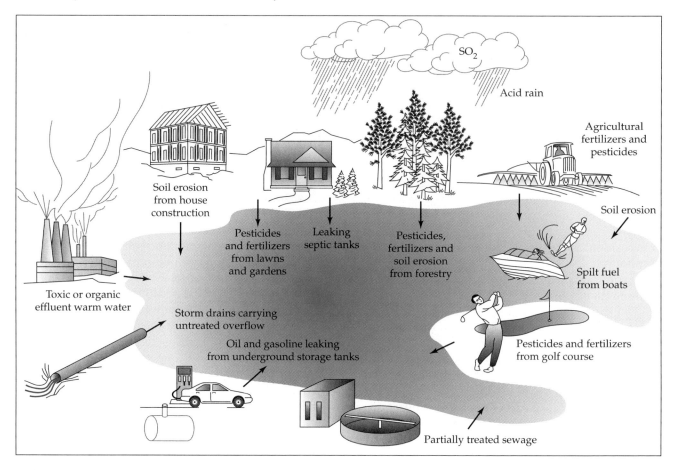

Figure 18.1 **Common sources of pollution to a temperate lake.** (a) Pollution associated with changes in land use; (b) the effect of increasing concentrations of fertilizers such as phosphate and nitrate on aquatic systems.

Chapter 7). The surface waters are rich in photosynthetic algae that produce oxygen and are separated from the hypolimnion by a strong temperature gradient such that the two water bodies do not mix. The dark waters of the hypolimnion have no oxygen production, only a high demand for oxygen by decomposers. Consequently, the bottom waters become deoxygenated. As oxygen depletion becomes increasingly severe, fish species die out. Those that demand the most oxygen, such as trout and bass, will die first, while carp and catfish will survive in a soupy, green pool. Ultimately, even those species will be lost.

Often this process of eutrophication is described as "poisoning" or "killing" the lake. These emotive descriptors are inaccurate; in biological terms the productivity of the system is increasing (so long as eutrophication is not so severe that it precludes photosynthesis). Nevertheless, the conversion of a "clean blue lake" into a smelly, green eyesore, albeit one teeming with microscopic life, is generally unpopular among human residents and visitors. Ultimately, eutrophication can continue to the point where all photosynthetic algae are lost, and the only living organisms are decomposers. At this extreme the system is truly dying.

Temperature and Eutrophication

An increase in the temperature of water can lead to profound physiological stress in local populations of organisms. The effects of eutrophication become more pronounced as water is warmed. Industrial effluent, or the outfalls of coolant water used in power stations, may be clean but warmer than the natural temperature of the lake or river (this is particularly true of northern temperate climates). Warm water will have three principal effects on a system.

First, the physiology of organisms is affected because temperature is likely to be an important param-

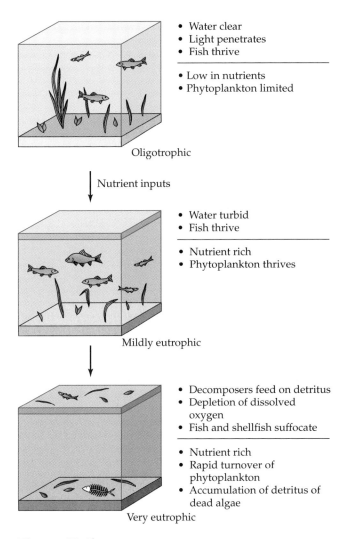

• Water clear
• Light penetrates
• Fish thrive

• Low in nutrients
• Phytoplankton limited

Oligotrophic

Nutrient inputs

• Water turbid
• Fish thrive

• Nutrient rich
• Phytoplankton thrives

Mildly eutrophic

• Decomposers feed on detritus
• Depletion of dissolved oxygen
• Fish and shellfish suffocate

• Nutrient rich
• Rapid turnover of phytoplankton
• Accumulation of detritus of dead algae

Very eutrophic

Figure 18.1b   (continued)

eter in their niche. Therefore, as temperature is increased, some organisms will pass from their zone of optimal living conditions into the zone of physiological stress. Similarly, some organisms will be favored by the warmer water. Consequently, a shift in species composition is expected as a result of the warm water inflow. This change may be relatively subtle and of little concern, but under more extreme conditions of warming, or where warming is coincident with pollution with organic material, the effect on the ecosystem will be more pronounced.

The second effect is that as temperature increases, biological and chemical processes will run faster, resulting in more-rapid rates of decomposition and hence an elevated biological oxygen demand (BOD). The third effect is that warm water cannot hold as much dissolved oxygen as cold water. Consequently,

in warm water, oxygen concentrations in the water will be low to begin with and will be reduced further by increased biological activity. Severely deoxygenated waters cannot support plants, fish, or amphibians, and to a casual observer this system will appear dead.

The effect of deoxygenation can be seen in rivers where there is a point source for eutrophying pollution, such as a sewage outfall. Immediately downstream of the outfall the river exhibits an **oxygen sag** as concentrations of dissolved oxygen are reduced by the BOD of decomposers (Figure 18.2). In this zone of oxygen depletion, all the fish may be lost from the system, and the only life will be some hardy plankton, insect larvae, and bacteria. The river water may be clear or turbid with the discharge, for deoxygenation does not necessarily cause a discoloration of the water. The sediment on the bottom of the channel may be brown or black and will appear to be bare. But remember, it is not dead; these systems are teeming with bacterial and algal life.

Reversing Eutrophication

Where there is a point source of nutrients, the fertilizing effects of the input wear off downstream as the pollution becomes more dilute. The distance that it takes for the fertilizing effects to diminish will depend on the flow of river water relative to the effluent input or the relative temperature of both bodies of water. If the effluent input to the river ceased, the system would recover relatively quickly. For most cases of eutrophication by pollution this is true: It is a reversible process. Studies of the eutrophication of several large bodies of water demonstrate that once pollution stops, nature will reestablish an equilibrium close to the unpolluted state. In Figure 18.3, the supply of phosphorus is compared in lakes before pollution, during maximum pollution, and after pollution has been restricted. Phosphorus is taken as a representative nutrient and provides a measure of the eutrophication of the lake. The phosphorus load is calibrated per square meter of lake surface area. This calibration makes allowance for large and small lakes, but it does not allow for the depth of the lake. The deeper the lake, the greater must be the supply of nutrients to eutrophy the system. Note also that the scale of each axis is logarithmic.

Lake Ontario was an oligotrophic lake before it was polluted. Pollution increased the input of total phosphorus to this lake from about 0.1 g/m^2 to about 1.1g/m^2, a tenfold increase. The pollution source was

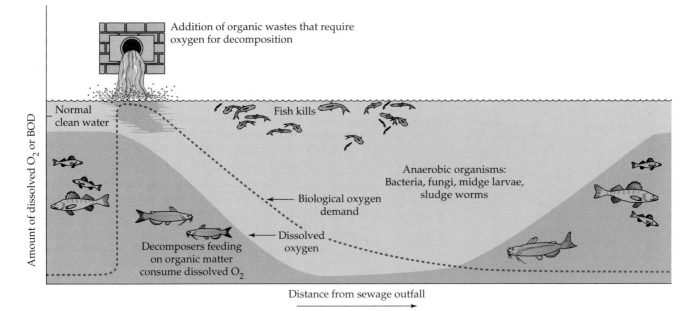

Figure 18.2 **Release of organic-rich effluent from a point source into a river results in a downstream oxygen sag.**

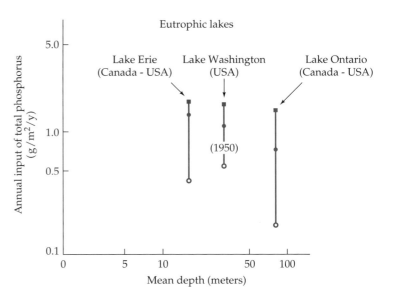

Figure 18.3 **Phosphorus concentrations in three lakes throughout a cycle of pollution and recovery after pollution stopped.** The open circles represent the phosphorus concentrations before the lakes were polluted; the closed squares represent the concentration of phosphorus when the lakes were most heavily polluted; the closed circles indicate the present phosphorus concentration now that pollution has been substantially reduced. (After W.H. Schlesinger, 1991.)

primarily industrial effluent, agricultural waste, and untreated human sewage. After legislation to control pollution was passed in 1965 and 1972, the input of nutrients fell. Lake Ontario is now marginally eutrophic, much cleaner than it was in the 1950s, and it will probably improve further. Lake Washington and Lake Erie have had similar histories, although they had a more-eutrophic starting point.

The truly encouraging aspect that emerges from this data set is that pollution leading to eutrophication is reversible, and nature takes care of the cleanup once the pollution is stopped. However, a lag period will occur between the termination of pollution and the lake's return to a natural state. This lag is the time it takes for the flushing of dissolved nutrients from the system or the burial of other nutrients. Living organisms and the layer of decomposing organic matter on the lake bed are rich in the nutrients provided by pollution. Before the lake can recover to a natural state, the nutrient level has to return to prepollution concentrations. Nutrients must therefore be lost from the system. They are removed as the nutrient-enriched muds are overlain by new layers of lake mud. As the nutrient-rich muds are buried beyond the range of burrowing worms and decomposers, the lake can return to its natural chemistry. Water flowing through the lake and out through a stream will flush some nutrients from the system and may speed the recovery.

Direct action by land managers can also speed the recovery process. The first step is to ensure that no new pollution is entering the system. Next, the process of recovery may be speeded by the oxygenation of the water. This can be achieved by bubbling air through the water or by creating a large fountain in the middle of the lake. In less extreme cases, biological solutions might be attempted through restocking the lake with snails and fish that feed on algal blooms (Figure 18.4). Dredging polluted muds from the bottom of the lake is a last resort, because the potential for environmental damage makes this worthwhile only when severe toxic pollutants have contaminated the lake. Any of these steps to restore an ecosystem is relatively ex-

pensive, particularly if attempted on the grand scale. In general, once the pollution has been stopped, nature is left to repair the system, a process that may take from years to centuries, according to the scale, severity, and nature of the pollution.

18.3 Biological Effects of Pollutants

It seems that all the things that taste really good kill us. Cream, butter, sugar, bacon, chocolate, and salt are all linked to heart disease or cancer. In truth, any food item taken in excess would do us harm; the same is true of most chemicals in the environment. A few, such as plutonium, are absolutely toxic even in minute quantities, but most can be absorbed at low concentrations and become dangerous only as a critical threshold is reached. That danger may be manifested

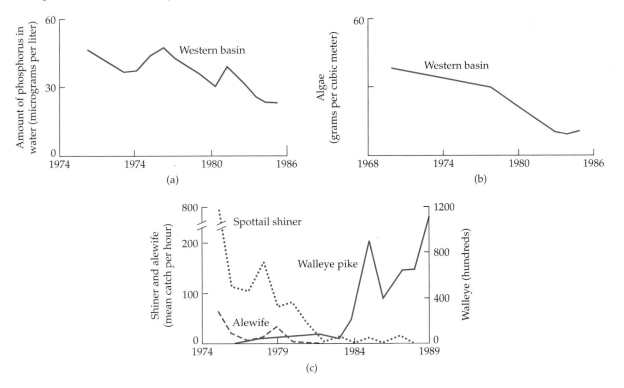

Figure 18.4   Changes in the ecology of the western basin of Lake Erie. (a) Changes in the phosphorus load between 1968 and 1986. High phosphorus promotes algal blooms that lower water quality and eliminate other species. (b) The gradual decline in algal abundance in Lake Erie. Populations of algae are regulated by zooplankton. When zooplankton numbers fall, algal populations explode and form algal blooms. (c) Predator-prey interaction of walleye pike (predator) and spottail shiner and alewife (both prey items) in Lake Erie. High population levels of spottail shiner and alewife (zooplankton predators) will depress the zooplankton population and allow algal populations to form blooms. The treatment of sewage and reduction of phosphate entering the lake improved water quality in the early 1970s. Once water quality had improved beyond a certain threshold, walleye pike flourished. As pike populations increased, they reduced populations of alewife and spottail shiner. Zooplankton benefited from this change, and consequently algal blooms were reduced, leading to a further improvement in water quality. Relative walleye abundance was calculated from catches of sport fishermen. Alewife and spottail shiner population estimates are from catch per hour by commercial trawlers. (After J.C. Makarewicz and P. Bertram, © 1991. American Institute of Biological Science.)

as death, but for most chemicals a lesser range of concentrations exist that do not kill, but inflict damage. Reproduction may be impaired, or exposure may induce long-term illness, such as cancers or lung disease. Toxicologists study the maximum concentration of a chemical that can be tolerated before it interferes with physiological processes. Here the concern is not primarily with the dose that kills, but with lesser concentrations that may still cause birth defects, disease, or pain and suffering. Although these estimates are generally applied to humans, they could also be applied to determining ecosystem damage from pollutants.

Biodegradable Pollutants

Examples of biodegradable pollutants can be found in both organic chemicals such as acetone or pyrethroid pesticides, or inorganic chemicals such as nitrous oxides, sulfuric acid, ozone, or chlorine. At low concentrations, these chemicals can be absorbed by ecosystems and will be degraded into less-harmful compounds. However, if the system is overloaded with any of them, organisms will start to exhibit signs of poisoning and ultimately will die. Because these are important industrial chemicals, trying to ban their emission is both impractical and economically undesirable. Therefore, we try to determine an optimum level of pollution, one that causes the least hardship to the general public, industry, and agriculture, without damaging the environment.

Nonbiodegradable Pollutants

Heavy metals, such as mercury, cadmium, and arsenic, and manufactured chemicals, such as PCBs and some pesticides, are examples of nonbiodegradable pollutants. These chemicals are highly toxic, so that low levels of exposure or low concentrations of these compounds are poisonous. Such chemicals are so foreign to living organisms that they are not metabolized and remain in the ecosystem basically unchanged. Worse than that, if eaten, they may be stored within the body. Each time chemicals such as PCBs, mercury, or dioxin are taken into the body they are added to the existing stock. If this accumulation continues, a toxic level is reached. The Romans were great poisoners, and they knew the toxic value of gradually administering poisons such as antimony or arsenic. Each meal was safe to eat, but the steady diet of a little poison time after time led to the death of the victim.

Even though nonbiodegradable pollutants may be relatively rare, they are stored in the bodies of an organism and passed on up the food chain in a process called **biological amplification** (also referred to as biological magnification). A predator absorbs all the stored pollutants in the hundreds or thousands of prey items that it eats, and each meal provides a dose of the toxin. The chemical is stored in the body of the predator, where the successive doses accumulate and become more concentrated. If the predator then falls prey to a larger carnivore, the entire dose of toxicity is taken to the next step in the food chain. Thus, the concentration of the toxin is amplified at each link in the food chain (Figure 18.5). Because heavy metals are toxic to many organisms at even small concentrations, it is not surprising to find that few plants and animals can withstand the environments created where these metals are mined. Small amounts of the mined material enter the soil water or the food chain and initiate the concentration of the poison. For several decades, environmentalists have warned that our rivers and coastal waters are becoming loaded with these substances.

A tragic example of this biological amplification received worldwide attention in the 1950s. Inhabitants of the fishing village of Minamata, Japan, were subjected to mercury poisoning that affected over 3500 people. Chronic illness, birth defects, even death were directly attributable to ingesting mercury. The Chisso Chemical Plant had been dumping mercury-rich waste into the bay—mercury that was thought to be in an inert chemical state when it was dumped. However, bacterial activity altered the metallic mercury to soluble, but no less deadly, methyl mercury. The mercury was incorporated into the food chain. The villagers, relying on fish and shellfish for their protein, were predators feeding high on a food chain that was amplifying mercury concentrations. The only solution to their pollution problem was to dredge the mercury-laden sediments from the bay.

Mercury poisoning continues as a threat, especially in the Amazon basin. A latter-day gold rush is taking place in remote areas of Amazonia, and, like the Californian gold rush of 1849, the result is a lawless frontier. More than 40,000 prospectors are working the river, clashing with and exterminating the local Indian peoples from areas previously set aside as tribal homelands. In addition to the social ills resulting from this gold fever, the gold extraction poses a serious pollution problem. Mercury is used to help separate gold from the river sands; the mix of gold and mercury is then heated to burn off the mercury and leave pure gold. The miners and their families operate as a cottage industry with no health safeguards, and they commonly suffer from mercury poisoning. The long-

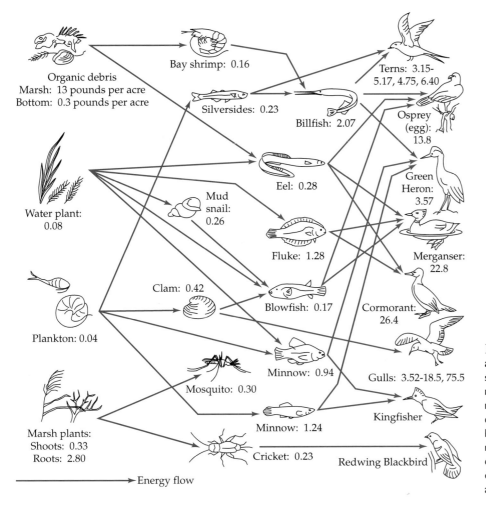

Figure 18.5 The biological amplification of DDT. DDT is stored in the bodies of plants and animals. The DDT becomes more and more concentrated along the food chain until it reaches toxic levels in the largest predators. Other organisms may also be affected by the lower concentrations of pesticide, but their demise is not so likely to stir public attention.

term effect of the mercury on the river downstream of the mining camps is unknown, but it is feared that it may locally threaten the unique fauna and flora of the river.

In addition to heavy metals, some organic chemicals also bioaccumulate (e.g., DDT, dioxin, and PCBs). These chemicals can poison humans outright or can cause cancers to form. Although banned since the mid-1970s in much of Europe and the United States, these chemicals are still widely used in other countries. In the United States, millions of tons of PCBs were released into the environment before being banned, and they are still present at dangerous levels in sediments of some water bodies, such as the Great Lakes, Chesapeake Bay, and Boston Harbor. This legacy of earlier pollution is likely to remain with us for decades to come because there is no natural process decomposing these chemicals to harmless forms. Eventually, the polluted sediments will be buried by other muds washed downstream, or they will be eroded and later deposited in the ocean deeps.

Field Experiments to Study the Effects of Pollution

Manipulations of natural habitats have been an important tool in ecological research. These studies commonly take the form of exclusion experiments or experimental manipulation of abiotic factors such as fire or nutrient load.

Exclusion experiments are designed to determine the impact that one species, or group of species, has on an ecosystem. At its simplest this can be a very simple experiment such as surrounding a small area of grassland with rabbit-proof fencing and removing any rabbits that are found within the enclosure. This has been done in many settings and generally results in a decrease in the species diversity of plants within the pen. In the absence of grazing pressure, the most competitive plant species will take over and displace the less competitive species. Notice that this is effectively the same experimental design and an equivalent outcome as described in the removal of *Pisaster* starfish from the rocky intertidal shoreline of Oregon (Chapter 12). The equivalent experiment in lakes is the removal of the top predator fish. Eliminating all the fish from the lake can be done with poisons such as Rotenone, but the ecological side effects are often too damaging for the experiment to be worthwhile. Consequently, most modern experiments are conducted in mesocosms within the lake. Each mesocosm is a mesh enclosure that may float or be attached to the bottom of the lake. The mesh is small enough to control the movement of the experimental organisms, but large enough to allow water to flow through the pen. The experimenter then adds the species to be studied, and has total control over the number of individuals, their age, sex ratios, etc. This kind of study allows the researcher to add, or exclude any member of a food web to determine how the community is shaped if that species is absent. An important aspect of any experimental design is to have controls in which everything is replicated, apart from the one variable being studied. With a mesocosm design, this is easily achieved by setting up additional mesocosms as controls.

The manipulation of abiotic factors is well demonstrated in experiments that add fertilizer to a system. Again controls are established to which no fertilizer is added, ensuring that the observed changes are indeed caused by the fertilization. In one remarkable experiment that has been running since 1856, 18 experimental grassland plots have been fertilized every year. Two control plots have received identical treatment except that they have not been fertilized. The number of plant species has been documented in each plot, and within the first 100 years the species diversity of the fertilized areas had fallen sharply compared with the control areas (Figure 18.6). The fertilization pro-

motes growth of species previously limited by the lack of the nutrient. Hence there is an increase in primary productivity, but these few species flourish at the expense of many others. Consequently, fertilization causes a reduction in the species diversity of the area. The same pattern of increased productivity, but reduced species diversity, is also observed in aquatic and forest settings.

18.4 Why Do We Pollute?

If uncontrolled pollution is so clearly bad for the environment, and bad for us, why does pollution continue? Almost all pollution could be contained by building stronger storage containers, by installing improved antipollution devices on smokestacks, by reducing vehicle use, and by ceasing to use agrochemicals. Industrial and agricultural production would be stifled, and the social cost would be exorbitant. The price of zero pollution is too expensive, but so too is the cost of zero regulation. Where there are no regulations on pollution, the cheapest way for industry, agriculture, or domestic users to vent waste chemicals is to dump them. Under these circumstances, chemicals will be poured into rivers or released into the atmosphere from chimneys. Vehicles will trail black smoke from their tailpipes, and garbage will be dumped on any convenient land. For an individual or a company, it will almost always be cheaper to pollute than to restrict pollution, thus pollution will be the norm until regulation prevents it.

The issue of pollution control has an ethical component: We do not have the right to befoul our planet. Ethical arguments may inflame passion, but they are often overshadowed by economic arguments. However, in recent decades there has been a general realization that pollution control also has a significant health component.

Arguably, the health of our population is one of the few factors that may take political priority over economic profit. Pollutant chemicals may induce cancers (carcinogens) or genetic abnormalities resulting in birth defects (mutagens). The health costs incurred will ultimately have a negative impact on industrial profits (Chapter 27).

Regulating these noxious chemicals so that they are maintained within "acceptable" levels becomes a priority for pollution regulators. Following events such as the burning of the Cuyahoga, the federal government passed a series of antipollution laws, such as the Clean Air Acts of 1963, 1967, and 1990 and the Clean Water Act of 1972 (Chapter 28). Regulation seeks to achieve a compromise between harm done

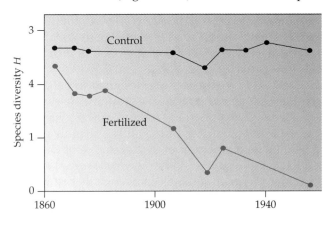

Figure 18.6 The effect of fertilization on species diversity. An experiment lasting more than 93 years tracked species diversity in fertilized and unfertilized grasslands at Rothampstead, England. In the unfertilized area there was no change in species diversity, but in the fertilized areas species diversity dropped markedly.

to the environment and our willingness to pay to prevent that harm.

18.5 Pesticides: Pollutants That We Need

Just as fertilizers can become pollutants, so too can pesticides. These chemicals are essential to maintaining agricultural productivity in many farm systems, yet they can become serious pollutants as they wash from fields or gardens. In the case of agricultural pesticides, or an antibiotic medicine, the goal is to kill a target organism as quickly as possible. The ability to develop agricultural pesticides in large quantities enabled crop breeders to develop strains of crops that were high-yield miracle crops, but vulnerable to pests. Vulnerability was ensured because the diversity of crops grown was greatly reduced in the Green Revolution, and the diversity of types of each crop was also greatly diminished. For instance, hundreds of local varieties of wet rice were abandoned in favor of one miracle variety. Hundreds of defensive mechanisms against pests were replaced by one. If a pest could attack this one variety, the entire planting would be at risk. Perhaps arrogantly, or just naively, it was thought that chemicals could be engineered to protect crops against disease. The reliance on pesticides was, and is, immense and is literally a matter of life and death. The crop must be protected in the field and after it has been harvested. Furthermore, the seeds for next year's planting must also be kept safe.

Before pesticides were introduced, it was estimated that approximately 30%–35% of the annual harvest was lost because of competition from weeds, insect damage, molds, rusts, and fungus. The pesticide industry is a vast international industry, costing farmers and governments billions of dollars each year. However, recent estimates of pest damage to crops are still around 30%–35% of the harvest. Of course, the remaining 70% of the harvest is much larger now than 50 years ago because crops improved during the Green Revolution. Indeed, it can be argued that these new crop strains would be hit much harder without pesticide inputs, and our increase in food production is intimately tied to chemical use. It is highly instructive to look at some of the attributes of pesticides, the history of their use, and some alternatives that may play a role in future agricultural development.

Defining a pest is highly subjective and depends entirely on circumstance. *Pest* and *weed* are really terms that we use for any organism that is living where we do not want it. In a rose garden, a seedling pine tree is a weed; but in a pine plantation, a rose is a weed. Pesticides can be subdivided, according to their targets, into herbicides, rodenticides, insecticides, fungicides, bactericides, and so on. Almost half of the pesticides used in the United States are weed killers designed to remove unwanted plants from croplands and lawns. Pesticides can also be grouped according to how the toxin works. **Sterilants** interfere with pest reproduction and hence eliminate the next generation. Chemicals that attack the nervous system of the organism and kill it almost instantly, as do most fly killers, are **neurotoxins**. With plants, there may be **contact poisons**, which kill the plant where they touch the leaf or stem; **systemic poisons**, which are absorbed through the roots and then poison the plant; or **soil sterilants**, which are broad-spectrum toxins that kill everything in the soil, including the target plant. The latter would be useless for agriculture or forestry, but would be appropriate for keeping weeds off railroad tracks, highways, or other areas where plants must be eradicated completely. Some toxins are long lasting and remain effective for several years; others are rapidly broken down into nontoxic chemicals and are harmless within a few days of application. Toxins may work specifically against a given organism or may be effective against a broad spectrum of plants or animals. Thus a wide diversity of chemical treatments exists, each being suitable for a different situation (although it could be argued that most are ideal for none).

Wartime often spurs research, and in the 1940s the chemical industry was galvanized with the search for effective pesticides to increase harvests and to prevent disease among the troops. After the war, agriculture reaped the benefits of that research, and pesticide use rose sharply in the 1940s and 1950s as part of the Green Revolution.

The First Warnings

In 1963, a slim book changed how society viewed science and agriculture. Rachel Carson's *Silent Spring* cast shadows of doubt over the use of pesticides and made *ecology* a household term. Carson was the original whistle-blower on the petrochemical industry. She exposed to the public evidence that pesticides were being overused, were having severe environmental consequences, and had the potential to become a health risk to humans. Her main arguments were as follows:

- Pesticides were not specific enough. Each application killed the target organism, but also killed others, some of which might be highly beneficial, such as honey bees and ladybugs.

- Chemical residues were becoming incorporated into food chains. Passed from one trophic level to the next, the concentration of toxin grew so that upper-level predators, which were not the target, were being killed. Humans were also at risk if they ate animals close to the top of these food chains, such as lake trout.

- Farm workers were often poorly informed about the potential danger to themselves in handling these powerful toxins. Insufficient training, inadequate protective clothing, and poor washing facilities led to contamination and disease (complaints that are still made by agricultural workers' unions).

- The overuse of chemicals was leading to the pollution of streams, rivers, and lakes, resulting in fish and bird kills. Contaminant chemicals were also feared to be reaching our groundwater drinking supplies.

None of these arguments could be refuted, so the pesticide industry defended itself by questioning the patriotism of whoever would believe such calumnies. Representatives of the chemical industry petitioned the publishers of *Silent Spring* to suppress the book on the grounds that limiting pesticide use would destabilize agriculture and help in a Soviet takeover of the world (the Cold War was indeed a strange time). Carson and her publisher were resolute, and the book was published.

Silent Spring and the ensuing public outcry drew clear lines between the agrichemical industry and the fledgling environmental movement. It took another 10 years for any kind of legislation regulating the use of pesticides to make its way through Congress. In 1972, the Federal Insecticide, Fungicide, and Rodenticide Control Act (FIFRA) was passed. Although some critics say that this legislation is poorly enforced, among the pesticides that it regulated was dichlorodiphenyltrichloroethane, better known as DDT.

DDT: A Fall from Grace

DDT was first synthesized in 1874 and belongs to a group of chemicals called organochlorides. In 1939, entomologist Paul Mueller discovered that DDT was a potent insecticide. As a powder or as a spray, it could kill fleas, lice, and mosquitoes upon contact. During World War II, DDT was distributed to allied troops to help them to clean lice from their uniforms, malarial mosquitoes from their quarters, and to generally reduce the load of biting insects around them. After the

war, widespread use of DDT virtually eliminated malaria, the disease that until that time had been the leading cause of death in tropical countries. In 1948, Mueller was awarded the Nobel Prize in medicine for his application of DDT as an insecticide. Twenty-four years later, it was a banned product in the United States.

In the 1950s and 1960s, DDT and the closely related chemical DDD were widely used on farms and in recreational areas to control pests. Some lakes, particularly those used for bathing and recreation, were routinely sprayed to spare the irritation caused by midges and mosquitoes. These insecticides were also in the vanguard of the "war on pests" declared under the Green Revolution. Crops were dusted regularly, sometimes excessively, with the result that DDT was washed from the fields into ditches and streams. Because organochlorides do not biodegrade, even though the DDT had left the field and was now in an aquatic system, it was still acting as a potent insecticide. The streams fed into lakes, perhaps far from the point of application, and began to alter the ecology of those systems.

In the 1960s, fish kills observed at popular fishing grounds and a noticeable decline in birds of prey alerted the public and scientists to a possible environmental problem. Water analysis revealed trace amounts of DDT, apparently nothing to worry about. Yet, upon analysis, it was found that the dead fish and hawks had high levels of DDT in their body fat. A further study of the hawks revealed that, in general, they were not dying of DDT poisoning; their numbers were declining because they were failing to reproduce. A side effect of the DDT exposure was that the eggs laid had very thin shells that cracked prematurely, offering the chick no chance of survival. Extinction is the inevitable consequence of a failure to reproduce. The hawk populations were severely threatened, and indeed some species, such as the peregrine falcon and the bald eagle, went extinct over large portions of their natural range.

DDT can pass quite freely through cell walls of plants and animals until it comes into contact with lipids (fat). DDT is a lipid-soluble chemical, and once incorporated into fat, it will stay there and will not be excreted. Gradually, with each meal, more and more DDT is absorbed into the fat; thus the organism becomes richer and richer in DDT. The first step in the food chain will always be a photosynthetic organism, and if photosynthetic organisms absorb water containing DDT, it will soon be found in all trophic levels

(Figure 18.5). Each prey item has been accumulating DDT, so each represents a concentrated package of that toxin. The more of these packages the predator eats, the more DDT it will acquire. Consequently, the many meals of prey items that themselves were concentrating DDT means that the fat of the predator will have much higher residues of DDT than does its prey. The jump in the pesticide load from one trophic level to the next is thus a function of biological amplification in the preceding trophic level and the many prey items that are eaten.

As with many issues relying on public concern to generate action, it took the decline of "desirable" species such as the bald eagle and trout to limit the use of DDT. Far more ecologically important organisms such as insect decomposers, vital to nutrient recycling, were also being killed, but it is impossible to motivate public concern over such critters. After 20 years without DDT, bald eagles and peregrine falcons are beginning to return to many of their former breeding grounds, and this cleanup is undoubtedly one of the great successes of the environmental movement.

DDT is just one example of many chemicals that can accumulate in food chains. Others include PCBs (solvents), dioxin (a bleaching chemical used in the paper industry), and TBT (an antifouling paint additive used in harbors, locks, and on boat hulls). Although the use of DDT has been banned in North America and much of Europe, it is still manufactured in those regions and shipped abroad to less-developed countries that do not have the same rigorous environmental regulations. About 25% of the pesticides (6 tons per hour) that are shipped abroad from the United States are chemicals banned for domestic use under FIFRA.

The Attributes of an Ideal Pesticide

Rachel Carson's attack on such broad-spectrum pesticides as DDT helps to identify the ideal attributes of a pesticide.

- It should be highly toxic to the intended target and harmless to other organisms. This feature ensures that if the pesticide is washed off the land, or is applied carelessly, it will not harm local wildlife, crops, or livestock.
- It should be quick acting and then rapidly decay to a harmless, inert chemical. The longer the pests attack, the greater the damage they cause. Therefore a rapid pest mortality following application of the chemical is sought. If the chemical remains active

after it has been applied to the field, there is the possibility of leaving poisonous residues on the crops or of poisoning livestock turned onto a field after harvesting. A further menace is caused by chemical accumulation in the soil. While each individual chemical may be a target-specific toxin, the brew of several pesticides together may have entirely different properties. It is not just the chemical as it is applied to the land, but also the products of its gradual decay that can contribute to a toxic chemical soup. For this reason, the ideal pesticide would quickly decompose into inert (that is nonreactive), nonpoisonous chemicals.

- Long-term exposure to the pesticide should not result in disease among humans.
- It should not encourage the development of genetic resistance in the pest.
- It should be inexpensive.

Lethal Doses of Pesticides

Toxicology is the science that studies the effects of poisons on organisms and the environment. The effect that a poison will have on an individual will depend on

the toxicity of the poison: effectively how poisonous the chemical is;

the dose received: how much of the poison was administered; and

the exposure to the toxin: the length of time that the toxin could have been absorbed.

An efficient pesticide is one that kills quickly and at relatively low concentrations. To find an optimum dose, scientists conduct experiments to determine the proportion of individuals killed by a given concentration of the pesticide within 48 hours. These are known as lethal dose experiments, and a 48-hour lethal dose experiment to find the concentration that kills 50% of the target organisms is abbreviated to 48-h LD_{50}. From the perspective of the farmer needing to control pests, this information is very useful because it gives a clear indication of the likely kill-rate of the pesticide. However, it must be realized that much lower concentrations would be troublesome if the pesticide washed off the land.

Ecologists must concern themselves not only with lethal effects, but also with sublethal effects. For example, if it is a pesticide causing the deformities in frogs in Minnesota, the effects are very damaging, even when sublethal (Chapter 8). Though frogs and fish are not the targets of the pesticide, they too may be damaged if exposed to it. A separate assessment of the 48-h LD_{50} may be conducted for these species to determine the concentration that would prove rapidly fatal to them. However, because the effect of the toxin is a function of both concentration and length of exposure, time becomes an important variable. A longer exposure to low concentrations may be as severely damaging as a brief (48 hour) exposure to high concentrations. Pesticide that has washed into a river may bathe organisms for weeks, if not permanently. Therefore, the long-term effects of exposure to the pesticide are more important than the short-term ones. Long-term exposure, even to very small concentrations of pesticide, may be enough to induce deformities or cause sterility. Furthermore, the cumulative effect of numerous pollutants may prove more deadly than exposure to a single chemical. In a study of the River Trent, in England, areas with varying concentrations of several toxins were compared. It was found that sites with less than 3% of the 48-h LD_{50} concentration of a range of toxins could support trout, whereas sites with 13%–20% could not. Sites with up to 20% of the 48-h LD_{50} concentration of toxins could support fish that are more tolerant of poor water quality than trout, but at concentrations greater than about 20% of the 48-h LD_{50}, all fish were lost from the river. Thus concentrations of a pesticide that appear to be "safe" in a laboratory experiment may prove to be lethal in the real world.

Pesticide Resistance

Just as some bacteria are becoming resistant to antibiotics, many agricultural pests are becoming resistant to the chemicals that are used to control them. Because most pest species are capable of sexual reproduction, a range of genotypes will exist within each generation. Pesticides are seldom capable of killing all the genotypes that they confront, and therefore some of the pest population will survive. These individuals, perhaps as few as 0.01% of the original population, are resistant to the pesticide, and the chances are good that their offspring (because both parents presumably survived the application) are also resistant genotypes. Thus, in just a few generations there can be a substantial recovery of a population decimated, but not exterminated, by the initial application of pesticide. Once that resistant population has become sufficiently large to inflict economic damage to crops, the farmer will need to use a new pesticide. On average, a new pesticide introduced to the market is highly effective

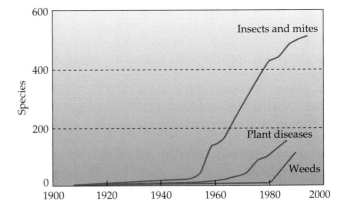

Figure 18.7 **The resistance of various pests has increased markedly since the onset of the Green Revolution.** Plants, pathogens, insects, and mites are increasingly difficult to control because genetic resistance is bred into the populations of pest species. Consequently, once-effective pesticides are now ineffective. (After P. Weber, 1994.)

for five to eight years before the target pest evolves significant resistance.

The race to stay ahead of pests becomes ever more difficult and expensive, and pesticide resistance is increasing among many organisms (Figure 18.7). By 1993, at least 520 species of insects and mites, 150 plant diseases, and 113 weed species demonstrated tolerance to pesticides that once controlled them. The answer to resistant pests is not to apply more pesticide or to apply it more often, although unfortunately that has been a common response.

18.6   Pesticide Alternatives

If, after 50 years of using pesticides, 30%–35% of potential harvests are still being lost, could there be a better way? In some cases, though probably not all, there may be. In natural systems, predation prevents the prey items from establishing huge populations. One agricultural method, **biological control**, seeks to mimic this relationship through using natural predators to control crop pests. Aphids and scale insects suck the sap from leaves and shoots. As they siphon off the products of photosynthesis, they deny the plant the ability to set a full yield of fruit. Because these insects can kill a tree or reduce the potential harvest, they are a major threat to orchard farmers. However, spraying insecticides against these enemies can result in much greater losses. The pesticide would kill not only the target pest but also the ladybugs that are the natural predator of these insects.

The ladybug population plays a vital role in keeping their prey populations at low densities. If the ladybugs are removed, then the natural regulation of the populations has been lost and the pest will increase in numbers. The farmer is then locked into what is known as the "pesticide treadmill," because without the predator the only option is to keep on applying pesticide. Another consequence of spraying the orchard may be to kill all the honey bees that pollinate the flowers. If the flowers are not pollinated by the insects they have to be pollinated by hand, an almost impossibly laborious process.

Using predators to control pests is not new and has had some spectacular successes. The two best-known cases are from Australia and California. In the late 1880s, Californian orchards were plagued by scale insects, and the entire citrus crop was threatened. Scientists determined that chemical sprays could not be used because they would kill pollinating bees; a biological control would have to be found. The scale insect was not native to California; it had been accidentally introduced from Australia, and so the scientists looked to Australia to find its natural predator. The introduction of 129 live "verdalia" ladybugs, a beetle that will feed as both adult and larva on scale insects, started on a single tree. After about four months, the tree was almost free of scale insects; within seven months the orchard was cleared of them; and after just one year the entire Californian orange crop had been saved.

The story doesn't end there. More than a hundred years later, the descendants of those ladybugs are still engaged in a deadly game of hide-and-seek. Local outbreaks of scale insects are located by the ladybugs, who raise as many young as they can on the scale insects. The ladybugs then disperse to hunt down the next possible food source. The scale insects are never completely eradicated, but neither are they getting out of control. The system is tightly regulated by the natural cycles of a predator-prey interaction.

In Australia, overgrazing by sheep laid the ground open to a cactus that escaped from ornamental gardens. The prickly pear cactus (*Opuntia inermis*) ran rampant over a vast area of Australia in the 1930s and 1940s. The loss to grazing land was a financial disaster. Biological control was attempted, and a moth whose larvae eat the prickly pear was introduced from Arizona. By the 1960s, the *Cactoblastis cactorum* moth had successfully eliminated the cactus from most of the rangelands, and it continues to maintain it at acceptably low levels to the present day.

A generalized representation of biological control (Figure 18.8) shows that the predator is very unlikely to eradicate the prey. Indeed, it is better that some of the pests survive, otherwise the predator would go extinct. So long as both prey and predator are present, an explosion of the pest population will be prevented. However, if the predator were to go extinct, the pest species could reinvade and cause damage to the har-

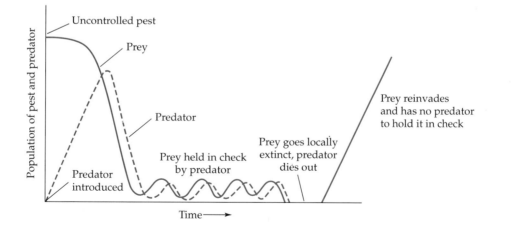

Figure 18.8 Predator-prey cycles for biological control. The prey (pest) population does not go to zero in a successful biological control situation. The predator and prey populations will be on different scales; there will be many more pests than predators. The diagram shows relative abundance of each through time. When low levels of pests are maintained through predation, some level of crop damage will ensue, but this may be economically preferable to paying for enough pesticide to control the pest completely. If the pest population were to go to zero, the predator would go extinct; it would starve. A reinvasion of the pest could lead to an infestation resulting in serious crop damage, hence it is better to maintain small populations of both pest and predator.

vest before a predator could be reintroduced. The farmer using biological control must therefore accept some crop damage. The desirability of biological control will depend on the economic threshold between pesticide cost and the crop loss. If the cost in crop loss due to pest damage is less than the expenditure needed to buy pesticide to completely exterminate the pest, the farmer is better off with biological control.

Although biological control has these shining success stories, on the whole it has not had widespread success. Part of the problem is that a number of requirements exist.

- The prey must have a potentially dominating predator.
- The prey must be arranged spatially and temporally to facilitate the maintenance of predator populations.
- The predator must be prey specific, otherwise it may become a pest in its own right.

18.7 Integrated Pest Management

The goal of integrated pest management (IPM) is to increase farm production while at the same time releasing the farmer from the expense and damage of the pesticide treadmill. Integrated pest management uses a variety of approaches to manage the field as an ecosystem.

Crop rotation ensures that in each growing season the land will support a different crop. This rotation denies a potential pest the opportunity to build its population generation after generation. This strategy is essentially one of pest avoidance and works on the principle of hiding your crop from a potential pest.

In some areas such avoidance strategies are not feasible, and a single crop is the most profitable and most productive way to farm. For instance, the wet rice paddies of Southeast Asia support several crops each year, and no other crop could provide a comparable amount of food. The Green Revolution an-

Ecology in Action

Biological Control of Rabbits

The European rabbit is native to Spain and has been introduced to provide food and fur in other locations. After the European settlement of Australia that started in 1788, Europeans repeatedly attempted to introduce rabbits. But it was the introduction of 12 rabbits to Southern Australia in 1859 that doomed Australia to battle the bunny. Within 50 years rabbits had invaded every habitable part of Australia (a continental area of about the same size as the lower 48 states of the United States). The rabbit population was competing and winning in a tussle for resources against both the native marsupials and also with introduced grazers such as cows and sheep. The rabbits could crop grass and herbs so short that cattle would starve, and the land was overgrazed. By the 1930s it was estimated that there were more than 600 million rabbits in Australia; something had to be done.

Hunting or trapping the rabbits was impractical because of the high reproductive rate of the rabbits. The fecundity of a rabbit is astonishing. A single female can have as many as 38 young in a good breeding season. Natural mortality takes care of about 80% of these rabbits, but even so, under favorable conditions rabbit populations can increase by a factor of 12 during a single year. To offset such massive fecundity, it is necessary

to kill about 85% of the rabbits. Due to their sheer numbers, and the vastness of Australia, biological control was the chosen remedy, and myxomatosis (rabbitpox) the tool. *Myxoma* is a relatively benign virus in South American rabbit populations. After extensive testing it was established that this disease was lethal to the European rabbit, but did not harm marsupials. Myxomatosis was introduced to Australian rabbits in 1951. The mortality rate of infected rabbits was 99.8% in the first year, and rabbit populations were decimated. However, as we would predict, in the following years, the disease was seen to be less and less lethal, as more rabbits survived infection. By the tenth year, only 7% of rabbits with myxomatosis died.

Despite attempts to reintroduce the disease, rabbit populations surged upward during the 1980s and 1990s. By 1995, Australia had 300 million rabbits, most of whom were resistant to myxomatosis. The rabbit was driving native marsupials to extinction and costing ranchers over $115 million in lost earnings each year. In a fresh onslaught against the furry foe, the Australian government experimented with another virus. Rabbit calicivirus causes rabbit hemorrhagic disease (RHD), a massive internal bleeding that kills rabbits within two days of infection. This virus was quarantined while it swer was to use miracle crop strains, increased fertilizer, and increased pesticide inputs. During the late 1960s and early 1970s, this approach proved to be expensive but successful. Governments then subsidized the use of fertilizers and pesticides in order to make the new technologies more affordable. One consequence of the subsidies was that pesticides became inexpensive and were used indiscriminately. An unforeseen effect of putting so much pesticide on the land was that it killed beneficial predators. Soon afterward, epidemics of insect pests, such as the brown rice hopper, were experienced for the first time. The hopper had become resistant to the pesticide, but its predator had not. For the first time, the hopper was not held in check by natural predation, and it ravaged the rice crop of Indonesia, destroying over 1 million tons of rice in 1977. A short-term solution was to plant rice hopper–resistant varieties of rice, but within a few years the hopper had changed once more to take advantage of the banquet of rice. In 1986, the president of Indonesia banned 57 out of 66

pesticides widely used on rice and phased out subsidies for the remaining pesticides. Pesticide consumption plummeted, and so too did the population of rice hoppers. Natural predators, wasps and spiders, built sufficient populations to control the next outbreak. In the years since 1986, Indonesia's consumption of pesticides has fallen 60%, the government has saved $120 million on subsidy payments, and the national rice harvest has increased by 15%. Other countries have yet to take such bold steps, but the example of Indonesia offers a concrete demonstration that IPM can be a cost-effective solution to pest problems.

Another example of IPM comes from the oil-palm plantations of Malaysia. Thousands of hectares of tropical forest have been converted to palm plantations in lowland Malaysia. The oil from the palm nuts is used for cooking and industrial use and represents a substantial source of income. Postharvest crop damage by rats has been a costly loss of income to planters, and rodenticides have proved expensive

was being investigated on a small island 5 kilometers from the Australian mainland. RHD killed about 80% of the rabbits on the island and apparently did not affect other mammals. Before the trial was complete and the virus was approved for release, rabbit calicivirus escaped from the experimental island and started to kill mainland rabbits. There is speculation as to whether this was as a result of flies carrying the virus to the mainland, or whether it was deliberately and illegally released by someone impatient for a reduction in rabbit populations. RHD proved to be highly effective, and in the first year after its release its lethality was as high as 95%, killing over 100 million rabbits. However, once again it is clear that the disease alone will not rid Australia of rabbits. A program of burrow gassing and poisoning will be needed to eliminate the resistant rabbit population.

The general lesson here is that biological control that relies on a pathogen to control a population will almost certainly fail, because resistant genotypes will dominate the population. A pathogen can be used to reduce a population to a level that can then be managed by other means. This outcome contrasts with a predator-prey interaction in which a dedicated predator can hold a prey (pest) population at low numbers. This begs the question: Why not introduce a predator to control rabbits? Unfortunately, we have introduced such preda-

tors. Cats and foxes have been introduced, but these predators are better equipped to kill native marsupials than they are to kill rabbits. The interaction of rabbits and these predators is discussed in Chapter 20. No predator is known that will specialize in rabbits and leave native marsupials alone.

Rabbits crowd around a waterhole in Australia. High rabbit densities and livestock provide a combination that leads to overgrazing and loss of native wildlife.

and ineffective. Recent success has been reported from plantations that have installed Barn Owl nest boxes. The Barn Owl (Figure 18.9) is found all around the world (it is the same species in Malaysia as in the United States and Europe), and the owls are a natural predator of rats. Owls were reintroduced to the plantations and have thrived. The rat population is declining, and money is saved on rodenticides.

In addition to the avoidance procedures and biological control, IPM may use sterile strains of pests that merge with the local pest population, mate with them, but produce no young. This reduces the fecundity of the native pests and can lead to a decline in pest burdens. Traps using chemical attractants, generally the sexual attractants (pheromones), can capture and eliminate pests. Alternatively, a trap crop may be used. In this procedure, a small area of the field is planted ahead of the main sowing. The crop sprouts, and, because it is the only set of shoots in the field, attracts all the pests. The trap crop is then hit with a concentrated dose of pesticide that eliminates the pest. This crop is then harvested and thrown away, because it probably carries toxic residues. The crop in the rest of the field sprouts later than the trap crop and grows in an environment from which the pests have been removed.

Figure 18.9 **Barn owls can be important predators, reducing the population of agricultural pests such as mice and rats.**

Summary

- Pollution does not necessarily kill an ecosystem.
- Pollutant chemicals are often ones that have beneficial uses in other settings.
- Pollutants can accumulate in the body (non-biodegradable) or be incorporated into natural systems (biodegradable). Either type may prove toxic.
- Pollution regulation requires the balancing of ethical, medical, and economic aspects.
- Natural systems can often repair the effects of pollution, so long as the pollution is stopped.
- Such repair of systems will take significant time, perhaps centuries, but can be hastened by environmental management.
- Pesticides are needed to maintain present agricultural outputs, but they pose a significant health and environmental hazard.

- Improved pesticides should aim to be target-specific, quick acting, rapidly biodegradable to harmless components, safe to humans, and inexpensive.
- Use of pesticides removes natural predators and allows surges of new pest species.
- Biological control is most effective when a highly specialized predator is employed to maintain low populations of a pest species.
- Introduced disease can temporarily reduce pest populations, but acquired resistance will lead to a pest population recovery. Disease needs to be one tool, not the only tool of biological control.
- Integrated pest management (IPM) uses biological control, pest avoidance, and limited pesticide inputs to reduce pest numbers.

Further Readings

Carson, R. 1963. *Silent Spring*. New York: Houghton-Mifflin.

Fry, W. E., and S. B. Goldwin. 1997. Resurgence of the Irish Potato Famine fungus. *BioScience* 47:363–372.

Jordan, T. E., and D. E. Weller. 1996. Human contributions to terrestrial nitrogen flux. *BioScience* 46:655–654.

Rosegrant, M., and R. Livernash. 1996. Growing more food, doing less damage. *Environment* 38 (7):6–11; 28–32.

Zimmerer, K. S. 1998. The ecogeography of Andean potatoes. *BioScience* 48:445–454.

Web Connections

On-line resources for this chapter are on the World Wide Web at:
http://www.prenhall.com/bush
(From this web-site, select Chapter 18.)

Aspects of Tropical Development

19

19.1 *Remote sensing and tropical forests*

19.2 *Misleading estimates of forest destruction*

19.3 *What are the factors driving deforestation?*

19.4 *Promoting the conservation of tropical rain forests*

19.5 *Sustainable agriculture in the forests*

19.6 *Before we blame it all on LDCs*

19.7 *Overview*

Because 50% to 80% of biodiversity is believed to lie in the tropics and the tropical nations have more land yet to be developed than do temperate countries, the tropics are central to concerns over the global loss of biodiversity. This chapter deals primarily with the causes underlying the loss of biodiversity in the tropics and some of the potential solutions.

The less-developed countries have the fastest-growing populations and the largest numbers of very poor people. They also control most of the globe's biodiversity. This combination alarms many conservationists. They foresee that tropical habitats, the great reservoir of planetary biodiversity, will be lost within a generation as exploitation mounts. It often appears that the more-developed countries show greater concern over the plight of the rain forests and coral reefs than over the fate of the poor. This perception, held by LDCs, surfaced at the Rio Earth Summit in 1992. Problems of high infant mortality, poor social infrastructure (hospitals, schools, sanitation, roads, and pensions), contracting economies, inflation, and illiteracy plague LDCs and must form a higher priority for their governments than environmental protection.

19.1 Remote Sensing and Tropical Forests

Determining the extent of human alteration to tropical habitats relies on eyewitness accounts or some kind of remotely sensed data. Many areas of tropical forest are almost impossible to reach, and eyewitnesses will tend to travel where they, and therefore settlers, can gain ready access. The investigator may then obtain a biased view that overestimates the extent of tropical forest destruction. Hence the most accurate estimates of tropical forest use are based on **remotely sensed** data, such as aerial photographs and satellite, sidescan radar, or infrared images.

Satellite images are increasingly important to science, and it is worth learning a little about how they are made. The most commonly used satellite images come from the *Landsat* Thematic Mapper (a sensor) carried on U.S. satellites or the French *SPOT* system. These satellites do not simply photograph Earth from space, they also make digital recordings of set wavelengths of energy reflected from Earth. For example, modern *Landsat* satellites record seven bands of energy that span the spectrum from ultraviolet to infrared energy.

Cloud cover is often a problem for conventional aerial photography and satellite images. Remote sensing can also be based on radar imagery, which is not affected by clouds and allows detailed representations of areas that carry an almost permanent cloud cover. These various sensors capture an image, but is it an image that can be readily interpreted? Humans cannot see radar, infrared, or ultraviolet energy. Consequently, these wavelengths must be translated into visible light if they are to be used.

In general, a satellite image (the picture output) is composed of three of the sets of wavelengths recorded. For example, a *Landsat* image has seven sets of wavelengths that could be used, but only three will be selected to form the image, such as ultraviolet, infrared, and blue light. The inputs of light are translated into wavelengths that can be seen by the human eye. Infrared might be expressed in shades of red, the ultraviolet in shades of green, and the blue in shades of blue. The result is an image that allows us to "see" in these normally invisible wavelengths. The image produced is known as a **false-color image**. At the end of the process we may obtain a detailed image, but can we trust it? For instance, if we look at a photograph of a forest, we see trees. But the remotely sensed data might not be picking up the reflected signal from trees but instead the wavelength reflected by water on the leaf surfaces. If the forest is wet when the data are collected, the imagery may show a forest as a lake. In order to determine that the patterns obtained are real and not artifactual, people must be sent in to check on, or **ground-truth**, a proportion of the sites described by the data. Our imagery of tropical forests is still incomplete, and although we can tell the difference between a pasture and a mature forest, the certainty of mapping accuracy declines as more subtle habitat separations are sought. At present, it is not possible to separate the range of secondary successional vegetation types that approach a mature forest; a mature forest is still confused in the imagery with late-successional regrowth. In short, despite huge increases in this kind of information technology we are still left to produce only educated guesses about the rate of forest destruction. The Worldwatch Institute estimated that by 1990 about 20% of tropical forests had been severely impacted by human actions.

19.2   Misleading Estimates of Forest Destruction

Conservation consultant Norman Myers's estimate is that about 240,000 km^2 of forest are lost each year, or about 46 hectares (135 acres) per minute. These first estimates were made in the late 1980s and led to predictions that many countries would have lost all their forests by 1990. Furthermore, some environmentalists extrapolate from this projection to suggest that within 35 years all the tropical forests will be gone. The first of these deadlines has passed, and the forests, though reduced, have not disappeared in the predicted areas, largely because of conservation initiatives.

The principal problem with the 35-year forest deadline is its generalism. The accuracy of the forest-loss estimate is also questionable. Let us take the latter first. One of the difficulties is in determining whether a "new" forest clearance is in fact new, or is the recutting of an area that has lain fallow long enough that it is difficult to separate from mature forest on the imagery. Mistakes of this sort could lead to an overestimate of forest destruction. Errors can also be made that underestimate the effect of forest clearance. For instance, the area affected by logging will not be just the area cut, but will also include some of the adjacent region. Often, the process of deforestation is a complex pattern of progressive fragmentation of the forest (Chapter 20). For now it is sufficient to realize that an area made up of a mosaic of forest fragments, covering perhaps half the total area of a landscape that was originally all forest, is likely to have lost many, perhaps 90%, of the species that demand deep-forest habitats.

The larger problem with the 35-year estimate of forest survival is that some forests are much more likely to disappear than others. The flood forests of the Amazon will still stand long after the coastal rain forests around São Paulo, Brazil, have been felled. These Atlantic rain forests are biologically and evolutionarily distinct from the Amazonian forests and are among the most species-rich habitats in the world, with recent estimates of 430 tree species per hectare. These forests are accessible and lie within an area of spreading urbanization and a human population of 100 million. Already, the Atlantic coastal forests have been reduced to less than 5% of their documented area at the turn of the century (Figure 19.1). Compare this with Amazonia as a whole, where an estimated 88% of the forest still stands as a relatively intact set

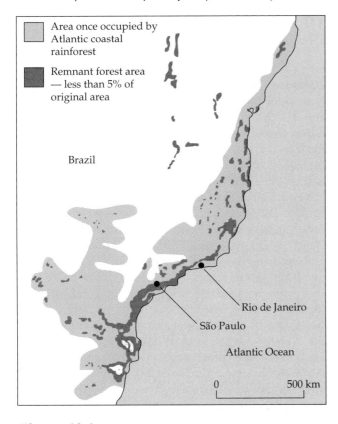

Figure 19.1   The destruction of the Atlantic coastal forest of Brazil. This is possibly the most species-rich habitat on Earth. Less than 5% of the original forest remains, and, apart from some small parks, the remainder will probably be felled in the near future. (After F. D. Por, 1992.)

On the one hand, the general statement of a 35-year expectancy for the survival of rain forests greatly overestimates forest destruction. The forces of exploitation will use this inaccuracy as a weapon to discredit the conservation movement. After 35 years, they will be able to point to large areas of real estate that remain forested and thereby claim that concerns regarding the loss of biodiversity are groundless. On the other hand, the estimate does not convey the full urgency of the fact that unique habitats rich in endemic species are being lost every year. For the most-endangered forests of the lowland Andes and the Atlantic seaboard, another decade of procrastination is likely to seal their doom.

19.3   What Are the Factors Driving Deforestation?

The short answer to the question "What is driving deforestation?" is "the expansion of human populations." However, the population growth of countries such as Mexico, Brazil, and the Philippines is not the direct cause of forest destruction. The direct cause is the extreme poverty of the vast majority of the population. Thus, the long answer involves population growth, economics, changes in land management, and the demand for hardwood. Human population growth is discussed in Chapter 16, the other topics in this chapter.

Changes in Land Management

The deforestation of large sections of Amazonia and other tropical lands is not necessarily a new phenomenon. For example, the first human populations in Amazonia entered the forests about 11,000 years ago and spread out along the major river systems. Earthworks, piles of shells, and other garbage dumped at the edge of settlements provide a record of human occupation. By 8000 years ago, these people were using pottery, and, by 6000 years ago, corn had been carried south from Mexico and was being cultivated in the Amazon basin. Field systems were being made by shifting agriculturalists, substantial forest areas around settlements would have been disturbed by fire, and the ecology of the area would have been influenced by hunting. About 2500 years ago, human populations seemed to increase rapidly, colonizing ever more remote areas of the Amazon, even areas far away from rivers. By the time the first Europeans penetrated the Amazon, they found almost continuous settlement

of habitats. Other areas that have already seen mass extinctions are the Andean foothill regions of Ecuador, Peru, and Colombia.

The transition of forest types up tropical mountains includes some of the most species-rich habitats on Earth. Some of these are also among the smallest habitat types, with the highest proportions of endemic species. Even neighboring valleys separated by a narrow ridge can support different assortments of species, some of which are found nowhere but in one particular valley. These habitats are so small, so rare, and so coveted for their timber and agricultural potential that they hold many of the species closest to extinction. Whole communities rich in endemic species are being lost at an unprecedented rate in the mountainous areas of South America, the Philippines, Indonesia, and Africa. In many areas, the greatest loss of biodiversity is associated with the deforestation of the premontane habitats at the edge of the great rain forest blocks rather than in the flat bottomlands of Amazonia or the Congo basin.

along the major rivers, and historian William Denevan estimates that the human population of Amazonia was in excess of 4 million and perhaps as many as 10 million people when the Europeans first arrived. A similar pattern of population settlement and expansion is seen in the Central American countries of Panama and Guatemala.

In the remote Darién province of Panama, tropical rain forests cloak the land as far as the eye can see. This region, the only section of the Americas not bridged by the Pan-American highway, is said to be impassable. In 1988, Paul Colinvaux, of the Smithsonian Tropical Research Institute, and I hiked close to the Colombian border to raise a sediment core from a previously unmapped lake, Lake Wodehouse. The forest surrounding the lake was the finest I had seen in Panama, with huge buttressed trees and a tremendous diversity of birds and mammals. The mud retrieved from the lake was analyzed for its fossil pollen content, and to my astonishment it revealed a 4000-year history of maize cultivation and forest burning, indicated by charcoal and the pollen and phytoliths (silica plant parts; identified by Dolores Piperno of the Smithsonian Tropical Research Institute) of maize (Figure 19.2). Almost as exciting as the discovery that this apparently pristine area had once supported farmland was evidence that all cultivation had stopped abruptly just over 300 years ago.

The history of European contact with New World populations is one of betrayal, religious coercion, and enslavement. But our most deadly gift was not the dogmatism of the Inquisitors and their minions but the diseases that they carried. Perhaps 98% of the native population succumbed to diseases against which they had no inherited defense (Chapter 26). The abandonment of agriculture evidenced in the Lake Wodehouse record would almost certainly have been due to this depopulation. That pattern would have been repeated countless times across South and Central America as small farming communities died out. A period of forest regrowth then took place to heal the scars inflicted by fire and axe. It can be argued, therefore, that forest disturbance is not a new phenomenon. In fact large expanses of apparently pristine forest throughout South and Central America may be only a few hundred years old.

Our view of "virgin" forest may be a romantic myth, engendered by the nineteenth-century biologists and expeditionaries. Undoubtedly, Henry Bates of England and Karl Friedrich von Martius of Germany, the great biologists who first described the Amazonian forests, traveled through luxuriant, mature rain forest. However, it was not necessarily a virgin forest. In many areas, it could have been the product of just two centuries of regrowth. The explorers were visiting the forests at a time when human population densities had reached their lowest level in at least 2000 years. The explorers took what they saw before them to be the norm, the "real" Amazonia; this view was a fully understandable example of Victorian-era now-centricism. However, if they had been able to visit those areas 1000 or 2000 years earlier they

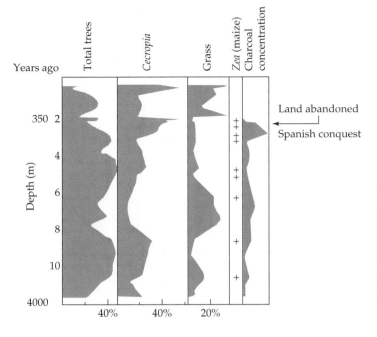

Figure 19.2   **The fossil pollen record from Lake Wodehouse, Panama.** Corn production and burning were continuous features of this landscape for 4000 years before the sudden cessation of human impact about 300 years ago. The abandonment of the land is probably related to disease introduced by European settlers. (After M.B. Bush and P.A. Colinvaux, 1994.)

might have found a much more exploited landscape. In any case, their accounts formulated our impression that humans have no place within the forest and that nature and agriculture in tropical forests are mutually incompatible. The archaeological data point to substantial populations of humans living for thousands of years in Amazonia, and clearly they did not eliminate all the wildlife, otherwise the Amazon forests would not have their splendid modern diversity of life.

Before I give the impression that it is quite all right to devastate tropical forest, I should make it clear that the disturbance inflicted by these past populations was caused by hunting and gathering, fishing, and shifting agriculture. These land uses were very different from modern ranching or slash-and-burn exploitation.

Shifting versus Slash–and–Burn Agriculture

Shifting cultivation has been documented as the "aboriginal" agricultural pattern in almost all agrarian tropical cultures. A patch of forest of 0.5 hectares is selected, the undergrowth and small trees are cleared with stone axes or machetes, and the area is then burned. Large trees may be killed by lighting fires around their base. An unshaded field with a coat of ash that is rich in minerals and plant nutrients results. At the onset of the wet season a mixture of crops is planted. The full range of crops needed by a family

will be grown, for in this system there is no, or very little, surplus available for trading. The clearing of the forest results in changed nutrient availability, particularly as phosphorus becomes insoluble, and results in declining harvests. As crop returns dwindle, it is time to clear a new patch of forest (Figure 19.3). Meanwhile, the original field is left to lie fallow. Each family may operate 15–30 such fields, working them in a strict rotation that leaves each field fallow for about 30 years. During this period, the process of secondary succession starts, and hence the organic content is restored to the soil. Woody species invade the open area and build biomass that will become the fertilizing residue of the next farming cycle. One six-person family can be sustained with about four hours of work a day on 6 hectares of good land or 20 hectares of poor soils.

Shifting agriculture was almost certainly the preconquest land use of the Americas. It is a sustainable practice that permits some coexistence of agriculture and wildlife. Trading of hunted game, fish, and any surplus crops or forest fruits may allow the farmer to barter for basic clothes, a machete, maybe even a dugout canoe. The moment that the farmer or his family decide they need a radio, an outboard motor, and other store-bought merchandise, this system will fail. It will also fail if human population densities grow to the point that each field must remain in cultivation rather than be allowed the lengthy fallow period.

The farming of the modern settlers, and those who seek to harvest enough to participate in a cash econ-

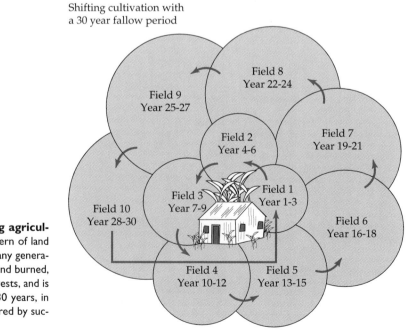

Shifting cultivation with a 30 year fallow period

Figure 19.3 Shifting agriculture. The indigenous pattern of land use can be sustained for many generations. Each field is cleared and burned, provides three or four harvests, and is then left fallow for about 30 years, in which time fertility is restored by succession.

omy, is slash-and-burn agriculture. This method can be practiced by displaced city dwellers who have become refugees from the city (willing or forcedly), borne into the forests on policies of agricultural expansionism. The process of clearance takes place along a front, steadily eroding the wall of forest. Crops are raised on the cleared ground until the land is exhausted, seldom more than five harvests. A lack of money puts expensive fertilizer and pesticides or improved seed stocks beyond the grasp of the farmer. Consequently, the land will be worked until nutrients are exhausted. A fallow period when the land is rested and ecological succession restores nutrients to the soil is another luxury that cannot be afforded. Money, essential to the survival of the family, can be made from the land by running cattle on the exhausted land. The cattle trample the soil, adding to soil erosion. Their grazing prevents succession from taking hold, and there is no soil recovery, only a steady degradation of the environment. The front of clearance then pushes deeper into the forest, and cattle are moved onto the vacant fields (Figure 19.4). Poverty locks the farmer into the slash-and-burn cycle that leads to the degradation of land and destruction of forest.

Logging and Forest Access

The impact of logging varies tremendously according to how it is carried out. A clear-cut, when all the trees are removed from a large area of land, lays the soil open to baking temperatures as the shading trees are removed (Chapter 14). Few soil organisms are able to withstand this extreme heating, and the process of decomposition will be greatly slowed.

Good drainage protects the land from erosion. If water does not soak into the soil, it may flow across the surface and wash away the topsoil. The rate at which water soaks into the ground can be expressed as the soil infiltration rate (distance that water soaks downward per unit time). In mature rain forest with undisturbed soils, infiltration rates are about 117 cm per hour. Under a regime of slash and burn, this drainage capacity is reduced to about 10 cm per hour. If machinery is used to remove the forest, the weight of giant backhoes and bulldozers will crush the soil, pressing out air spaces. Soils compacted in this way may have infiltration rates as low as 0.5 cm per hour. Lowered infiltration rates indicate a reduced water-holding capacity and an increased likelihood of soil erosion. Thus, where bulldozers have been used to re-

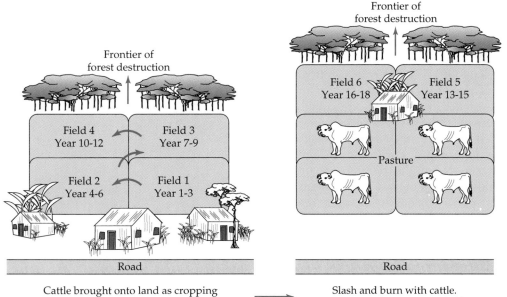

Figure 19.4   Slash-and-burn agriculture in the Amazon. Settlement often takes place along a river or road, with neighboring farms packing together. The fields are cleared, burned, and farmed until harvests decline, and then are given over to livestock. With no fallow period, the land does not recover fertility, and new fields must be cleared. The farmer is forced to move his house to stay close to the fields, and a frontier of forest destruction drives into the forest from the road. This is an unsustainable use of the land.

move trees, not only has the protective canopy been lost, but the soil character has been changed. Under such circumstances, the soil erosion may increase by a factor of 20 to 30 compared with the undisturbed system. In Sumatra, some experimental forestry programs have reinstated the traditional practice of using elephants to drag cut logs from the forest. The elephants cause much less damage both to the remaining trees and to the soil than do heavy machines.

We are familiar with scenes of devastation in which an entire forest has been felled in a clear-cut operation. An alternative to clear-cutting is the selective logging of forests, where only the most valuable trees are removed. Frequently, loggers will work to extract about 5%–8% of the trees from a forest. These cuts can form a sustainable forest industry if properly managed, although the long-term consequence of continually selecting a single species or group of species (such as mahoganies) will ultimately result in their local extinction. The key to the success of selective logging is the return period of the logger. Any period less than 30 years between cuts is virtually guaranteed to fail, and it may require considerably longer cycles for this method to be truly sustainable. Even a cutting program that removes only 5% of the trees can appear to be very damaging to the forest. In Malaysia, a study of selective logging revealed that for every tree intentionally felled, 16 or 17 more were damaged. Part of the problem is that vines bind the canopies of tropical forests. Although vines have thin trunks and constitute only 5% of the wood in a given forest area, their leaves may account for 45% of the photosynthetic area of the canopy. These vines moor one tree to another, and as one tree is felled, its neighbors are pulled down with it. This mass of falling debris snaps off other trees or damages their crowns as they crash to the ground. In a case study in Sumatra, rainforest ecologist Anthony Whitten recorded all the trees more than 30 cm in diameter before loggers moved in to remove 8% of the trees (Figure 19.5). When they left, more than half of the trees had been damaged. A wound where a branch has been ripped away or where the canopy is broken open is a site of potential infection. Many of the trees would recover, but certainly more than the intended 8% would be killed.

In a study of the effect of logging on mammal species diversity at Sekundur Reserve in Malaysia, the mature forest contained about 30 mammal species (Figure 19.6), most of which were bats and terrestrial rodents. A carefully controlled selective logging, defined as the removal of 8 to 10 trunks per hectare, resulted in almost no observed loss of native mammalian species. Four species of human-introduced

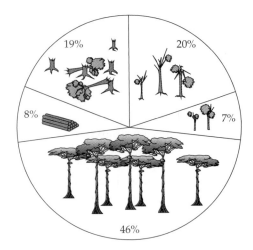

Figure 19.5 An 8% cut of mature rain forest results in damage to half the trees. A damaged tree is likely to be invaded by pathogens and may well die. (After A.J. Whitten et al.)

mammals, various types of rat, invaded the disturbed area, actually increasing the number of mammal species. However, as the clearance was intensified, there was a sudden drop in mammal diversity, so that of the initial 30 species, only 6 native species remained in the secondary forest. As the clearance was completed and cattle were run onto the land, only the introduced mammals remained.

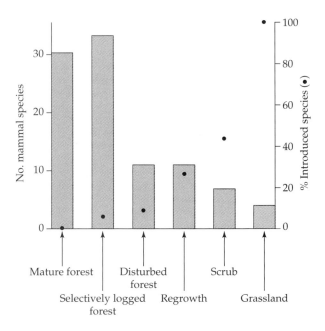

Figure 19.6 Changes in the diversity of introduced and indigenous mammals in progressively disturbed landscapes. Although species diversity is highest in the selectively cut forest, some of the mature-forest species will have been lost. They will have been replaced by species tolerant of disturbance and by introduced species, such as rats. (After A.J. Whitten et al.)

Logging in itself is seldom so damaging that nature cannot repatch the forest through the process of succession. The principal threat to forests from logging operations is that the areas logged are rapidly invaded by, and become home to, slash-and-burn agriculturalists. A logging track driven into a previously remote area provides instant access to the land-hungry settlers. Slash and burn will drive the forest farther and farther back from the original track. Satellite imagery of the Amazonian frontier of forest destruction looks like a mad ladderwork of initial logging tracks begetting myriad tangential tracks made by the settlers leading to the receding forest edge (see Color Plate 9).

Ranching

In the late 1980s, the U.S. fast-food industry received a lot of adverse publicity regarding its use of beef raised in the rain forests of South America. However, whether a particular hamburger chain does, or does not, use beef raised in the rain forest is largely immaterial to the fate of the forests. Ranching is a big business in many tropical countries and is closely tied to politics. The major landowners in many Latin American countries often belong to a small pool of elite families who form a ruling oligarchy. This powerful landowning lobby resists land reform, using force if necessary. Ranchers, miners, and farmers have fought those who would save the rain forests. According to Tyrone Miller, more than 1000 activists opposed to the continued rain forest destruction have been murdered in Latin American countries since 1985.

Government subsidies in some countries have encouraged the felling of forest to produce ranchland—policies aimed to bring a stronger economic and agricultural base to an emerging economy. The dilemma of short-term gain versus the degradation of the environment has proved to be a tough political problem, especially because the export of beef provides these countries with much-needed foreign capital. Furthermore, a substantial portion of the beef raised in the rainforest countries is used within their borders, and the protein from cheap beef is an essential part of many people's diets. It would be a brave politician who withdrew ranch subsidies. However, the subsidies promoted an uncontrolled destruction of the forest. It is estimated that about 30% of the Amazonian deforestation that took place between 1965 and 1983 was to provide grazing land.

Initially, on land cleared of rain forest, cattle densities rarely exceed 1 cow per hectare. After a few years of grazing, each cow needs 4 hectares. After 9 or 10 years, the land may no longer support cattle at all. At this stage, the topsoil has often been completely eroded, and ecosystem recovery, instead of taking a few decades, may take as much as 1000 years (Chapter 14). In the late 1980s, the Brazilian government undertook a bout of economic reform that included as a minor component removing ranch subsidies and enforcing the licensing of chainsaws. These brave reforms contributed to the rising rate of inflation, which reached more than 200% per year, and predictably the government fell.

The Demand for Tropical Hardwood

Tropical hardwood is cheap, so it is often the timber of choice in MDCs. In the rainforest countries, the labor is cheap, the trees are free, and subsidies are often available to drive new roads into an area. Logging operations can be highly profitable and represent an important source of income to the individuals involved. Timber is among the leading exports of LDCs. It brings in as much as cotton, twice as much as rubber, and nearly three times the revenue from cacao (*Theobroma cacao*, from which chocolate is made). However, the forestry practiced is not sustainable. The trees are effectively mined; compared with the number cut, very few are planted. A sign that the resource is not being used sustainably is that valuable trees such as mahogany, once common along the waterways of Amazonia, are now truly hard to find. In the past, timber was extracted from the riverside forests and great rafts of hardwood logs were floated downriver to markets or sawmills. Because the valuable timber has been cut from the river margins, the loggers are having to penetrate deeper into the forests in search of commercial-quality trees. This search for timber is driving the cycle of forest destruction. Construction of logging tracks promotes the arrival of settlers and the degradation of the land by slash and burn, preventing the regrowth of the forest, hence creating the need for further logging tracks to be driven deeper into the forest.

The real profit to be made from logging is not in selling a tree trunk, but in processing the wood. Sawmills convert the cheap logs into an expensive end product (such as high-grade lumber, veneer, or plywood). Indonesia, one of the major tropical timber producers, saws fewer than 30% of the trees cut for timber; the rest are shipped overseas as logs. In the 1980s, the demand for tropical hardwoods more than doubled (Figure 19.7), with Japan as the leading consumer, followed closely by the United States and Europe. Japan has forests that could provide timber, but as long as timber is relatively cheap on the world market, Japan will buy timber rather than deplete its own re-

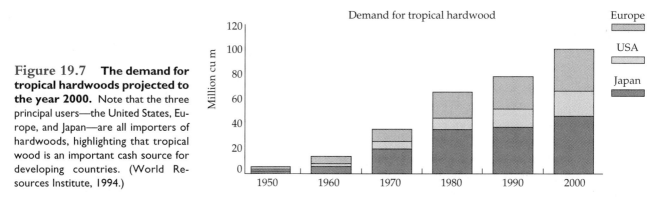

Figure 19.7   The demand for tropical hardwoods projected to the year 2000. Note that the three principal users—the United States, Europe, and Japan—are all importers of hardwoods, highlighting that tropical wood is an important cash source for developing countries. (World Resources Institute, 1994.)

sources. Much of the timber imported comes as raw logs, and Japan is the leading exporter of tropical hardwood plywoods. The forests of Indonesia, Malaysia, the Philippines, and Thailand have been the primary sources for Japanese timber imports, but as these forests are overexploited, new logging concessions are being sought in Africa and South America. The United States and Europe also practice this brand of imperialism, and it is the developed world that has the highest demand for timber.

19.4   Promoting the Conservation of Tropical Rain Forests

In 1987, the Brundtland Commission, an international research panel reporting to the United Nations, issued a report stating that the socioeconomic priorities of the LDCs must be attained before they can be expected to enforce environmental programs. This report promises to be one of the most influential policy documents of the late twentieth century. Previously, it had been assumed that if population growth could be controlled, the economies of the LDCs would improve and conservation projects would follow. The findings of the Brundtland Commission turn this assumption around to emphasize that economic development must precede population control.

However, this conclusion is based on the assumption that LDCs will follow a pattern similar to that of the MDCs through the demographic transition (Chapter 16). Even if they do, in the time that it takes for Brazil or the Philippines to make this progress, many of their unique habitats will have been lost. Few conservationists have confidence in a laissez-faire economic fix to the problems of biodiversity loss. Economics will have to be a part of the eventual solution, so the issue becomes one of revealing a

reason for LDCs to cherish their natural habitats in economic terms.

The Role of Ecotourism in Conservation

Tourism is a multibillion dollar international industry, and as travelers have become more exotic in their tastes, tours to investigate the wildlife of remote places have flourished. This ecotourism is seen by some as the likely savior of habitats from Alaska to Antarctica. The best-known areas developed explicitly for ecotourism were the Great Plains of East Africa. The first visitors from Europe were primarily big-game hunters, but lately the emphasis has shifted to animal watching and photography. The visitor can now explore parks such as the Serengeti by Jeep, hot-air balloon, or microlight aircraft. Similarly, accommodation ranges from luxurious lodges to campsites for those who want to live closer to the natural world they are visiting. An urge to see lions, elephants, zebras, and migrating herds of wildebeest is likely to be satisfied in one of these safaris. The appeal is evident.

A newer development in ecotourism is the rain forest tour, in which access is usually gained by canoe, foot, or occasionally on horseback. These are seldom luxury tours, and they are more demanding of the tourist. Also, the rain forest is very different from the grasslands of Africa for viewing wildlife. Most of the wildlife that a visitor will see will be insects, because most of the birds, monkeys, and other more appealing wildlife are high up in the canopy, where they can be heard but not seen. The traveler is not undertaking this kind of trip to obtain uninterrupted, close-up views of wildlife, but more to savor the flavor of the forest as an ecosystem. Whatever the goals of the visitors, ecotourists express a demand for natural areas and they spend money in order to fulfill their goal.

Ecotourism has the potential to bring employment and money into some poor rural communities. Under ideal circumstances, local arts and crafts can be sold as souvenirs, bed and board can be provided for travelers, and local guides can be paid to escort tourists through the reserve. All of these activities encourage the growth of the local economy. Monteverde, Costa Rica, is a rain forest reserve that is privately owned (started by American Quakers who moved to Monteverde). Tourists from all over the world pay a modest admission charge for the opportunity to walk the trails of the cloud forest. In the nearby village there flourishes a group of locally owned restaurants and boarding houses that are clearly benefiting from visitors to the forest. As money is injected into the local economy, the value of the reserve is reinforced and the local population will work to further its protection.

A less rosy view of ecotourism is that most of the money goes to tour companies in the national capital, that visitors are whisked in and out of the reserve areas in air-conditioned buses, and that they have little interaction with the local populace. The hotels established inside or on the edge of the park are owned by foreign or distant administrators, and relatively little money is spent in the local villages. This version of reality sees the jobs that are available to local people as low-paying service-related occupations. An example can be found in some of the ecotours that set out from Quito, Ecuador, to explore the Amazon, or from Chiang-Mai, Thailand, to visit the opium triangle. Commonly, each participant pays the tour operator more than $100 per day. However, the tours are more or less self-sufficient, carrying all their provisions in canoes or on horseback, and, although local villages are visited, little money changes hands with very little benefit felt by the rural population. Furthermore, the local traditions of land use and culture are compromised for the sake of wealthy tourists. Despite this controversy over monetary equity, once ecotourism is established in a country, there will be a very strong lobby to promote conservation so that this source of revenue continues.

Like any new venture, ecotours range from the highly professional to fly-by-night operations. It is likely that ecotours will become more regulated in the future to prevent the real dangers that can be faced. While on fieldwork in Amazonia, I met a group of ecotourists who had set out on a three-day guided tour; this was their seventh day in the forest. Their guide had misjudged the river conditions, and they had become stranded as river levels fell. With water too shallow for canoe passage, they were forced to hike back

through ankle-sucking mud to the nearest road. Hungry and ill-tempered after four days on foot, they were unlikely to return to that company for a second tour.

The ecotourism industry also needs to resolve problems that include pollution caused by sewage, refuse, and leaking motors and the social impact of tourists on local indigenous populations. A potential problem that has yet to be addressed is the disturbance to wildlife that tourism brings. Although tourists do not mean to harm the wildlife, they can have many subtle impacts. As a hotel set deep in the forest grows, so do its needs for staff and access roads. Soon a small village may develop to provide the infrastructure for the tourists who are trying to escape into the wilds. If the press of visitors becomes too great, the animals may be driven away from the center of tourist activity. Such disturbance often results in reproductive failure and population declines among the local wildlife. Such impacts could be lessened by setting a quota limiting the number of visitors to an area or by establishing a closed season to allow unimpeded reproduction of especially sensitive populations.

Debt-for-Nature Swaps

Many of the rain forest–owning nations are LDCs grappling with problems of an emergent economy, spiraling populations, and a deteriorating balance of trade. In order to finance their fledgling economies, these countries have borrowed money from other governments or private institutions. If development and money-earning potential do not follow swiftly after taking the loan, the LDC is faced with a growing debt as overdue interest mounts on the initial loan. Commonly, all the payments made by the LDC go toward paying off the interest while the principal, the actual amount borrowed, remains unpaid. Under such circumstances, the debt has no end. The financial drain on a poor country to repay its international debts can be intolerable and leads some countries to default entirely on repayments. Clearly, this system fails to work well for either the loaner, who is faced with a high risk of a bad debt, or the recipient, who is faced with a bottomless morass of debt.

In 1984, Tom Lovejoy of the World Wildlife Fund proposed a radical alternative to this impasse. A nongovernmental organization (NGO), such as Conservation International, The Nature Conservancy, or the World Wildlife Fund, would become a third party in the negotiations. This third party would purchase a fraction of the debt from the bank that had lent the money and would then be in a position to "forgive" the debt. In exchange for the cancellation of a portion of its debt, the debtor nation would take specific environmentally beneficial actions such as establishing and maintaining a rain forest nature reserve.

Banks benefit from this arrangement because it is better to take a 10% repayment on a debt than to receive nothing when the country defaults. The country benefits by losing some of its financial obligations, and conservationists in MDCs, for a relatively small investment, can save an endangered area of forest. In 1987, the first debt-for-nature swap led to the formation of the Beni Reserve in Bolivia.

Bolivia is a poor South American country with little industry and low wages. A typical family might earn $600 a year. Its human population of 7.5 million is growing at a rate of 2.9% and can be expected to double in the next 25 years. The Bolivian economy had become reliant on loans, and in 1987 Bolivia owed $5.7 billion. The lowlands of Bolivia edge into western Amazonia and contain some of the most diverse of all Amazonian forest regions. An area of 1.5 million hectares was guaranteed to be set aside as a nature reserve if a $650,000 debt could be purchased from a Swiss bank. In addition, the Bolivian government promised to provide $250,000 to help maintain the reserve. Conservation International, a U.S. charitable organization, raised $100,000 to buy the debt from the bank, and the purchase was completed in 1987.

This was a small, but important, first step because it established the precedent for a debt-for-nature swap. With hindsight, we can see that some mistakes were made. The agreement with the Bolivian government was not sufficiently explicit as to how the reserves would be protected, physically or legally, with the consequence that loggers have continued to extract high-value trees from the protected area. Negotiations as novel and complex as this are bound to require some learning by the participants. One hopes that those mistakes will not be repeated.

More troubling is the observation that the amounts of debt canceled are almost trivial compared with the overall debt problem. The biggest swap to date, a $50 million debt buyout on behalf of Costa Rica, resulted in only a 4% reduction in that country's debt. Compared with the benefits, which when seen on this scale appear small to the debtor nation, the issue of lost sovereignty may seem large.

That so many rain forest areas are sparsely inhabited does not mean that they are not prized possessions. A case in point is the disputed territory between Ecuador and Peru. In 1942, a brief and largely forgotten war between those two countries led to Peru's claiming a large portion of what until then had been Ecuadorian Amazonia. Most maps show the Peruvian version of history, except those from Ecuador, which almost double the size of the country by including an area of "disputed territory." To forestall a similar fate from befalling their far-flung reaches of Amazonia, the Brazilian government until very recently helped relocate settlers into the border regions with Peru and Bolivia, thus firmly claiming the ground as Brazilian. Into this nationalistic fervor for ownership of the remote rain forest we now cast conservationists. The offer to intercede and buy debt in return for nature reserves is seen by Brazil and other nations as being yet another way that the MDCs can try to influence the economic and social development of their country. The sacrifice of sovereignty, the power to do as you will with your land, is seen as too high a price to pay for a small easing in the burden of debt.

If you doubt that this jingoism is something felt in the United States, consider the public outcry in 1992 at the thought that a Japanese business venture was going to buy the trade concession for Yellowstone Park. That foreign money might finance and profit from the sale of hot dogs and sodas in one of our national parks was more than the American public could stomach, and the deal was squashed. Promising to forego any future development—such as logging or mineral wealth—of a million or more hectares of rain forest for the "international good" is a much larger invasion of sovereignty.

Sustainable Extractive Industries

If forest products can be harvested without damaging the ecosystem, these will provide a repeatable, sustainable harvest year after year. Rubber is one of the most important economic products from many forest areas. It is obtained by scarring the bark of the rubber tree so that it bleeds its latex-rich sap into collecting cups (Figure 19.8). The rubber trees are scattered through the forest, and the business of collecting the rubber is sustainable but extremely labor-intensive. In the 1930s, the Ford Motor Company, seeking to supply the growing demand for motor tires and rubber drive belts, gained the rights to establish rubber plantations on 1.5 million hectares of Amazonia. Pure stands of rubber were planted on 7500 hectares, but after just seven years the young trees were so ravaged by pests (remember you must be rare to survive in the tropics) and the soil erosion was so severe, that "Fordlandia" was abandoned. The rubber tappers made headline news again in 1988 when Chico Mendes, their

Figure 19.8 **Rubber is an example of a forest product that can be extracted without degrading biodiversity.** (a) The white latex that forms rubber oozes from the scars on the tree and is collected. (b) The latex is pressed into mats of raw rubber for transportation to market.

301

Ecology in Action

Can Extractive Industries Save the Rain Forest?

Ecologists have realized that an economic argument must be made if the tropical forests are to be saved. If a forest is viewed in strictly economic terms, its worth can be calculated in terms of the money that it generates. A typical hectare of lowland rain forest can generate money when it is felled and then farmed for cattle. This use would provide a one-time income of $1000 for the timber and an annual income of about $150 per hectare per year from the cattle. This use would provide $4000 over a 20-year period. Alternatively, if the forest was felled (a $1000 income) and then a plantation forest established, the farmed timber could be sold for about $3500 after 20 years of growth. Thus, after 20 years the forest provides an income of $4500 per hectare. Neither use promotes wildlife conservation because both involve habitat destruction. The more environmentally destructive use, cattle farming, provides a faster return on investment and is the one commonly adopted by poor, land-hungry settlers.

In 1989, Charles Peters of the Institute of Economic Botany excited interest with his observation that a single hectare of Peruvian forest contained 117 individuals of 12 species of plant that could yield economically valuable produce. He calculated that if the fruits and latex from one hectare of this forest were collected and marketed, they would be worth $697.79 per year. Even if 25% of the fruit were left in the forest to ensure regeneration, and applying a 5% discount rate (an economic tool to predict the value of future income to someone in the present, see Chapter 24 for an explanation), the income over 20 years would be $6330—almost half again as much as could be derived from the most-profitable traditional forestry use.

Apparently, the forests were worth more standing than felled. Peters and his coworkers suggested that the natural riches of the forest had simply been overlooked and that a systematic nondestructive harvesting would be economically optimal and conserve biodiversity. Could the solution to maintaining biodiversity be this easy?

Other ecologists criticize this work because it assumes that the forests produce predictable amounts of fruit year after year. Tropical forests have masting years in which a mass of fruit is produced (see Chapter 15), but very little fruit is formed for the next three or four years. Such a pattern provides a very erratic income, and worse, during the times of high fruit abundance, the market is likely to be glutted, and prices will fall. Another problem is that the calculations of worth made by Peters are based on fruits that are relatively rare. If the supply of fruit were to increase because of increased harvesting, the price would fall, leading to a reduction in the estimated income from the forest. Yet another difficulty lies in getting these highly perishable fruits to market and selling them to the consumer before they rot. Without a huge infrastructure of roads, refrigerated trucks, or boats, the market would remain local.

In an analysis of fruit production in Indonesian forests, Deborah Lawrence, a graduate student at Duke University, and her colleagues determined that villagers preferred to plant orchards of these fruit trees close to their homes, rather than search the forest for scattered individuals. It was not worth the effort to walk several kilometers to find a fruiting tree when one could be planted and grown.

Certainly, the natural forest products have been undervalued in the past, and they may play an important role in future multiuse strategies to save tropical forests. However, it seems unlikely that simply gathering nuts, berries, and rubber will provide the economic vehicle to safeguard biodiversity.

articulate and charismatic spokesman, was murdered by ranchers for advocating that large areas of Amazonia should be set aside as extractive reserves. An extractive reserve would allow the tapping of rubber and the harvesting of fruit and nuts, but preclude the use of the area for ranching or mineral and timber extraction. It has been suggested that more money can be made from harvesting the fruits of a forest than can be made from felling it for timber, and therefore extractive reserves could be a viable, biologically desirable alternative to logging or ranching. This idea has an intrinsic appeal to all conservationists. However, the assumptions that markets can be found for these products, and that they could be shipped economically, have been questioned by economists. Large-scale extractive reserves are included in the Brazilian government's development plans for Amazonia, but they have yet to be established.

Medicines from the Forest

Some years ago, I used to pass a billboard that was appealing for financial gifts to a national cancer charity. The sign read, "Please give generously: The cure for cancer doesn't grow on trees." But it almost certainly does. Many of the drugs that are commonly used to treat ailments from headaches to leukemia are derived from tropical plant or animal extracts. About 75% of the 3000 plants currently known to have some medicinal properties come from the rain forests. A few of these are shown in Table 19.1.

Many plants and animals defend themselves with chemicals that evolution has shaped to ward off bacterial, protozoan, and fungal attacks. It really isn't surprising, therefore, that with some modification these same chemicals are effective in repelling the fungi, bacteria, and protozoa that plague humans. Morphine and cocaine are both analgesics derived from tropical plants, and, though abused by some, they are important as the pharmaceutical base for an array of painkillers (it is rumored that cocaine was also used as an ingredient in the original formulation of Coca-Cola, but was replaced by a substitute compound in the 1920s). How different life would be without stimulants such as caffeine obtained from tea, chocolate, and coffee, all of which come from tropical forest plants. Each plant may have different chemicals in the leaf, bark, root, and flowers, so a single species may offer three or four chances of containing a useful compound.

Despite the wealth of drugs already isolated from tropical organisms, less than 4% of tropical plants have been analyzed for their pharmaceutical potential. Barely more than this proportion of tropical insects have even been described by science, let alone tested for chemical properties. Pharmaceutical researchers are scrambling to collect new products for testing in a wave of **chemical prospecting**. The indigenous witch doctors, shamans, and herbalists have a detailed knowledge of plants in their area—a knowledge that is unwritten but learned as an apprentice. Often this knowledge is held to be sacred and forms the basis of the indigenous social hierarchy. It is unreasonable to expect native peoples to divulge such information to outsiders, particularly as the proffered monetary reward is often going to be linked to increased Westernization and values that run counter to the native cultural beliefs. The average age of elder cultural leaders is estimated to be over 80 years old, and few have passed on their full experience to the younger (often more Westernized) generation. The extinction of these herbalists and their knowledge is a more imminent loss than the extinction of the beneficial forest species. The importance of this loss is apparent. When drug companies conduct an unguided search for useful chemical compounds, about 1% of the samples collected have some potential value. When they are guided by shamans and shown which plants to collect, their success rate increases to about 50%.

It is simplistic to think that cures for AIDS will be found in a brief foray into the forest. Chemical prospecting is a long and expensive business. Thousands and thousands of samples have to be collected, concentrated, and analyzed before a hint of a useful chemical is found. The company undertaking this search makes a huge financial investment in the faith that, ultimately, after years of trials it will be able to sell a new drug, recoup its expenses, and turn a profit for its patient investors. And here lies one of the stumbling blocks. Who should profit from the new drug?

Table 19.1   Some of the medicines that have been obtained from rain forest plant extracts.

Drug	Medicinal use	Source
Allantoin	Anti-bacterial	Blowfly larva
Cocaine	Analgesic	Coca bush
Cortisone	Anti-inflammatory	Mexican yam
Cytarabine	Leukemia	Sponge
Diosgenin	Birth control	Mexican yam
Erythromycin	Antibiotic	Bacterium
Morphine	Analgesic	Opium poppy
Quinine	Malaria	Chincona tree bark
Reserpine	Hypertension	*Rauwolfia* plant
Tetracycline	Antibiotic	Bacterium
Vinblastine	Hodgkin's disease and leukemia	Rosy periwinkle plant

The company who does much of the work or the individual or nation owning the forest where the source organism was located? This dilemma can be summarized as a friction between intellectual property and royalties. The developers of drugs claim that the idea, the intellectual property, is everything and that the drug was just a worthless leaf until it was developed. On the other hand, rain forest countries, and these are usually among the world's poorer nations, claim that royalties should be paid to the owner of the unique resource that provided the extract.

The case for the drug company needing to make a profit is obvious, otherwise it will not do the painstaking research and prolonged testing required to have a new drug approved. Perhaps the case for royalties to the tropical nation is less apparent. Consider the analogy that your family has a cure for headaches. The cure, knowledge of which has been passed from generation to generation, comes from a small flower that grows in your garden. One day you treat a visitor for his headache, and, stunned by the efficacy of the remedy, he asks for one of these healing plants, which you give him. Some years later, a photograph of your visitor arrives. He is lazing by a luxurious pool, with mansion and Ferrari in the background, and on the back of the photograph is scrawled, "Neat plant, made a fortune from it. Have a nice life." Wouldn't you feel just a little cheated? Far-fetched? Far from it. Many of the drugs so far "invented" for Western medicine were part of the herbalism of shamans and witch doctors. Surely the compounds have been improved, but the intellectual property may truly lie with the native cultures and not with the drug company.

Until now, the collection of potential drug extracts has not been restricted by the countries that own the forests. Official permits are sought and granted, but no financial remuneration by the drug company is made if a useful drug is discovered. In recent years, this extraction of resources from tropical countries without payment has begun to be branded as "intellectual colonialism." The 1992 Earth Summit was an international congress held in Rio de Janeiro, Brazil, to discuss major environmental issues. The meeting discussed a charter guaranteeing the maintenance of genetic biodiversity. But, because a compromise could not be reached between intellectual property and royalties, the charter was not enacted (Britain and the United States were the only nations that refused to sign the agreement). Perhaps more significant than all the treaties of the Earth Summit was another agreement signed in 1992. Merck, a multinational pharmaceutical company, negotiated chemical-prospecting

rights with the Costa Rican government organization (INBIO)—a public-private collaboration that aims to promote research and sustainable use of local biodiversity. Costa Rica has developed a strong environmental program, with a higher proportion of its land devoted to national parks than any developed nation, and it has been at the forefront of several conservation initiatives. It is also a country with exceptionally high biodiversity and reasonably accessible forest. Merck contracted to pay $1 million for the right to conduct chemical prospecting in Costa Rica and then to share a proportion of profits, purportedly between 1% and 3%, with the Costa Rican government. INBIO will supply Merck with 10,000 samples of rain forest species for laboratory testing. This agreement is not pure altruism on the part of Merck. They have built an improved relationship with the government of Costa Rica and are guaranteed access to the forests when future policy changes may restrict access to other companies. Furthermore, Merck can now lay claim to being the first pharmaceutical giant to become "rain forest friendly." The potential benefits of the good publicity could make this an advertising bargain!

The Rights of Indigenous Peoples

The logging of tropical forests and the expansion of European-style agriculture threatens not only the biodiversity of these regions, but also the lives of indigenous peoples. Humans evolved in the tropical plains of Africa and spread from there into other habitats. Some debate surrounds whether the first *Homo sapiens sapiens* actually used the African rain forests 100,000 years ago. However, strong evidence indicates that human populations have exploited these forests for at least the last 35,000 years. The aboriginal settlement of the Southeast Asian forests is similarly ancient, and the last forests to be occupied were those of the Americas about 11,000–12,000 years ago. The tribal nations that formed in all these areas have suffered or withdrawn from the influence of colonial invasion, of which the rise of post-Renaissance European powers was just the latest phase.

The rights of indigenous peoples are only just beginning to be recognized, but even now little protection is afforded them in the face of advancing "civilization." In the summer of 1993, a tribe of Yanomamö Indians was massacred by gold miners moving in to exploit part of the Yanomamö homeland in northwestern Brazil. A simple and ancient showdown—gold or people—and gold won. The Indians are caught in a catch-22, the eternal no-win situation: If the native peoples come to the city to negotiate for their futures and arrive in traditional dress, they are not taken seriously, because they are clearly "just savages." Conversely, if representatives of the Indian nations adopt Western dress codes, they are perceived to be a watered-down, Westernized version, the puppets of radicals, and are again ignored. A deep racism against most indigenous peoples prevents their being accorded the basic rights that the rest of society expects, and it certainly denies their claim to the moral ownership of the land.

19.5   Sustainable Agriculture in the Forests

The path to sustainability lies in controlling the forces that presently ravage the forests. More than half of the forest destruction is not done by large corporations but is carried out at the human scale of an individual trying to feed his or her family. Breaking from the slash-and-burn cycle requires the development of sustainable agricultural practices that are affordable and do not result in even greater poverty. With hindsight, we can easily predict that certain agricultural systems will fail. It is harder to assert which will succeed. Monocultures will be decimated by pests; wholesale forest removal will lead to soil erosion and increased drought stress and soil temperatures; "improved" strains of crops will often be unaffordable or require too many expensive artificial fertilizers and pesticides. The overriding problem in the tropics is the rate of nutrient cycling. Because so many nutrients are held in the biomass and so few in the soils, all but the most fertile floodplain sites and some volcanic ashes rich in minerals will prove to be poor farmland. Long rotations of fallow or successional periods must play a part in any successful agricultural or forestry program, but these represent a period of lost income to the farmer. These farmers are not thoughtless or stupid. If presented with a better alternative to their present poverty they would, with a little persuasion, adopt it. But, in truth, no one has yet formulated a farming strategy that can offer them improved, affordable returns.

19.6   Before We Blame it All on LDCs

Because of the sheer richness of tropical habitats in the LDCs, it is easy to forget that there are many species threatened with extinction in MDCs. In these countries,

the argument that poverty forces the loss of habitat, which leads to extinction, cannot hold. The two prime causes of extinction in MDCs will be habitat loss due to development and the introduction of exotic species.

Seldom is development in MDCs "essential" in the sense that the local human population cannot survive without it. However, a behavioral pattern was established during the early development of the landscape, when the rights to develop clearly lay with the landowner, and the right of the landscape to exist undamaged was subsumed in the need to generate prosperity. The continued development of a landscape in an MDC will bring further benefits to some, because the development may be to increase farm efficiency or to build new housing subdivisions. The construction of golf courses, marinas, theme parks, or other recreational areas may require clearing forests or eliminating wetlands. Such areas can provide new homes or vacation facilities that bring pleasure to thousands of people, and the improved welfare of the population has to be weighed against the cost to the environment. Where species are directly threatened with extinction as a result of land use, it is highly unlikely that the permission for such development would be granted. However, extinction events can be brought to a crisis point by a gradual decline in habitat quality because our actions increasingly fragment breeding grounds, feeding areas, or migration routes (Chapter 20). Conservationists are then left to try to protect the last populations of a species, when the real damage that led to the decline in numbers has been going on for years or decades.

Trying to change the balance of land-usage rights, to change from the right of an owner to develop to the obligation of an owner to conserve, is politically challenging. For more than two centuries in the United States, property rights have, with only a few limitations, been equated with the ability of the landowner to develop the land. Changing such rights to recognize a societal right that transcends the rights of the individual can be taken to strike at the heart of our philosophy of ownership, personal freedoms, and independence. Only in the most recent decades has there been any shift in the balance of property rights away from development and toward conservation. The legislation that most consistently promotes such a change in property rights is the Endangered Species Act of 1972, and not surprisingly, therefore, it has become the prime legislative target of prodevelopment lobbies (Chapter 28).

While habitat loss due to development in an MDC can be argued to be due to our insatiable appetite for an ever-higher level of personal wealth, a secondary cause of extinction is invasion of exotic species (Chapter 20). As a landscape becomes increasingly developed, species are introduced from other ecosystems, even from other continents; these introductions can become a major cause of extinction for local flora and fauna. Island ecosystems and the remaining patches of wild lands in MDCs are at most risk from invasion by introduced species.

19.7   Overview

It is easy to criticize LDCs for exploiting their forest resources, but MDCs became wealthy and powerful only by full use of their land area. Britain has a few scattered woodlands in western Scotland that might be natural; otherwise every inch of land has been used. In the United States, from sea to shining sea, we have felled our forest and plowed the prairies to build an economic giant, and our remaining national parks and wild lands are still being affected by development. We hardly are in a moral position to criticize Brazil, Cameroon, the Philippines, or other tropical nations where rain forests are being converted to farmland, for using their natural resources.

Tropical rain forests are most likely to be saved as a result of the economic development of LDCs. Tropical forest will not be saved by punitive sanctions or through massive handouts. So long as human populations continue to rise in LDCs, and so long as LDCs are economically weak, the forests will be harvested as their most immediate resource. Economic development would lead to a more equitable distribution of wealth, and, as the standards of living rise, there will be a decrease in population growth rates. The best path for curbing population growth is to promote the desire for small families. The education of the population in birth control, women's rights, and hygiene are inextricably linked to reducing infant mortality and, before long, the number of babies born (Chapter 16). The establishment of social services so that the

elderly are not wholly dependent on the young for their well-being is another prerequisite for population stabilization and saving the forests. Rather than being restricted, forestry should be developed so that it is sustainable. That means investing in forestry education and even in sawmills so that the profit from the extraction is kept in the country and used to elevate the domestic economy.

It is the right of every people to develop socially and economically. It is an outsider's hope that they will do so with some environmental sensitivity. Conservation is a luxury that those faced with grinding poverty cannot afford. Only when these countries develop a substantial middle class with some political power is it likely that enforceable legislation will be passed to protect forests. If the voters of such vast democracies as Brazil are persuaded that rain forests are an asset, not a hindrance to economic growth, the governments will be forced to act to conserve these habitats. Brazil has undergone what amounts to an intellectual revolution in the last few years, as a sudden concern for, and interest in, the rain forest developed. Suddenly, their soap operas are set in the rain forest and deal with environmental issues. The purchase of chainsaws without official permit has been prohibited and schoolchildren are receiving environmental education. Nighttime aerial photography of Amazonia shows that the number of fires set by farmers has been reduced over the last five years, perhaps a sign that forest destruction has slowed. There is still a long way to go, but these are some of the most hopeful signs on the conservation horizon. It remains to be seen whether other rain forest nations will follow the example set by Brazil.

In the MDCs, an increased awareness of the dangers of introducing exotic species and the need to control the spread of those that are already growing on our lands should be conservation priorities. Ranking alongside these is the need to protect endangered species, because many environmental ills can be assuaged with time, but lost biodiversity is lost forever.

Summary

- LDCs have the highest population growth rates, and some contain the areas with the highest biodiversity.

- Development is necessary in LDCs, and loss of natural habitats is the price paid.

- Rates of rain forest loss are difficult to measure, and simple predictions of when the forests will be gone are flawed.

- Some rain forests are much more endangered than others.

- Rain forests have long supported human activities, but the land use was different from that of modern slash-and-burn cultivation followed by cattle raising.

- Tropical soils can support sustainable agriculture, but they need a long fallow time (about 30 years) after every four or five harvests.

- Clear-cut forestry is extremely destructive to forest ecosystems.

- Selective logging can initially increase mammal diversity, although some habitat-sensitive species may be lost.

- The effect of repeated selective logging with 30-year fallow periods is unknown, but the method might be sustainable if well managed.

- In Amazonia, road building and ranching are two of the most harmful activities in areas of natural vegetation.

- Ecotourism is a huge potential source of revenue for areas rich in wildlife resources.

- Debt-for-nature swaps may provide a vehicle to protect the forests.

- The rain forests have provided many medicinal drugs, yet fewer than 4% of plants have been tested for their chemical properties.

- The medicinal knowledge of indigenous peoples is being lost faster than the forests.

- Problems relating to royalty payments (how much and to whom?) for medicines derived from tropical plants have yet to be solved.

Further Readings

Eden, M. J., 1990. *Ecology and Land Management in Amazonia.* New York: Belhaven Press.

Elisabetsky, E. 1990. The pharmacopeia from the forest. *Garden* (6):5–6.

Peters, C. M., A. H. Gentry, and R. O. Mendelsohn. 1989. Valuation of an Amazonian rainforest. *Nature* 339:656–657.

Lawrence, D., M. Leighton, and D. R. Peart. 1995. Availability and extraction of forest products in managed and primary forest around a Dayak village in West Kalimantan, Indonesia. *Conservation Biology* 9:76–88.

Rice, R. E., R. E. Gullison, and J. W. Reid. 1997. Can sustainable management save tropical forests? *Scientific American* 276 (4):44–49.

World Resources Institute. 1994. Women and sustainable development. In *World Resources 1994–95.* Oxford: Oxford University Press.

Web Connections

On-line resources for this chapter are on the World Wide Web at:
http://www.prenhall.com/bush
(From this web-site, select Chapter 19.)

Habitat Fragmentation and Extinction

20

E.O. Wilson of Harvard University has estimated that between 1000 and 50,000 species are going extinct each year. While this spread of numbers reflects a lot of uncertainty in the actual rate of extinction, it is important to remember that the number of species living on the planet is unknown (4 million to 100 million species being the commonly quoted range). Furthermore, the pace of extinctions caused by human activities vastly outstrips the rate of new species evolution. The great majority of extinctions are of small insects and microscopic organisms, life forms about which few people would become excited. However, ecologists are concerned about even these small creatures, because they may play an important, even if presently unidentified, role in the shaping of their ecosystem.

This modern wave of extinctions is primarily the result of the loss or degradation of suitable habitat. As humans develop Earth's resources for their own benefit (already 40% of the terrestrial primary productivity is used by humans), wildlife is squeezed into an ever-diminishing area. As that area is further reduced, the impact of one new development is more and more likely to affect the survival of an entire species. Humans hold the power of life or oblivion for thousands, perhaps millions, of species. An ethical concern is raised: Does our species have the right to induce a wave of species extinctions that is unparalleled since the mass extinctions of 65 million years ago? Must we accept the responsibility for causing the modern equivalent to the extinction of the dinosaurs? Or is this argument of catastrophic extinction mere hyperbole and overstatement that will prove groundless? To answer these questions scientifically, we must find out more about the factors that affect the stability and diversity of populations.

20.1 The Relationship Between Habitat Area and Species Diversity

The number of species within an area can be described in terms of a **species-area curve**. This curve tracks the expanding number of species found as the sampling area of a habitat increases. If all the plant species in 1 square meter of forest floor are counted, the total might be 5. If the search is then extended to cover 10 square meters, the species list might now number 15: the initial 5, plus 10 more. As the search area increases, so too will the number of species recorded, until all the common species have been recorded. Eventually, even the rarest species has been counted, and a numeric peak is reached. At that point, expanding the search area will not provide a single new species, so long as the search remains within one habitat type. On a larger scale, species-area data can be used to describe species diversity for entire islands or continents. The resultant curve of species number plotted against area provides a species-area curve (Figure 20.1). The number of species at the asymptote, the area required, and the steepness of the curve will vary from one set of habitats to another, but the basic shape, an initial rapid rise followed by a leveling, will be a recurrent feature regardless of habitat type. In most of these curves, about 90% of the species are documented after 50% of the habitat area has been surveyed; the importance of this trend will become evident later.

20.2 Lessons from Islands

Islands are isolated ecosystems, and by studying them, scientists can make many predictions regarding the fate of habitats fragmented by development. Island analogies have been used to describe processes that take place in fragmented habitats, and much of the theory of reserve design evolved from these studies. In 1963, Robert MacArthur, of Princeton University, and Edward O. Wilson of Harvard University proposed the theory of island biogeography to explain the number of species on an island. They related species diversity to the distance of the island from the mainland, the distance of the island from other islands, and the size of the island. That the number of species was affected by these factors was not a new observation, but it was the first time that the factors had been tied together in a comprehensive model. Their observations about island populations can be summarized as follows:

- Large islands support more species than do small islands. A large area is more likely to satisfy MVA or territory requirements (Figure 20.2).

- Islands far from the mainland support fewer species than islands close to the mainland (Figure 20.3). Remote islands such as Hawaii, the Galápagos, or Easter Island have few species because of the increasing difficulty of making the sea crossing.

- Islands that are part of an archipelago support more species than lone islands. The islands of an archipelago, such as the Aleutians or the Florida Keys, can serve as stepping stones. Species that could not colonize the farthest island from the mainland directly in a single flight can reach it by island hopping along the archipelago.

- Islands with high habitat diversity will support more species than an island of equal size with low habitat diversity. Because most of the species of a given habitat do not require large areas for survival, a diversity of habitats will maximize the number of (common) species that can be supported in an area (Figure 20.4).

- Extinction events are more likely the smaller the island population. Because island populations generally are small, they face an increased risk that they could be decimated by disease, drought, fire, or flood.

- The top predators of a mainland ecosystem are often missing from islands. The islands may not be large enough to support the viable populations of these predators. The lack of predators may also stem from the erratically fluctuating size of island prey populations.

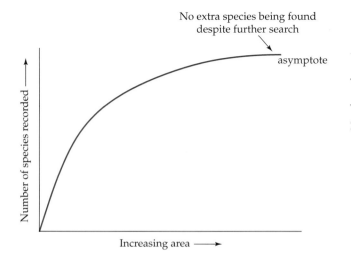

Figure 20.1 A species-area curve in which the number of species is plotted against habitat size.

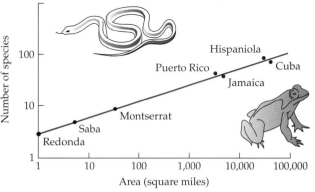

Figure 20.2 Larger islands support more species than do smaller islands. The data are for the number of reptile and amphibians species on Caribbean islands of different sizes. Note the logarithmic scales. (After R.H. MacArthur and E.O. Wilson: *Equilibrium Theory of Island Biogeography.* Copyright © 1967 by Princeton University Press. Reprinted by permission of Princeton University Press.)

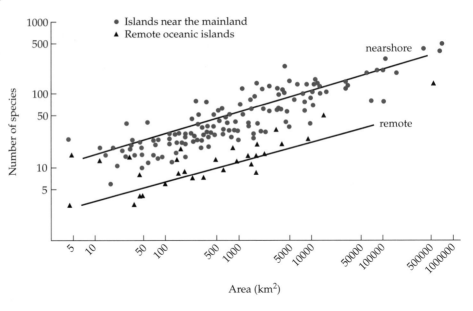

Figure 20.3   Distant islands support fewer species than nearshore islands. Data are for bird species on islands in warm oceans. Islands more than 300 km from the mainland. (After M. Williamson, 1981.)

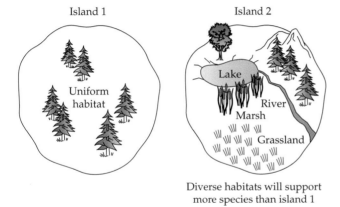

Figure 20.4   Given two islands of the same area, the one with the highest habitat diversity will support more species.

- Islands have an unusual balance of species. The species that live on an island and the lack of large predators often lead islands to have unusual species balances. Plants and animals that are rare on the mainland can, for lack of competition or predation, be common on islands.

- The sum of these factors leads to an overall realization that islands support fewer species per unit area of habitat than are found on the mainland.

MacArthur and Wilson's hypothesis determining the number of species that an island would hold was based on a mathematical model. Their model started with an empty island that gradually filled with species. They assumed that there was a pool of species on the mainland capable of colonizing the island. At first, any new arrival to the empty island was a new species for that island, so the rate of new species added (called immigration) was high. Because an increasing proportion of arriving species were already represented by individuals living on the island, the probability that a new arrival would represent an additional species declined. Consequently, as the island system matured, the immigration rate (number of new species arriving per unit time) plotted against the number of species in the mainland pool (or time) declined (Figure 20.5).

Conversely, extinction rates would be expected to increase as the island ecosystem matured. When there were no species on the island, there could be no extinctions. As the number of species on the island increased through time, some of them went extinct. MacArthur and Wilson viewed extinction as an outcome of natural ups and downs among fragile island populations (fragile because of small population size and the uncertainty of island life). All species were considered to have a similar and random probability of extinction.

Where the immigration curve intersected the extinction curve, for each species arriving on the island another species was going extinct. Overall, the number of species on the island would remain constant (equilibrial). However, because species were still arriving and others were going extinct, the species composition would be changing. Hence, this is not a static equilibrium in which everything is fixed, but a dynamic equilibrium in which a changing complex of species form the equilibrial number.

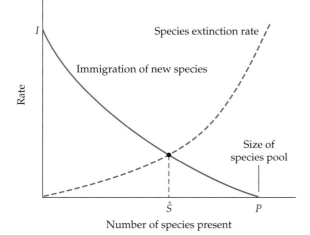

Figure 20.5 **The predicted curves for immigration to and extinction on an island, based on the equilibrium theory of island biogeography.** Where immigration and extinction intersect, a dynamic equilibrium is established for the number of species that will live on the island. (Data are from R.H. MacArthur and E.O. Wilson: *Equilibrium Theory of Island Biogeography.* Copyright © 1967 by Princeton University Press. Reprinted by permission of Princeton University Press.)

and Wilson, so too were extinction rates. In fact, the extinctions on Krakatau are not random; they can almost all be traced to successional change. MacArthur and Wilson's model is not consistent with the data from Krakatau. More important, the data suggest that an equilibrial model is unlikely to explain species abundance in most natural communities of plants and animals.

Because plants are the basis of food chains, so long as new species of plant are arriving, there is the possibility that whole new groups of animals can then occupy the islands. Therefore, equilibrium of animal diversity is most unlikely to occur before plants approach equilibrium. The forests of Krakatau are still rather monotonous, with only about 220 species of plants, 120 of which are trees. In an equivalent area of mainland forest, 1000 or more species of trees might be found. If it has taken 110 years for the first 120 species to arrive, and immigration rates are slowing down, then it may take many hundreds of years, perhaps thousands of years, for the plants to become equilibrial. In the meantime, the islands are affected by fresh eruptions, hurricanes, and fire, all of which are disturbances that might disrupt the succession. Consequently, my colleague Robert Whittaker of Oxford University and I predict that the islands are permanently nonequilibrial and will never reach their potential equilibrium. It is likely that this rule of thumb should be applied to other landscapes, and it is linked to the ideas of impermanence of communities discussed in Chapter 13.

20.3   Edge Effects and Habitat Fragmentation

Entering a forest on a summer's day, you brush past shrubs and vines before stepping into the cool, moist shade. Farther into the forest the shade is deeper, but here it is dappled with sun where shafts of light seep through chinks in the forest canopy. The golden rays of light are alive with motes of dust that hang almost motionless in the still air. How different this environment is from the hot, dry, windy conditions of the open field. Although the contrast is great, the transition from one to another is not abrupt and may be spread over 50–100 m. The transition from open ground to forest is marked by a set of environmental gradients: increased moisture both in the air (humidity) and in the soil; increased amounts of leaf litter and hence soil organic content; and decreased light avail-

ability, wind speed, temperature, noise, and pollution. This zone of transition is the forest edge. It offers a unique set of habitats and is often an area of considerable species richness. However, species diversity can be a misleading indicator of ecosystem health.

The plants of the deep forest are specialists that thrive under dark, cool, moist conditions, so not surprisingly many species fail to survive in either the open areas or the edge habitats. Some forest species, the least demanding in terms of habitat requirements, will be able to survive in the edge. Similarly, plants of the open ground will be unable to compete in the shaded conditions of the forest, but some will be able to live in the edge. Consequently, the edge of the forest will contain a mixture of both forest and open-ground plants. Animals are dependent on plants for food, so the edge will support a blend of forest and open-ground animals. The protection of the forest and the proximity of fields or grasslands for foraging will suit some animals—for example, deer, hares, and rabbits. The hunting conditions of the forest will attract animals that have their homes outside the forest—for example, cats and dogs. The edge supports a portion of the flora and fauna of both open ground and deep forest, and consequently, it may actually have a higher species diversity than either.

We noted that the species that can live under edge conditions are those with a broad environmental tolerance. These species are likely to be the species we are most familiar with, because so many of the habitats that humans create are edgelike: small woodlots on farms, hedgerows, gardens, and parklands. The species that are most threatened by our changes in land use are those with very narrowly defined niches and precise habitat requirements. Species with rigorous environmental demands are termed **indicator species**, because a healthy population of them in a habitat indicates that the ecosystem is meeting their requirements. The presence or absence of successfully reproducing indicator species therefore provides a quick estimate of the "quality" of an environment. Note that "successful reproduction" of the indicator species is the crucial component of this evaluation. Where the animal is a large predator, such as a bear, tiger, or Northern Spotted Owl, it is possible to find the indicator species hunting over a wide range of habitats that neighbor their breeding ground. While acceptable as a hunting ground, these habitats do not necessarily satisfy the niche requirements for a breeding ground. Thus, the Northern Spotted Owl can be found hunting in patches of regrowth (immature) forest, but it is able to reproduce reliably only in the old-growth (mature) forest.

If the edge effect is so important, the next logical question is, How far does it extend into the forest? The answer will vary according to the edge factor that one is considering. For example, wind speed, light, and humidity are probably affected within 100–300 m of the edge. In a study of edge effects in the temperate forests of Washington State's Olympic Peninsula, humidity was found to be reduced for 240 m into the forest. Similarly, in Amazonian rain forest, light-loving butterflies that normally do not live in the forest were found 300 m into the forest from the nearest edge. Small mammalian predators, such as cats, probably do not hunt much more than 1 km into the forest, whereas humans will often penetrate as much as 9 km on foot and as much as 15 km on horseback to hunt. Thus, the edge effect for seedling germination may be considerably narrower than for large tasty mammals. The edge effect would be insignificant in an area the size of Yellowstone National Park, but few of our parks are so large (Chapter 21), and most are profoundly affected by edge effects. These effects are extremely important to the ecology of progressively fragmented ecosystems. The smaller the habitat fragment, the greater will be the proportion that is influenced by the edge effect (Figure 20.7). In an extreme case, where the reserve is only 100 m across its narrowest axis, the whole area will be an edge. In such areas, no habitat is left for the indicator species, and it will be missing.

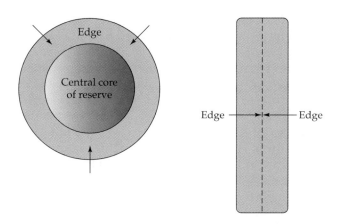

Figure 20.7 **The effects of edges on small nature reserves of equal area but different shape.** Note that long, thin reserves have a greater proportion of their area as an edge and that circular reserves will have more interior habitat.

20.4 Songbirds and Forest Fragmentation

During the past half century a decline in songbirds has been evident both in Europe and America. The species that are hardest hit are the migratory songbirds, such as warblers, vireos, tanagers, flycatchers, and thrushes. The American populations of these birds have wintering grounds in the subtropics and tropics and breeding grounds in temperate and boreal forest habitats. The long-distance migration made each spring and autumn takes a terrible energy toll on the birds, and such a life-style is tenable only if there are good feeding grounds in both winter and summer quarters. These birds do not have the energy reserves to make the long trek north or south without feeding; they are therefore also utterly dependent on finding feeding grounds en route. Thus, habitat quality and food availability over wide geographical areas are essential if these populations are to flourish. A further factor relating to the energetics of the long migration is that these species generally lay fewer eggs than the species that are resident. Resident songbirds can have clutch sizes of four to seven eggs (the European Robin was an example of a resident songbird; Chapter 10); migratory songbirds commonly lay only two or three eggs. In addition, the time taken for migration reduces the time available for nest building, so if there is a false start to their reproductive effort, all is lost for that year. Resident songbirds may have enough time for a second attempt.

The concern for the fate of migratory songbirds began in the 1980s when birders reported seeing fewer birds than usual. Studies in various parts of the United States confirmed that bird numbers had indeed fallen. It is estimated that the numbers of birds migrating each year has fallen by 75% since Europeans started to clear the North American forests, and by 50% since the 1960s (Table 20.1).

The first hypothesis raised to account for the decline was habitat loss in the United States, but conflicting results showed that such a simple explanation was not supported by all the facts. In some areas where forests were being lost, the bird populations were maintained. The next hypothesis was that it was the deforestation of the overwintering grounds in the tropical areas that was causing the decline. Again the data were inconclusive. Habitat loss seemed to be part of the puzzle, but it was not the whole answer. Work by David Wilcove provided an important insight. The birds that nest in fragmented habitats are subject to

Table 20.1 Population decline in migrant and nonmigrant birds in Rock Creek State Park, MD.[1]

| | Mean number of pairs sighted | | |
Species	1940s	1980s	% change
Migrants			
Red-eyed Vireo	41.5	5.8	-86.0
Ovenbird	38.8	3.3	-91.5
Acadian Flycatcher	21.5	0.1	-99.5
Wood Thrush	16.3	3.9	-76.1
Scarlet Tanager	7.3	3.5	-52.1
Yellow-throated Vireo	6.0	0	-100.0
Hooded Warbler	5.0	0	-100.0
Eastern Wood Peewee	5.5	2.8	-49.1
Black-and-white Warbler	3.0	0	-100.0
Nonmigrants			
Carolina Chickadee	5.0	4.3	-14.0
Tufted Titmouse	5.0	4.5	-10.0
Downy Woodpecker	3.5	3.0	-14.3
White-breasted Nuthatch	3.5	3.1	-11.4

[1] This single example is believed to be representative of many woodland areas in the eastern United States.
(*Source:* Data are from J. Terborgh, *Where Have All the Songbirds Gone*? Copyright © 1992 Princeton University Press. Reprinted by permission of Princeton University Press.)

much higher rates of predation by mammals that are adapted to edge conditions (such as raccoon and opossum) than are birds that nest deep in undisturbed forest. In 1992, Scott Robinson conducted detailed studies of fragmented woodlots in Illinois and showed that as many as 80% of the nests were suffering predation. Worse still, of the 20% that survived the attentions of raccoons and opossum, 76% of the nests were being parasitized by Brown-headed Cowbirds. Cowbirds lay their eggs in the nests of songbirds and rely on the unwitting host to raise the cowbird chick. The cowbird hatches earlier than the songbird chicks, grows quickly, and claims most of the food brought to the nest (Figure 20.8). The young of the songbirds starve. Under natural conditions, the threat posed by cowbirds is relatively small. However, the grain to be found around fields and feedlots has enabled cowbird populations to boom. A once insignificant foe of the songbirds is now, as a result of human actions, a major threat to their survival. Cowbirds feed in open fields, but they may fly up to 800 m in search of songbird nests. As forest patches decrease in size to less than

1.6 km in diameter, no nest is safe from parasitism. Costly programs to trap and kill cowbirds have been proposed as part of the strategy to save such endangered species as Kirtland's Warbler.

The answer to the decline in songbirds is now evident. The birds suffer from reduced habitat availability, and the pattern of habitat loss greatly increases predation and nest parasitism. In the Illinois woodlands, fewer than 7% of the songbirds in Robinson's study bred successfully. It will not take long for species' numbers to decline when experiencing a 93% failure in reproductive attempts.

In Europe, songbirds are in decline, although the reasons are somewhat different. Habitat fragmentation is still a major factor, but so too is hunting. The migratory songbird routes from northern Africa, the Middle East, and southern Europe, to their northern breeding grounds in northern Europe is made perilous by human predation. Each year flocks of migrating birds fly up the boot of Italy, where they are trapped or shot. Songbirds are considered a delicacy in Italy, and as many as 50 million become menu items each year.

Figure 20.8 A yellow warbler feeds a cowbird chick. Note how much bigger the cowbird chick is than both the "parent" or other chicks in the nest.

20.5 Metapopulations: Another Way to Think About Fragmented Populations

The overall populations of many plant and animal species are fragmented into small groups or **subpopulations**. For example, in an area where there are small ponds scattered across the landscape, we can census bullfrogs in each pond and assess a total population for the region. Each individual pond holds a recognizable subset of the overall population, and this is referred to as a subpopulation. It is most likely that a frog born in a small pond will find its mate within that pond. Therefore, the frogs within the subpopulation are reproductively isolated from other subpopulations. However, as we all know, "froggy will go a courtin." Frogs will hop from one pond to a neighboring one to find a mate. If frogs are hopping in all directions to find mates, we can assume that there is genetic exchange going on between all the subpopulations. The distance between subpopulations, the inhospitability of the intervening terrain, and the mobility of the organism will determine the amount of gene exchange taking place. Each frog hopping to a new pond is an immigrant to the new area and an emigrant from its old home. Richard Levins, of Harvard University,

proposed a new framework to describe the arrangement of these subpopulations in space and time. He suggested that all the subpopulations that are linked by immigration and emigration should be regarded as part of a larger **metapopulation**.

The metapopulation concept predicts that subpopulations are not all going to be equally successful. Subpopulations that exist in the best habitats will have such reproductive success that they would become overpopulated if the excess young did not emigrate. These individuals that leave the subpopulation will search for unoccupied or underoccupied habitats. At the other extreme, some of the subpopulations will inhabit poor sites and have such low reproductive success that there are insufficient young to replace the parents. Crucial to this concept is the understanding that individuals can find enough resources to survive in a location, but may not be able to find the resources to breed successfully. The resource shortfall may be caused by insufficient food to feed hungry young, as well as adults, or it could be that an essential habitat component is missing. For example, adult wood ducks can live in a wide variety of wetland settings, but they are hole-nesters and need large trees with cavities in which to breed. The felling of mature swamp forests led to a decline in wood ducks because it was the loss of nesting sites that prevented reproductive success (artificial nest boxes for wood ducks have proven successful in rebuilding their population). Thus, prior to the nestbox program, wood duck adults would have been seen in many locations, but only a few sites would actually have supported successfully breeding populations. In ecological terms, we can define reproductively successful subpopulations as having a fitness greater than 1, whereas the unsuccessful populations would have had a fitness of less than 1. The sites with a fitness greater than one supply emigrants to other subpopulations and hence are known as **source populations**. After each reproductive effort, the subpopulations of the source areas overflow, and their emigrants may establish new successful subpopulations in which case a new source area is formed. More likely, though, the emigrant will eke out an existence in marginal habitat with poor reproductive potential. It has then joined a subpopulation that is failing to replace itself through reproduction; that is, it has a fitness of less than 1. By definition, this subpopulation would go extinct unless it received immigrants from elsewhere. These areas that persist purely because they receive immigrants from source populations are termed **sink populations**.

Ecology in Action

Studying the Effects of Fragmenting a Rain Forest

The Biological Dynamics of Forest Fragments Project, brainchild of Smithsonian Institution ecologist Tom Lovejoy, was the largest biological experiment ever undertaken. Begun in 1979 and supported by the Brazilian government and the World Wildlife Fund, the experiment consisted of surveying, then fragmenting, a large area of Amazonia. Anticipating that the front of forest destruction would encroach on a vast area of mature forest, Lovejoy persuaded landowners to leave untouched square blocks of forest ranging from 1 hectare to 1000 hectares.

The first step in the experiment was to survey the diversity and population size of birds, mammals, and plants in the area before it was disturbed. Specimens of all the plants within the selected area were collected and identified. Individual trees were measured and tagged so that their future growth rates or mortality could be monitored. Similarly, inventories were made of animals by catching, identifying, and releasing them. The disturbance isolated the blocks of forest, and the biologists watched and waited.

The experiment is still running, and an immense database of forest change due to fragmentation is being amassed. The warmer, drier, brighter edge of the forest allowed some early successional species to colonize beneath the cover of tall trees. The seedlings of these colonizing species grew well, while the seedlings of deep-forest species foundered. Adult trees were also affected. Within 100 meters of the edge mortality rates were as high as 10% per year. These mortalities helped to break open the canopy and aid the immigration of gap-filling, early successional species. Although these mortalities were highest within a few meters of the forest edge, elevated rates of mortality were detectable for 100 m into the forest. If climatic changes are penetrating at least 100 m into the forest, then the edge effect must be at least this broad. A 1-ha forest block is 100 m on each side, so these areas were all "edge." Even the 10-ha block was 87% edge, the 100-ha block was 36% edge, and the 1000-ha block was 13% edge.

Because trees live a long time, they are relatively slow to respond to environmental change. Short-lived species, such a beetles provide a more dramatic example of the changes. Even the commonest forest beetles were strongly affected by the fragmentation. After 10 years of isolation the 1-ha fragments had lost 49.8% of common beetle species, the 10-ha forest blocks had lost 29.8%, and the 100-ha blocks had lost 13.8%. Other insect species lost included shade-loving butterflies.

In addition to the climatic changes, the fragmentation reduces the area available for hunting or for establishing territories. This is more likely to be a problem for army ants, mammals, and birds than for insects. Army ants forage in vast swarms across the forest floor, eating any animal in their path. The ant swarm is accompanied by a characteristic set of bird species that feed on the insects that jump or flutter to escape ahead of the advancing ants. These ant colonies need a forest territory of at least 10 hectares, so they failed to survive in the smallest forest fragments. Without the ants, the foraging niche of the birds was lost, and those species also went locally extinct. Thus, the loss of a species often has a cascade effect throughout the local food web.

A decade after the clearance, a decline in the habitat diversity of the remnant forests is becoming severe. Peccaries, the wild pigs of the Amazon, create mud wallows, puddles that are essential habitats for frogs and other amphibians. The peccaries fled from the smaller forest fragments, and, after just a few years without new puddles being formed and the old ones drying up, many amphibians were lost from the forests.

The longer-term effects on the ecosystems, such as the reproductive success of trees in the forest interior or subtle changes in nutrient cycling, will not become apparent for at least another decade. The full value of this experiment has yet to be realized, but already it has greatly increased our understanding of the effects of fragmentation on rainforest ecosystems.

Ronald Pulliam, working at the University of Georgia, estimated that for many natural metapopulations, as few as 10% of the population may live in source areas. The remaining 90% of the metapopulation live in sink areas and are totally dependent on the source areas for new recruits.

Even though source areas are the most important areas to metapopulation survival, sink areas are also valuable. Without sink areas, emigrants from source areas would die. The sink area provides a habitat in which these organisms can live. Only through having sink areas can a maximum metapopulation be maintained. A large metapopulation is important to maintaining genetic diversity. It should be noted that even though the net movement of individuals is from the source to the sink, there will be some migration in the other direction as individuals compete for space in the prime breeding grounds. This two-way movement ensures genetic exchange between all subpopulations and reduces the risk of **inbreeding**. Inbreeding occurs when closely related individuals mate with each other. Generally, this is most likely to happen when populations are very small (as is often the case on nature reserves) and leads to weaker offspring that are more prone to infertility, disease, or deformity. Another role of sink areas is that they contain individuals that could replenish the source population. Should the source population be eliminated, members from sink populations will flood back into the source area to fill vacated niches (remember the Maine Gunners experiment in Chapter 10).

A startling consequence of thinking about populations from the perspective of sink-source interactions is that areas supporting the most individuals are not necessarily the most important areas for the long-term continuance of the species. It is entirely possible that sink areas will have both larger populations and greater densities of individuals than source areas. Reproductive success, not adult distribution, is what matters in conservation biology. It follows, therefore, that not all subpopulations are of equal importance in terms of conservation. The loss of one among many sink populations from a metapopulation will have little effect on the well-being of the metapopulation. However, if the source population is the one that is lost, the entire metapopulation could go extinct.

A further realization is that the metapopulation relies not only on the areas that form breeding grounds for the various subpopulations, but also on the intervening migration routes. Without these routes there can be no replenishment of the sink areas, or gene flow from sinks back to the source areas. Under natural conditions, these migration routes would lie within areas of suitable (or least inhospitable) habitat. In Chapter 21 the importance of these migratory routes is discussed in the context of reserve management.

Contrasting Island Biogeographic Theory and Metapopulation Theory

Island biogeographic theory and metapopulation theory address different aspects of isolation. MacArthur and Wilson's equilibrium theory of island biogeography was based on the rates of immigration and extinction of species on islands and sought to explain the number of species that could occupy a given area. This theory did not attempt to explain population sizes or the genetic exchange between populations. More recent studies have emphasized

that species composition and number is affected by habitat diversity, the dispersal abilities of organisms, and local environmental history. These studies offer valuable insights into fluctuations in the number of species found within an ecosystem over time or following isolation.

Metapopulation theory takes one species at a time and seeks to explain the fate of linked subpopulations. Note that this is a hierarchical view in that a species may be comprised of many metapopulations, which in turn are comprised of subpopulations. Metapopulation theory describes the degree of isolation of fragmented populations, their population size, their local fitness, and their genetic exchange. As the science of conservation biology has grown, so the emphasis has shifted from island biogeography toward metapopulation dynamics. Metapopulation theory is a powerful tool to help prioritize areas to be set aside for conservation, and to assess the needs of endangered species.

20.6   The Threat Posed by Exotic Species

The three greatest threats to the conservation of many ecosystems are habitat loss, habitat fragmentation, and exotic species. Plants and animals in a landscape can be divided into those that occur there naturally, **native species**, and those that have been introduced as a result of human activities, **exotic species**. Many introductions of exotic species are accidental and can range from microscopic pathogens to large predators. A deadly example was the introduction of the protozoan parasite *Plasmodium* (which causes malaria) to America and the Caribbean. The malarial parasite was originally native to Africa but was inadvertently spread around the world by infected slaves and slavers. Other exotic species, such as false oat grass, a European grass that can reduce agricultural yields in the United States, may be economically damaging. Any cargo consignment of wood, fruit, or vegetables could carry insects or small organisms. Seeds can be inadvertently imported in soil, in clothing, or mixed in with seeds that are being deliberately traded. Larger animals, such as goats, pigs and horses, though often cherished by humans, can utterly change the height, texture, and species composition of local floras when introduced to a new setting. The result is that the native herbivores face new competition in addition to changed habitats. Even pets such as such as domestic cats and dogs can be immensely damaging to local faunas. These furry chums are ferocious predators of small animals. The addition of an exotic predator to islands, which typically have few mammal predators, can have disastrous consequences for native birds and mammals.

Why Are Exotic Species So Successful?

Up to now we have described the processes that keep populations in a local balance. Individual species are constrained by fitness, competition, predation, disease, and density-independent factors (Chapter 9). Why then should a new arrival, an exotic species, not be similarly constrained? The success of exotic species hinges upon a lack of shared evolutionary history with the organisms of the ecosystem that they are invading. In its natural setting, every species is burdened by disease and parasites that have evolved alongside it. Predation, ranging from lions picking off the young, to beetle grubs eating seeds, serves to limit population size. Other evolutionary relationships are also important: Prey evolve to hide or flee from local predators making them hard to catch, and plants develop defensive chemicals making them unpalatable. Among the native organisms, every time conditions are right for a population explosion, it is cut short by one of these long-term relationships.

Imagine being able to sidestep this evolutionary burden and enter a new realm, where the prey are defenseless, the plants are palatable, the parasites are fully engaged with your competitors, and there is no disease that infects you. This is the world of the exotic invader. Exotic species can build huge populations with relative impunity. Of course evolution would catch up with the exotics in the end, and they, too, would be subject to all the usual constraints on population. However, this evolutionary control may take thousands of years to develop, whereas human actions are introducing new exotics on an annual to decadal basis. In the interim between introduction and eventual control, the exotic species can radically alter the ecosystem that it entered. When this alteration leads to the extinction of species, the changes are irreversible and permanent.

Although to a conservationist any exotic species is unwelcome, fortunately not all exotic species seriously damage an ecosystem. Some species are introduced and barely maintain a toehold: They may fail to reproduce much of the time, and though not native, they are not a major pest. When you consider that almost all garden plants are exotic species, yet relatively few of them are truly a hazard to local ecosystems, it is clear that not all exotics run rampant. What is the common thread that links the exotic species that do become a major ecological hazard, such as kudzu vine, purple loosestrife, the Asian tiger mosquito, gypsy moth, rabbits, cats, and foxes?

All of these species are successful exotic species because of rapid reproduction, rapid dispersal, and the ability to cope with a wide array of habitats (Table 20.2). An indication of such flexibility is an original native distribution that is broad both in geographic extent and habitat types. Being able to take advantage of new food items is a distinct advantage for predators. The gypsy moth that has been eating its way across the Midwest for two decades seems capable of digesting almost any species of broadleaved tree in its path. Similarly, a domestic cat, although a wonderful pet,

Table 20.2 Some general characteristics of exotic species likely to become invasive pests and of communities most susceptible to invasion. Note that this list is neither complete nor without exceptions.

Characteristics of successful invaders	Characteristics of invadable communities
High fecundity	Habitat disturbed by humans
Short generation time	Early successional
Good disperser	Climate similar to original native habitat
Opportunistic	Absence of predators/pathogens on invading species
Highly adaptable to new conditions	Absence of predators in evolutionary history
Broad geographic native range	Introduction of fire
Broad habitat tolerance	
Broad diet	
Human commensalist	

(*Source:* After Lodge, 1993.)

is a fierce predator of almost all small mammals and birds. This ability to base its diet on whatever is available is an important attribute for an exotic predator, because a new setting may lack any familiar plants and animals. If the species is a human commensalist, that is, it is not perceived to be a threat to humans but is well adapted to living among us, it stands an excellent chance of colonizing areas as they are invaded by humans. Even organisms that we regard as pests, such as mosquitoes, fleas, and rats, that thrive around us but are harmful, may prove so difficult to eradicate that they effectively become human commensalists.

Turning the question around, we can ask which are the habitats most open to invasion by exotic species? The short answer is, those habitats most impacted by humans. Human occupation of an area produces a characteristic pattern of soil churned for fields and gardens, increased fire frequency, early successional vegetation stands, and eutrophied water bodies. These landscapes are repeated the world over as humans (the most invasive of all exotic species) colonize an area. Consequently, the niches that we offer exotic species are broadly similar. Burning land to clear vegetation and remove dangerous animals has been a common characteristic of human settlement. The new fire regime will result in local extinctions of fire-intolerant species, leaving vacant niches that exotic fire-tolerant species occupy. Species within ecosystems that do not burn regularly, have few adaptations to cope with fire and will be most adversely affected by its introduction.

Exotic species will be at their most competitive when colonizing an area with a similar climate to that of their original native range and when they do not share an evolutionary history with the area's native inhabitants. The importance of evolutionary history is especially evident on islands where the native plants and animals may have evolved in isolation. As islands frequently lack large predators, the birds may evolve to be flightless and to lack a fear of predators. The Dodo, a giant flightless pigeon, evolved on the island of Mauritius. There were no predators, and thus the Dodo did not need to fly. Once freed of the need to become airborne, the Dodo could invest energy in building a larger, more energetically efficient body. Similarly, in the absence of predators, there is no evolutionary benefit to being watchful or skittish. When eighteenth century sailors discovered Mauritius, they found the Dodo to be so "tame" that they were able to walk up to them and club them. The Dodo was not tame; it was just naive. Dodos had no evolutionary programming to make them run away from another large animal, for the sailors were the first predators encountered in the Dodo's abbreviated evolutionary history. The naivety of island animals makes them easy targets for exotic predators such as cats and foxes in New Zealand, the Galapagos, Hawaii, and other islands around the world.

Two examples of the influence of exotic species on native systems are considered in more detail: The effect on native mammals of deliberately introducing cats and foxes to Australia (cats were introduced to control rats and foxes to provide hunting), and the effects of an accidental introduction of zebra mussels to the Great Lakes.

Australia's Struggling Mammals

Almost half of the mammals known to have gone extinct this century lived exclusively in Australia. Of 300 Australian mammal species, 19 (6.3%) have disappeared forever and another 10 species have such seriously declining populations that they are threatened with extinction. To understand the cause of extinctions it is necessary to consider the contrasting impacts that Aborigines and Europeans have had on Australia.

The first human inhabitants of Australia, the Aborigines, arrived about 40,000 to 50,000 years ago. Aborigines changed the face of Australia through the use of fire. Australia had long been a land of natural fires, but with the arrival of Aborigines fire was used to drive game and provide tender new growth that would increase the carrying capacity of prey such as wallabies and kangaroos. Aborigines were so skilled in their use of fire that anthropologists call this style of land management "fire-stick farming." The fires were small and carefully set, and areas were allowed to undergo successional recovery before being burned again. Often the period between fires would exceed 30 years. The result was a landscape with a mosaic of different-aged successional stands of vegetation. Each successional stage offered different niches for plants and animals and, therefore, characteristic opportunities for hunting and gathering. This initial change in land management coincided with a wave of mammalian extinctions about 40,000 years ago, but the remaining fauna were well suited to the new conditions and persisted until a fresh wave of extinctions began in the late 1800s.

In the 200 years since the first Europeans colonized Australia, the landscape of the continent has been transformed again. Old-growth forests, whose massive trees survived the fire management used by Aborigines, fell to saw and ax. In both these and drier regions, the land was burned every 3 to 5 years instead of every 30 to 50 years. Consequently, the mosaic of successional stages was replaced with a uniform landscape of the just burned. The new grasslands were suitable for introduced cattle and sheep and unwanted grazers such as rabbits. Equally unpopular with the farmers, red and grey kangaroos and some species of wallabies thrived in the new habitat. However, populations of many smaller marsupial herbivores dwindled.

For these small marsupials, worse was yet to come. Introduced species not only elevated levels of competition, but also of predation. It is important to note that the marsupials that have gone extinct, and those most threatened with extinction, are generally small (weighing between 30 grams and 5500 grams), do not use burrows, and have slow reproductive rates (Figure 20.9). These characteristics suggest vulnerability to predators such as cats and foxes and an inability to recover if subjected to an intensified predatory pressure. Both cats and foxes are now common as wild animals in Australia, though both are exotic species introduced by European settlers.

The presence of the predators alone may not be enough to result in marsupial extinction. Indeed, ecological theory predicts that predator-prey interactions do not usually result in a prey species being driven to extinction (Chapter 12). An interesting twist is added when the predator-prey relationship is supplemented by the presence of an exotic prey species. For example, on MacQuarie Island, a local native bird population, the MacQuarie Island Parakeet, coexisted for more than 60 years with an exotic predator: the cat. However, within 20 years of the introduction of rabbits to the island, the parakeets were extinct. It is unlikely that the rabbits were such direct or strong competitors with the parakeets that they drove the parakeets to extinction. It is more likely that with the arrival of rabbits, cat populations increased rapidly. Rabbits, especially young ones, are suitable prey for cats, and the increased prey availability allowed the

cat population to swell. The extra cats resulted in extra predation on the parakeets. No matter how rare the parakeet became, the cat population was supported by the availability of rabbits. The previously balanced predator-prey relationship between cat and parakeet was upset. Such an imbalance in which the normal predator-prey relationship breaks down as predators become unusually abundant is **hyperpredation**.

Hyperpredation may also be part of the cause of the decline in small marsupials. In this case, the ready abundance of prey in the form of rabbits supports large predator populations. If the cats and foxes prefer the marsupials or find them easier to catch than rabbits—perhaps because the marsupials do not dart down burrows—marsupial populations could be driven to zero. Eradicating rabbits has been a goal in Australia since the early 1900s, but has been unsuccessful to date. Attention has now been turned to removing the predators. It is important to note that controlling both the rabbits and their predators needs to be carried out in parallel. Otherwise, if the rabbits are eliminated, but the predators are untouched, the exotic predators will increase their hunting pressure on native mammals. Similarly, if the rabbits are untouched and the predator populations reduced, the rabbit populations will increase in size as they are freed from predation.

The most recent strategy is the introduction of a new virus that causes rabbit hemorrhagic disease (Chapter 18). The disease may seriously reduce the rabbit population, at least temporarily. Fox populations are being attacked by an experimental poisoning program. So far, the program has been locally successful in western Australia where a recovery of native marsupials is evident (Figure 20.10). Populations of quoll, numbat, and rock-wallaby increased rapidly following local elimination of foxes. However, hyperpredation is not the only, or even the most fundamental cause of marsupial extinction. Controlling predator numbers is a positive first step, but more far-reaching changes in land management will be needed

Figure 20.9 Australian marsupials threatened with extinction. (a) numbat, (b) rock-wallaby, and (c) quoll. All of these native mammals fall prey to foxes, an exotic predator.

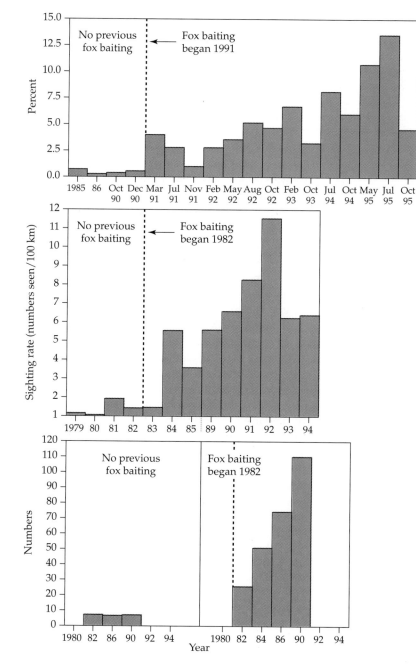

Figure 20.10 Increases in Australian mammal populations following fox poisoning. Quoll, numbat, and rock-wallaby all increased in abundance once fox populations were reduced or eliminated by poisoning. (a) Percentage of live-mammal traps containing quoll before and after poisoned baits were used to reduce fox populations in Batalling Forest, W. Australia. (b) Numbat sightings in Dryandra Woodland, W. Australia, before and after fox baiting. (c) Rock-wallaby sightings in adjacent forests, one in which foxes were eliminated, the other where there was no eradication program, near Kellerberrin, W. Australia. (After C. Bailey, 1996.)

to re-establish populations of threatened marsupials. A vital component is to restore the mosaic landscape produced by fire-stick farming and to safeguard the remaining old-growth forests.

Mussel-Bound but Helpless

A native of Russian lakes, the zebra mussel spread (Figure 20.11) throughout Europe as an exotic species in the 1960s and 1970s. The invasion of the Americas started in 1985 or 1986 when zebra mussels, probably carried from Russia in the bilges of ships, were accidentally flushed into the Great Lakes. This small mollusk now poses a threat to the local ecology, power producers, boat owners, and sport fisheries.

The first living zebra mussel in North America was collected on June 1, 1988, close to Detroit. By 1994, the zebra mussel had spread throughout the Great Lakes and entered eight major river systems: St. Lawrence, Hudson, Mississippi, Ohio, Illinois, Tennessee, Susquehanna, and Arkansas. In 1999, the

Figure 20.11 Zebra mussels attached to freshwater clams. The zebra mussel arrived in the United States in the 1980s. The mussel forms vast colonies that damage water intakes and foul any hard underwater surface.

zebra mussel was to be found from New Orleans to Detroit, and as far west as Nebraska (Figure 20.12).

The zebra mussel is a clamlike mollusk that grows to be about 25 mm long. It can reproduce when less than a year old, and each female can produce 1 million eggs a year. The mussels float free as larvae and attach to almost any flat or hard underwater surface, where they mature. Arriving in a new continent, the mussels were virtually free from pathogens, parasites, and predators, and the Great Lakes provided nearly ideal conditions for rapid population growth.

The mussels are colonial, and as many as 500,000 individuals may be attached as a solid mass on each square meter of substrate area. The mussels encrust and clog the hinges of lock gates, the intake ducts of power stations, and the rudders of ships. Tubes that carry water, such as drains or intake pipes for industry,

can have a 70% reduction in their effective cross-sectional area because of clogging mussels. Removing the pests is expensive. In 1990, it cost Detroit Edison more than $500,000 to clear their intake pipes of zebra mussels, and it is estimated that the cost to industry, shipping, and sport fishing could be $5 billion in the Great Lakes region alone by the year 2000.

The mussel is carried to new bodies of water by the natural flow of water, as larvae on the feathers of migrating wildfowl, in bait buckets, and on trailered boats. Unless a new and presently unknown pathogen arrives, the expansion of the mussel is likely to continue. Ecologists have tried to determine the potential expansion of the geographical range of the mussel. Experiments have been conducted to define its tolerance to ranges of salinity, pH, calcium (needed to build its shell), and temperature. To date, it appears that most freshwater systems in North America with a pH of 7.4 or greater are vulnerable to invasion. Even areas once thought to be too warm for the mussel, such as the Southwest and Florida, are at risk. Experimenters have observed that the mussel can acclimate to new temperatures after a few generations. With most waters vulnerable to invasion, the next issue to be faced is, What would be the effect of invasion?

The mussels are filter-feeders. They draw in water, extract algae, and excrete their waste as a sediment. An individual mussel can filter 1 liter of water every 24 hours, and a typical colony covering one square meter of substrate could filter 180 million liters per year. When the mussels establish vast colonies, they can reduce the algal population and hence the primary productivity of a lake. The nutrients in the algae are either absorbed by the mussel or drop to the bottom in their feces. Effectively, the mussels are transferring

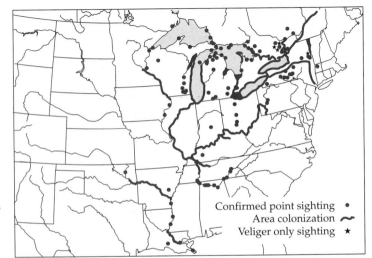

Figure 20.12 A range map of the Zebra mussel in the United States for February, 1998. (After Cornell Seagrant project.)

Confirmed point sighting •
Area colonization 〜
Veliger only sighting ★

nutrients from the water column to the sediment. The consequence of reducing the algal population of the lake is to erode the base of food chains, hence reducing the fish population of the lake. The mussels are already threatening to reduce biodiversity in some lakes and rivers.

The negative effects of the zebra mussel are already apparent in the Great Lakes, but the presence of the mussel does have some positive aspects. The vast filtration capacity has been a major factor in returning Lake Erie to an oligotrophic system, after a century of eutrophication by sewage and fertilizer discharge. In the first decade of the mussels' living in Lake Erie, the vertical distance that light could penetrate the lake (an index of water visibility) had increased from a few centimeters to almost 10 m. Zebra mussels are sensitive to pollution and can be an aquatic version of the canary in the coal mine. In Europe, changes in the pumping behavior of the mollusk are found to be a reliable indicator of the presence of chemicals such as organochlorines, radioactive compounds, and toxic metals. If only they could be controlled, zebra mussels could be a tool to help in the fight against pollution and eutrophication of our lakes and rivers, but as yet the mussel remains more of a threat than a benefit.

Control of the mussel by pesticides and other conventional means has been tried and found to be ineffective or unacceptably damaging to other organisms. One possible avenue, at least for industry and boat owners, has been suggested by biologist Susan Fisher of Ohio State University. Fisher found that the mussels avoid settling on areas where even a small amount of potassium is released into the water. Industrial locations threatened by the mussels could perhaps be protected by a steady trickle of potassium pumped into the water at potential attachment sites. Similarly, boats, buoys, and piers may benefit from being treated with a paint containing potassium. However, natural habitats will not receive such treatment, and it is likely that within a few years the zebra mussel will be found in most freshwater habitats, and even some moderately saline ones, in the United States and southern Canada.

20.7 Extinction or Crying Wolf?

The estimates of species extinctions quoted by E. O. Wilson and other ecologists are generally based on observations of island populations. If two islands with populations that have reached a dynamic equilibrium are compared, one half the size of the other, the smaller island will support about 90% of the species found on the larger island. Thus, the species-area curve for a

habitat is used to describe the number of extinctions associated with reduced habitat area. If habitat area were halved, a 10% extinction would be predicted when the populations approach equilibrium. Critics of these statistics point out that they do not hold up to scrutiny. A commonly cited instance is that of the forests of the eastern United States. There, habitat has been almost completely disturbed, yet there has been no catastrophic extinction. Almost all of the eastern forests have been felled in the past 300 years (Figure 20.13), and of all the bird species that inhabited those forests only four species, the Carolina Parakeet,

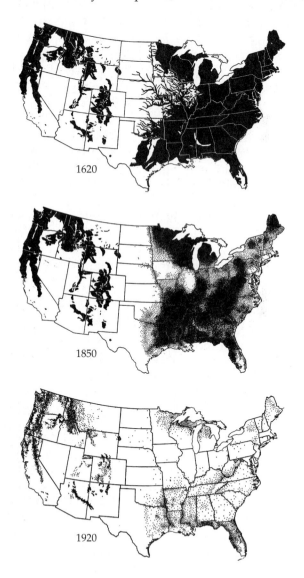

Figure 20.13 **Sketch map showing the area of uncut forest in the eastern United States in 1620, 1850, and 1920.** Note that this map is not a map of total forest cover because it does not show the areas of successional regrowth. (Data are from W.B. Meyer, 1995.)

the Ivory-Billed Woodpecker, Bachman's Warbler, and the Passenger Pigeon have gone extinct. These statistics on birds are used to suggest that conservationists and ecologists are "crying wolf." However, the conclusion that conservationists are exaggerating the probability of extinctions rests on four untrue assumptions: all the bird species could live only in old-growth stands; all the bird species extinctions have already occurred (some populations, such as the Red-Cockaded Woodpecker and the songbirds remain insecure); there is no regrowth of forests that can support bird species; and birds are as sensitive to extinction as other organisms.

Some forest-dwelling birds are restricted to the deep undisturbed forest and cannot tolerate any disturbance. These species go locally extinct when their forest is disturbed. Another group of woodland bird species is apparently indifferent to disturbance and can thrive in woodland edges, parks, and gardens. Probably the largest group of woodland bird species consists of species that are adversely affected by the disturbance but do not go extinct. Their populations decline not suddenly, but gradually, and they may or may not stabilize at a lower population size.

The spatial scale of the logging of eastern U.S. forests is well established; only 3% of forest remains uncut. However, a temporal scale is also important because, although forest habitats were disturbed, they were not all cut at the same time. The forestry created a mosaic landscape of cut, recovering, and uncut forest. The recovering forest provided successional habitats, culminating in mature forest regrowth. If you had been able to stand and watch the process of logging and a 200-year recovery at a single point, you would have seen the local extinction and subsequent reinvasion of many species. The species were not regionally extinct; they had small populations in tracts of forest that had not yet been cut. Gradually, as the succession proceeded, almost all of them returned. Thus, when the last of the old forests were cut, the reservoir of species diversity now lay in the mature regrowth forests. Not all species could make this series of relocations, and some went extinct.

When Europeans first entered the land, almost all of the area was covered in forest. Despite the fact that 97% of this forest was cut over the next 300 years, forest cover at any one time has always been greater than 20%. In recent decades, the forest coverage has actually increased to about 30%, although most of it is in the relatively early stages of succession.

A second temporal-scale factor is that the original model of species loss versus habitat reduction was for "equilibrial" populations. Earlier, we discussed reasons

why systems seldom, if ever, attain equilibrium (Chapters 13 and 14), and these forest bird populations provide a further example. Even though the cutting of the original forests has stopped and forest area is actually increasing, the effects of the original disturbance have yet to be fully realized. The populations of songbirds are still falling as a result of ecological changes associated with habitat fragmentation. Their population decline is slow but steady rather than a sudden extinction event. It is unknown how many of these populations are stable at a new, lower level and how many will go extinct. That the forest is still changing successionally may help some of these bird populations to recover. Thus, the system is not equilibrial, so there is a problem in checking the predictions of the island-based model. Although the full extinction effects resulting from past land use are not yet evident, it is complacent and inaccurate to dismiss the link between extinction and habitat fragmentation.

Because there are relatively few species of birds compared with insects, a single bird species extinction may indicate the loss of hundreds, perhaps thousands, of insect species. A further factor that favors the survival of birds is that they are highly mobile; they can avoid being caught within a burning forest or can flee the bulldozer. Fire will drive birds out of a section of forest, but with a few exceptions it will not kill many birds. The same fire will incinerate plants, insects, reptiles, and mammals that lie in its path, for these are usually too slow to flee. Therefore, a closer study of populations should be made before asserting that all is well simply because there is unexpectedly little evidence of extinction among birds.

Given that deforestation will continue, is it possible to be moderately optimistic that extinctions can be avoided if the exploitation is carefully controlled? The answer is "to some extent," but rare or endemic species with very local distributions such as the Northern Spotted Owl would be lost.

Many species will survive deforestation, providing they can retreat into mature habitats and then recolonize the maturing successional system. However, if keystone species are lost, the repercussions on the entire system will be profound. The most serious flaw in the argument that nature is a self-healing system is that people who use that argument often forget that nature cannot heal itself when land use precludes regrowth and the recovery of the ecosystem. If ecosystem functions are destroyed, with the result that the land is thoroughly degraded, the successional sequence that might maintain wildlife is delayed or corrupted.

The possibility that conservationists need to be rabble-rousers to gain the ear of media and the public should not be ignored. Some of their claims are exaggerated, but to put it simply, if the niche of an animal or plant disappears, that organism will go extinct. Niches are being destroyed as development replaces natural habitats. Extinctions do occur as each kilometer of tropical forest is felled; most are of species science has never even described, but some are of spectacular large animals such as the Javan tiger (declared extinct in 1994). If extinctions have not been as extreme as conservationists predicted, it does not indicate a lack of a problem, but rather that there is still time to safeguard the majority of Earth's biological diversity.

Summary

- Habitat destruction, degradation, and exotic species are the greatest causes of modern species extinction.

- Islands provide a simple system in which the effects of isolation can be studied.

- With two areas of a habitat, differing only in size, the larger will hold more species.

- Equilibrial models of island systems are best applied to very simple island systems. In more complex systems nonequilibrial models are more informative.

- Fragmentation of habitats reduces the quality of remaining habitat and leads to loss of biodiversity.

- Fragments, particularly small, irregularly shaped areas, are strongly affected by edge effects, reducing the effective size of the core habitat.

- Populations in edges experience high rates of predators, parasitism, and competition from invasive and exotic species.

- Metapopulation theory treats populations as isolated subpopulations connected by migration events.

- Metapopulation models allow the genetic and population effects of isolation to be examined (aspects not covered by equilibrium theories of island biogeography).

- Some subpopulations are sources and some are sinks. Sources are disproportionately important to the surival of a species.

- Exotic species, when released from predation and pathogens, have a competitive advantage over native species.

- Exotic species are often human commensalists and are adapted to living in the disturbed landscapes created by human occupation.

Further Readings

Ehrlich, P. R., and E. O. Wilson. 1985. Biodiversity studies: Science and policy. *Science* 253:758–762.

Lodge, D. M. 1993. Species invasions and deletions: Community effects and responses to climate and habitat change. P. M. Kareiva, J. G. Kingsolver, and R. B. Huey, eds. In *Biotic Interactions and Global Change*. Sunderland, MA: Sinauer Associates.

Mann, C. C. 1985. Are ecologists crying wolf? *Science* 253:736–738.

Meffe, G. K., and C. R. Carroll. 1994. *Principles of Conservation Biology*. Sunderland, MA: Sinauer Associates.

Smith, A. P., and D. G. Quinn. 1996. Patterns and causes of extinction and decline in Australian conilurine rodents. *Biological Conservation* 77:243–267.

Whittaker, R. J. 1998. *Island Biogeography, Ecology, Evolution and the Light-house Keeper's Cat*. Oxford: Oxford University Press.

Web Connections

On-line resources for this chapter are on the World Wide Web at:
http://www.prenhall.com/bush
(From this web-site, select Chapter 20.)

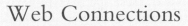

Working to Save Biodiversity

21

That biodiversity should be maintained has become an article of faith among conservationists, but an honest evaluation of why we should save every "insignificant" bug and weed is far from easy. Natural systems frequently include apparent redundancy; that is, many organisms seem to have overlapping roles within the ecosystem. For example, in a single rainforest tree there lived 47 species of ant. Would it really matter if there were only 46? If this is true, why does it matter if some species go extinct?

21.1 What Is So Good about Biodiversity?

One of the common answers has been that all nature is interwoven and that if any piece is removed it will start a chain reaction of collapse that will ultimately doom us all. Clearly, this alarmism is nonsense. There have been thousands of extinctions this century and yet ecosystems have survived. Another answer that is essentially the same line of argument but couched in more scientific terms is that food webs with lots of species on each trophic level are more robust than ecosystems that have species-poor food webs. High biodiversity provides ecological stability and is, therefore, a good thing. Although this is true in broad terms, this answer is intellectually dishonest. Removing a single species, even several species, from a complex

food web with hundreds of members may have no real ecological effect. Perhaps a better way to express these sentiments is that while we can describe an organism, map its DNA, and deduce its evolutionary history, we still do not know its importance to an ecosystem. Rather than guess and be wrong, ecologists would prefer to ensure the survival of every organism just in case it does play a pivotal role.

An assertion that is accurate, though in danger of being overworked, is that high biodiversity ensures us a supply of new genetic and pharmaceutical material. Less than 5% of described organisms have been investigated for their potential importance as sources of medicinal chemicals. That many of our drugs are derived from naturally occurring chemical compounds (Chapter 19) underscores the potential importance of any organism. The genetic pool present in wild species is similarly important. Centuries of selective breeding has reduced the genetic diversity among crops and livestock. The genes that were lost in this process may have held the key to overcoming a new disease or changed climatic conditions. By maintaining the wild stocks of that species and its close relatives, a reservoir of genes is available for future manipulation. Perhaps one of these species will have a gene that can be reintroduced and prove valuable. As we develop the ability to engineer organisms, the possibilities of transferring genes between organisms seems almost endless. Even a bacterium may hold a gene that could be beneficial when introduced to a cow or a corn plant or that could prevent disease among humans (Chapter 29). These arguments establish that any organism *could* be valuable, they certainly do not make the case that all organisms *are* valuable.

Another justification for maintaining biodiversity is that we cannot predict which species will be needed for biological control. As we are forced to deal with increasing resistance in pathogens and pest species, or as more areas are overrun with exotic species, the use of biological controls will grow. The organisms that are most successful as control agents are often insects, fungi, and bacteria. Prior to being needed, these otherwise insignificant organisms would never make the list of nature's most desirable species. However, as with the potential for medicines, relatively few species are likely to be useful as control agents.

Perhaps the most honest argument is also the most subjective. As far as we know, humans are the first creatures to have morality, conscience, a sense of history, and the ability to think critically. We have stepped beyond the role of a normal species on this planet and have assumed dominion over nature. This leads to two basic mind-sets. In one, Earth and all life on it are available for our use and well-being; we can save the useful species and allow the others to go extinct. The fallacy of this position is that it presupposes that we can tell useful from useless species, that we can artificially control ecosystem functions and thereby maintain ecosystem values, and that this would improve our standard of living.

One example that demonstrates our lack of competence at manipulating ecosystems and emphasizes our lack of ecological foresight is the current crisis facing many fruit farmers. The pesticides used to control insect pests on the crops and fruit trees have also killed the pollinators. Without the myriad bees, wasps, beetles, flies, bats, and birds that carry pollen from male to female flowers, the female flower is never fertilized and no fruit forms. Pollination by hand or machine is impossibly expensive. Orchard yields and profits have suffered because an ecological service, the free pollination of millions of flowers by bees, was taken for granted.

The other mind-set is that we are one species among many, albeit a very powerful one, and that it is our moral obligation to use our creativity and intellect to sustain life on the planet and not destroy it. This does not force us to hug every tree, but we should be able to assert that nature has both an underestimated economic value (Chapter 27) and an intrinsic value. The intrinsic value of nature is different for each one of us,

but includes spiritual, religious, and aesthetic appreciation of the living world. Assuming the 70% of Americans who declare themselves to be environmentalists feel some connection to the natural world, then the wholesale extinction of species could be likened to the desecration of churches, the burning of books, or the shredding of the Mona Lisa.

If it is unacceptable for our generation to preside over a rapid decline in global biodiversity, some remedial action is needed. Commonly, two basic strategies are proposed to ensure the future survival of endangered populations. One is to establish protected areas of habitat within which, it is hoped, wildlife populations will stabilize and the other is to boost the population of rare species through breeding programs.

21.2   Why Have Nature Reserves?

To protect and manage biodiversity are lofty goals and require large amounts of money and land. Conservation, therefore, is going to be strongly affected by politics and economics. Many times conservationists must accept a compromise or the least damaging solution to conflicts over land use. Consequently, land that is set aside for "reserves" may not always be a nature reserve in the classic sense of an area where natural systems and organisms are totally protected. The four most common reasons to form a reserve are to maintain "nature" in an urban area, to maintain a particular ecosystem function, to save a particular species, and to save a habitat or ecosystem.

In the past, most reserves have been established to protect a single organism, such as the bison, spotted-owl, or the Florida yew. Protecting the animals by legislation and setting aside land for their feeding and breeding will be key parts of a program to turn a population away from extinction. The habitat may be modified and managed to improve the conditions for the target organism, but in general most of the natural ecosystem functions will need to be maintained. This kind of protection is called **single-species management**. Protecting large animals that require a lot of land to support their population results in the incidental protection of many smaller creatures that need the same habitat type. These large, land-demanding species are termed **umbrella species**; their conservation provides an umbrella of protection over other organisms. Sometimes the target organism is a member of what may be thought of as the charismatic megafauna. These are "cute" animals that the public will pay to protect, such as giant pandas, wolves, tigers, and elephants. When one organism is being used as a vehicle to conserve a suite of less well-loved species, it is known as a **flagship species**.

Conservation could also be aimed at a **keystone species** (Chapter 12). These animals and plants exert an influence disproportional to their biomass upon the rest of the ecosystem. If their population changes, either up or down, the nature of the local community is changed. For example, in systems where wolves are the keystone species, the loss of wolves can lead to such an increase in deer populations that they overgraze an area and seriously affect regeneration of seedlings. With different plant species surviving the grazing pressure, the community enters a new successional trajectory. Thus the presence of a few animals can shape the entire community.

Saving an entire habitat or ecosystem breaks away from single-species management. Instead of targeting conservation to secure a particular species, the goal of this style of conservation is to maintain all the natural ecosystem functions. Here, managers may have to step in to manipulate the system to replicate natural features that have been affected by human activity. For example, where fire suppression has stopped an area from burning, controlled burning of the habitat may be done to simulate natural periodicities and intensities of fire. Similarly, weirs may have to be installed on streams to replicate the natural hydrology in areas where water extraction has lowered water tables. This style of management may create the most "natural" conditions, but its critics say that it does not pay enough attention to the fate of individual species.

21.3 The Population Needed for Survival

Determining the minimum number of individuals needed to ensure the survival of a species has been an important issue for conservation biology. One individual, even one pair, does not constitute a breeding population. The goal of conservation is to maintain a **minimum viable population (MVP)** of even the rarest species. Alternative definitions of MVP are that the population is large enough to retain 90% of the genetic diversity of a species for 200 years, or that the population is sufficient to guarantee, with 95% confidence, the survival of the species for 100 years. It would be ideal to have a magic number that is the MVP for all species, but there is none.

As populations increase in size, their chance of extinction lessens, and the risks posed by different causes of extinction diminish. When there are fewer than 10 pairs of an organism, there is a very strong probability that chance events such as an unusual ratio of the sexes, or plain bad luck, will lead to their extinction within a few generations. Populations with more than 20 but fewer than 50 individuals have a better chance of survival, but there are so few individuals that close relatives breed. Such matings produce offspring that may have lowered fitness, resulting in **inbreeding depression**. Mating events between individuals more closely related than cousins are likely to cause a decline in the genetic quality of the group. Rare, damaging genes, that had previously remained hidden become more abundant. In the terminology of genetics, there is a decline in heterozygosity among the herd (all individuals are becoming more genetically alike), and the benefits of genetic diversity are lost. Inbred individuals have increased vulnerability to disease, are likely to be small, and may have low fertility or even gross deformities.

Similarly, at very small population sizes, especially if there are fewer than 20 reproductive individuals, random genetic changes become increasingly important and start to affect the genetic composition of a group. The tendency is that as a result of chance, an allele (a form of a gene) is not passed from one generation to the next. The loss of the allele is permanent and lowers the genetic diversity of that generation. This process of random genetic change is independent of natural selection; thus, when populations are small, there can be significant shifts toward poor genotypes purely as a result of chance. This random change of genetic composition from one generation to the next is **genetic drift**.

At populations of greater than 50 individuals, the genetic decline in herd quality is much slower, and at populations greater than 500, the population is genetically stable, showing no loss of heterozygosity. These observations led to the 50/500 rule that suggested that if populations could be maintained above 50 individuals they would be safe in the short term, but for longer-term success populations of at least 500 would be needed. So far we have only considered the population as if it were a herd that existed independently of fire, hard winters, disease, and other stresses. Even a genetically stable population might still be vulnerable to demographic slumps in which outside events lower numbers drastically and lead to extinction. To allow for continued evolution of a species, and for it to persist in perpetuity (as opposed to the 100 years required by the MVP definition), populations of between 2500 and 5000 are a minimum. Before we adopt any one of these values as an MVP, remember the case of the Passenger Pigeon (Chapter 13), where many thousands of birds were needed in a single flock to trigger breeding. For that species, it was minimum flock size that was all important. For our purposes, we will accept the MVP of 500 individuals, recognizing that this is truly a minimum number.

Genetic Bottlenecks

A **genetic bottleneck** occurs when a population loses much of its genetic diversity as a result of a population decline. If a population is reduced to just a few survivors, those that remain will not contain the full complement of genetic diversity of the initial population.

Imagine you and your nine closest friends are the only survivors of the human race; your group cannot contain the full genetic diversity of humans, but it probably does contain almost all the diversity of the immediate population in which you live. So long as both sexes are represented in the group, you could form the core of breeding individuals to rebuild the human race. After many generations, there might once again be thousands of humans on the planet, and their genetic diversity would still be high. However, if that initial group had all been members of the same family, or, more importantly, if the population was held at a very small size for more than a few generations, there would be a profound, lasting loss of diversity. In Figure 21.1a a population crashes, but quickly rebuilds its numbers. The loss of genetic diversity is

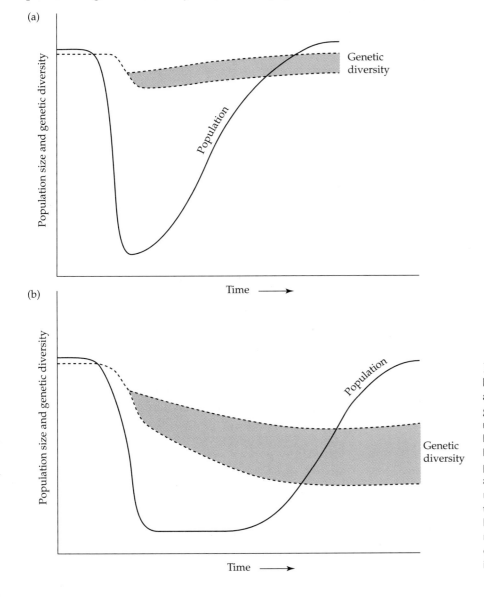

Figure 21.1   **Trends in population size and genetic diversity as a population passes through a genetic bottleneck.** The loss of genetic diversity is a result of a fall in numbers of individuals and is exacerbated by inbreeding and genetic drift if the population remains small for more than a few generations. The shaded area represents the range of genetic diversity that might result if survivors are unrelated (upper boundary), or are closely related (lower boundary), to each other. Diversity slowly increases after its lowest point because of evolution.

relatively small and would depend on chance factors and also the degree of relatedness among survivors. In Figure 21.1b the population crashes and stays low for many generations. While the population is low, genetic drift and inbreeding depression progressively reduce genetic diversity with each generation. As the population size increases, there is no further loss of diversity due to genetic drift. This population has passed through a genetic bottleneck. Again chance and the relatedness of the initial survivors will provide a wide variation in the genetic diversity of the recovered population. Note the difference between the outcomes of Figures 21.1a and b. In 21.1a both the population and the genetic diversity recover close to their starting point. If you did not know that there had been a past population crash it would be hard to detect on the basis of genetics. In Figure 21.1b, although the population has recovered in terms of size, it is genetically impoverished. A genetic comparison of this population with others would reveal much lower genetic diversity. A rule of thumb is that to induce a genetic bottleneck there must be fewer than 10 to 15 individuals and their population must stay low for more than two or three generations.

Examples of animals with very low genetic diversity include the cheetah, which may have gone through one or more bottlenecks (though this is not certain) caused by climate change at the end of the last ice age and more recently by hunting. A more certain example is the lions of the Ngorongoro Crater of Tanzania. In 1962, a plague of biting flies killed most of the lions, leaving only 9 females and 1 male. This population is largely cut off from other lion populations, and despite the migration of a few males from the neighboring Serengeti population, the Ngorongoro lions passed through a genetic bottleneck. Recently, the population has risen to about 125 individuals, but genetic diversity is lower than in prides of lions in the nearby Serengeti.

21.4 The Area Needed for Survival

Sustaining an MVP requires a matching set of resources, such as nest sites, food, and water. These requirements define the **minimum viable area (MVA)** and will vary considerably from one species to the next. As a rule of thumb, the larger the organism, the greater the resource demand because of the energy requirements of the animal. These larger animals would be expected to have a larger feeding area or territory than smaller animals. The total hunting and living area occupied by an individual is its **home range**, which is commonly related to its body weight and its feeding patterns. If a comparison is made between equally sized animals that graze (eat leaves and stems) and ones that hunt for seeds, the grazing animals need only 25% of the home range of the seed-eaters. Remarkably, grazers also need about 25% of the home range of carnivores. The explanation is that both seeds and meaty prey items are relatively rare compared with leaves. Clearly, both diet and body size will play a role in determining the minimum viable area of a species.

The flow of energy and biomass through food chains can be used to determine the relative resource or space requirements of an individual at each trophic level (Figure 21.2). Each herbivore in the huge population on the second trophic level has very low resource requirements, whereas the top carnivore with its high energetic requirements, but relatively rare prey, requires huge hunting areas to support even a single individual. Apart from this general energy statement, the ecology of a particular organism will determine its MVA. For instance, is the animal territorial? If it is, how big is one territory? How large an area is needed for 250 territories? If the predator to be protected is migratory, the MVA will have to include the entire migratory region. If the predator is not migratory, its prey for critical periods of the year may be; in that case, the MVA must include the migratory range of the prey. Thus, the whole ecosystem must be studied to determine whether a population of a given area will be stable if it is isolated.

Conservation biologist Christine Schonewald-Cox provides an approximation of the area needed to support minimum viable populations of small herbivores, large herbivores, and large carnivores. Her data elegantly support the prediction that large reserves contain larger populations of animals and that the populations of animals will be determined by their different

Figure 21.2 Resource requirements of individuals within a food chain. Resource needs per animal at each trophic level can be calculated by inverting the Eltonian pyramid of numbers (Chapter 6).

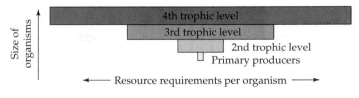

feeding strategies (Figure 21.3). If it is assumed that a minimum viable population of 500 individuals is required, a very rough approximation of the required minimum viable area for each feeding strategy can be read from the graph. The small herbivores would need approximately 100 km², the large herbivores would need 10,000 km², and the large carnivores would need about 1 million km². Half of the world's reserves and protected areas are areas less than 100 km². On this basis very few reserves could offer areas large enough even for birds and small mammals, let alone large predators. Existing reserves are too small and too few for truly effective conservation, but before we sink too deep into gloom and despondency, consider that these statistics are based on single reserves, and it may be that the presence of other reserves can have an important effect.

The Presence of Other Reserves

Island biogeographic theory and metapopulation theory (Chapter 20) encourage us to think of movement of species between habitat islands. If there is movement of individuals between reserves (either naturally or artificially) each individual reserve does not have to be large enough to sustain an MVP. As the landscape becomes more fragmented, the decision must be made as to how many nature reserves can be set aside and where they should be located. In the 1970s a debate emerged over whether it was better to have a single large or several small (SLOSS) reserves. Under normal circumstances a large reserve is preferable to a small one because it will support a greater diversity of species, will have larger populations of each species, and will have the best chance of sustaining a species

with large resource requirements. The prediction that reserves of at least 20,000 km² would be needed to prevent the loss of top carnivores is reinforced by a study conducted in 14 western North American national parks. The study examined the effects of relaxation (the species lost following isolation) in parks of various size. A strong correlation was found between park size and the number of local mammal extinctions (Figure 21.4). Within the first 43 to 94 years after the park was established, as many as seven species of mammal were lost from a single park. Parks of fewer than 10,000 km² were especially susceptible to extinctions. It is important to note that the full effects of relaxation may take longer than the 43 to 94 years since these parks were formed and that more extinctions are likely in the future.

Island biogeographic theory provides a rule of thumb that 10% of the species with breeding populations in an area will go extinct if the area of intact habitat is halved. Furthermore, the species lost are probably the ones that are most demanding in terms of habitat requirements and are, therefore, the rarest ones. Thus, although the diversity has declined by only 10%, the species most in need of protection have been lost.

The disadvantage of having a single large reserve is that if you put all your eggs in the proverbial one basket, a natural disaster or a disease epidemic could seriously affect populations in the area, possibly resulting in the extinction of some species. Saving several smaller areas has a greater probability that one or more of the populations will survive such a disaster. Also, the small areas will be spread over a wider geographical area and therefore, as a group, may capture more species diversity. This suggestion is supported

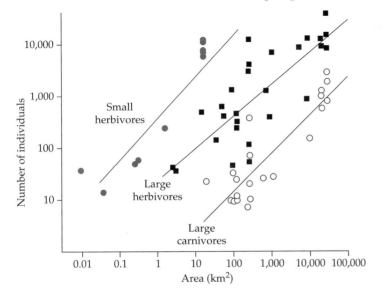

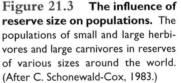

Figure 21.3   The influence of reserve size on populations. The populations of small and large herbivores and large carnivores in reserves of various sizes around the world. (After C. Schonewald-Cox, 1983.)

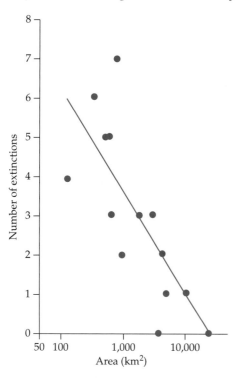

Figure 21.4   Mammal extinctions in North American reserves. The number of naturally occurring mammal extinctions in reserves that occurred within the first 43–94 years after the reserve was established. Larger reserves suffered fewest extinctions, smaller reserves suffered most. (After Newmark, 1987.)

by the observation that in similar habitats the difference in actual species combinations is directly related to increasing distance from the starting point. For instance, the mixed oak forests of eastern Connecticut will be more similar to those of western Connecticut than to those of Georgia or Ohio. There is much to be said, therefore, for scattering reserves throughout the range of a habitat type, rather than having one centrally located superreserve. A trade-off is clear between the optimal size and optimal number of reserves. Again, no magic solution is likely to be forthcoming. The answer to the SLOSS debate is that it was irrelevant. In truth, conservationists do not have the luxury of picking and choosing which reserves they want; they take whatever land they can get.

A more meaningful way to approach reserve design is to look at existing reserves and determine whether they can achieve their goal of protecting biodiversity—in almost all instances the answer is "no"—and then consider ways to improve the situation. The basic problem is that most reserves are too small and too isolated to maintain minimum viable populations of large mammals. One way of attempting to reconcile problems associated with small reserve size and

future climate change is to integrate reserves into the broader landscape and link reserves into a web of protection.

Cores and Buffers

Because ecosystems are linked to other ecosystems, or a small habitat patch is part of a larger ecosystem, it is important to take into account surrounding land use when siting a nature reserve. For example, if the goal is to save river otters, there is little point in trying to establish vibrant otter populations if the reserve is sited downstream from a polluting industry, say a paper mill, steelworks, or gold mine. Even large reserve areas can be subject to outside influences from riverborne pollution, acid deposition, flood management schemes, or excessive recreational use. It is necessary to safeguard the ecological functions that contribute to the protected landscape. For example, if the park contains a large wetland area that is flooded each year as a local river washes over its banks, the local ecosystem receives an input of nutrients from the flood water. These nutrients provide fertile soils, but the lingering flood waters result in anaerobic soil conditions that may last several months. Such floodplains are filled with wetland species adapted to waterlogged conditions. If the annual flooding ceased, the habitat that the reserve was created to protect would change. Preserving the habitat, therefore, requires protecting the target area and enough of the river's drainage basin to ensure that the correct regime of flooding continues. The upstream presence of industry polluting the river, a hydroelectric dam evening out the flow rate of the river, or extraction of water for agriculture could all adversely influence the wetland. Thus, the siting of the reserve must become part of a larger scale of planning in which land use around the reserve is developed in such a way that it does not damage the reserve.

Multiple-use modules, seek to integrate reserves into the overall landscape through defining appropriate land uses in concentric areas around the core of the reserve. At this larger landscape scale, the actual reserve area is viewed as the **core** area of the conservation effort. Within the core area access is strictly limited. The absolute priority is for the maintenance of the ecosystem. Many conservationists would like to see humans excluded completely from the core area. However, such prohibitions are unpopular and expensive to enforce. Core areas generally have some hiking trails, and there may be backwoods "primitive" campsites, but that would be the limit of its development. There would be no vehicle access, no sewer systems, or piped water supply. Surrounding the core is the **inner buffer** area, the role of which is to prevent edge effects from causing the deterioration of the reserve. The buffer zone is to be kept as pristine as possible close to the core, with more disturbance permitted with greater distance. Close to the core, nature trails, backpacking, and horse riding would be appropriate uses. A little further from the core, interpretive centers and car parking would provide access for visitors. Surveys of park users in Britain revealed that the vast majority do not stray more than a few hundred meters from where they park their car. Visitors are therefore likely to be concentrated in a few locations, and if uncontrolled, they could damage the area, inadvertently ruining what they have come to experience. Interpretative trails, boardwalks, and guided tours all serve to limit the disturbance caused by visitors and at the same time help them to enjoy and learn from their "natural experience." The buffer area will generally have more sophisticated campgrounds that can handle vehicles. The **outer buffer** has less emphasis on wildlife protection, but it should serve as a barrier to the worst forms of pollution and to urbanization. Forestry, golf courses, agroforestry, public gardens, or residential development with large lot sizes could all be suitable land uses for the outer buffer area. Neither a golf course nor a commercial forest will be managed as a reserve, but both may reduce noise and atmospheric pollution (so long as pesticides and fertilizers do not stray into the reserve) and may help to maintain ecosystem functions within the core area. These green areas will also increase the hunting area of owls and hawks, so they may effectively increase the size of the reserve and allow the existence of higher populations of predators.

Yellowstone National Park and the surrounding national forests demonstrate the use of buffer zones to increase the habitat area of the core reserve (Figure 21.5). The distinction between a national park and a national forest is an important one. National parks in the United States were established for the protection of wildlife in a setting that may be visited by people. Cutting trees, picking flowers, and hunting any creature within their boundaries are forbidden. National forests are for-profit timber and mining areas in which hunting may be allowed. Although each segment of national forest will be disturbed according to a logging regime, until that time comes, it will be used by wildlife. These lands, though not secure in the long term, can provide good hunting and breeding habitats in the short term. One result is that animals requiring huge territories, such as cougar and bears, will

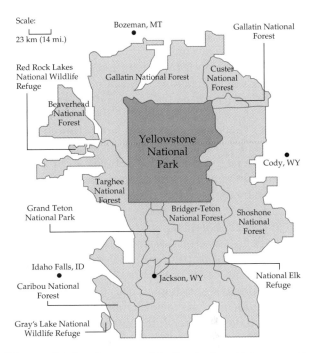

Figure 21.5   A core and buffer design. Yellowstone National Park (core and inner buffer) and the adjacent national forests (outer buffer) are an example of how planning buffer areas can increase the effective reserve size.

wander between the national park and the national forests. Hence, surrounding a national park with national forests greatly increases the effective size of the reserve beyond its core area. For species that are protected from hunting by law, the habitats offered by national forests are beneficial. However, for game animals that can be hunted, like the black bear, the national forests can prove to be extremely dangerous. A black bear that has its lair within a national park may wander across the jurisdictional boundary into a national forest, where there is a strong probability that during the hunting season it will be shot. In these cases, a buffer that permits hunting can actively reduce the number of animals that the reserve supports.

Nodes and Corridors

Ideally, a nature reserve that seeks to protect an area of high biodiversity will maintain that diversity in the future. The idea that a nature reserve can have a rigidly defined boundary assumes that there will be no change in the future area occupied by the community. As we learn more about how communities change in terms of both geography and species composition, we realize that species will need to respond to environmental change through migration. Reed Noss, a conservation biologist at Oregon State University, suggests that the areas of high biodiversity should be thought of as nodes. The nodes will not be constant through time and are expected to migrate. In Noss's vision of reserve design, the nodes are identified and then linked by corridors of wild lands that would allow species to move freely between nodes. If the node migrates, then the species will be able to establish populations in the corridors. This design has many appealing features. One is that a particular node can be smaller than the minimum viable area for the population of a given species, so long as the sum of area among the nodes connected by the corridors does encompass the MVA.

Clearly, the concept of nodes and corridors fits well with the concept of metapopulations (Chapter 20) because both are based on the premise of continual range extension and contraction of species, and gene flow between subpopulations. However, there are also some important distinctions. One is that many species do not conform to the metapopulation concept, and they may have relatively little interaction with other populations. In this case the connectivity of nodes and corridors produces a single continuous habitat rather than a landscape of isolates. The roles of the corridors are also subtly different. In the metapopulation model it is assumed that new habitat suitable for colonization will always be present, therefore, the connecting corridors need only be wide enough to facilitate migration. In Noss's model, the corridors may have to become temporary homes for the population if a node is no longer habitable (we must assume that all other land is developed and unavailable for colonization). Therefore, the corridors are more than just a highway between nodes; they must be suitable for the longer-term survival of individuals. In general this will require the corridors to be wider in Noss's scheme than in the basic metapopulation model.

The value of these corridors will depend on their breadth and length and on the demands of the species using them. Some corridors may be needed only to connect nearby woodlots; others, ideally, would provide a flyway for birds migrating, for instance, from Central America to Canada. In general, the longer the corridor, the broader it must be. Although humans can walk along a narrow line such as a footpath, most animals follow more random tracks as they cross the countryside. Thus, what may appear to be an adequate corridor, say a few tens of meters of forest on either side of a river, may prove inadequate for the movement of wildlife locally or to serve as a long-distance migration corridor. Guidelines for successful corridor widths have yet to be established, but studies have shown that wolves in Alaska need corridors in excess

of 12–22 km wide, while a bobcat needs corridors at least 2.5 km wide.

Some states, such as Florida, are designing ambitious plans to link their state park and forest systems with a loose web of natural and seminatural habitats (Figure 21.6). Such schemes are expensive. Often, the land comprising the corridors must be bought or the owners must be compensated for its use as a wildlife area. It is hoped that these corridors will allow the migration of animals, hence increasing gene flow and reducing the chance of a population's dying out as a result of inbreeding.

Some wildlife corridors need to be only as long as the width of a road. Roads fragment landscapes and may become a barrier to dispersal, both because they are an alien surface and also because animals may be killed crossing the road. In Florida, the construction of a four-lane highway, "Alligator Alley," to carry rapidly moving traffic east to west across the Everglades, posed a potential danger to wildlife. The population of the Florida panther was bisected by the road, and it was a real threat that these and other rare animals would become roadkills. A high chain-link fence was erected to prevent animals from jumping out in front of traffic, but this safety barrier could have caused the populations on either side of the road to become reproductively isolated. Isolation was prevented by tunnels beneath the road, permitting the safe passage of animals from one side of the road to the other. Video cameras installed in these tunnels have shown that they are used by a wide variety of animals, especially deer. In this instance, short wildlife corridors have proved highly successful.

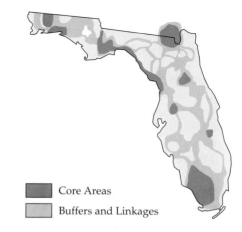

Figure 21.6   Proposed wildlife corridors in Florida. The corridors would allow a greater movement of populations and reduce the likelihood of inbreeding. (After R.F. Noss, 1991.)

Although in nature subpopulations would be linked through migration, in a landscape modified by human activity the benefits of connected subpopulations must be balanced by some largely unknown risks. While connectedness of subpopulations promotes gene flow, it also promotes the distribution of pathogenic organisms, parasites, and predators. For example, if a disease epidemic wipes out an isolated subpopulation, other isolated subpopulations may be safe. However, if the subpopulations are connected, it is more likely that the entire metapopulation will be exposed to the disease. Exotic species are a significant wildlife threat (Chapter 20). Their invasions are not curbed by predators or pathogens, and frequently the only thing that prevents a population explosion of the exotic species is a barrier to dispersal. When subpopulations are linked by corridors of similar habitat, there are no barriers, and the newcomer can spread through the entire system.

Another criticism of long corridors is that there has been little work to ensure that most of the animals that enter a corridor emerge safely at the other end. The animals may be exposed to increased dangers of predation, poaching, and being run over. If this is the case, then the corridor is simply sucking individuals out of the subpopulation, decreasing the probability that the metapopulation will survive. Overall, corridors are believed to be beneficial to wildlife conservation, but further research will be needed to verify that they are.

21.5 The Management of Reserves

Although setting aside areas for conservation is praiseworthy indeed, if these are to last for future generations, the job of maintaining them has only just begun. The management, particularly of small "island" areas, must anticipate and offset adversity for the protected populations. Setting up a nature reserve is similar to starting a business; a clear goal is needed (the why behind the establishment of the reserve), as is a plan to achieve that goal. Monitoring the development of the reserve determines what progress is being made toward the goal, and action must be taken to rectify the situation if the goal is not being met. A realistic management plan is absolutely essential for the successful establishment of reserves.

Often the area that can be set aside for protection is not large enough to support all the functions of a natural ecosystem, and management is needed to replicate those lost functions. Reinstating ecosystem functions may require active management to remove exotic species, conduct controlled burns, improve the hydrology, or reforest the area with native species. All these actions must be documented in the management plan. Another facet of the management plan is how to implement protection of the reserve. Such protection may need armed guards, such as those accompanying the remaining wild black rhinoceros (one guard with each animal). For small reserves, a fence and signs may be all that are required. If the park is protected by the law, then it may be the police or the army who ensure that laws are upheld. Whatever means is pursued, there is some cost to the citizens. If the tax base or the agency who owns the land cannot afford the protection that is called for in the management plan, the park becomes vulnerable to exploitation. Hunting, logging, farming, and urbanization may all encroach on the reserve area if nothing is done to prevent them from doing so. Such areas, which are indicated to have the status of a park on a map but for which there is no real protection, are reserves in name only. Such hypothetical parks are termed **paper parks**. The problems in protecting these parks may range from over-ambitious conservation programs, in which the management budget is spread too thinly, to governmental corruption.

People in Reserves

The rights of indigenous peoples have long been neglected and are only just beginning to be recognized. Among these is the right to continue traditional hunting and farming practices within a designated reserve. Hostility can be generated when parks are established through government edict without consideration for the needs of people who already live on that land. The hostility may be manifested in poaching, illegal logging and burning, or running livestock into the reserve area. Sometimes a confrontation between government and the local population is almost unavoidable, especially when the local ways of life are responsible for the demise of the species to be protected. Under these circumstances, paying the local people to change their hunting or farming practice may allow the target population to recover. As a last resort, the decision may be made to relocate the human population, in which case new lands must be found for them.

Increasingly it is realized that obtaining a consensus among the affected population to support the establishment of a park is the ideal solution. The first step in reaching this consensus is to consult the local people about their needs. Small adjustments in park boundaries at the planning stage may reduce later conflicts. For example, shifting the park boundary so that it does not enclose ancestral burial grounds may smooth the way to better relations. Almost certainly, the project will need an extensive education program informing the local people why that particular place was chosen for the reserve. Successful projects include some benefit for the local population, perhaps through employment as park rangers or guides.

One way to provide for indigenous populations is to establish large tracts of real estate as cultural reserves. Tribal peoples not only live within these areas but also govern them. Management plans for the Brazilian Amazon call for the establishment of more than 20 million hectares of such homelands. However, significant problems exist with this solution. The indigenous peoples of Brazil already have designated lands and legal protection, at least in theory. In practice, however, their property rights continue to be violated. An example is the massacre of the Yanomamö Indians who protested the invasion of their homelands by gold miners in 1993 (Chapter 12). On the other side of the issue is a concern expressed by some conservationists for the future use of these lands. The general assumption is that the indigenous people will maintain the area in a natural state, but this is an assumption, not a guarantee. Kent Redford of the Nature Conservancy points out that the social structure of the Amazonian tribes is likely to change and become more Westernized. The old ways of hunting and gathering and shifting cultivation are likely to be succeeded by more intensive land use. The tribes that control these lands could convert them to agriculture, pasture, or any use of their choosing; that is their prerogative. When a "sustainable" use of the forest is chosen, the criteria of sustainability used by the indigenous people may prove to be different from those of conservationists. For example, the indigenous people may be attempting to gain a sustainable harvest of Brazil nuts, which can be sold as a cash crop. The management to maximize the harvest of Brazil nuts would be favorable to some, but not all, of the original forest wildlife. Redford points out that conservationists cannot count on these lands as areas that will necessarily be wildlife sanctuaries in the future.

Co-ownership and Natural Boundaries

Nature is almost without sharp boundaries. Yet throughout history, land has been carved into territorial blocks by humans. Pieces of property range in size from individual gardens to whole countries, and these areas are subject to strict demarcation and defense. Very seldom do the boundaries conform to the pattern set by local ecosystem boundaries, and the importance of this difference will become apparent. Sometimes the land was apportioned according to a straight line on a map, such as the 49th parallel separating the United States and Canada. Rivers, such as the Jordan River separating Israel from Jordan, are convenient property boundaries and form more than 200 national boundaries worldwide. The popularity of rivers as national boundaries is that they have two defined sides and present obstacles to military invasion. Rivers have become natural choices for human-imposed boundaries on a landscape. For this reason, nature reserves are often bound by rivers. The advantage of rivers as boundaries is that the edge of the reserve is easily mapped and identified, but this demarcation of a boundary also engenders multiple management problems. Rivers are natural divides on a geopolitical landscape, but they are the natural middles of an ecological one. Rivers provide particular challenges to reserve managers. Some of these are as follows:

- A watershed describes the entire geographic region that drains water into a river (Chapter 6). As water drains into the river from either side, the river becomes the center of the drainage feature. If a watershed has multiple land uses, those upstream or on the opposite bank of a reserve area bound by a river could have a deleterious impact.

- As the number of partial watersheds affecting the reserve is increased, the number of landowners who need to be included in the management plan swells.

- The river (often the primary route for travel) provides multiple points of uncontrolled access to the reserve, inviting poaching and illicit logging (Figure 21.7a).

Instead of having rivers to provide boundaries, reserves should be bounded by their watershed divides. The advantages of this boundary are that a single checkpoint can control river access to the reserve (Figure 21.7b) and that the number of landowners who affect the reserve is minimized.

Where the river forms a property divide, it is common to find two different types of land use on opposite banks. Similarly, the upstream and downstream uses of land may be quite different. Determining a coherent management plan for an entire watershed that

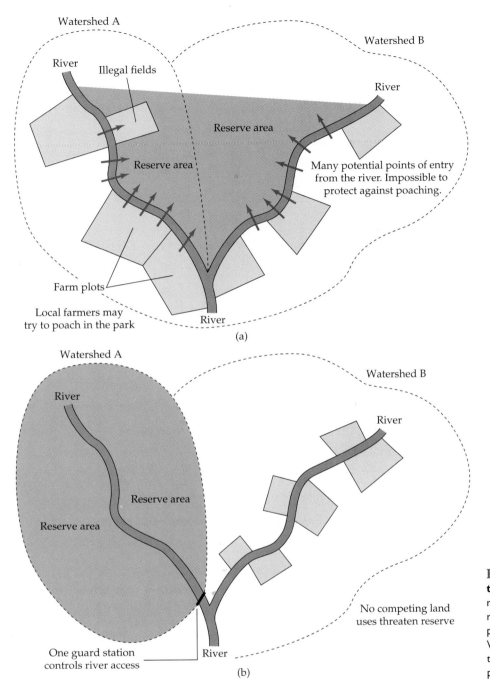

Watershed A

River Illegal fields

Watershed B

River

Reserve area

Many potential points of entry
from the river. Impossible to
protect against poaching.

Reserve area

Farm plots

Local farmers may
try to poach in the park

River

(a)

Watershed A

Watershed B

River

River

Reserve area

Reserve area

No competing land
uses threaten reserve

One guard station
controls river access

River

(b)

**Figure 21.7 Alternative ways
to site a nature reserve.** (a) With
rivers forming the boundary of a nature
reserve, there is unlimited access for
poachers who arrive in boats. (b)
When the watershed is used to define
the park boundary, a single river access
point can be guarded.

is not completely contained within a reserve may require the cooperation of diverse landowners who have different ideals of land usage (Figure 21.8). Each party that will be affected by land-use plans is termed a **stakeholder**, and each needs to be consulted and represented in the decision-making process. This does not necessarily mean that everyone will be happy with the outcome, because the desires of individual stakeholders may not reflect the common opinion (or the law). Consequently, even when consensus has been reached, there may still be the need to enforce a policy. Because the park boundary may also be a national boundary, the negotiations can involve international dispute resolution and diplomacy.

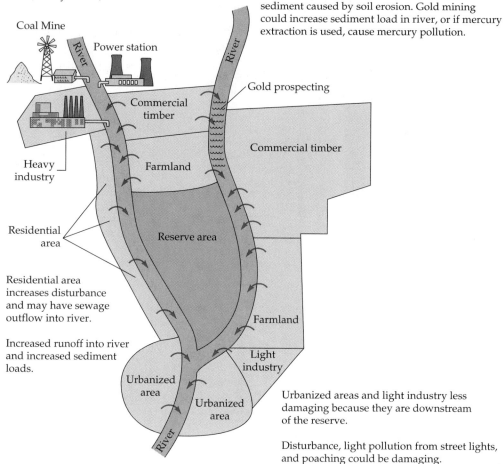

Industries could pollute river with acids, heavy metals, or warm water.

Commercial timber and farmland could pollute river with pesticides and fertilizer; also increase sediment caused by soil erosion. Gold mining could increase sediment load in river, or if mercury extraction is used, cause mercury pollution.

Coal Mine

Power station

Gold prospecting

Commercial timber

Commercial timber

Heavy industry

Farmland

Residential area

Reserve area

Residential area increases disturbance and may have sewage outflow into river.

Increased runoff into river and increased sediment loads.

Farmland

Light industry

Urbanized area

Urbanized area

Urbanized areas and light industry less damaging because they are downstream of the reserve.

Disturbance, light pollution from street lights, and poaching could be damaging.

Figure 21.8   A landscape of diverse land use and numerous stakeholders who could affect the success of the reserve area.

21.6   Restoration Ecology: The Next Thrust of Conservation?

As world populations grow, agricultural lands are lost to desertification and wilderness areas are consumed by development. Meanwhile, the science of **restoration ecology** is emerging. The goal of restoration ecology is to take a degraded landscape and return it to its original condition so that it may support a full range of indigenous wildlife. However, the wider application of this knowledge would allow the restoration of farmland, pastures, forestry, wetlands, or fisheries. The reinstatement of wildlife and the resurrection of farmland are both important to the preservation of natural systems. Taking degraded land and converting it into a wilderness area has the clear advantage of increasing the habitat area available for wildlife, inherently increasing the likelihood of maintaining the minimum viable area for some species. The benefit of restoring agricultural land is equally important, though perhaps less obvious. The reason that the land was degraded is often poor past agricultural practices. Let us take the example of slash-and-burn agriculture in a tropical forest. The land had yielded several harvests before being turned over to pasture. Cattle prevented the successional recovery of the land and augmented soil erosion, finally leaving just an infertile subsoil at the surface. The farmer has now moved on and will be cutting new fields from mature forest areas. The cycle of destruction continues. If the original fields could be returned to productivity, they would be better for the farmer as they are almost certainly nearer to the original access road or river than are present fields.

Ecology in Action

Rebuilding a Prairie

Prairies rich in grasses and wildflowers once stretched from Illinois to the Rockies. These were the natural pastures of huge herds of bison, deer, elk, and antelope, and the herbs truly were "as high as an elephant's eye." The wetter areas of the prairies made good cropland, and the drier areas provided good grazing land for domestic sheep and cattle. Prairies were plowed, overgrazed, or reseeded, until less than 1% of natural prairie remains. The black-footed ferret, a small predatory mammal, is reliant on natural prairie habitats and has become one of our most endangered species. The effort to save the prairie ecosystem has been going on for more than 60 years and is one of the first examples of restoration ecology in the United States.

The start of the conservation effort can be traced to President Roosevelt's attempt to reduce the effects of the Great Depression of the 1920s and 1930s through the New Deal. Public money was poured into ambitious road, bridge, and dam construction projects to provide employment and jump-start the economy. One of these initiatives was to form the Civilian Conservation Corps. Unmarried men aged 18–25 were provided work in forestry, dam construction, and fire fighting. They were paid $30 a month, of which $25 had to be sent home to their families. Among these public works projects was an ambitious plan to restore an area of prairie at Madison, Wisconsin.

In 1936, work began, transporting prairie soil and lifting sods of the native grasses from natural prairie remnants and bringing them to the Arboretum of Wisconsin University. This work was done carefully to avoid damaging the source areas. The sods contained the roots and seeds of a wide diversity of plants that were native to a natural prairie in Wisconsin. The sods were planted and nurtured until the plants they contained became established in the Curtis Prairie. However, the species composition of the newly formed prairie was not as diverse as that of the source areas. The sods introduced from the prairie fragments held all the desired diversity, but many of these plants failed to establish populations in the new system. Worse still,

agricultural weeds appeared to be outcompeting the native species.

Ecologists investigated how the constructed prairie was different from the natural one. The most significant factor proved to be fire. Natural prairies burn every few years, and all the native species are fire tolerant. The created prairie was so carefully manicured that it had not been allowed to burn. Experimentation revealed that a single burn eliminated almost all the weed species and triggered the germination of buried seeds. Carefully planned burns proved to be the missing ingredient in the path to successful restoration.

Although the feat of reconstructing a prairie is impressive, what makes Curtis Prairie even more of an achievement is that for the last 40 years the project has been conducted entirely by volunteers. The size of the restored prairie has increased, and the local community has been involved in every aspect of its formation. The organizers have gone out of their way to make the restoration process of Curtis Prairie colorful and interesting. The prairie is burned, mowed (to simulate grazing by bison), and even trampled (to simulate the passage of a bison herd) by volunteers. This project inspired numerous other prairie restoration programs, some of them even larger, such as the 155-hectare restored prairie at the Fermi National Accelerator Laboratory outside Chicago. Volunteers sowed the seeds of more than 70 species of native prairie plants in the early 1970s. This prairie now supports 125 species of plants, and 224 bird species have been sighted. The organizers of this project expect the soils of the prairie to continue to mature and the wildlife to gradually undergo successional adjustments toward a natural prairie ecosystem.

A controversial project that has been proposed is to restore a "buffalo commons" to about 380,000 km^2 of low-grade farmland in the Dakotas, Texas, Wyoming, and Nebraska. This would be prairie restoration on a truly massive scale, way beyond anything so far attempted, but the potential gains lie in ecotourism, low-intensity (high-quality) grazing, and wildlife management.

Furthermore, the farmer would no longer need to invade mature forest; this reduction in cutting the forest would be a major step toward local conservation of biodiversity. However, the process of restoring fertility to degraded land is hugely challenging, because it requires knowledge, patience, and finance—three

assets that are seldom available on the frontier of forest destruction.

Let us consider what has happened to degraded land from an ecological point of view. The degradation can have many causes and consequences: overgrazing leading to erosion and the loss of topsoil; mine waste chemically poisoning an ecosystem with toxins such as copper; canalization of rivers reducing floodwaters and causing wetlands to dry out. In all these cases, the ecosystem functions have been damaged, perhaps to the point where the land lies barren or has low primary productivity.

Ecological succession is nature's way to heal rifts in the forest, but for it to work the forest must be allowed to recover. The force that perturbed the system and led to the degradation must have stopped its disruption. For example, if cattle are herded onto abandoned croplands, their presence prevents the successional processes that would lead to the establishment of dense vegetation, changing microclimatic conditions, and improvement in soil quality. As the pasture declines in quality, overgrazing increases in intensity and any new green growth is eaten. The grazing must stop before succession can get started.

So long as the driving force behind the disruption has stopped, ecological succession would eventually lead to a recovery. However, it is not guaranteed that the system that emerges from the recovery would be one that is desirable. In many instances, the land surrounding the affected area has been heavily disturbed, its natural vegetation has gone and has been replaced by crops, weeds, and other invasive species. The local seed source for the natural colonists of a successional sequence no longer exists. The successional area would receive only the seeds of weeds and other introduced species. Such a circumstance increases the likelihood that the successional area would become home to crop pests, possibly threatening the harvests of nearby farms. In cases where the land has been poisoned, it may take many centuries for the toxins to be washed from the system.

Active management speeds and directs the recovery of the land. Topsoil imported from another site could provide an inoculum of decomposers and seeds. Revegetating the area with carefully selected trees and shrubs, characteristic of early-successional landscapes could speed the recovery. All the while, the water table, pathogens, soil fertility, and pests must be monitored, because this artificially hastened recovery will remain extremely fragile until nature takes over from people as the instrument of succession. Clearly, this process is labor intensive and will test the limits of our understanding of ecosystems. An analogy is that of an engine that needs to be repaired. With even a little knowledge, we may be able to conclude that the engine isn't running because a wire is loose or the

engine has run out of fuel. It takes a different level of knowledge to start with some steel, plastic, glass, and aluminum and construct a Ferrari. In ecosystem restoration, the ecologist is often starting "from scratch" and dealing with systems in which the processes are imperfectly understood and in which the roles of species yet to be described by science are unknown.

Ecosystem restoration is a very young science, with most projects taking place on a relatively small scale. However, in areas such as South and Central America, eastern Africa, the Great Plains of the United States, and vast expanses of India and China, the potential gain from recovering degraded lands is immense. As we run short of areas to cultivate new farmland, to grow new forests, or to preserve biodiversity, what could be more logical than to make better use of abandoned or under-used lands?

21.7 Crawling from the Brink of Extinction

The Dodo, Tasmanian wolf, Passenger Pigeon, Carolina Parakeet, and Javan tiger met oblivion. These and thousands of other species have recently and prematurely toppled into the abyss of extinction, driven there by human hunting and land-use changes. But just because a species starts to decline in numbers does not necessarily mean that it can no longer coexist with humans. The American bison, alligator, gray wolf, and Bald Eagle are recent success stories in conservation biology. Active measures were taken to prevent the decline of those species: Hunting was banned, pesticide use was restricted, and their habitats were protected.

The key to the success of programs to conserve or reintroduce species is the continued existence of the organism's niche. Here, niche is used in the sense of an *n*-dimensional set of habitat characteristics (Chapter 9). All the facets of the niche must exist, including the relationship with predators and competitors. The invasion of a species, say, an unpalatable Asian grass invading a U.S. prairie and replacing a palatable native grass, would change the food availability and consequently change the niche of a grazing mammal. Similarly, if an island that had no predatory mammals gains a human population who introduce rats and cats, the niche of the native fauna will be changed radically. In other words, whatever caused the initial decline in the population of the endangered species must not be only halted, but reversed. Successful conservation relies on the potential for the existence of a larger, more stable population of the species. Such a recovery will require effective protection and a commitment to the well-being of the species so that the population decline is not repeated.

The job of the conservation biologist is to establish the correct niche of the animal, including the provision of the minimum viable area for a stable population. Such action may mean reclaiming land that has been cleared and developed. Consider the political and local opposition to preventing the development of an area because it is home to an endangered species. There are many examples of the conflicts that arise when a proposed change in land use threatens to eliminate an endangered species: The urbanization of coastal scrub leads to the loss of the California Gnatcatcher (a bird); the damming of streams in Tennessee for hydroelectric projects could eliminate the Tennessee darter (a fish); or the logging of ancient forest causes the decline of the Northern Spotted Owl. Now consider the uproar if land that has *already* been developed is taken for use as a wilderness so that a minimum viable population of an endangered species can be established. Such a solution is rarely politically acceptable.

In a generally law-abiding society where the law is enforced, protective legislation of populations has proved successful. However, where the rewards for successful poaching outweigh the likely penalty, or where legal enforcement is weakened by corruption, the law alone will not be sufficient to protect a species. Legislation has failed to protect the rhinoceros populations of East Africa. Rhinoceros horn (actually not true horn but formed from matted hair) is treasured as a medicinal compound in Asian markets, and despite genuine efforts to protect the animals, poaching continues to diminish their numbers. As rhinoceros become rarer, the scarcity drives up the black-market price of the horn, increasing the profitability of poaching. One attempted solution has been for the park rangers to cut the horns off the living rhinoceros, so that there is no profit to be made from killing them. The horn is somewhat like a long fingernail in that much of it can be removed without discomfort to the animal, but the base of the horn is underlain by the sensitive "quick." Therefore when the animals are dehorned, the base of the horn is left. However, even these dehorned rhinoceros are vulnerable to poaching. The remaining stub of horn still carries a market value, or perhaps the animals are

killed before it is seen that they have been dehorned. Black rhinoceros numbers have dwindled from about 30,000 animals in the 1970s to fewer than 3000 in the 1990s. It is widely believed that extinction is unavoidable. Contrast this decline, where a combination of poverty and disregard for the law have been powerful factors, with two success stories: the American bison and the gray wolf.

Reclaiming the Plains for the American Bison

The American bison is perhaps the best-known case of rescuing an animal from impending extinction, through using a captive breeding program. The Great Plains once supported 50 to 60 million bison, but just a few decades of unprecedented human hunting pressure decimated the herds. In the years of 1871 and 1872, 8.5 million bison were shot, and many of the carcasses were left to rot. In 1874, public outcry led to the passage of protective legislation, but President Ulysses Grant vetoed the measure, claiming that the supply of bison meat and reduction in herd size were needed to pursue the "Indian Wars." Finally, in 1889, with the wild population reduced to about 150 animals and a total population of fewer than 1000, the bison received protection. Captive herds were raised and released, adding to the natural recovery of those left in the wild. As a result of their protection, the population of bison is approximately 200,000, and they are so abundant that they can be used as a food source for humans once more. The closely related European bison was not so lucky; the last wild one was shot in Bialowieza Forest, Poland, on February 9, 1921.

Reintroducing Gray Wolves

The gray wolf (Figure 21.9) is another of the conservation movement's great success stories. Once the top carnivore from Virginia to the west coast and from Alaska to Mexico, the wolf was hunted close to extinction in the lower 48 states by settlers and ranchers. In the late 19th century, bounty programs were initiated offering financial rewards for each wolf killed. As late as 1965 a hunter could still earn $50 for each wolf carcass. Even within national parks the wolf was not safe. In 1914, Congress approved funds to eradicate wolves from Yellowstone, a program that was not terminated until 1935. Under the onslaught of shooting, trapping, and poisoning, the wolf population of the lower 48 states fell as low as 600 individuals. With

Figure 21.9 **The gray wolf.** The gray wolf was hunted to extinction throughout much of its former range. Reintroduction is being attempted in Yellowstone National Park and central Idaho.

the passage of the Endangered Species Act of 1973, the gray wolf was listed as endangered, giving it a high degree of protection and forcing government authorities to formulate plans to conserve and rebuild wolf populations. One way to increase wolf populations was to reintroduce wolves to parts of their former range. The reintroductions also needed to take into account the ecology of wolves. These are pack animals that hunt over an area of between 80 km^2 and 1500 km^2 according to food availability. Not everyone wants to have a wolf living close by because they would be capable of killing pets, stock, or even people. Therefore, the wolves had to be reintroduced to areas as remote from humans as possible. Only a few areas could provide sufficient separation of people and domesticated animals from wolves. After years of planning the two sites selected for the reintroduction were Yellowstone National Park and U.S. Forest Service lands in central Idaho.

Wolves captured in Alberta and British Columbia were transported to the release points where they were kept in large enclosures until they had become familiarized with their surroundings. Between January and March of 1995, 15 young adult wolves were released in Idaho and 14 wolves comprising three family groups were released into Yellowstone. If all went well, the wolves would breed and attain stable

populations by 2002, at which time they could be removed from the endangered species list.

The first real problem for the reintroduction program came when, on February 5, 1996, one of the Yellowstone wolves was shot on a sheep ranch. This incident provided more ammunition for the local cattle and sheep farmers who were opposed to the reintroductions. A series of legal actions were undertaken that culminated in a U.S. District Court ruling in December 1997 that the wolves of Yellowstone should be captured and removed. The removal of the wolves would almost certainly result in their death, and the judge amended his ruling to state that the wolves should not be removed if it meant killing them. The courts have yet to decide on the fate of the wolves, although the court of public opinion is in favor of the reintroduction. In a 1998 poll sponsored by the National Wildlife Federation, 63% of recipients stated they were in favor of wolves being reintroduced to Yellowstone, and this number increased to an 84% approval if removal of the wolves meant killing them.

In the meantime, the wolves are thriving. The first wolf cubs were sighted in May of 1996, with at least three litters being born. The reintroduction has been a spectacular success; Yellowstone and Central Idaho currently support an estimated 163 wolves and their offspring. Elsewhere in the United States wolf populations have gained in numbers as their persecution has ended. Montana, Minnesota, Wisconsin, and Michigan all have healthy wolf populations, and there are populations, of unknown size in the Dakotas, Wyoming, and Washington. The conservation of this species has been so successful that it is 1 of just 11 species ever removed from the endangered species list as "recovered."

Breeding and Releasing Endangered Species

For a rare species, each newborn is a vital potential addition to the breeding group, but under natural conditions, the survival rate of individual offspring is low. Through the process of natural selection, most will die before reproducing. As the populations of these rare creatures continue to decline, a point may come where individuals are so scattered that males fail to find females and reproduction fails, or a single piece of misfortune—a flood, an epidemic of pneumonia, or a cyclical decline in prey populations—could wipe out the remaining individuals. Is it better then to stand aside and watch the population disappear, or should action be taken to save it?

One solution is to capture the wild animals, gathering them into a single protected group in a zoo where they will be nurtured and protected. Under these conditions the young born within the group stand a greater chance of survival, and the population should increase in size. When the population has grown, animals can once again be released into the wild. In a less extreme form, some individuals of the natural population are left in place, while others are placed in captivity, and, as they reproduce, their young are reintroduced to the wild. These kinds of zoo-based breeding programs are known as *ex situ* (out of the natural setting) conservation. Currently, programs are in place for the captive breeding of many species, including the California Condor, giant panda, and black-footed ferret (Figure 21.10).

The reproductive success of these programs can be dramatic, particularly among cats, which breed freely in captivity. However, if the initial pool of breeding animals is small, then inbreeding depression and genetic drift can adversely affect the new generations. The captive rearing of black-footed ferrets began with just 7 individuals, and although more than 200 individuals were bred from this group they

(a)

(b)

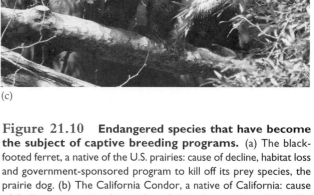

(c)

Figure 21.10   Endangered species that have become the subject of captive breeding programs. (a) The black-footed ferret, a native of the U.S. prairies: cause of decline, habitat loss and government-sponsored program to kill off its prey species, the prairie dog. (b) The California Condor, a native of California: cause of decline, persecution by ranchers, habitat loss, and possibly pollution. (c) The giant panda, a native of montane bamboo groves in China: cause of decline, habitat loss.

manifested signs of declining genetic quality such as internal bleeding, webbed feet, and kinked tails. These genetic problems can be avoided by starting with a larger number of captive individuals and by swapping individuals between zoos. In most cases, reintroducing the animals into the wild has proved harder than breeding them. The behavior of captive animals is different from that of wild ones: they may lose the ability to hunt and climb trees, or worse, they may lose their fear of humans. With careful forethought, most of these problems can be averted, but there remains the problem of reversing the process that has led to the initial population decline. Unless the goal is simply to maintain a zoo population, the crucial issue is habitat protection for the endangered species. To succeed in the wild, the zoo-raised individuals must be released into an area large enough, and of appropriate quality, to support a minimum viable population.

Saving the Cute and Cuddly

A criticism of the *ex situ* conservation movement is that it concentrates on cute, large, furry animals; ones that can be anthropomorphized or regarded as overgrown pets. Beetles, plants, phytoplankton, and fungi do not receive the same attention, yet they may be equally as important in their ecosystem. The reason for this discrimination is strictly economic. *Ex situ* conservation is hugely expensive, and of the thousands of mammal species on the planet, present zoo facilities could deal with only 200 to 300 species to be maintained at a population size of 500 individuals. The public are likely to support efforts to save a spectacular, photogenic animal such as the giant panda, but they would not respond to the plight of an endangered snail, snake, or spider with the same generosity. This reality usually leaves the 30 million non-mammal species to fend for themselves.

Ex situ programs are likely to be the only way to keep some species from going extinct, but it is likely that most of these species will never leave the protection of the zoo. If spectacular animals are to remain in the wild, or the majority of Earth's biodiversity is to be maintained, the goal must be *in situ* (within the natural environment) conservation. Nature reserves and protected areas are the best hope for maintaining wildlife.

Summary

- Biodiversity is important as a source of genetic material and new drugs, increased ecosystem stability, and for ethical and moral reasons.

- Nature reserves may be designed to protect single species, groups of species, or ecosystem functions. Each will require different management plans and strategies.

- Active management is likely to be necessary to maintain the ecosystem functions of small reserves.

- The success of management must be evaluated and practices modified to improve outcomes.

- Estimates for minimum viable population range between 500 and 5000 individuals, according to criteria.

- Size and isolation of reserves are important factors to be addressed in the design of the reserve.

- Nodes and corridors would allow the migration of species as conditions change.

- Surrounding a core area with buffer zones of complementary land use can increase the effective size of the reserve.

- Parks and reserves are only as good as the actual protection from poachers or development that they afford.

- Involving stakeholders in management decisions is critical to the success of reserve areas.

- Captive breeding programs can boost wild populations only if the niche for an MVP still exists.

- *Ex situ* conservation is expensive and can help only a few species.

Further Readings

Flather, C. H., M. S. Knowles, and I. A. Kendall. 1998. Threatened and endangered species geography. *BioScience* 48:365–376.

Leopold, A. 1949. *A Sand County Almanac and Sketches Here and There.* New York: Oxford University Press.

Mladenoff, D. J., R. G. Haight, T. A. Sickley, and A. P. Wydeven. 1997. Causes and implications of species restoration in altered ecosystems. *BioScience* 47:21–31.

Packard, S., and C. F. Mutel, eds. 1997. *The Tallgrass Restoration Handbook, for Prairies, Savannas and Woodlands.* Washington, DC: Island Press.

Rosenberg, D. K., B. R. Noon, and E. C. Meadows. 1997. Biological corridors: Form, function, and efficacy. *BioScience* 47:677–687.

Soulé, M. E., ed. 1986. *Conservation Biology: The Science of Scarcity and Diversity.* Sunderland, MA: Sinauer Associates.

Stein, B. A., and S. R. Flack. 1997. *1997 Species Report Card: The State of U.S. Plants and Animals.* Arlington, VA: The Nature Conservancy.

Wilcove, D. S., D. Rothstein, J. Dubow, A. Phillips, and E. Losos. 1998. Quantifying threats to imperiled species in the United States. *BioScience* 48:607–615.

Web Connections

On-line resources for this chapter are on the World Wide Web at:
http://www.prenhall.com/bush
(From this web-site, select Chapter 21.)

Atmosphere, Air Pollution, and Ozone

22

For the last 400 million years, the time that life has existed on land, the composition of the atmosphere has been remarkably constant. Because air has been a constant factor, there has been no evolutionary need to waste energy checking the quality of air in each breath we take. Hence, we cannot smell any of the major gases in the air. Since we have started to industrialize and burn fossil fuels, the air we breathe has been changing in quality. The major constituents remain the same, but important changes in trace gases or in new atmospheric components, such as benzene, dioxin, or CFCs (chlorinated fluorocarbons), become cause for concern. Many kinds of pollution exist as a result of chemical releases from industry, energy production, transportation, or forest clearance. In this chapter the relationship between ozone and other chemicals is investigated.

22.1 The Composition of the Atmosphere

Nitrogen (N) is the most common atmospheric gas, forming about 78% of normal air. The next most common gases are oxygen, at about 21%, and argon and carbon dioxide combined at almost 1%. Table 22.1 summarizes the important components of air.

Nitrogen is virtually inert, and it is the very stability of this gas that allows life as we know it. The balance between nitrogen and the much more volatile gas, oxygen, is a fine one. If oxygen were the primary gas in the atmosphere, Earth would be an inferno. Physicist James Lovelock has suggested that if oxygen concentrations were to rise above 25%, even damp organic material would burn freely. Fires would spring

Table 22.1 The ten most common components of dry clean air.

Gas	Chemical notation	Volume in air (%)
Nitrogen	N_2	78.08
Oxygen	O_2	20.94
Argon	Ar	0.934
Carbon dioxide	CO_2	0.035
Neon	Ne	0.00182
Helium	He	0.00052
Methane	CH_4	0.00015
Krypton	Kr	0.00011
Hydrogen	H_2	0.00005
Nitrous oxide	N_2O	0.00005

(*Source:* Data are from W.H. Schlesinger, 1991.)

up with the slightest provocation and would burn until all the potential fuel was exhausted. By way of contrast, if oxygen levels were to fall below 15%, even the driest material would not burn. If these assertions are true, either an enriched or a depleted oxygen concentration in the atmosphere would profoundly change evolutionary pressures.

An increase in fire frequency would lead to much higher rates of habitat disturbance, and fire resistance would become a principal evolutionary determinant of survival. Where habitats burn frequently under present conditions, physiological adaptations to fire resistance are evident; for example, trees grow a thick corky bark to insulate the precious conducting vessels of the trunk. However, in rain forests and other areas that do not normally burn, an increase in fire frequency would have a considerable effect on species abundances.

Conversely, lowered oxygen concentrations would greatly affect the metabolism of many animals, particularly endothermic vertebrates and others that have high oxygen demands. Evolution could undoubtedly compensate for lower oxygen availability, but some significant changes in physiology would be needed.

At present, our atmospheric concentrations of gases are fortuitously regulated by various chemical cycles. New oxygen release from photosynthesis is balanced by the action of reducing minerals produced by volcanic activity or exposed by erosion, by sulfur bacteria that metabolize oxygen, and by respiration (Chapter 3). Robert Berner and Donald Canfield of Yale University have produced computer simulations of past atmospheric environments. Their simulations indicate that this regulation has not always been maintained. They suggest that during the Carboniferous period, about 300 million years ago, oxygen concentrations may have been substantially higher than modern values, perhaps between 25% and 35% of the atmosphere (Figure 22.1). The computer model then "speculates" that oxygen fell to modern levels, or a little lower, around 250 million years ago, increased from 22% to 27% during the Cretaceous age (130–70 million years ago), and then started a gradual decline to modern levels. Berner investigated the chemical composition of air bubbles trapped in fossil amber of the Cretaceous age and found that atmospheric oxygen concentrations may have been as high as 30% at that time. Many ecologists question these data, citing the possibility that wildfires would have jeopardized the survival of terrestrial life. These doubters suggest that some chemical change may have acted to increase oxygen concentrations in the gas bubbles over the last 100 million years and that the computer simulations, as acknowledged by Berner and Canfield, are only first-order approximations. This is an exciting debate, because it challenges assumptions about the constancy of our atmosphere and clearly demonstrates the need for collaboration between atmospheric chemists and evolutionary biologists. Further research is needed for a resolution. In the meantime, the results suggesting elevated oxygen concentrations in the past must be viewed as being highly conjectural.

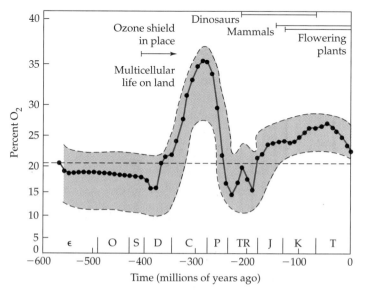

Figure 22.1 **Best estimates of changes in oxygen concentrations in the last 300 million years.** The concentrations of oxygen shown should be treated as only a rough guide; the real concentrations could have lain anywhere within the shaded error zone. These are the results of a computer model and are an interesting but speculative scenario that is disputed by some scientists. (After R.A. Berner and D.E. Canfield, *American Journal of Science*. Reprinted with the permission of *American Journal of Science*.)

22.2   Layers in the Atmosphere

The atmosphere is thinnest—that is, the concentration of gas molecules is lowest—at the transition from "outer space" to the **thermosphere** (Figure 22.2). The thermosphere contains ionizing gases that are subject to high levels of solar energy; however, the molecules are so widely spaced that atmospheric pressure is close to zero. Light energy is converted to heat only when it is stored, and storage requires *something* to absorb it. In the thermosphere, an astronaut taking a space walk in a pressurized suit would heat up rapidly, yet the surrounding space might feel cold because there are so few molecules storing energy.

The next lower layer of the atmosphere is the **mesosphere**, another zone of very low atmospheric pressure. At about 50 km above the surface of Earth, there is a rapid increase in atmospheric temperature, marking the transition from the mesosphere to the **stratosphere** (Figure 22.2). This region of the atmosphere contains some water vapor and high concentrations of ozone. Ozone concentrations in the stratosphere are about 1000 times greater than in any other atmospheric layer,

and this region of the atmosphere is referred to as the **ozone shield**. Because ozone molecules can trap heat effectively, the stratosphere is relatively warm. Beneath the stratosphere lies the lowest level of the atmosphere, the **troposphere**. In some areas, at the boundary of the stratosphere and the troposphere, high winds flow from west to east. These are the **jet streams**. The jet streams follow the discontinuity between the stratospheric and tropospheric air masses and play a major role in determining the location of polar air masses and therefore of weather systems on Earth (Chapter 4).

The troposphere makes contact with the surface of Earth and has the greatest concentrations of most gases, the highest pressures, and hence the highest sensible heat. It contains about 80% of all the atmospheric chemicals (measured by mass) and is the section of the atmosphere that life inhabits. Most of the weather that determines storms, droughts, floods, and consequently the success or failure of harvests, is generated within the troposphere. This is not only the layer that affects us most directly, but also the layer most changed by human actions such as pollution and deforestation.

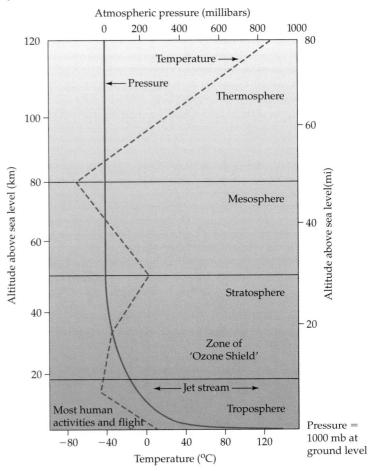

Figure 22.2   The atmosphere has layers that are defined by their relative pressure and temperature. Space is cold, not because it lacks solar energy, but because there is nothing to absorb light and store it as heat. Temperatures increase toward the planet surface because concentrations of atmospheric gases increase. The stratosphere is relatively warm because, within this layer, 99% of incoming ultraviolet light is absorbed by ozone (O_3).

22.3   Air Pollution

When calculating the risk to human health caused by air pollution, both the concentration of the pollutant (more concentrated means greater health risk) and the level of exposure must be considered. Most Americans spend the majority of their days and nights indoors. Therefore, indoor air pollution is potentially the most serious health threat, because exposure to these pollutants is greatest.

By far the most serious source of indoor pollution comes from cigarette smoke, followed by exposure to asbestos dust (which can cause lung disease), and solvents such as those found in new carpets or plywoods (which can cause asthma and other respiratory ailments). Another potentially dangerous indoor pollutant is radon-222. Radon-222 is a radioactive gas released from decaying uranium-238. Uranium is most frequently found in volcanic rock types (particularly granites) and in some sedimentary deposits rich in phosphates. Consequently, areas with severe radon problems are found on the volcanic rocks that form the Appalachians and New England and in phosphate-rich sediments in Florida.

Radon is constantly entering the atmosphere and does so in such small concentrations that it poses no threat to us. However, in a mine shaft, or in a house with a poorly ventilated basement, it is possible for radon to leak out of the ground and to be trapped. If radon concentrations become high, and if the space is used on a regular basis, this exposure to radon gas can pose a significant cancer threat.

Most of the studies on radon have been conducted on coal miners who have experienced high concentrations and long exposures to the gas. In these populations, exposure to radon is found to increase significantly the probability of developing lung cancer. In the general population, the risk from radon is less, but the U.S. Environmental Protection Agency (EPA) suggests that between 8000 and 20,000 Americans die from radon-induced cancer each year. Despite this warning and extensive publicity campaigns to have homes and workplaces tested for radon, the public remains largely unconcerned about it.

Of the annual deaths predicted by the EPA, only about 200 would be nonsmokers. Apparently, radon is most deadly when combined with smoking. Part of the casual attitude may be because it is smokers who are most at risk. Those who smoke presumably accept the risk of getting lung cancer, and when they contract it, smoking rather than radon is blamed.

In most homes, radon can be reduced to harmless levels simply by ventilating the basement. Although indoor pollution is a serious human health issue, the various forms of outdoor air pollution are likely to have greater effects on the natural world.

22.4   Our Love–Hate Relationship with Ozone

Although holes in the ozone layer have received a lot of media attention, and it has become the general perception that ozone is beneficial, this is not wholly true. On the one hand, ozone is absolutely essential to the continued existence of life; on the other, it is a highly destructive gas that damages plants, increases the rate of any oxidative process, such as the rusting of metal, and is a major contributor to photochemical smog. The key to this paradox of our love-hate relationship with ozone lies in where the ozone is found in the atmosphere. The ozone that we need is **stratospheric ozone**, but **tropospheric ozone** is a problem gas.

What Is Ozone?

Ozone (O_3) is a molecule that contains three atoms of oxygen (Figure 22.3), rather than two, as in the oxygen molecule (O_2). Two oxygen atoms can bind together strongly by what are called *covalent bonds*. However, if a third oxygen atom is present in the molecule, the covalent bonds cannot form; thus, the atoms are held in their configuration by electrical attraction. This electrovalent bond is much weaker than the covalent bond; hence, the molecule is always susceptible to having one of its oxygen atoms stripped by an "oxygen-hungry" molecule. The reactivity of a chemical is in large part determined by the strength of the bonds between the atoms. The weak bonds of the ozone molecule make it an even more highly reactive gas than oxygen. Because of its reactivity, ozone can change the body chemistry of plants or animals that it enters and is consequently a highly toxic substance.

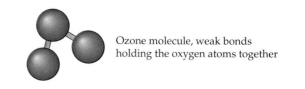

Ozone molecule, weak bonds holding the oxygen atoms together

Figure 22.3   The ozone molecule contains three atoms of oxygen that are weakly bonded to each other.

22.5 Tropospheric Ozone: The Hate Relationship

The fogs for which London was famous were the product of cold, damp air mixing with smoke from coal fires. The result was a yellow-green opacity that hung in the streets so that, even during the day, visibility was only an arm's length. Known by Londoners as "pea-soupers" because of their color and density, these fogs were better described as **industrial smog**. The Industrial Revolution of the late 1800s was built on coal and the cheap energy it provided in the form of steam power or electricity, and coal was burned for heat in homes. Coal is generally a dirty energy source, producing a lot of smoke and acid chemicals such as sulfur dioxide (SO_2). Much of the smoke was released close to ground level, where it mixed with humid air to form smog. The sulfur gave the smoky clouds a greenish yellow tint, creating the "pea-soup" effect. Thus, it was the smoke particles and emissions from thousands of domestic hearths and large coal-burning power stations that turned a fog into a smog.

The last of the great London smogs occurred in combination with a serious outbreak of influenza. Thousands of people suffered in an epidemic of bronchitis and pneumonia, conditions especially dangerous to the elderly or those in poor housing. In the wake of the smogs of the 1950s, clean air regulations were passed in London, forbidding the burning of coal in domestic fireplaces. Such regulations have made industrial smog pollution a matter of history in many Western cities. In China and parts of eastern Europe where there is still a heavy reliance on coal, industrial smog remains a significant environmental health hazard.

In most countries, **photochemical smog** has largely replaced industrial smog as an urban pollutant. In the cities of the developed world, as reliance on coal has declined, combustion of oil has increased. Gasoline and oil both produce nitrogen oxides, NO_x (NO and NO_2), as they are burned. As a result of tailpipe emissions, cities with dense traffic are likely to accumulate high concentrations of NO_x in the lower 100 m to 200 m of the atmosphere. It is the reaction of these gases with other chemical pollutants and sunlight that forms photochemical smog. The chemical process that leads from NO_x to the production of ground-level ozone is powered by ultraviolet (UV) radiation.

Ultraviolet energy from sunlight pries oxygen atoms away from oxygen-rich molecules. This process is called *dissociation* and describes the splitting of an oxygen molecule into two single atoms of oxygen. The UV energy can also split NO_2 into NO and a free oxygen atom. The single oxygen atoms are highly reactive and are likely to combine with an oxygen molecule (O_2) to form a molecule of ozone (O_3). The mixture of ozone, NO_x, dust and particles of carbon, and organic chemicals (often the products of partial combustion of gasoline or diesel fuel) make up photochemical smog (Figure 22.4). This smog is visible from the vantage point of a landing aircraft as a reddish brown haze hanging above many cities: The characteristic color comes from its nitrogen dioxide (NO_2) content. Nitrogen dioxide also combines with water vapor to produce nitric acid, making the smog acidic. The smog irritates people's eyes and respiratory tracts and is particularly dangerous to those who suffer from asthma or respiratory disorders.

Some cities have worse smog problems than others; the difference is partly a function of geography. Cities with dense populations and heavy traffic will produce large quantities of these pollutant chemicals. A city lying within a bowl of hills will have less wind to flush pollutants away, and in the still air of summer the photochemical smog can become an acute problem. Rain tends to clean the air of some NO_x, so dry climates are more prone to smog than are wet ones. Also, high temperatures promote the chemical reactions that lead to smog formation. Thus, a large city such as Los Angeles, Denver, Bangkok, or Mexico City, built in a valley with a hot, dry or seasonal climate will be especially smog prone. In these cities, the ozone and acids of photochemical smog become a health hazard to humans. The ozone also results in increased rates of oxidation (absorption of oxygen molecules by a surface), weathering buildings and promoting the decay of anything that will rust.

Temperature inversions can promote the formation of photochemical smog because they form a pool of still air close to the ground. Under normal conditions, cities generate warm air. Fumes from exhaust or from chimneys are often warm relative to the natural, or ambient, air. The principal surfaces of our cities—concrete, pavement, metal, and asphalt—are effective radiators of heat energy. Thus, it is normal to find that the air temperature of a city is a few degrees warmer than that of the surrounding countryside.

Warmed city air rises and carries with it many of the pollutants created in the city. However, under certain conditions, particularly in dry valleys such as Los Angeles or within the bowl of hills that flank Denver, the rising of the polluted air may be prevented by a cap of warmer air. Such a cap can form as a result of the arrival of an air mass that is warmer than the air in the city, such as warm air blowing in from the ocean.

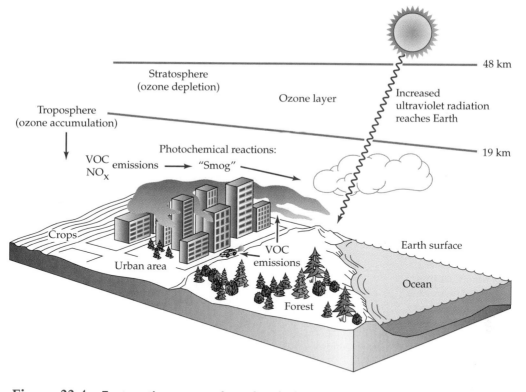

Figure 22.4   Factors that create photochemical smog. NO_x and volatile organic compounds (VOCs) react in the presence of ultraviolet light to form ozone and smog.

Alternatively, the city air can be chilled by cold, dense air descending into the valley and pooling in the valley bottom. When the cold air is overlain by warmer air, a **temperature inversion** has formed (Figure 22.5). Often the cool air may be held for several hours, in which time pollution becomes more and more noticeable within the city. Occasionally, the inversion may last for several days, during which time local pollutants are building up and may reach levels dangerous to human health.

It is not just cities that suffer from excessive tropospheric ozone. Unusually high levels of ozone have been documented above the forests of the Amazon basin. Here, traffic cannot be to blame. The trees produce hydrocarbons that are highly reactive and can, in the presence of UV energy, dissociate NO_2 to release an oxygen atom that is then free to form an ozone molecule. Alternatively, carbon monoxide (CO) can be a part of a fairly complex chemical pathway that leads, through the splitting of NO_2, to ozone production. The burning that accompanies deforestation releases large quantities of carbon monoxide into the atmosphere, so in Amazonia there is both a natural and a human-induced element to the generation of high ozone concentrations.

Clean air close to the ground generally has an ozone concentration of about 20 ppb (parts per billion). Some forested areas of the United States look and smell clean, but they are being affected by increasing levels of ozone. These areas are prone to high ozone concentrations because of their natural production of hydrocarbons. But for high concentrations of ozone to form, there also has to be a high concentration of NO_2. This NO_2, which is broken down by UV energy, is being supplied by fossil fuel combustion in neighboring metropolitan areas. Ozone concentrations in excess of 80 ppb are now commonly recorded in the forests of the eastern United States (Figure 22.6).

The importance of increasing ozone concentrations even in areas with sparse human populations is made clear through the experimental growth of crops at different ozone levels. The experimental data show that plant growth is inhibited as ozone concentrations increase (Figure 22.7). The "normal" growth of crops is determined by measuring the weight of a typical plant after a set time at ozone concentrations of 20 ppb. The weight of this plant can then be compared with the weight of plants grown at elevated concentrations of ozone. It can be seen that, as atmospheric concentrations

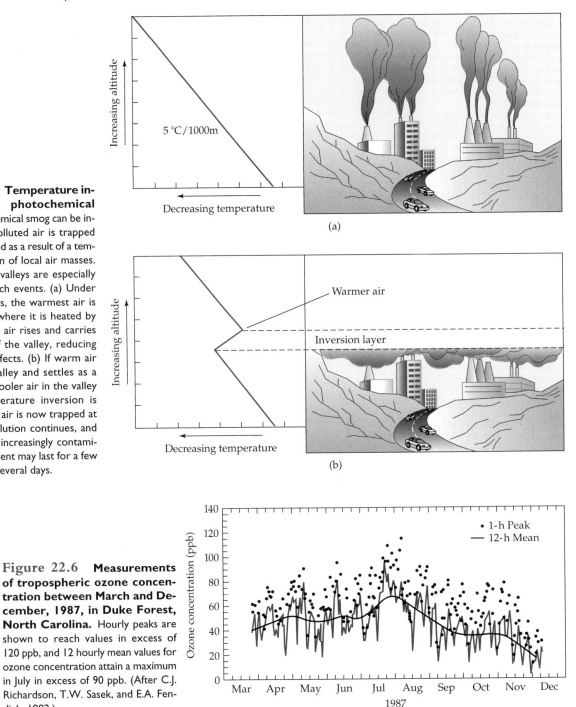

Figure 22.5 Temperature inversions and photochemical smog. Photochemical smog can be intensified when polluted air is trapped close to the ground as a result of a temperature inversion of local air masses. Cities located in valleys are especially susceptible to such events. (a) Under normal conditions, the warmest air is at ground level, where it is heated by Earth. The warm air rises and carries pollutants out of the valley, reducing local pollution effects. (b) If warm air flows into the valley and settles as a layer on top of cooler air in the valley bottom, a temperature inversion is formed. Polluted air is now trapped at ground level, pollution continues, and the air becomes increasingly contaminated. Such an event may last for a few hours, even for several days.

Figure 22.6 Measurements of tropospheric ozone concentration between March and December, 1987, in Duke Forest, North Carolina. Hourly peaks are shown to reach values in excess of 120 ppb, and 12 hourly mean values for ozone concentration attain a maximum in July in excess of 90 ppb. (After C.J. Richardson, T.W. Sasek, and E.A. Fendick, 1992.)

of ozone rose above 80 ppb, all of the crops in the experiment started to show decreased yields. The poor growth of the plants reflects the absorbance of ozone by the plant through the stomata. Inside the plant, the ozone then reacts with, and degrades, chlorophyll. The loss of chlorophyll reduces the photosynthetic efficiency, hence lowering the productivity of the plant.

The consequence on a larger scale is that in ozone-rich areas, crop and timber yields will fall. Given that 80 ppm of tropospheric ozone in the eastern United States is routinely documented, the point at which our tropospheric ozone pollution could have a significant detrimental effect on wildlife and farms is being approached.

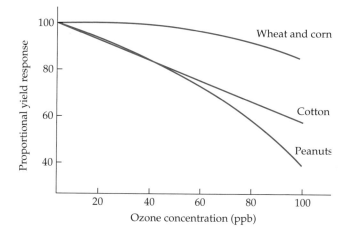

Figure 22.7   The effect of increasing atmospheric ozone concentrations on yields of crops raised in growth chambers. As ozone concentrations increase above 80 ppb, all the crops tested are seen to provide poorer yields. (After W.W. Heck and A.S. Heagle, 1970.)

22.6   Stratospheric Ozone: The Love Relationship

It was not the presence of oxygen in the atmosphere that allowed life to leave the oceans and become terrestrial but the presence of stratospheric ozone. Many organisms can live without oxygen, so the biological significance of developing an oxygen-rich atmosphere lay not in the oxygen itself but in a secondary chemical reaction that could take place only in the presence of oxygen. High in the atmosphere, oxygen molecules were transformed by ultraviolet light from their normal configuration of O_2 to ozone (O_3). The oxygen molecules high in the stratosphere are broken down through dissociation, resulting in the absorption of UV light in the wavelengths 180–240 nm.

$$O_2 + \text{ultraviolet light} \rightarrow O + O$$
$$(180\text{–}240 \text{ nm})$$

The free oxygen then joins an O_2 molecule to form ozone.

$$O + O_2 \rightarrow O_3$$

The ozone layer formed by this process is not uniformly distributed around the globe; it is thickest in the stratosphere above the poles and thinnest at the equator. If ozone were forming constantly, with none being lost, there would be a continual accumulation of this gas in the stratosphere. However, since atmospheric concentrations stabilized at about 21% oxygen, it is likely that the concentration of ozone has remained constant. Therefore, there must be a natural

process of ozone degradation to maintain the balance. A little ozone leaks downward into the troposphere, but much more is broken back down to oxygen as a result of another photochemical reaction. In this reaction, UV radiation in the spectrum 200–320 nm (a slightly longer wavelength of UV than is absorbed in the creation of ozone) provides the energy.

$$O_3 + \text{ultraviolet light} \rightarrow O_2 + O$$
$$(200\text{–}320 \text{ nm})$$

The single oxygen atom can then pull an oxygen free from an ozone molecule to form two oxygen molecules.

$$O + O_3 \rightarrow O_2 + O_2$$

The balancing reactions of ozone production and breakdown absorb incoming UV radiation. Ultraviolet radiation arriving in the atmosphere may be divided into three categories, according to wavelength. UV-A has the longest wavelength and UV-C the shortest. Ozone absorbs more than 99% of the UV-C wavelengths, about half of the lower-energy UV-B, and little of the lowest-energy (relatively harmless) UV-A wavelengths. If the equilibrium of the reactions that absorb UV radiation is perturbed, a change in the influx of ultraviolet radiation to the surface of Earth is experienced.

The Importance of the Ozone Shield

Life was able to evolve in the oceans before there was an ozone shield because water reflects ultraviolet light. However, if any life form had tried to colonize the land, exposure to UV light would have played havoc with its DNA. DNA absorbs UV light, especially between wavelengths of 280 and 320 nm, and UV radiation disrupts DNA replication to the point of reproductive failure or death. Consequently, an ozone shield was a prerequisite to the colonization of land. Once the ozone shield was in place 400 million years ago, life moved out of the ocean and onto the land.

Relatively small increases in UV radiation can cause mutations during the replication process that can result in the production of a cancerous cell. Because the cells most exposed to UV radiation are the skin cells, one of the consequences of prolonged exposure to UV wavelengths is skin cancer. Elevated levels of UV radiation are also associated with increased incidence of eye cataracts, a clouding of the iris that leads to blindness.

It is not just humans that are affected by increased levels of radiation. These conditions can induce eye cancer in other mammals, and high dosages of UV radiation, as experienced on some mountaintops, can lead to reproductive failure in birds and lizards.

Perhaps of most concern is the damage caused by UV light to photosynthetic systems, for this damage has the potential to reduce the primary productivity, hence threaten the entire food web, of an ecosystem.

Stratospheric Ozone Depletion

In the 1930s, Du Pont Industries developed a new supercoolant that was nonflammable, noncorrosive, relatively inexpensive, and apparently chemically stable. The chemical, a chlorofluorocarbon, was patented by Du Pont and given the trade name of Freon. Freon was an almost ideal chemical, or so it seemed, for use as the coolant in air conditioners and refrigerators. Chlorofluorocarbons (CFCs) were also widely used as the dispersant gas in pressurized spray cans of paint, insecticide, and hair spray. A further use for these chemicals was to provide the bubbles in polystyrene plastic foam, better known in the United States by the Dow Chemical trade name Styrofoam.

In 1972, research teams investigating the effect of the space shuttle's exhaust gases concluded that chlorine released as hydrogen chloride (HCl) by the shuttle as it swept through the stratosphere could be a damaging pollutant. The research demonstrated that chlorine could destroy ozone. In 1974, the first warnings of previously unforeseen chemical reactions involving CFCs were sounded by Sherwood Roland and Mario Molina. These chemists, working at the University of California at Irvine, raised the hypothesis that CFCs, chemicals containing chlorine, would rise into the stratosphere, where they would degrade ozone. After an initial flurry of interest, little attention was paid to this problem until, in 1985, the British Antarctic Atmosphere Survey demonstrated a strong depletion of the Antarctic ozone shield.

The South Pole receives the least radiant energy of any place on Earth, and after the long night of the Southern Hemisphere winter, the water in the stratosphere above Antarctica exists as tiny ice crystals. CFCs and ozone molecules are held on the surface of the ice crystals. With the first warming of spring (September and October) in Antarctica, the ice clouds in the stratosphere return to water vapor and the CFCs and ozone are freed. A stock of CFCs has been accumulated during the winter months, and their liberation leads to the sudden degradation of the ozone. The chemical process involved is simple. The CFC started as a very stable chemical configuration, but in the presence of ultraviolet light the bonding of the molecule is altered to leave one of the chlorine atoms only weakly attached. As the modified CFC encounters an ozone molecule (which is also weakly bonded), a chlorine

atom breaks from the CFC and strips an oxygen atom from the ozone molecule (Figure 22.8).

$$Cl^- + O_3 \rightarrow ClO + O_2$$

The chlorine monoxide that results from this reaction is still a highly reactive chemical, and as it encounters another ozone molecule it will strip one of the oxygen atoms.

$$ClO + O_3 \rightarrow ClO_2 + O_2$$

The chlorine dioxide is then broken down by ultraviolet light to leave a free chlorine atom and an oxygen molecule.

$$ClO_2 + \text{ultraviolet light} \rightarrow Cl^- + O_2$$

Thus, through this cycle, the single chlorine atom released by the CFC has degraded two ozone molecules and is still free to go through yet another cycle. It is

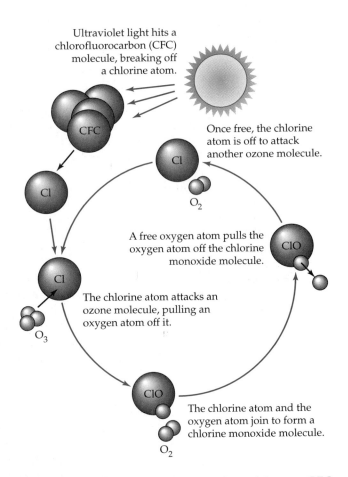

Figure 22.8 A chlorine atom released from a CFC (chlorofluorocarbon) strips an oxygen atom from an ozone molecule. Each chlorine atom may repeat this cycle as many as 50,000 times, breaking down 100,000 ozone molecules.

Labels in figure:
Ultraviolet light hits a chlorofluorocarbon (CFC) molecule, breaking off a chlorine atom.

Once free, the chlorine atom is off to attack another ozone molecule.

A free oxygen atom pulls the oxygen atom off the chlorine monoxide molecule.

The chlorine atom attacks an ozone molecule, pulling an oxygen atom off it.

The chlorine atom and the oxygen atom join to form a chlorine monoxide molecule.

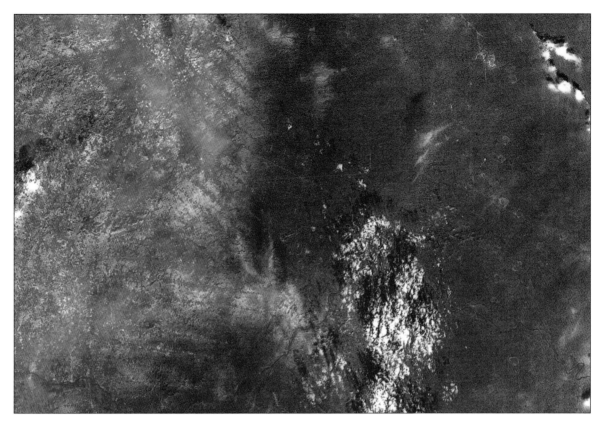

Plate 3

A false color satellite image showing the destruction of rainforest in Brazilian Amazonia. The red is undisturbed forest while the blue is disturbed areas. Note how the ladderwork of roads (straight blue lines) penetrates the forest. Settlers move in along the new roads producing a frontier of destruction (pink) that is advancing from left to right.

Plate 4

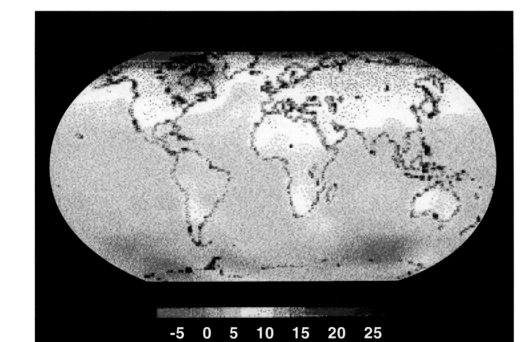

Temperature changes with a doubled concentration of atmospheric CO_2. The predicted temperature change that will accompany a doubling of atmospheric carbon dioxide concentrations according to the Geophysical Fluid Dynamics Laboratory model (this is a fairly typical result for other models as well). Red indicates a warming of as much as 2.5°C, while green is unchanged and blue is a cooling of 1-2°C. Note that it is the highest latitudes that are predicted to change the most and that land masses will experience greater temperature change than oceans.

(Data from the Geophysical Fluid Dynamics Laboratory.)

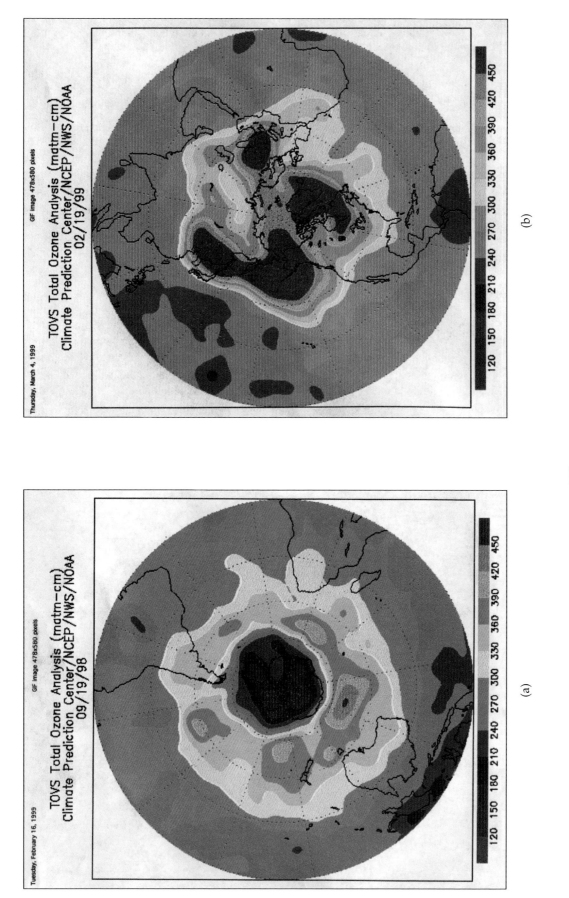

Plate 5

False color satellite data showing the seasonal depletion of stratospheric ozone. Ozone concentrations are in Dobson Units (DU) over (a) the Antarctic in September 1998, and (b) the Arctic in February 1999. Ozone concentrations lower than 220 DU represent a "hole" in the ozone shield. *(Data are from the National Oceanographic and Atmospheric Administration.)*

estimated that a single chlorine atom could pass through 50,000 such cycles, resulting in the breakdown of 100,000 ozone molecules. The numbers are mind-boggling, especially when it is considered that a single cup of expanded-bead polystyrene contains more than 1 billion molecules of CFCs.

Other chemicals are also implicated in ozone depletion. Hydrochlorofluorocarbons (HCFCs), which have been introduced in place of CFCs, also deplete ozone, but at about 2% to 10% the rate of CFCs. Other chlorine-based chemicals such as methyl chloride can also lose a chlorine atom and degrade ozone. The closely related chemical, bromine, is another ozone-degrader and may account for as much as 5–10% of ozone depletion.

Although these chlorine- and bromine-based compounds are generally coolants and solvents, agricultural chemicals also play a role in ozone depletion. When nitrogen-based fertilizers such as nitrate and ammonium are applied to the land, not all the nitrogen is absorbed by the plants. More than half of the nitrogen is released to the air as nitrous oxide (N_2O) as microbes break the fertilizer down to simpler chemicals (nitrification and denitrification; Chapter 6). The nitrous oxide rises into the stratosphere where it degrades ozone. Although in urban areas, the majority of N_2O emissions are from tailpipes, globally, agricultural emissions are about 5 times those of tailpipes. If nitrous oxides were to double in atmospheric concentration, it is estimated that they would account for a 10% increase in ozone depletion and a 20% increase in UV radiation penetrating to Earth's surface.

Ozone Holes over the Antarctic

The hole in the Antarctic ozone shield, though first described in 1985, dates back at least to the 1960s. For more than two decades the seasonal depletion of ozone had been documented, but because ozone concentrations were thought to be constant, computer programs were written to ignore any measured fluctuations. Satellite data from the mid-1980s revealed that each spring a vast area above Antarctica lost about half of its ozone. Ozone concentrations are measured in Dobson Units, and values of 400 DU to 600 DU are considered normal. The area of the ozone "hole" is taken to be areas with less than 220 DU. Since the 1980s, the Antarctic ozone hole has grown, and each year in the mid- to late 1990s the maximum extent of the depletion area has exceeded 20 million square kilometers—about three times the area of the continental United States.

The lowest measured concentration of stratospheric ozone was recorded in the Antarctic spring of 1993.

On October 12, 1993, the ozone shield was thinned by about 70% compared with normal concentrations. The concentration of ozone at the South Pole fell to 96 DU, while ozone concentrations in the midlatitudes were about 350–400 DU. This event is thought to have been atypically severe and enhanced by the presence of chlorine gases and sulfuric acid droplets emitted during the eruption of Mount Pinatubo in 1991. The sulfuric acid does not degrade the ozone, but the droplets may provide the reaction surfaces on which chlorines can break down ozone both faster and at lower temperatures than on a water droplet. The chemicals released by the Mount Pinatubo eruption have been washed out of the atmosphere, and by 1994 the thinning of the ozone layer returned to its 1991 pre-eruption level (see Color Plate 5). However, this thinning still represents a 40% weakening of the ozone shield compared with its natural state.

The depletion is also changing the vertical structure of Earth's ozone shield. Prior to the depletion of the ozone layer, the highest concentrations of ozone occurred about 16 km above Earth's surface (Figure 22.9). The thinning of the ozone layer is not uniform at all elevations in the atmosphere and is most pronounced where the polar vortex concentrates ice crystals. Not only has the depletion become more severe during the 1990s, but it has spread vertically. In 1986 maximum depletion occurred between 14 km and 18 km elevation (Figure 22.9), whereas in 1997, the region of peak depletion extended from 14 km to 20 km elevation.

To determine the magnitude of an ozone depletion event it is necessary to factor in both the geographic area of the depletion and the number of days that the depletion lasted. Figure 22.10 documents an *integrated area* for the ozone hole. The integrated area is the sum of daily measurements of the area of the ozone hole. This method allows both the area and the duration of the event to be incorporated into a single statistic. The steady growth of the ozone hole in the 1980s and 1990s is consistent with our understanding of the action of ozone destroying chemicals in the atmosphere.

The Annual Break Up

In December, changes in polar wind patterns cause the Antarctic ozone hole to break up. The break up is in some ways the most dangerous time for humans. Because Antarctica is largely uninhabited, there is little human exposure to elevated ultraviolet radiation so long as the ozone hole lies above the South Pole. However, in late November and December the ozone hole breaks up, sending discrete parcels of ozone-depleted air away from the pole. As these stratospheric

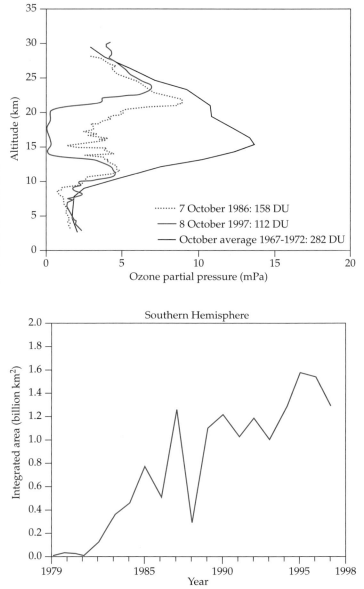

Figure 22.9   Changes in the vertical abundance of ozone over the Antarctic. The 1967–1971 ozone concentrations are taken to be the natural state. A progressive decline in ozone concentrations (expressed as partial pressure) is evident in the 1986 and 1997 data sets, and the zone of depletion is seen to be expanding vertically.

.... 7 October 1986: 158 DU

——— 8 October 1997: 112 DU

——— October average 1967-1972: 282 DU

Figure 22.10   The seasonally integrated ozone hole area for Antarctica 1979–1997. Integrated values are based on the sum of daily measurements of the area with less than 220 DU of ozone for the duration of the depletion event.

air masses drift across continents, they offer windows for the penetration of ultraviolet light. During the early Australian summer, the UV radiation levels can rise as much as 20%, and this increase may be responsible for the high incidence of skin cancer in Australians. One estimate suggests that 7 million Australians (in a nation of only 18 million people) will develop skin cancer as a result of this exposure.

Ozone Thinning over the Arctic

In 1988, a similar, though smaller, thinning of the ozone layer was discovered in the Arctic during April and May. A 9%–20% ozone depletion became evident and since then has been observed each spring. On February 10, 1993, the lowest ozone concentrations yet observed, 230–240 DU, were documented over the North Pole. These low values may be related to exceptionally low stratospheric temperatures in 1992 and 1993. However, the low temperatures should not be assumed to be related to human pollution; they are more likely a vestigial effect of the eruption of Mount Pinatubo in 1991. Thus, maximum and minimum ozone concentrations are not especially important, but long trends of declining ozone are indeed worrisome. It is unlikely that the ozone depletion in the Northern Hemisphere will be as severe as that over the Antarctic. The configuration of the southern continents and the

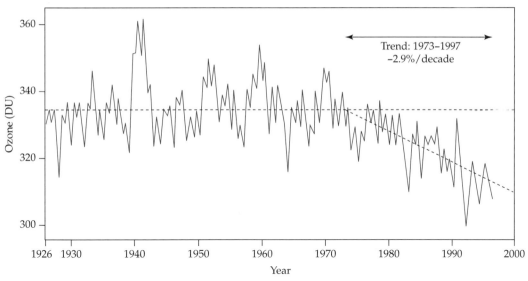

Ozone at Arosa, Switzerland, since 1926

Trend: 1973–1997
−2.9%/decade

Figure 22.11 **The concentration of stratospheric ozone over Arosa, Switzerland, 1926–1993.** The rhythmic rise and fall of ozone concentrations probably reflects variability in the output of solar energy. Between 1926 and 1973, despite the rhythmic fluctuations, the overall ozone concentrations remained constant. Since 1973, there has been a new trend, a decline in ozone concentrations of 2.9% per decade.

strong polar vortex, the deep depression that traps air over Antarctica each winter, makes this area more vulnerable to ozone depletion than the Arctic. Even so, as the Northern Hemispheric polar vortex weakens in the late spring, bubbles of stratospheric air with reduced ozone concentrations drift away across parts of northern Europe and North America. To date, it is northwestern Europe that has experienced the highest UV inputs due to this ozone depletion. The local media in parts of Britain, the Netherlands, and Germany now include a daily "UV count" and advise fair-skinned people to stay out of the sun on days of exceptionally high UV radiation. A long-term data set monitoring ozone concentrations in the Swiss Alps reveals a significant downward trend in ozone concentrations since 1973 (Figure 22.11).

Also evident in the Swiss data is the natural variation in ozone concentrations through time. The cause of this variation is not completely understood, but may relate to fluctuations in the sun's output of energy, which would affect the rate of stratospheric ozone formation. Nevertheless, it is evident that the decline in ozone concentrations is a separate phenomenon.

Globally, it appears that about 4% to 5% of Earth's ozone has been degraded since the early 1980s. If the trend of ozone loss continues unabated and is due to our pollution of the atmosphere, it is predicted that the ozone shield will be depleted by a further 10% by the year 2050. The consequences of such a depletion

are unknown, but a panel of international scientists convening in 1991 estimated that a 10% ozone depletion would result in an additional 300 million cases of skin cancer and 1.6 million cataracts in the last decade of the century.

22.7 Protecting the Ozone Layer

News that the ozone layer was damaged caught public attention, and the international response to limit ozone depletion can be seen as a success story. Curtailing emissions of CFCs and other chemicals that release chlorine or its chemical neighbor, bromine, took considerable retooling in industry, but substantial progress in this direction has been made.

Before any political mandates were laid down, industry had realized the good publicity that could be gained from "ozone-friendly" products. The general public started to purchase spray cans that used non-CFC propellants, and by the late 1980s most of these products were labeled as CFC-free. This was a case of consumer demand providing industry with an incentive to change.

In 1987, the Montreal Protocol was established, an agreement that established goals to reduce the emissions of ozone-depleting chemicals. The agreement has been progressively strengthened at subsequent meetings, both in the number of signatories, which has risen from 24 to 92 nations, and in the schedule of emissions reduction. The 1992 phase of this agreement led to signatory nations phasing out CFC-11 and CFC-12, the worst offenders among the chemicals, by 1996. Whether a global ban on CFCs can be realized will depend on policies adopted by developing nations with burgeoning populations and huge potential demands for refrigeration and air conditioning. At present these countries, such as China and India, still use CFCs, and the potential exists for a significant increase in CFC production as a larger proportion of their populations can afford refrigerators. The onus will fall on the developed nations to encourage the use of more modern technology so that the work of the Montreal Protocol is not undone.

Although adherence to the Montreal Protocol has greatly reduced CFC emissions, the chemicals used in place of CFCs are not as benign as once thought. As CFCs were phased out they were replaced by "ozone-friendly" HCFCs. But, as noted above, HCFCs still degrade the ozone layer. HCFCs will be phased out by 2030. HCFCs are gradually being replaced by hydrofluorocarbons (HFCs), which are believed to be truly ozone friendly. However, the chemical structure of HFCs, like CFCs and HCFCs, trap heat and will be potent greenhouse gases (Chapter 23).

Other chlorine- and bromine-based chemicals are also under scrutiny in the United States. Methyl chloride, a solvent, was phased out in 1996, and methyl bromide, an agricultural pesticide, will be phased out in 2010 in MDCs, and the amount used in LDCs will be frozen at 1995–1998 levels by 2002. Within the United States, methyl bromide will no longer be used after 2001.

Controlling nitrous oxides from automobile exhaust emissions has been addressed by the Clean Air Act of 1990 (Chapters 23 and 28), however, agricultural emissions of N_2O are likely to increase as agriculture becomes increasingly reliant on fertilizer inputs. Agricultural N_2O outputs are not large enough to cause or maintain a depleted ozone layer, but they will slow its recovery.

The Future of the Ozone Layer

The ozone-friendly sales pitch has paid dividends for air-conditioner, freezer, and refrigerator manufacturers as CFC substitutes are introduced. Because of the response of the buying public and industry, goals for CFC reduction have actually been exceeded in many countries. Global CFC production peaked at 1.25 million tons in 1988, but by 1994, production levels had fallen to about 0.3 million tons. If this trend continues, and nations adhere to the Montreal Protocol (a major assumption given that China and India are not signatories), the ozone layer will continue to be depleted until about the year 2000. CFCs remain active in the stratosphere for 40 to 200 years, meaning that it takes that long for the ultraviolet radiation to modify the CFC molecule to release the chlorine atom. Thus, although the output of CFCs into the stratosphere has been reduced, the ones already there have yet to decay and release the damaging chlorine. Fortunately the residence time of HCFCs is only about 10 to 20 years. After bottoming out around the year 2000, stratospheric ozone concentrations should start to increase, taking 50 to 100 years to regain their natural concentrations.

....

Summary

- The proportions of the gases in the atmosphere are not fixed absolutely, but fluctuate within a fairly narrow range through time.
- Cigarette smoke is the most serious indoor air pollutant.
- Tropospheric ozone contributes to photochemical smog.
- The presence of a stratospheric ozone shield allowed plants and animals to colonize the land about 400 million years ago.
- Stratospheric ozone protects us from damaging ultraviolet radiation. Thinning of the ozone layer increases the risk of skin cancer, eye cataracts, and faulty DNA replication.
- Chlorine- (CFCs, HCFCs, methyl chloride), bromine- (methyl bromide), and nitrogen-(NO_2) based compounds cause the breakdown of stratospheric ozone.
- The ozone layer thins by about 70% above Antarctica between September and October and by about 40% above the Arctic between February and May.
- The Montreal Protocol and subsequent international agreements have caused a marked reduction in CFC emissions since 1987. However, not all countries are signatories to the agreement.
- The residence time of CFCs in the atmosphere is about 70 years. Recovery of the ozone layer will be delayed until CFCs already released are lost from our atmosphere.

Further Readings

French, H. F. 1997. Learning from the ozone experience. In *State of the World 1997*. Edited by L. Brown, C. Flavin, and H. French. Washington, DC: Worldwatch Institute.

Mosier, A. R., et al. 1998. Assessing and mitigating N_2O emissions from agricultural soils. *Climate Change* 40:7–38.

Retallack, S. 1997. God protect us from those who "protect the skies." *The Ecologist* 27(5):188–191.

Rowland, F. S. 1989. Chlorofluorocarbons and the depletion of stratospheric ozone. *American Scientist* 77:36–45.

Soroos, M. S. 1998. Preserving the atmosphere as a global commons. *Environment* 40(2):7–13, 32–35.

Web Connections

On-line resources for this chapter are on the World Wide Web at:
http://www.prenhall.com/bush
(From this web-site, select Chapter 22.)

Climate Change and Global Warming

23.1 *The greenhouse effect*

23.2 *Carbon dioxide concentration through time*

23.3 *Climatic triggers*

23.4 *Human actions and climate change*

23.5 *Computer simulations of a warmer world*

23.6 *The potential effects of a 2.5°C warming*

23.7 *Carbon sequestration: A new way to think about a tree*

23.8 *Global warming: A risk to be ignored?*

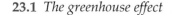

Although a single statistic for a hot summer or a cold winter means nothing, as a series of such events develop it is worth looking for a cause. Thirteen of the nineteen years between 1980 and 1998 were among the hottest on record since measurements began in 1881. The hottest years ever recorded were 1997 and 1998. A progressive warming trend is evident for almost two decades, with the only respite in the 1990s due to the cooling influence of dust ejected during the volcanic eruption of Mount Pinatubo in 1991. Is this a convincing trend toward unprecedented warming, or just a brief naturally warm climatic cycle?

Global climate change resulting from human activities is one of the most contentious topics in environmentalism, ecology, and government. The prime source of power that humans generate comes from burning **fossil fuels**. Fossil fuels (coal, oil, and natural gas) are organic material, and whenever they are burned, carbon dioxide (CO_2) is released into the atmosphere. The fear is that pollution by common anthropogenic (created by humans) pollutants such as carbon dioxide, methane, and ozone may provide the spark for unparalleled increases in global temperature. It is probable that the planet has been warmed since the more-developed countries started their industrialization about 150 years ago. Accurate records of atmospheric carbon dioxide have been kept for the last

114 years, and in that time, global temperatures have risen about 0.4°–0.7°C (Figure 23.1). The apparently close correlation between temperature and carbon dioxide emissions supports the argument that industrial emissions, in particular gases released by the burning of fossil fuels, will lead to further climatic change over the next century. However, the argument of future warming is necessarily based on a set of predictions. How good are those predictions? Should they be ignored?

23.1 The Greenhouse Effect

On the basis of our distance from the sun, Earth would be expected to have a temperature about 33°C cooler than present. An average temperature for the planet would then be about -18°C, and our planet would be a frozen chunk of rock coated in ice. These temperatures have never been experienced on Earth, because in the earliest times the fusion of planetary fragments left Earth radiating heat. By the time Earth had cooled enough for water to exist in liquid form on the surface, a primitive atmosphere was already in place, and this insulated Earth from the full effects of heat loss.

Our atmosphere allows much of the incoming shortwave radiation from the sun to strike Earth. This energy is principally high-quality light energy in the ultraviolet and visible spectra, with lesser inputs of

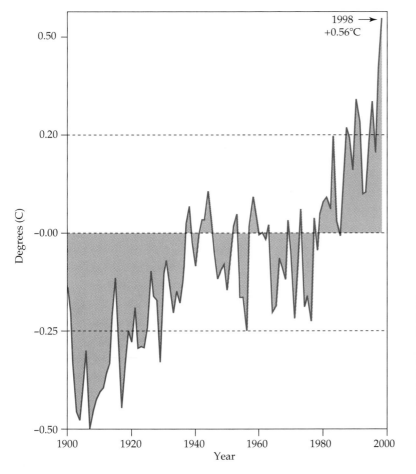

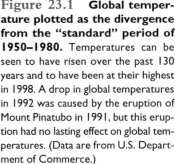

Figure 23.1 Global temperature plotted as the divergence from the "standard" period of 1950–1980. Temperatures can be seen to have risen over the past 130 years and to have been at their highest in 1998. A drop in global temperatures in 1992 was caused by the eruption of Mount Pinatubo in 1991, but this eruption had no lasting effect on global temperatures. (Data are from U.S. Department of Commerce.)

longer wavelength energy, such as heat (Figure 23.2). On first striking the atmosphere, and then the planet surface, much of the energy is absorbed temporarily. The absorbed energy undergoes a change of energy status, and, following the laws of thermodynamics, the energy undergoes a downward transfer in quality, such as longer or more erratic wavelengths. Much of the energy is transformed from light to heat. The energy absorbed and held as heat during the day is irradiated back into space at night. However, the fact that not all the heat is lost to the atmosphere has led to the comparison of the atmosphere to a greenhouse.

The glass of a greenhouse allows light to enter. The light falling on plants and the soil is stored as heat and reradiated, but the heat is trapped inside the glass, causing the greenhouse to become warmer than the outside air. Similarly, the atmosphere is highly permeable to shortwave radiation and allows the passage of light. Much of the energy transformed to heat is trapped by atmospheric gases—particularly water vapor, carbon dioxide, and nitrous oxide (N_2O)—that prevent this radiated heat from being lost into space. Thus, the atmosphere operates in much the same way

as a greenhouse: It lets light in but traps heat. The warming of Earth attributable to atmospheric gases is termed the **greenhouse effect**.

The natural thermal insulation of the greenhouse gases raises the average global temperature from $-18°C$ to $15°C$. The greenhouse effect is thought to be increased as concentrations of water vapor and carbon dioxide rise or to be lessened as they decline. As air warms, its capacity to hold water vapor increases. If atmospheric pollution causes air temperatures to rise, the water-holding capacity of the air will increase, and it is this additional water vapor that would be the primary greenhouse gas of global warming. It is also this water vapor, in particular its potential to form clouds, that is proving to be one of the biggest hurdles to providing accurate models of future warming trends.

23.2 Carbon Dioxide Concentration Through Time

Bubbles of gas trapped in glaciers provide strong evidence for changes in carbon dioxide concentrations

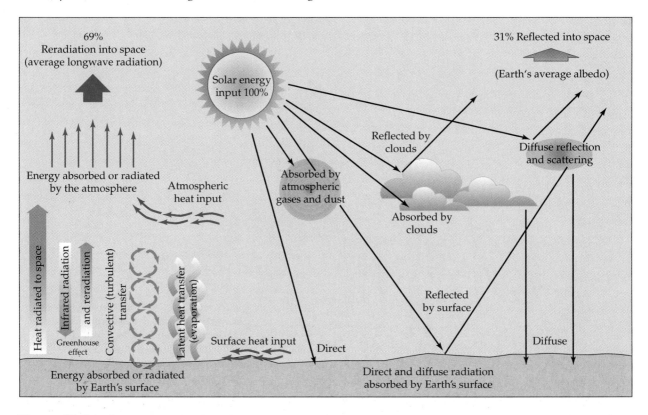

Figure 23.2   The atmospheric heat budget. Solar energy enters the atmosphere as light energy. About half of this is either absorbed and stored as heat energy or reflected back into space. Light reaching the surface of Earth may be stored as heat energy before being reradiated to the atmosphere, where some of it may be absorbed. The heat stored in the atmosphere warms Earth. The absorption of energy and its storage as heat by the atmosphere is called the greenhouse effect.

through time. This chemical is a rare (0.03%) atmospheric gas, but its presence is essential for successful photosynthesis. Its rarity also makes it more vulnerable to a halving or doubling of concentration than an abundant gas such as oxygen. For example, it has been estimated that if all the trees on the planet were burned, the oxygen concentration in the atmosphere would be reduced by about 300 ppm (parts per million), changing the concentration of oxygen from 209,480 ppm to 209,180 ppm; a change that would barely be noticed. An increase of 300 ppm in carbon dioxide would double its concentration and thereby would potentially increase photosynthetic productivity and elevate global temperatures. On the other hand, a decrease of 300 ppm would eliminate carbon dioxide from the atmosphere, photosynthesis would fail, all plants would die, and food chains would collapse. The overall atmospheric composition has varied relatively little in the last 400 million years. We can assume that variations in the concentration of a gas that is rare, but vital, are more likely to have disrupted ecosystems and driven evolution than small variations in an abundant gas.

The fossil record and computer simulations suggest that high temperatures and carbon dioxide concentrations have been correlated for more than 100 million years. There has been a general downward trend in carbon dioxide concentrations during the last 100 million years, with peak prehistoric values as much as twice those of today. One hundred million years ago, reptiles similar to alligators were able to survive in the polar regions, and remains of dinosaurs have recently been discovered in the Antarctic. These records indicate that Earth was perhaps 10°C warmer when carbon dioxide was at its maximum atmospheric concentration than it is today. For the last several million years, carbon dioxide concentrations have stabilized to fluctuate within a range of about 180 ppm (an atmospheric concentration of 0.018%) and 270 ppm (0.027%). These fluctuations match the waxing and waning of ice ages.

A detailed image of carbon dioxide concentrations over the last 100,000 years has been obtained from air bubbles trapped in ice caps. In areas such as Antarctica, Greenland, and the high peaks of the Andes, some of the snow that falls never melts. Air trapped between the snowflakes becomes pressed into bubbles as the annual snowfall turns into a layer of ice. Each individual bubble is a time capsule that captured the air and preserved it. The difference between the chemistry of the bubbles from oldest to youngest in the ice cap provides a record of changes in the composition of the atmosphere through time. Ice cores raised from these ice caps can provide an annual record of atmospheric change going back more than 100,000 years. Such studies reveal that the interglacial period that preceded the last ice age had carbon dioxide concentrations approximately equal to those of modern times. It is believed that this period was about 1°–2°C warmer than the present. However, during the ice ages, when the atmospheric concentration of carbon dioxide was halved (Figure 23.3), global temperatures slumped by 6°C (about 9°C, in mainland North America). With the start of the present interglacial, about 10,000 years ago, the carbon dioxide concentrations rose rapidly to about 260 ppm, where they remained until the start of the Industrial Revolution.

That the previous times of very high carbon dioxide concentrations allowed very warm climates and that, during ice ages, both carbon dioxide concentra-

tions and temperatures were low are indicators of the potential power of greenhouse gases. But before leaping to conclusions, remember the link between temperature and carbon dioxide is one of correlation. No causal link has been shown. It is possible that carbon dioxide and temperature oscillations were both influenced independently by a third factor. Thus, the carbon dioxide could track the temperature change without actually causing it.

As a result of carbon dioxide released by the burning of fossil fuels, atmospheric concentrations of this gas have risen 30%, to about 365 ppm, in the last 200 years. For many scientists, it is not the actual change in carbon dioxide concentration that is most worrisome but the rate of the change. The only comparable rate of increase within the last 100,000 years took place at the transition between ice age and post-ice age climates: a period of rapid and profound climatic warming that caused the mass migration of species to cooler locations.

One way to look at industrial emissions is that, for the last 200 years, humans have conducted a global-scale "experiment." If carbon dioxide and the other gases are causally related to climate change, we can predict that increasing the greenhouse gases will precede a warming. Human actions have increased the concentration of the gases, and if the relationship is simple, there should be a quantitative correlation between the carbon dioxide added and increased temperature.

Two problems face this experiment. First, there is only one Earth, so there is no control for the experiment. Ideally, there would be another planet Earth that had unchanged greenhouse gas concentrations. Any difference between the climates of the two Earths would be attributable to the change in atmosphere. Without the control, it is impossible to tell for certain which observed changes would have happened anyway. Second, the dynamics of climate and atmosphere are anything but simple.

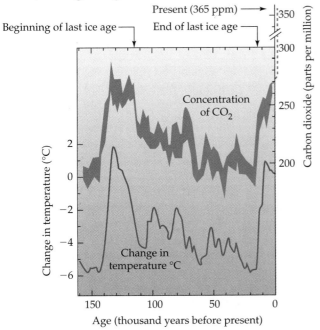

Figure 23.3   Atmospheric carbon dioxide concentration and global temperature have risen and fallen synchronously for the last 150,000 years. The carbon dioxide is measured from bubbles trapped in an ice core from Antarctica. Note that this record terminates when the measured carbon dioxide concentration is about 280 ppm, shortly after the start of the Industrial Revolution. Current atmospheric carbon dioxide levels are about 365 ppm, so we now have higher carbon dioxide concentrations than at any time in the past 200,000 years (shown as a dotted line).

23.3   Climatic Triggers

The climate of Earth has a complex history and continues to change, sometimes subtly and sometimes dramatically. In the past 2.5 million years, there have been 22 glacial (ice age) and interglacial episodes (Chapter 13). Recent evidence suggests that the flip-flop of climate that kicks Earth out of a warm interglacial period into a cold glacial may take as little as three to five years. To estimate the probability that we face another such change in the near future, or to find

+feedback

if we have warmed our planet beyond the bounds of natural climate variation, requires a study of past and present climates.

Variations in Earth's wobbles, tilt, and elliptical orbit around the sun will bring about changes in the received input of energy to the planet. These natural rhythms, the **Milankovitch cycles**, appear to underlie the gross structure of glacial and interglacial periods of the last 2.5 million years (Chapter 13). With overlapping rhythms of about 22,000, 40,000, and 100,000 years, these cycles may set the scene for long-term changes in climate, but they cannot be responsible for short-term climatic variation. These shorter variations have been found to be a part of both our modern world and also of the past ice ages. Recent evidence from long ice cores suggests that simply to view ice ages as switching on and off, and the world as going from cold to warm and back to cold again, is a gross oversimplification.

Researchers working on the deep-ocean sediments demonstrated that, during the last interglacial, a supposedly warm period, there were "cold events" of the same magnitude as full-glacial cooling (Figure 13.3). These cold periods appear to last only a few decades, thus they do not function at the scale of changes in astronomical cycles, and they are too fast and too profound to be caused by changes in atmospheric composition. A satisfactory explanation for them has yet to be found. As climate change on this scale has never been witnessed, computer models are used to estimate what could cause rapid environmental change. The models suggest that such rapid pulses in climate could be due to feedback loops of cooling caused by an increase in the amount of light reflected by Earth and changes in ocean circulation patterns. Because these are the suspected causes of profound short-term climate change, particular attention is paid to how they work and how they might be affected by pollution.

Albedo and Temperature Regulation

Objects that appear dark do so because they absorb much of the light that falls on them. The energy is stored as heat, with the result that a dark object warms quickly in sunlight. Conversely, pale objects reflect most of the light that hits them and thus remain cool. The reflectance from white or shiny objects, such as the surface of the ocean, ice caps, clouds, or droplets of water on leaves, results in the scattering of light back into the atmosphere. This reflectance is termed **albedo** and is an extremely important factor in establishing global heat budgets. The albedo of our planet

is not constant through time and will change according to climatic conditions.

Albedo may change when polar ice caps expand during glacial periods. For example, during ice ages vast reflective surfaces of ice cover much of Canada, northern Europe, and a much increased Antarctic ice cap. Under these conditions, a high proportion of the sun's energy falling on the planet is reflected directly back into space. This reflected energy does not add to the heating of Earth. An expansion of glaciers that increases albedo could set in motion a rapid cooling of the planet by means of a positive feedback mechanism. The colder it gets, the more ice there is; more ice leads to more albedo, resulting in more reflected energy; with a reduced energy input, the planet gets even colder, and the cycle continues. Conversely as the ice sheets melt, albedo is reduced, and the planet will lock into a heating cycle.

At present, glaciers are contracting and are probably close to their lowest ice volume in the past 2 million years. This observation suggests that we are in a heating cycle. Reduced ice mass leads to lower albedo, but the effect may be somewhat offset by increased cloudiness. Northern hemispheric cloudiness appears to be increasing, probably as a result of industrial emissions. However, estimates of cloud cover are based on a short run of data (from 1965 to 1993) and provide no information on the density and reflectiveness of the clouds or their exact location in relation to cities.

Changes in Sea Ice

In 1997, an analysis by William de la Mere of the Australian Antarctic Division provided a remarkable insight into long-term shrinking of the Antarctic ice mass. The data are based on the location of whale kills by the British and Norwegian whaling fleets. Whale hunters knew that in the Southern Hemisphere summer (October to April) the large whales were found at the edge of the Antarctic ice. Because these whales were hunted along the ice margin, their kill sites reflect the approximate position of the edge of the pack ice in summer. Between 1931 and 1987 the species killed and its precise location were recorded for every kill made by the British and Norwegian whaling fleets. In the 1920s and 1930s the target species were blue whales and humpback whales (Figure 11.14), whales that hugged the ice margin. By the 1940s blue whale stocks were too low to be worth exploiting and hunting for fin and humpback whales intensified. Between 1930 and 1954 there is no significant difference in the latitude of the whale kills, and it is evident that the ice margin lay at approximately 61° S to 62° S

(Figure 23.4). From the last few records of the 1950s, the pack-ice margin appears to be retreating southward, but statistically these data points do not form a distinct group. By the end of the 1950s, all the large whale species close to the ice pack were too scarce to be worth hunting and the hunt switched to the next largest whale, the sei whale. Sei whales do not swim close to the ice margin and so for the next decade we have no information from whale kills about the location of the summer pack ice. As sei whale populations were overexploited and the hunt centered on minke whales, the whalers once again prowled the edge of the pack ice. The kills recorded from 1972 onward are located between 64° S and 66° S; clearly poleward of the kills made in the 1930s to 1950s. These data suggest that between the mid-1950s and the mid-1970s the Antarctic pack ice retreated about 2.8° S, representing a loss of ice area of 25%. Such a strong change in the extent of pack-ice suggests that the Antarctic ocean had warmed during the period of this study. Furthermore, the loss of that huge reflective area will increase the amount of solar energy absorbed by the planet and increase the probability of further warming.

Ocean Circulation

Ocean currents are great transporters of heat, and they modify the climates of adjacent landmasses. Cold water currents, such as the Labrador current that flows south past Newfoundland, chill a region and often bring fog and rain. Conversely, warm water currents, such as the Gulf Stream that warms western Europe, can raise annual temperatures, particularly alleviating the cold of winter. If these currents were perturbed from their present course, profound climatic change could follow. Whether these heat transports can be shut down midway through an interglacial is unknown. However, there have been times in the past when the ocean currents have been switched on and off. About 11,000 years ago, the normal pattern of ocean circulation was apparently interrupted for about 500 years. Climatic warming at the end of the last ice age caused the melting of the great ice sheets. Oceanographer Wally Broecker of the Lamont-Doherty Observatory in New Jersey has proposed that huge discharges of cold, fresh meltwater poured down the Mississippi and St. Lawrence Rivers. This cold water so changed the ocean circulation that the currents transferring heat northward from the subtropics to high latitudes were interrupted. Without this source of warmth, the northern glaciers ceased to melt and started to expand southward again. This cold episode, known as the Younger Dryas event, lasted for 500 years and is documented in northern Europe, North America, and perhaps as far away as China and Argentina. About 10,500 years ago, the warming of the planet reached a critical point where, despite meltwater flow, the warm currents were reinstated and the present interglacial started in earnest.

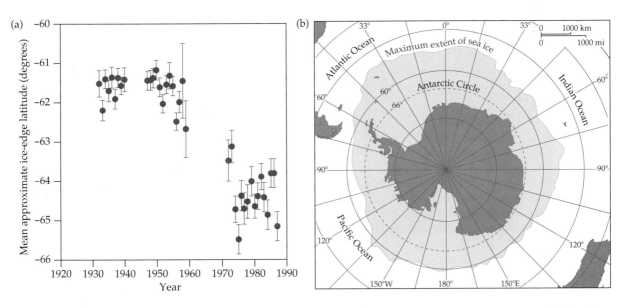

Figure 23.4 **Changes in the summer margin of Antarctic sea ice based on whaling records 1931–1987.** (a) Data points represent the mean latitude of whales killed at the Antarctic ice margin. Between the 1950s and the 1970s, the killing grounds moved poleward by 2.8° of latitude. These data indicate that the area of Antarctic sea ice shrank by about 25% in the middle decades of this century. (b) The maximum historical coverage of sea ice. Note that the area of ice is now reduced.

Recent research has suggested that, because of deforestation and extraction of water from subterranean aquifers, the flow of freshwater into the oceans is increasing. However, these inputs are too puny to affect the circulation of ocean currents. It seems unlikely that human action could directly influence this powerful system. Therefore, if there is to be a major human-induced climate change, it is more likely to come from a change in albedo than from ocean circulation.

Climate and Chaos

Chaos is an absence of order. In a truly chaotic system, no pattern exists. In climate systems, there is often an underlying pattern (on the scale of hundreds of years), but the year-to-year, even decade-to-decade, variability is chaotic. Although our knowledge of Earth's atmospheric system is steadily improving, our ability to predict change in climate seems as remote as ever.

Simple ideas of glacial and interglacial cycles have given way to a view in which the shorter-term climatic variation (periods of tens of years) can be almost as profound as that of the long-term (periods of hundreds or thousands of years). An example comes from our own interglacial. Between the 1400s and the mid-1800s, Europe and North America experienced the "Little Ice Age." In the 1800s, some winters were so cold that the adventurous could walk on the ice from Staten Island to Manhattan. This was a time of expanding glaciers, harsh winters, shortened growing seasons, and frequent crop failures. The "Little Ice Age" has become a feature of textbooks that document this 500-year period as being 1°–2°C cooler than present. However, climatologists now note that, within this period, there were climatic cycles almost as warm as the present and other times of exceptional cooling, notably in the early 1600s and early 1800s. Although overall the Little Ice Age was a period of cooler temperatures, it is evident that, had tabloid newspapers been printed, we might still have seen headlines like "Hottest July on Record," "Driest Summer Ever!", and "Cold Kills!"

The warm phase that undoubtedly exists in the twentieth century could be a natural oscillation in global climate, which would be followed by a cooler period. One clue that might indicate whether this warming is natural is the date it started. A natural cycle could have started anywhere, but a human-induced cycle is most likely to have started with the large-scale burning of fossil fuels in the late 1800s.

A problem of the best time period for study now becomes apparent. The period from 1700 to present indicates that the present high temperatures are part of a general, probably natural, warming. No distinct rise in temperature is attributable to industrial pollution. Yet, if we look at the record for the last thousand years, it appears that "normal" temperatures are somewhat cooler than those of the twentieth century. The rapid change in temperature observed at the beginning of the century is, in this scenario, likely to have been caused by human activities.

Which is the appropriate time period to study? Clearly, this simple comparison of past temperatures is not going to provide a definitive answer regarding human alteration of global climates. But it does allow us to make the following basic observations:

- Weather and climate are more chaotic than once thought, and little attention needs to be paid to record temperatures, record rainfalls, and the like, unless they prove to be sustained.

- Temperatures are now higher than at any time in the past 1000 years.

- Temperatures appear to have increased sharply over the past 100 years, although they cannot be said to have exceeded "natural" bounds.

23.4 Human Actions and Climate Change

Apart from the possibility of inducing a nuclear winter as a result of atomic warfare, the most likely way that humans could change the climate is through air pollution. The greenhouse effect is the natural process whereby tropospheric gases trap heat. If we add to the concentrations of those gases, we could effectively double-glaze the greenhouse. Double-glazing does not prevent light from entering your home, but it prevents heat from escaping; enriching the troposphere with these greenhouse gases has exactly the same effect.

The principal anthropogenic sources of these greenhouse gases are the burning of organic material—such as oil, coal, gas, or their derivatives—and the burning that accompanies deforestation. The carbon cycle can now be redrawn to include human activities. The "sinks" of carbon remain the same, but there are more sources (Figure 23.5). After a century of increasing global carbon emissions, the output of carbon has remained more or less constant at about 7–8 billion tons per year between 1989 and 1998. This release, primarily from burning fossil fuels and deforestation, represents about two to three times as much carbon as can be absorbed by natural systems, resulting in a steady increase in the concentration of atmospheric carbon

The Global Carbon Cycle

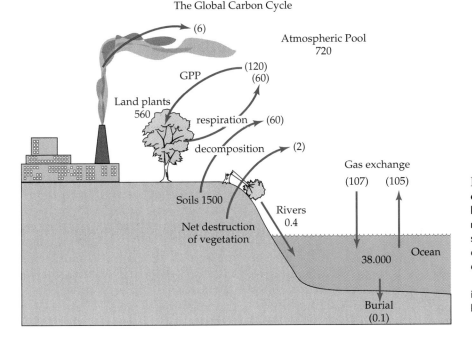

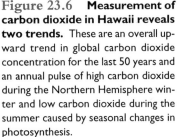

Figure 23.5   **The carbon cycle expanded to show the effects of human activities.** More carbon is now being released than is being absorbed, so there is an accumulation of carbon dioxide in the atmosphere. All carbon values are expressed in units of 10^{15} grams of carbon, while fluxes are in grams of carbon per year. Fluxes are bracketed. (After W.H. Schlesinger 1991.)

dioxide. Other lesser sources of carbon dioxide include the oxidation and respiration of organic material once held as peat in wetlands that have been drained.

Over the past century fossil fuel use has increased rapidly (see Figure 25.1). The product of combusting any carbon-based fuel is carbon dioxide. The best data set showing how the increased use of fossil fuels has affected atmospheric carbon dioxide concentrations comes from Hawaii (Figure 23.6). Far from local industrial contamination, the carbon dioxide concentrations above Hawaii document two distinct records to produce this upward-sloping sawtooth diagram. The Northern Hemisphere appears to have an annual flux

in carbon dioxide concentrations, a very subtle change that causes the teeth on the curve. Each summer, the concentration of carbon dioxide falls slightly, and each winter it rises, responding to seasonal demand for carbon dioxide by photosynthesis. During the winter, when there is less photosynthesis, carbon dioxide intake is reduced, and the output of carbon dioxide through the respiration of organic matter continues, effectively raising carbon dioxide concentrations. As the leaves appear in the northern temperate forests in the spring, the demand for carbon dioxide rises; consequently, the carbon dioxide concentrations fall. These rhythms are constant; witness how the peaks and

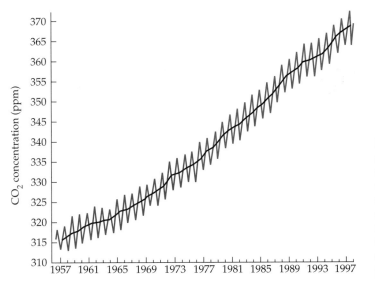

Figure 23.6   **Measurement of carbon dioxide in Hawaii reveals two trends.** These are an overall upward trend in global carbon dioxide concentration for the last 50 years and an annual pulse of high carbon dioxide during the Northern Hemisphere winter and low carbon dioxide during the summer caused by seasonal changes in photosynthesis.

troughs vary uniformly about the average line. It is the average concentration that is climbing, and this climb is a direct reflection of carbon dioxide pumped into the atmosphere. This excess carbon dioxide has not been used up, has not been absorbed by the natural sinks, and thus has been added to the global greenhouse. In the past 100 years, atmospheric carbon dioxide concentrations have risen from 280 ppm to about 365 ppm.

Other greenhouse gases have also increased in concentration, and it is important to note that some of these are disproportionately effective at trapping heat (Table 23.1). Because of their structure, some molecules can absorb more heat than others, and if we take carbon dioxide as representing an absorbance efficiency of 1, it is apparent that methane, nitrous oxides, and chlorofluorocarbons (CFCs) are much more potent greenhouse gases. The residence time indicates how long the chemical remains active in the atmosphere. This value also indicates the time it would take for the atmosphere to return to a natural state for that gas if pollution were to stop. Despite the heat-trapping potential of such gases as CFCs, the sheer volume of carbon dioxide makes it the most important single greenhouse agent produced by human activities. In addition, natural water vapor remains the principal agent of the greenhouse effect. Alone, carbon dioxide and the other pollutant gases would cause a very small increase in atmospheric temperature. However, a delicate balance exists between atmospheric temperature and evaporation from the ocean surface. The small temperature change caused by the pollutants increases the water-holding capacity of the air and results in increased evaporation from the ocean surface. Atmospheric water vapor is a very effective greenhouse gas, and it is these molecules that store most of the heat.

Table 23.1 Atmospheric residence time and relative warming effect of greenhouse gases.

Gas	Residence time* (years)	Effect†
Carbon dioxide	500	1
Methane	7–10	25
Nitrous oxides	150	230
Chlorofluorocarbons	60–400	10,000–20,000
Hydrofluorocarbons	260	10,000

*The period that an average molecule will remain chemically active in the atmosphere.
†The estimated lifetime heat-trapping potential of each molecule relative to a single molecule of carbon dioxide.

Thus the pollutants provide a small tweak to the atmospheric system that sets in motion a shift to a new equilibrium.

Not Everyone Believes in Global Warming

Two principal objections can be made against the claim that human activities cause global warming. The first is that satellite-based measurements of Earth's temperature show no warming in the last 25 years despite land-based measurements that suggest increasing temperatures. Some scientists are concerned that the land-based temperature measurements are biased toward warmth because so many are taken near cities, which are known to emit heat. The global scale data gathered from satellites should provide an unbiased estimate of temperature. Satellites have been gathering these data since the mid-1970s, and from these measurements it appears that Earth has a stable, or even cooling surface temperature. However, in 1998 the satellite data was found to have a methodological error. The satellite data did not include a correction for aging satellites gradually falling out of their orbit. The satellites underestimate temperature change by about 0.005°C to 0.007°C per year; equivalent to 0.5°C to 0.7°C per century. Once this correction was applied to the data set, the satellite and land-based temperature estimates both agree on the amount of global warming experienced this century.

The second objection is that the great "global experiment" of adding greenhouse gases to the atmosphere has demonstrated that there is no direct linkage between carbon dioxide concentrations and temperature. Proponents of this objection note that the observed change in temperature of 0.4°–0.7°C is less than the expected values of 1.2°C. They suggest that if changed carbon dioxide concentrations do not bring about temperature change, then the anthropogenic enrichment of greenhouse gases is a nonissue. This argument has been welcomed by industries who are, or whose products are, the primary contributors of greenhouse gases (electrical utilities, oil and coal mining, and auto manufacturers). These industries have vigorously opposed regulation of emissions of greenhouse gases from industrial sources, on the basis that the assumption that atmospheric pollution leads to global warming is unfounded.

Where Is the "Missing Heat?"

If global warming is a reality, an explanation should be sought for the discrepancy between the observed and expected data. It is possible that an unknown process is delaying the response of climate to atmospheric changes. Perhaps the atmosphere and temperature are not in equilibrium, and there is an increasing time-lag effect as greenhouse gases become enriched. In other words, this hypothesis suggests that the heating will come, it just takes time for it to become evident. This is a possibility, but it is not good science to settle on an explanation that invokes unknown causal factors for a phenomenon.

Another possibility is that Earth has just passed through three decades of natural cooling and that this cooling has masked the heating signal caused by an enhanced greenhouse effect. Again, such an argument is weak, because no independent evidence exists to confirm that such a cold period happened.

Volcanic activity may have played a part in reducing global temperatures and thereby reduced the overall warming trend. Violent eruptions can inject clouds of dust into the stratosphere that block incoming sunlight and reduce the energy input to the planet (Figure 23.7). However, these events provide a temporary cooling effect and cannot account for a long-term repression of the influence of increased concentrations of greenhouse gases.

One hypothesis to explain the missing heat problem is that the full impacts of atmospheric pollution are being offset by another factor. In other words, while the increased greenhouse gas concentrations warm the planet, another pollutant is cooling the planet. Observations to support this hypothesis come from a number of sources.

- Computer models using carbon dioxide concentrations to predict changes in atmospheric temperature are less accurate for the Northern than for the Southern Hemisphere.
- There is more industry in the Northern than in the Southern Hemisphere.
- Nighttime temperatures have risen faster than daytime ones over the past decades.
- Cloudiness, in particular the presence of low stratiform clouds, has increased in recent years around industrial centers.

The possible explanation that threads these apparently disparate observations together is that pollutants are causing the increase in cloudiness and that this cloudiness is moderating the warming. Because more pollutant chemicals are released in the Northern than

Figure 23.7   Volcanic dust from Mount Pinatubo in the streets of Olongapo, Philippines. The eruption of Mount Pinatubo blasted dust particles high into the atmosphere and lowered global temperatures by 0.5°–1°C for a year after the eruption.

in the Southern Hemisphere, if one of these chemicals is reducing global warming, it might be expected to have a greater effect in the north than in the south. Which of the thousands of chemicals released by industrialized nations could be reducing the effect of increased concentrations of greenhouse gases?

The chemicals that have received most attention are particulate aerosols, such as nitrogen oxides (NO_x) and sulfur dioxide (SO_2). The proposed mechanism is that these molecules and tiny particles of soot in the troposphere attract water molecules. The pollutants form the nucleus around which a water droplet forms. If there are enough of these **condensation nuclei**, a cloud will form. These clouds appear shiny from space, so they have high albedo. They reflect sunlight away from Earth during the day and thus reduce the energy input to the planet. The reduced energy input almost balances the warming potential of the strengthened greenhouse effect, so daytime temperatures remain nearly constant. At night, the clouds are still there, and the longwave radiation of stored heat emitted from Earth cannot escape into space. Conse-quently, nights become warmer. This hypothesis provides an elegant explanation of the apparent lack of global warming and fits the known facts regarding diurnal temperature variations. Furthermore, the observation that the 1980s was a period of increasing cloudiness is also consistent with this hypothesis.

Computer simulations of temperature change in response to increased greenhouse gas concentrations that include the effect of aerosols provide the best predictive tool of future climate change (Figure 23.8). The modeled and observed temperatures are seen to be closely correlated and suggest an overall warming this century of about 0.7°C.

If it is true that the effect of increasing greenhouse gas concentrations is being offset by other pollution, it raises the possibility that emissions of carbon dioxide do not matter so long as emissions of sulfur dioxide and nitrous oxides keep pace. Perhaps if sulfur dioxide and nitrous oxides were not damaging pollutants in their own right, this gamble of offsetting one kind of pollution with another would gain support. However, both of these pollutants are heavily implicated

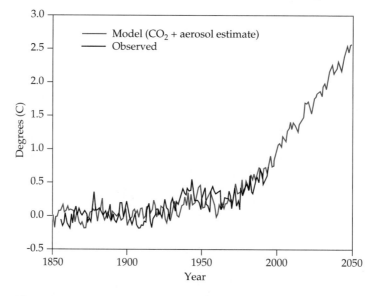

Figure 23.8 Comparison of recorded and predicted temperatures from computer simulations of climate (global circulation models = GCM) since 1850. This model includes the cooling influence of atmospheric aerosols. Modeled and observed temperatures are plotted as the departure from the 1880–1920 mean temperature. The closer the agreement of model and observed values, the greater the reliance that can be placed upon the modeled predictions. During the 1980s and early 1990s predicted temperatures were consistently higher than modeled temperatures. Some of the departures from the modeled data coincide with major volcanic eruptions. Other pollutants may be mitigating temperature effects, the world may just have gone through a cool period, or the predictive models may be flawed. (Data are from Geophysical Fluid Dynamics Laboratory web-site.: http://www.gfdl.gov)

in the increasing acidification of lakes, rivers, and soils (Chapter 24). Moreover, most scientists would be skeptical that humans understand atmospheric chemistry well enough to attempt such a manipulation without disastrous consequences.

23.5   Computer Simulations of a Warmer World

Predictions of how pollutants may affect the climate of Earth are essential if wise regulations and policies are to be formulated. Because there is no precedent, there is no experience that can guide decision making. The best approximations of what the future will hold are derived from computer simulations. These programs attempt to model the interaction between energy inputs, the atmosphere, land and oceans, and the greenhouse effect. Consequently, simulating future climatic change is immensely complex; the number of variables in these models is enormous. The models seek to provide a three-dimensional representation of the temperature characteristics of the land, ocean, and atmosphere. To do this, the models include variables such as the geographical variation in concentrations of atmospheric gases, ocean salinity, temperature of land and sea, albedo, cloudiness, winds, and seasonal differences. The globe is divided into grid squares, or cells, of 2° of latitude and longitude, and nine vertical cells model the atmosphere above each grid cell on Earth's surface.

A starting value for each of the variables is entered for each cell. The cells then have to be made interactive, so that if one value changes in one cell, the cells around it change in a realistic manner. For example, if the ocean temperature is rising, the ocean will heat the atmosphere cells above it, and they will start to hold more moisture, which evaporates from the ocean surface, and ocean salinity will increase. The increased water vapor load over the oceans may result in increased precipitation as the air mass moves onto a continental region.

The ability to run such models has often been a function of available computer power and research funds. A major improvement took place when the models made the ocean and atmosphere interactive, and the latest modification was to make cloudiness more realistic. The models form the "best guess" of what the future may bring, but they also demonstrate our lack of understanding of how pollution affects the world.

The veracity of the predictions cannot be checked until those events occur. The only way to check the accuracy of these models is to run them "backward" and compare the model predictions against past temperature records. If all the variables are plugged into the model for a past climate, does the model make an accurate prediction of the events that followed? The models can be checked against a variety of time periods, using historical records to provide the "observed" values for relatively recent times. A longer perspective can be gained by checking the model predictions against the fossil record. The models have been tested backward over a time span of the last 2 million years. Each new version of these computer simulations has improved enormously, but they are still far from perfect. For instance, when run backward, the models still fail to produce glaciers during the last ice age. The modelers themselves do not make extravagant claims for the accuracy of their predictions, but state that each model result is just one possible future scenario.

A feature that is consistent from one model to the next is that the Arctic and high northern latitudes will be the most affected by global warming (Color Plate 4). However, the actual degree and rate of warming vary wildly according to the assumptions of the model. In the 1980s, predictions of an average global warming of 5°C to 10°C resulting from a doubling of atmospheric carbon dioxide were common. More recently, these estimates have generally been downgraded to a range of values from 1°C to 5°C. Indeed, such variability exists within a single model according to the way that clouds are represented. In the last year or two, climate modelers have realized that clouds are hugely complex systems that differ in their thermal properties. They can vary in their density, height from the ground, duration, and loading with particulates (dust, smoke, or chemical pollutants); all of these factors will cause albedo to vary. Changes in albedo affect the amount of incoming radiation that reaches the planet surface and the insulation of reflected heat released from Earth. Realistic modeling of clouds would represent a major breakthrough in the predictive power of the models.

Despite all these uncertainties, most scientists believe that global warming will result from an enhanced greenhouse effect. A reasonably conservative estimate of the impact would be a rise in average temperature of between 1°C and 2.5°C as carbon dioxide is doubled.

23.6   The Potential Effects of a 2.5°C Warming

Global warming is not proven, but it is anticipated, and the preventive treatment is likely to be costly. The total control of greenhouse gas emissions would require a reshaping of industry and land use. Is this too high a price to pay? In the absence of a crystal ball, the best data available are considered and then a decision is made as to whether the threat warrants a costly remedy. For this analysis, it is assumed that carbon dioxide doubles to 700 ppm late in the twenty-first century and that such an increase is likely to generate a 2.5°C increase in global temperature.

Agricultural Impact

In some areas, such as Alaska, Siberia, and Asia, a warming would improve agricultural yields because it would lengthen the growing season. A poleward shift of cultivable lands would be expected as Earth warmed. In Asia, this movement could lead to improved or sustained crop yields because the soils to the north of the present agricultural zone are potentially fertile. However, in North America, a northward expansion of the crop belt would be limited by the poor soils of the Canadian Shield. This shift from the fertile soils of the Midwest toward the thin soils of the shield could lead to a significant reduction in agricultural output from the world's leading area for the production of cereal crops. In drier areas, such as North Africa and parts of the southwestern United States, a reduction in precipitation and an increase in evaporation rates would convert marginal agricultural land to true desert. Globally, it is estimated that land lost to agriculture would outweigh land gained, and the cultivable area would be reduced by about 20%.

One suggestion has been that increased carbon dioxide levels in the atmosphere would promote plant growth. Studies conducted in laboratory settings showed that if plants were grown in twice the normal concentration of atmospheric carbon dioxide and were well supplied with water and nutrients, they did indeed grow larger and faster. These results, it was claimed, demonstrated that under future conditions increased agricultural productivity would counter any loss of cropland, and global warming would have a net beneficial effect on food supply. Such claims have yet to be substantiated, and they are currently the subject of intense debate.

The counterargument is that the experiment was not an accurate reflection of the natural world or of a real agricultural system. The plants were so well supplied with water, light, and nutrients that the improved growth was the result of increased nitrogen uptake promoting photosynthetic production. The extra carbon dioxide in the artificial atmosphere prob-

ably helped the plant growth, but from these data it cannot be stated, with certainty, that it was the prime cause of the improved growth rates. When plants were grown in an enriched carbon dioxide atmosphere, but under natural soil conditions (without additional fertilizer and water), results were mixed. Some species showed improved growth, while others did not respond to the extra carbon dioxide. In other experiments carried out over longer periods of time, it was found that the benefits of growing in double the normal atmospheric concentration of carbon dioxide were not sustained. After an initial growth spurt, the plants established an equilibrium photosynthetic rate that was almost the same as that achieved by plants grown under normal conditions.

This debate will continue as new research emerges, but it seems likely that elevated carbon dioxide alone will not materially affect crop productivity. If water and nutrients are in good supply, as would be the case in most farms in more-developed countries, an improvement in yield might be expected for some crops. Plant physiologist Boyd Strain of Duke University suggests that under good growing conditions, plants using the C3 photosynthetic pathway (the one used by most plants; see Chapter 3) will stand to gain more from an atmosphere enriched in carbon dioxide than would plants using the rarer C4 pathway (Figure 23.9a and b). The C4 photosynthetic process enables a plant to absorb carbon dioxide rapidly and reduce the amount of water lost as it opens its stomata to absorb carbon dioxide. However, this C4 mechanism is energy expensive: The plant expends a lot of energy to absorb the carbon dioxide. Thus, for C4 plants, increasing carbon dioxide concentrations would not be of much advantage; their yield would not be expected to increase much.

Around the world, crops are a mixture of C3 plants, such as wheat, oats, potatoes, and rice, and C4 plants, such as corn, sorghum, and cotton. Under normal circumstances, the C4 pathway offers a significant competitive advantage when water is scarce; thus it is most commonly found in desert or semidesert plants. If a general drying accompanies global warming, it may be that the superior water-use efficiency of C4 plants will enable them to grow when a C3 crop would fail completely. But for farmers who cannot afford irrigation or fertilizers, there may be little benefit from increased concentrations of carbon dioxide alone. Similarly, trees in commercial timber plantations where little fertilizer is added, or in natural systems, may show little increase in growth rates. Evidence for this came from the measurement of growth of loblolly pine

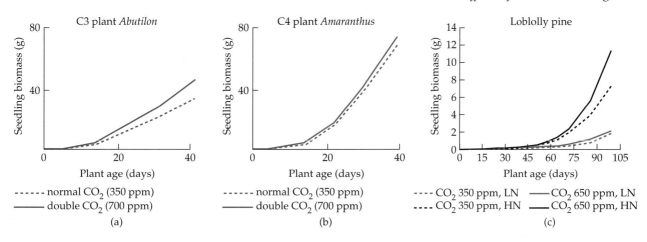

Figure 23.9   (a) Plant (*Abutilon* sp.) grown under ambient and doubled carbon dioxide, supplied with abundant fertilizer, light, and water. Note the marked improvement of growth in the C3 plant (as shown by the distance separating the two lines) when grown in an enriched CO_2 atmosphere. (b) Growth response of a C4 plant grown under ambient and doubled carbon dioxide, supplied with abundant fertilizer, light, and water. Note that the C4 plant is barely affected by the change in carbon dioxide concentration. (Data for (a) and (b) are from J.K. Dippery, D.T. Tissue, R.B. Thomas, and B.R. Strain, 1995.) (c) Growth response of the loblolly pine seedlings (*Pinus taeda*), a C3 plant grown under ambient and doubled carbon dioxide, but given a restricted supply of nutrients and water (replicating natural or nonintensive agricultural conditions). The seedlings that were fertilized and those that were both fertilized (high nitrogen = HN) and grown in enriched CO_2 grew faster than other seedlings. When nitrogen was not supplemented (low nitrogen = LN), increased levels of CO_2 did not affect plant growth. (Data for (c) are from K.L. Griffin, R.B. Thomas, and B.R. Strain, 1993.)

seedlings. When fertilized and exposed to high concentrations of carbon dioxide the trees showed enhanced growth rates compared with the control (normal carbon dioxide, fertilizer added). When the experiment was repeated with and without fertilizer, it was found that after a short-lived increase in growth rate, doubling carbon dioxide concentrations did not increase growth rates unless fertilizer was added (Figure 23.9c). In other words, the growth enhancement attributable to elevated concentrations of carbon dioxide is quickly overridden as another resource becomes limiting.

In summary, there may be a slight overall crop benefit accruing from elevated carbon dioxide, but only a very modest one, and probably not enough to compensate for the loss of agricultural land and spread of arid regions.

Nonagricultural Changes in the Distribution of Species

During the warming that followed the last ice age, species migrated poleward and upslope. Each followed an independent migration route, giving rise to the observation that communities are ephemeral through time (Chapter 13). The advent of global warming would again cause species to move poleward, and once more that movement would lead to the reshaping of communities. The elevation at which

frost is recorded is very significant for plant distributions, and under constant climatic conditions it would be unchanging. Since 1970, this height has moved upslope at an average of 1.5 m per year (consistent with 0.7°C warming per century).

Broad observations are possible regarding the expansion and contraction of biomes. The higher latitudes will experience greater changes than the lower latitudes, and this difference will translate into a contraction of tundra areas and an expansion of boreal forests. Deserts will spread as many areas become drier.

The precise outcome of species migrations, in terms of the composition of new communities, is impossible to predict. Figure 23.10a and b illustrates predictions for the future ranges of beech and sugar maple trees in eastern North America given a doubling of atmospheric carbon dioxide. The new ranges represent the present realized niche of the trees in terms of precipitation and temperature. The data suggest that there would be a northerly migration, but when two different climate models are compared (Figure 23.10b and c), it is apparent that considerable discrepancies exist between the predictions. Neither group of scientists responsible for these models would claim perfect predictive power; this is not a clash of scientific opinion, but rather a demonstration that small differences in the way the data are modeled can produce strikingly different outcomes.

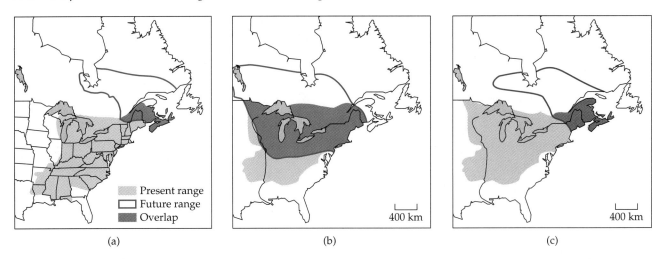

(a) (b) (c)

Figure 23.10 Changing distributions of trees with global warming. The modern distribution of (a) beech and (b) sugar maple in the eastern forest of North America and a prediction of where beech may be found given a doubling of atmospheric carbon dioxide. (c) Modern and predicted distribution of sugar maple given a doubling of carbon dioxide using the United Kingdom Meteorological Office (U.K.M.O.) model. (Data are from M.G. Davis and C. Zabinski, 1992. In *Global Warming and Biodiversity*. R. Peters and T. Lovejoy (eds.), © Yale University Press.)

Figure 23.10b shows a straight northward shift of the climate type (and hence the tree), whereas the model shown in Figure 23.10c suggests a strong northeastward movement. The implication of this latter model is that the Great Lakes region would be significantly drier than at present.

Beyond the difficulties of the climatic prediction, it should be recognized that the predicted distributions are also based on some assumptions of constancy. For example, they assume that changes in temperature and precipitation would be more important than changes in seasonality. Also, it is assumed that the interspecific relationships are constant. For example, it is assumed that an existing predator or pathogen would not suddenly decimate the beech population over a portion of

its new range. When such uncertainty exists as to what will happen to a single species, ecologists cannot predict the structure of future communities.

Such changes in distribution have implications for our ability to provide protected areas for the conservation of biodiversity. Let us simplify this reassortment to consider just two species that are the object of conservation on a nature reserve (Figure 23.11). The ranges of the two species overlap under present conditions, but as global warming takes effect, species A, the species with the more northerly distribution, dies out in most of the southern edge of its range. Meanwhile species B is only slightly affected by the new conditions and shows a small northward expansion. The nature reserve established to protect both species

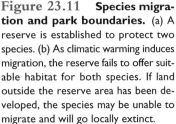

Figure 23.11 Species migration and park boundaries. (a) A reserve is established to protect two species. (b) As climatic warming induces migration, the reserve fails to offer suitable habitat for both species. If land outside the reserve area has been developed, the species may be unable to migrate and will go locally extinct.

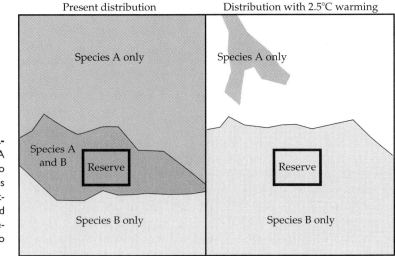

when their ranges overlapped is now failing to protect species A. The continued existence of species A depends upon the availability of habitat in its new range. If all of its potential habitat has been replaced by urban or agricultural development—and this is likely because the only protected area is the nature reserve—species A will be unable to migrate and will go locally extinct. Thus, climate change offers another reason to ensure that protected areas are spaced so that species can flow through them as they are forced to migrate (Chapter 21).

Increasingly Erratic Climates

In addition to a general warming trend, the global climate models predict that the climate will become more erratic. Seasonality—that is, the difference between winter and summer or between wet season and dry season—is likely to become more pronounced. Increased heating of the land and water is likely to generate increased convectional activity, and this could lead to increased storminess. The movement of the intertropical convergence zone (see Chapter 4) would be amplified, and therefore hurricane activity would increase. All of these observations are consistent with the perceived trends in climatic instability over the past decades.

However, some caution is needed before concluding that the climate would be truly transformed. Heightened awareness of the potential for climate change means that better climate records are kept now than ever before, and the apparent increase in variability may be a function of improved data gathering. Equally, just as Earth's climate may naturally cycle between warm and cold times, it may also have cycles of stability and instability. Perhaps the shift to less-stable climates is natural and not a product of atmospheric pollution. The scientific jury is still out.

Changes in Sea Level

Another consequence of a 2.5°C warming would be a rise in sea level. Ocean water would expand as its temperature rises, and the result would be rising sea levels. The increased temperatures will also result in a partial melting of the Antarctic and Arctic ice caps, and the released water will flow into the oceans and further raise sea levels. Note that sea level is only influenced by melting ice when that ice lay on dry ground. When floating ice melts it does not increase sea level, just as when a drink has ice poking up above the rim of a glass it does not overflow when the ice melts. The weight of the ice has already displaced an

equivalent volume of water, and thus when the ice melts there is no net change in water level. So far this century's sea level appears to have risen about 15 cm. The net rise in sea level attributable to a doubling of CO_2 is estimated to be between 0.5 m and 1.5 m.

Such a gain in sea level sounds trivial, but it would cause billions of dollars of damage. Many of our major cities are built close to sea level, or even below it, and are protected from flooding by elaborate systems of levees, flood barrages, and dikes. Even a modest rise could overwhelm these sea defenses. In the United States, cities such as Boston, Baltimore, New York, Charleston, Miami, New Orleans, Jacksonville, and Los Angeles would be threatened by even a 0.5 m rise in sea level. In other areas, substantial parts of the Netherlands, Denmark, and Bangladesh would be submerged, and centers such as London, Copenhagen, Venice, Bangkok, and Jakarta would face severe flooding. Clearly, the wealthy countries would build new sea defenses, the cost of which would run into billions of dollars. In the poorer areas of the world, such as Bangladesh, it is likely that adequate sea defenses would not be built and that a high price would be paid in human suffering.

23.7   Carbon Sequestration: A New Way to Think About a Tree

It has been suggested that one solution to offset the industrial emission of carbon dioxide is to plant forests. As the trees grow, they absorb carbon dioxide through photosynthesis and incorporate a portion of this carbon into their body structure. If a sufficient area could be planted, there might be no net increase in global carbon dioxide despite continued industrial emissions. If the total carbon absorption of an average tree is calculated, it is possible to determine the area of forest that would need to be planted to achieve this balance. Given present industrial emissions, the planting would need to cover an area the equivalent of Australia. To gain a significant increase in carbon storage this would have to be land that could support forest but that is presently barren or under pasture. Such a scheme does not seem realistic as a one-step solution to our atmospheric carbon problem; its implementation would be costly, and enough land might not be available. Another difficulty is that once trees are fully grown their net uptake of carbon is greatly reduced. When trees die and

Ecology in Action

Ecologists Monitor Boreal Fires and Climate Change

Ecologists are monitoring the size, intensity, and frequency of fires in boreal forest regions to gain an understanding of future risks of forest fire. Wildfires can cost lives, homes, and thousands of dollars in lost timber production. Yet they are a natural part of many ecosystems. One prediction of global climate change is that northern forests will burn increasingly often.

Because many boreal forests are in remote areas, gaining information on the frequency and extent of modern fire events calls for aerial photography or satellite imagery. Fires show up in the heat-sensitive infrared bands of the satellite imagery, and recently burned areas will have a different reflectance of light and heat energy than mature forest. The early successional herbs and sapling trees appear smooth from space, whereas the mature forest appears rough and craggy. Thus, the surface roughness can provide information about the age of the forest. Fire is the primary cause of disturbance in these remote areas, so the different roughnesses can provide an index of when each section of the forest last burned. A different approach is to use satellite images to compile a map of recently burned regions, and from this map compute the area burned each year. When this information is pieced together, a prediction can be made of how often fire will revisit an area.

The truth of these predictions is checked by visiting a representative portion of the sites as part of a "ground-truthing" operation. To the knowing eye, trees carry the scars of past fires, and these signs provide a valuable, independent tool to determine the fire history of an area. When a boreal forest tree is cut, the stump reveals both growth rings and fire scars. When a fire runs across the floor of a forest, it will scar the bark and the active con-

Scarred pine trunk. Fourteen years of rapid growth before one-fourth of the trunk was lost due to fire. Narrower growth rings indicate slow growth for the last 12 years.

rot, the carbon is returned to the atmosphere via respiration. Consequently tree planting is not truly ridding the environment of excessive carbon dioxide but is only sequestering it until a later date. Defenders of the carbon sequestration scheme note that if the timber is harvested and made into furniture, or if the logs are sunk into ocean trenches (where they would not rot), respiration of the dead wood could be avoided and the carbon would not find its way back into the atmosphere. Alternatively, they suggest that we look on carbon sequestration as a way to buy time. The carbon of trees destined to die and rot will be held for the life span of the tree, which might provide 60–100 years in which to scale back carbon emissions gradually.

Although a global-scale carbon sequestration project seems unlikely, the thought process behind it provides us with a new set of values in forestry.

Standing timber can now be seen not only as a potential wood harvest but also as stored carbon. The costs of global warming, if and when it takes hold, may prove to be high. The stored carbon is not contributing to global warming, or to those costs, and therefore it has a market value—the worth of the damage that is avoided by having the carbon stored (Chapter 27).

23.8 Global Warming: A Risk to Be Ignored?

The exact consequence of global warming is not proven, although the above threats are considered real by most scientists. A ban on the emission of greenhouse gases is wholly unrealistic; it would be immensely costly, even if it were practicable. We are harnessed ducting vessels that lie just beneath the bark. If the tree survives, new growth rings will resume, burying the scar inside the tree (see photo). Thus, by cutting a few trees in a forest and counting the number of rings between fire scars, and since the last fire scar, the timing of past fire events is determined.

Values of 80 to 200 years for the return periodicity of fire would be typical in much of Canada, although some areas appear to have cycles as long as 5000 years between fires. It is critically important for land managers that the understanding of fire in these ecosystems moves beyond a purely historical perspective and that the factors that cause variability in fire frequency are identified. It is necessary to analyze soil moisture at different times of the year, and soil moisture will be a function of temperature, precipitation, slope, soil type, and aspect (north- or south-facing). In addition, the time between fires affects the amount of accumulated combustible materials on the forest floor—the longer the intervals between fires, the more there is—and the type of plants. All of these factors play a role in determining the probable return period of a fire. If any one of them were to vary, such as a decrease in soil moisture due to reduced precipitation, a change in fire frequency would be expected.

Climate models predict that as carbon dioxide concentrations increase, summer temperatures in the boreal regions of Alaska, Canada, Europe, and Asia may increase by 4°–8°C. With such summer temperatures, the surface of the land will dry faster, a process compounded by a predicted reduction in precipitation. The new conditions will increase the probability of forest fires. Through computer simulations, ecologists can predict the frequency of fire, its rate and direction of spread, and the conditions that are most likely to immediately precede a forest fire. It is hoped that these models will provide fire crews and inhabitants with warning before a major fire event.

While the human consequences of more accurate models may be measured in lives saved, the ecological benefit is a better understanding of probable consequences for wildlife populations. The migration of species in response to the warming will provide one source of disturbance. Furthermore, the increased frequency of fire will also be a destabilizing influence on these forest systems. Because the succesional sequence takes several hundred years to reach maturity, it is apparent that with migration and increased fire frequency there will be a shift toward earlier successional forests. The loss of mature forest habitats and replacement by younger successional forests and novel community assemblages will provide a different range of niches for wildlife. Too little is understood of the individual ecology of species to make firm predictions of what will happen, but the models demonstrate that populations of species of mature forest systems should be monitored closely because they are clearly at risk.

to carbon dioxide production until we can replace all our energy supply with nuclear, solar, tidal, or wind power. A total ban, even a substantial reduction, of carbon dioxide emissions would require massive changes in industrial processes. Consequently, introducing a regulation limiting carbon dioxide emissions is politically impossible without decisive, unequivocal evidence of global warming. On this basis, many scientists would advocate taking a moderate stance on limiting greenhouse gas emissions.

It is relatively cost-effective to reduce emissions by the first 50%; thereafter pollution control becomes more and more expensive, until it is vastly expensive to remove the last 10% (Figure 23.12). The potential costs of full-scale global warming are so high that it is prudent to take action to limit them. However, given the uncertainties in predicting future climate change, it might be an overreaction to advocate setting economically crippling limits on greenhouse gas emissions. To date, carbon dioxide and methane have not been the target of regulation in North America or Europe, although voluntary measures that industry could adopt have been suggested by the U.S. government. These include the instigation of car-pooling programs for industries with more than 500 employees and nonmandatory goals for carbon dioxide emissions.

In 1997, at an international conference on global warming in Kyoto, Japan, it was proposed that the more developed countries should stabilize their CO_2 emissions at 1990 levels by the year 2010. To achieve this goal, CO_2 emissions must be reduced by about 14% of those projected for the year 2000. Recognizing that industrial growth is needed to raise standards of living in the less developed countries, these nations are not required to reduce their emissions. The conference proposal carries no legal weight until it is ratified

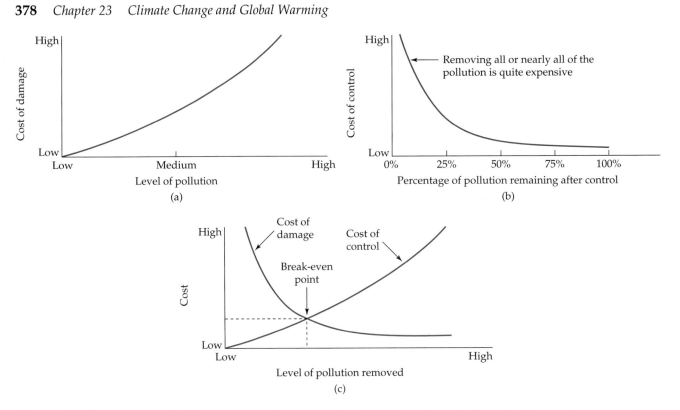

Figure 23.12 **The cost of pollution control.** Removing the first 50% of the pollution load is relatively inexpensive, but a total prevention of pollution is very costly.

by the governments of member nations. Fears that the required emissions reductions will place U.S. industry at a competitive disadvantage have slowed the adoption of this treaty.

From this whole research arena two facts emerge: First, the atmosphere is being enriched with heat-absorbing gases; second, until there is conclusive proof of global warming, the public and politicians would rather gamble that any detrimental effects of this pollution will be mild than change their behavior.

This sets the scene for a refinement of the global experiment. We have the hypothesis: Atmospheric pollution with heat-absorbing gases will cause global temperatures to rise. We have testable predictions: Sea level will rise, deserts will spread, climate will become more erratic, and some new areas will become cultivable. Within our lifetime, we should have a definitive answer refuting or accepting the hypothesis.

Summary

- A natural greenhouse effect raises global temperature by 33°C and makes Earth habitable.
- Increasing the concentrations of greenhouse gases in the atmosphere will bring about climate change.
- The principal greenhouse gas is water vapor.
- The detectable change in global temperature this century is about +0.5°C.
- The release of particulate aerosols may be mitigating the effects of global warming.

- Principal consequences of global warming are destabilization of climates, rising sea levels, increased desertification, poleward migration of species, and loss of agricultural areas.
- Reducing carbon dioxide emissions is an early goal of environmentalists, but it is politically sensitive.
- Carbon sequestration could delay some effects of global warming until other solutions are found.

Further Readings

Bolin, B. 1998. The Kyoto negotiations on climate change: A science perspective. *Science*. 279:330–331.

Karl, T. R., N. Nicholls, and J. Gregory. 1997. The coming climate. *Scientific American*. 276 (5):78–83.

Mazza, P. 1998. The invisible hand: Is global warming driving El Niño? *Sierra*. 83 (3):68–72; 92–95.

deLa Mere, W. K. 1997. Abrupt mid-twentieth-century decline in Antarctic sea-ice extent from whaling records. *Science* 389:57–60.

Schneider, D. 1997. The rising seas. *Scientific American*. 276 (3):112–117.

Web Connections

On-line resources for this chapter are on the World Wide Web at:
http://www.prenhall.com/bush
(From this web-site, select Chapter 23.)

How Does Acid Deposition Affect Ecosystems?

24

The term **acid rain** was first used by Robert Angus Smith as he investigated the air chemistry of Britain's industrial heartland in the 1850s. Cotton mills and heavy industry powered by coal spewed smoke over the city streets. Smith demonstrated that the factory emissions dirtied the air with soot and changed the chemistry of the rain, making it acidic. In the late 1950s, signs were detected that pollution resulting in increased atmospheric acidity could adversely affect forests and fisheries far from the industrial source. This trend was first noted in Europe and Canada; it was not demonstrated in the United States until the early 1970s. Though this form of pollution is commonly known as acid rain, a better term is **acid deposition**, because the acidity can be delivered as a gas or as a dust, as well as in the rain.

Controversy still surrounds the actual effect that acid deposition has wrought on our lakes, rivers, and forests. On one side of the debate, environmentalists have been successful in making acid rain an issue of national and international concern. Consequently, federal and state legislatures have passed laws that limit acid pollution. However, in the early 1990s, a lobby emerged arguing that tree and fish mortalities are not the result of acid imbalance in the system but are due to insect damage, disease cycles, and other natural processes. To investigate the worthiness of these two arguments, we must determine how changes in acidity could affect natural systems.

24.1 Acidity: Definition and Sources

What Is Acidity?

An **ion** is an atom that has either lost or gained one or more electrons. Because the electron carries a negative electrical charge, its departure or arrival will affect the charge of the ion. For instance, the hydrogen atom (H) may lose an electron (e^-) to form a positively charged hydrogen ion (H^+).

$$H \rightarrow H^+ + e^-$$

Similarly, a hydroxide molecule can gain an electron to become negatively charged.

$$OH + e^- \rightarrow OH^-$$

In pure water, hydrogen and hydroxide ions have generally combined to form H_2O, which carries no charge, but about 1 in every 10 million hydrogen ions is not bonded and floats free. The abundance of free hydrogen ions in pure water is taken as our standard for neutrality. If there are more free hydrogen ions than in the standard, the substance is an acid; if there are fewer, the substance is an alkali, or is basic. Because the number of free hydrogen ions varies enormously from one substance to another, the scale of acidity, the **pH** of a substance, is counted on a logarithmic scale. Thus, for a change of 1 pH unit there has been a 10-fold change in the number of hydrogen ions.

The pH index is expressed as the *negative* logarithm of the number of free hydrogen ions. Therefore, the *lower* the value on the pH scale, the *greater* the number of free hydrogen ions, and the more acidic the substance (Figure 24.1).

The Acidity of Natural Systems

Humans are mostly water. It bathes our tissues, fills our cells, and acts as a solvent to move nutrients around our bodies. Water is similarly important to most life forms; it is the medium in which many essential chemical reactions take place. The pH of local water can have a direct effect on the pH of an organism's body fluids and thus can affect the speed of chemical reactions within the body. Hence, the great majority of organisms are highly susceptible to changes in the pH of their surroundings or water supply.

Most natural systems are slightly acidic; even "clean" rain typically has a pH of about 5.6. This acidity is due to atmospheric water vapor combining with naturally occurring carbon dioxide molecules to form a weak solution of **carbonic acid**. Aquatic and terrestrial life has evolved to deal with this mild acidity as the normal input of water to a system. However, the pH of the water is usually modified before it is absorbed by the roots of plants. The rocks that provide the mineral component of a soil may themselves be acid or alkali, a property that will be reflected in the soil. If the rocks are pH-neutral, or so resistant to erosion that they provide very little mineral input, the soils will be acid.

Soil acidity will also be affected by the proportion of decomposing organic material it contains. As organic matter is broken down, it releases acids. Consequently, soils that are very rich in organic matter tend

to have a low pH. Examples of soils acidified by organic material are peats of wetlands and forest soils carpeted by mats of pine or spruce needles.

As rain percolates down through an acidic soil, it absorbs the free hydrogen ions that make the soil acidic. The soil water is then more acid than the rain, and the plant roots will be surrounded by water of pH 4 or even lower. Where the geology is alkaline, as in chalk or limestone rich in carbonates, the soil water will be deficient in free hydrogen ions and rich in hydroxide ions (OH^-). Consequently, the soil and its soil water may have a pH as high as 8 or 9, providing alkaline living conditions for all the organisms of that area. Thus, even though the clean rainwater falling on their leaves is pH 5.6, plants must cope with a supply of water that can be considerably more acidic or alkaline.

Temperature, water availability, and soil acidity are prime examples of environmental gradients (Chapter 9). Given the pH range of natural systems, from the acid bath of a lake beside a volcanic vent to the alkaline conditions of a limestone grassland, it is most unlikely that any of the same species would be found at both extremes. Most species can thrive under only a narrow range of pH conditions. Therefore, community composition will be sensitive to relatively small shifts in acidity, just as they are to changes in any other environmental gradient. Species turnover, where one species is replaced by another that occupies a very similar niche, may result in sharp transitions as soil acidity changes. Consequently, if the acidity of the water entering the system is modified, there may be a considerable reassortment of species within the communities; the changes may be so profound that they warrant renaming the community.

Sources and Cycles of Atmospheric Acidity

The primary pollutants causing the deposition of acid are **sulfur dioxide (SO_2), nitric oxide (NO)**, and **nitrogen dioxide (NO_2)**, and to a lesser extent **chloride (Cl^-)**. The two oxides of nitrogen co-occur to such an extent that they are usually grouped together as **(NO_x)**. These acidifying chemicals can become attached to smoke and dust particles and fall as a dry acid dust close to the source that emits them, or they can stay in the atmosphere as gases. Sulfur dioxide and NO_x are gases that oxidize—that is, they gain oxygen atoms—in the atmosphere and then dissolve in water droplets to form **sulfuric acid (H_2SO_4)** and **nitric acid (HNO_3)**, respectively. These acids are in the rain, snow, and sleet that falls from polluted air.

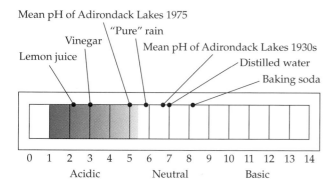

Figure 24.1 The pH scale. This is a logarithmic scale in which each unit change in pH represents a 10-fold change in the number of free hydrogen ions. The more free hydrogen ions there are, the lower the pH, and the more acid the chemical.

Sulfur dioxide and NO_x are produced by volcanoes and so have been natural components of the atmosphere since the earliest times. Indeed, it was atmospheric sulfur that was the substrate metabolized in the first photosynthetic reactions 3.5 billion years ago. Sulfur dioxide and NO_x are rare gases in the natural atmosphere, but human activities in the past century have greatly increased their concentrations, primarily through the burning of fossil fuels.

Nitrogen oxides are released by the combustion of gasoline, and, despite the technology of catalytic converters, mufflers, and tailpipes, some escape to the atmosphere. Tailpipe emissions from vehicles are now the largest single source of these gases.

Emissions of sulfur dioxide are strongly associated with the burning of coal. Coal powered the industrial expansion of the United States and Europe in the 1800s and early 1900s, and it was once commonly used as a fuel for domestic heating. However, the manifest dirtiness of coal as a power source and the increasing availability of natural gas led to a change in consumption. In most industrialized nations, the burning of coal, and hence the production of sulfur dioxide, is now primarily associated with electricity generation and the smelting of iron ore.

The amount of sulfur released for each ton of coal burned will depend on both the pollution-control technology and the type of coal. Coal is the fossilized remains of swamp plants that died during the Carboniferous period, a geologic era that ended 286 million years ago. The swamps were rich in decomposer bacteria that produce hydrogen sulfide (rotten egg gas; H_2S). Despite the activity of the bacteria, plants collapsing into the swamps were not completely decayed, so an organic peat mat accumulated. Through time, these peats were transformed into coal. During the fossilization process, sulfur from the corpses of plants and some of the bacterial hydrogen sulfide bonded to the carbon molecules that were to become coal. A typical sulfur content for coal used by the U.S. power industry is about 2.2%. However, some coal, known as **low sulfur coal**, has a sulfur content less than 1%. The sulfur remains locked to the carbon until it is burned, when it is released as sulfur dioxide.

Sulfur dioxide is not a particularly reactive gas, but it is soluble in water, so it does not accumulate in the atmosphere. Over the course of 7–14 days, the sulfur dioxide undergoes atmospheric oxidation and dissolution in water vapor. The next rainstorm washes the sulfur dioxide out of the atmosphere, and it falls to the ground as a dilute solution of sulfuric acid.

24.2   The Effect of Acid Deposition on Terrestrial Systems

In some areas, the soil contains enough calcium ions, an alkali, to buffer any potential increase in acidity. If the buffering action is strong enough, the soil pH may be unchanged despite the presence of acid deposition. The buffering comes about because the clay minerals in soils have a mild negative electrical charge, so they attract and hold the positive ions. The strength of this negative charge is termed the **cation exchange capacity (CEC)**. The critical word here is *exchange* because, as the balance of ions within the soil changes, the actual ions held on the surface of the clay may be swapped. A hierarchy of replacement exists among the ions: Aluminum is the hardest ion to dislodge, then comes hydrogen, and then calcium on down to sodium.

$$Al^{3+} > H^+ > Ca^{2+} > Mg^{2+} > K^+ > NH^{4+} > Na^+$$
easier to displace →

Thus, when hydrogen ions are added to the soil, they can replace calcium on the surface of the clays. The absorption of the hydrogen ion by the clay effectively removes it from the pool of reactive soil chemicals, and its potential for creating acidity has been reduced.

The above hierarchy of CEC assumes that all the ions are about equally abundant, but the order of the system breaks down if one of them, for instance H^+, is present in much greater concentrations than the others. If there is insufficient calcium in the soil to act as a buffer and the soils are already acid, increased inputs of H^+ ions can cause the release of aluminum even though it is higher in the exchange hierarchy. Soils that lack the calcium ions to buffer changes in pH will be vulnerable to damage by acid rain. Such calcium-poor soils cover about 70% of the eastern United States.

An example of hydrogen replacing aluminum and its consequences has been documented recently in the northeastern United States, where one of the common clay minerals is gibbsite, an oxide of aluminum (Al_2O_3). When soil acidity increases, there is a massive surge in the number of free H^+ ions, and the Al^{3+} of the gibbsite is replaced by H^+. This replacement has a huge impact on the local ecosystem because aluminum can be a potent poison. So long as the aluminum remains bonded to the oxygen as Al_2O_3, it is harmless. But when it is released as Al^{3+}, it is poisonous. The free aluminum ions, released by the acidification of the soil, are highly toxic to many plants and animals. Thus, increasing the acidity of the soil can release toxins that kill trees and, if washed into aquatic systems, poison rivers and lakes.

The data in Table 24.1 provide us with an insight into the rate of decomposition in two forests at two different times. Between 1966 and 1983, Norwegian soils became markedly more acidic, often changing by one full pH unit. Although the rates of decomposition, and hence nutrient cycling, are not measured directly, much information can be inferred from the concentrations of organic matter, nitrogen, and phosphorus in the soil.

The rate of decomposition can be thought of as the time it takes for a nutrient held in the leaf on a tree to pass through the cycle of leaf death, breakdown into its composite chemicals by decomposers, and reabsorption by the plant to make a new leaf (Chapter 6). The faster the rate of decomposition the less organic material is left in the soil at the end of a year. Thus, relatively low numbers in Table 24.1 indicate relatively fast rates of decomposition.

Leaf litter was broken down and recycled fastest in the beech forest, leaving less organic matter to accumulate in the soil. In both forests between 1966 and 1983, it is evident that organic matter and nutrients were accumulating in the soil. These data are important because they indicate that the rates of decomposition and the nutrient cycling slowed down. Because it is taking longer for nutrients to be made available for plant uptake, it is likely that the plants will become nutrient-limited and the net primary productivity (NPP) of the system will be reduced.

Death of decomposers, poisoned by excess aluminum in the soil or simply by the increased acidity could explain the slow rate of decomposition. A further factor limiting plant growth would be if the mycorrhizal fungi that live symbiotically on the roots of trees have been poisoned (see Chapter 6); their death

would inhibit the absorption of nutrients by the roots. Another effect of increased soil acidity is that some essential nutrient chemicals, such as phosphorus, are precipitated as insoluble compounds that cannot be absorbed by plants. For these various reasons, the accumulation of nutrients in acidified soils is a sure sign that productivity is falling.

Trees tend to enhance the negative effects of acid rain on a soil. Although most plants take in almost all their water through their roots, some water is absorbed through the leaf surface. Rain, dew, and mist all provide moisture that in unpolluted systems will usually have a pH of around 5.6. As the atmospheric moisture becomes more acidic, so the water absorbed through the leaf surface will have a low pH. Acid within the leaf will degrade chlorophyll and reduce the photosynthetic capacity of the plant. The acidity is buffered internally by the plant, using calcium as an alkali to neutralize the acid. Plants achieve this buffering by pumping hydrogen ions out through their roots in exchange for calcium. Thus, the plants are drawing in an alkali and pumping out an acid, making the soil conditions yet more acidic (Figure 24.2). There may come a point when hydrogen ions are sufficiently common in the soil, and calcium is exhausted; then aluminum ions are released from the surface of clays and trees will be poisoned by absorbing the aluminum ions through their roots. Aluminum is an important chemical in the acid deposition equation, but it is by no means the only toxin that plays a role in determining the fate of forests and aquatic systems. Other potential toxins, such as cadmium, lead, copper, zinc, and arsenic, can all be released as the acidity increases.

At this stage, a good ecological question to ask is, Can we predict which ecosystems will be most affected by acid deposition? A general answer will be, Wherever acid deposition falls on soils that cannot buffer the effects of a low pH input. More specifically,

Table 24.1 Effect of acid rain on the storage of organic matter and nutrients in Norwegian forest soils.

Type of forest	Year	Organic matter (kg/ha)	Nitrogen (kg/ha)	Phosphorus (kg/ha)
Beech forest	1966	29,600	809	52
	1983	48,100	1269	71
Spruce forest	1966	49,000	960	53
	1983	96,300	2034	101

(*Source:* C.O. Tamm, 1989.)

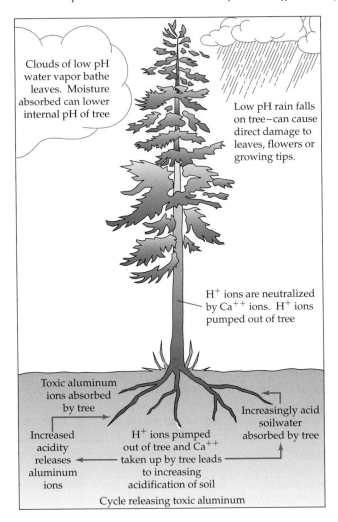

Clouds of low pH water vapor bathe leaves. Moisture absorbed can lower internal pH of tree

Low pH rain falls on tree–can cause direct damage to leaves, flowers or growing tips.

H⁺ ions are neutralized by Ca⁺⁺ ions. H⁺ ions pumped out of tree

H^+ ions are neutralized by Ca^{++} ions. H^+ ions pumped out of tree

Toxic aluminum ions absorbed by tree

Increasingly acid soilwater absorbed by tree

Increased acidity releases aluminum ions

H^+ ions pumped out of tree and Ca^{++} taken up by tree leads to increasing acidification of soil

Cycle releasing toxic aluminum

Figure 24.2 Trees can increase the acidity of local soils as a result of their internal chemical buffering. Calcium ions are absorbed through the roots while hydrogen ions are pumped out into the soil. The increased soil acidity frees aluminum ions, which are toxic. The aluminum ions are then absorbed by the trees, leading to tree mortality.

we find that many of the areas hardest hit by acid rain are forests on hilltops. Such populations appear to be especially vulnerable, and their increased mortality rates may be due to one or more of the following factors:

- Low clouds are more likely to form around high ground than in the lowlands, and if the clouds are acidified, the trees are being bathed in acid water vapor with a pH as low as 3.5.

- The humidity of this cloudy environment can also promote the growth of some acid-loving mosses that kill the mycorrhizal fungi on the tree roots, and hence reduce the nutrient uptake of the trees.

- Soils on hilltops tend to be thin and more leached than those in the lowlands, so they are less likely to have available calcium for buffering acidity unless the bedrock is a calcium-rich chalk or limestone.

- The interaction between hydrocarbons, nitrogen dioxide, and UV radiation produces tropospheric ozone (Chapter 22). The combination of ozone and atmospheric acids provides a brew that is especially toxic to evergreen trees.

Acid or Aphids?

Mount Mitchell, North Carolina, is the highest point in the mountains of the eastern United States, and it is one of the areas where tree mortalities have been attributed to the combination of high ozone concentrations and acid deposition (Figure 24.3). Ironically, the trees themselves may be releasing terpenes, highly reactive hydrocarbons that combine with nitrogen dioxide (NO_2) to form ozone (Figure 24.4). The atmospheric acids are borne in on westerly winds from the industrial areas of Tennessee. Almost all the red spruce and Fraser fir above 2000 m elevation on Mount Mitchell are losing their needles, and about half of them are dead. The acidity alone may not be enough to account for the death of the trees, nor are the concentrations of ozone necessarily fatal. However, recent studies demonstrate that pollutant chemicals can interact to produce a synergism of destruction. Experiments show that plants exposed to low levels of ozone or sulfur dioxide pollution appear unaffected. But if the plants are exposed to the same levels of both pollutants, there is a marked increase in mortality rates.

Attacks of aphids and other insects have been offered as an alternative explanation, particularly by the industrial lobby, for widespread tree mortality in the eastern forests. Indeed, in the past few years there has been an outbreak of aphids and gypsy moth caterpillars on the trees of the Appalachian forests. Aphids suck the photosynthetic products from their hosts and have undoubtedly killed large numbers of trees. Similarly, the damage caused by gypsy moth caterpillars has resulted in the defoliation of whole hillsides. Nevertheless, the decline of evergreen trees in the eastern United States cannot be accounted for solely in terms of increased predation. The death of the trees is more widespread than the insect infestations, and it follows a pattern that is consistent with areas of high ozone concentrations, significant acid deposition, and the lack of a natural chemical buffer. If the health of the forests is being reduced by chemical attrition, the insects may merely be hastening the demise of an already weakened host.

Figure 24.3   Dead coniferous trees around the summit of Mount Mitchell, North Carolina. The trees were killed by acid deposition from low-level clouds.

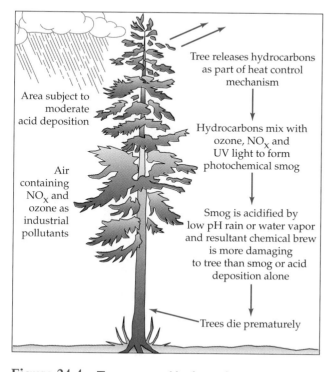

Area subject to moderate acid deposition

Air containing NO_x and ozone as industrial pollutants

Tree releases hydrocarbons as part of heat control mechanism

Hydrocarbons mix with ozone, NO_x and UV light to form photochemical smog

Smog is acidified by low pH rain or water vapor and resultant chemical brew is more damaging to tree than smog or acid deposition alone

Trees die prematurely

Figure 24.4   Terpenes and hydrocarbons are released by some trees. These chemicals can contribute to a form of natural photochemical smog. This smog becomes harmful to the plants when it is combined with clouds that are acidic.

24.3   The Effect of Acid Deposition on Aquatic Systems

The oceans are too vast to have been affected by acid deposition, but life in lakes and rivers is especially sensitive to changes in pH. It was the loss of fish from Scandinavian lakes that triggered some of the initial concern over acid deposition. By the 1970s, fish stocks had been entirely eliminated from most lakes in southern Norway, representing an important commercial and recreational loss.

Just as on land, aquatic ecosystems are affected directly by the increase in acidity and indirectly by the release of toxins as pH falls. The toxins released by the acidity may be washed into the system as water drains from the land or freed from the sediment in the bottom of the lake. The poisoning of the food chain starts with the photosynthetic algae, and once the food base is undermined the whole system is threatened with collapse. The toxins also work directly on invertebrates and on the fish of higher trophic levels. For example, as acidity increases, rising aluminum concentrations irritate the gills of brook trout. The trout secretes mucus, which is a response to cleanse the gills of an irritant chemical or parasite. In this instance, the aluminum irritation continues and so much mucus is produced that the gills become clogged and the trout die of asphyxiation.

Severe acid pollution resulting in water with a pH of less than 4.5 kills adult trout, and, between pH 5 and 4.5, reproduction will be inhibited. Female fish may fail to ovulate, or, if the eggs are viable, the hatchlings are often deformed or fail to develop (Figure 24.5). The young prove susceptible to changes in pH even at levels that are still tolerable for adults. Thus, it is possible to have a population of aging trout with no replacements in the next generation.

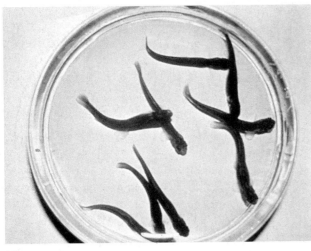

(a)

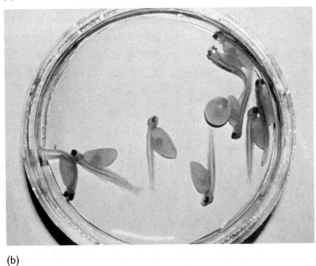

(b)

Figure 24.5 Young trout exposed to water of (a) pH 5, which they can tolerate, and (b) pH 4.5, which is lethal to them.

The effects of decreasing pH on fish populations was documented in George Lake, Ontario. The acid rain originates from the industrial heartland of the United States, and in just 12 years, between 1961 and 1973, it caused the pH of George Lake to fall from 6.5 to 5.0. The progressive decline in fish populations (Figure 24.6) started with the failure of reproduction among walleye, an acid-intolerant species. The fish were failing to reproduce, but the survival of some adults explained the time lag of seven years between the last reproduction and the last catch of the adult fish. As pH fell from 5.7 to 5.0, there was a decline in the abundance of the more acid-tolerant white sucker fish and an even greater decline in their biomass, indicating that the average size of fish in the lake was also declining. When the lake was so acid that it was pH 5, the remaining fish showed deformities attributable to a lack of calcium. Fish need calcium to grow healthy bones and for successful ovulation.

Increasing acidity can have a dramatic effect on all trophic levels in a food chain. Although fish such as yellow perch and lake trout can survive a pH as low as 4.5, many of their prey items, including mayflies and hatchling fish, cannot tolerate water with a pH lower than 5 (Figure 24.7). In studies of invertebrates (animals without backbones, such as insects, mollusks, and crustaceans) in New England streams, it was found that a change in pH from 5.5 to 5 brought about a halving of the species diversity. A similar study in England revealed a considerable variation of species diversity in unpolluted streams (pH 6–7), but invertebrate diversity declined markedly as pH decreased (Figure 24.8). Most of the species that still survive in acidified water are struggling to do so, and their populations are reduced. The food chain has been weakened, and if top-level predators are not killed outright by the changed chemistry, they may starve. The decline in photosynthetic productivity as a lake turns acidic can result in a once moderately eutrophic lake looking blue. The blue appearance of the lake is due to a lack of life rather than to a lack of nutrients, but because we equate a blue lake with a clean lake, their appearance does not generate immediate concern.

Acid deposition has the most effect on lakes in areas that have acid geologies, where there are no buffering alkaline soil chemicals, and in areas that have a significant period of freezing temperatures. During winter, acids arrive with each snowfall and are stored in the snowpack until the spring. With the first warmth of spring, the snowpack melts, and meltwater

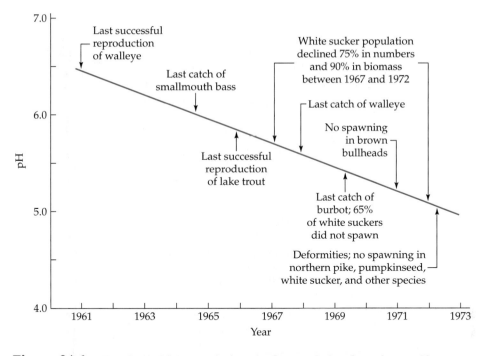

Figure 24.6 **The fate of fish populations in George Lake, Ontario, as pH was reduced by acid deposition.** Failure of reproduction is followed by the loss of the species. As pH continues to fall, the remaining fish increasingly suffer from deformities due to calcium deficiency. (After R. Beamish, 1976.)

laden with acidity pours into rivers and lakes. The first 30% of the meltwater carries almost all of the acid with it, and commonly has a pH between 3 and 3.5. Such pulses of acidity, though of short duration, are lethal to most organisms.

It is important to note, as an aside, the difference between the toxicity of acid pollution—which kills outright, prevents reproduction, and causes mortality at all levels in the food chain—and the toxicity of chemicals that bioaccumulate (Chapter 18). Bioaccumulation affects the highest trophic levels the most, often leaving the phytoplankton and zooplankton relatively untouched. Conversely, acidity and aluminum ions, although potentially toxic to plants and animals, do not accumulate through the food chain.

24.4 Acid Transport and Buffered Systems

Dry deposition of acid pollutants will generally be close to their source of origin; they can be a severe, but essentially local, problem. The poisonous effects of the acid on local vegetation are generally so obvious that it is possible to identify the producer—an important

point, if pollution regulations need to be enforced. The acidic gases that dissolve in atmospheric water present a different set of problems. Industries emitting these gases may install tall smokestacks that help the emissions to be caught by the wind and dispersed. This is an example of the maxim "Dilution is the solution to pollution." Certainly, when a small volume of acid emission is mixed with a larger volume of air, it is no longer possible to see a distinct cloud of chemical gas. And when the point source of pollution is raised 100 meters into the air, the local consequences of the pollution are much reduced. However, another saw is perhaps more accurate: "What goes up must come down."

Most areas of the industrialized world have a prevailing wind direction, a predictable path followed by the wind most days out of the year. In the United States, high-pressure cells form in the Northwest, swing south and east across the country until about the longitude of the Dakotas, and then drive toward the eastern or northeastern coast. These pressure cells are following the course of the jet stream, a powerful belt of moving air that flows west to east across the continent, marking the boundary between the upper level of the troposphere and the lower stratosphere (Chapter 22). Because air will move from an area of

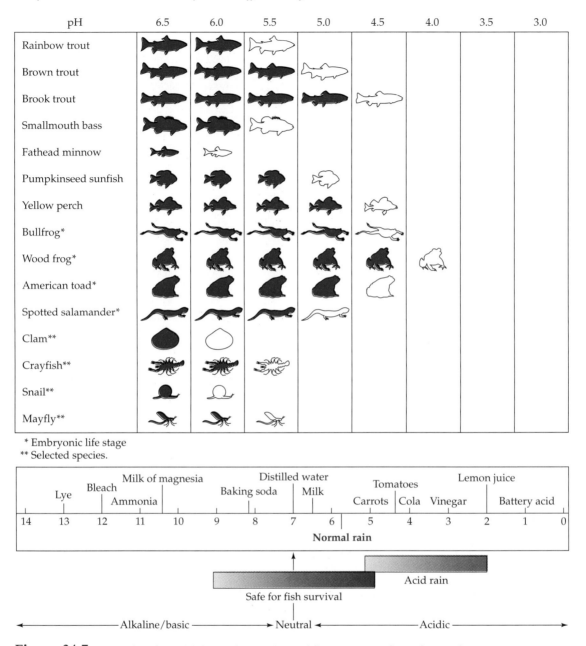

Figure 24.7   A scale of sensitivity to increasing acidity among selected aquatic creatures. The approximate pH level at which the organism dies is indicated by an organism icon in white.

high pressure toward one of low pressure, the wind direction mirrors the eastward procession of the pressure systems. In the eastern United States, most winds blow from the west or northwest and so are termed "westerlies." Thus, the jet stream determines the paths of the pressure systems, the pressure systems determine the winds, and the pollution travels on the wind. Although the gases, particulates, and steam from a smokestack quickly become invisible, they, and pollutants from other smokestacks, stream with the wind.

It can take up to two weeks for the sulfur dioxide emitted from a smokestack to become incorporated in rain as sulfuric acid. In those two weeks, the gas can be blown a long distance from the pollution source. Consequently, the pollution emitted from the industrial heartland of the Midwest falls in a band from Tennessee all the way to Labrador (Figure 24.9).

In Europe, where the prevailing winds are also from the west, the industrial centers of Britain and Germany emit acid gases that fall on Sweden and

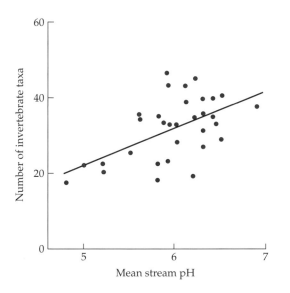

Figure 24.8 The species diversity of invertebrates in streams within the Ashdown Forest, southern England, plotted against the pH of the stream water. As the pH decreases—the water becomes more acidic—species diversity declines. (After C.R. Townsend, A.G. Hildrew, and J.E. Francis, 1983.)

Norway. It is estimated that only about 45% of the sulfur dioxide produced by British industry falls within Britain, making it a net exporter of pollution. Sweden, on the other hand, receives about 500% of the sulfur dioxide that Britain produces, making it a net importer. Some of the pollution arriving in Sweden emanates from Britain, some from Germany, and some from the countries of eastern Europe. By the time the pollution arrives in Sweden, it is impossible to apportion blame to a particular country, let alone to an industry. It is easy to understand the moral outrage of the unfortunate pollution recipients at finding their forests and fisheries spoiled by a third party. However, it is also understandable that the polluters do not want to accept responsibility for damage to another country that cannot be proved to be their fault alone. The potential for conflict over pollution restitution is not restricted to Europe. Throughout the early 1980s, the responsibility for acid deposition on Canadian forests was a source of tension between the governments of Canada and the United States. The United States was clearly the exporter of some of the pollution, but the administration of President Reagan would not accept total responsibility. These tensions also exist at a state level, and a full resolution has yet to be reached between the states of New England (importers) and those of the Midwest (exporters).

24.5 Solutions to the Acid Deposition Problem

Limiting the burning of coal or ensuring that acid-forming chemicals are "scrubbed" from emissions would stop the problem at its source. However, such a simple solution is unlikely to happen because coal is such an important fuel commodity, providing 28% of the power generated globally. Although the economic recession of the early 1990s and the restructuring of eastern European industry have led to local declines in coal usage, the global consumption is more or less steady. The principal users of coal are now countries attempting a rapid industrial expansion. China derives 75% of its power from coal and is presently the largest single user, accounting for 25% of all the coal that is burned globally. India is the fourth largest consumer of coal. Both these nations rely on their domestic coal reserves to supply relatively inexpensive electricity. Developing nations, particularly those governed by totalitarian regimes, generally have poor pollution control records. Their priority must lie in basic social and economic development; environmentalism is perceived as a luxury that comes later. If the developed nations are truly concerned about reducing acid emissions, it is highly likely that they will have to subsidize clean energy programs in the developing world.

The effects of acid deposition are not all on a global scale, and local legislation could make a substantial difference in the quality of fisheries and forests. In the United States, the Clean Air Act of 1990 stipulated that sulfur dioxide emissions would be cut from the 1990 level of 17 million metric tons to 8.9 million tons by the year 2000. By 1995, emissions had been reduced below 6 million tons. Emissions of NO_x were regulated for the first time by this act. The new target outputs of these gases were met, reducing 5 million metric tons of NO_x emissions to less than 3.5 million tons by 1992.

Cities would be major beneficiaries of reduced acid deposition. Acid damages buildings made of acid-soluble materials, such as marble, sandstone, and limestone, and the Council on Environmental Quality estimates that air pollution causes about $10 billion in architectural damage to U.S. buildings each year. The acidity in the rain also speeds the rusting of metal and etches the finish on paint, leading to the premature aging of cars; this damage costs the public billions of dollars in depreciation.

"Band-aid solutions" are possible. Cars can be given a protective coat that inhibits acid damage, and

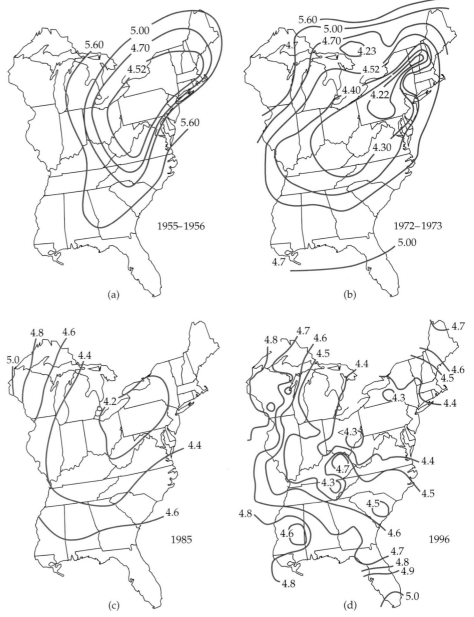

Figure 24.9   The acidity of rain across the eastern United States between 1955 and 1996. Between 1955 and 1985 there was a steady decline in the pH of rain falling in the eastern United States. Clean air legislation has reduced sulfur emissions and the 1996 data indicate a slight increase in the pH of precipitation across many areas. (Data are from the National Atmospheric Deposition Program/National Trends Network.)

adding lime to the soil or to lakes helps offset the effects of increasing acidity. These solutions are expensive and merely mask some of the most obvious symptoms without treating the disease. Each year, Sweden spends approximately $40 million adding lime to lakes in order to preserve conditions that can support fish. If this practice were extended to the United States the annual costs would be about $8 billion to deacidify our lakes. The regulations imposed by the 1990 Clean Air Act have already gone a long way to stabilizing the decline in water quality caused by acid deposition in the United States. Indeed, the first signs of rising pH values are apparent. Note in

Figure 24.9 that the area receiving precipitation with a pH of 4.3 or less is much smaller now than in 1984. That such a change is evident so soon after the reductions in SO_2 went into effect is encouraging and demonstrates that environmental improvement on a grand scale is possible.

Emissions Trading: A New Solution to Pollution?

The proposed emissions reductions of the Clean Air Act of 1990 were particularly worrying to the energy industry. Coal- and oil-burning utilities (power companies)

account for 70% of the sulfur dioxide production in the United States and are the largest individual polluters. Not all utilities could reduce emissions by 50% within 10 years, because doing so might require a complete overhaul of their manufacturing process. The goal of the legislation is to improve air quality, not to drive industry out of business. Rather than set impossible targets, the act sets an overall air quality goal and recognizes that some industries will be overachievers and some underachievers in the quest for reduced pollution.

The act establishes a system that penalizes polluting industries and rewards clean industries. On the basis of previous sulfur dioxide emissions, each company is issued an allowance for the amount of pollution that it may produce in the following year. The novel part of this scheme is that unused portions of the allowance may be banked, traded, or sold.

When a company has an unused pollution allowance that it holds against a future need (either a future expansion of the company or a decrease in the pollution allowances allocated), it can bank the allowance. Alternatively, a company can trade an unused portion of its pollution quota to offset pollution from another plant owned by the same company. The third option is for the unused pollution allowance to be sold through a commodities market to a company in a "dirty" industry.

This sale of allowances allows industries to continue to pollute in the short term. Thus, the heavy sulfur dioxide polluters have the choice of paying to pollute or reducing their emissions to comply with the standards set by the act. An important part of this process is that the money used to purchase a pollution permit does not go to the government as a tax but is paid as a reward to the "clean" company. Ultimately, all companies will have to comply with the new emissions standards, but the allowance system gives the firms a breathing space to develop and install cleaner technologies.

A small but important point is that these allowances are not permits because they do not confer any rights to the industry to maintain these levels of pollution in the future. Because the government issues the allowances, it can control the overall amount of pollution that will be released each year. Through time, the yearly number of allowances issued by the government will be decreased, bringing an improvement in air quality. If this scheme works as it is intended, the allowance system will ensure that the nation's industries will not exceed a total pollution budget. The effect of the act has been to significantly reduce SO_2 pollution both at a national and a regional scale. Note the

sharp drop in SO_2 emissions between 1990 and 1995 across the industrial heartland of Ohio, Illinois, Indiana, and Michigan (Figure 24.10).

Critics of the system suggest that instead of a general cleanup, ghettos of heavy industries will form pollution black spots by buying pollution allowances. Further criticisms center on the charges made by the brokers who run the actual allowance exchange system. Some estimates suggest that the brokerage fees would act as a strong disincentive for any firm to try to sell its excess allowances.

The Early Pollution Credit Sales

The firms that wish to purchase extra pollution allowances can either buy them in private sales or bid for them in an annual auction. The allowances are auctioned in units of 1 metric ton of sulfur dioxide. In other words, by purchasing one allowance, the company can release one additional metric ton of sulfur dioxide into the atmosphere. The first auction of sulfur dioxide allowances was held in 1993. It was anticipated that the allowances would sell for about $400 per metric ton; the actual price was somewhat lower at $320 per ton. In March 1996, the auction price for a one metric ton sulfur dioxide allowance had fallen to just $69 (although this did rise to $100 in 1997). With a price as low as $69, it is evident that the demand for the allowances cannot be high. If the industries are complying with the law, they must be finding other ways to meet the new air quality standards.

In 1994, of the 260 utilities with the highest sulfur dioxide emissions, only 24 utilities (9%) took part in the bidding, and only 10 (4%) successfully bought allowances. Clearly, with such a low participation, the utilities have chosen a different strategy to comply with the law. About 55% have switched to low-sulfur coal, and about 16% have installed scrubbers, devices that remove pollutants from emissions before release. About 5% have converted to natural gas and 3% have been shut down. A further 18% appear to have done nothing. Either they were allocated too many allowances initially and did not need to reduce emissions, or they prefer to pay fines on the excess pollution they create.

The Electric Power Research Institute estimates that alternative ways to reduce sulfur dioxide emissions range in cost from $330 to $1147 per metric ton per year, much more than the $69 for a pollution allowance. That more utilities have not opted for the trading program may express a lack of faith in this innovative approach. Traditionally, the utilities industry has been deeply conservative, and it may take some

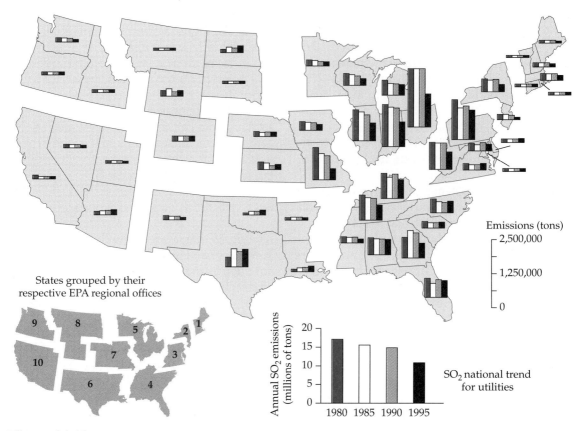

Figure 24.10   Emissions of sulfur dioxide across the United States 1980–1995. The Clean Air Act of 1990 led to a marked drop in SO₂ emissions in the industrial heartland of the midwestern states. (Data from the Environmental Protection Agency.)

time for this new policy to be seen as a viable way to do business. Utilities may also be worried that buying allowances could result in negative publicity such as, "Local utility spends thousands to continue to pollute our air."

Perhaps the most important outcome of the SO_2 trading program is that it signals a change in the way that the federal government deals with industrial polluters. Rather than simply telling companies what to do and when to do it, the permit system offers more flexibility for industries to find a way to meet a pollution reduction goal that is right for them. Permit trading may become a standard way of buffering industry from the worst impacts of compliance with environmental regulations. Indeed, the Kyoto conference on global warming (Chapter 23) proposed a trading program for CO_2 in which nations could buy and sell emissons permits during a compliance period.

Summary

- Unpolluted rainwater typically has a pH of about 5.6.
- Rainfall across much of the eastern United States is now pH 4.2–4.5.
- Sulfur dioxide and NO_x are the principal chemicals that cause acid deposition.
- Coal burning and vehicle emissions are the largest contributors to acid deposition.

- Increased rain acidity can damage plants, animals, buildings, and cars.
- Plants can be poisoned by aluminum released from acidified soils.
- Local soils and bedrock, if rich in alkalis, can buffer the effects of acid deposition.
- Interaction of acids with other pollutants can produce added toxicity.

- The Clean Air Act of 1990 and international agreements have started to address air quality and acid deposition.

- The sulfur dioxide allowance trading program has started slowly, but the SO_2 emissions targets of the Clean Air Act have been achieved.

- The permit trading program demonstrated that initial industry estimates of the cost of abating pollution were overstated.

Further Readings

Kowalok, M. E. 1993. Common Threads. *Environment* 35 (4):13–200; 35–38.

McCormick, J. 1998. Acid pollution: The international community's continuing struggle. *Environment* 40 (3):17–20; 40–45.

Hutterman, A. et al. 1994. *Effects of Acid Rain on Forest Processes*. New York: John Wiley.

Munton, D. 1998. Dispelling the myths of the acid rain story. *Environment* 40 (6):4–7; 27–34.

Web Connections

On-line resources for this chapter are on the World Wide Web at:
http://www.prenhall.com/bush
(From this web-site, select Chapter 24.)

PART

FOUR

Ecology and Society

The Use and Supply of Energy

25

Energy is a central theme that connects apparently disparate parts of ecology and environmentalism. Without the sun, plants would be without energy, and without plants, animals would starve. It is the need for energy that drives predator-prey cycles and limits reproductive effort. Our demand for metabolic energy underlies all agricultural development and its environmental consequences. We may think of these, the power of sunlight and metabolism, as "natural" sources of energy, whereas energy production in commercial power plants is a "supplemental" source of energy. Yet the division between the two is fuzzy. For instance, we need "artificial" energy in the form of gasoline and diesel fuel to power farm machinery and oil byproducts to synthesize fertilizers and pesticides. The manufacture of farm chemicals in huge industrial plants requires electricity, as does the farmer's house or the store where we purchase produce. Thus, we find that even providing natural sources of energy for metabolism requires large supplements of artificial energy. Without these inputs of energy, farm production would plummet and we would be denied the food used for our metabolic energy. Energy availability and the well-being of human populations are thus linked. The health of human populations is of interest to ecologists in its own right, but so too are the consequences of human actions as populations grow. Energy generation is also of interest to ecologists because of pollution associated with power production and economic development of oil-owning countries that have high biodiversity.

25.1 Power Plants Do Not Make Energy

Although we routinely speak of "energy production," to create energy is an impossibility. Just like physical matter, energy can be neither created nor destroyed; it can only be converted from one form to another (First Law of Thermodynamics; Chapter 3). The initial input of energy is almost always solar energy that is converted to another form. For example, solar energy is trapped by plants and stored as chemical energy. The plant dies, fossilizes, and turns into coal. The coal is mined, burned, and converted to heat energy. The heat generates steam, which is used to turn turbines (kinetic energy) that generate electricity. With each transformation of energy, some energy escapes. Just as in a food chain, this energy is not "lost" in a real sense; it just has not been captured by the next step of the procedure. This waste energy may be vented as steam, hot water, noise, or chemical residues. The proportion of energy that is successfully converted into electricity, compared with the initial potential energy of the source fuel (coal, oil, gas), is generally around 30%.

25.2   A Brief History of Energy Use in the United States

Throughout much of this century, the demand for energy grew steadily, although the means of meeting that demand changed as one technology succeeded another. Indeed, there can be few better ways to chart the development of the United States over the last 150 years than to look at patterns of energy consumption (Figure 25.1).

In the mid-1800s the principal fuel for heating and cooking was wood; energy consumption was low, as a largely agrarian society relied on horses and mules for transport and farm power. The economic transition to an industrialized nation is energy expensive as factories and houses replace fields. Fewer farmers must support a growing urban population, while food and industrial commodities must be shipped by road, rail, or river. The whole infrastructure of a nation must be built. Consequently, as the United States made that transition during the late 1800s, energy consumption rose sharply. Coal powered this industrial revolution and was not overtaken in terms of importance until after World War II, when oil became the prime energy source.

The postwar economic boom increased the demand for fuel as did the intensified agricultural practices em-

ployed in the Green Revolution. The attainment of the "American dream" by unprecedented numbers of Americans during the 1950s and 1960s was built upon easy access to energy. The economy boomed, farm and industrial output grew, and changes required energy. Cars became larger, air conditioning and central heating became the norm rather than the exception, and electrical appliances ranging from freezers to hair dryers came within the grasp of the majority. This same period saw the growth of another oil-based industry that changed the face of consumerism: the manufacture of plastics. The demand for energy surged upward.

For the first 70 years of the century, the demand for electricity had grown about 8% each year. In other words, the capacity to generate power had to be doubled every nine years. The United States, like most of the developed nations of western Europe, was becoming increasingly reliant on imported oil to fuel power stations and millions of internal combustion engines. A true turning point in the politics of energy came in 1973 when the Middle Eastern nations banded together to form the powerful Organization of Petroleum Exporting Countries (OPEC). Until that time, the Arab nations of the Middle East had negotiated independent deals to sell their oil. By organizing this cartel, they could eliminate cost-cutting competition and charge higher prices.

In 1973, a religious and resource-driven war pitched Jew against Arab as Israel fought Egypt and Syria.

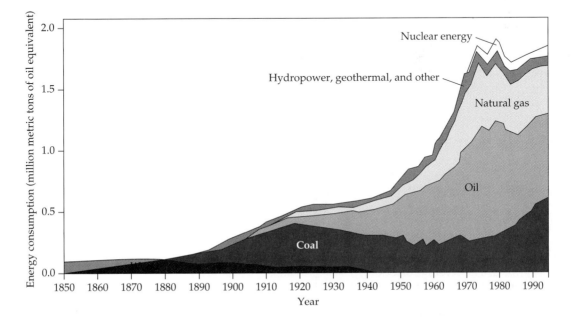

Figure 25.1   Changes in energy sources and demand in the United States since 1850. (Data is from F.T. MacKenzie, 1998.)

Figure 25.2 **Motorists wait in line for precious gasoline.** The OPEC oil embargo of 1973–1974 caused gasoline shortages in the United States and underlined the vulnerability of a nation that is dependent upon imported energy sources.

American support for Israel in the 18-day war so angered the Arab nations that an OPEC oil embargo was imposed on the United States. The embargo lasted until March 1974 and demonstrated how vulnerable the U.S. economy had become to changes in the politics of oil distribution. Before 1973, the United States had been relying on OPEC oil for 30% of its consumption. As OPEC restricted the sale of oil to the United States, there developed a shortage of oil and gasoline, and gas stations hosted long lines of motorists waiting to fill their tanks (Figure 25.2). Despite

higher prices at the gas pump, the United States continued to increase its imports as soon as the supply was resumed. Between 1973 and 1981, the price of oil rose from about $3 to $35 per barrel (Figure 25.3), imports hit a maximum of 48% of all oil used in the United States, and two-thirds of the imported oil came from OPEC nations. However, in the late 1990s a demand for cash in oil-producing nations has led to high rates of extraction and a glut of oil on the world market. The overproduction has resulted in prices as low as $11 per barrel. In real terms, that is, relative to income, gasoline is now cheaper for motorists in the United States than ever before.

In the latter half of the twentieth century, the United States has gone from the world's greatest oil exporter to the greatest importer. Subsidized oil prices, and a near refusal to impose a tax on gasoline, provide Americans with the cheapest oil and gasoline of any more-developed country (MDC) and the least incentive to become more fuel efficient. Although the overall demand for energy in the United States reached an approximate peak in the early 1970s, the United States remains a gasoline-dependent society. Since the mid-1970s, energy demand has grown at just 3% each year, reflecting a decreasing population growth rate, the first attempts to conserve energy, and the near saturation of the domestic market with major electrical appliances. However, oil imports alone cost the United States $80 billion in 1990, an annual debt of about $360 per person that must be paid or balanced by exports.

25.3 Nuclear Power: Fallacy of a Dream Foretold

Could nuclear power be the answer to our energy dilemma? The energy planners of the 1950s and 1960s

Figure 25.3 **Average price of a barrel of crude oil between 1973 and 1998.** Oil prices are standardized to 1996 dollars. (Data are from U.S. Department of Commerce and U.S. Department of Energy.)

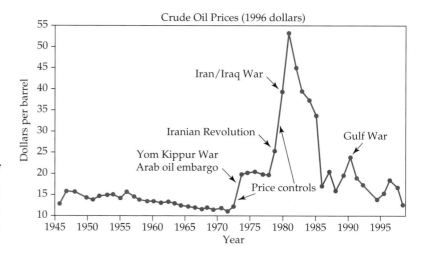

thought so. Put yourself in their role and look at the demand for energy between 1940 and 1970 (Figure 25.1). Energy consumption was skyrocketing, while fossil fuels were known to be finite resources, and on the horizon was the promise of nuclear power. It is hardly surprising that so much hope, anticipation, faith, and finance were invested in the nuclear energy industry. The nuclear dream foretold of electricity that was so cheap to produce that it could be virtually given away. A cleaner, safer landscape, free of the smoking chimneys of conventional power stations was envisioned. The peacetime production of power had the added benefit of producing weapons-grade plutonium[239] for use in Cold War nuclear arsenals, guaranteeing peace through power.

Nuclear energy is produced by fracturing the nucleus of uranium. Uranium comes in a variety of isotopic forms; the most common is ^{238}U, but the rarer ^{235}U isotope that makes up about 0.3% of uranium is the form required for nuclear fission. The technology to produce nuclear power safely is available. Layers of safeguards can be installed to make sure that the system is thoroughly controllable, yet the reliability of nuclear reactors remains a cause for concern. The problems usually have a human origin. Occasionally the probems are design flaws, but more often it is a building or operating fault. Specifications demanded for the construction of the containment chamber, the portion of the reactor that prevents radioactive leakage, may be unrealistic for the scale of construction and insufficient after exposure to the elements. In some cases, contractors trying to cut corners have been responsible for delays and costly reconstruction. Worse, the shortcuts may not be discovered until they cause a leak of radioactive material. The extreme environment close to the reactor core also presents a unique set of engineering problems. To date, all the reactors built have involved delays and cost overruns. When they do come into production, the reactors function well below their designed optimal operating efficiency; hence, energy from these nuclear sources remains relatively expensive.

The idealism that met the nuclear era began to evaporate as cheap nuclear energy did not materialize and as a series of accidents eroded public confidence in reactor safety:

• 1957: A tank of radioactive waste exploded at the Kyshtym reactor in the Soviet Union. Secrecy shrouded this incident; this was believed to have been the principal site where plutonium was manufactured for the Soviet weapons program. The number of fatalities is not known, but several hundred square kilometers remain uninhabitable, and 30 villages no longer appear on maps. It is thought that this may have been the most severe of all nuclear accidents.

• 1957: A fire in the British Windscale reprocessing plant that takes uranium waste and converts it to plutonium resulted in radioactive gases being vented. Thirty-three local inhabitants died prematurely of cancer. The plant was renamed Sellafield and continues to operate.

• 1979: A coolant failure resulted in a partial meltdown of the Three-Mile Island reactor in Pennsylvania. One hundred forty-four thousand people were evacuated; no fatalities have been documented.

• 1986: Explosions and fire destroyed a Soviet reactor at Chernobyl. A plume of radioactive gas contaminated much of northwestern Europe. Many of the 135,000 people who were evacuated will not be able to return to their homes. The official death toll stands at 36, though this does not include premature cancer victims. As many as 4 million people may have been exposed to radiation from this accident (Figure 25.4).

During the 1970s, public backing for the initiation of new projects waned (Figure 25.5), and the proportion of the population favoring the building of new nuclear power stations declined to less than 20%. Consequently, all of the 13 nuclear power stations planned and approved in the United States since 1975 were canceled before construction began. Despite its problems, the nuclear industry still provides 22% of all power used in the United States, 50% of power used in Sweden, and 75% of power in France. While the trend in much of Europe, Canada, and the United States has been to cut back and decommission nuclear plants, Japan, France, and South Korea are expanding their nuclear programs.

Although it may be argued that nuclear power is an environmentally clean source of energy, mining uranium results in considerable environmental damage. Some of the largest known deposits lie within areas that are environmentally sensitive. Australia supplies much of the world's uranium, but the richest ore deposits lie within the pristine wilderness of Kakadu National Park, located in the Northern Territories. For three decades the mining industry has sought to expand its Ranger Mine that operates in Kakadu. An open-pit mine of the sort proposed is an immense hole in the ground, surrounded by piles of mine waste. Miners have to be housed, and this leads

Figure 25.4 The Chernobyl power plant that exploded in 1986. The head firefighter, Lt. Colonel Leonid Telyatnikov, points to the damaged Chernobyl reactor.

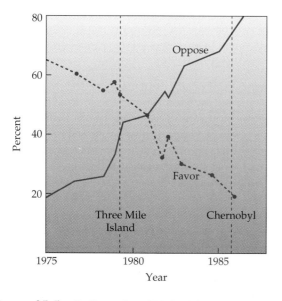

Figure 25.5 Poll results of United States public opinion on building more nuclear power plants between 1975 and 1986. (Data are from World Watch Institute, 1987.)

to the establishment of a new town, complete with roads, airstrips, and sewage treatment. The proposed development lies in the heart of an area that is so culturally and environmentally important that it was awarded the status of a World Heritage Area. Local aboriginal populations have a 50,000-year history of occupation of this land, and the land itself is an integral part of their mysticism and religion. Despite the protests, the Ranger mine is very likely to be expanded.

The supply of uranium^{235} is limited, and it is estimated that known economically viable reserves will support the nuclear industry for about another 40 years. Although uranium is the staple fuel of most "conventional" nuclear reactors and plutonium^{239} is a waste product, fast-breeder reactors use plutonium^{239} as their fuel. These reactors are considered by nuclear proponents to be much safer than other conventional fission reactors. However, if a problem develops with their liquid sodium coolant, the chain of failure leading to an explosion might be measured in seconds rather than minutes. The only fast-breeder reactor to come on-line, France's "Superphenix," has proved to be expensive to build, run, and maintain. Plans to build similar reactors in Germany and Britain have been shelved until such technology can be shown to be economically viable. Japan is presently negotiating with France to purchase plutonium^{239} to power its fast-breeder program. The prospect of this incredibly toxic material being shipped around the world from France to Japan is causing considerable disquiet in environmental circles.

The greatest problem to be overcome by the nuclear industry is the treatment of radioactive waste. Plutonium and cesium are highly radioactive and toxic chemicals; a few grams of either is sufficient to poison an entire city. These waste chemicals of the nuclear industry maintain their radioactivity for 10,000 years; thus, their disposal must be done in such a way that they cannot leak or be stolen for illegal use. Plutonium and cesium are too toxic to risk dumping in the oceans or launching into space. No way is known to detoxify them through chemical procedures, and the only feasible, though imperfect, solution is to bury them underground in specially constructed chambers. At Yucca Mountain, Nevada, a site has been prepared to store such waste. By 1998, after spending $2 billion dollars and 11 years of research and preparation, storage has not yet begun. Indeed, after 40 years of nuclear power production all the high-level waste ever produced has either been used for military purposes or has been placed in temporary storage; none has been stored permanently. Burying such deadly material raises concerns over the long-term safety of the dump sites and the morality of passing a potential radioactive nightmare to future generations. Given our limited understanding of how our Earth works, can we be sure that the storage area will be safe from earthquakes, a glacier, or flooding for more than 10,000 years? Given our understanding of humans, can we be sure that these sites will be protected and maintained until the wastes are safe?

It is a popular cry in environmental circles that the nuclear dream has faded, that the pious hopes of the 1950s are now reduced to decommissioned reactors and waste dumps of toxic nuclear material. Perhaps it was a dream that just came too early, one that will recur and that we will be able to accomplish in the future. Nuclear technology will continue to develop, as will the safeguards that prevent leakage of radioactive materials. Improved construction techniques and more sophisticated robotic control operations may pave a safer nuclear path in the future. The specter of a mounting pile of nuclear waste remains the most troubling problem to overcome. No chemical technique to detoxify the waste has been conceived. It is likely, however, that as other energy prices rise and nuclear power offers a cheap alternative, we will become more accepting of the idea that we can produce and store highly radioactive waste. Theoretically, nuclear power remains a potentially clean and plentiful source of electricity, and it is likely that we will see the resurrection of the nuclear dream.

25.4   Our Future Stocks of Energy

Fossil fuel resources such as coal, gas, and oil are not infinite because we are using them at a rate faster than they are forming. Oil exploration continues, but it is widely believed that we have already found about 90% of the crude oil on the planet. So far almost half (800 billion barrels) of Earth's known and expected reserves of conventional crude oil have been extracted, with about another 1000 billion barrels remaining. If the approximate quantity of oil left to be extracted is known, and future demand is estimated, it is possible to compute the time that these resources will run out. Assuming rates of present usage, conventional crude oil is expected to last for about 50 years, natural gas for 60 years, and coal for several hundred years. If this were true, society would change out of all recognition within our lifetime as we are forced to replace all our major fuels.

Economists argue that these numbers are likely to be as overly pessimistic as the 1 billion human population maximum predicted by Malthus back in 1798. The reasons for inaccurate predictions are:

- We do not know the full reserves available on the planet and therefore cannot say when they would be depleted.
- Presently known deposits that are uneconomic to tap may become economically viable as technology changes or prices rise.
- Consumption will decline as prices increase, slowing our use of the resource.
- Substitute energy sources will become economic as energy prices rise.
- Improved technology may make energy conversion much more efficient.
- Conservation measures by users may reduce demand for power.
- Recycling waste oil could prolong reserve life.

In 1998, Colin Campbell and Jean Laherere, both experienced petroleum geologists suggested that while we may not have to be concerned about when the last drop of oil is used, oil prices will soon rise. They estimate that we have about another 10 years of cheap oil. After about 2010, global oil production will start to fall as reserves in major oil fields became harder to access (Figure 25.6). If demand for oil is not met from another source, oil prices will rise sharply.

The world has vast reserves of "unconventional oil" that are held as tar sands and oil-rich sludge. At present the technology does not exist to extract oil from these deposits in a cost-effective manner, but with the incentive of rising oil prices these alternative oil sources are likely to be exploited. The political ramifications of accessing these deposits would be huge. The OPEC nations will recede in importance as their conventional oils are exhausted, and countries with the large reserves of unconventional deposits, Venezuela,

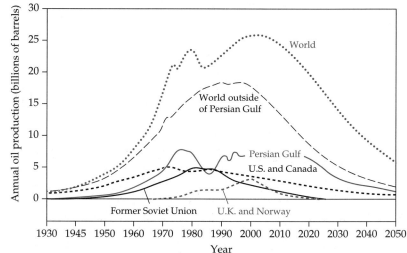

Figure 25.6 The end of cheap oil around 2010? The lower the production, the higher the price of oil. Analysts predict that falling outputs of oil, due to exhausting the best deposits, will start to push oil prices upward after the year 2010. (After C. Campbell and J. Laherrere, 1998.)

the former Soviet Union, and Canada, would become the energy brokers of the future.

American economist Julian Simon takes an extreme stance and suggests that technology will triumph and there will be no future power shortage. He suggests that because of a combination of human ingenuity and new energy-source discoveries, our energy supplies will prove to be near infinite. Perhaps the most likely future trend lies between the harrowing predictions of an energy-starved world and Simon's utopian vision of human initiative.

25.5 Energy and Pollution

Pollutant gases are released whenever energy is derived from burning fossil fuel. Burning any fossil fuel releases gases that contribute to global warming; some are also sources of particulates, acid deposition, and the chemical constituents of photochemical smog. Coal is the most abundant fossil fuel, yet it is being replaced as the fuel of choice in the more-developed countries (MDCs) wherever possible. Although a cheap and plentiful energy source, coal produces large quantities of pollution as it is combusted. In the United States, 900 million tons of coal are burned each year, releasing 18 million tons of sulfur dioxide (SO_2), 5 million tons of nitrogen oxides (NO_x), and 4 million tons of dust into the atmosphere. The NO_x released is a potential greenhouse gas, and both nitrous oxide and sulfur dioxide cause acid deposition. Of pollution released within the United States, coal burned by power stations accounts for 75% of sulfur dioxide, 33% of nitrous oxide, and 50% of industrial carbon dioxide production. Restricting emissions even to these levels, re-

quires 40% of the cost of building a coal-burning power plant to be spent on pollution control. Furthermore, 35% of the daily operating costs are spent on emission reduction. With investment, coal can provide energy and cause little pollution, by switching to low-sulfur coal and by installing scrubbers that wash dust and polluting chemicals from the gases emitted. The technology to reduce the pollution exists and, as time goes by, that technology may become more affordable.

However, not every country is so concerned with pollution regulation, and industrial expansion remains their immediate goal. China is the world leader in coal production, obtaining 75% of its energy from coal, and it is projected that between 1994 and 2002 China will use coal to increase its electricity generating capacity by 40%.

Other fossil fuels "burn cleaner" than coal, producing fewer pollutants, but carbon dioxide, hydrocarbons, and nitrous oxides are all released whenever fuels are burned. A further source of environmental damage associated with fossil fuel and uranium extraction is the local pollution that results from the mining of coal and uranium or from drilling for oil and gas. For mines, this pollution usually takes the form of spoil heaps where the waste rock is piled and the open pits are dug to extract coal or uranium.

The oil and gas industry tends to leave fewer scars on the landscape, but spills of oil along pipelines or from tankers can be major sources of pollution. One proposal has been to build power stations that burn wood. This can work only where wood is relatively cheap; consequently, these schemes have been targeted for tropical countries such as Belize. Carbon dioxide would be given off as the wood is burned, but a larger concern is the deforestation that might accompany such a project. If the plantations established to supply the power station cannot supply enough wood, there would be a strong incentive to fell mature rain forest to supply cheap firewood.

The cleanest source of commercially viable energy is hydroelectric power. A dam is built to hold back a body of water, and the pressure of water pushing out through a narrow outlet and falling through turbines is used to generate the electricity. However, hydroelectric projects are not free of environmental impact. The principal problems associated with hydroelectric projects are the loss of land flooded by the reservoir and downstream ecological changes that result from damming the river. The land to be flooded by the dam is often either a wetland or good farmland. Wetlands are already greatly reduced in area, and consequently a relatively high proportion of their floras and faunas are endangered species. A lake and a wetland are very different habitats, so the presence of a reservoir will not compensate the displaced wildlife: Wetland species are likely to go extinct locally. If the land is being farmed, the local human population must be uprooted, compensated, and resettled. Resettlement programs often lead to the deforestation of fresh areas of land or overcrowding on existing farmlands.

Damming a river allows the downstream flow of water to be regulated. The huge seasonal pulses of water that would have brought flooding are now absorbed and regulated by the reservoir. The flow of the river becomes more predictable, but the disruption of natural flood events interrupts the supply of sediments to floodplains or the flooding of seasonal wetlands. The muds carried in a river make fertile soils, and the farmland adjacent to great rivers such as the Nile or the Mississippi owe much of their productivity to mud deposited by flood waters. If the floods are prevented by the dam, then the fertility of the land may decline. Instead of being deposited on the land, the mud now settles out in the still waters of the reservoir, especially against the wall of the dam. In time, this mud would fill the reservoir, and to maintain the volume and depth of water in the reservoir requires expensive dredging operations. Between the regulation of flow that might prevent flooding and the loss of a nutrient input, the ecology of natural wetland areas downstream from a dam is altered completely. As the wetland niches have been changed, so a modification of species composition will follow. Thus, rare or endemic species, both at the site of the reservoir and in the entire downstream drainage, are threatened with extinction as a result of loss of habitat.

25.6 "Alternative" Energy Sources

Solar power is the oldest energy source on the planet, and its importance as a natural source of power became the center of religious observance among peoples as varied as Aztecs and Druids. Springs heated naturally by geothermal energy became spas in Roman times, and surplus hot water was used to heat the Romans' bathhouses. Wind power has driven ships since the invention of the sail, and windmills and waterwheels were the power plants of great civilizations from the Chinese and Etruscans of 6000 years ago to Renaissance Europe. Much of the European industrial revolution was built on water power, and it was not until steam power became affordable and reliable in the mid-1800s that fossil fuels played a large part in energy supply. The return to these once-important sources of power is now regarded as the search for "alternative" or unconventional energy.

In the wake of rising oil prices associated with the formation of OPEC, the developed nations rushed to find cheap energy from wind, solar, wave, or geothermal power. These were infant technologies still far from commercial viability, and, like all new endeavors, needed a phase of experimentation to develop the appropriate machinery and computer resources. Research and development funding was always a fraction of that invested in nuclear and conventional power sources, and, as would be expected, success was mixed.

Wind Power

In the 1970s, wind power was the alternative energy source nearest to economic viability. However, because the prototypes were based on helicopter technology adapted for power generation, rather than a purpose-designed windmill, results were disappointing. The original plan was to make huge windmills with rotors more than 90 m (the length of a football field) in diameter, but these proved vastly expensive. Problems also lay in the erratic nature of the wind; power production from each windmill could not be predicted. When series of windmills were linked to feed into a common power network, there were difficulties in creating a concerted power output. Because of these perceived difficulties with the technology, federal funding for wind power research in the United States was cut 90% in the 1980s.

Wind power remains the "new" energy-generating system most likely to make a significant contribution to energy supply. It offers cheap, clean, sustainable power and does not add to global warming. The momentum in developing wind as a source of power has shifted from the United States to Europe (Figure 25.7). Denmark currently produces about 2% of its power from wind, while the United States produces less than 1%. Britain estimates that wind could supply about 20% of its power requirement and that 10% of the power demand of the European Union could be met from this source using existing technology. The development of new technology is proceeding rapidly, and the European Union countries are investing 10 times as much money as the United States in the development of wind power. Most of the early problems have now been resolved, and there remain only two principal barriers to a widespread introduction of wind power: wind availability and public acceptance.

The windmills need to be laid out in farms, each with 10 to several hundred machines, depending on their size (Figure 25.8). For the United States, it is estimated that it would require about 0.6% of the land area of the 48 contiguous states, primarily in the Midwest, to generate 20% of the national energy requirement. Often, the only land available for such an enterprise is fairly remote and may conflict with maintaining the area as a wildlife refuge or for recreational use. The windmills look modernistic and, when planted in a wild setting, take on a somewhat surreal appearance. The aesthetics of a wild area are undoubtedly changed by their inclusion in the landscape. However, aesthetic values can change with time. The citizens of Paris were enraged at the appearance of the Eiffel Tower on their skyline in 1889, but it is now a

Figure 25.8 **A field of windmills in Palm Springs, California.**

natural treasure. Although windmills are not desirable in all locations, most people would be happier living close to a windmill than to a nuclear reactor or a coal-burning power plant.

Solar Power

There are a number of ways to harness solar energy for the production of heat or electricity. The most common manufactured solar power source, and probably the most likely to be developed in the future, is the photovoltaic, or solar, cell. Each photovoltaic cell is small, and many must be used together to form a solar panel that produces from 30 to 100 watts of energy. The panels can be used singly or in clusters to provide energy on a small scale, or great arrays of shiny cells can be connected so that their energy is united in a central power plant (Figure 25.9). The cost of solar power has tumbled, with costs reduced by more than 90% between 1975 and 1993. However, power stations using photovoltaic power to generate electricity remain a very expensive source of electricity; each kilowatt-hour costs 30 cents, about five times the price as fossil fuel. Photovoltaic cells are economically viable

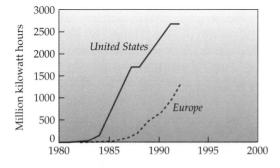

Figure 25.7 **Wind power–generating capacity in the United States and Europe, 1981 to 1992.** Although the United States has the largest wind-generating capacity in the world, the European countries have expanded their programs in the 1980s and 1990s while the U.S. program has slowed. (Data are from L. Brown, H. Kane, and D.M. Roodman, 1994.)

Figure 25.9 **The solar collectors for the Odeillo solar-powered electricity-generating plant, France.**

when used in appliances that have low energy needs and are relatively costly. Consequently, solar-powered watches, calculators, and signs are becoming more and more common. Solar power becomes more affordable for large-scale energy generation if it can be combined with natural gas to provide energy at 8 cents per kilowatt-hour. Advocates of solar technology predict that the next generation of parabolic dish reflectors could produce electricity for 6 cents per kilowatt-hour early in the twenty-first century. Even so, this form of energy would require that large areas be set aside for power generation; about 20,000 square kilometers of solar plant would be needed to supply 10% of the U.S. energy demand. Some estimates suggest that by 2030 about 20% of energy demand can be met from solar power. The reality of this prediction will depend on continued development, because solar power remains an uneconomic means of large-scale energy production. The U.S. federal budget for solar energy research and development was cut 76% between 1981 and 1990, at a time when other governments, such as those of Germany, Spain, Switzerland, and Japan, as much as tripled their investment in this area.

Geothermal Power

Tapping the heat energy of Earth's interior could provide a sustainable source of energy. Local geological conditions are the primary determinant of the role that

geothermal power can play because source areas are not evenly distributed. At present, these schemes have proved successful and cost effective. Because the trapped heat often co-occurs with sulfur dioxide gas, mildly radioactive rock, and toxic metals such as mercury, air and groundwater pollution are potential problems that must be overcome. The world's largest consumer of geothermal power is northern California, where 6% of the energy comes from this source. The U.S. Department of Energy estimates that in the future geothermal energy production could increase by a factor of 30.

25.7 Meeting Future Energy Demand

The pattern of energy supply was one of decreasing costs and increasing efficiency throughout the first part of the century. In 1892, each kilowatt-hour of electricity cost $4 in today's currency. By 1930, this amount of energy cost 60 cents, and by 1970 the cost was about 7 cents per kilowatt-hour for most conventional power sources. Part of the decline in cost was attributable to more efficient power stations (Figure 25.10), although by the 1960s, power station efficiency had reached an apparent asymptote of about 33%. The 67% of wasted energy cannot be converted to electricity; through the process of entropy it degrades to heat energy. Im-

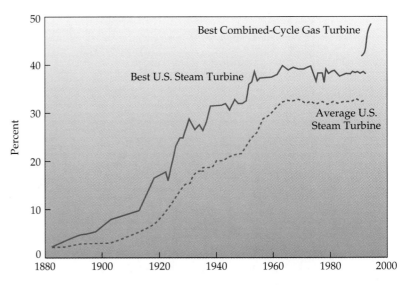

Figure 25.10 The efficiency of electricity generation from fossil fuels, 1880–1992. (Data are from C. Flavin and N. Lenssen, 1994.)

proving generating efficiencies beyond this value of 33% has proved to be enormously difficult.

A new generation of natural gas–powered generators called *combined-cycle plants* promises improved efficiencies by using the "waste" heat generated from the gas turbine to power a second turbine. One such plant installed in 1993 by General Electric in South Korea is 50% efficient, and a new design announced in 1993 is predicted to be 53% efficient.

In the United States, a quiet revolution in energy supply is taking place with the establishment of small independent energy producers. A change in the law (the Public Utilities Regulatory Policies Act of 1992) allowed independent suppliers to build competing plants to use renewable fuels or to **cogenerate** heat.

Cogeneration is the use of the "waste" heat produced by conventional generators to create steam, which can power machinery, heat a building, or provide hot water. Already, large industry and even small enterprises, such as hotels, fast-food restaurants, and hospitals, are using cogenerated heat. The efficiency of such systems can be as high as 90%, and they offer the advantage of cutting energy bills by 15%–30%. In 1993, for the first time, independent energy suppliers added more output capacity than large utility companies (Figure 25.11). One vision of the future is that small high-efficiency generators could supply energy needs of individual apartment buildings or whole neighborhoods. Given present technology, however, they would still use nonrenewable fuels.

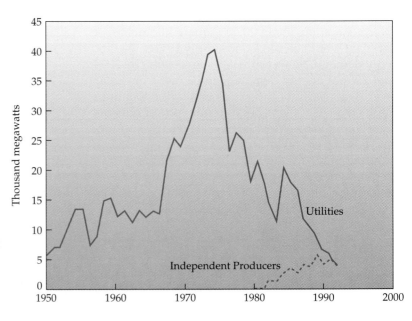

Figure 25.11 Rise of the independents. Additional generating capacity of electric utilities and independent power companies 1950–1992. (Data are from C. Flavin and N. Lenssen, 1994.)

25.8 Energy Conservation and Efficiency

Considering the future of the energy industries brings to mind the advice on how to become rich: It's not what you earn, it's what you spend. It seems that a similar message applies to the power-generating industry. The United States is the most profligate energy consumer, with each person, on average, using more than twice as much energy as in any other nation, including the industrial giants of Europe and Japan. Americans currently use 300 times as much energy per person per day as do people in some less-developed countries such as Malawi and Nepal.

There are some signs that North Americans may be becoming wiser energy consumers. Since the early 1970s, the U.S. and Canadian demand for power has slackened considerably, not because of reduced industrial output or diminished power requirements in the home, but because energy conservation began to become a design feature. The average mileage per gallon of automobiles rose from about 13 miles per gallon in 1970 to about 28 miles per gallon in the 1980s. Improved window design and insulation of walls and attic spaces increased heat efficiency, as did new designs of furnaces, heat pumps, and air conditioners. Many industries re-evaluated their usage of energy and found ways both to cut costs and increase competitiveness. The result was a stabilization of U.S. energy demands despite a growing population and economy.

An important aspect of energy conservation has come about through **demand-side management (DSM)**. DSM is the apparently paradoxical statement that it could be better for both consumer and utility if energy demand were lowered. Conventional economics would argue that it is always best to sell more and to do so at the maximum price that the market will bear. However, DSM indicates that improving the efficiency of homes and industry costs less than building the new generating plants needed if wasteful energy use persists. At first, this change in market structure was fiercely resisted by the utilities, and it has taken a stick-and-carrot approach by government to begin the changes.

The stick came with regulatory agencies' ordering utilities to encourage energy conservation. Advice slips advocating energy conservation measures were enclosed with fuel bills sent to consumers. These notices advised consumers that they could save money by setting home thermostats to moderate settings, reducing water temperature in hotwater tanks, regularly servicing appliances, and improving wall, roof, and window insulation. In some states, cash rebates were offered for the installation of energy-saving compact fluorescent lamps. Subsidized loans were made available on weatherization and industrial refits leading to energy savings. It is estimated that the major utility companies have invested $2 billion in DSM in the last few years. The carrot offered by regulating agencies came in the form of incentives paid to utilities compensating them for lost revenue.

As a result of DSM, the utilities are modifying their role in the marketplace by ceasing simply to generate electricity and starting to become full-service companies selling energy-efficient appliances. DSM appears to work. To generate a kilowatt-hour of energy costs the utility about 7 cents, but the same amount of energy can be saved for just 2.1 cents. Over the next decades, some U.S. utilities plan to cut their generating capacity, relying on conservation measures to lower demand. Even the largest utility, Pacific Gas and Electric, plans to meet 75% of its increased energy demand through conservation measures.

Another way to think about energy efficiency is with reference to the pollution caused as fossil fuels are burned. One index is to compare industrial output against carbon emissions to provide what is termed the carbon efficiency. In recent years, **carbon efficiency** has been rising, indicating that for each kilogram of product, less carbon is being released, primarily as carbon monoxide or carbon dioxide, to the atmosphere. This result may be due to improved industrial techniques, or it may reflect industry's switching to nonfossil fuel energy sources. Energy sources that increase carbon efficiency are wind, hydroelectric, solar, or nuclear power because they release no carbon to the atmosphere. Since 1979, U.S. industry has reaped about seven times the energy benefit through conservation measures as it has from exploiting new energy sources. However, there is still room for improvement. U.S. industry lags behind its leading economic rivals, Japan and the former West Germany, in terms of energy efficiency. Those countries devote only about 5%–6% of their gross national product (GNP) to purchase energy, whereas the United States spends about 11% of its GNP on energy. This difference is passed from producer to consumer in the form of higher prices, placing U.S. exports at a disadvantage compared with more energy-efficient competitors.

One of the benefits of urban renewal is the energy conservation that can be achieved through the remodeling of older houses. About 40% of heat lost from a home in a cool climate is through cracks around windows, doors, foundations, and walls. Another 20% of heat is lost through the roof. Insulating these areas and fitting double-glazing to windows can reduce these losses by about 70%. Such improvements make the home a more pleasant living space and save money on fuel bills. Enforcing construction codes that ensure energy efficiency on new building is another way to work toward more energy-efficient cities.

Mass transit by train, bus, or subway is much more energy efficient than use of cars and is already an integral part of European and Asian city life. However, in the United States it is commonly used only in the larger cities, and in many areas public transport suffers from a bad public image. If mass transit fails to offer a realistic alternative to routine use of cars, the American love affair with the six- and eight-cylinder engine may become the object of **sin taxes**. When an individual's actions lead to a cost to society as a whole, rather than banning that action outright, society can impose a financial penalty: a sin tax. Powerful car engines consume more fuel than smaller engines and also produce more pollution, both of which are costs to society. Sin taxes allow anyone wishing to purchase a gas-guzzler the freedom to do so, but they impose a penalty for that indulgence. Sin taxes on cars, cigarettes, CFCs, and alcohol are already in place in many states.

If short-term costs of energy spiral upward, the next real efforts at energy conservation may deny the need for an expanded power industry. It has been estimated that improving energy efficiency could save the equivalent of 60 billion barrels of oil by the year 2020, whereas all known natural gas and oil deposits might yield 18 billion barrels in the same time period. In the long run, it seems likely that social and economic imperatives, such as congested streets, pollution, and rising energy costs, will cause Americans to reevaluate attitudes toward public transport and inefficient industrial and domestic use of energy. Power, from one source or another, will continue to be available. But to avoid shortages, we may have to be more restrained in our consumption than we have been for much of this century.

25.9   Energy and Development

The industrialization of the United States, Canada, and Europe was founded on energy derived from coal. The provision of affordable energy lay at the heart of the social transitions that resulted in reduced family size and improved standards of living. The developing nations of today also need cheap and plentiful energy.

Coal could fulfill this need, but it is in the best interests of the MDCs to promote other energy sources in the less-developed countries (LDCs) because coal combustion produces so much pollution. If the LDCs were to power their industrialization with coal, the efforts of the MDCs to reduce global and regional pollutants, such as carbon dioxide and sulfur dioxide emissions, would be undermined. At the same time, the MDCs want to see the LDCs develop for a number of reasons. One humanitarian motivation is that a strong infrastructure reduces the chance of famine and disaster. Political motivations are based on the assumption that as nations develop they are more likely to become democracies, less likely to have repressive regimes, and more likely to be politically stable. A third set of reasons that should encourage the MDCs to promote development in LDCs is environmental. It is predicted that reduced birth rates will accompany development, and these, in turn, will reduce the need to expand slash-and-burn cultivation into the forests. As prosperity increases, more farmers will be able to afford to intensify and improve their agricultural practices. The result will be that more food will be produced from the same acreage, again reducing the need to deforest the land. If the key to this development is the availability of cheap energy, and MDCs want it to be "clean" energy, then it is in their best interest to export modern energy technology to the LDCs.

An expensive lesson learned by utilities in MDCs is that having the ability to produce more energy than is actually needed results in financial losses. Many LDCs are currently building power-production facilities, and one role for MDCs is to advise them on how to avoid becoming locked into costly programs that would lead to the overproduction of power. Another role is for MDCs to assist in the research and development of tailor-made technologies that meet the needs of LDCs. Urban populations can be supplied with electricity from large power stations, but small-scale local utilities are needed to supply the needs of remote villages. Such technologies should be low-maintenance and affordable, using fuel sources that are inexpensive or locally produced—for example, wood, leafy biomass, or oil.

The history of MDCs assisting in power generation in LDCs is littered with cases of vast projects—say, flooding large areas to provide hydroelectric power—that were poorly located and proved too expensive and too environmentally destructive. If MDCs are to encourage conservation of biodiversity and natural resources in LDCs, a substantial contribution can be made by financing utilities, but the emphasis should be on small-scale, locally managed programs.

Summary

- The cheapest commercial energy sources are from fossil fuels (coal, oil, and gas).
- Fossil fuels are a finite resource and are likely to become more expensive as reserves are depleted.
- Predictions of "times until exhaustion" for fuel sources based on present use rates are unlikely to be accurate.
- Nuclear power has the potential to be a clean, efficient source of power, but mining uranium is environmentally damaging.
- No prudent way has yet been found to dispose of the waste. After 40 years of commercial nuclear power production none of the waste has been disposed of permanently.

- All forms of fossil fuel power production contribute to global warming.
- Coal-powered power stations are the largest source of sulfur dioxide, the leading cause of acid deposition.
- "Alternative" energy sources are not free of environmental costs; they too may pollute, be unsightly, or have significant pollution associated with their manufacture.
- Conservation of energy may prove more profitable than building new power stations.

Further Readings

Ahearne, J. F. 1993. The future of nuclear power. *American Scientist* 81:24–35.

Anderson, R. N. 1998. Oil production in the 21st Century. *Scientific American* 278 (3):86–91.

Bormann, F. H. 1978. An inseparable linkage: Conservation of natural ecosystems and the conservation of fossil energy. *BioScience* 26:754–759.

Campbell, C. J., and J. H. Laherrere. 1998. The end of cheap oil. *Scientific American* 278 (3):78–83.

Reisner, M. 1998. Coming undammed. *Audubon* 100 (5):58–65.

Web Connections

On-line resources for this chapter are on the World Wide Web at:
http://www.prenhall.com/bush
(From this web-site, select Chapter 25.)

Human Disease: Evolutionary and Ecological Perspectives

26.1 *Battling malaria: Nearly a success story*

26.2 *Drug resistance and diseases that haunt us*

26.3 *Evolutionary thoughts about virulence*

26.4 *The ecological perspective*

26.5 *The emergence of new diseases*

26.6 *Human immunodeficiency virus*

Imagine the consequences of a disease that is untreatable, from which there is no recovery; one spread with every cough and sneeze; a disease with no visible symptoms until the final hours; a disease that kills within 24 hours of infection. All of these attributes exist in different viral and bacterial diseases, yet no single disease encompasses them all. Some diseases, such as cholera, pneumonic plague, smallpox, influenza, and yellow fever, come close in terms of their potential to devastate a human population, but our species has yet to succumb to a superdisease. Why hasn't a supervirus or superbacterium evolved?

In the last 100 years, medicine and social improvements have conquered many once-fatal illnesses such as typhoid, polio, and tuberculosis. Smallpox, once a devastating killer, has been eradicated completely. Our control over these and other familiar ailments has made the inability of science to find a cure for acquired immunodeficiency syndrome (AIDS) all the more shocking. If the history and future of human society are to be understood, the effects of disease, not just AIDS, should be considered from an ecological viewpoint.

Every organism is a potential host, a rich resource waiting to be tapped by the appropriate parasite. We can think of ourselves as a set of vacant niches ripe for invasion and occupation by other organisms. We provide shelter (our bodies), a central heating system that can incubate their young at a constant and predictable temperature (36.8°C), and a rich source of food (the

products of our digestion or cell metabolism). We are such an attractive proposition to potential pathogens that, like other animals, we have had to evolve a sophisticated set of defensive mechanisms.

- Our skin, so long as it is intact, is an impermeable barrier to most would-be pathogens.
- Everything we eat is subjected to an acid bath in our stomach. This helps start the digestive process, but will also kill almost all viruses and bacteria as they enter our bodies.
- We have white blood cells that prowl our bloodstream searching out foreign bodies.
- Our lymphatic system siphons off invaders that may be roaming our intercellular spaces. Lymph bathes these regions, and its flow sweeps up bacteria, viruses, or foreign proteins and carries them to the lymph nodes, where white blood cells attack them.

If a bacterium, virus, or other pathogen does gain a foothold in our bodies, we set up a series of defensive procedures. If the infection is mild, a first response may be to change our body temperature. Our immune system causes the temperature to increase by 1° or 2°C and, in so doing, radically changes the environment of the invading pathogen. Under the warmer conditions, pathogen reproduction slows, giving our body time to respond in other ways to clean up the intrusion. If the initial temperature rise is ineffectual, our body temperature may rise by a few degrees more. At this stage, we will be running a strong fever. Clearly,

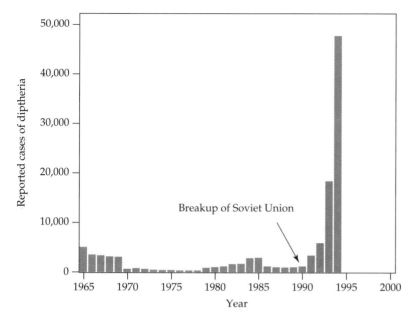

Breakup of Soviet Union

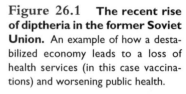

Figure 26.1   The recent rise of diptheria in the former Soviet Union. An example of how a destabilized economy leads to a loss of health services (in this case vaccinations) and worsening public health.

a severe fever is debilitating and can so weaken the sufferer as to cause death, but remember that a mild fever may be the best defense against diseases such as the common cold.

Another of the responses of the body might be to clear pathogens from the respiratory system. An increased secretion of mucus prevents the pathogens from adhering to the inner walls of the lungs, sinus, or nose. As we blow our noses or cough up phlegm, we are ridding our bodies of huge numbers of pathogens caught in the mucus. If we have an intestinal infection, we may suffer from diarrhea, and again this is our bodies' attempt to flush pathogens from our system. Thus, our classic symptoms of cough, runny nose, fever, and diarrhea are all defensive mechanisms hard at work to control an infection.

Acquiring Resistance

It has long been known that we can catch some diseases only once. Measles, chickenpox, rubella, and smallpox all produced severe disease reactions with first infection, and the viruses were so abundant that first infection usually happened in the first few years of life. Those who survived these "childhood" diseases could be re-exposed to the disease but would not catch it again. In combating the disease the first time, the host's immune system produces antibodies. The antibodies combine with the proteins of the invading cell (antigens), effectively disabling the pathogen. If the host survives the disease, blood cells called memory B lymphocytes are created that carry the "response code" to produce fresh antibodies should a reinvasion

of the pathogen occur. The subsequent production of the antibody is immediate; the host is immune to reinfection. Thus, most adults have developed resistance to the common diseases of their surroundings.

Vaccination programs make use of the body's ability to create antibodies. A vaccine is a mild, or closely related, form of the pathogen that is deliberately introduced to a human body. The person's immune system overwhelms the pathogen and creates antibodies. When later exposed to the real disease the vaccinated person's immune system will respond quickly and forcefully, preventing the disease from getting a foothold. Vaccination was the key to the eradication of the smallpox virus and has greatly reduced the incidence of measles and diptheria. However, until a disease is completely eradicated the vaccination program must remain in place. One effect of social and economic chaos in the former Soviet Union during the 1990s was that the vaccination programs were not continued. The result was that diptheria, a disease that had been persistent at very low levels, made a troubling recovery (Figure 26.1).

26.1   Battling Malaria: Nearly a Success Story

In the 1950s, it seemed that malaria was to become a threat of the past, a medical curiosity that once plagued the southern United States and many tropical regions. However, in the last 20 years, the malarial protozoan, *Plasmodium*, has made a spectacular and highly unwelcome recovery, and some strains of it are alarmingly resistant to almost every drug.

There are several species of malarial pathogens that belong to the genus *Plasmodium*. These single-celled creatures are injected into a human by the bite of the *Anopheles* mosquito. Within the human host, vast numbers of the pathogen reproduce asexually in the liver and bloodstream (Figure 26.2). A mosquito biting an infected human withdraws blood infested with *Plasmodium*. The *Plasmodium* reproduce sexually within the mosquito and are injected with the mosquito's saliva as it bites the next human.

The First Attempts to Control Malaria

The nineteenth-century belief was that yellow fever and malaria were ills spread by bad night air and that the immoral were most susceptible. In 1897, Ronald Ross, a physician and amateur microbiologist, dissected an *Anopheles* mosquito after it had fed on a malarial patient and found that the *Plasmodium* had been transferred to the mosquito and was growing. From his clinical observations and research, Ross constructed the infection route for malaria. But this researcher, working in a remote Indian province, a far-flung corner of the British Empire, was largely ignored, and his claims that malaria was mosquito-borne were thought fantastic. It took another decade before mosquito eradication became the main plank of malarial control. Drugs to prevent or treat the disease came later, and, with the combination of drugs and the mosquito-control program, malaria seemed to be receding as a threat to human health.

Genetic Resistance and the Return of Malaria

All pathogens reproduce and, one way or another, produce a population with a wide range of slightly differing genetic forms, or genotypes. A medicine or antibiotic targeted at that organism is a chemical so toxic to the pathogen that it will kill almost all of these genotypes. But some do not die. There always exists the possibility of individuals with a configuration of genes that will not be killed, those belonging to a resistant genotype. By continuing to apply the toxin, we are driving the evolution of the population toward the resistant strain, and with each generation the proportion of resistant individuals in the population will grow. The proportion of resistant genotypes may take a long time to become serious. Conversely, it may be that the resistance trait is inherited by all the offspring, in which case the toxin will be wholly ineffective the next time it is applied.

Resistance to preventive drugs and the antibiotics used to cure malaria is rising. Several species of malarial *Plasmodium* exist, the most dangerous being *Plasmodium falsiparum*. These protozoans show a remarkable ability to mutate against our drugs. Considering that an infected patient typically has 100 billion of the parasites in circulation, even a drug that was 99.999% effective would still leave a million parasites alive! Thus, the chances that a rare genotype will survive become quite good. Political instability in nations with a

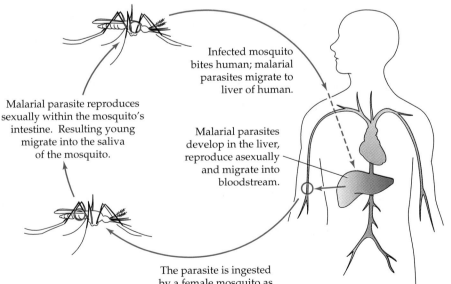

Malarial parasite reproduces sexually within the mosquito's intestine. Resulting young migrate into the saliva of the mosquito.

Infected mosquito bites human; malarial parasites migrate to liver of human.

Malarial parasites develop in the liver, reproduce asexually and migrate into bloodstream.

The parasite is ingested by a female mosquito as she bites and collects blood.

Figure 26.2   The life cycle of the malarial *Plasmodium*.

malarial problem adds to the difficulty of trying to instigate a comprehensive campaign against the disease. Currently, malaria is primarily a disease of developing countries and is most severe where social strife, poverty, and civil war have led to a collapse of medical programs. Furthermore, local populations may regard Western medicine with skepticism and its practitioners as profiteers. Humans enjoyed an initial success with the introduction of pesticides to control the mosquitos and more effective drugs, but the malaria parasites are now evolving toward increasingly untreatable strains. Current estimates are that malaria affects about 300 million people worldwide and that each year it is responsible for 2 to 4 million deaths.

26.2   Drug Resistance and Diseases That Haunt Us

The malarial *Plasmodium* is far from the only pathogen that exhibits drug resistance. Disease-causing bacteria are proving to be worthy adversaries for researchers. The first antibiotic drugs were introduced in 1940, but some *Staphylococcus* strains were exhibiting drug resistance as early as 1944. We are now faced with some (thankfully rare) organisms that are resistant to the full range of antibiotics. The tuberculosis (TB) bacterium (*Mycobacterium tuberculosis*), thought to have been thoroughly beaten in the 1950s and 1960s, is now returning, and the new strains are resistant to all 11 antibiotics that were once effective treatments. A study of 50 species of disease-causing bacteria revealed that only two species, lyme disease and syphilis, could be treated with a single antibiotic. By the late 1980s some streptococcal and pneumococcal infections were resistant to every antibiotic except vancomycin. This drug was the safety net that saved lives when all other drugs failed. In 1989 only 0.3% of infections were vancomycin resistant, but by 1993 this number had soared to 13%.

Most of us think of antibiotics as pills to treat human ailments, but they have wider applications than that. One of the most important uses of antibiotics is as disinfectants within hospitals. A worrying trend is the increasing resistance of bacteria to antibiotics used to sterilize operating rooms and wards. It is now estimated that 1 in 20 patients admitted to a hospital will acquire a bacterial infection while they are there.

The ability of bacteria to combat antibiotics comes from their peculiarly unselective approach to swapping DNA. Bacteria can reproduce asexually, and that

is how they build populations very rapidly. However, they can also "conjugate." Two individuals fuse, swap portions of genetic material, and then separate to produce two genetically novel offspring. In some versions of conjugation, the process is driven by packets of DNA, called plasmids, that float free within the bacterium. Great masses of bacteria can clump as they exchange plasmids, and, in a break with our "normal" view of reproduction, unrelated species of bacteria can take part in the trading of genes. Thus, if one particular bacterium develops a gene for resistance to a drug, there is a strong possibility that neighboring bacteria, even different species, will acquire that resistance in the near future. In recent research, bacteria have been shown to mutate to defend against a poison (antibiotic) that they have never encountered, the mutation being triggered by the trading of genetic material with a bacterium that has survived.

Viruses present a different set of problems. Viruses are not truly living organisms, and they cannot be treated by antibiotics, although other kinds of drugs can be effective. If viruses are not alive and cannot reproduce sexually, how can they have so much variability? Viruses invade the cells of their host, capture DNA, and then reconfigure it to their needs. Because the DNA molecule is a double helix, it can be unzipped down the middle, and each half can then be rebuilt to produce two exact replicas of the initial molecule. This principle is followed in all living organisms; they all replicate DNA. In many organisms, including humans, any errors that occur in the copying procedure can be repaired quickly; hence our DNA replication is precise.

Many of our viral diseases, such as influenza, measles, AIDS, and mumps are caused by retroviruses that use a sloppy approach to copying DNA. Retroviruses start with a strand of RNA that forms a template for making DNA. This procedure is the reverse of the normal condition, in which RNA is made from copying DNA; hence the virus is called a retrovirus. The RNA is like an unzipped half of a DNA molecule, and it attracts the nitrogen bases that form the base pairs of DNA (cytosine, guanine, thymine, adenine). The "zipped" pairs always occur in the combinations cytosine with guanine and adenine with thymine, but in RNA, the base ultracil pairs with adenine (Figure 26.3). The RNA of a retrovirus forms a template for a mirror image of the DNA molecule. The appropriate bases are attracted and attach to the RNA template to form a chain that is RNA on one side and DNA on the other. The DNA strand then unzips from the RNA and attracts a matching set of bases to complete the double helix.

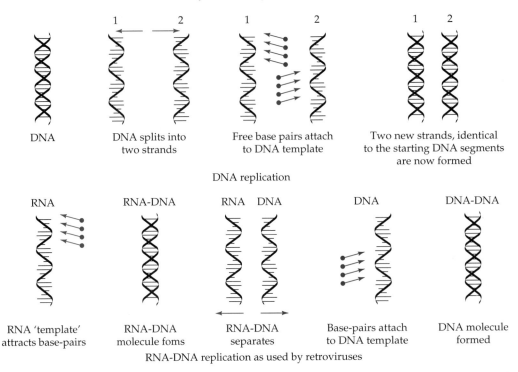

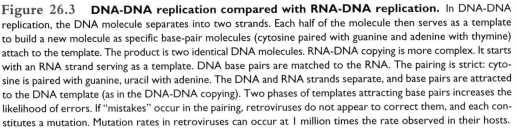

RNA-DNA replication as used by retroviruses

Figure 26.3   DNA-DNA replication compared with RNA-DNA replication. In DNA-DNA replication, the DNA molecule separates into two strands. Each half of the molecule then serves as a template to build a new molecule as specific base-pair molecules (cytosine paired with guanine and adenine with thymine) attach to the template. The product is two identical DNA molecules. RNA-DNA copying is more complex. It starts with an RNA strand serving as a template. DNA base pairs are matched to the RNA. The pairing is strict: cytosine is paired with guanine, uracil with adenine. The DNA and RNA strands separate, and base pairs are attracted to the DNA template (as in the DNA-DNA copying). Two phases of templates attracting base pairs increases the likelihood of errors. If "mistakes" occur in the pairing, retroviruses do not appear to correct them, and each constitutes a mutation. Mutation rates in retroviruses can occur at 1 million times the rate observed in their hosts.

Many errors can and do occur in this replication, and to date no evidence has been found of any corrective mechanism. Viruses may be mutating a million times as fast as their host. The mutation rate of retroviruses is so high that they are regarded not as a simple clonal lineage but as a swarm of genotypes that deviate to a greater or lesser extent from an average "type." The benefit to the virus is that any host immune response can be overcome by the pathogen swarm; one of them will be able to pick the newly installed defensive lock. This battle between host and pathogen underlies the Red Queen hypothesis (Chapter 10) and is clearly a fundamental condition of life.

Medical researchers striving to find new drugs and pathogens rallying to form resistant populations are replaying a variant of the Red Queen hypothesis: racing and racing and getting nowhere. It is as well to remember that viruses and bacteria have been on the planet for more than 3.5 billion years. Bacteria have proved themselves to be remarkably resilient to change, and, as environmental writer Mark Caldwell observed: "Bacteria are far more experienced in the drug industry than humans are. They have been manufacturing antibiotics—and finding ways to counter them—for eons."

26.3   Evolutionary Thoughts about Virulence

Pathogens invade our bodies to take advantage of our products of metabolism. The faster they draw on our supplies, the more we notice their presence. If the infestation is severe, it may lead to the collapse of our system and death. Virulence, the capacity of a pathogen to overcome our bodies' defenses, is often tied to the rate at which a pathogen reproduces inside the host body; hence, it is proportional to the drain of host resources. A traditional view of virulence is that most diseases start as virulent outbreaks with high rates of lethality and, through the process of natural selection, gradually become more benign. If a disease is too damaging, the host dies before the pathogen is transferred to a new victim. Hosts infected with benign strains are more likely to remain active long enough to pass the virus to a new victim. Consequently, over time there is a selective advantage to developing more and more benign strains. Recently, an alternative view has been proposed that links disease virulence to modes of transmission.

Paul Ewald of Amherst College observes that a pathogen that requires direct transmission—that is, an infectious disease transferred from host to host without going through an intermediary—relies on the host's being mobile long enough to pass the disease around. The virus of the common cold is passed through contact with the mucus and nasal secretions of the infected; it is very widespread, but it does not kill. The projectile blast of a sneeze or contact with a hand that has wiped a runny nose can serve to pass the virus to a new host. If the pathogen fails to reach a host, it degrades rapidly, lasting just a few hours before its outer protein coat breaks down. A virus that depends on this means of transmission should not be too damaging to the host. If the pathogen were highly virulent and caused death rapidly, it would stand little chance of genetic immortality.

In contrast, smallpox is an example of a highly virulent virus that can survive outside the body for many days, even years. The smallpox virus can lie on clothing or blankets carried from the host to a new victim. Pathogens transmitted by a vector organism, such as a mosquito or tick, can also afford to be highly virulent. The immobility of a dying host may actually increase the probability that the host will be bitten by a mosquito. Furthermore, the rapid reproduction of pathogens ensures that each drop of blood sucked from the victim will carry a wealth of potential colonists of a new host. The virus gains no advantage from maintaining a low level of infection under these circumstances; virulent strains predominate among diseases with this mode of transmission, and often the victim dies a few days after infection.

Waterborne diseases may be similarly virulent. In a comparison of bacterial diseases that affect the human digestive tract, Ewald demonstrated that the forms of cholera and dysentery that are spread through contamination of water by excrement are the most virulent (Figure 26.4). *Vibrio cholerae* is the bacterium that causes the "classical" devastating cholera of refugee camps where there is heavy contamination of drinking water (Figure 26.5). Virulence is maintained by a close-packed human population relying on few water sources, all of which support cholera populations.

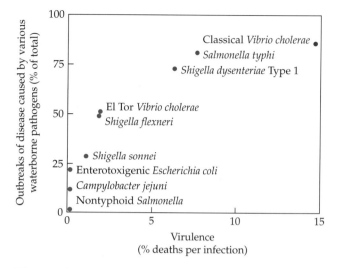

Figure 26.4 **Difference in virulence between waterborne pathogens.** Virulence is measured as deaths per 1000 infections plotted against the outbreak of disease. (After P.W. Ewald, 1993.)

Billions of cholera bacteria are flushed from the victim's body with each wave of vomiting or diarrhea. These bacteria may be washed by rain into the water supply, or relatives tending the sick may wash themselves or contaminated clothes in the water. Under these circumstances, the most virulent strains of cholera can maintain viable populations despite killing their victim within 24 hours of infection. Breaking this cycle calls for the human population to disperse or for hygiene to improve. As contamination declines, the less virulent strain of the cholera bacterium, "El Tor" *Vibrio cholerae*, outcompetes the virulent strain, and the percent of deaths per infection is reduced from 15% to 3%. Thus, purifying a water supply not only excludes many pathogens but selects against the most virulent strains.

The relationship between different strains of a pathogen is competitive; each host infected by the benign strain may deny that individual to the virulent one. Thus, maintaining infection of nonvirulent pathogens not only provides a level of immunity but also controls the numbers of the virulent population. The widespread occurrence of benign strains may prove to be one of our best defenses against virulent cholera and dysentery epidemics. For this reason, attempts to completely eradicate a disease may backfire on us. Instead of dealing with a widespread but benign form of the disease, we might end up having to deal with new outbreaks of virulent strains no longer held in check by competition.

Figure 26.5   Rwandan refugees fill water bottles and bathe beside a decomposing body. Overcrowding and poor sanitation in a refugee camp provide a continual threat of virulent cholera outbreaks.

Virulence and New Diseases

Completely new diseases are rare; most outbreaks are from old diseases with a slight mutant twist that lends extra virulence. However, the possibility exists that a completely new killer virus could suddenly appear. Some examples from the animal kingdom have raised alarm over the possibility that a new supervirus may emerge and wipe out humanity. In April 1983, a new virus was detected on a chicken farm in Pennsylvania. At first, the virus was benign, causing little damage to the chickens, but in October of that year, a highly virulent mutant strain appeared and within two months 17 million chickens were dead. All chickens that could have come into contact with those infected were destroyed, and the outbreak was stopped. The containment of this pathogen was so rapid that relatively little can be learned from this outbreak of how a new disease might move through a human population.

A deadly outbreak of disease that does provide some insights on the way a novel pathogen surges through a host population is provided by myxomatosis (rabbitpox). Form an inital lethality of 99.8% the virulence of the virus quickly declined to a lethality of about 7% (Chapter 18). This was one of the cases that prompted the assertion that viruses evolve to become less virulent. However, that is essentially a non-Darwinian argument because it does not take into account individual competitiveness among viral lineages. The viruses do not develop a common will to become nicer to their host, it is just that an optimum virulence (neither the most virulent nor the most benign) will emerge through natural selection. The decline in myxomatosis lethality probably reflects a lower density of nonimmune hosts and therefore a slowing of the infection cycle.

In a human parallel, smallpox, measles, influenza, and cholera were spread by Europeans wherever they went. The effects of these diseases were most pronounced in Central and South America. The arrival of the Conquistadors also introduced waves of disease that swept through indigenous American populations between A.D. 1500 and 1600. Population demographers estimate that, before European contact, the population of South America was between 4 and 10 million, but within 100 years of the arrival of Europeans repeated disease epidemics reduced these populations to about 100,000 (a mortality of about 90%–98%). The Americas were conquered not militarily by the Conquistadors but biologically by the pathogens they carried. The vulnerability of humans without disease resistance was comparable to that of rabbits without resistance to the myxomatosis virus (though it should be noted that the humans were hit by an armada of ailments, not a single disease). The same combination of increased resistance and a relatively sparse population ended the epidemics among the indigenous population.

More recently, outbreaks of killer diseases have caught the public imagination. In 1993, hantavirus flared up, leaving 16 dead in the southwestern United States. Hantavirus is a disease of rodents, passed to humans when they inhale dust contaminated by the urine of an infected deer mouse. The new form was massively virulent, too virulent to be sustained, and it subsided rapidly (see the box titled Ecologists Search for a Pattern in Hantavirus Outbreaks).

Many potentially fatal diseases, such as chickenpox, measles, and rubella, have attained an equilibrial state of moderate virulence, making them more of a nuisance than a true danger. The message from these examples is that a new disease can be extremely destructive, but it is unlikely to maintain extreme virulence for long.

26.4   The Ecological Perspective

Studying the evolutionary aspects of disease provides information on virulence and the maintenance of infection cycles. However, an ecological view reveals information on the abundance and distribution of pathogens and how these can be changed by human actions. Changes in population densities, either of people or of pathogen vectors, can increase or decrease the transmission of disease. If human actions, such as clearing forest, living in cities, planting crop monocultures, damming streams, polluting the air, or introducing livestock to new areas are advantageous to a pathogen, a new assault by disease may follow. Clearly, these ideas are related to the evolutionary argument, because as population densities increase, so do the opportunities for pathogen virulence.

Changes in Human Population Density

As human populations increase in density they become susceptible to diseases that previously could not be sustained. The size of the population required by the pathogen will depend upon the virulence, the mortality rate, the mode of transmission, and the proportion of the population that can gain resistance. For instance, the measles virus is thought to require city sizes of at least 300,000 to survive. At lower population concentrations, the number of previously uninfected people will be too small for the disease to survive. The first evidence of diseases' causing the abandonment of urban centers may be the cultural declines of the Harappan region of India and Pakistan (1800 B.C.) and of Mesopotamia (2000 B.C.). Historian William McNeill suggests that cholera may have driven the population from the cities. Cholera requires high population densities and the possibility of fecal contamination of drinking water. The growing urbanization of these two cultures and their rudimentary sewer systems may have provided ideal conditions for a cholera epidemic. If the inhabitants fled the stricken city, the exodus would have reduced the population density and so resulted in decreased bacterial virulence.

Changes in Vector Population Density

The recent conversion of rain forests to agricultural land has changed the species mix. Some species have gone extinct; others, particularly those that favor disturbed conditions, have thrived. Standing pools of water are often associated with human development. Soil compaction, which prevents drainage, may result in the formation of temporary pools. Trash and stored material can also catch water that becomes a breeding ground for mosquitoes and midges. An example of development's providing the vehicle for increased disease was the occurrence of an epidemic of Oropouche virus. In 1960, a road was constructed to connect the Amazonian city of Belém to the Brazilian capital, Brasília. Land-hungry settlers colonized the land on either side of the highway and grew cacao (*Theobroma cacao*, chocolate). The cacao beans are formed within a husk, which is discarded at harvest. The husks piled up around the villages. Rainwater pooled in the husks and provided an ideal breeding ground for a rare forest midge. The midge was the vector for Oropouche virus, previously only known in sloths. Midge populations exploded with the abundance of breeding grounds and 11,000 humans were infected by the virus, which created flulike symptoms. The turning point in this example was the dynamic population growth of the midge, as a result of conditions brought on by changed land use.

Changes in the Landscape

As forest is replaced with endless vistas of a single crop or as rivers are dammed to form reservoirs, new opportunities are created for pathogens as well as for their vectors. Schistosomiasis, a debilitating disease, has spread from tropical Africa to most tropical regions and is thought to affect about 200 million people worldwide.

The pathogen is a flatworm, or fluke, of the genus *Schistosoma*. The free-swimming fluke larvae bore through the skin of humans wading in infested water and then migrate to the blood vessels of the host, matures, and lay eggs. (Figure 26.6). Some of the eggs produced by the fluke are passed out of the body in our feces. For the schistosome lifecycle to be completed, the eggs must enter an aquatic habitat. A free-swimming larva hatches from the egg and seeks out its next host: an aquatic snail. The larvae reproduce asexually within the snail and leave it to seek a human host. Humans wading in the water, as they wash, plant rice, or fish, risk infection from the fluke boring through the soles of their feet, and the cycle continues. The host snails favor sluggish water and lakes, conditions that are ably met by irrigation projects. As streams are dammed to regulate their flow and provide a dry-season water supply, the habitat is changed from being unsuitable to ideal for the snails. Consequently, as the land is developed the disease can spread. This disease provides a fine example of how changes in land use can promote a deadly disease.

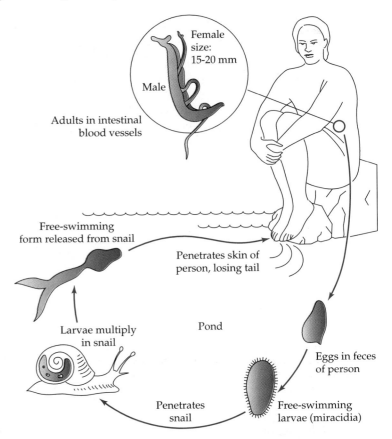

Figure 26.6   The infection cycle of *Schistosoma*.

Climate Change and the Spread of Disease

Climatic fluctuations have been shown to correlate with outbreaks of malaria, dengue fever, pneumonic plague, hantavirus and cholera, and climate change is expected to bring these diseases to new areas. Disease may enter a new area because of two principal factors, the acceleration of infection cycles, and the arrival of disease vectors.

The infection cycle is the time taken for the disease to move from one host, through a vector, to another host (Figure 26.2). The faster this cycle runs, the greater the chance that an epidemic can start. Therefore, increasing the speed of the infection cycle may increase the number of infected hosts, and hence promote disease virulence. Climate can affect the infection cycle, particularly when the vectors are ectothermic, such as mosquitoes and snails. Warm temperatures increase the metabolic rate of an ectothermic creature. A raised metabolic rate allows the mosquito to become more active, increases the number of hours it can feed, and thereby increases the number of times that it bites. Together, these factors increase the probability that it will acquire blood tainted with a pathogen such as malaria or dengue fever. The higher temperature also hastens

the reproductive rate of pathogens within the mosquito. In a study of dengue fever, it was found that an increase in ambient temperature from 30°C to 35°C allowed the infection cycle to be completed in 7 days rather than 12 days. Such an increase in the rate of the infection cycle translates into a tripling of disease transmission.

The climate models do not predict a 5°C increase in temperature for most tropical and subtropical areas as a result of greenhouse effects (Color Plate 4), but note the increase in soil temperature (which provides a good indicator of average air temperature) in areas that have been cleared of tropical forest (Figure 14.7). Thus as we discuss climate change, do not forget that local land-use changes supplement the global patterns described in Chapter 23.

The migration of disease vectors could cause the spread of tropical diseases to the United States. If the climate warms significantly over the next 100 years, plant and animal populations will expand poleward and upslope (Chapter 13 and 23), and diseases and their vectors are likely to show similar patterns of migration.

Pim Martens and colleagues at the National Institute of Public Health and the Environment in The Netherlands, have integrated data from the United Kingdom Meteorological Office climate model and the known habitat needs of malarial disease vectors. The climatic assumptions of Martens' model are that a doubling of atmospheric CO_2 induces a 3°C increase in temperature by the year 2100. Given this temperature, malaria could expand its geographic range by about 30%. Under the new climatic conditions, parts of the United States, Canada, northern Europe, and Australia would provide suitable habitat for *Plasmodium falciparum*, a particularly deadly form of malaria (Figure 26.7a, b). In general areas that already have malaria will see slightly increased transmission rates, but it is the adjacent areas somewhat poleward, or up-slope, that will see the greatest increase in the threat of malaria. This does not mean that Vancouver will have more malaria than Zaire, but it does mean that people presently safe from malaria will face an increased risk of contracting the disease. The model predicts 80 million new cases of malaria per year, about 1 million of

which would be expected to die; effectively increasing malarial deaths by 50% compared with the present.

Note that the central and southern regions of the United States are already potentially malarial, but that malaria is uncommon there. A few years ago we would have said that malaria had been virtually eliminated in the United States, with the exception of travelers bringing it back from foreign countries. This level of eradication was achieved through mosquito control and by having screens on our windows. The screens reduced the number of bites achieved by the mosquitoes and so broke the infection cycle. Outbreaks of locally transmitted malaria have recently been reported in California, Florida, Georgia, Michigan, New Jersey, New York, and Texas. These occurrences are consistent with the prediction that as the climate becomes warmer with wetter summers, there will be an increase in the potential for malarial transmission.

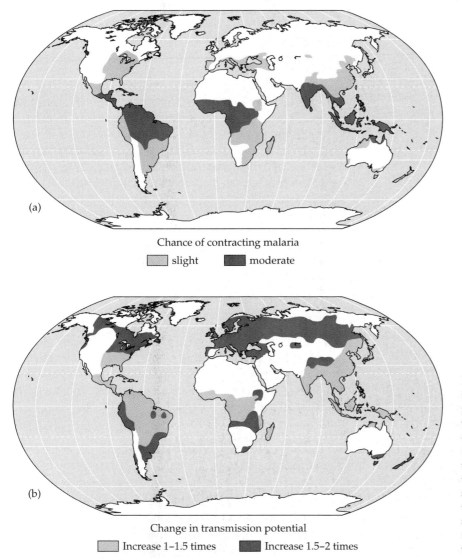

Chance of contracting malaria

slight   moderate

Change in transmission potential

Increase 1–1.5 times   Increase 1.5–2 times

Figure 26.7 **Estimated changes in the transmission of *Plasmodium falciparum* malaria resulting from climate change.** (a) Chance of transmission based on present climate. (b) The increased probability of malarial transmission given a 3°C increase in temperature. Virtually all moist temperate and tropical regions would be potentially malarial regions. Where malaria already exists conditions would be improved for its transmission. (Data from P. Martens, 1997.)

Climate change may also increase the number of vector generations that can reproduce each season. For example, the larvae of mosquitoes that carry malaria need water temperatures in excess of 16°C for successful development, and warmer temperatures mean that ponds and puddles will provide breeding habitat for more months of the year; hence, there will be more disease transmission.

The recent El Niño cycles have provided some additional insights into the effects of warming Earth. El Niño brings warm wet winters to Mexico and California (Figure 4.16). Even though the El Niño does not last long enough for all members of a community to migrate in response to changes in climatic conditions, highly motile organisms such as mosquitoes can occupy new ranges. In Central and South America, the upslope movement of the mosquito, *Aedes aegypti*, brought dengue fever to the mountains of Mexico and Colombia. Because this mosquito cannot tolerate temperatures lower than 10°C, dengue fever is normally limited to elevations below 1000 m elevation. However, during the lengthy El Niño of 1990–1995, local warming allowed an upslope expansion of mosquito populations, and dengue fever claimed victims at elevations of 1700 m in Mexico and 2200 m in Colombia. Rwanda provides a good example of the influence of temperature on malarial outbreaks. In 1987, Rwanda experienced temperatures that averaged 1°C greater than normal and a 337% increase in cases of malaria, with many of the additional cases coming from mountainous regions.

It is not only temperature that affects disease outbreaks. Increased rains and storminess are part of the predictions of a hotter world. Warm air increases evaporation rates, and with wetter air there is the probability that gentle rains are replaced by downpours. Intense rains lead to flooding and ponding of water, which provide ideal conditions for mosquito reproduction. However, it is not only mosquito-borne diseases that can be promoted by wet weather. The 1990 El Niño brought rains to the Four Corners region of the Southwest, leading to the outbreak of hantavirus which is spread by mice.

The same El Niño event is blamed for an outbreak of cholera. Rita Colwell, President of the University of Maryland Biotechnology Institute, and colleagues have suggested that the 1991 cholera epidemic that infected more than half a million people in 19 South and Central American countries was induced by an El Niño event. During the El Niño that began in 1990, ocean-dwelling crustaceans, called copepods, built huge populations in the warm waters off the coast of Peru. The copepods were a natural host of the cholera-causing bacterium, *Vibrio cholerae*, and an important component of local food chains. As a result of predation on the copepods, the *Vibrio cholerae* spread through the food chain until it was present in seafood eaten by humans.

A further factor to consider is that if droughts, floods, storms, and other disruptive climatic phenomena lead to people being forced from their homes or having an inadequate food supply, these people will be made much more vulnerable to all kinds of disease.

Contact with Exotic Diseases

Human evolution started in Africa, and, because pathogens coevolve with their hosts, Africa accumulated a greater diversity of organisms that use humans than any other continent. Because many of our pathogens require an intermediary host to complete their life cycle, as humans left their evolutionary homeland, they left behind those disease-bearing organisms. The first humans to enter the Americas about 13,000 years ago found a land free of human-specific pathogens; it was that same freedom from disease that was to ensure the destruction of their descendants some 11,500 years later when they came in contact with the Conquistadors.

When two populations come into contact for the first time, the potential exists for an exchange of novel diseases that can be disastrous. Eighteenth- and nineteenth-century Africa was justifiably known as the "white man's grave." Traders, slavers, scientists, missionaries, and imperialists came to inglorious ends as they perished of malaria, yellow fever, dysentery, typhoid, or sleeping sickness. Exposure to these diseases, to which the interlopers had no resistance, often was rapidly fatal. McNeill raises the fascinating argument that the presence of so much endemic disease prevented the wholesale colonization of Africa by Europeans, so Africa, apart from the very south, though often subject to foreign rule, remained culturally African. How different world history would have been had a similar battery of diseases awaited the Conquistadors in the Americas.

26.5 The Emergence of New Diseases

A finite chance exists that a pathogen will cross over from a nonhuman species into humans. Such events are rare, but when they occur the consequences can be extremely damaging. As human densities increase in previously remote areas, we come into contact with new species of animals, which may transmit disease to us directly or indirectly. AIDS is probably an example of direct disease transmission in that one of the most likely scenarios for the first advent of this virus in humans is that it was transmitted by a monkey bite (see below). An example of indirect transmission is influenza.

The influenza epidemic of 1918 was the most catastrophic outbreak of human disease in the last two centuries. This strain of influenza crossed into the human population from pigs, and in the space of one year it killed 20 million people, leaving more dead than World War I, which ended November 11, 1918. The virus was extremely virulent, but the social upheaval of the four preceding years, destruction of housing, reduced food supply, overloaded medical system, worry, and fear may all have made populations especially vulnerable. In 1976, an outbreak of this same "swine influenza" was documented at Fort Dix, New Jersey, but this time the disease was quickly brought under control with good medical care; only one death resulted.

Other outbreaks of new forms of influenza were experienced in 1957 and 1968, each representing a new combination of viral characters. When such pandemic outbreaks occur, a wave of an identifiably new virus sweeps across the globe. The source is a reassortment of two previously separate viral strains that combine to form a new virus. In all of the documented outbreaks, the new viral genes came from a combination of avian and human pathogens, probably via pigs. The pigs can be infected with viruses from both ducks and humans. Inside the pig, a chance reassortment of some viral genes can result in a new virus. A minute, but distinct, possibility exists that the new virus can infect humans and cause a serious disease. All the pandemics of this century have started in China: Is this more than coincidence? It has been suggested that the Chinese style of animal husbandry, in which ducks (many of which have almost permanent symptomless influenza), pigs, and humans live in proximity, may encourage the interspecific transfer of influenza viruses.

The bouts of flu that hit us each winter are usually mutant strains of a well-known influenza virus and are usually the relatively benign version of the last virus that caused pandemic infection. Thus, our present influenza viruses are derivative strains of the 1968 pandemic virus. The most recent virus to form a pandemic that has crossed from animals into humans is HIV.

26.6 Human Immunodeficiency Virus

Human immunodeficiency virus (HIV) causes the disease AIDS, and, although the disease has arisen in the period of medical documentation, its history is only partially established. AIDS was first identified in 1978, but this was not the "start" of the disease. In Zaire, in 1959, a blood sample was taken from a man dying from what, in hindsight, appears to have been AIDS. The blood has since tested HIV-positive. It should be noted, however, that tests for HIV produce false positives, and stored blood samples are more prone to these than are fresh ones. The same year, 1959, a seaman died in Britain, and, again with hindsight, the clinical description of his death matches that of an AIDS fatality. Assuming that these were AIDS fatalities, it is likely that HIV had been in the human population for at least 10 years before those deaths (see below), and that it may have been with us, but undocumented, for some time longer than that. So where did HIV come from?

Two fundamentally different human immunodeficiency viruses are presently known: the highly virulent HIV-1 and the less virulent HIV-2. Both viruses ultimately degrade the immune system and lead inexorably to death. HIV-2 can be almost benign in the sense that the immune system is barely affected and that the incubation period between initial infection and the onset of AIDS can be many years. HIV-1 and HIV-2 both appear to have come from monkeys (simians), because the most closely related viruses are the simian immunodeficiency viruses (SIVs). Studies of the molecular contents of HIV and SIVs that are found in different monkeys and apes suggest a theory covering the pattern and timing of transmittal. Plenty of room exists for error in these estimates, and scientists are by no means unified on the interpreted history of these viruses, but it appears likely that HIV-2 diverged first and has been in humans for about 150 years. HIV-1 appears to have been more recent, perhaps in the human population only since about the 1950s. Although the same basic rules apply to HIV-2, most of the following discussion deals with the infection and spread of the more virulent HIV-1 strain.

HIV infection comes about through the transfer of body fluids (blood, semen, or lymph) from an infected (HIV-positive) person. It is believed that HIV-2, and possibly HIV-1, are derived from SIV. The means of transfer is unknown, but once inside the human body the SIV mutated to form HIV.

Ecology in Action

Ecologists Search for a Pattern in Hantavirus Outbreaks

In August 1993, a wave of mysterious deaths plagued the Four Corners region of the southwestern United States. The sickness was soon recorded in 15 states. The victims suffered flulike symptoms of fever, nausea, and vomiting; fluid flowed into their lungs, and some died of respiratory distress. Antibiotics were ineffective, and 27 of the first 45 cases proved fatal. Fortunately, the disease was not infectious between humans, and this limitation greatly suppressed its spread through human populations. If humans could not spread the sickness, what was the mechanism of disease dispersal?

The disease was identified as a new form of hantavirus, a disease that was well documented in China but not previously known in the United States. Medical investigators found that the disease was spread by deer mice (see photo). The virus is transmitted to us from the dust associated with mice nests and mouse urine and droppings. Humans breathing the dusty air of a house or barn used by deer mice run the risk of contracting hantavirus.

Perhaps the most important ecological question is to determine what caused the outbreak of the virus in 1993 and what would be the trajectory of future infections? There is no evidence to suggest that this was a new virus. Why then did it suddenly become a problem to humans

The deer mouse, a vector for hantavirus.

After entering a host, HIV appears to lie virtually dormant for several years. But it is not a true dormancy; the viral population is undergoing change throughout this time. HIV has especially high rates of mutation, and, as the virus replicates, a swarm of genotypes is emerging. Some of the swarm have the genetic capacity to breed faster than others, and all are attacked by the host immune system. The proportions of fast and slow breeders gradually changes as time goes on. Predation by the immune system keeps the virus in check, but the fast breeders will outcompete the slow breeders. Gradually the fast breeders take over and place more and more strain on the immune system. To maintain its population, the virus parasitizes more and more of the proteins being made by the cell. This drain may start as 1%–2% of manufactured protein and accelerate to almost 50% in the final stages of the disease. As HIV is draining the cells that it infests and straining the immune system, it is also attacking the immune system itself. As the HIV population increases in the host, all the time selecting for faster-breeding (more virulent) strains, the host is weakened and made more vulnerable to other diseases. A critical point is reached when the immune system is overwhelmed: The host may fall sick with respiratory diseases such as tuberculosis, develop skin cancer, and experience retinal decay. With the collapse of the immune system the population of the fast-breeding viral strain spirals upward. The phase in which the host falls acutely ill is clinically diagnosed in 1993? What changed in the ecology of either the humans, the mice, or the virus at that time? If humans had not changed their behavior greatly in 1993, then the search narrows to the mice or the virus. Because the link between the disease and the humans is the vector, the ecology of the mouse is a good place to start the search.

By chance, ecologists were already monitoring the deer mouse population of the Four Corners region. The study had started before the disease was documented and was undertaken because the mouse population had increased 10-fold between May 1992 and May 1993. In their investigation of the mouse population explosion, the ecologists were able to demonstrate the links in the chain of events that led to the disease outbreak.

Basic ecological observations of deer mice showed that they feed primarily on piñon nuts and grasshoppers and that food availability is the biggest limiting factor to their population. What limits seed production in the dry windy region of Four Corners? The answer is rainfall. For 10 years, the region had been gripped by a drought so severe that deer mouse populations fell to such a low level that most of the animals that prey on deer mice had been eliminated. The El Niño (Chapter 4) event in the spring of 1992 brought heavy rains to the Four Corners region and resulted in a bumper crop of piñon nuts. With increased food abundance and no predation, the mouse population flourished. This population explosion could have had two important consequences. First, the extra mice would search out new nest sites, so they would move into homes and barns. Consequently, human contact with the mouse and the virus-laden dust would be greatly increased. A second consequence of increased mouse population size is that it could have increased the virulence of the virus, just as denser human populations allow cholera to increase in virulence.

Cleaning homes and barns to reduce human contact was a priority in limiting human contact with the deer mice. However, even without this measure, the lethality of the virus to humans would be expected to fade. By October 1993, the deer mouse population had collapsed back to near normal levels. With the decline of mouse populations, human contact would be reduced and viral virulence would decline. Consequently, the disease would be expected to recede back to obscurity. The 1993 outbreak of hantavirus killed a total of 37 people in 15 states before it simply petered out.

Deer mice are found throughout Canada, Alaska, and the United States, with the exception of the East Coast and the Southeast. Researchers are now attempting to determine if it is possible, or likely, that another bout of hantavirus could flare up at a new location. An El Niño event that brings rain to this region increases the danger of another outbreak. While there is little that can be done about El Niño, learning more about the deer mice may be important. A study of deer mouse populations across the country is being made to determine which regions have mice that are carriers of the virus. Throughout their range, deer mice are captured in humane traps that do not injure the animal. Workers protected from the virus by breathing apparatus and protective suits empty the traps regularly and collect blood samples from the mice. The mice are then tagged and released. These studies allow researchers not only to calculate the numbers of deer mice in different habitats and at different geographical locations, but also to estimate the proportion that carry the hantavirus.

This case study underlines how even subtle climate change or the extermination of a predator that keeps natural populations in check can have dire consequences for humans.

as AIDS. A typical time from the initial infection with HIV to the development of full-blown AIDS is about 10 years. Thereafter, unless the person is treated with the drug AZT, life expectancy is generally less than two years. AZT is not a cure for AIDS; at best, it will retard the effects of the disease and delay death.

From an evolutionary standpoint, the increasingly virulent HIV strains will win the competition within the host, but the most virulent strains will not be passed on. As a host shows symptoms of AIDS, the chance of sexual acceptance by an uninfected partner is reduced. Also, in the latter stages, when the virus attains maximum virulence, the host is too weak to be interested in sex. These ultravirulent strains, therefore, play almost no part in the transfer of HIV. An excep-

tion is found among IV (intravenous) drug users, as the need for a "fix" may last until their dying breath. The needle used for their injection can become infected with ultravirulent strains, and IV users, therefore, stand an increased risk of being infected by virulent HIV strains. The hypothesis that infection attributable to IV drug use will result in a shorter time from infection to full-blown AIDS appears to have been substantiated by research.

The virulence of HIV-1 is dependent on sexual promiscuity or the frequent sharing of needles. Frequent partner swapping does not just increase the chance of contracting the disease, it also affects the genetics of the pathogen. The greater the number of sexual contacts, the greater the chance that a "late" or virulent form of the pathogen will be passed on. Thus, in societies where an individual can have many sexual partners in a relatively short space of time, there will be a tendency for the more virulent HIV strains to be favored. Promiscuity increases both the chance of an individual's becoming infected and the deadliness of the disease. Some experimental evidence supports the hypothesis that host behavior affects pathogen virulence. In experiments conducted on mice that have been genetically altered to suffer AIDS symptoms, the virulence of HIV-2 was seen to be low until conditions of more frequent partner swapping were replicated: Then the virulence increased markedly. It remains to be seen whether HIV-1 would mutate to become as benign as HIV-2 if social changes reduced the rates of new partner infection.

Geographic Patterns in AIDS Transmission

Prevention programs and good health services have contained the spread of HIV in North America and Europe. Rates of HIV infection in the industrialized nations have yet to exceed 1% of the population. But Europe and North America are not the areas most severely affected by AIDS. Health studies in some African and Asian countries reveal alarming levels of HIV infection, with new infections occurring at twice the rate previously believed. Each day 16,000 people become infected with HIV, almost 6 million new cases per year. Levels of HIV infection are estimated to be

about 2% in the general African population but much higher in some urban areas. The World Health Organization estimates that, in the teeming cities of Uganda, Nigeria, Kenya, and Rwanda, HIV infection rates are as high as 60% among female prostitutes, 20% among pregnant women, and 30% among the reproductive population (Figure 26.8). But the disease is spreading out from the cities, and in Zimbabwe one in four people have HIV. A relatively recent realization is that Asia is almost as severely affected as Africa. One in 2 prostitutes in Cambodia test positive for HIV, as do 6% of soldiers. In India, though the infection rate is less than 2% of the total population, this still amounts to 4 million people, the largest pool of HIV cases that any nation has to deal with. It is estimated that almost 90% of all people infected with HIV live in Asia and sub-Saharan Africa (Figure 26.9). These statistics reveal immediately two fundamental differences between the AIDS epidemic in Africa and that in America. First, AIDS is far more prevalent in these African countries than in the United States. Second, in the United States, about 90% of HIV victims are in the high-risk groups of nonmonogamous homosexual men and IV drug users. In Africa and increasingly in Asia, it is a disease that is now thoroughly rooted in the general population.

The spread of AIDS in African countries such as Zimbabwe, Uganda, Malawi, and Kenya is in large part due to the economic structuring of those countries. Rural poor migrate to the city to find work. The men often leave their families in the villages and, alone

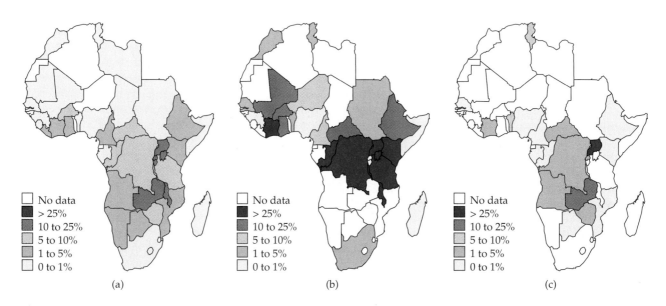

Figure 26.8 Minimum estimates of HIV-1 infection in Africa in 1991. (a) Among sexually active populations in cities, (b) female prostitutes (primarily in cities), and (c) sexually active populations in rural settings. (After R.M. Anderson, R.M. May, M.C. Boily, G.P. Garnett, and J.T. Rowley. Reprinted with permission from *Nature* © 1991 MacMillan Magazines Limited.)

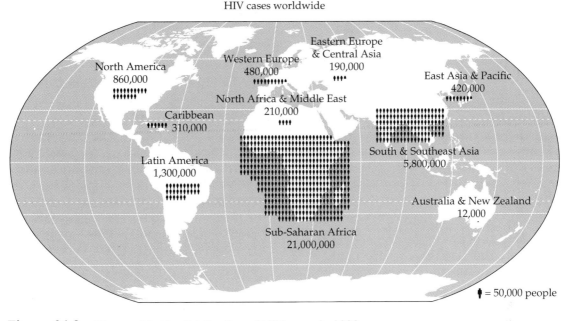

HIV cases worldwide

Figure 26.9   The worldwide distribution of HIV cases in 1998.

in the city, make use of prostitutes. For lack of employment, urban women frequently become part- or full-time prostitutes, and in the poor sections of the city they may have more than 1000 clients each year. The conditions for acquiring and passing on highly virulent forms of HIV are nearly ideal. The disease is then spread back to the rural areas when men who have been visiting prostitutes return to their wives and girlfriends. Another factor that facilitates the spread of HIV is the rapid increase in "bush taxis," minibuses, and trucks that bring even remote areas into contact with the major cities, thereby greatly increasing the mobility of the populace and their pathogens.

It has also been argued that the spread of HIV-1 into the heterosexual "low risk" population is a function of the time that the disease has been present in Africa. If that argument is true, the highest rates of infection should be found where HIV-1 first entered the human population. However, the rates of infection in western Africa, where it is believed that HIV first occurred, are not as high as in eastern Africa. Western African countries, such as Senegal, which are predominantly Muslim, and in which societal convention is rigidly opposed to extramarital sex, show relatively low rates of infection with HIV-1. While neighboring countries have double digit infection rates, in Senegal, condom distribution and sex education has kept HIV infections below 2%. In these areas, HIV-2, the less virulent strain, is more abundant than HIV-1. It seems that behavioral patterns are more important in determining the level of HIV

in a population than is the time since the onset of the epidemic.

Evidence of the effectiveness of modifying behavior in restricting the spread of HIV is seen among the male homosexual population in the United States. Since the identification of HIV as a sexually transmitted disease, the behavior of gay men has generally become more conservative, with reduced rates of partner swapping and unprotected sex. Consequently, the rate of new infection has fallen sharply in this group over the past 10 years. A group that shows no such behavioral adaptation is the IV drug users, and their rates of infection continue to climb. The campaign for "safe sex" has been of proven worth, but a determined effort to target IV drug users for AIDS prevention is needed. That IV drug use is illegal, and therefore is often denied or hidden, makes it hard to contact the people most at risk. Further, the craving for drugs defies rationality, and absurdly dangerous risks are ignored for the sake of the fix. The sharing of needles is often concentrated in the inner city areas among urban poor drug users (richer folk may have their own drug paraphernalia). AIDS shows the greatest chance of breaking out of the traditional high-risk groups into the general population where there is direct contact with these IV users.

A sobering thought is that approximately 5–9 million people worldwide have already developed AIDS and that half of them are dead. In 1998, the World Health Organization estimated that 30 million people had been infected with HIV, and given present medical knowledge, will surely die prematurely.

Summary

- Natural defenses such as antibodies, coughing, and vomiting protect us from a constant barrage of pathogens.
- Our genetic individuality provides a defense against pathogens.
- Disease pathogens are continually evolving to get around our natural and artificial (medicinal) defenses.
- Forms of disease resistant to medicines and vectors resistant to pesticides are resulting in the return of diseases once thought to be "conquered."
- Social upheaval such as warfare or economic recession can interrupt disease prevention programs and allow disease to return to a population.
- The virulence of a pathogen is related to its mode of transmission.
- The virulence of a pathogen is also related to the density of the host population and the frequency of transmission from one host to another.
- Ecological changes, such as changed land use, can cause increased occurrence of previously rare diseases.
- The contact between two populations that have spent some time evolving in isolation is a time when potentially lethal pathogens can be swapped.
- Climatic change shifts the ranges of disease vectors, enlarging the area where the disease is present, hence increasing the human populations threatened by the disease.
- HIV is most virulent among populations with the highest transmission rates—that is, the more often needles or sexual partners are swapped.
- HIV in Africa has moved into the general population.

Further Readings

Brown, K. S. 1996. Do disease cycles follow changes in weather? *BioScience* 46:479–481.

Epstein, P. R., et al. 1998. Biological and physical signs of climate change: Focus on mosquito-borne diseases. *Bulletin of the American Meteorological Society.* 79:409–417.

Martens, P. 1998. *Health and Climate Change: Modelling the Impacts of Global Warming and Ozone Depletion.* London: Earthscan Publications Ltd.

Pinkerton, S. D. and P. R. Abramson. 1997. Condoms and the prevention of AIDS. *American Scientist* 85:364–373.

Platt, A. E. 1996. Infecting ourselves: How environmental and social disruptions trigger disease. *Worldwatch Paper 129.* Washington, DC: Worldwatch Institute.

Wills, C. 1996. *Yellow Fever Black Goddess.* Reading, MA: Addison Wesley.

Web Connections

On-line resources for this chapter are on the World Wide Web at:
http://www.prenhall.com/bush
(From this web-site, select Chapter 26.)

Environmental Economics

27

conomics is the science that seeks to identify the most efficient allocation of scarce resources, such as raw materials, labor, production facilities, and investment capital. Because no individual, firm, or nation has all the resources it needs, decisions must be made to optimize the use of those available. Economics is the tool used in that decision-making process, and, consequently, it affects all of us. Our standard of living, employment prospects, and environment are all shaped by economic policies instigated by government, industry, or lending institutions. Taxes are spent, development choices are made, and money is granted to one project and not another, on the basis of economic projections.

For more than two centuries, economists have sought to increase the predictive accuracy of models that describe what would happen as any industrial or consumer factor changed. The traditional economic model features a feedback loop between consumers and industry (Figure 27.1a) that is supplied by raw materials. If consumers demand more, additional goods will be produced and more raw materials will be consumed. Traditional economics totals all the costs incurred in the production of a good and compares them with the revenue from the sale of the good. Production of a good often results in the creation of byproducts or waste materials of no value to the pro-

ducer. Examples would be waste steam from a boiler, carbon dioxide from burning fuel, and heat from the production of electricity. Any of these byproducts that is not assigned a value in the costs of production is termed an **externality**. One definition of an externality is that it is an unaccounted product, or factor, in a production process. Many forms of pollution are externalities. In traditional economics, pollution is merely the venting of a waste product; it carries no financial loss or gain, and, when the waste material leaves the facility, it leaves the economic world. With the advent of the environmental movement, many economists re-evaluated the traditional model and found that externalities can have a significant impact (good or bad) on the functioning of traditional economic markets (Figure 27.1b). They suggested that externalities should be included within the main economic model and that, through their inclusion, economics becomes a more powerful predictive tool. This broader discipline that fuses ecology, environmental science, and economics has become known as *environmental economics*. Questions that had previously lain in the realm of philosophers are now also in the domain of the environmental economist. How much is a species worth? Can we place a monetary value on clean air or on an unpolluted river? Should we include contentment in our index of economic well-being?

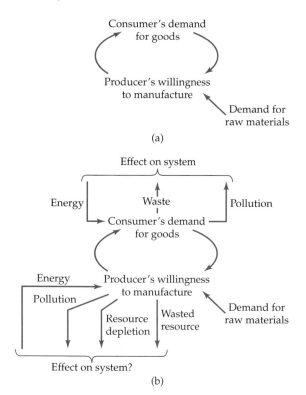

(a)

(b)

Figure 27.1   A positive feedback loop between consumers and industry. As consumers demand more, industry supplies more. (a) In the traditional economic view, the supply of materials, such as iron ore, oil, or coal, is the only reliance on the natural world. (b) In environmental economics, externalities such as water pollution, depletion of timber stocks, or poor air quality are included as negative feedbacks. As consumers demand more, producers produce more, thereby increasing pollution and hence lowering the quality of life.

27.1   Traditional Economics and Market Values

In the most basic of economic models, it is assumed that when an item is scarce it will have a price (only things that are ubiquitously abundant, such as air, will not have a price) and that the scarcer the commodity, the higher the price will be. The actual price will be determined by the opposing forces of supply and demand. If demand is great and supply is small, as in the case of diamonds, Van Gogh paintings, and out-of-season sweet corn, the price will be high.

The trading relationship of producer and consumer can be described by **supply and demand** curves (Figure 27.2a). The supply curve represents the quantity of the commodity that the supplier is willing to produce at a given price. The demand curve represents the number of purchases that can be expected at each

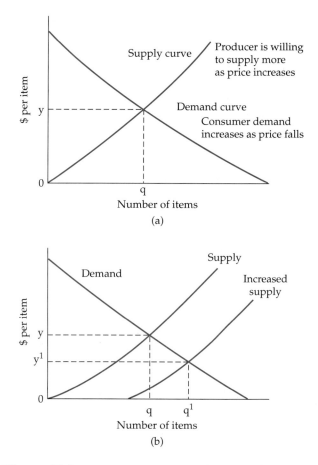

Figure 27.2   A simple model of supply and demand for a good. (a) Supply and demand curves for a good. As demand increases, the price that the consumer is willing to pay increases. As price increases, the supplier is willing to produce more. The intersection of the two lines is the market price of the good. y is the sale price of the good, and q is the quantity supplied to the market. (b) Increased supply moves the equilibrium from y to y¹. At y¹ the lower price leads to increased demand q¹.

price. The supply and demand curves intersect where the price matches consumers' willingness to buy with supplier's willingness to sell.

The price of a good is not necessarily fixed through time, so the supply and demand curves can move in response to changed market conditions. The price for sweet corn will be high in the spring, when corn is scarce, but it will tumble in the late summer. As the harvest comes in from the fields, the sweet corn supply increases dramatically, and, in order to unload the glut of corn, the producers have to drop their price. The supply curve shifts to the right (Figure 27.2b), and a new equilibrium price is found where the demand curve crosses the new supply curve. To know whether the producer is making a profit, we need to know the costs of production.

Estimating Costs

Production costs are determined by totaling the cost of raw materials, the wages paid to labor in the production process (this includes wages of manufacturers, distributors, and sellers), and the costs of the capital facilities used, such as industrial plant, machinery, transportation, and shopping mall. If this total cost is divided by the number of items produced, a cost per item is obtained. However, this estimate ignores the cost of pollution caused by the industrial process and may not reflect the depletion of reserves of raw materials, such as oil or timber. The costs of these raw materials should be reflected in higher prices as the materials become progressively scarcer. Paradoxically, oil is as cheap now as at any time in the last two decades, despite our use of billions of barrels of irreplaceable crude oil each year. Timber is also underpriced, because in many areas it is treated as a one-time crop, and no effort is made to replant or produce a sustainable harvest. The nonsustainable use of these natural resources is like a going-out-of-business sale: The merchandise is being cleared at bargain prices in order to gain a quick cash return.

Environmental economists maintain that realistic costs for pollution and the depletion of resources are real production costs and that they should be included within the cost calculation of the commodity. However, the problem of determining the real cost is not an easy one to solve if neither the raw material nor the pollution has a realistic price. Perhaps a price for oil that would reflect a truer cost to society of using a nonrenewable resource would be to determine how much the amount of energy the oil yields would cost to generate using a sustainable energy source such as wind power. For timber, a realistic price would include the cost of converting the industry from one that is purely extractive to one that farms timber. Trying to estimate the cost of pollution is far more complex; it would depend on the type, magnitude, and frequency of the pollution. In some instances of pollution, human health effects would be one criterion that could be used to determine societal costs.

In 1984, the Union Carbide chemical plant in Bhopal, India, exploded, releasing a deadly gas cloud. The Indian government estimates that about 2600 people died immediately, and a further 2500 died from aftereffects of the explosion. Estimates for those injured or blinded by the toxic cloud range from 10,000 to 200,000. How should a disaster on this scale be valued? One way would be to add the cost of the destroyed factory and the compensation paid by the company. The cleanup of the disaster cost about $100 million, and, in 1989, the Supreme Court of India ordered Union Carbide to pay a total of $470 million in compensation to the victims of the accident. The total cost of the accident on this basis was about $570 million. On this basis, the average compensation payment to each victim, assuming 200,000 claims, would be $2350 each. Had that accident occurred in the United States, would the valuation have been the same? In the United States, a human life is generally valued by the courts at between $2 million and $8 million, at least 1000 times as high as the compensation paid to the victims of Bhopal. Thus, estimates of the cost of the accident are heavily influenced by the local valuation of life.

Most pollution is not as devastating in its consequence as the Bhopal explosion. Indeed, it is rare that anyone dies directly and immediately from pollution, although many people may suffer increased likelihood of respiratory disease, cancer, or allergies. The costs of such pollution have long been ignored, but to do so may not be in the company's best interest. Long-term ailments among workers may result in lost workdays due to ill health. A sickly workforce will be unreliable and will have low productivity. Other costs may include increased medical expenditures that will drive up insurance rates and an increased turnover of employees as people choose to leave a polluted environment (their leaving results in training costs for new staff). Worsened industrial relations may result if the workforce is discontent with their general living conditions and blames them on the firm. Any of these factors will increase the costs incurred by the company, and hence reduce its profits. Thus, even though usually ignored in economic accounting, pollution can result in very real costs to the producer.

As we look further afield, we see that not all pollution effects should be measured in terms of human health. Pollution will have an impact on ecosystems, and we should evaluate those costs. Some pollution, such as acid deposition or ground-level ozone generated by factories, will result in reduced crop and forest yields. Such damage is not taken into account as an industrial cost, although the farmer or forester could put a dollar amount on the lost harvest. If the damage is wrought on a national park or other area that is not being used for commercial gain, then there is no crop involved and no money is lost. But does this mean that there is no cost?

Traditional economic theory deals only with commodities that have a price, which brings us to the concept of market value. Goods and services—such as a house, an apple, a haircut, or a log—all have a **market value** and hence a price, but there are many goods that have no price, yet are prized. Examples of goods without a market price are religious and political freedom, happiness, love, or a breathtaking view (Figure 27.3). Does this really mean that they have no value? Millions of people have died fighting for their religious beliefs or for freedom, causes for which they were prepared to pay the ultimate price. Where a good has a worth but not a price, it represents a **non–market value**.

Non–market-valued goods are frequently omitted from traditional economic calculations. Similarly, the costs of externalities such as pollution, noise, or resource depletion are negative because they cause damage or are unpleasant to live with. That these externalities have the potential to be of economic significance is immediately apparent. Other externalities, such as a beautiful view or waste steam that can be harnessed for cogeneration, are positive. A positive externality such as a view of the ocean is clearly going to increase the value of a piece of property. Hence, positive externalities are also significant economic factors. Many environmental qualities such as clean air, wilderness, and biodiversity need to be treated as non–market values. If we are to establish the cost of damaging a wilderness or a beautiful scene, we need to determine how much it was worth in its pristine state, or, in economic terms, What is the benefit of zero disturbance?

Figure 27.3 How much is a beautiful view or the chance to walk in a wilderness worth?

Estimating Benefits

In a traditional economic analysis, earnings are entered into the **benefits** column. However, the preceding argument has led us to the point where it is clear that we derive benefit from things even though we never gain income from them. Wonderful views, the smell of clean air, or a glimpse of a bear all bring us pleasure, but should these feelings count as economic benefits? Many people would say that they work five days a week so that they can enjoy two days of recreation on the weekend and two weeks of annual vacation. Thus, their prime benefit in terms of quality of life comes from their recreation not from their production. The places to participate in this recreation are goods with non–market values. One of the goals of environmental economics has been to develop ways that these non–market values can be quantified into dollar amounts so that they can be incorporated into economic decision making.

The crux of this argument is that future decisions regarding land use will be made on the basis of economics and that the costs and benefits of competing uses will be compared. If wildlife, soil quality, hydrology, clean water, and use of an area for fishing or hunting are not given a monetary value, they will not enter these calculations; effectively they would be given a $0 value.

Developing a sound economic basis to determine the dollar value of non–market values has proved both difficult and contentious. The backbone of these estimates is carefully conducted surveys of the public intended to elicit the value they place on a particular good, such as a particular wilderness area, a Bald Eagle, or a clean beach. Sometimes, the method to determine a value relies on direct questions as part of a long questionnaire. At other times, calculations can reveal an otherwise hidden set of values. For instance, if there are two beaches, one dirty and close to the city and one more distant but cleaner, visitors to the distant beach can provide information regarding the extra cost that they have incurred in coming to the cleaner area. These data can be elicited in a questionnaire asking why they came to that beach, how far they have traveled, and their travel expenses. The additional cost incurred by the visitor to the clean beach represents a minimum value that they were prepared to pay to avoid pollution.

Analyses of hundreds or thousands of such questionnaires yield an average value. This value can be multiplied by the number of visitors that use the beach each year to provide an estimated value to society for the cleanliness of the beach. In the jargon of environmental economics this is a **use value**, because each visitor was reporting his valuation of something that he was using (the clean beach). We may also hold **non–use values**; an example would be the donation of time and money by volunteers publicizing the plight of rain forests. Some of these people hope to travel to the forests, so they have a vested interest (use value) in the preservation of those habitats. Perhaps they want to go there to do research, to vacation and watch monkeys, or to hunt. All of these are "uses" of the forest that could motivate someone to work to save it as a resource. Yet many of the donors and volunteers never expect to visit a rain forest. The reward for their efforts lies simply in the knowledge that this spectacular ecosystem exists. Theirs is a non–use valuation of the forest.

The calculated worth of many environmental features can be assessed to include both use and non–use values. What is the value of a grizzly bear or a blue whale? The traditional way of valuing a species considers all the products that can be obtained from the corpse. The whale can be boiled down to produce fat and oil, while the bear can make a fine hearth rug. Hunters would pay to shoot the bear, so that is another value that could be assessed. However, trying to value a bear or a whale as a living creature, and one that stays that way, is a different proposition. A controversial estimate of worth, but probably the best available, is based on our **willingness to pay** to keep the animal. Surveys of the public are used to elicit how much the creature is worth to us.

It is noticeable from the valuation of wildlife (Table 27.1) that large mammals are generally the most highly prized species. Among the birds, the Bald Eagle is a national symbol, and its high profile guarantees a higher valuation than that of the rarer Whooping Crane. We clearly respond to cute and cuddly creatures, to those that we perceive to be intelligent, and to those that receive media attention.

Willingness to pay is one of several different kinds of contingent valuation methods (CVMs). A CVM is a survey based on questionnaires to determine the value that the public places on the existence or use of a natural asset (such as water quality in a river, air quality, or a scenic landscape). CVMs can be used to establish the full cost to society of a change in land use. For example, in these questionnaires, the proposed land use and the existing land use are contrasted, and the reader is provided with an unbiased account of the pros and cons of the proposed change. That the survey is not slanted to provoke a particular outcome is essential if the result is to provide a valuation of the existing habitat.

A contingent valuation study was used to determine the reparation damages assessed against Exxon after the Exxon *Valdez* oil spill in Alaska. Exxon challenged the accuracy of the valuation of habitat damage, and, to resolve the issue, a special panel of disinterested experts was convened by the National Oceanic and Atmospheric Administration (NOAA). After researching the claims of both sides, this panel stated that "contingent valuation can produce estimates reliable enough to be a starting point for a judicial process of damage assessment." This expert advice has set an important legal precedent, validating the use of CVM analyses in assessing damages against polluters.

However, contingent valuations are not without their critics, even within the environmental movement. A common objection is that we should not put an actual dollar value on wildlife (or natural settings) and that an ethical argument is better. This contention rests on the recognition that it is our absolute duty to maintain the planet's biodiversity and that reducing a 4 billion year evolutionary history to a line on an account sheet is unethical. In this valuation, all species of plants and animals are priceless because they are irreplaceable.

An alternative argument is that we have no idea which plant or animal may contain the chemical that provides a cure for cancer or some other disease. Therefore, our assessment of worth based on our existing knowledge may prove to be a gross underestimation. An example would be the *Taxus* tree of the Pacific rain forests (Figure 27.4). Until recently, this tree would have been considered virtually worthless in monetary terms. It was too small for timber production and appeared to be just an insignificant member of the forest understory. However, the discovery that a chemical purified from the tree can be used to treat cancer suddenly changed the valuation of the

Table 27.1   "Willingness to pay" estimates for endangered species.

Species	U.S. $ per person per year
Grizzly bear	18.5
Bald Eagle	12.4
Blue whale	9.3
Bighorn sheep	8.3
Bottlenose dolphins	7.0
Endemic shiners (fish)	4.5
Whooping Crane	1.2

(Data are from D.W. Pearce, 1993.)

Figure 27.4 Harvesting the bark of the yew tree (*Taxus* sp.) of the old-growth forests of the Pacific Northwest. The bark provides Taxall, a drug found to be effective against cancer, but cutting the bark kills the tree. What is the worth of this tree?

species. This insignificant species rapidly became the center of research activity and has the potential to spawn millions of dollars in drug sales.

A further danger of ascribing monetary values to wildlife is that the dollar value may be used in isolation in a decision process. The value of a shiner (small fish of rivers in the southeastern United States) may be estimated as only $4.50, so would a compensation equivalent to $5.00 for each person in the United States more than offset the extinction of the species? One elegant argument that highlights the problem of considering species on an individual basis is to liken an ecosystem to an aircraft held together by rivets. In this analogy, each species is a rivet holding down its small part of the ecosystem. Some of these rivets, such as the ones holding on the wings, would clearly be important; others may seem less significant. Given that we are uncertain of the importance of some of the rivets, how many can we safely remove? As we taxi out onto the runway, how many rivets do we wish to see in place? This analogy is appealing, as it suggests that species diversity is linked to ecosystem stability (*Note:* Ecosystem stability does not cause diversity, but if di-

versity is lost, ecosystem functions can be destabilized). The loss of a single rivet does not cause the plane to crash, but if enough rivets are removed some failure of the structure is anticipated.

If it is unwise to assign values to individual species, perhaps it is possible to value ecosystems through their functions (Chapter 6). Ecosystems may serve to regulate water flow, prevent floods, and enhance water quality, or they may sequester carbon and help prevent global warming (Chapter 23). David Pearce, an environmental economist at the University of London, investigated the value of an area of forest purely in terms of its ability to hold carbon. Trees are composed of carbon, and the forest can be thought of as a sink that stores carbon. Carbon will cycle between trees, decomposers, the atmosphere, and back to trees again, but so long as the area remains forested, each tree is a carbon sink. If the forest is removed by cutting and burning, the carbon is released as carbon dioxide, either from fire or by the work of decomposers. No trees remain on the land, so the carbon dioxide released into the atmosphere is not being balanced by photosynthetic uptake. The carbon dioxide will add to the global warming problem, and, because it is predicted that global warming will have a negative economic impact, it should be possible to estimate the cost of each metric ton of carbon dioxide emitted.

To calculate the cost of each ton of carbon dioxide released to the atmosphere, we first have to establish the likely costs of global warming. A number of models attempt to do this, and they vary according to the predicted rate of warming and the categories of costs considered. No one can be certain of the influence of global warming over the next century, although the range of global temperature change will probably lie between 0.5°C and 4°C, with an average warming of about 2°C. An estimate of costs produced by economist William Cline assumed a 2.5°C increase in temperature and suggested a total cost of $53.4 billion (expressed in 1988 U.S. dollars; Table 27.2).

Pearce argues that we should not be satisfied with this global budget. It is more useful to calculate the cost per ton of carbon dioxide emitted, because this helps us in a benefit-cost analysis of pollution. A crude estimate of costs can be obtained by calculating the metric tons of carbon dioxide required to double atmospheric concentrations (the predicted release for the next century), and then dividing the total cost of the global warming by the metric tons of carbon dioxide liberated. From this calculation, Pearce estimates the cost of carbon dioxide pollution to be about $20 per metric ton of carbon dioxide emitted.

Table 27.2 Global economic costs due to a doubling of carbon dioxide.

Problem	Cost in 1988 U.S. $ (billions)
Primary sector damage (agriculture, forestry, etc.)	18.1
Other economic sectors (damage to industry)	16.6
Sea-level rise (flooding of cities)	6.1
Loss of well-being (increased mortality due to drought, health effects, migration of populations to farmable areas)	5.4
Ecosystem loss (based on monetary valuation of ecosystems)	3.5
Pollution (air and water quality changes due to greenhouse gases)	3.0
Natural hazards (such as increased storm activity)	0.7
Total	53.4

(*Source:* J.A. Dixon, 1994.)

The $20 per metric ton cost can be used in a cost analysis of the carbon dioxide liberated when a hectare of tropical forest is burned. In a moist tropical forest, about 280 metric tons of carbon are sequestered in the vegetation. The initial clearance (if done by fire) will release much of this carbon into the atmosphere, and over the next 10–20 years much of the remaining ash and debris will decay, releasing carbon dioxide. But new vegetation will grow, and this will sequester carbon. If the forest is replaced by a grassland, about 80 tons of carbon per hectare will be held in the grass. Thus, the burning of a hectare of forest results in a net release of 200 metric tons of carbon (if this were a dry forest or parkland habitat, the figure would be about 100 metric tons). Therefore, Pearce argues, the real cost of clearing that land should include $4000 (200 metric tons × $20) for global-warming damage. A hectare of potential pastureland on the margin of the Amazonian forest, forest threatened with imminent destruction, can be purchased for between $30 and $300. A logical conclusion then would be that it is better to buy the forest as a carbon reserve, regardless of its wildlife conservation value, than to allow it to be burned.

In a first attempt to estimate a total worth for the services rendered by nature, Robert Costanza of the University of Maryland and his colleagues obtained values ranging from $16 trillion to $54 trillion per year, with an average value of $33 trillion per year. The authors emphasize that they believe these to be minimum estimates of natural services. They estimated decomposition and the cycling of nutrients (worth $17 trillion per year) to be the most important single service performed by nature. The estimate of the value of decompositional cycles was based on how much it

would cost us to achieve this amount of nutrient cycling artificially. In other words, if nature shut down, but we still needed to turn dead trees back into nutrients or to filter nutrients out of water flowing from land into a river, how much would we have to pay to engineer a solution? This study can be criticised because it is based on willingness to pay calculations and produces values that are higher than the sum of all GNPs. Willingness to pay assumes the ability to pay, and so obtaining values greater than the amount of money available suggests that the value of services has been overestimated. The exact numbers produced in this study are certainly open to criticism, but it was a first brave attempt to demonstrate that the services that we often take for granted would be inordinately expensive for us to simulate (even if we could). Indeed, perhaps natural services, which we conventionally value at $0, are worth about as much as all the goods and services produced by humans each year.

27.2 Benefit–Cost Analysis: A Two–Edged Sword

Benefit-cost analysis asks, Is it worth it? Add up all the benefits that accrue from a project and divide by the total costs. If the resulting number is greater than 1, the project produces an overall benefit; if the number is smaller than 1, costs outweigh benefits.

If the only benefits and costs considered are the monetary values of traditional economics, the benefit-cost analysis will underestimate both environmental benefits and the costs imposed by pollution and resource depletion. If it is the environmental lobby advocating the use of a benefit-cost analysis it is likely to include costs of externalities, both positive and negative. Thus, while both the industrial and the environmental lobbies may advocate benefit-cost analysis, the data to be included would vary radically according to who is making the estimate. Consequently, when a benefit-cost analysis is presented, it is essential to determine how that statistic has been calculated.

A further factor that is of critical importance and is likely to vary between developers and conservationists is the **discount rate**. The idea behind discounting is that, even without inflation, something is worth more to us now than it will be in 10 years. For instance, if I were to offer you a check for $1000 would you rather have it now or next year? You would want it now. Therefore, in your mind, a check for $1000 to be cashed next year is worth less than the one that can be cashed immediately. If I then offered you either a check for $1000 dollars next year or $700 now (I am guaranteed to pay either check), you would probably choose to wait. Similarly, if the choice were between $999 now and $1000 next year, you would take the money now. To an economist, you have defined the outer limits of the discount value of the check; it lies somewhere between $700 and $999. Through some careful questioning, we might find that the value of a $1000 check to be cashed next year is, in your mind, really worth $950 (or 95% of its face value). The discount amount is the difference between your value of the check and the face value; in this case it is 5%. Because this difference was 5% in one year, economists assume that the discount rate will be similar into the more distant

future, with a rate of 5% per year. This is just like compound interest on a savings account, except that the multiplier is [(100−discount rate)/100], producing a negative interest rate.

The discount rate is then used to reduce costs and benefits that will occur in the future to **present day values**. Thus, a cost of $1000 that occurs immediately shows on the balance sheet as $1000, but a long-term benefit, for example $1000 in saved medical expenses 10 years from now (assuming a compound 5% discount rate), would be listed as a benefit of $599. The 5% discount rate is a low one, and some calculations are based on a 10% discount rate. If this rate were applied in the above example, the benefit would be $349 (Table 27.3).

The discount rate makes immediate costs and benefits monetarily more important than benefits that show up later. This effect is more pronounced the higher the discount rate. Most environmental arguments are based on the belief that there are long-term benefits to not polluting, to saving biodiversity and habitats, or to maintaining health. However, the full benefits will not be measured within a 5-year, or even a 50-year period. On the other hand, most developers have high initial costs, because they purchase land or build a facility and need a return within the next few years. Inevitably, developers are concerned with short-term profits. Discount rates are inherently biased toward a short-term planning horizon and, therefore, will work against most environmental arguments. The setting of the discount rate is frequently somewhat arbitrary, and environmentalists would favor a rate as

Table 27.3   The discounted value of $1000 over a 10-year period with 5% and 10% discount rates.

Remaining benefit	Value at discount rate of 5%	10%
	1000.00	1000.00
after 1 year	950.00	900.00
after 2 years	902.50	810.00
after 3 years	857.38	729.00
after 4 years	814.51	656.10
after 5 years	773.78	590.49
after 6 years	735.09	531.44
after 7 years	698.34	478.30
after 8 years	663.42	430.47
after 9 years	630.25	387.42
after 10 years	598.74	348.68

close to zero as possible, while pro-industry lobbies are likely to want the highest possible discount rate to be applied in a benefit-cost analysis.

Benefit-cost analysis has the potential to become a political weapon. A proposal instigated by the anti-regulatory lobbies would require that federal activities should proceed only if they have a favorable benefit-cost ratio. Because this legislation is designed to reduce regulation of pollution and the protection of habitats, the benefit-cost ratio would be calculated on the basis of traditional economics. The calculation of this ratio would ignore the non–market values of pollution, biodiversity, and ecosystem health, treating them as externalities, and have a relatively high (10%) discount rate.

Such conservative benefit-cost analyses will frequently justify the exploitation of the environment, when an approach that features both environmental and developmental goals may lead to more balanced development. An example came from the Canadian pulp paper industry. The process of converting timber to paper pulp involves the use of acids and bleaches that often are flushed into local rivers. The untreated effluent from paper mills is highly toxic to river life, and local rivers are often unsightly and devoid of fish. Pulp paper mills need not be so environmentally damaging; pretreatment of the effluent, chemically and biologically, can render it virtually harmless. But such treatment adds to production costs. The powerful Canadian pulp paper industry had long fought federal pollution regulations, presenting benefit-cost analysis to support their position and suggesting that regulation would cost jobs in the timber and paper industry. This view proved shortsighted when the European Union, a major customer for Canadian paper, banned imports of paper that was not produced in an "environmentally benign" process. The resultant loss of sales cost the Canadian industry thousands of jobs. The traditional benefit-cost analysis that they had made of pollution control was a marked failure because it did not take into account a strong consumer preference for "clean" industries.

27.3   How Much Are We Prepared to Pay to Prevent Pollution?

A progressive decline in return is evident for each dollar invested in cleaning up pollution. The first remedial steps are cheap, but the later stages of removing

every trace of pollutants is very expensive. If a benefit-cost analysis of the pollution (including externalities) is calculated the result will show the overall costs to society of the pollution versus the costs of cleanup (Figure 27.5a). From these data it should be possible to identify the optimal investment in pollution control. Clearly, this point will vary through time as social attitudes change. As desire to protect the environment grows, so the intersect of the two lines will shift to the right, indicating a willingness to spend more on pollution control (Figure 27.5b). Conversely, if popular

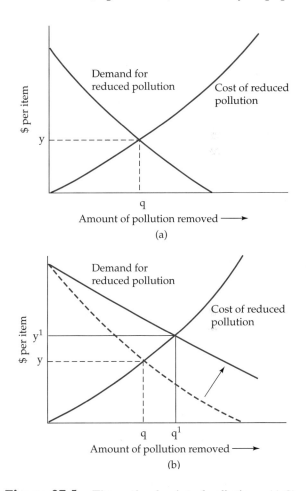

Figure 27.5   The optimal point of pollution. (a) Cleaning up pollution is increasingly expensive the cleaner the environment becomes. The demand for pollution regulation is highest when the pollution is obvious (highest) and lowest when the environment seems to be clean. The intersect of these two lines provides the price that society is willing to pay (y) to curb pollution and the level of acceptable pollution (q). (b) If societal values change to demand more pollution regulation the demand curve will shift to the right. This shift indicates that consumers are willing to pay more (y^1) to decrease pollution levels, that is, to increase the amount of pollution cleaned up (q^1).

(continued)

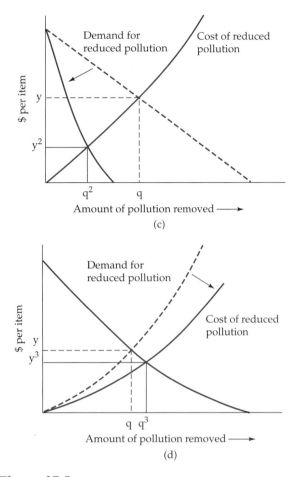

Figure 27.5 *(continued)*
(c) If society demands economic growth and deregulation of polluters—that is, they are willing to spend less on pollution control (y^2) and will tolerate more pollution (q^2)—the demand curve will shift to the left. (d) If technology improves and pollution control is cheaper, consumers will pay less (y^3) for a cleaner environment (q^3).

opinion shifts toward wanting unfettered economic growth, the optimal spending will shift to the left (Figure 27.5c). Also, as technologies change, it may become cheaper to clean up pollution or to limit pollution emissions. We would expect the new technology to increase the demand for clean air and water at the new equilibrium position (Figure 27.5d). Our pollution control is, therefore, a function of how much we are prepared to pay for a clean environment.

The payment to support pollution control measures may take the form of taxes on pollution-producing behavior, such as taxes on cigarettes, gasoline, and diesel fuel. A portion of taxes raised by the government is spent to monitor and enforce compliance with pollution codes and on subsidies to encourage dirty industries to install cleaner technology. For most goods, the cost of pollution control incurred by industry is passed on as higher prices to the consumer.

27.4 The Environmental Industry: Economic Drag or Stimulus?

The bumper sticker reads "Save a Logger, Kill an Owl." Once more it appears that economics is to be pitted against environmentalism. Many of the regulations introduced to protect the environment or limit pollution are seen to rob workers of their jobs. A widely held belief is that environmental regulation limits industrial growth and costs jobs (Figure 27.6). Let us examine the evidence for and against this assertion.

Anecdotal evidence, the experiences of individual companies and industries, is often used to provide an insight into the costs of environmental regulation. For instance, the logging lobby argues that the protection of old-growth forests in the Pacific Northwest has cost between 20,000 and 140,000 jobs. The American Petroleum Institution attributes the loss of 400,000 jobs in the chemicals industry to environmental regulations, and, according to the Motor Vehicle Manufacturing Association, 300,000 jobs have been lost in the automotive industry. A somewhat smaller set of numbers has been claimed for jobs *created* by the new environmental consciousness. Local recycling programs provide employment across the country, an example being 14,000 jobs created in California in 1992, and it is estimated that the enforcement of the 1990 Clean Air Act will generate 60,000 new jobs. Arguments can be made for both sides in this debate, but anecdotal evidence can be biased in many ways to support the slant of the writer. These case-specific studies should be treated with caution because they can be used to "prove" anything.

The underlying cause of job losses blamed on environmental regulation sometimes reflects a downsizing of the workforce accompanied by increased automation. In these cases, claims that layoffs are induced by compliance with environmental regulations may be a smoke screen covering the restructuring of companies. A company's admission that it is laying off workers to facilitate a policy to become "leaner and meaner" could result in strikes and stoppages. If government regulations can be blamed, then the situation is seen as inevitable, and the company can maintain good worker relations. At times, compliance with regulations and restructuring are inextricably

Figure 27.6   Protesters worried that environmental regulation will cost them their jobs.

linked. For example, a paper manufacturer confronted with new pollution regulations is forced to install new devices to control effluents. One way to meet regulations is to install an additional stage in the production line to clean up the effluent before it is discharged. Alternatively, it may be a better investment to replace old technology with new, cleaner machines that do not produce the pollution in the first place. It is likely that the new machines will be more automated, so workers will be laid off. Compliance with the regulations has initiated (though not mandated) a restructuring of the company, leading to a smaller workforce and reduced operating costs.

Another example of restructuring comes from the forests of the Pacific Northwest. Because of long-term depletion of regional timber stocks, logging corporations were faced with the prospect of laying off workers even before the issue of protecting the Northern Spotted Owl was raised (Chapter 11). Job losses were the inevitable outcome of a half-century of mismanagement of timber reserves. Environmental legislation protecting old-growth forest on federal lands has accelerated an industrial contraction, but it is not the root cause for the collapse. Automation and modernization of timber mills will lead to as many as 25% of workers' being laid off in some mills regardless of environmental restrictions on logging.

There is no doubt that some job losses are directly due to increased environmental regulation. However, this number may be smaller than some pro-industry lobbies would like us to believe. An analysis by the U.S. Bureau of Statistics in 1988 concluded that job loss and environmental regulations are not causally related. Their data suggest that 99.9% of jobs lost in the United States are due to factors other than environmental regulation.

The meteoric increase in public awareness over environmental issues since the 1960s has provided a whole new sector in the U.S. and world economy. In the period from the 1960s to 1992, spending in the United States to protect the environment grew three times faster than the GNP. The environmental industry is currently worth between $60 billion and $170 billion annually (according to how the industry is defined) and represents about 2.8% of our GNP. It can be argued that this is a booming section of our economy, a generator of jobs with 600% growth over the last 20 years. The growth is anticipated to continue, so expenditure on the environmental industry could increase by 50%, even allowing for inflation, by the year 2005.

Economic analyst Roger Bezdek suggests that environmental regulations do not diminish the wealth of a nation but redistribute wealth from polluters to pollution controllers and abaters and firms that do not generate pollution. Many of the jobs that are lost because of new regulations are in heavy industry, whereas many of the jobs created are for workers with a training in law, management, or environmental sciences. The jobs lost are not directly equivalent to those gained: The worker laid off from the production line in Toledo does not benefit from the appointment of an attorney in New York. Indeed, the new opportunities tend to cluster in administrative centers, such as state and federal capitals, or in the head office of a company. As the more-developed countries proceed toward a postindustrial society, employment opportunities will shift from blue-collar to white-collar jobs. Environmental regulation is one factor within this larger framework that pushes society toward a new balance.

27.5   A Digression on the Meaning of Sustainability

The movement of more-developed countries toward the postindustrial society does not necessarily indicate that their use of resources is wise. The realization that "business as usual" will result in the depletion of such vital factors as soil fertility, hydrological balance, and fossil fuel deposits has triggered a stampede toward **sustainable development**. Sustainable development has become a goal that is almost universally advocated. However, it has not only proved unobtainable but has also been the source of considerable confusion, because *sustainable* has a very different meaning to its various users.

Economists pride themselves on practicing a science that optimizes the allocation of limited resources; ecologists revel in understanding how limiting resources regulate plant and animal populations. Both disciplines see that it is the rarest factor in the equation that is the most important. Given this commonality, you might expect economists and ecologists to understand each other's models, but their world views are fundamentally different.

Sustainability has been adopted by both camps to describe an ideal use of resources, but this apparently unifying term means different things to each. To environmentalists and ecologists, sustainability means a use of resources that will allow the indefinite survival of ecosystems. Ecosystems comprise both the wildlife and the natural processes, such as nutrient cycling, soil drainage, and climate, of an area. Implicit in the persistence of ecosystems is the notion that species will not be driven to extinction, the balance of individual populations will be maintained, and natural resources, such as mature forests, fertile soils, and stocks of fossil fuels, will be managed so that they are not exhausted. However, given the pressures of modern society, these pure goals appear unattainable, and some compromise must be found. Most ecologists would agree that overly preservationist stances that would limit the economic development of less-developed countries are morally questionable and unlikely to achieve conservation goals.

In contrast to the ecological view, economists see sustainability as an allocation of resources that leaves future generations no worse off than present generations. It is a founding assumption of economics that all resources have substitutes and that human ingenuity will identify them. This view permits one resource to be substituted for another so long as it does not lead to economic decline. For instance, while the standing timber of an old-growth forest in Oregon is a potential resource, so too is a fertile field. Thus, it is legitimate to fell all the forests so long as the land provides a lasting agricultural, or urban, return of equal worth. It would also be legitimate to harvest the trees in order to produce wealth that results in investment, for this would provide a monetary asset usable by future generations. Similarly, it is legitimate to harvest all the cod from the seas if an alternative protein source at an equivalent price can be identified (the extinction of cod as a species is of no economic concern). Nonrenewable resources, such as oil, can be exploited to the last drop so long as an affordable, alternative energy source, be it solar, nuclear fusion, or ethanol, is identified.

Economic theory was developed independently of ecological thinking, and it comes as a cruel surprise to some economists that, in nature, resources are limiting and are not substitutable. Plants need phosphate to grow; nitrate, potassium, or water cannot be used as a substitute. Over millions of years, the giant panda has evolved a digestive tract to feed on bamboo; it cannot survive on oak leaves or pasture grasses. Ecosystems are by definition substitutable; that is, a forest ecosystem that is logged for farmland will be replaced by an agricultural ecosystem. This does not mean, however, that the replacement is a sustainable (in the economic sense) substitute for the first. Deforestation leads to the establishment of a new ecosystem. However, an eroding landscape with ruptured nutrient cycling and an altered microclimate that supports a few invasive species is not an adequate substitute for the original mature forest. It is clear that the meanings of sustainability used by economists and ecologists have yet to be reconciled and that a better fusion of the ideas of the two disciplines will be needed if sustainable development is to become anyone's reality.

Now another distinction needs to be made, and that is between *economic development*, the improvement of the well-being of the population, and *economic growth*, the increase in the size of the country's economy. Development can be achieved through improved education, improved health care, and individual empowerment. Growth requires exploiting natural resources for industrial and agricultural expansion. By definition, economic growth will leave progressively fewer resources for nature. Thus, while economic development can be sustainable, economic growth cannot.

27.6 How Do We Evaluate Development?

In the 1940s, a new means emerged to compare the wealth generated each year by a country. This measure, the gross national product (GNP), has been adopted as the basic criterion evaluating the prosperity and development of nations. GNP is calculated by adding up annual governmental expenditure, consumer expenditure, and investment. The GNP can then be divided by the number of people in the country to obtain the average per capita (per person) earnings for that year. The advantages of GNP are that it provides a standardized method of comparing economic factors in nations of different sizes and different stages of development, regardless of the national currency. Importantly, it is a single statistic that can be incorporated into political debate.

Development, however, is poorly represented by this simple measure. Financial development is certainly an important component of development, but the GNP statistic yields no information on the distribution of wealth and the social development of the nation. Every country has its rich and poor, but in wealthy countries there is a block of the population that occupies the middle ground. In the poorest countries, there is no middle class and a "wealth gap" divides the rich ruling elite from the poverty-ridden masses. In countries such as Brazil, where this hierarchy of ownership has yet to break down, more than 80% of the wealth may be owned by just 8% of the population. In richer countries, the rich are still astonishingly wealthy, but the middle class emerges to control a significant proportion of the nation's assets, and the proportion of the population in poverty is reduced.

As per capita income increases, it is likely that basic life-sustaining needs will be met and that standards of sanitation, education, and health care will improve. The population is then in a position to enter a series of positive feedbacks: As they become healthier, they have reduced infant mortality; that reduction is likely to translate into a decreased total fertility rate. The lowered birth rate relieves some of the pressure on an emerging economy; more resources can be invested in each child, so their education is improved, and they have better opportunities and, therefore, health as

adults. None of this cycle of improvement can be determined with assurance from the GNP statistic.

Another problem that environmental economics has highlighted in the calculation of GNP is that it does not consider externalities. Much of the natural wealth of a country, such as soil fertility, forests, fish stocks, and oil deposits, is not included in the calculation of GNP. Only the portion of the resources that is extracted is valued. The lumber cut during a year is part of the GNP, but the area of forest remaining, an area that may be growing or shrinking, does not figure into the accounting. Thus, countries can appear to have expanding economies. But if this development is based on a steady depletion of natural resources, the GNP provides a false sense of prosperity. Many of the developing countries are relying on rates of extraction that cannot be sustained, and as much as half of their apparent growth may be due to this overuse of natural resources. Economic growth based on practices that deplete these resources is like selling off the family silver to pay the bills. All appears well because the bills have been paid, but the silver is gone and the bills will arrive again next month.

An example comes from a study of the economy of Indonesia by Robert Repetto of the World Bank and his colleagues (Figure 27.7). The Gross Domestic Product (GDP) is similar to GNP except that it excludes earnings from foreign investments. The GDP shows a doubling of income during the period in question. (*Note:* This is not a per capita estimate and so has not been corrected for population growth.) The increase in Indonesian earnings was largely stimulated by the discovery and exploitation of oil fields and the use of forest resources. Oil is a nonrenewable resource, so some cost allocation should be made for the fact that some of the oil has been used up. Forest clearance could be viewed as a crop that is harvested time and again; however, the extraction methods used often result in a one-time yield of large high-quality timber. No thought is given to repeat harvests, and care is not taken to prevent soil erosion and depletion of the resource. Consequently, this kind of exploitive forest use may not be repeatable in an economic time frame and can be considered as a cost. Taking these factors into account, Repetto recalculated Indonesian economic growth to obtain a new value, the "net domestic product," a more realistic assessment of real economic advance. The average annual growth rate of 7% documented by Indonesia between 1971 and 1984 was reduced to a more modest 5.5%. Because the Indonesian population was growing at a rate of between 2% and 3% throughout this period, a calculation of the

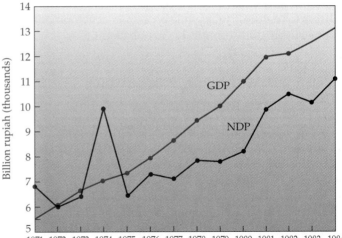

Figure 27.7   The increase in gross domestic product (GDP) for Indonesia since 1965. Data are plotted in "constant" 1973 rupiah to compensate for any change caused by inflation. If depletion of natural resources as a result of exploitation of nonrenewable resources is subtracted from the GDP to provide the net domestic product (NDP), the apparent strong growth of GDP is reduced. (Data are from World Resources Institute, 1989.)

annual growth per capita in net domestic product would fall from 5.5% to about 3%.

Another way in which traditional calculations of GNP overestimate economic well-being is by including all expenditure as if it were equally good. Money generated by the construction of new homes, art, food, or education all add materially to the quality of life and deserve a place in the wealth estimate of a nation. However, expenditure on cleaning up a nuclear accident, pollution control, and medical costs incurred as a result of smoking are also included in the GNP. Unless an oil spill reduces tourism or has some similar impact on the local economy, it is not counted as a negative within the calculation of GNP. The fees of the company cleaning up an oil spill, or federal compensation for damage resulting from the spill, become part of the profits of private companies and so are counted as earnings. The cleanup costs accompanying the environmental disaster of the wreck of the Exxon *Valdez* oil tanker appear in the GNP as pure profit. Thus, a country that spends a large proportion of its available income masking its industrial and social ills appears to be more profitable than a country that has sustainable clean industry, that does not have disasters, that has a healthier population with lower medical needs, but that maintains the same level of output. The second of these two countries would probably be a more desirable place to live and be more developed in many ways, but it will have a lower GNP.

Many of the countries of western Europe, Canada, and the United States are starting to develop an accounting procedure to produce a revised set of national accounts. These alternative estimates of national finances are called **satellite accounts**.

One such set of accounts is the **index of sustainable economic welfare (I.S.E.W.)**, which takes into account the costs of pollution, depreciation of natural assets, loss of ecosystems, and other factors that contribute to, or detract from, our well-being. If the I.S.E.W. is compared with the GNP for the United States over the past 40 years (Figure 27.8), the United States is seen to pass through times of growth, stagnation, and even economic decline. The GNP suggests a steady growth of about 2% per annum, while the late 1970s and 1980s are shown to have been a time of 0.8% per annum decline in the I.S.E.W. This downturn is largely attributable to estimated environmental degradation. One problem with the I.S.E.W. is that it requires a huge statistical data base that is simply unavailable for many nations.

At present, no two nations are using the same methods to create satellite accounts, so the results are not comparable. The United Nations is developing a standardized way of presenting these accounts, although

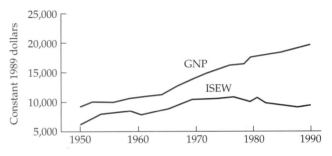

Figure 27.8   An estimate of the index of sustainable economic welfare (I.S.E.W.) for the United States compared with GNP since 1950. GNP appears to overestimate our economic welfare throughout much of this period. (Data are from J. Dixon. Copyright © Wadsworth Publishing, Inc.)

it is recommending that they be kept separate from the traditional estimates of GNP.

It is widely anticipated that within the next decade, satellite accounts will be applied regularly to estimates of national wealth. However, the environmental community needs to remember that these valuations must be believable, and they must be able to convince traditional economists that sustainable (in the ecological sense) development is the wisest, not an eccentric, course to follow. Wild assertions about the value of an average forest tree (one estimate suggests a worth of $190,000) are unlikely to achieve the desired goal of a revision of benefit-cost thinking.

27.7 Global Budgets and Local Accounts

Some pollutants such as carbon dioxide and CFCs, while produced through local forest burning or factory emissions, have global effects; these are termed **global pollutants**. A realization to emerge from this statement is that global pollutants produced anywhere in the world will affect us and that it is, therefore, in our interest to try to prevent that pollution from taking place.

An example is found in the global regulation of CFC emissions. (Air conditioners and refrigerators commonly use CFCs as refrigerants; as these leak from the appliances, they destroy the stratospheric ozone layer; Chapter 20). Since 1987, more than 100 nations have undertaken to reduce or eliminate their CFC emissions. These nations have imposed taxes or air quality standards on their industries to ensure compliance with the new goals. Consequently, many companies have invested heavily in reducing CFC emissions, a process that has been very successful but also very expensive. However, some nations that are major CFC emitters have not signed these protocols to limit their pollution. One such nation is China. The Chinese object that the alternatives to CFCs are very expensive and that they cannot impose these costs on emerging industries. The continued, even expanding, use of CFCs in China threatens to undo the work of the 100 signatory nations to reduce the concentrations of these chemicals. How can nations such as China be encouraged to use the more expensive CFC substitutes? Should the "stick" of economic sanctions or the "carrot" of subsidies be used? Most economists see sanctions as a last resort and purport that it is ultimately economically more profitable for all involved to encourage than to punish. One way that the CFC problem can be resolved is by calculating the amount that a nation such as China would spend on CFCs to install refrigeration over a given period. Then calculate the cost of the same amount of refrigeration capacity using non-CFC coolants. The difference between the two costs is the amount that it would cost China to become "ozone-friendly." That amount of money can then be offered to China as international aid on condition that they eliminate CFC production. The Chinese would then be in a position to install non-CFC coolants at no loss, and the donors of the money would be assured that their previous investment in non-CFC technologies is not undone. This is a "win-win" situation, in which both recipient and donor achieve their goals and the environment is protected.

One way to demonstrate the cost of environmental change is through estimates of damage caused by natural disasters. Storms, droughts, fires, floods, and freezes all carry a monetary toll in terms of lost production, destruction of property, and deaths that can be estimated. 1998 had more weather-related disasters than any previous year. The most severe problems were faced in China, where flooding killed 3000 people and forced 230 million people from their homes. Even allowing for inflation, all the damage incurred in the 1980s ($82.7 billion), was less than the damage caused in 1998 (Figure 27.9). The great unanswered questions is; What proportion of this change and, therefore the costs, is attributable to human activities? If, as is suspected, global climate change due to pollution is affecting the strength and frequency of disastrous events (Chapter 23), not only do these data suggest that worse is yet to come, but they also help to provide a dollar value for the potential benefits of combating climate change. Such accounting provides another way to attribute a cost to pollution.

A more encouraging realization that emerges from seeing pollution in a global perspective is less obvious. For a global pollutant such as carbon dioxide, some local pollution may be unavoidable as industry is established or a new power station comes on-line. However, that emission of carbon can be offset by a compensating action that leads to the absorption of carbon from the atmosphere anywhere in the world. Some countries, such as Norway, which have heavy investments in the oil industry, may choose to finance the maintenance of "carbon reserves" in forested countries in exchange for their continued sale of oil—which will be combusted and liberate carbon dioxide. Thus the polluting country can be seen to be environmentally conscious.

In an example from the United States, Applied Energy Services (A.E.S.) is installing an 18,000 MW

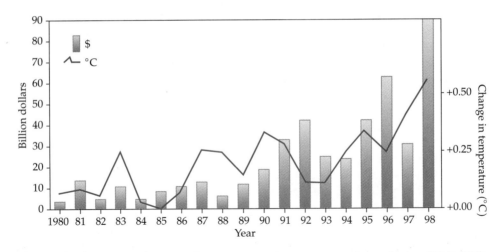

Figure 27.9   Costs of natural disasters and global temperature change 1980–1998. Disaster costs include infrastructure and crop losses, but do not include health costs, damage to natural resources, or long-term environmental damage. Global temperature change is expressed as deviations from the 1950–1980 mean temperature. (Data are from the Worldwatch Institute and the National Oceanic and Atmospheric Administration.)

(megawatt) coal-fired power station in Uncasville, Connecticut. To offset potential global warming effects, A.E.S. has undertaken to plant enough trees in Guatemala and Paraguay to sequester 15 million tons of carbon. This project helps with the public image of the power station and also serves to inject capital into the regions where the tree planting will be carried out. The resulting employment of 40,000 local families will lead to a substantial growth in the local economies. A.E.S. is putting in $2 million of the total $20 million costs, $1.2 million is being provided by the government of Guatemala, and the balance is coming from such organizations as CARE, the U.S. Peace Corps, and from US-AID. It is evident that the various organizations concerned participate for different motives. A.E.S. benefits by obtaining planning approval for its new power station and from improved public relations. The Guatemalan government is investing $1.2 million in return for about $18.8 million dollars in foreign investment. The other partners are participating in an environmentally desirable program that provides benefits on a local and a global scale. This is another example of a win-win situation, even if the price paid is a polluting power station in Connecticut. It should be emphasized that this kind of compensatory project can work only when the pollutant is global. Pollutants, such as pesticides, heavy metals, and fertilizers, which have local or regional effects, must be treated locally.

Summary

- Traditional economics does not take into account the costs of externalities such as pollution, ill-health, loss of biodiversity, or soil erosion that can result from industry, agriculture, or forestry.
- Environmental economics seeks to ascribe values to externalities and include them in calculations.
- In traditional economics, if a commodity is not traded (has a market value) it has zero value.
- Goods can be valuable but have no market value. Environmental economics is the science of ascribing non–market values.
- High discount rates promote short-term resource exploitation.
- Environmental regulations have contributed to a contraction of jobs in heavy industry.

- Whether environmental regulations benefit or hurt the economy is not proven.
- Sustainability means different things to ecologists and economists.
- Sustainability used in an ecological sense may be consistent with economic development, but cannot be reconciled with economic growth.
- Gross national product is an imperfect tool for comparing the economic or environmental welfare of nations.

Further Readings

Costanza, R., et al. 1997. The value of the world's ecosystem services and natural capital. *Nature* 387:253–260.

Pearce, D. 1998. Auditing the Earth. *Environment* 40(2): 23–28.*

Portney, P. R. 1998. Counting the cost: The growing role of economics in environmental decision making. *Environment* 40 (2):15–18 and 36–38.

Tietenberg, T. 1997. *Environmental and Natural Resource Economics*. 4th ed. New York: HarperCollins Publishers, Inc.

Web Connections

On-line resources for this chapter are on the World Wide Web at:
http://www.prenhall.com/bush
(From this web-site, select Chapter 27.)

*Pearce (1998) is a commentary on Costanza et al. (1997) and includes a rebuttal by Costanza.

Environmental Legislation and Policy

28

28.1 *Common law and the environment*
28.2 *Protection under statutory law*
28.3 *A pocket history of environmental legislation*

28.4 *Focus on five pieces of legislation*
28.5 *Property rights versus environmental legislation*

Environmentalism and industry are often uncomfortable bedfellows, although each may make placatory noises regarding the other. Environmentalists do not wish to be seen as out of touch with financial reality, and industrialists would avoid being labeled as insensitive defilers of the planet. A 1990 Gallup poll indicated that 76% of Americans consider themselves to be environmentalists, at least to the extent that they want to promote the well-being of wildlife and to live in a healthy environment. Many of these people would also be in favor of economic expansion, full employment, and a capitalist economy. Voters do not polarize cleanly into environmentalists or industrialists, and politicians are mindful of maintaining a balance between the two. Each side is represented by powerful lobbies and would lead us to believe that no compromise could, or should, be made. Nevertheless, common sense and a long-term view of the economic situation often result in a workable, albeit informal, compromise, minimizing the number of times that irreconcilable conflicts between wildlife advocates and developers emerge. However, the rival interests can become almost diametrically opposed when it comes to the formal regulation of industry or the status of endangered organisms. Nowhere does the environment versus industry standoff come into sharper focus than in environmental legislation.

The U.S. Constitution guarantees that citizens can find legal redress through two separate channels. **Common law** is built on legal precedent and custom, owing much of its original form to the British legal system. The other branch is **statutory law**, the edicts passed by national or state assemblies that form our written laws. Laws are designed to prevent individuals, corporations, or government from committing acts that are antisocial or a nuisance to the public. Laws and regulations are sometimes seen to be government's meddling in the lives of private individuals, but, as we shall see, some level of environmental regulation is necessary.

28.1 Common Law and the Environment

Common law is applied where statutory law does not exist, and it is therefore likely to be used as new, or unique, problems occur. Common law is the only recourse for protection before the legislature has formulated statutory law to deal with a situation. Some environmental problems that could be tackled by common law would be to prevent the pollution of land by a neighbor's activities, to prevent hunting on private land, and to seek damages for medical ailments resulting from nearby toxic chemical usage. The first step in the legal process would be to seek an **injunction** to stop the activities of the offending party. The case can then be pressed to claim financial **damages** as reimbursement for the harm.

Three areas where common law can be prosecuted are **trespass**, **nuisance**, and **negligence**. Trespass is the invasion of land owned by you and does not require a proof of damage. Someone who appears on your land carrying a chainsaw or a shotgun could be prosecuted for trespass, even before they have cut a tree or shot a bird. However, most environmental cases are less tangible, particularly if they are related to pollution. For almost 200 years, "private nuisance" laws were the most common vehicle for pollution prosecution. Perhaps the first such prosecution was a private nuisance case that came to an English court in 1611. A pigsty had been built next to William Alred's property, and he gained the court's protection from the frightful stench that had invaded his land and home. This case from the British courts was one of many that formed the initial backbone of American legal precedent. Gradually, the U.S. legal system developed its own traditions, often mirroring the development of the country. Court decisions made during the Industrial Revolution reflected a stance that industry must be allowed to prosper, no matter the cost. In 1927, without clear and concise codes to guide the courts in claims for restitution against polluters, Justice Learned Hand of the U.S. Supreme Court commented that this whole area of "nuisance law" was in "great confusion."

Environmental nuisances can take the form of disturbance by noise or pollution that results in health problems. For example, toxic waste that induces cancer is a nuisance. Nuisance law could also be used to sue the pilot of a crop-dusting plane who flies across your land dribbling carcinogenic (cancer-inducing) pesticide. In this instance, the presence of the agent, the pesticide, is an invasion of your land and therefore constitutes the nuisance.

Negligence is also grounds for a common law suit. If our hapless crop-dusting pilot was having a particularly bad day, he or she might have gone on to spray pesticide on a worker in the target field. The spray went onto the correct place, but either the farmer or the pilot may be deemed negligent in not taking all reasonable precautions to ensure the safety of the worker.

Prosecution under common law is generally most successful when cases are simple and the offending party is an individual or a small company. Taking on a large industrial corporation is akin to the legend of David and Goliath, except that this time David's slingshot is loaded with marshmallows instead of pebbles. Such legal actions can be so costly that rather than a single individual attempting the fight, a collective effort of individuals with a similar complaint, or **class action**, is brought against the offender.

The ultimate prosecutors of common law are ambulance-chasing attorneys. While these practitioners have a somewhat tarnished reputation, the stereotype of their art should be studied by environmental groups seeking redress, because they perform exactly as the law demands. First, the injured party must look as pathetic as possible. Use anything, a wheelchair, neck brace, crutches, or bandages, to gain the sympathy of the court. Leave judge and jury in no doubt that a grievous injury has been inflicted. Next, with trembling finger, point to the accused and paint the picture of the uncaring inebriate who has savagely damaged your client. Last, convince the court that this injury has denied quality of life and potential for financial independence.

This sequence can be broken down to establishing injury, guilt, and damages: the essentials of a common law prosecution. In the case of a traffic accident, establishing these is relatively straightforward, but environmental cases can become very complex. For example, let us consider the documentation that has to be provided to carry a suit against a polluter on health grounds. Even if we believe that the pollutant chemicals are resulting in elevated rates of cancer in the local community, suing the polluter for damages is not as simple as it sounds. First, it has to be shown that exceptional harm was done to the plaintiff rather than to the general public. It is not possible to bring a common law suit on the basis that the effluent contains carcinogens and may cause cancer in the future: The victims must already have the disease. Furthermore, an unusually high incidence of the disease must then be demonstrated in the area affected by the pollutant. This is the establishment of injury.

The establishment of guilt requires the identification of the carcinogen that is directly responsible for the cancer. The presence of this substance in the effluent has to be proved. Measuring the amount of the toxin in the modern effluent may be sufficient to establish the presence of the carcinogen; however, it is possible that the pollution content of the discharged effluent has changed through time. Because the disease must already be present to meet the establishment of injury, the pollution must have taken place at some time in the past. In the meantime, industrial change may have altered the present composition of the effluent, resulting in the absence of the carcinogen. If the chemical can be traced to the effluent of the present or the past, the onus still rests on the plaintiff to prove that it came from that particular source of effluent. Industries frequently congregate, and a successful prosecution has to determine which effluent source was *the* one that caused the cancer. Otherwise, a legitimate defense would be that the chemical came from a neighboring chemical plant, agricultural pesticide residues, or the cigarette you tried when you were 12 years old. A successful prosecution such as this requires expert sampling, laboratory research, considerable legal advice and representation, and a large financial outlay. A large company can delay procedures at every turn so that the process may take years to complete. In practice, such prosecution is beyond the resources and patience of most individuals.

Because of the difficulties of pursuing this kind of prosecution, the damages are often decided in out-of-court settlements, in which the company tacitly admits guilt, or the appearance of guilt, by paying restitution. This procedure short circuits the proceedings by eliminating the court case. Attorneys from each side show their strength, and the defendant may then choose to make compensation before the matter goes to trial. Both sides stand to benefit because a settlement may reduce the turmoil, stress, and financial hardship that plague both sides in a lengthy court case. From the defendant's point of view, avoiding the courtroom reduces adverse publicity as well as costs of defense. Furthermore, the compensatory payment becomes a matter of negotiation rather than an amount imposed by the judge.

Common law is such a difficult means of attaining recompense or of preventing environmentally damaging acts that the importance of statutory law is greatly increased. In practice, statutory law is the only practical way to regulate polluters and those who would destroy wildlife habitats. Both industrialists and environmentalists realize the importance of statutory law, so environmental legislation becomes a fiercely fought turf war, in which neither side willingly yields an inch.

28.2 Protection under Statutory Law

Most legislation is triggered by a dramatic event documented by the media. An explosion in a power plant or a factory, the pollution of a river flowing through a national park, or the contamination of an aquifer with pesticide residues could all attract enough attention to require legislation. Our elected representatives are also human, and they will respond to reports of a disaster with the same shock as the community at large. Their attention will be heightened when an event of primarily local importance is taking place within their district. For other events, they will respond to pressure from their constituents (Figure 28.1).

Once the legislature becomes involved, whether at state or federal level, they may form a committee to investigate the issue. Scientific, economic, and social implications of imposing new regulations will be considered. Researchers will review the literature for information and form advisory statements for the elected officials. If hearings are held, advocates for and against changed regulations will be called for expert testimony. Throughout these proceedings, the media, the public, and **special interest groups** represented by lobbyists will be pressing their own case. Special interest groups play an important part in decision making

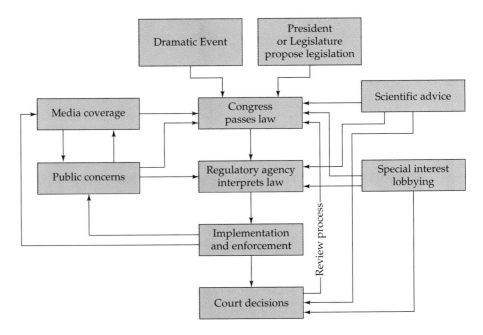

Figure 28.1 The flow of information that leads to environmental legislation. Note that lobbyists can exert influence at all stages of the process. Additional biases that are not shown may be due to overall social standards, campaign donations to legislators, or the way that an incident is reported. (Data are from J. Dixon. Copyright © Wadsworth Publishing, Inc.)

as they represent large blocks of voters who may make elective choices on single issues, such as abortion (Pro-Choice or Pro-Life), civil liberties (American Civil Liberties Union), environment (Sierra Club), or gun regulation (National Rifle Association). Some of these groups may also make substantial campaign contributions to individual delegates, and are, therefore, in a powerful position to sway voting by government representatives. As new legislation is debated, it will evolve in response to these various pressures. Passage of a bill to become a federal act requires congressional and presidential approval, but even then the process is not over. The new act may provide regulatory power to a government agency, but it is the nature of the law to be revised. Mindful of this, the special interest groups will continue to lobby to have an existing law amended to their favor.

28.3 A Pocket History of Environmental Legislation

Until the 1960s, it was believed that the amounts of chemicals being added to the environment were so small, relative to the overall size of the biosphere, that they would have no lasting consequence. As the repercussions of human actions on our surroundings were realized, so the first environmental laws were introduced. The Clean Air Acts of 1963 and 1965, though largely ineffective, were landmark pieces of legislation because they were the first regulations imposed to protect the general environment. In the wake of Rachel Carson's book *Silent Spring* and the discovery of pesticide buildup in food chains, federal and state regulations hailed down on potential polluters in the 1960s and 1970s.

It is a common misconception that the strongest environmental laws are passed by liberals, while conservatives restrict such legislation. Conservatism as represented by the Republican party in the United States has changed since 1970. The Reagan administration of the 1980s represented a particular brand of conservatism that sought to minimize government regulation, especially of industry. The modern Republicans are also essentially of this school of thought. The Democrats, while doing very little to further environmental protection in this same period, have become, by default, the party with a stronger environmental agenda. However, it is important to note that these political divides have not always been the case. Probably the most far-reaching piece of U.S. environmental legislation came from

the office of a conservative Republican, President Richard M. Nixon.

The National Environmental Policy Act (NEPA) was passed in 1969 and led directly to the establishment of the Environmental Protection Agency (EPA) in 1970. The EPA was given the mandate "to prevent and eliminate damage to the environment and biosphere," not a small undertaking. The EPA was empowered to enforce government regulations such as the Clean Air Acts. A wave of legislation was enacted—a total of nine acts—that gave the EPA powers of environmental watchdog over pollution, emissions standards, and waste disposal issues. A further requirement of NEPA was that all federal agencies prepare an environmental impact statement before instigating any major new program. The EPA was given the task of being the ultimate arbiter, determining whether these statements were accurate and the likely impact acceptable.

When President Ronald Reagan came to office in a landslide victory in 1980, one of his stated goals was to reduce the regulatory powers of the EPA. This he achieved by changing the management of the agency and cutting the regulation enforcement budget. For several years, the watchdog was muzzled. While the American public may have wanted a reduction in federal involvement in their daily lives, unharnessed pollution was unpopular. Consequently, in 1982, the EPA was once again turned loose on polluters, albeit as a rather subdued watchdog.

28.4 Focus on Five Pieces of Legislation

It is impractical to describe all the environmental legislation that has been passed, but five pieces have become, or are likely to become, increasingly discussed: the Mining Act of 1872; the Clean Water Act of 1972; the Endangered Species Act of 1973; the Comprehensive Environmental Response, Compensation, and Liability Act of 1980; and the Clean Air Act of 1990.

The Mining Act of 1872 and the "Wise Use" Movement

The Mining Act of 1872 was established to facilitate the development of federally owned lands in the western United States. Mine prospectors could stake an area for just $2.50 per acre and claim title to it, providing they undertook some mining activity in pursuit of valuable minerals. The miners could then keep the profits and own the land. This act is still on the statute books and has been the mainstay of efforts to increase mining activities on such public lands as national parks. Consequently, environmental groups have petitioned government to repeal the act. This tension between mining consortia and environmentalists has been growing and led to the **"Sagebrush Rebellion."** This was an antienvironmentalist movement that started in the 1970s, largely financed by mining companies espousing the interests of ranchers, miners, and loggers. One of their goals, not surprisingly, was to open all federal land (including national parks and wilderness areas) to mining. The movement gained brief acceptance within the Reagan administration becaue one of its leaders, James Watt, was appointed to President Reagan's cabinet as Secretary of the Interior. His stay was short-lived, lasting just two years; he was ousted from the inner circles of Capitol Hill. A resurgent interest in environmental matters in the late 1980s encouraged President Bush to describe himself as the "Environment President" during his 1988 election campaign, and his Clean Air Act of 1990 lived up to this billing. However, his vice presidential appointee, Dan Quayle, was not proenvironment; his opinions were evident from his role as chair of the "Council for Competitiveness." This was a conservative think tank that aimed to optimize the nation's economic growth, was bent on deregulation, and sought to undo the power vested in the Endangered Species Act and the EPA. **Wise Use** became the slogan of this and other antienvironmentalist groups. *Wise use* does not refer to a cautious development of natural resources but to the dogma that it is wiser to use resources than to let them lie in the ground.

Like the Sagebrush Rebellion, the Wise Use movement is closely tied to the mining corporations, plus industrial groups who stand to gain from deregulation, such as the makers of off-road vehicles and logging companies. One of the most influential of these groups is the CDFE (Center for the Defense of Free Enterprise), which published a Wise Use Agenda (Table 28.1).

The agenda goes on to recommend the opening of wilderness areas to dirt bikes and snowmobiles and the denial of access to wildlife researchers. The Wise Use movement has a potentially valuable role to play. The presence of an opposition lobby will often spur greater accuracy and documentation and could result in the strengthening of the environmental case. The environmental movement has been, and still is, guilty of making appeals on an emotional rather than a scientific basis. However, the Wise Use groups present a

Table 28.1   Highlights and interpretation of the Wise Use Agenda.

- National parks should be opened to multiple uses, including mining.
- Some endangered species should be identified as being "nonadaptive," the implication being that these cannot, and should not, be saved because they cannot live within human landscapes.
- The Forest Service should start a campaign to explain to the public "the wise commodity use of the national forests." For this statement read "the establishment of mines and the felling of old growth."
- A Global Warming Prevention Act should identify decaying wood as a major contributor of atmospheric carbon dioxide. The aim would be to "convert in a systematic manner all decaying and oxygen-using forest growth . . . into stands of oxygen-producing, carbon dioxide–absorbing young trees." This is another statement advocating the immediate logging of old-growth forests.
- The mineral and oil resources of the Arctic National Wildlife Refuge should be made available for utilization immediately.
- Obstructionist liability should be passed as law. Such a law would require anyone objecting to the commercial development of public lands to post a bond equal to potential profits before the matter is considered. For example, some individual, or an environmentalist group, objecting to the exploitation of oil and minerals within the Arctic National Wildlife Refuge would have to put up billions of dollars, (industry's potential profits) as a bond. This measure would prevent effective protest against any major industrial development of public lands.

misleading image to the public. They claim to be grass roots organizations, but they are generally funded by mining consortia and big business, and they masquerade under misleading "Earth-friendly" names. Examples are the U.S. Council for Energy Awareness (nuclear power industry), Wilderness Impact Research Foundation (mining and logging interests), Californians for Food Safety (chemical companies and processed-food producers), and the National Wetlands Coalition (real estate developers and oil and gas industry). The names imply a commitment to conservation, which is the opposite of the truth.

The Clean Water Act of 1972

In the early 1970s, many of the nation's rivers were heavily polluted with industrial effluent. These waters were unfit for swimming, and the fish that could survive in them were inedible. Chemical wastes rich in mercury, arsenic, and PCBs were concentrated in the fish through bioamplification, making them unsafe for human consumption. Congress enacted a truly ambitious piece of legislation with the stated goal of making all U.S. waters (lakes, rivers, estuaries, and coastlines) swimmable and fishable by 1983.

The Clean Water Act of 1972 was complex and covered topics as diverse as oil spills, the discharge of point-source pollution, and the destruction of wetlands. National pollution limits were set, which required potential polluters to use the best available technology to reduce their discharges of toxic wastes. The Environmental Protection Agency was charged with enforcing these regulations and bringing offenders to court.

The deadline for making all waters swimmable and fishable has long since passed, and many of the nation's waters, though cleaner than in 1972, are still polluted. This should not be taken as an indication that the act failed; rather our understanding of pollution sources has improved. When the act was initiated the most significant polluters of inland waters were point source polluters such as papermills, steelworks, and chemical plants (Chapter 16). The pollution from these industries can be pinpointed and monitored because they have an identifiable pipe or outfall that discharges pollutants. Such point source–polluting industries have been forced to reduce pollutant emissions with the consequence that their share of the pollutant load has been greatly reduced. The water still is not clean, partly because point source pollution has not been eliminated but also because of the effect of non–point source pollution. In 1972, non–point source pollution was a relatively minor component of the total pollution load, but as the point source polluters have reduced their effluent, the proportion of the pollution attributable to non–point sources has risen. Non–point source pollution, such as the artificial fertilizer that is carried in urban waste water, is difficult to regulate. Because an origin for the pollution cannot be identified, it is hard to quantify the amount of pollution produced from an individual farm, lawn, or parking lot; hence it is almost impossible to apportion responsibility. The means to deal with non–point source pollution is an active area of research in environmental economics and law and may prompt changes in the law in the next few years.

In another section of the Clean Water Act, section 404, the Army Corps of Engineers is given responsibility (subject to Environmental Protection Agency approval) for issuance of permits to allow the dredging and filling of navigable waters. This may seem an innocuous and trivial aspect of a far-reaching law, but it has become a highly contentious piece of legislation. The conflicts arise over the interpretation of "navigable waters." Most of us would assume that these were waters plied by boats, but not so in a legal sense. Test cases have shown that these waters include wetlands, even ones that have no standing water on the surface! The courts have accepted that no firm boundary can be established where a navigable waterway ends and dry land begins, and wetlands at that transition have been accepted as part of the waterway. Consequently, wetlands can be neither deepened nor filled without permission.

This provision is used to regulate the filling of wetlands for commercial or municipal development. Once the ground surface is raised above the level of the water table, the land can be used for development, say, building shopping centers, apartment complexes, or industrial sites. No specific law exists to protect wetlands from development, even though it has been widely accepted that these ecosystems should be prioritized for protection. Permission must be gained to convert a wetland, but relatively few applications to convert wetlands to other uses are denied. One reason is that there are plenty of ways of developing a wetland that are not included within this law, such as drainage, logging, and conversion to agricultural land. "Swamp Gas," a wetland conservation lobby, reports that each year about 70,000 conversion projects in wetlands are attempted in the United States, of which only about 15,000 are required to obtain permits from the Corps of Engineers, and of these only about 500 applications are denied.

While the need for some regulation of wetland development is recognized by most parties, considerable disagreement exists as to the definition of what constitutes a wetland and the level to which government intervention is legitimate. The proconservation lobby will strive to expand the role of this legislation, while the antiregulators will seek to redefine wetlands to greatly reduce the land area affected by this act. This is another section of the law that is likely to be reviewed in the near future.

The Endangered Species Act of 1973

In 1973, the Endangered Species Act was passed in an attempt to prevent extinctions of rare species in the United States. (Note that this was passed under the Republican administration of President Richard Nixon.) In 1976, the Committee on the Status of Endangered Wildlife in Canada (COSEWIC) was established to identify protection programs for rare species

Ecology in Action

Ecologists and Environmental Impact Surveys

Although many ecologists are employed in colleges and universities, the majority work in industry. One of the employment markets for ecologists that has grown rapidly since the 1980s has been the field of **environmental consulting**. An example of the role of ecologists in these consultancies is in the preparation of **environmental assessments** (E.A.s) and **environmental impact statements** (E.I.S.s). Federal and state laws require any project involving government lands, government spending, or large-scale projects carried out by the private sector to be preceded by an E.A. These surveys require expert knowledge and are often completed by ecologists working for environmental consultancy firms. Also, with the enforcement of CERCLA, any landowner is liable for toxic or hazardous waste on their land. Consequently, a purchaser should always obtain an independent assessment to determine if there is the slightest possibility that a previous owner has stored or spilled chemicals on the land.

Carrying out the E.A. requires that an ecologist visit the areas in question to determine if it is the breeding ground of any endangered or threatened species. To make the determination, the ecologist must inventory the plants and animals of all the affected areas during times of the year when it is likely that an endangered species would be present. The survey also takes into account the impact of the proposed project on areas that could be affected indirectly by the development. For instance, a proposed project to straighten a section of river to make boat access easier would affect not only the river but also adjacent wetlands and downstream areas. An E.A. would also locate and delineate any wetlands (Chapter 8) and determine if they were of sufficient size that they qualified for protection (this size varies according to state law).

The search for hazardous or toxic waste would normally include an on-site search for abandoned 55 gallon drums or other chemical containers, staining of the ground by oil, or the presence of undocumented underground storage tanks. The E.A. would also include a search of databases to determine if the site was ever documented to have storage tanks. If some chemical dumping was found, a second phase of the E.A. would be required to determine the severity of the problem. The resolution of the problems comes under the E.I.S.

Among the requirements of an E.I.S. is a description and comparative assessment of alternatives (such as other routes for a new road or an alternative location for a shopping center), a description of the environment that will be affected by the proposed development, and a full analysis of the environmental consequences of both the proposed action and the alternatives.

The final E.I.S. must include all the information gained on the target site and on the alternative sites. Under an E.I.S., mitigation required for the destruction of wetland or threatened species habitats is laid out in detail. Similarly, the treatment of dumped or leaked chemicals is detailed in an E.I.S. If the chemical containers are still intact and there has been no leakage, the chemicals can be removed by hauling them to a handling facility. If, on the other hand, the chemicals have permeated the soil and entered the groundwater, the landowner will be required to remove all the contaminated soil, drill test wells to monitor removal, and possibly cleanse the local aquifer. Removing chemicals from an aquifer requires the water to be pumped to the surface, decontaminated, and then pumped back down as clean water into the aquifer. Such cleanups can take 10 years or more and cost millions of dollars.

Clearly, these surveys require considerable fieldwork and the study of site plans and local maps; but, above all, they require experience. Each decision made by the environmental consultant could cost someone large amounts of money, and, therefore, the scientist must be prepared to present authenticated data and defend her or his conclusions to a court.

in that country. The nomenclature adopted for the conservation efforts establishes two separate categories of organisms in need of protection: those deemed endangered and those that are threatened. The act defined an **endangered species** as "any species which is in danger of extinction throughout all or a significant portion of its range." The lesser status of **threatened species** is defined as "any species which is likely to become an endangered species within the foreseeable future throughout all or a significant part of its range."

In 1973, the U.S. endangered list covered 92 species, and by 1998, this number had grown to 1138 (animals and 669 plants), with more than a thousand listed as threatened. Another 4000 species have been nominated as deserving endangered status, and about one-tenth of those are expected to receive full protection. In Canada, the number of species listed as endangered and threatened are 46 and 50, respectively. The lower number of Canadian species reflects the relatively low species diversity of tundra and boreal forest ecosystems and the fact that vast areas of these ecosystems remain relatively intact.

The Endangered Species Act does not attempt to save the last individual or the last pair of a species, for that would be futile. If a population becomes too small, there is very little chance that the species can survive. The act is attempting to preserve viable breeding populations of species that are presently showing a downward trajectory in their numbers. The act is not foolishly aiming to forestall natural extinctions; most extinctions in recent history are induced by human activity. Our actions change the world so rapidly that nature cannot keep up and species are lost as their habitat is changed. Recognizing this fact, the Endangered Species Act specifically forbids federal agencies from undertaking any action that is "likely to jeopardize the continued existence of any endangered or threatened species or result in the destruction or adverse modification" of their habitat. This clause is very clear. No federal agency can undertake activities that could lead to the extinction of any species. This description includes allowing the development of federal land that is the home to an endangered or threatened species, financing a development that could lead to an extinction event, or permitting the hunting of a species to extinction. A further provision of the Endangered Species Act charges federal and state agencies with responsibility to restore the habitat of endangered or threatened species if it has been damaged on federal or state lands. The law also requires that the Secretary of the Interior prepare a species recovery plan for each endangered species. This plan should include estimates of minimum viable areas for populations (Chapter 21), habitat requirements, and ways to achieve a rebound in the population number so that the animal can be removed from the endangered species list.

The implementation of the Endangered Species Act, as it was originally formulated, was to be decided on the best available scientific and commercial knowledge. In other words, expert testimony from scientists would be sought, and if recommendations consistently indicated that a species should be protected, the protection would become law. The cost of that protection to industry, landowners, or the community was not a consideration. The reason was that neither Congress, nor scientists, nor developers can put a precise value on a species (Chapter 27). Even the most apparently insignificant species could provide a chemical that would be a valuable cancer cure or become a new food additive and thus have immense commercial value. Consequently, the Endangered Species Act determines the worth of a species to be "incalculable," and on that basis the species could always be worth more than a field of wheat, a shopping center, or a hydroelectric project.

After some initial confrontations in which the Endangered Species Act held up multimillion dollar development projects, the wording of the law was changed. In its present state, the law states that a critical habitat must be protected, "taking into consideration the economic impact, and any other relevant impact . . ." This clause provides a legal loophole, so projects deemed essential for national security or the well-being of a community can override the Endangered Species Act if no workable alternative exists. The committee that makes this decision has the potential to condemn a species to extinction and hence is known as the "God Squad." In over 15 years, the God Squad has met only three times, and it has yet to grant complete exemption from the Endangered Species Act.

Where the act is most likely to impinge on the actions of the private landowner is the clause that prohibits the "taking" (hunting, trapping, collecting, harming, or disturbing) of threatened or endangered species. For example, if an endangered marsh plant is identified on a piece of land, the drainage of the marsh would harm the plant and would be unlawful. However, in the saga of the Northern Spotted Owl and forest cutting in the Pacific Northwest, the cutting of old-growth forest, the habitat of the owl, is a legislative issue only on public lands. On private lands, there is effectively no protection for owl habitat. Despite the "taking" clause, the primary importance of the act deals with protection of species on public lands or on lands affected by federal spending.

The act also contains clauses that are linked to international efforts to reduce extinctions. Globally, perhaps as many as 1000 species per year are being driven to extinction by human activities. However, many of these are insects or plants, species for which there is no international record keeping. Public concern is greater for mammals, birds, reptiles, and amphibians than for bugs. Consequently, these larger animals are the only ones consistently documented. The global status of rare species is published by the International Union for Conservation of Nature and Natural Resources (I.U.C.N.). In their 1989 "Red Data" book among those listed as endangered are 2214 bird species, 1666 species of mammals, 858 reptile species, and 129 amphibian species. Imports of these species into the United States are banned under the Endangered Species Act. Despite legal protection, species such as white rhinoceros and African elephants face imminent extinction by poaching. Nevertheless, for most species, it is the loss of breeding habitat and feeding grounds that is the critical issue.

The Endangered Species Act of 1973 must be reviewed periodically because the original act was effective for only 10 years. The 1983 renewal was delayed until 1988. The act remained in effect, but new funding levels and closure of loopholes were debated for five years. A basic issue that must be addressed with each renewal of the act is: Has it been effective? Since 1973, only eight endangered species have recovered sufficiently to be removed from the list. In what is hoped to be a proconservation publicity coup, the Bald Eagle was downgraded from an endangered species to a threatened species (Figure 28.2). Since the 1960s, the population of the Bald Eagle has increased from 800 in the lower 48 states to about 8000 in 1994. The recovery of the eagles is beyond dispute, but the cause is contested. Environmentalists claimed that the eagles demonstrate the success of the habitat preservation clauses of the Endangered Species Act. The antiregulatory lobby argues that this recovery is due not to habitat protection but to changes in pesticide use (specifically the ban on DDT) and hunting. Naturally, the protection of the eagles from bioamplifying pesticides and from bullets has helped their populations to recover. However, if their breeding territories are not protected, then the benefit of other measures is canceled.

Are populations of species that have not been removed from the list at least increasing? Of the 912 species listed between 1973 and 1994, the populations of 364 (40%) were considered to be stable or improving in 1994, while the population decline of another 33% had slowed. Seven of the listed species have gone extinct, and a further 14 are thought to have gone extinct (note this is only among species listed as endangered in the United States). Ecologists would interpret these numbers cautiously. The law certainly appears to be helping in the recovery of species, but as population recovery is measured over decades, it is too early to determine how many of these species can be saved from extinction.

Because the Endangered Species Act has become the focus of some antiregulatory groups, it is perhaps as well to see the effects of this act in perspective. Some foresight and planning can almost always avert outright clashes between developers and conservationists. According to a study conducted by the World Wildlife Fund, in the period between 1987 and 1991, of almost 75,000 projects to be undertaken by federal agencies, only 19 were canceled because of the Endangered Species Act. Nevertheless, the inconveniences caused to companies and citizens who find their plans stymied or delayed by the presence of an organism of which they have never heard is a real problem. Talk shows on radio and television are potent publicity tools, and statutes are amended and written by politicians who pay close attention to public opinion. Even when the Endangered Species Act clearly provides protection for an organism, environmentalists can be made to appear foolish and uncaring in the eyes of the public. Many endangered species are not charismatic furry animals, but are flies, mollusks, or weedy-looking plants. Making the case for their survival as being more important than the retirement plans of a hard-working couple, a farmer's livelihood, or the jobs offered by

Figure 28.2 Bald Eagle fishing. Is the recovery of Bald Eagles a major success for the Endangered Species Act or the result of pesticide regulation?

a new mall, inevitably leads to conflict. If the Endangered Species Act is to be maintained, environmentalists will have to use the act's powers responsibly, not as a tool to prevent development.

Overall, the act does not appear to be a major obstacle to thoughtful development, and its success is reflected in general popularity. A 1992 public opinion poll of Americans revealed that 66% were in favor of the act and only 11% were against it.

The Comprehensive Environmental Response, Compensation, and Liability Act of 1980

By definition, no one wants waste, and it is fundamental to our way of thinking that we do not like to pay for what we don't want. Therefore, it is easy to see that no one wants to pay large sums of money to dispose of waste. From an economic perspective, waste material constitutes an externality, so the economic benefits of wise disposal of this material are considered to be of little interest. Until the 1970s, industry, municipalities, and agriculture paid little attention to the disposal of potentially toxic waste materials. These chemicals were frequently dumped or buried; "out of sight, out of mind" was the ruling philosophy. As the number of dump sites increased, the probability that they would adversely affect nearby inhabitants also increased. Horror stories of seeping chemicals resulting in miscarriages, birth defects, cancer, respiratory disease, and lowered disease resistance were at first ignored. However, these accounts were substantiated, and the effects of pollution weeping from the ironically named Love Canal spurred national debate and government action.

The Love Canal community lived on land owned by the City of Niagara, New York (Figure 28.3). In 1953, the city had bought the land for $1 from the Hooker Chemical and Plastics Corporation, who had been using the abandoned canal as a chemical waste dump since 1942. By 1953, the site contained 19,000 metric tons of hazardous waste, much of it buried in steel drums. Before the sale, Hooker Chemical had made an attempt to seal the old dump site by covering it with a cap of clay. This procedure was standard at the time, but it proved to be ineffective in preventing the leakage of chemicals as the steel drums corroded. Chemicals leaked onto the surface of the ground and into the basements of adjacent houses. More than 80 toxic chemicals, some carcinogenic, were identified in the ooze from the site. Elevated rates of birth defects and respiratory disease were reported by local residents, and, after years of dispute, more than 1000 fam-

Figure 28.3   Toxic waste and Love Canal. Tim Moriarty, a resident of Love Canal for 35 years, is forced to leave his home because of toxic waste seeping into his basement.

ilies were relocated. This sad story gained national attention, but who were the guilty parties? Who should pay for the cleanup of the site? Should it be the chemical company who caused the pollution or the city who built houses and a school on the land?

The first attempt to control the disposal of toxic waste was the Resource Conservation and Recovery Act (RCRA) of 1976. RCRA set down the standards that companies must follow to lawfully dispose of hazardous waste. For the first time, the law stipulated that a record must be kept of what happened to hazardous chemicals from the time they were made to the time of their disposal. This type of tracking is known as **cradle to grave** tracking and should reveal the amount of a given chemical that is "lost" along the way. The tracking discouraged illegal dumping or release of toxic materials. RCRA did not deal with the dumps of hazardous waste that built up before 1976 or with accidental spills. Thus, it did not apply to Love Canal.

In 1980, the Comprehensive Environmental Response, Compensation and Liability Act (CERCLA) was passed in an attempt to make polluters pay for the mess that they make (or have already made). This law has become better known as the **Superfund** legislation, after the pool of money established to instigate these cleanups. The law has two principal aims: to penalize polluters (through fines or imprisonment) who release or threaten to release hazardous chemicals into the environment and to force polluters to pay for the cleanup of sites polluted with hazardous chemicals. The legislation defined more than 700 substances as constituting hazardous waste and established a bureaucracy to list and prioritize sites for cleanup.

Superfund is grounded in the **polluter pays** principal that seeks to penalize those responsible for the initial pollution. Once a site is identified, the responsibility for the cleanup is taken by the EPA, and the EPA then bills the guilty party. The liability also extends to landowners who allowed pollutants to be dumped on their land or to the landowner of an area from which pollutants are presently escaping. The act establishes the liability of both present and past polluters. Thus, after an initial startup funding, the money raised by Superfund prosecutions was meant to pay for the cleanup. As the law was envisioned, the taxpayer would not have to foot a substantial bill.

The fairness of the Superfund legislation has been criticized repeatedly. It is evident that the present owners of the land should be liable if chemicals are leaking from their land and endangering local populations. It is also essential that past owners of the land be held accountable for pollutants that they left behind; otherwise, a company could dump chemicals on the land, sell the property, and be immune from prosecution while an innocent party is buying land that holds hidden pollutants. For example, land could be rented by a chemical company that illegally buries chemical waste. Before the lease expires, the chemical company covers its tracks, hiding the dump site. The landowner may know nothing of this activity and proceeds to sell the land. The buyer of the land is innocently and unknowingly inheriting a chemical dump but, nevertheless, would be liable for a portion of the cleanup cost.

There is no guarantee that land purchasers are exempt from Superfund prosecution; however, courts are less likely to assign liability if the buyer can show that a thorough effort was made to determine that the site was free of hazardous waste and that she had no prior knowledge of hazardous waste associated with the land. A new industry of site inspections has developed, undertaking detailed site assessments, including soil and runoff water tests to search for hidden wastes. Such surveys have become routine before large real estate sales, especially when the land may have had a past commercial use.

The polluter-pays approach seems to be a reasonable and equitable way to address the environmental menace of dumped chemicals. But 15 years and $13 billion later, cleanups have started at fewer than 200 sites, and only a handful have been completed. Most of the money (perhaps as much as 80%) spent on Superfund has been absorbed by legal and administrative fees. The sticking point has been proving culpability. The standard defense of many companies accused of dumping chemicals has been a bout of finger pointing to draw other potentially guilty parties into the dispute. As the guilt or innocence of these other companies is debated, legal costs mount and the pollution remains. This problem has proved more or less intractable, and many see this well-intentioned and necessary piece of law as one of the gravest failures of environmental legislation. However, it should not be forgotten that pollution cleanup and site restoration are only two of the original goals of the act. Another goal of the act, to reduce illegal or careless dumping of chemicals through inflicting massive penalties on would-be polluters, appears to have been successful. Companies dare not break the rules of CERCLA, and, so far, fewer toxic dump sites are created now than in the past.

The need for some form of Superfund legislation is not in doubt, but the range of sites that it covers, the burden of proof to establish buyer innocence, and the way that the law is implemented are likely to be reviewed in the near future.

The Clean Air Act of 1990

The Clean Air Act of 1990 is the most stringent regulation of atmospheric emissions yet passed in the United States, and estimates for the cost of its implementation range from $4 billion to $60 billion. While the reduction of chemical outputs responsible for acid rain (Chapter 22) and stratospheric ozone depletion (Chapter 20) would be anticipated in any modern clean air package, an innovative section of the act allows the trading of pollution (see Chapter 24). Table 28.2 itemizes the key components of the act.

The initial goals of the Clean Air Act of 1990 (phase 1) have largely been met. Phase 2 of the act goes into effect in the year 2000 and further reduces allowable levels of pollutant gases and dust-sized particulates. However, the act does not address the release of

Table 28.2 Principal provisions of the Clean Air Act of 1990.

- Reduce CFC emissions 20% by 1993 and 50% by 1999 (see updated reductions under post–Montreal protocols: Chapter 20).
- Reduce emissions of 189 toxic chemicals by 90% between 1995 and 2003.
- Meet federal emission standards for ozone between 1993 and 1995. This provision applies to 87 of the smoggiest cities. Eight cities have been granted until between 2005 and 2007 to conform, and Los Angeles must meet more stringent state standards by 2010.
- Reduce hydrocarbons by 35% and nitrogen oxides (NOx) by 60% for all cars by 1994. Even more strict emissions standards will go into effect in 2003.
- Reduce particulate matter from large diesel trucks and buses by 98% by 1998.
- Require oil companies to sell cleaner-burning fuel in the nine worst-polluted cities by 1995.
- Reduce sulfur dioxide emissions from coal-burning power stations by almost 50% by the year 2000, or by 2005 if they switch to "clean coal technologies." Concurrently reduce NOx emissions by 2.1 million metric tons.

carbon dioxide and other greenhouse gases, nor does the act deal with indoor air pollution.

28.5 Property Rights Versus Environmental Legislation

Denying a landowner the full use of his or her land because of a new governmental regulation might be deemed **regulatory takings**. For example, a 1 km strip of beachside property that is 60 m wide is bought for $2 million by a developer. Planning permission is sought for a set of condominiums that would bring a profit of $5 million. In the meantime, the state government passes a regulation forbidding new construction within 90 m of the ocean. The 90-m-wide no-building zone is justified on the basis that houses constructed on the front row of dunes increase coastal erosion and are at considerable risk from storm damage. Such building regulations are common on areas of eroding beaches and generally represent the amount of land expected to erode in 30 years (the length of many mortgages). Regardless of the good intention of the regulation, the property owner finds that he or she cannot build on the land. The developer is now the proud owner of a $2 million piece of unusable real estate that will have a much lower resale value. Who, if anyone, is at fault? And, if the state government is at

fault, is it liable for the purchase price of the land or the potential profit from the land when purchased?

With the increase in the number of environmental regulations, land-use and takings cases could arise in many ways. Any of the following might prevent full use of your land: having an endangered species on your land, being denied a dredge and fill permit for a wetland, having zoning restrictions placed on the kind of building allowed, or having emissions to air or water regulated.

Chief Justice Rehnquist has argued that the Constitution provides protection against regulatory takings in the Fifth and Fourteenth Amendments. The Fifth Amendment of the U.S. Constitution states "nor shall private property be taken for public use without just compensation." The Fourteenth Amendment does not contain such a clause but has been interpreted widely to support this rule of law. Together these amendments provide the platform for takings cases in which the plaintiff (alleged victim) demands relief from regulations or government action that threatens their constitutional property rights. This area of the written law has not changed since the signing of the Constitution, but in recent years it has been given a new latitude.

The original interpretation of these amendments consistently assigned a takings only when there was an actual physical invasion of property. In other words, compensation would be given only if the government or its agents took possession of your land (the beachfront developer would have received no compensation). Even within this narrow construct, government could still avoid a takings if it could show that its action was motivated by a need to forestall "harmful or noxious" actions.

During the 1980s, new appointments made by Presidents Reagan and Bush to the U.S. Supreme Court created a Court more sympathetic to the property rights of the individual. In a series of cases, with opinions written by Chief Justice Rehnquist and Justice Scalia, the court started to include economic loss in addition to physical invasion as a grounds for takings. The *Lucas* case in South Carolina was a turning point in takings legislation. This was the first case where a takings was deemed to have occurred solely on the basis of economic loss. Lucas was a beachfront developer who had bought two parcels of land for $975,000 in 1986, with the intention of building on them. Subsequently, the South Carolina legislature passed a law banning building close to eroding beaches. Lucas, who now could not develop his land, argued that all economic worth of the land was gone. Consequently, he sued the local government for regulatory takings. In

1992, the case made its way to the U.S. Supreme Court, which, in a landmark decision, declared that a takings had indeed occurred and awarded Lucas $1.2 million compensation.

The law is there to protect everyone and to ensure that constitutional rights are maintained, and to this extent compensation for a regulatory takings should be enforced. However, it is almost impossible to conceive of any law designed to improve the environment of the general public that would not result in an economic loss to someone. Since the Lucas case, the U.S. Supreme Court has awarded takings for a partial loss of economic worth. Environmentalists fear that regulations will become unenforceable, either because of legal battles and fees or as a result of financially crippling settlements. Fear of huge financial losses could lead a local government to back down when a violation is spotted rather than confront the wrongdoer.

In 1994, the Republicans proposed a "Contract with America" in which a takings could be awarded for a loss of only 10% of the value of property. It is widely believed that had this legislation passed, regulatory activity by local government and the EPA would have been severely restricted (the object of the proposal), and much of the statutory protection granted to the environment over the last 30 years would have been undone.

Summary

- Legislation polarizes industry and conservationists.
- It is difficult to pursue environmental issues using common law.
- A common law case must prove the identity of the offender, that actual injury was caused, and that it was the actions of the accused that resulted in the injury.
- Statutory law is needed to protect the environment or workers from the possible excesses of industry.
- Statutory law can empower regulatory agencies to oversee industry and force compliance with the law.

- Statutory law is enacted by elected officials and so is subject to lobbying and political maneuvering.
- Passage of the most significant environmental legislation has not been the exclusive domain of either political party.
- The "Wise Use" movement is backed by developers and industrial interests but masquerades under proenvironment titles.
- Most environmental legislation is reviewed every 5 to 10 years, giving politicians the chance to strengthen or weaken it.
- Regulatory takings are now based on economic rather than on physical occupation criteria.

Further Readings

Easter-Pilcher, A. 1996. Implementing the Endangered Species Act. *BioScience* 46:355–363.

Hearne, S. 1996. Tracking toxics: Chemical use and the public's "Right-to-Know." *Environment* 38(6):4–9 and 28–34.

Hoffman, A. 1995. An uneasy rebirth at Love Canal. *Environment* 37(2):4–9 and 25–31.

Luoma, J. R. 1992. Eco-Backlash. *Wildlife Conservation* (6):27–36.

Ritter, D., M. Russell, R. L. Reisenweber, and T. P. Grumbly. 1995. How clean is clean? *Environment* 37(2):10–15 and 32–34.

Wilkinson. T. 1998. *Science Under Siege: The Politician's War on Nature and Truth.* New York: Johnson Books.

Web Connections

On-line resources for this chapter are on the World Wide Web at:
http://www.prenhall.com/bush
(From this web-site, select Chapter 28.)

Peering into the Future

29

History is littered with people who tried to predict the future, yet no one has been able to do it successfully. Human society has always been volatile. Seemingly all-powerful empires come and go. Civilizations rise and fall. Some are brought down by internal feuding and warfare, and some, as in the case of the Mayans and Easter Islanders, by environmental change. The pace of change is accelerating as new technology changes all aspects of our lives, and it is the direct and indirect effects of this technological revolution that make the future so hard to predict.

29.1   Is the End Nigh?

Forecasts by the pundits of doom and gloom may be part of the basic human psyche. Apocalyptic visions are found in all societies, be they the result of angering gods, the product of unrestrained warfare, or natural disasters. Perhaps the predictions of societal collapse and starvation based on environmental failure are just a new twist to an old set of fears. If they were, we would have little to worry about. However, while predictions that "the end of the world is nigh" have in the past been based on dreams, local catastrophes, or ambiguous signs, there is no doubt that humans are changing the global environment that they inhabit.

In 1798, Thomas Malthus predicted imminent famine as population growth outstripped food supply.

His vision of the future failed to anticipate new crops, new farming practices, and mechanization. Perhaps it is useful to think of Malthus as being simply ahead of his time. A 200-year error in his calculations may seem a long time to us, but it is the evolutionary blink of an eye. For more than 99% of human evolutionary history our population was more or less stable. It is only in the last 0.25% of that history that our population has soared. If human population growth were to continue at its present rate, the 50 billion mark (the human carrying capacity of the planet; Chapter 16) would be passed when you are about 60 years old. It is reasonably certain that population growth will level off before this number, but it is important to realize that within your adult life you will witness a new phenomenon, the global slowing of human population growth.

In the spirit of Malthus, a more recent forecast of doom and gloom was made by a group of eminent scientists who became known as the Club of Rome. In 1972, they predicted that human populations would crash in the twenty-second century. Their scenario was simple: Humans are using finite resources, such as oil, gas, and forests, in an unsustainable manner. Agricultural practices are heavily dependent on oil for powering machinery and for fertilizers and pesticides produced by the petrochemical industry. As oil resources are finally depleted, agricultural production will slump; famine and a general food shortage will

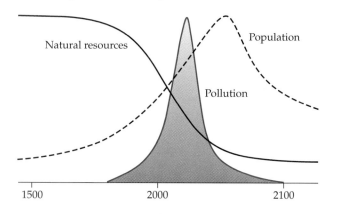

Figure 29.1 **The predictions of a computer model generated by the Club of Rome.** As population increases, pollution increases and natural resources are used up. A critical point is reached when the natural resources are spread too thinly, pollution creates problems of toxicity, harvests fail, and human population collapses. This model's principal weakness is that it makes no allowance for new technologies. (After D. Meadows, J. Randers, and W. Behrens, 1972.)

lead to a crash in world population. The whole process will be exacerbated by increasing pollution and warfare as competition for limiting resources intensifies (Figure 29.1). Some scientists suggest that this whole scenario has already been played out in miniature on the world's most isolated island, Easter Island.

29.2 The Lesson of Easter Island

The tale of Easter Island brings together some of the threads of island life, the arrival and growth of a human population, and habitat loss. The story might have been an epic of human initiative overcoming adversity and people coexisting peaceably with island creatures. But the real story is one of warfare, cannibalism, exotic species, and extinction.

Easter Island lies roughly midway between Chile and Polynesia, a volcanic mount battered by the storms and high seas of the South Pacific Ocean (Figure 29.2). More than 3500 km from mainland Chile and more than 2200 km from Pitcairn Island (the nearest inhabited land), Easter Island is almost impossibly remote. And yet, blown by storms and drifting on currents, seafaring Polynesians in enormous dugout canoes found their way to the island about A.D. 400. They found a forested island with large palm trees but few animals that could provide food. The island had no mammals living on it, but the seas contained fish, and on land there were birds' eggs. The settlers also carried a particular variety of Polynesian rat with them, which they ate as a protein source during the many days at sea. Either deliberately or accidentally, the rat was introduced to the islands; an act that would have unforeseen repercussions.

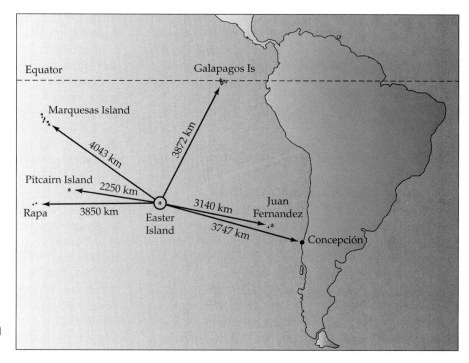

Figure 29.2 **The location and isolation of Easter Island.**

This culture existed in isolation for more than a thousand years. Their lack of contact with other peoples is indicated by their legend that all the other lands of Earth had sunk beneath the waves and that they were the only survivors. Their culture and beliefs were rooted in Polynesian tradition, but as society evolved so too did a fresh set of legends and ceremony. The religious centerpiece of their society was a reverence of ancestors, and great statues, probably representing their forebears, were carved from the island's volcanic rock. Some of these statues were more than 6 m high with a hat, or topknot, of a different carved rock 2 m high (Figure 29.3). The statues are highly stylized people with long sloping faces. For many years, the statues were thought to have been "blind," but excavations in 1978 revealed that the statues had eyes of white coral surrounding a vivid red iris made from volcanic rock. All faced the sea, staring at a world that had disappeared beneath the waves. The carving, transportation, and erection of these giant statues was no small undertaking, and it is probable that the islanders specialized into professions of masons, farmers, and fishermen. At the height of their culture, about 10,000 lived on Easter Island.

In 1992, British archaeologist Paul Bahn reviewed the island's archaeological records, which included many remains and carvings that were hidden in subterranean caves. His fellow researchers John Flenley and Sarah King reconstructed the forest history using fossil pollen records held in the mud of crater lakes. Both lines of evidence document the rise and fall of this island civilization. Before it was settled, Easter Island supported an endemic palm that can be identified from its pollen and its nuts. The palm became important to the islanders, because it was the only tree large enough for the construction of dugout canoes. These large canoes enabled the settlers to fish for sharks and other large fish, a task impossibly dangerous in a small canoe. The seagoing canoes also allowed

Figure 29.3 **A line of statues on Easter Island.**

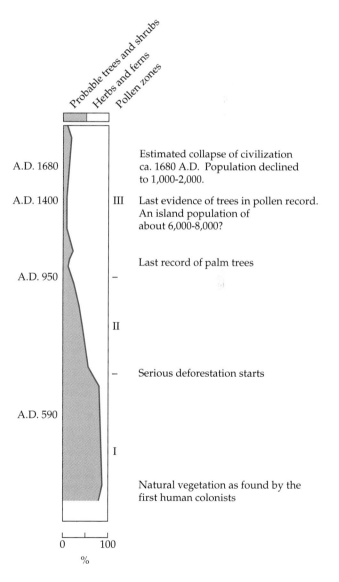

Figure 29.4 **An annotated summary diagram of the pollen record from Easter Island.** (After F. Bahn and J.R. Flenley, 1992.)

the islanders to visit bird colonies on uninhabited islands up to 400 km distant. There, they could collect bird eggs, which became a valuable protein source. The palms were also important to the Easter Island culture because they were the only tree large enough to provide the wooden rollers and hoists to move the statues from the quarries to the clifftops where they were erected. Despite their importance economically and socially, the palms were cut. Unlike an oak tree, a palm tree, once cut, dies. The palm population was much reduced by the logging, but the final factor sealing the doom of the palms may have been the rat that the initial settlers had carried as a protein source during their journey. Palm nuts dating to about A.D. 1500 found in caves all show tooth marks left by the mouse-sized Polynesian rat. The rats appear to have bred freely and established large populations on Easter Island, and they may have become a serious predator of nuts, preventing the reproduction of the palm. The pollen diagram from Easter Island shows the islands to have been forested for more than 30,000 years before the arrival of humans (Figure 29.4). Between A.D. 400 and 1400, there is a steady decline in tree pollen, and by the time of first European contact in the late 1600s the islands were virtually treeless.

The islanders, despite an advanced artistic culture, knew nothing of the soil erosion that would follow deforestation or the consequent loss of fertility in their fields. When the last palms were cut, the islanders could no longer build canoes big enough for deep-water fishing or for harvesting eggs from distant islands. Warfare driven by famine raged in the late 1500s and 1600s; weapons suddenly appear in the archaeological record, and families lived in hidden subterranean caves. Cannibalism, perhaps of prisoners of war, is suggested by the discovery of human bones that bear the marks of being butchered. The population of Easter Islanders fell from about 10,000 people to fewer than 1000 in the space of 200 years (Figure 29.5). Easter Island is not a unique example of island deforestation leading to societal collapse. The same sequence of overexploitation of natural resources, starvation, and warfare is evident in the archaeological record of Pitcairn and Henderson Islands. The difference is that, on those islands, the human population died out completely about six generations after felling the last tree.

The "doom and gloom" contingent find it hard not to see a parallel between Easter Island and the whole Earth. They say that Earth supports the only known life in the universe, yet humans practice many of the follies of the Easter Islanders. Like them, we ignore the consequences of erosion, a burgeoning human population, and the ever-present threat of war. And we appear to be content with ignoring those factors . . . until it is too late.

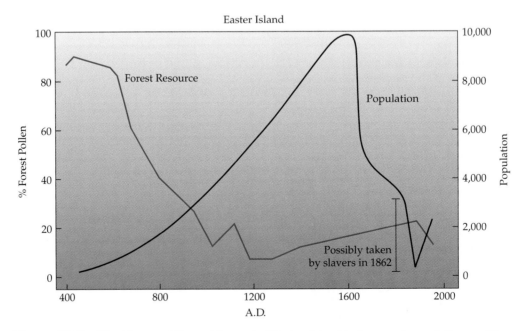

Figure 29.5   The history of the settlement of Easter Island in terms of population and forest resources. (After P. Bahn and J.R. Flenley, 1992.)

29.3 Human Initiative Saves the Day?

It is a maxim of historians that we never learn from history and that we repeat mistakes made by earlier societies. However, a fundamental difference does exist between the world predicted by the Club of Rome, which essentially predicts a repeat performance of the Easter Island tragedy, and our future. Easter Islanders lacked the knowledge and raw materials to change their technology, we do not. Technology has so far found a way to deal with our problems, providing more food and water, improving our living conditions and health, offsetting the direst predictions of Malthus. However, these improvements are not "free," they carry a cost in terms of pollution, resource depletion, and societal change. Contrary to media frenzies over isolated tragedies, our twentieth century society is safer than that of any preceeding time. Of course there wil be social change and increasing regulation, but there is no suggestion that we are about to collapse into anarchy. Similarly, although we can see severe problems with fisheries and the destruction of pristine habitat, these do not threaten the survival of humanity. The real lesson to learn from Easter Island is that *humans lack an environmental failsafe switch in their brain.* Evolution has not prepared us to make environmental decisions on a large scale; it will have to be a conscious, not an inherited, decision to use our limiting resources responsibly.

Malthus and the Club of Rome made their dire forecasts on the basis of "business as usual conditions" and do not (because they cannot) make allowance for future changes in technology or social choices. The current prediction that 8 billion people can be supported with some acceptable standard of living assumes present technology. That is an estimate grounded in reality, even though there is plenty of room for error. Predictions based on hypothetical future technologies have no such basis in reality. If we were to say that, by the year 2050, we expect technology to double our capacity to support humans, allowing the existence of a population of 16 billion, this estimate may or may not be true. Which technology is it that could double population and yet leave us with acceptable biodiversity, energy, water, and food availability? Will those people all be added to the teeming cities of the developing world (as would be the present trend), or will they be spread evenly across the surface of the globe? We cannot include technological improvements in our predictions because, unless grounded by knowledge, they are pure fantasy.

Thus, with hindsight we may be dismissive of the past attempts to predict the future, but as we attempt to peer into our own future our predictions are equally flawed. Two hundred years ago Thomas Malthus was able to look at a simple equation of logarithmic population growth and arithmetic increase in food production and predict a future shortfall that would limit population growth. We still have an exponentially growing population, and we have concerns regarding our ability to match that growth with food production. However, we have added whole new tiers of uncertainty, each of which could radically alter the future. We now have the potential for globally destructive warfare, human-induced changes in climate, an increasing load of pollutants, AIDS, higher demands in terms of standards of living, and a depleted resource base. On the other side of the equation, never has technology been surging ahead so rapidly, and never has the world population been as healthy, as educated, or, arguably, as environmentally aware.

Perhaps of all the technology that is on the horizon, none offers a better example of potential benefits and perils than genetic manipulation.

29.4 The Rise of Genetic Manipulation

Genetic manipulation of organisms has been a standard tool of medical science for more than a decade, but public acceptance of "genetically engineered" foods has been slower. The Green Revolution of the 1950s and 1960s relied on "miracle crops," all of which were hybrids. It is likely that at the forefront of the next Green Revolution will be genetically manipulated varieties. Genetic engineering will pose new moral dilemmas. How far should we change a species? Is it legitimate to produce legless cows for leaner meat or chickens that cannot walk because their breast is so large? Can a devout Hindu eat a bowl of engineered rice that contains a gene from a cow? These, along with the moral questions posed by our ability to screen a human fetus for genetic disorders, hair color, sex, or predisposition to disease, generate a new subject that has been dubbed **genethics**.

Manipulation versus Hybridization

Hybridization is the mating of dissimilar forms of the same species (usually) to produce offspring that, one hopes, contain the best characters of both parents. Occasionally, different, but closely related species are crossed to produce hybrid young. For example, a tangerine crossed with an orange produces a tangelo. However, most crop improvement consists of crossing two different strains of the same species. Perhaps one strain has a high yield but has the genetic trait to be flavorless, while the other has the traits for good flavor but relatively low yield. When these two strains are crossed, a proportion will be flavorless and have a low yield; these are discarded. A proportion of that same group of offspring will be like their parents and these too are discarded, but some will be flavorful and have high yield. To derive a marketable crop of seed that has these good characters is a time-consuming and costly business. One benefit to the seed producer is that the improved variety, an **F-1 hybrid**, although growing strongly and setting large quantities of seed or fruit, will be sterile, or produce weak offspring. Consequently, the farmer has to buy fresh F-1 seed from the grower each year, so the seed-producing industry stays in business.

Traits such as flavor, seed size, disease susceptibility, plant height, animal fat content, protein content, and drought resistance are controlled by genes. Since 1973, it has been possible to **sequence** genes to identify which portion of a chromosome holds a particular gene. Once identified, the length of DNA that forms the desired gene can be cut out. The cut portion of the DNA can be lifted out of that chromosome, cloned, and inserted into the chromosome of a completely different species. This process is exact, and once the desirable gene has been identified, the development time to produce a new crop or animal variety would be much more rapid than from hybridization.

Ideally, if a gene for drought resistance could be isolated from a desert plant it could be transferred to corn. The resultant corn plant would then have an increased tolerance to drought. Because this is a genuine genetic transformation, the seeds from the modified corn would contain the drought-resistance gene. One of the great differences between genetically engineered crops and the "miracle crops" resulting from hybridization is that the gene modification is built into the lineage of the engineered crop. Frequently, because the new plant is not a hybrid, it can set viable seed for planting the next year. In other words, the farmer would have to buy only one generation of "super-corn," and thereafter the seed from each harvest could be planted and still provide drought-resistant corn.

Hybrids can be raised only from two reproductively compatible varieties, such as different varieties of peanut, and this requirement limits the range of traits available to the seed grower. In genetic engineering,

we manipulate the basic building blocks of life, DNA. Because the same genetic code is shared by all organisms, a gene that codes for a particular protein in a bacterium, will produce exactly the same protein if it is inserted into your DNA sequence. Thus, the gene for growth regulation could be extracted from a cow and put into a chicken or indeed into a peanut. That is not to say that we will soon have chickens and peanuts the size of cows, but it might result in a chicken that reaches its optimal size faster than before, or a peanut with improved disease resistance. This ability to transfer genes from completely unrelated species is what makes genetic engineering such a powerful tool.

The difficulties of developing genetically altered crops are immense. From a technical standpoint, identifying the gene that provides a desirable trait requires many months of intense laboratory activity. Even then, the desired trait may prove to be linked to an unwanted trait. The same gene may confer both rapid growth and mildew susceptibility; one is inextricably linked to the other. If a gene is imported from another species, it may bring about unforeseen, even undetected, changes in the chemistry of the crop. While genetic engineering is theoretically possible and has a huge potential for success, it is still a science in its infancy.

Weighing Benefits and Costs

Between 1988 and 1994 more than 2000 field trials of 44 genetically engineered plant species were conducted. Most of these trials were to include genes for insect resistance, disease resistance, herbicide tolerance, and stress tolerance. The benefit of a crop that is insect or disease resistant is obvious, perhaps herbicide and tolerance are not so apparent. Keeping a crop weed-free reduces competition and so enhances yield. Farmers expend a lot of time and energy to keep their fields clean of weeds. When weeds are closely related to the crop plant it becomes difficult to find a conventional herbicide that will kill the weed but not the crop. Imagine how much easier it would be to simply spray the field with a herbicide that kills all the weeds, but leaves the crop unharmed. Already, experimental strains of 19 of our most common crops have been genetically modified to be resistant to at least one herbicide. Stress tolerance refers to the plant's ability to withstand environmental extremes such as heat, drought, salt, or nutrient deficiency. One of the problems facing this line of research is that an organism's ability to withstand a particular extreme, say frost, is not controlled by a single gene. The more genes that have to be manipulated, the more complex, costly, and inherently failure-prone, is the experiment.

Perhaps, that genetic manipulation to provide stress tolerance has lagged behind the development of herbicide-tolerant lines reflects economic rather than scientific criteria. Potentially, new races of rice, manioc, sweet-potato, and corn could be produced that would be extremely drought resistant, or that do not need expensive fertilizers. Furthermore, genetically altered seed, unlike seeds of hybrids, will grow to produce plants that bear fully viable seed. The farmer can keep a portion of the yield and sow it for the next crop. There is no longer a need to buy seed every year. Such crops would be invaluable to farmers in LDCs. However, because these farmers are too poor to pay for the transgenic seed there is no economic incentive to develop the crop. The research and high-tech manipulation of genes required to produce new crops and the field testing and the licensing of each version of a crop costs many millions of dollars. Only the large business empires will be in a position to produce these crops. Remember also, the corporations producing the transgenic crops are often the same as the ones producing fertilizer and pesticides. To date, therefore, most genetic development has not been directed toward crops that would reduce sales of fertilizer and pesticide, but toward herbicide-tolerant crops. The benefit to the agribusiness is clear; they are selling a package of plant and chemical. This argument that agribusiness will only develop transgenic crops that will increase the sale of their other products has an undeniable logic, but it may be assuming the worst. Our knowledge of genetic engineering is not sufficiently advanced to produce the perfect crop. By degrees, gene manipulation will improve crops, but these are likely still to be genetic improvements of hybridized strains. It is hybrid seed that is sold to the farmer. As this seed will grow into plants that set sterile or poor seed, the agribusiness companies will still be in a position to sell seed each year.

Environmentalist Jeremy Rifkin opposes genetic manipulation on the grounds that the transfer of genetic material may bring entirely unforeseen results. The gene might escape into wild populations through hybridization of the crop with a closely related weed. If the hybrid is fertile, the gene for herbicide resistance could be passed back into the weed population. Within a few years, the herbicide would be ineffective at controlling weeds, thus increasing weed problems rather than reducing them. Another argument is that pesticides are only effective if they are used rarely, but massively. If pests survive the onslaught of pesticide,

resistance is bred into the population. If we take the gene that codes for the best natural defenses against pests and implant it into all our crops it is a certain recipe for the evolution of pests to become resistant.

Market Acceptance of Engineered Foods

Might public apprehension regarding the wholesomeness of genetically altered foods reduce their commercial viability? Most gene changes would be rather subtle, perhaps replacing a gene that shuts down the production of a particular enzyme or hormone with one that allows the production of that chemical to continue. If the changes are that subtle, geneticists have argued, most of the genetic alterations will be no more profound than are achieved using hybridization; they are just quicker to produce and more exact. Consumers are not advised when they are buying hybrid vegetables (almost everything you eat comes from a hybrid), nor are they warned if the genetic modification is deemed minor. The regulation stipulates that produce must be labeled as genetically engineered only if it carries a gene from a grossly different species, such as a chicken gene inserted into a lettuce, or if it could produce a potentially toxic chemical.

In 1993, the first genetically altered produce, the *Flavr Savr* tomato, was introduced to stores in the Midwest. In Mexico and California, tomatoes are picked when they are still green and are trucked east. The tomatoes arrive in the store blemish-free after thousands of miles of shipping, but they are tasteless compared with those bought in a local farmers' market. The *Flavr Savr* tomato has a genetic alteration that delays the rotting of the tomato. This gene allows the tomato to stay on the vine an extra three to five days, in which time the fruit reddens and accumulates the sugars and acids that provide flavor. The new tomato variety has the benefit of flavorfulness and resistance to damage while being shipped. The trial marketing appeared to be successful, in the sense that the public bought it. However, as a genetically engineered product it failed because its ability to survive transport was no better than a regular tomato. But the important message was that the American public did not shy away from buying genetically altered produce.

No one seriously fears the attack of the killer mutant tomato, but there is genuine concern that human health could be adversely affected as we increase our intake of new proteins and hormones. Perhaps the transplanted gene carries a hidden trait that causes an unexpected chemical change in the plant or animal.

The new chemistry might cause an allergic response in some people who eat it. As more and more consumers are exposed to genetically engineered produce, we can expect a small number of people to fall ill, even die, as a result of an adverse reaction to something they ate. Should this be the signal that we have gone down the wrong track and that genetic engineering should be abandoned? Perhaps we can learn from another historical example.

When canned goods were first introduced in the 1800s, they revolutionized the grocery market. For the first time, out-of-season produce could be kept on shelves without having been pickled, smoked, or dried. In 1844, a state-of-the-art expedition led by the English explorer Sir John Franklin set off to find the Northwest Passage. The expedition was well-funded, and the group took the latest in food technology: a complete diet of canned meat, vegetables, and fruit. Franklin was last seen in 1845, and subsequent searches established that he and almost all of his party died in 1847. In the icy Arctic, the bodies were still well preserved when found by an expedition in the 1980s. The recovered bodies were subjected to an autopsy. The autopsy revealed that they had been poisoned. The culprit was not a maniac expedition member or bad food, but the cans that had been sealed with a bead of lead. Living on nothing but canned goods month after month, the expeditioners had gradually accumulated lethal doses of lead in their bodies.

Canned goods are no longer sealed with lead, and technological improvement has made canning a thoroughly safe way to store perishable food. Consider how different our agricultural, military, exploratory, and consumer history would had been if the early setbacks in canning had led to its abandonment. Genetic engineering represents a brave new world of technology, and with it there will be some risk of mishap. The consumer must be protected by adequate regulatory safeguards, but if mistakes are made, we should learn from those errors, not abandon the whole array of techniques.

29.5 Another Clone? There's More Where This Came From

In 1996, a remarkable experiment conducted by Ian Wilmut and colleagues of the Roslin Institute in Scotland, resulted in the birth of a cloned sheep. A **clone** is an exact genetic copy of the mother and, apart from any differences resulting from her fetal environment, should be physically identical to her mother. Wilmut took a cell from the udder of a Finn Dorset sheep (a sheep with a white face), and extracted the nucleus that held a complete genetic code for this type of sheep. Mature cells are metabolically active, and for Wilmut's experiment he needed the cell in a placid state. Wilmut achieved this by starving the extracted nucleus for several days (Figure 29.6). The nucleus lapsed into a quiet state, in which its genetic machinery came to a temporary halt; a state suspected to be similar to that of a newly fertilized nucleus in a fetal cell. The next step was to remove the nucleus from the egg of a different breed of sheep, a Scottish Blackface. Without the nucleus the egg contained no genetic information. The nucleus from the Finn Dorset sheep was then inserted into the "empty" egg of the Scottish Blackface to form a transgenic cell (a cell containing the nucleus of another individual). It is important to realize that this cell was not fertilized. An egg is haploid, having half the full complement of chromosomes, and would normally receive the other half of the genetic material from sperm during fertilization. However, in this case, the nucleus came from an adult body cell, rather than a gamete, and had a full complement of chromosomes. Because the transgenic egg was genetically complete, it could develop without fertilization. Unlike a normal pregnancy in which the young has half of the genetic material from each parent, this embryo contained only genes from its Finn Dorset mother. The egg was implanted into the uterus of a Scottish Blackface sheep where it developed. After 148 days of pregnancy, a 6.6 kg, perfectly formed, Finn Dorset lamb was born. The lamb looked nothing like her surrogate Scottish Blackface mother (Figure 29.7), but was genetically identical to her Finn Dorset mother. With the exception of Mary's little lamb, "Dolly" became the first of her species to achieve stardom.

The prospect of cloning our finest cattle, combined with the ability to manipulate individual genes, is giving rise to an emerging science that has become known as "Pharming." Pharming has the potential to revolutionize the livestock industry, changing the quantity and quality of meat, eggs, and milk. Consequently, Dolly's birth saw a huge injection of money to promote translating this research achievement into increased food production.

The birth of Dolly provoked public concern that scientists were about to clone people. The first objection to human cloning is that we would produce not people, but monsters. Cloning remains a very inexact science. On the path to producing Dolly, Wilmut

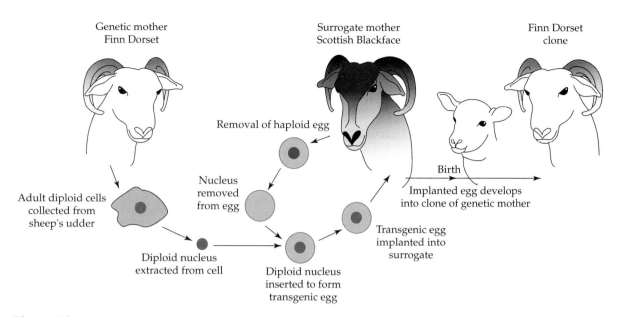

Genetic mother
Finn Dorset

Surrogate mother
Scottish Blackface

Finn Dorset
clone

Removal of haploid egg

Nucleus
removed
from egg

Adult diploid cells
collected from
sheep's udder

Birth

Implanted egg develops
into clone of genetic mother

Transgenic egg
implanted into
surrogate

Diploid nucleus
extracted from cell

Diploid nucleus
inserted to form
transgenic egg

Figure 29.6   Cloning Dolly. Cells are taken from the udder of a Finn Dorset sheep. The cells are cultured and then induced into a placid state. The nucleus is removed from the cells. An egg is taken from a Scottish Blackface sheep and its nucleus is removed, emptying the egg of its genetic material. The nucleus from the Finn Dorset cell is inserted into the "empty" egg. The egg, which has never been fertilized, is placed back into the Scottish Blackface sheep where it matures into a Finn Dorset lamb that is genetically identical to its Finn Dorset mother.

Figure 29.7   Dolly the cloned sheep with her "mother." The nucleus of an udder cell from a Finn Dorset sheep was transplanted into the cell of a Scottish Blackface sheep from which the nucleus had been removed. The transgenic cell was implanted into the womb of a Scottish Blackface sheep. The cell developed into Dolly, the first cloned mammal.

swapped 277 nuclei, but only one developed to maturity. Since Dolly was born, Holstein calves have been cloned, and in 1998, Ryozo Yanagimachi of the University of Hawaii, raised three generations of cloned mice, ending with more than 60 genetically identical individuals. However, attempts to clone animals such as pigs, which are developmentally more similar to humans than calves, mice, and lambs, have been unsuccessful. Even among species where cloning is possible, many of the embryos fail to develop, dying within the first weeks of the pregnancy. Of the few that survive to birth, many are born with such gross defects that they cannot survive. Such technical obstacles will be overcome in the next few years, but for the present, the chance of cloning a fully functional human is very small (and almost certainly beyond the ability of the few scientists who claim that it is their goal).

The second objection to human cloning is that it threatens the essence of our humanity—our individuality. Each of us is unique, with emotions, hopes, and thoughts that belong to no one else. The idea of entering a room filled with copies of ourselves is the stuff of nightmares. A science fiction movie from the 1980s, *The Boys from Brazil*, starred Gregory Peck as a mad scientist cloning a whole army of Hitlers who would ultimately conquer the world. The movie raises an important question: Would a clone be an exact copy of the parent? If we were to clone Michael Jordan, would we automatically get an equally great basketball player? Would a clone of Mother Teresa be saintly? And would Hitler's clone be a terrifying dictator? Perhaps we can allay some of these fears by realizing that the probable answer is "No" to all of these questions. The developmental environment of an individual, from the moment of conception until death, shapes personality, and even physique. The development of a fetus is heavily influenced by environmental conditions in the womb. A different womb, at a different time, would produce a different baby; therefore, history is unlikely to repeat itself even with clones.

Medical Benefits of Genetic Manipulation and Cloning

The potential medical benefits from gene manipulations and transgenic cell biology are huge. We could avert cancer by correcting a lifetime's accumulation of imperfectly copied DNA. We could cure genetically based disease by swapping a bad gene for a good one. Organ donation, with all its inherent problems, could be obviated by growing a heart, kidney, or liver, from

your own cells. A new eye could be grown and then implanted to replace one lost in an accident.

However, there are ethical questions to face. How far should we pursue a flawless body? It can certainly be argued that many of our greatest artists and thinkers have overcome tremendous physical adversity; perhaps that is not a coincidence. By removing our flaws are we making ourselves less human, less inspired? One of the most politically powerful opponents of these new branches of science comes from an unexpected quarter, the "Right to Life" movement. Where does the process of cloning begin? Is it when the nucleus is implanted into the cell? This would be a close equivalent to the moment of conception, and some of the more extreme members of the "Right to Life" movement suggest that this practice should be banned. They argue that if the cell only becomes a clone when it is implanted into a surrogate mother, was the cell not "alive" before? As a society we must decide whether the rights of a transgenic cell outweigh the potential benefits of this new branch of medicine.

Although most of us recoil in horror at the thought of cloning humans, perhaps we can imagine some circumstances where cloning a whole human would be acceptable. Take the instance of a couple who have one child, but can have no more. Their child is killed by a drunk driver. Would it be immoral to clone a cell from the body of the child and allow the parents the opportunity to renew their parenthood?

29.6   Future Choices

Perhaps every generation feels that they stand at a crossroads in history as they face the challenges of their time. Previous generations have decided the fates of nations and of empires. Your generation will face decisions that are truly global, decisions that have been postponed by modern leaders because they were hard to make, or where the data were equivocal.

The hardest decisions will be those that condemn humans to poverty and suffering. Hopefully, human population growth will slow as a result of economic and social development and no overt decision will have to be made to limit the reproductive freedom of individuals; but this possibility cannot be ruled out. Inequalities of wealth and opportunity are likely to become more exaggerated as some countries fail to control their population growth. Almost by definition, these will be the poorest countries where famine and hardship are constant companions. Is international aid the appropriate solution for overpopulation in areas facing desertification? And if so, in what form? Should countries with stable populations close their borders to further immigration thereby protecting their natural resources and quality of life? The choices that will shape our world will come from social choice rather than edict, and the environmentally literate (you) will need to voice your opinion. New balances must be found between biodiversity and economic growth, pollution and wilderness, property rights and regulation, and alternative and conventional energy use. Certainly, to an ecologist, there has never been a generation with more power to protect and conserve or to exploit and eradicate. Please choose wisely.

Summary

- The Club of Rome predicted that overexploitation of resources and pollution would lead to a crash in human population in the twenty-second century.
- On Easter Island a once vibrant society was reduced to warfare and cannibalism after they overexploited their natural resources.
- Humans do not automatically pull back from the brink of ecological disaster, they have to decide to do so.
- Changing technology makes predicting future environmental trends virtually impossible.
- Genetic manipulation has the potential to revolutionize agriculture, but economic considerations may overshadow the potential for applying the technology to LDCs.
- Genetically "improved" species are fertile, unlike hybrids.
- The ability to clone individuals or the organs of individuals opens up entire new fields of medicine and animal husbandry.
- Genetic engineering raises new ethical and moral dilemmas.
- Your generation will make global decisions that shape the future of life on Earth.

Further Readings

Bahn, P., and J. R. Flenley. 1992. *Easter Island, Earth Island*. London: Thames and Hudson.

Gallopin, G. C., and P. Raskin. 1998. Windows on the future: Global scenarios and sustainability. *Environment* 40 (3):6–11 and 26–31.

Meadows, D., J. Randers, and W. Behrens. 1972. *The Limits to Growth. Universe Books*. Report of the Club of Rome. New York.

Snow, A. A., and P. M. Palma. 1997. Commercialization of transgenic plants: Potential ecological risks. *BioScience* 47:86–96.

Wilmutt, I., A. E. Schnieke, J. McWhir, A. J. Kind, and K. H. S. Campbell. 1997. Viable offspring derived from fetal and adult mammalian cells. *Nature* 385:810–813.

Web Connections

On-line resources for this chapter are on the World Wide Web at:
http://www.prenhall.com/bush
(From this web-site, select Chapter 29.)

Glossary

abiotic factors   Environmental influences produced other than by living organisms, for example, temperature, humidity, pH, and other physical and chemical influences. Contrast *biotic factors*.

absolute poverty   The lack of sufficient income in cash or exchange items for meeting the most basic human needs for food, clothing, and shelter.

acid   Any compound that releases hydrogen ions when dissolved in water. Also, a water solution that contains a surplus of hydrogen ions.

acid deposition   Any form of acid precipitation including the fallout of dry acid particles. See *acid precipitation*.

acid dew   Acidic dew; results from water vapor condensing on dry acid fallout.

acid fallout   Molecules of acid formed from reactions involving nitrogen and sulfur oxides and water vapor settling out of the atmosphere without additional water.

acid precipitation   Includes acid rain, acid fog, acid snow, and any other form of precipitation that is more acidic than normal (i.e., less than pH 5.6). Excess acidity is derived from certain air pollutants, namely sulfur dioxide and oxides of nitrogen.

acid rain   See *acid precipitation*.

adaptation   A change in an organism's structure or habits that produces better adjustment to the environment; a genetically determined characteristic that enhances the ability of an organism to cope with its environment.

adaptive radiation   Evolutionary diversification of species derived from a common ancestor into a variety of ecological roles.

adsorption   The process of chemicals (ions or molecules) sticking to the surface of other materials.

aerenchyma   Large air-filled cells that allow rapid diffusion of oxygen within wetland plants.

aerobic   Pertaining to organisms or processes that require the presence of oxygen.

age class   Individuals in a population of a particular age.

age structure   Within a population, proportions of people who are old, middle-aged, young adults, and children.

agricultural pollution   Contamination of the environment with liquid and solid wastes from all types of farming, including pesticides, fertilizers, runoff from feedlots, erosion and dust from plowing, animal manure and carcasses, and crop residues and debris.

agroforestry   Production of tree crops in a manner similar to agriculture. Also, production of trees along with regular crops.

air   The mixture of gases, namely 78% nitrogen, 21% oxygen, and 0.035% carbon dioxide, making up the atmosphere. Water vapor and various pollutants may also be present.

air pollution   The presence of contaminants in the air in concentrations that overcome the normal dispersive ability of the air and that interfere directly or indirectly with human health, safety, or comfort or with the full use and enjoyment of property.

alga, pl. **algae**   Any of numerous kinds of photosynthetic plants that live and reproduce entirely immersed in water. Many species, the planktonic forms, exist as single or small groups of cells that float freely in the water. Other species, the "seaweeds," may be large and attached.

algal bloom   A proliferation of living algae in a lake, stream, or pond.

allele   One of two or more alternative forms of a gene located at a single point (locus) on a chromosome.

alpha diversity   Diversity of species within a habitat.

alternative farming   Farming methods designed to minimize the use of agricultural chemicals.

altruism   In an evolutionary sense, enhancement of the fitness of an unrelated individual by acts that reduce the evolutionary fitness of the altruistic individual.

ambient standards   Levels of air pollutants set by law as the maximum tolerated in order to maintain environmental and human health.

anaerobic   Pertaining to organisms or processes that occur in the absence of oxygen or to any thing (such as soil) or condition that has no oxygen.

annuals   Plants that grow from seed, flower, set seed, and die, thus completing their life cycle in a single year.

anthropogenic   Referring to pollutants and other forms of impacts on natural environments that can be traced to human activities.

aquaculture   Farming of aquatic or marine systems; rearing of organisms such as fish, algae, or shellfish under controlled conditions.

aquiclude   A layer of impermeable rock that bounds an aquifer.

aquifer   A layer of rock, sand, or gravel through which water can pass; an underground bed or layer of earth, gravel, or porous stone that contains water; the place in the ground where groundwater is naturally stored.

asexual reproduction   Reproduction that is completed by one individual. The young are genetically identical to the "mother."

assimilate   To incorporate into the natural working or functioning of the system as, for example, natural organic wastes are assimilated (broken down and incorporated) into the nutrient cycles of the ecosystem.

assimilation efficiency   The percentage of energy ingested in food that is assimilated into the protoplasm of an organism.

asymptote   A straight-line plateau forming the highest values of a data set.

atom   The fundamental unit of all elements.

autotroph   An organism that obtains energy from the sun and materials from inorganic sources. Contrast *heterotroph*.

average GNP   The gross national product of a country divided by its total population.

bacterium, pl. **bacteria**   Any of numerous kinds of microscopic organisms that exist as simple, single cells that multiply by simple division. Along with fungi, they constitute the decomposer component of ecosystems. A few species cause disease.

base   Any compound that releases hydroxyl (OH^-) ions when dissolved in water. A solution that contains a surplus of OH^2 ions.

Batesian mimicry   Resemblance of an edible (mimic) species to an unpalatable (model) species to deceive predators.

benefit-cost   Used to describe an analysis and/or comparison of the value benefits, in contrast to the costs of any particular action or project.

best management practice   Farm management practices that best serve to reduce soil and nutrient runoff and subsequent pollution.

beta diversity   Diversity of species between habitats.

biennials   Plants that flower, set seed, and die in their second year.

biodegradable   Capable of being decomposed quickly by the action of microorganisms.

biodiversity   The diversity of living things found in the natural world. The concept usually refers to the different species but also includes ecosystems and the genetic diversity within a given species.

biogeochemical cycle   The passage of a chemical element (such as nitrogen, carbon, or sulfur) from the environment into organic substances and back into the environment.

biogeography   The branch of biology that deals with the geographic distribution of plants and animals.

biological amplification   The concentration of a substance as it "moves up" the food chain from consumer to consumer.

biological control   Control of a pest population by introduction of predatory, parasitic, or disease-causing organisms.

biological oxygen demand (BOD)   The amount of oxygen that will be absorbed or "demanded" as wastes are being digested or oxidized. This includes both biological and chemical processes. Potential impacts of wastes are commonly measured in terms of the BOD.

biomass   Dry weight of living material in all or part of an organism, population, or community; commonly expressed as weight per unit area, biomass density.

biomass pyramid   Refers to the structure that is obtained when the respective biomasses of producers, herbivores, and carnivores in an ecosystem are compared. Producers have the largest biomass, followed by herbivores and then carnivores.

biome   A major terrestrial climax community; a major ecological zone or region corresponding to a climatic zone or region; a major community of plants and animals associated with a stable environmental life zone or region (for example, northern coniferous forest, Great Plains, tundra).

biosphere   The whole Earth ecosystem.

biota   All the living organisms occurring within a certain area or region.

biotic factors   Environmental influences caused by living organisms. Contrast *abiotic factors*.

biotrophic parasite   A parasite that thrives only as long as its host stays alive.

birth control   Any means, natural or artificial, that may be used to reduce the number of live births.

boreal   Occurring in the temperate and subtemperate zones of the Northern Hemisphere.

bottom-up control   A system in which the number of individuals in each trophic level is limited by primary productivity. Contrast *top-down control*.

broad-spectrum pesticides   Chemical pesticides that kill a wide range of pests. They also kill a wide range of nonpest and beneficial species; therefore, they may lead to environmental upsets and resurgences. The opposite of narrow-spectrum pesticides and biorational pesticides.

buffer   A substance that will maintain the pH of a solution by reacting with the excess acid. Limestone is a natural buffer that helps to maintain water and soil at a pH near neutral.

buffering capacity   Refers to the amount of acid that may be neutralized by a given amount of buffer.

bycatch   Organisms accidentally caught and killed during commercial fishing operation.

calorie   A fundamental unit of energy. The amount of heat required to raise the temperature of 1 gram of water 1 degree Celsius. All forms of energy can be converted to heat and measured in calories. Calories used in connection with food are kilocalories, or "big" calories (Calories), the amount of heat required to raise the temperature of 1 liter of water 1 degree Celsius.

carbon-14   A radioactive isotope of carbon (atomic weight 14) that can be used for dating organic materials.

carbon monoxide   A highly poisonous gas, the molecules of which consist of a carbon atom with one oxygen attached. Not to be confused with nonpoisonous carbon dioxide, a natural gas in the atmosphere.

carcinogen   A chemical or physical agent capable of causing cancer.

carcinogenic   Having the property of causing cancer, at least in animals and by implication in humans.

carnivore   An animal (or plant) that eats other animals. Contrast *herbivore*.

carrying capacity   The amount of animal or plant life (or industry) that can be supported indefinitely on available resources; the number of individuals that the resources of a habitat can support. Also called biological carrying capacity.

cation   A molecule with a positive electrical charge.

cation exchange capacity   The readiness (often of a clay mineral) to swap cations for other ions (particularly hydrogen ions).

cell respiration   The chemical process that occurs in all living cells, wherein organic compounds are broken down to release energy required for life processes. Higher plants and animals require oxygen for the process as well and release carbon dioxide and water as waste products, but certain microorganisms do not require oxygen. See *anaerobic respiration*.

cellulose   The organic macromolecule that is the prime constituent of plant cell walls and hence the major molecule in wood, wood products, and cotton. It is composed of glucose molecules, but since it cannot be digested by humans, its dietary value is only as fiber, bulk, or roughage.

CFCs   See *chlorofluorocarbons*.

chemical buffer   A substance that can neutralize potential changes in pH.

chemical energy   The potential energy that is contained in certain chemicals; most importantly, the energy contained in organic compounds such as food and fuels, which may be released through respiration or burning.

chlorofluorocarbons (CFCs)   Synthetic organic molecules that contain one or more of both chlorine and fluorine atoms.

chlorophyll   The green pigment in plants responsible for absorbing the light energy required for photosynthesis.

Circumpolar Current   Cold-water current that encircles Antarctica.

Clean Air Act of 1970   Amended in 1977 and 1990, this is the foundation of U.S. air pollution control efforts.

Clean Water Act of 1972   The cornerstone federal legislation addressing water pollution.

clear-cutting   The practice of cutting all trees in an area, regardless of species, size, quality, or age.

climate A general description of the average temperature and rainfall conditions of a region over the course of a year.

climax community A community capable of indefinite self-perpetuation under given climatic and edaphic conditions. The term mature community is now preferred.

clone A lineage of individuals reproduced asexually.

closed system An ecosystem in which nutrients are retained completely.

coevolution Development of genetically determined traits in two species to facilitate some interaction, usually mutually beneficial.

coexistence Occurrence of two or more species in the same habitat; usually applied to potentially competing species.

cogeneration The joint production of useful heat and electricity. For example, furnaces may be replaced with gas turbogenerators that produce electricity, while the hot exhaust still serves as a heat source. An important avenue of conservation, it effectively avoids the waste of heat that normally occurs at centralized power plants.

cohort Those members of a population that are of the same age, usually in years or generations.

command-and-control strategy The basic strategy behind most air and water pollution public policy. It involves setting limits on pollutant levels and specifying control technologies that must be used to accomplish those limits.

commensalism Relationship between two species that benefits one without affecting the other.

common law Laws built on legal procedure. Contrast *statutory law*.

commons A resource used by everyone but owned by no one, for example, fresh air.

community A group of populations of plants and animals in a given place; used in a broad sense to refer to ecological units of various sizes and degrees of integration.

competition The interaction that occurs when organisms of the same or different species use a common resource that is in short supply or when they harm one another in seeking a common resource.

competitive exclusion principle The hypothesis, based on theoretical considerations and laboratory experiments, that two or more species cannot coexist and use a single resource that is scarce, relative to demand for it.

competitor species A species adapted to long life, slow growth, and reproduction, capable of defending itself and competing for resources. Contrast *opportunist species*.

compound Any substance (gas, liquid, or solid) that is made up of two or more different kinds of atoms bonded together. Contrast *element*.

Comprehensive Environmental Response, Compensation, and Liability Act of 1980 See *Superfund*.

condensation The collecting together of molecules from the vapor state to form the liquid state, as, for example, water vapor condenses on a cold surface and forms droplets. Contrast *evaporation*.

conservation The management of a resource in such a way as to ensure that it will continue to provide maximum benefit to humans over the long run. Conservation may include various degrees of use or protection, depending on what is necessary to maintain the resource over the long run. *Energy:* Saving energy. It not only entails cutting back on use of heating, air conditioning, lighting, transportation, and so on but also entails increasing the efficiency of energy use, that is, developing and instigating means of doing the same jobs (e.g., transporting people) with less energy.

consumer An organism that obtains its energy from the organic materials of other organisms, living or dead. Contrast *producer*.

contact poison A poison that kills the organism on contact.

continental drift The movement of the continents from their original positions as parts of a common landmass to their present locations.

continental island An island that is near to and geologically part of a continent, for example, the British Isles or Trinidad.

continental shelf The shallow part of the seafloor immediately adjacent to a continent.

continental slope Ocean bed connecting continental shelf to abyssal depths.

contour farming The practice of cultivating land along the contours, across rather than up and down slopes. In combination with strip cropping, it reduces water erosion.

contraceptive Any device or drug that is designed to allow sexual intercourse but prevent pregnancies.

control group The group in an experiment that is the same as and is treated like the experimental group in every way except for the particular factor being tested. Only by comparison with a control group can one gain specific information concerning the effect of any test factor.

controlled experiment An experiment with adequate control groups. See *control group*.

convection currents Wind or water currents promoted by the fact that warming causes expansion, decreases density, and thus causes the warmer air or water to rise. Conversely, the cooler air or water sinks.

cooling tower A massive tower designed to dissipate waste heat from a power plant (or other industrial process) into the atmosphere.

coral bleaching The loss of color from a coral as it expels its zooxanthellae—usually a stress response.

coriolis effect The deflection of large bodies of water or air to the right in the Northern Hemisphere and to the left in the Southern Hemisphere.

cost-benefit ratio/benefit-cost ratio The value of the benefits to be gained from a project divided by the costs of the project. If the ratio is greater than 1, the project is economically justified; if less than 1, it is not economically justified.

cost-effective Pertaining to a project or procedure that produces economic returns or benefits that are significantly greater than the costs.

courtship Any behavioral interaction between individuals of opposite sexes that facilitates mating.

covalent bond A chemical bond between two atoms, formed by sharing a pair of electrons between the two atoms. Atoms of all organic compounds are joined by covalent bonds.

critical number The minimum number of individuals of a given species that is required to maintain a healthy, viable population of the species. If a population falls below its critical number, its extinction will almost certainly occur.

crop rotation The farming practice of planting the same field with a different crop each year to prevent nutrient depletion.

crude birth rate Number of births per 1000 individuals per year.

crude death rate Number of deaths per 1000 individuals per year.

cultural change Any modification of characteristics specific to a population that is transmitted by learning rather than by genetic mechanisms.

damages Financial reparation ordered to be paid by the accuser to the plaintiff in a common law decision.

DDT 1,1,1-trichloro-2,2-bis(p-chlorophenyl) ethane; the first of the modern chlorinated-hydrocarbon insecticides.

debt crisis Refers to the fact that many less-developed nations are so heavily in debt that they may not be able to meet their financial obligations, e.g., interest payments. Failing to meet such obligations could have severe economic impact on the entire world.

debt-for-nature swap Trading portions of the foreign debt that Third World countries owe us for their saving portions of their natural environment, e.g., tropical rain forests.

decomposers Consumers, especially bacteria and fungi, that change their organic food into mineral nutrients.

deforestation Removal of trees from an area without adequate replanting.

degrade To lower the quality or usefulness.

demographic transition The transition from a condition of high birth rate and high death rate through a period of declining death rate but continuing high birth rate, then finally to low birth rate and low death rate. This transition may result from economic development.

demography Population trends (growth, movement, development, and so on).

denitrification Enzymatic reduction by bacteria of nitrates to nitrogen gas.

density The number of individuals of a species per unit area.

density-dependent Having an influence on individuals that varies with the number of individuals per unit area in the population.

density-independent Having an influence on individuals that does not vary with the number of individuals per unit area in the population.

deoxyribonucleic acid See *DNA.*

desert A region receiving very small amounts of precipitation or where (for example, ice caps) the moisture present is unavailable to vegetation.

desertification Declining productivity of land due to mismanagement. Overgrazing and overcultivation, allowing erosion and salinization, are the major causes.

detritus feeders Organisms, such as termites, fungi, and bacteria, that obtain their nutrients and energy mainly by feeding on dead organic matter.

detritus The dead organic matter, such as fallen leaves, twigs, and other plant and animal wastes, that exists in any ecosystem.

developed countries Industrialized countries, United States, Canada, Western Europe, Japan, Australia, and New Zealand, in which the gross national product exceeds $7000 per capita.

developing countries All free-market countries in which the gross national product is less than $7000 per capita.

dimorphism The occurrence of two forms of individuals in a population.

dinoflagellate Microscopic algae. Some species have been implicated in the formation of red tides and fish kills.

dioxin A synthetic, organic chemical of the chlorinated hydrocarbon class. It is one of the most toxic compounds known to humans, having many harmful effects, including induction of cancer and birth defects, even in extremely minute concentrations. It has become a widespread environmental pollutant because of the use of certain herbicides that contain dioxin as a contaminant.

direct competition Exclusion of individuals from resources by other individuals, by aggressive behavior, or by use of toxins.

discount rate A multiplication factor to calculate the present value of costs or benefits experienced in the future.

dispersal Movement of organisms away from the place of birth or from centers of population density.

dissolved oxygen (DO) Oxygen gas molecules (O_2) dissolved in water. Fish and other aquatic organisms are dependent on dissolved oxygen for respiration. Therefore concentration of dissolved oxygen is a measure of water quality.

distribution The area or areas (taken together) where a species lives and reproduces.

DNA (deoxyribonucleic acid) The natural organic macromolecule that carries the genetic or hereditary information for virtually all organisms.

dominance The influence or control exerted by one or more species in a community as a result of their greater number, coverage, or size.

ecological Pertaining to the living environment.

ecological efficiency The percentage of energy in biomass produced by one trophic level that is incorporated into biomass by the next highest trophic level.

ecological fitness The number of a parent's young that live to reproduce; divided by two if sexual reproduction is involved.

ecological impact The total effect of an environmental change, whether natural or human.

ecological succession See *succession.*

ecology The branch of science dealing with the relationships of living things to one another and to their environment.

ecosystem A biotic community and its interaction with the abiotic environment.

ecosystem function The attributes of an ecosystem that determine the living conditions of its inhabitants and those of neighboring ecosystems.

ecosystem values The attributes that humans value in the ecosystem.

ecotone The transition zone between two diverse communities.

ecotourism The enterprises involved in promoting tourism of unusual or interesting ecological sites.

ectotherm An organism primarily heated by external sources.

edaphic Pertaining to soil.

El Niño A reduction in atmospheric pressure differentials across the Pacific that sets in motion a chain reaction of climate change. Occurs every 2–7 years.

El Niño Southern Oscillation (ENSO) Flip-flopping pressure systems in the South Pacific that trigger short-lived global changes in climate.

electrons Fundamental atomic particles that have a negative electrical charge but virtually no mass. They surround the nuclei of atoms and thus balance the positive charge of protons in the nucleus. A flow of electrons in a wire is synonymous with an electrical current.

element A substance that is made up of one and only one distinct kind of atom. Contrast *compound.*

emergent A tree with a canopy that forms above the general upper most continuous canopy.

emigration The movement of organisms out of a population.

endangered species A species whose total population is declining to relatively low levels throughout its range, such that, if the trend continues, will result in extinction.

Endangered Species Act The federal legislation that mandates protection of species and their habitats that are determined to be in danger of extinction.

endemic An organism that is native to a particular region.

endotherm An organism primarily heated by its own metabolic activity.

energy The ability to do work. Common forms of energy are light, heat, electricity, motion, and chemical bond energy inherent in compounds, such as sugar, gasoline, and other fuels.

energy resource A natural supply of energy available for use, for example, Earth's internal heat, fossil fuels, hydropower, nuclear energy, solar energy, and wind.

enrichment   The addition of nutrients to an ecosystem, for example, the addition of nitrogen to waterways by agricultural runoff.

entropy   Increasing disorder that reduces the capacity of energy to do work.

environment   All the biotic and abiotic factors that affect an individual organism at any one point in its life cycle.

environmental gradient   The possible range of a factor, such as light, temperature, or moisture.

environmental impact   Effects on the natural environment caused by human actions. Includes indirect effects through pollution, for example, as well as direct effects, such as cutting down trees.

environmental impact statement   A study of the probable environmental impacts of a development project. The National Environmental Policy Act of 1969 (NEPA) requires such studies prior to proceeding with any project receiving federal funding.

environmentalism   Embodies the assumptions that what we generally view as natural resources are products of the natural environment and can be maintained only insofar as the natural environment is maintained.

environmental movement   Refers to the upwelling of public awareness and citizen action regarding environmental issues that occurred during the 1960s.

EPA   U.S. Environmental Protection Agency. The federal agency responsible for control of all forms of pollution and other kinds of environmental degradation.

epidemiology   The study of disease in populations or groups.

epilimnion   The upper layer of water in a lake, usually warm and containing high levels of dissolved oxygen.

epiphyte   A plant that lives on another plant but uses it only for support, drawing its water and nutrients from natural runoff and the air.

equilibrium   A condition of balance, such as that between immigration and emigration or between birth rates and death rates in a population of fixed size.

erosion   The process in which soil particles are carried away by wind or water. Erosion moves the smaller soil particles first and, hence, degrades the soil to a coarser, sandier, stonier texture.

estuary   A bay open to the ocean at one end that receives freshwater from a river at the other. Hence, mixing of fresh- and saltwater occurs (the water appears brackish).

eugenics   The science of improving individuals (especially humans) physically and intellectually through controlling hereditary factors.

eukaryote   An organism in which each cell has a distinct nucleus.

eutrophic   Refers to a body of water, characterized by nutrient-rich water supporting abundant growth of algae and/or other aquatic plants at the surface. Deep water has little or no dissolved oxygen.

eutrophication   The process of becoming eutrophic.

evaporation   Molecules leaving the liquid state and entering the vapor or gaseous state as, for example, water evaporates to form water vapor.

evapotranspiration   The sum of the water lost from the land by evaporation and plant transpiration.

evolution   The theory that all species now on Earth are descended from ancestral species through a process of gradual change brought about by natural selection.

exotic species   A species introduced to a geographical area where it is not native.

exploitation efficiency   The proportion of available prey consumed by predators.

exponential growth   The steepest phase in a growth curve, where the curve is described by an equation containing a mathematical exponent.

exponential rate of increase   The rate at which a population is growing at a particular instant, expressed as a proportional increase per unit time.

externality   Any effect of a business process not included in the usual calculations of profit and loss. Pollution of air or water is an example of a *negative* externality—one that imposes a cost on society that is not paid for by the business itself.

extinct   Of a species, no longer represented by living individuals.

extractive reserves   As now established in Brazil, forest lands that are protected for native peoples and others who harvest natural products of the forests, such as latex and Brazil nuts.

famine   A severe shortage of food accompanied by a significant increase in the local or regional death rate.

fecal   Refers to the solid excretory wastes of animals. Consists of undigested material passing through the gut and bacteria that have begun to feed on it.

fecundity   The potential of an organism to produce living offspring; the number of eggs or sperm.

fertility   The number of eggs produced by a female that are fertilized.

fertilizer   Material applied to plants or soil to supply plant nutrients, most commonly nitrogen, phosphorus, and potassium but may include others. Organic fertilizer is natural organic material, such as manure, which releases nutrients as it breaks down. Inorganic fertilizer, also called chemical fertilizer, is a mixture of one or more necessary nutrients in inorganic chemical form.

First law of thermodynamics   The fact based on irrefutable observations that energy is never created or destroyed but may be converted from one form to another, e.g., electricity to light. See also *Second law of thermodynamics*.

fishery   Fish species being exploited or a limited marine area containing commercially valuable fish.

fission   Splitting or division; nuclear fission is the splitting of the nuclei of the atoms of certain elements into lighter nuclei and is accompanied by the release of relatively large amounts of energy.

fitness   See *ecological fitness*.

food chain   Figure of speech describing the dependence for food of organisms upon others, in a series beginning with plants and ending with the largest carnivores.

food web   The combination of all the feeding relationships that exist in an ecosystem.

forest   A region that, because it receives sufficient average annual precipitation (usually 75 cm [30 in.] or more), supports trees and small vegetation.

fossil fuels   Coal, oil, and natural gas, so called because they are derived from the fossil remains of ancient plant and animal life.

founder effect   The principle that a population started by a small number of colonists will contain only a fraction of the genetic variation of the parent population.

freshwater   Water that has a salt content of less than 0.01% (100 parts per million).

fundamental niche   The niche, as determined by the maximum tolerable combination of environmental conditions, of a species, in the absence of competition. Contrast *realized niche*.

fungicide   A chemical used to kill fungus.

fusion   The combination of two atoms into a single atom as a result of a collision, usually accompanied by the release of energy.

gene   A unit of genetic information.

gene flow   The exchange of genetic traits between populations by movement of individuals, gametes, or spores.

gene pool The sum total of all the genes that exist among all the individuals of a species.

generation time The time between the birth of a parent and the birth of its offspring.

genetic bottleneck Loss of genetic diversity as a result of sustained inbreeding within a small population.

genetic drift Change in gene frequency caused solely by chance, usually unidirectional and more important in small populations.

genetic engineering The artificial transfer of specific genes from one organism to another.

genotype The genetic constitution of an organism or a species in contrast to its observable characteristics. Contrast *phenotype*.

genus The taxonomic category above the species and below the family; a group of species believed to have descended from a common direct ancestor.

geometric rate of increase The factor by which the size of a population changes over a specified period. Contrast *exponential rate of increase*.

geometric series A series in which each number is obtained by multiplying the previous one by one factor, e.g., 1, 3, 9, 27, etc.

geothermal Refers to the naturally hot interior of Earth. The heat is maintained by naturally occurring nuclear reactions in Earth's interior.

geothermal energy Useful energy derived from the naturally hot interior of Earth.

glacial drift The crushed rock and clays deposited by a melting glacier or ice sheet.

glacial epoch The Pleistocene Epoch, the earlier of the two epochs comprising the Quaternary period, characterized by the extensive glaciation of regions now free from ice.

global pollutants Chemicals that are dispersed throughout the stratosphere.

global stability Ability to withstand perturbations of a large magnitude without being affected. Contrast *local stability*.

global warming The term given to the possibility that Earth's atmosphere is gradually warming because of the greenhouse effect of carbon dioxide and other gases. Global warming is thought by many to be the most serious global environmental issue facing our society. See also *greenhouse effect* and *greenhouse gases*.

glucose A simple sugar, the major product of photosynthesis. Serves as the basic building block for cellulose and starches and as the major "fuel" for the release of energy through cell respiration in both plants and animals.

GPP See *gross primary productivity*.

grassland A region with sufficient average annual precipitation (25–75 cm [10–30 inches]) to support grass but not trees.

greenhouse effect The heating effect of the atmosphere upon Earth, particularly as carbon dioxide concentration rises, caused by its ready admission of light waves but its slower release of the heat they generate on striking the ground.

greenhouse gases Gases in the atmosphere that absorb infrared energy and contribute to the air temperature. These gases are like a heat blanket and are important in insulating Earth's surface. They include carbon dioxide, water vapor, methane, nitrous oxide, and chlorofluorocarbons and other halocarbons.

Green Revolution Refers to the development and introduction of new varieties of wheat and rice (mainly) that increased yields per acre dramatically in some countries.

Gross National Product per capita The total value of all goods and services exchanged in a year in a country, divided by its population. A common indicator for the average level of development and standard of living for a country.

gross primary productivity (GPP) Production before respiration losses are subtracted; photosynthetic production for plants and metabolizable production for animals.

groundwater Water that has accumulated in the ground, completely filling and saturating all pores and spaces in rock and/or soil. Groundwater is free to move more or less readily. It is the reservoir for springs and wells and is replenished by infiltration of surface water.

guild A group of species that uses a set of resources in a similar or complimentary way.

Gulf Stream Current of warm water flowing across the North Atlantic from the Caribbean to western Europe.

habitat The sum of the environmental conditions where an organism, population, or community lives; the place where an organism normally lives; the environment in which the life needs of an organism are supplied.

Hadley cells Air masses that rise over equatorial regions, flow poleward, and descend 20°–30° north or south of the equator, before returning to the equator, close to the ground.

harem A group of females controlled by one male.

herbicide A chemical used to kill or inhibit the growth of undesired plants.

herbivore An organism that eats plants. Contrast *carnivore*.

heredity Genetic transmission of traits from parents to offspring.

heterogenic organisms Organisms that can switch between asexual and sexual reproduction.

heterotroph An organism that obtains energy and materials from other organisms. Contrast *autotroph*.

hierarchy A rank order; the pecking order, leadership, or dominance patterns among the members of a population.

home range The area in which an individual member of a population roams and carries on all of its activities.

horizon (of soil) Levels within a soil profile that differ structurally and chemically. Generally divided into A, B, C, E, and O horizons.

host The organism that furnishes food, shelter, or other benefits to an organism of another species.

host-specific When insects, fungal diseases, and other parasites are unable to attack species other than their specific host.

humidity The amount of water vapor in the air. See also *relative humidity*.

humus A dark brown or black, soft, spongy residue of organic matter that remains after the bulk of dead leaves, wood, or other organic matter has decomposed. Humus does oxidize but relatively slowly. It is extremely valuable in enhancing physical and chemical properties of soil.

hunter-gatherers Early humans who survived by hunting and gathering seeds, nuts, berries, and other edible things from the natural environment.

hybridization Breeding (crossing) of individuals from genetically different strains, populations, or, sometimes, species.

hydrarch succession The gradual sedimentation and filling of a lake until it is solid ground supporting terrestrial vegetation.

hydric soil Soil that is so frequently saturated with water that the upper layers are anaerobic.

hydrocarbons *Chemistry:* Natural or synthetic organic substances that are composed mainly of carbon and hydrogen. Crude oil, fuels from crude oil, coal, animal fats, and vegetable oils are examples. *Pollution:* A wide variety of relatively small carbon-hydrogen molecules resulting from incomplete burning of fuel and emitted into the atmosphere. See *volatile organic compounds*.

hydroelectric dam   A dam and associated reservoir used to produce electrical power by letting the high-pressure water behind the dam flow through and drive a turbogenerator.

hydroelectric power   Electrical power that is produced from hydroelectric dams or, in some cases, natural waterfalls.

hydrogen ions   Hydrogen atoms that have lost their electrons. Chemical symbol, H^1.

hydrological indicators   Clues that indicate whether a site floods or has a water table near the soil surface.

hydrological cycle   The movement of water from points of evaporation through the atmosphere, through precipitation, and through or over the ground, returning to points of evaporation.

hydroperiod   The duration, and timing of events that saturate the soil with water.

hydrothermal vent   Fissure in the sea floor through which heated water rich in dissolved chemicals flows into the ocean.

hypolimnion   The layer of cold, dense water at the bottom of a lake.

hypothesis   An educated guess concerning the cause behind an observed phenomenon that is then subjected to experimental tests to prove its accuracy or inaccuracy.

immigration   The movement of individuals into a population.

inbreeding   A mating system in which adults mate with relatives more often than would be expected by chance.

inbreeding depression   Loss of genetic diversity that results from close relatives mating. Promotes occurrence of deleterious recessive genes.

Index of Social and Economic Welfare (ISEW)   Calculation of national well-being taking into account diverse social, economic, and environmental factors.

indirect competition   Exploitation of a resource by one individual that reduces the availability of that resource to others.

industrialized agriculture   Using fertilizer, irrigation, pesticides, and energy from fossil fuels to produce large quantities of crops and livestock with minimal labor for domestic and foreign sale.

industrial smog   The grayish mixture of moisture, soot, and sulfurous compounds that occurs in local areas where industries are concentrated and coal is the primary energy source.

infant mortality   The number of babies that die before age 1.

infiltration/infiltrate   The process of water soaking down into soil as opposed to its running off the surface.

infiltration rate   Amount of water per unit time, per unit area that soaks down into a soil.

infrared radiation   Radiation of somewhat longer wavelengths than red light, the longest wavelengths of the visible spectrum. Such radiation manifests itself as heat.

infrastructure   The sewer and water systems, roadways, bridges, and other facilities that underlie the functioning of a city and that are owned, operated, and maintained by the city.

injunction   A court order to stop a practice.

innate capacity for increase *r*   Measure of the rate of increase of a population under controlled conditions (also referred to as intrinsic rate of increase).

inorganic compounds or molecules   *Classical definition:* All things such as air, water, minerals, and metals that are neither living organisms nor products uniquely produced by living things. *Chemical definition:* All chemical compounds or molecules that do not contain carbon atoms as an integral part of their molecular structure. Contrast *organic.*

insecticide   Any chemical used to kill insects.

integrated pest management (IPM)   Two or more methods of pest control carefully integrated into an overall program designed to avoid economic loss from pests. The objective is to minimize the use of environmentally hazardous, synthetic chemicals. Such chemicals may be used in IPM but only as a last resort to prevent significant economic losses.

interplanting   Rows of different crops grown together to reduce crop loss to predators.

interspecific   Between species; between individuals of different species.

intertropical convergence zone (ITCZ)   Region of warm, wet rising air where the trade winds meet.

intraspecific   Within a species; between individuals of the same species.

inversion   See *temperature inversion.*

ion   An atom or group of atoms that has lost or gained one or more electrons and, consequently, has acquired a positive or negative charge. Ions are designated by 1 or 2 superscripts following the chemical symbol.

ionic bond   The bond formed by the attraction between a positive and a negative ion.

IPM   See *integrated pest management.*

irrigation   Any method of artificially adding water to crops.

isolating mechanism   Any condition, for example, a genetically determined difference or a mechanical or geographical separation, that prevents gene flow between two populations.

isotope   A form of an element in which the atoms have more (or less) than the usual number of neutrons. Isotopes of a given element have identical chemical properties, but they differ in mass (weight) as a result of the additional (or lesser) neutrons. Many isotopes are unstable and give off radioactive radiation. See *radioactive decay, radioactive emissions,* and *radioactive materials.*

iteroparous   An organism that reproduces multiple times. Contrast *semelparous.*

kettle lake   Lake formed when a block of ice melted and left a depression in the surrounding clays.

keystone species   A species with a disproportionately large impact on the ecosystem it inhabits relative to to its biomass.

kinetic energy   The energy inherent in motion or movement, including molecular movement (heat) and movement of waves, hence, radiation including light.

La Niña   The opposite phase of the El Niño Southern Oscillation to El Niño. Trade winds are strengthened as the atmospheric pressure differential across the Pacific is heightened.

larva, pl. **larvae,** adj. **larval**   A free-living immature form that occurs in the life cycle of many organisms and that is structurally distinct from the adult. For example, caterpillars are the larval stage of moths and butterflies.

Law of conservation of energy   See *First law of thermodynamics.*

Law of conservation of matter   In chemical reactions, atoms are neither created, nor changed, nor destroyed. They are only rearranged.

LDC   Less-developed country, typically with low GNP, high population growth, low literacy, and low industrialization.

leachate   The mixture of water and materials that are leaching.

leaching   The process by which soluble materials in the soil, such as nutrients, pesticides, or contaminants, are washed into a lower layer of soil or are dissolved and carried away by water.

life cycle   The various stages of life, progressing from the adult of one generation to the adult of the next.

life table   Tabulation presenting complete data on the mortality schedule of a population.

limiting factor   A factor primarily responsible for limiting the growth and/or reproduction of an organism or a population. The limiting factor may be a physical factor, such as temperature or light, a chemical factor, such as a particular nutrient, or a biological factor, such as a competing species. The limiting factor may differ at different times and places.

limiting resource   The nutrient or substance that is in shortest supply in relation to an organism's demand for it and among all the resources it needs.

Lindemann efficiency   The proportion of energy available to a predator that is converted to the predator's body mass.

linkage   Occurrence of two loci on the same chromosome; functional linkage occurs when two loci do not segregate independently at meiosis.

litter   In an ecosystem, the natural cover of dead leaves, twigs, and other dead plant material. This natural litter is subject to rapid decomposition and recycling in the ecosystem, whereas human litter, such as bottles, cans, and plastics, is not.

littoral   Shallows of a lake or sea that allow penetration of light to the bed of the water body. Region of submerged plant growth.

loam   A solid consisting of a mixture of about 40% sand, 40% silt, and 20% clay.

locus   The site on a chromosome occupied by a specific gene.

logistic equation   A model of population growth described by a symmetrical S-shaped curve with an upper asymptote.

malnutrition   The lack of essential nutrients such as vitamins, minerals, and amino acids. Malnutrition ranges from mild to severe and life-threatening.

Malthusian theory of population   The theory of English economist and religious leader Thomas Malthus that populations increase geometrically (2, 4, 8, 16) while food supply increases arithmetically (1, 2, 3, 4), leading to the conclusion that humans are doomed to overpopulation, misery, and poverty, and that population levels will be reduced by disease, famine, and war.

mangrove   Tropical and subtropical estuarine communities composed of trees and shrubs.

market value   The price of a commodity determined by trade. Contrast *non–market value.*

masting   Synchronized fruiting of a species.

mature community   A community (of plants generally) in which the rate of change is driven by external change (climate change) rather than succession.

maximum sustainable yield (MSY)   Largest proportion of the population that can be harvested indefinitely.

MDC   More-developed country, typically with high GNP, low population growth, high literacy, and a strong economy.

meander   A loop in a river where the length of the water course is at least twice the shortest distance across the neck of the loop.

megafauna   Large animals, generally > 50 kg.

mesotrophic   An organism of moderate fertility.

metabolism   The sum of all the chemical reactions that occur in an organism.

metapopulation   Subpopulation that is part of a larger population and is connected to other subpopulations by migration.

methane   A gas, CH_4. It is the primary constituent of natural gas. It is also produced as a product of fermentation by microbes. Methane from ruminant animals is thought to be responsible for the rise in atmospheric methane, of concern because methane is one of the greenhouse gases.

microclimate   The actual conditions experienced by an organism in its particular location. Due to numerous factors such as shading, drainage, and sheltering, the microclimate may be quite distinct from the overall climate. Often used to refer to climate beneath a vegetation canopy.

Milankovitch cycles   Variations in Earth's orbit around, and geometry to, the sun.

mimicry   When one species copies another in order to gain a selective advantage.

mineral   Any hard, brittle, stonelike material that occurs naturally in the earth's crust. All consist of various combinations of positive and negative ions held together by ionic bonds. Pure minerals, or crystals, are one specific combination of elements. Common rocks are composed of mixtures of two or more minerals. See also *ore.*

mineralization   The breakdown of protein to produce ammonium.

mining of aquifers   When withdrawal of water from an aquifer exceeds the rate of recharge.

mitigation   Compensatory activity or payment to offset damage.

mitigation bank   Fund into which mitigation payments are made.

molecule   A specific union of two or more atoms held together by covalent bonds. The smallest unit of a compound that still has the characteristics of that compound.

monoculture   Cultivation of a single crop to the exclusion of all other species on a piece of land.

Montreal Protocol   An agreement made in 1987 by a large group of nations to cut back the production of chlorofluorocarbons that damage the ozone shield. A 1992 amendment called for signatory nations to cease production of CFC 11 and 12 by 1996. Less-developed nations have until 2004 to reduce CFC emissions.

moraine   Rock debris pushed in front of, or at the side of, an advancing glacier.

Müllerian mimicry   Mutual resemblance of two or more conspicuously marked, distasteful species to reinforce predator avoidance.

mutagenic   Causing mutations.

mutant   An organism with a changed characteristic resulting from a genetic change.

mutation   A change in the genetic makeup of an organism resulting from a chemical change in its DNA.

mutualism   An interaction between two species in which both benefit from the association.

mycorrhizae   The mycelia of certain fungi that grow symbiotically with the roots of some plants and provide for additional nutrient uptake.

National Environmental Policy Act of 1969 (NEPA)   Led to the establishment of the Environmental Protection Agency. It also required federal authorities to conduct environmental impact studies for projects involving federal funds or lands.

national forests   Administered by the National Forest Service, these are public forest and woodlands that are managed for multiple uses, such as logging, mineral exploitation, livestock grazing, and recreation.

national parks   Administered by the National Park Service, national parks are lands and coastal areas of great scenic, ecological, or historical importance. They are managed with the dual goals of protection and providing public access.

native species   A naturally occurring species of that community. Contrast *exotic species.*

natural (to describe a substance or factor)   Occurring or produced as a normal part of nature apart from any activity or intervention of humans. Opposite of artificial, synthetic, human-made, or caused by humans.

natural enemies   All of the predators and/or parasites that may feed on a given organism. Organisms used to control a specific pest through predation or parasitism.

natural resources As applied to natural ecosystems and species, this term indicates that they are expected to be of economic value and may be exploited. Likewise the term applies to particular segments of ecosystems such as air, water, soil, and minerals.

natural selection The natural process by which the organisms best adapted to their environment survive and those less well adapted are eliminated.

necrotrophic parasite Parasite that continues to feed on its host after host has died.

NEPA See *National Environmental Policy Act.*

neritic Marine environments that extend from the intertidal zone to the edge of the continental shelf.

net primary productivity (NPP) Production after respiration losses are subtracted.

net production efficiency Percentage of assimilated energy that is incorporated into growth and reproduction.

net reproductive rate *R* The number of offspring a female can be expected to bear during her lifetime, for species with clearly defined discrete generations.

neurotoxin A chemical that damages the nervous system of an animal.

neutron A fundamental atomic particle found in the nuclei of atoms (except hydrogen) and having one unit of atomic mass but no electrical charge.

niche The total of all the relationships that bear on how an organism copes with both biotic and abiotic factors it faces.

nitrate A salt of nitric acid; a compound containing the radical NO_3; biologically, the final form of nitrogen from the oxidation of organic nitrogen compounds.

nitric acid (HNO^3) One of the acids in acid rain. Formed by reactions between nitrogen oxides and the water vapor in the atmosphere.

nitrogen cycle The biogeochemical processes that move nitrogen from the atmosphere into and through its various organic chemical forms and back to the atmosphere.

nitrogen fixation/nitrogen fixing The process of chemically converting nitrogen gas (N^2) from the air into compounds such as nitrates (NO_3^2) or ammonia (NH_3) that can be used by plants in building amino acids and other nitrogen-containing organic molecules.

nitrous oxide A gas, N_2O. Nitrous oxide comes from biomass burning, fossil fuel burning, and the use of chemical fertilizers, and is of concern because in the troposphere it is a greenhouse gas and in the stratosphere it contributes to ozone destruction.

non–market value A commodity that has a value, but one that cannot be traded, such as liberty.

non–use value A value ascribed to something even though we may never use it. Contrast *use value.*

nonbiodegradable Not able to be consumed and/or broken down by biological organisms. Nonbiodegradable substances include plastics, aluminum, and many chemicals used in industry and agriculture. Particularly dangerous are nonbiodegradable chemicals that are also toxic and tend to accumulate in organisms.

non–point sources (of pollution) Pollution from general runoff of sediments, fertilizer, pesticides, and other materials from farms and urban areas as opposed to pollution from specific discharges. Contrast *point sources.*

nonrenewable resource A resource available in a fixed amount (such as minerals and oil), not replaceable after use.

nuclear power Electrical power that is produced by using a nuclear reactor to boil water and produce steam which, in turn, drives a turbogenerator.

nuclear winter A pronounced global cooling that would occur as a result of a large-scale nuclear conflict. It is based on theoretical projections concerning the amount of dust and smoke that would be ejected into the atmosphere and the resulting decrease in solar radiation.

nucleic acids The class of natural organic macromolecules that function in the storage and transfer of genetic information.

nucleus *Physics:* The central core of atoms, which is made up of neutrons and protons. Electrons surround the nucleus. *Biology:* The large body contained in most living cells that contains the genes or hereditary material, DNA.

nutrient *Plant:* An essential element in a particular ion or molecule that can be absorbed and used by the plant. For example, carbon, hydrogen, nitrogen, and phosphorus are essential elements; carbon dioxide, water, nitrate (NO_3^-), and phosphate (PO_4^{3-}) are respective nutrients. *Animal:* Materials such as protein, vitamins, and minerals that are required for growth, maintenance, and repair of the body and also materials such as carbohydrates that are required for energy.

nutrient cycles The repeated pathway of particular nutrients or elements from the environment through one or more organisms back to the environment. Nutrient cycles include the carbon cycle, the nitrogen cycle, the phosphorus cycle, and so on.

nutrient sink A habitat that releases fewer nutrients than it receives.

nutrient source A habitat that releases more nutrients than flow into it.

ocean conveyor belt Circulation of water throughout the world's oceans driven by subtle differences in temperature and salinity.

oligotrophic Low in nutrients and organisms; low in productivity.

omnivore An organism whose diet includes both plant and animal foods.

OPEC Organization of Petroleum Exporting Countries.

open system An ecosystem that loses nutrients (generally washed away) from nutrient cycles. Contrast *closed system.*

opportunist species An organism adapted to rapid growth and reproduction with little ability to withstand competition. Contrast *competitor species.*

optimal foraging theory Animals that hunt and feed in such a way as to maximize their net energy gain.

optimal range With respect to any particular factor or combination of factors, the maximum variation that still supports optimal or near-optimal growth of the species in question.

optimum The condition or amount of any factor or combination of factors that will produce the best result. For example, the amount of heat, light, moisture, nutrients, and so on that will produce the best growth. Either more or less than the optimum is not as good.

organic *Classic definition:* All living things and products that are uniquely produced by living things, such as wood, leather, and sugar. *Chemical definition:* All chemical compounds or molecules, natural or synthetic, that contain carbon atoms as an integral part of their structure. Contrast *inorganic.*

organic compounds or molecules Chemical compounds or molecules, the structure of which is based on bonded carbon atoms with hydrogen atoms attached.

organic gardening or farming Gardening or farming without the use of inorganic fertilizers, synthetic pesticides, or other human-made materials.

overgrazing The phenomenon of animals grazing in greater numbers than the land can support in the long run. There may be a temporary economic gain in the short run, but the grassland (or other ecosystem) is destroyed, and its ability to support life in the long run is vastly diminished.

oxidation A chemical reaction process that generally involves breakdown through combining with oxygen. Both burning and cellular respiration are examples of oxidation. In both cases, organic matter is combined with oxygen and broken down to carbon dioxide and water.

oxygen sag Reduction in dissolved oxygen close to a point source pollution, causing increased BOD.

ozone A gas, O_3, that is a pollutant in the lower atmosphere but necessary to screen out ultraviolet radiation in the upper atmosphere. May also be used for disinfecting water.

ozone hole First discovered over the Antarctic, this is a region of stratospheric air that is severely depleted of its normal levels of ozone during the Antarctic spring because of CFCs from anthropogenic sources.

ozone shield The layer of ozone gas (O_3) in the stratosphere that screens out harmful ultraviolet radiation from the sun.

paleoecology The study of past ecosystems.

PANS (peroxyacetylnitrates) A group of compounds present in photochemical smog that are extremely toxic to plants and irritating to eyes, nose, and throat membranes of humans.

parallel evolution Evolution of two separate lineages to produce species that are similar in habit and appearance.

parasite The organism that benefits in an interspecific interaction in which one organism benefits and the other is harmed.

parasitoid A specialized parasite that is usually fatal to its host and therefore might be considered a predator rather than a classic parasite.

parent material The rock material, the weathering and gradual breakdown of which is the source of the mineral portion of soil.

parthenogenesis Reproduction without fertilization by male gametes, usually involving the formation of diploid eggs whose development is initiated spontaneously.

particulate matter In air pollution, solid particles and liquid droplets, as opposed to material uniformly dispersed among the air molecules.

parts per million (ppm) A frequently used expression of concentration. It is the number of units of one substance present in a million units of another. For example, 1 g of phosphate dissolved in 1 million grams (= 1 ton) of water would be a concentration of 1 ppm.

pathogen, adj. **pathogenic** An organism, usually a microbe, that is capable of causing disease.

PCBs Polychlorinated biphenyls, a family of chemicals similar in structure to DDT.

peat Soil type largely composed of partly decomposed organic material.

pelagic Pertaining to the upper layers of the open ocean.

per capita rate of population growth *r* Rate of population growth per individual; used for species with overlapping, nondiscrete generations.

percolation The process of water seeping down through cracks and pores in soil or rock.

perennials Plants that survive and grow year after year as opposed to annuals. Contrast *annuals, biennials*.

permafrost A permanently frozen layer of soil underlying arctic biomes.

persistence Of pesticides, the length of time they remain in the soil or on crops after being applied.

pesticide A chemical used to kill pests. Pesticides are further categorized according to the pests they are designed to kill—for example, herbicides to kill plants, insecticides to kill insects, fungicides to kill fungi, and so on.

pesticide treadmill Refers to the fact that use of chemical pesticides simply creates a vicious cycle of "needing morepesticides" to overcome developing resistance and secondary outbreaks caused by the pesticide applications.

petrochemical A chemical made from petroleum (crude oil) as a basic raw material. Petrochemicals include plastics, synthetic fibers, synthetic rubber, and most other synthetic organic chemicals.

pH Scale used to designate the acidity or basicity (alkalinity) of solutions or soil. pH 7 is neutral; values decreasing from 7 indicate increasing acidity; values increasing from 7 indicate increasing basicity. Each unit from 7 indicates a 10-fold increase over the preceding unit.

phosphate An ion composed of a phosphorus atom with 4 oxygen atoms attached, PO_4^{3-}. It is an important plant nutrient. In natural waters, it is frequently the limiting factor. Therefore, additions of phosphate to natural water are frequently responsible for algal blooms.

photic zone The surface zone of a body of water that is penetrated by sunlight.

photochemical smog The brownish haze that frequently forms on otherwise clear sunny days over large cities with significant amounts of automobile traffic. It results largely from sunlight-driven chemical reactions among nitrogen oxides and hydrocarbons, both of which come primarily from auto exhausts.

photosynthesis Synthesis, with the aid of chlorophyll and with light as the energy source, of carbohydrates from carbon dioxide and water, with oxygen as a byproduct.

photosynthetic efficiency The proportion of solar energy falling on the area occupied by a plant that is converted to body mass or respired.

photosynthetic organism An organism capable of carrying on photosynthesis. Contrast *nonphotosynthetic*.

photovoltaic cells Devices that convert light energy into an electrical current.

phylum A basic taxonomic group.

phytoplankton The plant community in marine and freshwater situations, containing many species of algae and diatoms that float free in the water.

plate tectonics The study of the global-scale movements of Earth's crust that have resulted in continental drift and the creation of many geological formations.

Pleistocene A geologic epoch, characterized by alternating glacial and interglacial stages, that ended about 10,000 years ago, lasted for 2 million years.

pneumatophores Breather roots through which wetland plants absorb oxygen.

point sources (of pollution) Pollutants coming from specific points such as discharges from factory drains or outlets from sewage treatment plants. Contrast *nonpoint sources*.

pollutant Any natural or artificial substance that enters the ecosystem in such quantities that it does harm to the ecosystem; any introduced substance that makes a resource unfit for a specific purpose.

pollution   Contamination of air, water, or soil with undesirable amounts of material or heat. The material may be a natural substance, such as phosphate, in excessive quantities, or it may be in very small quantities of a synthetic compound, such as dioxin, that is exceedingly toxic.

polyculture   The growing of two or more species together. Contrast *monoculture*.

polygamy   A mating system in which a male pairs with more than one female at a time (polygyny) or a female pairs with more than one male (polyandry).

polygyny   See *polygamy*.

population   A group within a single species, the individuals of which can and do freely interbreed. Breeding between populations of the same species is less common because of differences in location, culture, nationality, and so on.

population density   The number of individuals of a species per unit of area.

population momentum   Refers to the fact that a rapidly growing human population may be expected to grow for 50–60 years after replacement fertility (2.1) is reached because of increasing numbers entering reproductive age.

potential energy   The ability to do work that is stored in some chemical or physical state. For example, gasoline is a form of potential energy; the ability to do work is stored in the chemical state and is released as the fuel is burned in an engine.

ppm   See *parts per million*.

precipitation   Any form of moisture condensing in the air and depositing on the ground.

predator–prey relationship   A feeding relationship existing between two kinds of animals. The predator is the animal feeding on the prey. Such relationships are frequently instrumental in controlling populations of herbivores. Sometimes also used to refer to a plant–herbivore association.

prey   In a feeding relationship, the animal that is killed and eaten by another.

primary consumer   An organism such as a rabbit or deer that feeds more or less exclusively on green plants or their products, such as seeds and nuts. Synonym: *herbivore*.

primary producers   In an ecosystem, those organisms, mostly green plants, that use light energy to construct their organic constituents from inorganic compounds.

primary productivity   Production by autotrophs, normally green plants.

primary succession   See *succession*.

productivity   Amount of energy (or material) formed by an individual, population, or community in a specific time period. See also *gross primary productivity* and *net primary productivity*.

prokaryote   An organism that lacks a distinct nucleus.

protein   The class of organic macromolecules that is the major structural component of all animal tissues and that functions as enzymes in both plants and animals.

proton   Fundamental atomic particle with a positive charge, found in the nuclei of atoms. The number of protons present equals the atomic number and is distinct for each element.

protozoan, pl. protozoa   Any of a large group of microscopic organisms that consist of a single, relatively large complex cell or in some cases small groups of cells. All have some means of movement. Amoebae and paramecia are examples.

punctuated equilibrium   A model that depicts macroevolution as taking place in the form of short periods of rapid speciation, alternating with long periods of relative stasis.

qualitative   Refers to issues involving purity.

qualitative defense   Chemical defense that is highly toxic even in small quantities.

quantitative   Refers to issues involving numbers.

quantitative defense   Chemical defense that acts in proportion to its consumption.

radon   A radioactive gas produced by natural processes in the earth that is known to seep into buildings. It can be a major hazard within homes and is a known carcinogen.

rain shadow   The low-rainfall region that exists on the leeward (downwind) side of mountain ranges. It is the result of the mountain range's causing the precipitation of moisture on the windward side.

***r* and *K* strategies**   Alternative expressions of traits that determine fertility and survivorship to favor rapid population growth at low population density *r* or competitive ability at densities near the carrying capacity *K*.

range of tolerance   The range of conditions within which an organism or population can survive and reproduce, for example, the range from the highest to lowest temperature that can be tolerated. Within the range of tolerance is the optimum, or best, condition.

realized niche   The portion of a species' fundamental niche that it can occupy in the presence of competition.

recharge area   With reference to groundwater, the area over which infiltration and resupply of a given aquifer occurs.

recruitment   Addition, by reproduction, of new individuals to a population.

recycling   The practice of processing wastes and using them as raw material for new products as, for example, scrap iron is remelted and made into new iron products. Compare *reuse*.

Red Queen hypothesis   The idea, named for the character in Lewis Carroll's *Through the Looking Glass*, that a species must continually evolve just to keep pace with environmental change and with other species, let alone to get ahead in the coevolutionary struggle.

red tide   Algal bloom, probably caused by high nutrient availability.

reducing   The loss of oxygen from a molecule (to the environment or another reaction surface).

refugia   Isolated areas of habitat in an otherwise unsuitable environment.

regulatory takings   Financial loss to an individual or company, incurred as a result of changes in government regulation.

relative humidity   The percentage of moisture in the air compared with how much the air can hold at the given temperature.

relaxation   Loss of species diversity following habitat isolation.

renewable energy   Energy sources, namely solar, wind, and geothermal, that will not be depleted by use.

renewable resources   Biological resources, such as trees, that may be renewed by reproduction and regrowth. Conservation to prevent overcutting and protection of the environment are still required, however.

reproductive rate   The rate at which offspring, eggs, or seed are produced.

reserves (of a mineral resource)   The amount remaining in the earth that can be exploited, using current technologies and at current prices. Usually given as *proven reserves*, those that have been positively identified, and *estimated reserves*, those that have not yet been discovered but that are presumed to exist.

resource   A substance or object that is required by an organism for normal maintenance, growth, and reproduction and that is finite.

Resources Conservation and Recovery Act of 1976 (RCRA)   Cornerstone legislation to control indiscriminate land disposal of hazardous wastes.

respiration The complex series of chemical reactions in all organisms by which stored energy is made available for use and produces carbon dioxide and water as byproducts.

restoration ecology The branch of ecology devoted to restoring degraded and altered ecosystems to their natural state.

runoff Water entering rivers, lakes, reservoirs, or the ocean from land surfaces.

Sagebrush Rebellion Movement to oppose environmental legislation affecting grazing and mining rights on federally owned lands in the western United States. The start of the Wise-Use movement.

sand Mineral particles 0.2–2.0 mm in diameter.

satellite accounts Calculations of a nation's environmental, social, and economic well-being. A broader view than traditional measures, such as Gross National Product.

scientific fact An observation regarding some object or phenomenon that has been repeated and confirmed by the scientific community.

scientific method The methodology by which scientific information is generated. Involves observing, formulating specific questions and hypotheses regarding the question's answer, then testing the hypotheses through experimentation.

Second law of thermodynamics The fact based on irrefutable observations that in every energy conversion (e.g., electricity to light) some of the energy is converted to heat, and some heat always escapes from the system because it always moves toward a cooler place. Therefore, in every energy conversion, a portion of energy is lost. Therefore, since energy cannot be created (first law), the functioning of any system requires an energy input.

secondary consumer An organism, such as a fox or coyote, that feeds more or less exclusively on other animals that feed on plants.

secondary succession See *succession.*

sediment Soil particles, namely sand, silt, and clay, being carried by flowing water. The same material after it has been deposited. Because of different rates of settling, deposits are generally pure sand, silt, or clay.

sedimentation The filling in of lakes, reservoirs, stream channels, and so on with soil particles, mainly sand and silt. The soil particles come from erosion, which generally results from poor or inadequate soil conservation practices in connection with agriculture, mining, and/or development. Also called siltation.

seed bank Live seeds that are lying dormant within the soil. If conditions change, seeds could germinate.

selection pressure (evolution) An environmental factor that results in individuals with certain traits, which are not the norm for the population, surviving and reproducing more than the rest of the population. This results in a shift in the genetic makeup of the population. For example, the presence of insecticides provides a selection pressure toward increasing pesticide resistance in the pest population.

semelparous Organism that reproduces once then dies. Contrast *iteroparous.*

sewage The organic waste and wastewater generated by residential and commercial establishments.

sex ratio The proportion of males to females in a population.

sexual reproduction Reproduction involving segregation and recombination of chromosomes such that the offspring bear some combination of genetic traits from the parents. Contrast with *asexual reproduction,* where all the offspring are exact genetic copies of the parent.

sexual selection Selection by one sex for specific characteristics in individuals of the opposite sex, usually exercised through courtship behavior.

shadow pricing In cost-benefit analysis, a technique used to estimate benefits where normal economic analysis is ineffective. For example, people could be asked how much they might be willing to pay monthly to achieve some improvement in their environment.

silt Soil particles between the size of sand particles and clay particles, namely particles 0.002–0.2 mm in diameter.

sinkhole A large hole resulting from the collapse of an underground cavern.

sink population A subpopulation with an average fitness < 1.

slash-and-burn agriculture The practice, commonly exercised throughout tropical regions, of cutting and burning jungle vegetation to make room for agriculture. The process is highly destructive of soil humus and may lead to rapid degradation of soil.

smog See *industrial smog* and *photochemical smog.*

soil A dynamic system involving three components: mineral particles, detritus, and soil organisms feeding on the detritus.

soil erosion The loss of soil caused by particles being carried away by wind and/or water.

soil fertility Soil's ability to support plant growth, often refers specifically to the presence of proper amounts of nutrients. The soil's ability to fulfill all the other needs of plants is also involved.

soil profile A description of the different, naturally formed layers within a soil.

soil respiration Gas exchange resulting from the decomposition of organic material in the soil.

soil structure The phenomenon of soil particles (sand, silt, and clay) being loosely stuck together to form larger clumps and aggregates, generally with considerable air spaces in between. Structure enhances infiltration and aeration. It develops as organisms feed on organic matter in and on the soil.

soil texture The relative size of the mineral particles that make up the soil. Generally defined in terms of the sand, silt, and clay content.

solar cells See *photovoltaic cells.*

solar energy Energy derived from the sun. Includes direct solar energy (the use of sunlight directly for heating and/or production of electricity) and indirect solar energy (the use of wind, which results from the solar heating of the atmosphere, and biological materials such as wood, which result from photosynthesis).

solubility The degree to which a substance will dissolve and enter into solution.

solution A mixture of molecules (or ions) of one material in another. Most commonly, molecules of air and/or ions of various minerals in water. For example, seawater contains salt in solution.

source population A subpopulation with an average fitness > 1.

specialization (evolution) The phenomenon of species becoming increasingly adapted to exploit one particular niche but thereby becoming less able to exploit other niches.

speciation The evolutionary process whereby populations of a single species separate and, through being exposed to differing forces of natural selection, gradually develop into distinct species.

species (both singular and plural) Organisms forming a natural population or group of populations that transmit specific characteristics from parent to offspring; a group of organisms reproductively isolated from similar organisms and usually producing infertile offspring when crossed with them.

species diversity   The number of species in a community or region.

stability   Absence of fluctuations in populations; ability to withstand perturbations without large changes in composition.

standards (air or water quality)   Set by the federal or state governments, the maximum levels of various pollutants that are to be legally tolerated. If levels go above the standards, various actions may be taken.

standing biomass   That portion of the biomass (population) that is not available for consumption but that must be conserved to maintain the productive potential of the population; the total number of living organisms.

standing crop   See *standing biomass.*

starvation   The failure to get enough calories to meet energy needs over a prolonged period of time. It results in a wasting away of body tissues until death occurs.

statutory law   Laws enacted by local, state, or federal government, also by international bodies, such as the European Parliament.

sterilant   Chemical that kills all organisms.

stomata, sing. stoma   Microscopic pores in leaves, mostly in the undersurface, that allow the passage of carbon dioxide and oxygen into and out of the leaf and that also permit the loss of water vapor from the leaf.

stratosphere   The layer of Earth's atmosphere between 16 and 47 km above the surface that contains the ozone shield. This layer mixes only slowly; pollutants that enter may remain for long periods of time. See also *troposphere.*

stream bank erosion   The eating away of the stream bank by flowing water. It is greatly aggravated by flooding.

strip cropping   The practice of growing crops in strips, alternating with grass (hay) at right angles to prevailing winds or slopes in order to reduce erosion.

subsistence farming   Farming that meets the food needs of the farmers and their families but little more. It involves hand labor and is practiced extensively in the LDCs.

succession (ecological or natural)   The gradual, or sometimes rapid, change in species that occupy a given area, with some species invading and becoming more numerous while others decline in population and disappear. Succession is caused by a change in one or more abiotic or biotic factors benefiting some species but at the expense of others. *Primary succession:* The gradual establishment, through a series of stages, of a mature ecosystem in an area that has not been occupied before, e.g., a rock face. *Secondary succession:* The reestablishment, through a series of stages, of a mature ecosystem in an area from which it was previously cleared.

sulfur dioxide (SO$_2$)   A major air pollutant, this toxic gas is formed as a result of burning sulfur. The major sources are burning coal (coal-burning power plants) that contains some sulfur and refining metal ores (smelters) that contain sulfur.

sulfuric acid (H$_2$SO$_4$)   The major constituent of acid precipitation. Formed as a result of sulfur dioxide emissions reacting with water vapor in the atmosphere. See also *sulfur dioxide.*

Superfund   The popular name for the Comprehensive Environmental Response, Compensation, and Liability Act of 1980. This cornerstone legislation provides the mechanism and funding for the cleanup of potentially dangerous hazardous waste sites; aimed at protecting groundwater from leakage from abandoned chemical waste sites.

surface water   Includes all bodies of water, lakes, rivers, ponds, and so on that are on the surface of the earth, in contrast to groundwater, which lies below the surface.

survivorship   The proportion of individuals in a specified group alive at the beginning of an interval (e.g., a five-year period) who survive to the end of the period.

suspension   With reference to materials contained in or being carried by water, materials kept "afloat" only by the water's agitation that settle as the water becomes quiet.

sustainable agriculture   Agriculture that maintains the integrity of soil and water resources such that it can be continued indefinitely. Much of modern agriculture is depleting these resources and is, hence, not sustainable.

sustainable development   Development that provides people with a better life without sacrificing or depleting resources or causing environmental impacts that will undercut future generations.

symbiosis   In a broad sense, the living together of two or more organisms of different species; in a narrow sense, synonymous with mutualism.

synergism/synergistic effect/synergistic interactions   The phenomenon in which two factors acting together have a very much greater effect than would be indicated by the sum of their effects separately—as, for example, modest doses of certain drugs in combination with modest doses of alcohol may be fatal.

taxonomy   The science of identification and classification of organisms according to evolutionary relationships.

tectonic plates   Huge slabs of rock that make up Earth's crust; can be continental or oceanic.

temperature inversion   The weather phenomenon in which a layer of warm air overlies cooler air near the ground and prevents the rising and dispersion of air pollutants.

territory   Any area defined by one or more individuals and protected against intrusion by others of the same or different species.

theory   A conceptual formulation that provides a rational explanation or framework for numerous related observations.

thermal equator   The latitude receiving maximum solar energy at a particular time.

thermal pollution   The addition of abnormal and undesirable amounts of heat to air or water. It is most significant with respect to discharging waste heat from electric generating plants, especially nuclear power plants, into bodies of water.

thermocline   The thin transitional zone in a lake or ocean that separates the epilimnion from the hypolimnion.

thermoregulation   The control of body temperature.

threatened species   A species not yet endangered but whose population levels are low enough to cause concern.

time lag   Delay in response to a change.

top-down control   An ecosystem in which the number of organisms in any trophic level is determined by the abundance of predators. Compare *bottom-up control.*

topsoil   The top few inches of soil, rich in organic matter and plant nutrients; the A and O horizons together.

total fertility rate   The average number of children that would be born alive to each woman during her total reproductive years, assuming she follows the average fertility at each age.

trace elements   Those essential elements that are needed in only very small or trace amounts.

trade-offs   The things that are given up to get or achieve something else that is valued.

trait (genetic)   Any physical or behavioral characteristic or talent that an individual is born with.

transpiration   The loss of water vapor from plants. Evaporation of water from cells within the leaves and exiting through stomata.

treeline   The uppermost altitudinal limit of forest vegetation.

trophic level   Feeding level with respect to the primary source of energy. Green plants are at the first trophic level, primary consumers at the second, secondary consumers at the third, and so on.

troposphere   The layer of Earth's atmosphere from the surface to about 16 km in altitude. The *tropopause* is the boundary between the troposphere and the stratosphere above. This layer is well mixed and is the site and source of our weather, as well as the primary recipient of air pollutants. See also *stratosphere.*

tundra   Level or undulating treeless land, characteristic of Arctic regions and high altitudes, having permanently frozen subsoil.

turbine   A sophisticated "paddle wheel" driven at a very high speed by steam, water, or exhaust gases from combustion.

ultraviolet (UV) radiation   Radiation similar to light but with wavelengths slightly shorter than violet light and with more energy. The greater energy causes it to severely burn and otherwise damage biological tissues.

undernutrition   A form of hunger where there is a lack of adequate food energy as measured in calories. Starvation is the most severe form of undernutrition.

upwelling   The process whereby, as a result of wind patterns, nutrient-rich bottom waters rise to the surface of the ocean.

urban   Pertaining to city areas.

use value   A value for a commodity based upon our desire and ability to use it. Contrast *non–use value.*

vector   An organism (often an insect) that transmits a pathogen (for example, a virus, bacterium, protozoan, or fungus) acquired from one host to another.

verification   The checking of observations and/or experiments by others to determine their accuracy.

vicariant speciation   The evolution of a new species, as a result of an existing population being split into two or more parts, each of which is genetically isolated.

vitamin   A specific organic molecule that is required by the body in small amounts but that cannot be made by the body and therefore must be present in the diet.

volatile organic compounds (VOCs)   A category of major air pollutants present in the air in vapor state, including fragments of hydrocarbon fuels from incomplete combustion and evaporated organic compounds, such as paint solvents, gasoline, and cleaning solutions. They are major factors in the formation of photochemical smog.

water table   The upper surface of groundwater. It rises and falls with the amount of groundwater.

water vapor   Water molecules in the gaseous state.

water-holding capacity (soil)   The property of a soil relating to its ability to hold water so that it will be available to plants.

waterlogged soil   Soil in which all airspaces are filled with water.

waterlogging   The total saturation of soil with water, which results in plant roots not being able to get air and dying as a result.

watershed   The total land area that drains directly or indirectly into a particular stream or river. The watershed is generally named for the stream or river into which it drains.

watershed management   Controlling activities in a watershed to maintain water quality in the reservoir.

weather   Short-term, highly variable conditions determining day-to-day local atmospheric conditions.

weathering   The gradual breakdown of rock into smaller and smaller particles, caused by natural chemical, physical, and biological factors.

wetlands   Areas that are at least seasonally wet and are flooded at more or less regular intervals. Especially, marshy areas along coasts that are regularly flooded by tide.

wilderness   Undisturbed area, as it was before human-made changes.

wildlife management   Humans attempting to provide balance for wildlife in whatever ways possible, after the natural balance has been upset.

wind farms   Arrays of numerous, modestly sized wind turbines for the purpose of producing electrical power.

wise-use movement   A supposed grass-roots movement opposing environmental regulation. Heavily supported by mining, forestry, chemical, and automotive industries.

work (physics)   Any change in motion or state of matter. Any such change requires the expenditure of energy.

world view   A set of assumptions that a person holds, regarding the world and how it works.

zones of stress   Regions where a species finds conditions tolerable but suboptimal; where a species survives but under stress.

zooplankton   The animal community, predominantly single-celled animals, that floats free in marine and freshwater environments, moving passively with the currents.

zooxanthellae   Dinoflagellates and other algae that lives as symbionts within coral.

References

Abramovitz, J. N. 1996. Imperiled Waters, Impoverished Future: The Decline of Freshwater Ecosystems. *Worldwatch Paper* 128. Washington, DC: Worldwatch Institute.

Ahearne, J. F. 1993. The future of nuclear power. *American Scientist* 81:24–35.

Alvarez, L. W., W. Alvarez, F. Asaro, and H. V. Michel. 1980. Extraterrestrial Cause for the Cretaceous-Tertiary Extinction. *Science* 208:1095–1108.

Anderson, R. M., et al. 1991. The spread of HIV-1 in Africa: sexual contact patterns and the predicted demographic impact of AIDS. *Nature* 352:581–589.

Anderson, R. N. 1998. Oil production in the 21st century. *Scientific American* 278 (3):86–91.

Bahn, P., and J. R. Flenley. 1992. *Easter Island, Earth Island.* London: Thames and Hudson.

Batt, B. D. J., et al. 1989. The Use of Prairie Potholes by North American Ducks. In *Northern Prairie Wetlands*, A. G. Van Der Valk, ed. Ames, IA: Iowa State University Press. Pp. 204–227.

Beamish, R. 1976. Acidification of lakes in Canada by acid precipitation and resulting effects on fishes. *Water Air Soil Pollution* 6:5401–5514.

Begon, M., J. L. Harper, and C. R. Townsend. 1986. *Ecology.* Cambridge, MA: Blackwell Scientific Publications.

Berger, J. 1990. *Environmental Restoration.* Covelo, CA: Earth Island Press.

Berner, R. A., and D. F. Canfield. 1989. A new model for atmospheric oxygen over phanerozoic time. *American Journal of Science* 289:333–361.

Berner, R.A., and G.P. Landis, 1988. Gas bubbles in fossil amber as possible indicators of the major gas composition of ancient air. *Science* 239:1406–1409.

Billings, W. D. 1938. The structure and development of old field short-leaf pine stands and certain associated physical properties of the soil. *Ecological Monographs* 8:437–499.

Blum, U., and W. W. Heck. 1980. Effects of acute ozone exposures on snap bean *(Phaseolus vulgaris)* at various stages of its life cycle. *Enviromental and Experimental Botany* 20:73–86.

Boersma, P. D. 1977. An ecological and behavioral study of the Galapagos penguin. *The Living Bird* 15:43–93.

Bolin, B. 1998. The Kyoto negotiations on climate change: a science perspective. *Science* 279:330–331.

Bond, G., et al. 1997. A pervasive millennial-scale cycle in North Atlantic Holocene and glacial climates. *Science* 278:1257–1266.

Bormann, F. H. 1978. An inseparable linkage: conservation of natural ecosystems and the conservation of fossil energy. *Bioscience* 26:754–759.

Brasseur, G. 1987. The endangered ozone layer. *Environment* 29 (January/February).

Braun, E. L. 1955. The phytogeography of unglaciated Eastern United States and its interpretation. *Botanical Review* 21:297–375.

Brodie, E. D., Jr., and E. D. Brodie III. 1980. Differential avoidance of mimetic salamanders by free-ranging birds. *Science* 208:181–182.

Broecker, W. S., and G. H. Denton. 1989. The role of ocean-atmosphere reorganizations in glacial cycles. *Geochimica et Cosmochimica Acta* 53:2465–2501.

Brown, K. S. 1996. Do disease cycles follow changes in weather? *Bioscience* 46:479–481.

Brown, L. E. 1997. Facing the prospect of food scarcity. In *State of the World 1997*, L. E. Brown, C. Flavin, and H. French, eds. Washington, DC: Worldwatch Institute. Pp. 23–41.

Brown, L. R., H. Kane, and D. M. Roodman. 1994. *Vital Signs.* Washington, DC: Worldwatch Institute.

Buchmann, S. L., and G. P. Nabhan. 1996. *The Forgotten Pollinators.* Washington, DC: Island Press.

Bush, M. B., and P. A. Colinvaux. 1994. A paleoecological perspective of tropical forest disturbance: records from Darien, Panama. *Ecology* 75:1761–1768.

Bush, M. B., and R. J. Whittaker. 1991. Krakatau: colonization patterns and hierarchies. *Journal of Biogeography* 18:341–356.

Bush, M. B., and R. J. Whittaker. 1993. Non-equilibration in island theory of Krakatau. *Journal of Biogeography* 20:453–457.

Bush, M. B. 1994. Amazonian speciation: A necessarily complex model. *Journal of Biogeography* 21:5–17.

Bush, M. B. 1996. Amazonian conservation in a changing world. *Biological Conservation* 76:219–228.

Bush, M. B., D. R. Piperno, and P. A. Colinvaux. 1989. A 6000 year old history of Amazonian maize cultivation. *Nature* 340:302–303.

Buvinic, M., and M. Lycette. 1988. Women, poverty, and development in the third world. In *Strengthening the Poor: What Have We Learned*, J.P. Lewis, ed. Pp. 153–154. Washington, DC: Overseas Development Council and Transaction Books.

Caldwell, M. 1994. Blessed with resistance. *Discover Magazine* 1 (1):46–8.

Caldwell, M. 1994. Prokaryotes at the gate. *Discover Magazine* 14 (8):45–50.

Campbell, C. J., and J. H. Laherrere. 1998. The end of cheap oil. *Scientific American* 278 (3):78–83.

Carson, R. 1963. *Silent Spring.* New York: Houghton-Mifflin.

Christopherson, R. 1994. *Geosystems: An Introduction to Physical Geography.* Upper Saddle River, NJ: Prentice Hall.

Christopherson, R. 1995. *Elemental Geosystems.* Upper Saddle River, NJ: Prentice Hall.

Clements, F. E., 1916. *Plant Succession: An Analysis of the Development of Vegetation.* Publication 242. Washington, DC: Carnegie Institute of Washington.

Cleveland, D. A., D. Soleri, and S. E. Smith. 1994. Do folk crop varieties have a role in sustainable agriculture? *Bioscience* 44:740–751.

Cline, W. 1992. *The Economics of Global Warming.* Washington, DC: Institute of International Economics.

Cole, K. L. 1990. Late Quaternary vegetation gradients through the Grand Canyon. In *Packrat Middens—the Last 40,000 Years of Biotic Change*, J. Betancourt, T. R. Van Devender, and P. S. Martin, eds. Tucson, Arizona: University of Arizona Press. Pp. 240–258.

Colinvaux, P. A. 1975. *Why Big Fierce Animals are Rare.* Princeton, NJ: Princeton University Press.

Colinvaux, P. A. 1980. The *Fates of Nations.* Englewood Cliffs, NJ: Simon and Schuster.

Colinvaux, P. A. 1987. Amazonian diversity in the light of the paleoecological record. *Quaternary Science Reviews* 6:93–114.

Colinvaux, P. A. 1992. *Ecology 2.* New York: John Wiley and Sons.

Colinvaux, P. A., and B. D. Barnett. 1979. Lindemann and the ecological efficiency of wolves. *American Naturalist* 114:707–718.

Collins, M. 1990. The *Last Rainforests: A World Conservation Atlas.* New York: Oxford University Press.

Committee on Global Change. 1988. *Toward an Understanding of Global Change.* Washington, DC: National Academy Press.

Connell, J. H. 1978. Diversity in tropical rain forests and coral reefs. *Science* 199:1302–1310.

Conway, G. R., and E. B. Barbier. 1990. *After the Green Revolution.* London: Earthscan Publications, Ltd.

Cook, R. E. 1969. Variation in species diversity of North American birds. *Systematic Zoology* 18:63–84.

Costanza, R., et al. 1997. The value of the world's ecosystem services and natural capital. *Nature* 387:253–260.

Cox, C. B., I. N. Healey, and P. B. Moore. 1976. *Biogeography: An Ecological and Evolutionary Approach.* 2nd ed. Oxford, England: Blackwell Scientific Publications.

Cunningham, W. P., and B. W. Saigo. *Environmental Science.* 3rd ed. Dubuque, IA: William Brown Publishers.

Currie, D. J., and V. Paquin. 1987. Large scale biogeographical patterns of species richness. *Nature* 329:326–327.

Daly, M., and M. Wilson. 1983. *Sex, Evolution and Behavior.* 2nd ed. Belmont, CA: Wadsworth.

Darwin, C. R. 1859. *On the Origin of Species by Natural Selection.* London: Murray.

Davies, N. D. 1977. *Behavioural Ecology.* Oxford, U.K: Blackwell Science.

Davis, M. B., 1982. Holocene vegetational history of the eastern United States. In *Late Quaternary Environments of the United States,* H. E. Wright, Jr., ed. Minneapolis, MN: University of Minnesota Press. Pp. 166–181.

Davis, M. B. and C. Zabinski. 1992. Changes in geographic range resulting from greenhouse warming: effects on biodiversity in forests. In *Global Warming and Biodiversity,* R. Peters and T. Lovejoy, eds. New Haven, CT: Yale University Press. Pp. 297–308.

De Duve, C. 1995. The Beginnings of Life on Earth. *American Scientist* 83:428–437.

De Candolle, A. P. A. 1874. *Constitution Dans le Régne Végétal de Groupes Physiologiques Aplicables à la Géographie Ancienne et Moderne.* Geneva, Switzerland: Archives des Sciences Physiques et Naturelles.

Deevey, E. S., Jr. 1947. Life tables for natural populations of animals. *Quarterly Review of Biology* 22:283–314.

Deevey, E. S., Jr. 1951. Life in the depths of a pond. *Scientific American* 185 (10):68–72.

De la Mere, W. K. 1997. Abrupt mid-twentieth-century decline in Antarctic sea-ice extent from whaling records. *Science* 389:57–60.

Diamond, J. M., and E. Mayr. 1976. Species-area relation for birds of the Solomon Archipelago. *Proceedings of the National Academy of Science, USA. Natn. Acad. Sci., USA* 73:262–266.

Dippery, J. K., et al. 1995. Effects of low and elevated CO_2 on C3 and C4 annuals. *Oecologia* 101:13–30.

Dixon, J. A. 1994. *Environment Dissemination Notes,* February 4, 1994. Washington, DC: The World Bank.

Dodd, A. P. 1940. The *Biological Campaign against Prickly Pear.* Commonwealth Prickly Pear Board. Government Printer. Brisbane, Australia.

Dunning, J. A., and W. W. Heck. 1977. Response of bean and tobacco to ozone effect and light intensity, temperature and reference humidity. *Journal of the Air Pollution Control Association* 27:882–886.

Durning, A. 1992. *How Much Is Enough? The Consumer Society and the Future of the Earth.* New York: W.W. Norton.

Ehrlich, A., and P. Ehrlich. 1992. The population explosion: why isn't everyone as scared as we are? *The Amicus Journal* 12 (4):22–29.

Ehrlich, P. R., and J. Roughgarden. 1987. *The Science of Ecology.* New York: Macmillan.

Ehrlich, P. R., and E. O. Wilson. 1985. Biodiversity studies: science and policy. *Science* 253:758–762.

Elton, C., and M. Nicholson. 1942. The ten year cycle in numbers of the lynx in Canada. *Journal of Animal Ecology.* 11:215–244.

Emiliani, C., and N. J. Shackleton. 1974. The Brunhes epoch: isotopic paleotemperatures and geochronology. *Science* 183:511–514.

Emmel, T. C., and G. T. Austin. 1990. The tropical rain forest butterfly fauna of Rondonia, Brazil: species diversity and conservation. *Tropical Lepidoptera* 1:1–12.

Epstein, P. R., et al. 1998. Biological and physical signs of climate change: focus on mosquito-borne diseases. *Bulletin of the American Meteorological Society.* 79:409–417.

Erwin, D., J. Valentine, and D. Jablonski. 1997. The origin of animal body plans. *American Scientist* 85:126–137.

Erwin, T. L. 1982. Tropical Forests: their richness in coleoptera and other arthropod species. *Coleopterist's Bulletin* 36:74–75.

Erwin, T. L. 1991. How many species are there?: revisited. *Conservation Biology* 5:330–333.

Ewald, P. W. 1993. The evolution of virulence. *Scientific American* 268 (4):86–93.

Ewald, P. W. 1994. *Evolution of Infectious Diseases.* Oxford: Oxford University Press.

Fisher, S. W., et al. 1991. Molluscidal activity of potassium to the zebra mussel *Dreissenia polymorpha:* toxicity and mode of action. *Aquatic Toxicology* 20:219–234.

Flather, C. H., M. S. Knowles, and I. A. Kendall. 1998. Threatened and endangered species geography. *Bioscience* 48:365–376.

Flavin, C., and N. Lenssen. 1994. Powering the Future: Blueprint for a Sustainable Electricity Industry. *Worldwatch Paper* 119. Washington, DC: Worldwatch Institute.

Flenley, J. R., and S. M. King. 1984. Late Quaternary pollen records from Easter Island. *Nature* 307:47–50.

Flenley, J. R., et al. 1991. The late Quaternary vegetational and climatic history of Easter Island. *Journal of Quaternary Science* 6:85–115.

Food and Agriculture Organization of the United Nations (FAO). 1993. Agrostat PC, on Diskette. Rome, FAO.

Foreman, D. 1995. Making the world safe for large predators and other living organisms. *Sierra Club Magazine* October:52.

Freed, S. A., and R. S. Freed. 1985. One son is no sons. *Natural History* 10:12–13.

French, H. F. 1997. Learning from the ozone experience. In *State of the World 1997,* L. Brown, C. Flavin, and H. French, eds. Washington, DC: Worldwatch Institute. Pp. 151–171.

Garson, P. J., W. K. Pleszcynska, and C. H. Holm. 1981. The "polygyny threshold" model: a reassessment. *Canadian Journal of Zoology* 59:902–910.

Gause, G. F. 1934. The *Struggle for Existence.* Baltimore: Williams and Wilkins.

Gentry, A. H. 1988. Tree species richness of upper Amazonian forests. *Proceedings of the National Academy of Science* 85:156–159.

Gilman, K. 1982. Nature conservation in wetlands: two small fen basins in western Britain. In *Ecosystem Dynamics in Freshwater Wetlands and Shallow Water Bodies*, D. D. Logofet and N. K. Luckyanov, eds. Volume 1. SCOPE and UNEP Workshop, Moscow, USSR. Center of International Projects. Pp. 290–310.

Gould, S. J. 1977. *Ever Since Darwin*. London: Penguin Books.

Gould, S. J. 1989. *Wonderful Life*. New York: W.W. Norton.

Grant, B., D. F. Owen, and C. A. Clarke. 1995. Decline of melanic moths. *Nature* 373:565.

Grant, P. R. 1986. *Ecology and Evolution of Darwin's Finches*. Princeton University Press: Princeton.

Greenland Ice-Core Project Members. 1993. Climate instability during the last interglacial period. Recorded in the GRIP Ice Core. *Nature* 364:203–207.

Grime, J. P., and D. W. Jeffery. 1965. Seedling establishment in vertical gradients of sunlight. *Journal of Ecology* 53:621–642.

Haffer, J. 1969. Speciation in Amazonian forest birds. *Science* 165:131–137.

Haines-Young, R., and J. Petch. 1986. *Physical Geography: Its Nature and Methods*. London: Harper and Row.

Halpert, M. S., et al. 1993. Fifth Annual Climate Assessment. Washington, DC: U.S. Department of Commerce.

Hardin, G. 1960. The competitive exclusion principle. *Science* 131:1292–1297.

Hardin, G. 1986. Cultural carrying capacity: a biological approach to human problems. *Bioscience* 36:599–606.

Hardin, G. 1968. The tragedy of the commons. *Science* 162:1243–1248.

Heck, W. W., and A. S. Heagle. 1970. Measurements of photochemical air pollution with a sensitive monitoring plant. *Journal of the Air Pollution Control Association* 20:97–99.

Heinselman, M. L. 1963. Forest sites, bog processes and peatland types in the glacial Lake Agassiz region, Minnesota. *Ecological Monographs* 33:327–374.

Hensley, M. M., and J. R. Cope. 1951. Further data on the removal and repopulation of the breeding birds in a spruce-fir forest community. *Auk* 68:483–493.

Hillis-Colinvaux, L. 1980. The genus *Halimeda:* primary producer in coral reefs. *Advances in Marine Biology* 17:1–327.

Hoffman, A. 1995. An uneasy rebirth at Love Canal. *Environment* 37 (2):4–9 and 25–31.

Horn, H. S. 1971. The adaptive geometry of trees. *Monographs in Population Biology 3*. Princeton, NJ: Princeton University Press.

Horne, A. J., and C. R. Goldman. 1994. *Limnology*. New York: McGraw-Hill.

Hutchinson, G. E. 1959. Homage to Santa Rosalia, or why are there so many kinds of animals? *American Naturalist* 93:145–159.

Hutterman, A., et al. 1994. Effects of Acid Rain on Forest Processes. Wiley Series in Ecological and Applied Microbiology. John Wiley and Sons: New York.

Jacobson, J. L. 1991. India's Misconceived Family Plan. *Worldwatch* (6):18–25.

Janzen, D. H. 1970. Herbivores and the number of tree species in tropical forests. *American Naturalist* 104:501–528.

Johnston, D. W., and E. P. Odum. 1956. Breeding bird populations in relation to plant succession of the piedmont of Georgia. *Ecology* 37:50–62.

Junk, W. J. 1982. Amazonian floodplains: their ecology, present and potential use. In *Ecosystem Dynamics in Freshwater Wetlands and Shallow Water Bodies*, D. D. Logofet and N. K. Luckyanov, eds. Volume 1. SCOPE and UNEP Workshop, Moscow, USSR. Center of International Projects. Pp. 98–126.

Karl, T. R., N. Nicholls, and J. Gregory. 1997. The coming climate. *Scientific American*. 276 (5):78–83

Karr, J. R. 1982. Avian extinction on Barro Colorado Island, Panama: a reassessment. *American Naturalist* 119:220–239.

Keith, L. B. 1983. Role of food in hare population cycles. *Oikos* 40:385–395.

Kerr, R. A. 1996. A new dawn for sun-climate links? *Science* 271:1360–1361.

Kerr, R. A. 1997. A new driver for the Atlantic's moods and Europe's weather? *Science* 275:754–755.

Kowalok, M. E. 1993. Common threads: research lessons from acid rain, ozone depletion, and global warming. *Environment* (6):12–20 and 35–38.

Krebs, C. J. 1978. A review of the Chitty hypothesis of population regulation. *Canadian Journal of Zoology* 56:2463–2480.

Kricher, J. C. 1997. *A Neotropical Companion*, 2nd ed. Princeton, NJ: Princeton University Press.

Kubasek, N. K., and G. S. Silverman. 1994. *Environmental Law*. Upper Saddle River, NJ: Prentice Hall.

Lack, D. 1947. *Darwin's Finches*. Cambridge, U.K.: Cambridge University Press.

Lack, D. 1954. The *Natural Regulation of Animal Numbers*. New York: Oxford University Press.

Larick, R., and R. L. Ciochon. 1996. The African emergence and early Asian dispersals of the genus *Homo. American Scientist* 84:538–551.

Lawrence, D. C., M. Leighton, and D. R. Peart. 1995. Availability and extraction of forest products in managed and primary forest around a Dayak village in west Kalimantan, Indonesia. *Conservation Biology* 9:76–88.

Laws, E. A. 1993. *Aquatic Pollution*. New York: John Wiley and Sons.

Leigh, E. G., Jr., A. S. Rand, and D. M. Windsor, eds. 1982. The *Ecology of a Tropical Forest; Seasonal Rhythms and Long-Term Changes*. Washington, DC: Smithsonian Institution Press.

Lewin, R., 1983. What killed the giant mammals? *Science* 221:1036–1037.

Likens, G. E., et al. 1970. Effects of cutting and herbicide treatment on nutrient budgets in the Hubbard Brook Watershed Ecosystem. *Ecological Monographs* 40:23–47.

Lindeman, R. L. 1942. The trophic dynamic aspects of ecology. *Ecology* 23:399–418.

Lodge, D. M. 1993. Species invasions and deletions: community effects and responses to climate and habitat change. In *Biotic Interactions and Global Change*. P. M. Karieva, J. G. Kingsolver, and R. B. Huey, eds. Sunderland, MA: Sinauer Associates, Inc. Pp. 367–387.

Lovejoy, T. E., et al. 1986. Edge and other effects of isolation on Amazon forest fragments. In *Conservation Biology: The Science of Scarcity and Diversity*, M. Soulé, ed. Sunderland, MA: Sinauer Associates, Inc. Pp. 257–285.

Luoma, J. R. 1992. Eco-backlash. *Wildlife Conservation* (6):27–36.

Mabberley, D. J. 1992. *Tropical Rain Forest Ecology*. 2nd ed. London: Blackie.

MacArthur, R. H., and E. O. Wilson. 1967. The theory of island biogeography. *Princeton Monographs in Population Biology 1*. Princeton, NJ: Princeton University Press.

MacArthur, R. H., and R. Levins. 1967. The limiting similarity, convergence and divergence of coexisting species. *American Naturalist* 101:337–385.

MacArthur, R. H. 1958. Population ecology of some warblers of northeastern coniferous forests. *Ecology* 39:599–619.

MacKenzie, F. 1997. *Our Changing Planet*. Upper Saddle River, NJ: Prentice Hall.

MacLulich, D. A. 1937. Fluctuations in the number of the varying hare *(Lepus americanus)*. *University of Toronto Studies in Biology Series* 43:1–136.

Makarewicz, J. C., and P. Bertram. 1991. Evidence for the restoration of the Lake Erie ecosystem. *Bioscience* 41:216–223.

Malthus, R. T. 1798. *An Essay on the Principle of Population as It Affects the Future Improvement of Society.* London: Johnson.

Martens, P. 1998. *Health and Climate Change: Modelling the Impacts of Global Warming and Ozone Depletion.* London: Earthscan Publications, Ltd.

Martin, P. S., and R. G. Klein, eds. 1984. *Quaternary Extinctions: A Prehistoric Revolution.* Tucson, AZ: University of Tucson Press.

May, R. M. 1983. Parasitic infections as regulators of animal populations. *American Scientist* 71:36–45.

Mazza, P. 1998. The invisible hand: is global warming driving El Niño? *Sierra* 83 (3):68–72 and 92–95.

McCormick, J. 1998. Acid pollution: the international community's continuing struggle. *Environment* 40 (3):17–20 and 40–45.

McCullough, D. 1977. *The Path between the Seas.* New York: Simon and Schuster.

McKnight, T. 1996. *Physical Geography.* Upper Saddle River, NJ: Prentice Hall.

Meadows, D., J. Randers, and W. Behrens. 1972. *The Limits to Growth.* Report of the Club of Rome. New York: Universe Books.

Meffe, G. K., and C. R. Carroll. 1994. *Principles of Conservation Biology.* Sunderland, MA: Sinauer Associates, Inc.

Melillo, J. M., J. D. Aber, and J. F. Muratore. 1982. Nitrogen and lignin control of hardwood leaf litter decomposition dynamics. *Ecology* 63:621–626.

Meyer, W. B. 1995. Past and present land use and land cover in the USA. *Consequences: the Nature and Implications of Environmental Change* 1:24–33.

Miller, G. T., Jr. 1996. *Living in the Environment.* 9th ed. Belmont, CA: Wadsworth Publishing Company.

Minnich, R. A. 1983. Fire mosaics in southern California and northern Baja California. *Science* 219:1287–1294.

Mitsch, W. J., et al. 1979. *Environmental Observations of a Riparian Ecosystem During Flood Season.* Research Report 142, Illinois University, Water Resources Center, Urbana, Illinois.

Mitsch, W. J., and J. G. Gosselinck. 1993. *Wetlands.* 2nd ed. New York: Van Nostrand Reinhold.

Morgan, M. D., and P. M. Pauley. 1996. *Meteorology: The Atmosphere and the Science of Weather.* 5th ed. Upper Saddle River, NJ: Prentice Hall.

Morris, S. C. 1989. Burgess Shale Faunas and the Cambrian Explosion. *Science* 246:339–346.

Morse, S. S. 1993. *Emerging Viruses.* Oxford: Oxford University Press.

Mosier, A. R., et al. 1998. Assessing and mitigating N_2O emissions from agricultural soils. *Climate Change* 40:7–38.

Munton, D. 1998. Dispelling the myths of the acid rain story. *Environment* 40 (6):4–7 and 27–34.

Murie, A. 1944. The Wolves of Mount McKinley. *Fauna of the National Parks of the United States,* Fauna Series 5. Washington, DC: Government Printing Office.

Myers, N. 1989. *Deforestation Rates in Tropical Countries and Their Climate Implications.* Washington, DC: Friends of the Earth.

Nebel, A., and E. Wright. 1995. *Introduction to Environmental Science.* Upper Saddle River, NJ: Prentice Hall.

Newmark, W. D. 1987. Animal species vanishing from U.S. parks. *International Wildlife* 17:1–25.

Nilsson, A. 1992. *Greenhouse Earth.* New York: John Wiley and Sons.

Nixon, S. W., and C. A. Oviatt. 1973. Ecology of a New England salt marsh. *Ecological Monographs* 43:463–498.

Noss, R. F., and L. D. Harris. 1986. Nodes networks and MUMs: preserving diversity at all scales. *Environmental Management* 10:299–309.

Noss, R. F. 1991. Landscape connectivity: different functions at different scales. In *Landscape Linkages and Biodiversity,* W. E. Hudson, ed. Washington, DC: Island Press. Pp. 27–39.

O'Neill, R. V., et al. 1986. A hierarchical concept of ecosystems. *Monographs in Population Biology 23.* Princeton, NJ: Princeton University Press.

Odum, E. P. 1971. *Fundamentals of Ecology.* 3rd ed. Philadelphia, PA: Saunders.

Ogden, J. G., III. 1966. Forest history of Ohio: radiocarbon dates and pollen stratigraphy of Silver Lake, Ohio. *Ohio Journal of Science* 66:387–400.

Paine, R. T. 1966. Food web complexity and species diversity. *American Naturalist* 100:65–75.

Pauly, D., et al. 1998. Fishing down marine food webs. *Science* 279:860–863.

Pearce, D. W. 1993. *Economic Values and the Natural World.* Cambridge, MA: Massachusetts Institute of Technology Press.

Pearce, D. W. 1998. Auditing the earth. *Environment* 40 (2):23–28.

Pearce, D. W., and R. K. Turner. 1990. *Economics of Natural Resources and the Environment.* Baltimore, MD: Johns Hopkins University Press.

Pearl, R. 1928. *The Rate of Living.* New York: Knopf.

Pechmann, J. H. K., et al. 1991. Declining amphibian populations: the problem of separating human impacts from natural fluctuations. *Science* 253:892–895.

Percival, R. V., et al. 1992. *Environmental Regulation.* Boston: Little, Brown and Co.

Perrins, C. M. 1965. Population fluctuations and clutch size in the great tit, *Parus major. Journal of Animal Ecology* 34:601–47.

Peters, C. M., A. H. Gentry, and R. O. Mendelsohn. 1989. Valuation of an Amazonian rainforest. *Nature* 339:655–656.

Pianka, E. R. 1970. On *r*- and *K*-Selection. *American Naturalist* 104:592–597.

Pielou, E. C. 1991. *After the Ice-Age: The Return of Life to Glaciated North America.* Chicago: University of Chicago Press.

Pinkerton, S. D., and P. R. Abramson. 1997. Condoms and the prevention of Aids. *American Scientist* 85:364–373.

Platt, A. E. 1996. Infecting Ourselves: How Environmental and Social Disruptions Trigger Disease. *Worldwatch Paper* 129. Washington, DC: Worldwatch Institute.

Por, F. D. 1992. *Sooretama—the Atlantic rain forest of Brazil.* The Hague, the Netherlands: SPB Academic Publishing.

Portney, P. R. 1998. Counting the cost: the growing role of economics in environmental decisionmaking. *Environment* 40 (2):15–18 and 36–38.

Pough, F. H. 1988. Mimicry of vertebrates: Are the rules different? *American Naturalist* 131(Suppl.): S67–S102.

Primack, R. B. 1993. *Essentials of Conservation Biology.* Sunderland, MA: Sinauer Associates, Inc.

Pulliam, H. R. 1988. Sources, sinks, and population regulation. *American Naturalist* 132:757–785.

Raup, D. M. 1991. *Extinction: Bad Genes or Bad Luck.* New York: W.W. Norton.

Redford, K. H., and A. MacLean Stearman. 1993. Forest dwelling native Amazonians and the conservation of biodiversity: interests in common or in collision? *Conservation Biology* 7:248–255.

Repetto, R. 1986. *World Enough and Time: Successful Strategies for Resource Management.* New Haven, CT: Yale University Press.

Repetto, R. 1991. What Progress in Natural Resource Accounting? Testimony before the Joint Economic Committee, U.S. Govt. Oct. 10, 1991.

Retallack, S. 1997. God protect us from those who "protect the skies." *The Ecologist* 27 (5):188–191.

Rice, R. E., R. E. Gullison, and J. W. Reid. 1997. Can sustainable management save tropical forests? *Scientific American* 276 (4):44–49.

Richards, O. W., and N. Waloff. 1954. Studies on the biology and population dynamics of British grasshoppers. *Anit-Locust Bulletin* 17:1–182.

Richardson, C. 1994. Ecological functions and human values in wetlands: a framework for assessing forestry impacts. *Wetlands* 14:1–9.

Richardson, D. J., T. W. Sasek, and E. A. Fendick. 1992. Implications of physiological responses to chronic air pollution for forest decline in the southeastern United States. *Environmental Toxicology and Chemisty* 11:1105–1114.

Robinson, S. K. 1992. Population dynamics of breeding neotropical migrants to a fragmented Illinois landscape. In *Ecology and Conservation of Neotropical Migrant Landbirds,* J. M. Hagan and D. W. Johnston, eds. Washington, DC: Smithsonian Institution Press. Pp. 408–418.

Rogers, R. A. 1995. *The Oceans Are Emptying.* New York: Black Rose Books.

Rubenstein, J. M. 1996. *The Cultural Landscape: An Introduction To Human Geography.* Upper Saddle River, NJ: Prentice Hall.

Sagan, C. 1980. *Cosmos.* New York: Ballantine Books.

Schaefer, M. D. 1957. Some considerations of population dynamics and economics in relation to the management of marine fisheries. *Journal of the Fisheries Research Board of Canada* 14:669–681.

Schidlowski, M. The atmosphere. In *The Handbook of Environmental Chemistry,* Vol. 1, Pt. A, The Natural Environment and the Biochemical Cycles. O. Hutzinger ed. New York: Springer-Verlag. Pp. 1–16.

Schlesinger, W. H. 1991. *Biogeochemistry: An Analysis of Global Change.* San Diego: Academic Press.

Schneider, D. 1997. The rising seas. *Scientific American* 276 (3):112–117.

Schonewald-Cox, C. M. 1983. Conclusions: guidelines to management: a beginning attempt. In *Genetics and Conservation: A Reference for Managing Wild Animal and Plant Populations,* C. M. Schonewald-Cox, S. M. Chambers, B. Macbryde, and L. Thomas, eds. Menlo Park, CA: Benjamin Cummings. Pp. 414–445.

Schulz, J. P. 1960. *Ecological Studies on Rainforests in Northern Sunnam.* Amsterdam: North-Holland.

Shafer, C. L. 1990. *Nature Reserves. Island Theory and Conservation Practice.* Washington, DC: Smithsonian Institution Press.

Shaffer, M. L. 1981. Minimum population sizes for species conservation. *Bioscience* 31:131–134.

Simon, J. 1981. *The Ultimate Resource.* Princeton, NJ: Princeton University Press.

Slud, P. 1976. Geographic and climatic relationships of avifaunas with special reference to comparative distribution in the neotropics. *Smithsonian Contributions to Zoology* 212:1–149.

Smith, A. P., and D. G. Quinn. 1996. Patterns and causes of extinction and decline in Australian conilurine rodents. *Biological Conservation* 77:243–267.

Smith, V. K. 1993. Nonmarket valuation of environmental resources: an interpretive appraisal. *Land Economics* 69:1–26.

Snow, A. A., and P. M. Palma. 1997. Commercialization of transgenic plants: potential ecological risks. *Bioscience* 47:86–96.

Soulé, M. 1985. What is conservation biology? *Bioscience* 35:727–734.

Stafford, J. 1971. Heron Populations of England and Wales 1928–1970. *Bird Study* 18:218–221.

Starr, H. F. 1982. *Ecosystem Hierarchies.* Princeton, NJ: Princeton University Press.

Stein, B. A., and S. R. Flack. 1997. *1997 Species Report Card: The State of U.S. Plants and Animals.* Arlington, VA: The Nature Conservancy.

Stewart, R. E., and J. W. Aldrich. 1951. Removal and repopulation of breeding birds in a spruce-fir forest community. *Auk* 68:471–482.

Stiling, P. 1996. *Ecology: Theory and Applications.* Upper Saddle River, NJ: Prentice Hall.

Strain, B. 1987. Measurements of tropospheric ozone concentration between March and December 1987 in Duke Forest, North Carolina. Unpublished Data. Duke University.

Tallis, J. H. 1992. *Plant Community History: Long-term Changes in Plant Distribution and Diversity.* New York: Chapman & Hall.

Tamm, C. O. 1989. Comparative and experimental approaches to the study of acid deposition effects on soils as substrate for forest growth. *Ambio* 18:181–191.

Taylor, R. H. 1979. How the Macquarie Island Parakeet became extinct. *New Zealand Journal of Ecology* 2:42–45.

Terborgh, J. 1988. The big things that run the world—a sequel to E.O. Wilson. *Conservation Biology* 2:403–403.

Terborgh, J. W. 1974. Preservation of natural diversity: the problem of extinction prone species. *Bioscience* 24:715–722.

Terborgh, J. 1992. *Where Have All the Songbirds Gone?* Princeton, NJ: Princeton University Press.

Tietenberg, T. 1997. *Environmental and Natural Resource Economics.* 4th ed. New York: HarperCollins Publishers, Inc.

Tilman, D. 1988. Dynamics and structure of plant communities. *Princeton Monographs in Population Biology* 26. Princeton, NJ: Princeton University Press.

Tissue, D. T., R. B. Thomas, and B. R. Strain. 1996. Growth and photosynthesis of loblolly pine *(Pinus taeda)* after exposure to elevated CO_2 for 19 months in the field. *Tree Physiology* 16:49–59.

Townsend, C. R., A. G. Hildrew, and J. E. Francis. 1983. Community structure in some southern English streams: the influence of physicochemical factors. *Freshwater Biology* 13:521–544.

Transeau, E. N. 1926. The accumulation of energy by plants. *Ohio Journal of Science* 26:1–10.

United Nations (U.N.) Population Division. 1992. *Long-Range World Population Projections: Two Centuries of World Population Growth, 1950–2150.* New York: U.N.

United Nations Children's Fund (UNICEF). 1993. The *State of the World's Children, 1993.* New York: UNICEF.

United Nations Educational, Scientific and Cultural Organization (UNESCO). 1990. *Compendium of Statistics on Illiteracy, 1990.* Paris: UNESCO.

Valiela, I. 1995. *Marine Ecological Processes.* New York: Springer-Verlag.

Van-Valen, L. 1973. A new evolutionary law. *Evolutionary Theory* 1:1–30.

Vig, N. J., and M. E. Kraft. 1994. *Environmental Policy in the 1990s.* 2nd ed. Washington, DC: CQ Press.

Vitousek, P. M., et al. 1986. Human appropriation of the products of photosynthesis. *Bioscience* 36:368–373.

Webb, T. III. 1988. Eastern North America. In *Vegetational History,* B. Huntley and T. Webb III, eds. Dordrecht, the Netherlands: Kluwer. Pp. 385–414.

Weber, P. 1994. Net loss: fish, jobs and the marine environment. *Worldwatch Paper* 120. Washington, DC: Worldwatch Institute.

Weber, P. 1994. Resistance to pesticides growing. In *Vital Signs,* L. R. Brown, H. Kane, and D. M. Roodman, eds. Washington, DC: Worldwatch Institute. Pp. 92–93.

Wegener, A. 1929. *The Origin of Continents and Oceans.* New York: Dover Publications. (Translation of German Original).

Weiner, J. 1994. *The Beak of the Finch.* New York and Cape, London: Knopf.

Werner, E. E., and D. J. Hall. 1974. Optimal foraging and the size selection of prey by the bluegill sunfish, *Lepomis macrochirus. Ecology* 55:1042–1052.

Werner, E. E., et al. 1983. An experimental test of the effects of predation risk on habitat use in fish. *Ecology* 64:1540–1550.

White, T. D., G. Suwa, and B. Asfaw. 1994. *Australopithecus ramidus,* a new species of early hominid from Aramis, Ethiopia. *Nature* 371:306–312.

Whittaker, R. H., and G. E. Likens. 1973. Carbon in the biota. In *Carbon and the Biosphere,* G. M. Woodwell and E. V. Pecan, eds. Washington, DC: National Technical Information Service. Pp. 281–302.

Whittaker, R. H. 1975. *Communities and Ecosystems.* 2nd ed. New York: Macmillan.

Whittaker, R. J., M. B. Bush, and K. Richards. 1989. Plant recolonization and vegetation succession on the Krakatau Islands, Indonesia. *Ecological Monographs* 59:59–123.

Whitten, A. J., et al. The Effect of Mechanized Exploration on the Tropical Rain Forest Exosystems of the Island of Pagai Selatan, Western Sumatra. Sumatra Barat Report No. 293. Sunderland, MA: Sinauer Associates, Inc.

Wilcove, D. S. 1985. Nest predation in forest tracts and the decline of migratory songbirds. *Ecology* 66:1211–1214.

Wilcove, D. S., et al. 1998. Quantifying threats to imperiled species in the United States. *Bioscience* 48:607–615.

Williamson, M. 1981. *Island Populations.* Oxford: Oxford University Press.

Willis, E. O. 1974. Populations and local extinctions of birds on Barro Colorado Island, Panama. *Ecological Monographs* 44:153–169.

Willis, E. O. 1979. The composition of avian communities in remanent woodlots in Southern Brazil. *Papéis Avulsos De Zoologia,* 33:1–25.

Wilmutt, I., et al. 1997. Viable offspring derived from fetal and adult mammalian cells. *Nature* 385:810–813.

Wilson, E. O. 1987. The arboreal ant fauna of Peruvian Amazon forests: a first assessment. *Biotropica* 19:245–251.

Wilson, E. O. 1992. *The Diversity of Life.* New York: W.W. Norton.

Woldegabriel, G., et al. 1994. Ecological and temporal placement of early Pliocene hominids at Aramis, Ethiopia. *Nature* 371:330–333.

Wood, B. J. 1971. Development of integrated control programs for pests of tropical perennial crops in Malaysia. In *A.A.A.S. Symposium on Biological Control,* C. B. Huffaker, ed. Boston: Plenum Press. Pp. 422–457.

World Bank. 1993. *Investing in Health.* World Development Report, 1993. New York: Oxford University Press.

World Bank. 1994. *Making Development Sustainable.* Washington, DC: The World Bank.

World Commission on Environment and Development (Brundtland Commission). 1987. *Our Common Future.* London: Oxford University Press.

World Resources Institute. 1989. *Wasting Assets: Natural Resources in the National Income Accounts.* Washington, DC: World Resources Institute.

World Resources Institute. 1994. *World Resources 1994–95.* Oxford: Oxford University Press.

World Resources Institute. 1998. *World Resources 1997–98.* Oxford: Oxford University Press.

Photo Credits

Index

489

Hantavirus, 416, 418, 420, 422–23
Hardin, Garret, 164
Hardwoods, 216–17
  tropical, demand for, 297–98
Harem, 154–55
Harmful algal bloom, 93–94
Hazardous waste, 453–54
HCFC (hydrochlorofluorocarbon), 355, 358
Heat, transport by oceans, 48–49
Heat budget, atmospheric, 362
Heavy metals, 273, 278
Henderson Island, 460
Herbicide, 281
Herbicide-resistant crops, 462
Herbivores, 36, 74, 330
  pyramids of power, 176–77
Heterogonic life-style, 146–47
Heterotrophs, 30
Heterozygote, 144, 328
HFC (hydrofluorocarbon), 358, 368
Histosol, 84
HIV. *See* Human immunodeficiency virus
Holdfast, 96
Home range, 330
Homelands, 336
Homes, energy efficiency of, 407–8
Hominids, 4, 242–43
Homozygote, 144
Hong Kong, algal blooms in, 94
Hubbard Brook, New Hampshire, 78–80, 85, 105, 213
Hudson Bay, 58
Hudson's Bay Trading Company, pelts submitted to, 180–81
Human carrying capacity, 245–46
Human cloning, 464–66
Human disease, 410–25
  AIDS, 421–25
  climate change and, 418–20
  defensive mechanisms against, 410–11
  drug-resistant pathogens, 413–14
  ecological perspective on, 417–20
  hantavirus outbreaks, 422–23
  human population density and, 417
  introduced by European settlers to New World, 293, 416, 420
  virulence of pathogens, 414–16
Human immunodeficiency virus (HIV), 421–45
  transmission of, 423–24
  ultravirulent strains of, 423
Human population. *See also* Demographic transition
  demographics, 246–47
  density and disease, 417
  population policy, 247
Human population growth, 306, 457, 466
  agriculture and, 244, 260–64
  asset or bane, 250–51
  consumerism and, 255
  exponential, 244–46
  first age of reproduction and, 251–52
  limiting, 250–54
  possibilities for, 245–46
Humans
  development of agriculture, 244
  entry into North America, 202–3
  evolution of, 241–43
  from hunter-gatherer to urban dweller, 243–44
  nutritional requirements of, 258–60
Humpback whale, 364
Humpbacked whale, 169
Hunter-gatherer, 243–44, 319
Hunter-gatherer-fisher, 231
Hunting, 333, 341. *See also* Predator
Hunting pressure, 4–5
  extinction of ice-age mammals by humans, 202–4
  on Passenger Pigeon, 204, 206
  sexual dimorphism and, 156
Hutchinson, G. Evelyn, 136–38

Hybridization, genetic manipulation vs., 461–62
Hydrarch succession, 99–100
Hydric soil, 123
Hydrochlorofluorocarbon. *See* HCFC
Hydroelectric power, 403
Hydrofluorocarbon. *See* HFC
Hydrologic indicator, wetland, 124
Hydrologic regulator, wetlands as, 112–13
Hydrological cycle, 89–90
Hydroperiod, of wetland, 109–10
Hydrothermal vent, 9
Hyena, 178
Hyperpredation, 320
Hypolimnion, 101–4, 273–74
Hypothesis, developing and testing of, 2–3

Ice, 101–2
Ice age, 49–50, 72, 82, 97, 190–96, 202, 362
  orbital wobble and, 191–93
  in tropics, 237–39
Ice cap, 191
  melting of, 375
Ice sheet, 49–50, 97, 193, 364
Immigration, 132, 248, 310–12, 315
Immune system, 411
  in AIDS, 422
Imported oil, 397–98
in situ conservation, 344
Inbreeding, 317
Inbreeding depression, 328, 330, 343
Inceptisol, 84
Index of sustainable economic welfare (ISEW), 440
Indicator species, 312–13
Indigenous peoples
  herbalists, 303–4
  in nature reserves, 335–36
  rights of, 304–5
Indonesia
  economic growth in, 439–40
  rice crop in, 287
Indoor air pollution, 349
Industrial chemicals, 278
Industrial culture, 248
Industrial revolution, 397
Industrial smog, 350
Industry, impact of environmental regulation on, 436–38
Infant mortality, 251, 439
Infant survival, 251
Infection cycle, 419

Infiltration, 96
Infiltration rate, 85, 295
Influenza, 416, 421
Infrastructure, 397, 408
Injunction, 444
Inner buffer area, 333
Inorganic fertilizer, 261
Insecticide, 281
Insects
  biodiversity of, 227
  counting in tropical forest, 227–28
  effect of acid deposition on, 386, 388
  pest pressure in tropics, 236–37
Inshore fishery, 159
Instar, 129
Integrated pest management (IPM), 270, 286–88
"Intellectual colonialism," 304
Interbreeding population, 27
Interglacial period, 49–50, 190, 193–94, 363–64, 366
Intermediate disturbance hypothesis, 220–21
International debt, debt-for-nature swaps, 299–300
Interplanting, 270
Interpretative center, 333
Interspecific competition, 133–34, 139, 220
Intertidal zone, 67–68
Intertropical convergence zone (ITCZ), 44–45, 375

Intraspecific competition, 132
Intravenous drug users, 423–25
Introductions. *See* Exotic species
Invertebrates, effect of acid deposition on, 386, 388–89
Ion, 380
IPM. *See* Integrated pest management
Iridium, 28
Irrigation, 263–64, 417
ISEW. *See* Index of sustainable economic welfare
Island biogeography, 309–12
  equilibrium theory of, 311–12
  metapopulation theory vs., 317
Islands
  in archipelago, 309
  distance from mainland, 309–10
  evolution on, 140
  habitat diversity on, 309–10
  immigration to and extinction on, 310–12
  lack of predators, 309
  lessons from, 309–12
  speciation on, 27–28
Isle Royale (Lake Superior), moose and wolf populations on, 38
Isotope
  defined, 191
  ocean histories and, 191
ITCZ. *See* Intertropical convergence zone
Iteroparous species, 149
Ivory-Billed Woodpecker, 324

Janzen, Daniel, 237
Javan tiger, 325
Jet stream, 45–46, 52–53, 348, 387–88
Job loss, blamed on environmental regulation, 436–37
J-shaped growth curve, 244–45

Kelp, 96
Kettle lake, 97
Keystone species, 184, 234, 324, 328
Kilimanjaro, 67
Kissimmee River, 120–21
Krakatau, Indonesia, 210–13, 311–12
Kudzu, 318
Kwashiorkor, 259
Kyshtyn reactor, Soviet Union, 399

La Niña, 53
Labrador current, 365
Ladybugs, as biological control agents, 284–85
Laherere, Jean, 401
Lake, 89
  acid deposition in, 385–88
  disappearance of, 99–100
  eutrophication of, 273
  formation of, 97–99
  freezing of, 101
  kettle, 97
  nutrients in, 100–103
  oxbow, 97–98
  oxygen in, 102, 275
  seasonal changes in, 102–5
  sources of pollution to, 274
  stratification cycle, consequences for wildlife, 103–5
  temperature of, 101–2, 274–75
  variability of, 100–101
Lake Erie, 276–77
  zebra mussels in, 323
Lake Ontario, 275–76
Lake Washington, 276
Lake Wodehouse, 293
Land bridges, 21
Land management, of tropical forests, 292–94
Land use, 341, 417, 430–31, 455
Landstat Thermatic Mapper, 290–91
Land-usage rights, 305
Lateral recharge, 96
Latitudinal gradient, 67
Laurentide ice sheet, 193–94

Common Conversions

METRIC TO ENGLISH

Metric Measure	Multiply by	English Equivalent
Length		
Centimeters (cm)	0.3937	Inches (in.)
Meters (m)	3.2808	Feet (ft)
Meters (m)	1.0936	Yards (yd)
Kilometers (km)	0.6214	Miles (mi)
Nautical mile	1.15	Statute mile
Area		
Square centimeters (cm²)	0.155	Square inches (in.²)
Square meters (m²)	10.7639	Square feet (ft²)
Square meter (m²)	1.1960	Square yards (yd²)
Square kilometers (km²)	0.3831	Square miles (mi²)
Hectare (ha) (10,000 m²)	2.4710	Acres (a)
Volume		
Cubic centimeters (cm³)	0.06	Cubic inches (in.³)
Cubic meters (m³)	35.30	Cubic feet (ft³)
Cubic meters (m³)	1.3079	Cubic yards (yd³)
Cubic kilometers (km³)	0.24	Cubic miles (mi³)
Liters (L)	1.0567	Quarts (qt), U.S.
Liters (L)	0.88	Quarts (qt), Imperial
Liters (L)	0.26	Gallons (gal), U.S.
Liters (L)	0.22	Gallons (gal), Imperial
Mass		
Grams (g)	0.03527	Ounces (oz)
Kilograms (kg)	2.2046	Pounds (lb)
Metric ton (tonne) (t)	1.10	Short ton (tn), U.S.
Velocity		
Meters/second (mps)	2.24	Miles/hour (mph)
Kilometers/hour (kmph)	0.62	Miles/hour (mph)
Knots (kn) (nautical mph)	1.15	Miles/hour (mph)
Temperature		
Degrees Celsius (°C)	1.80 (then add 32)	Degrees Fahrenheit (°F)
Celsius degree (C°)	1.80	Fahrenheit degree (F°)

Additional water measurements:

Gallon (Imperial)	1.201	Gallon (U.S.)
Gallons (gal)	0.000003	Acre-feet

1 cubic foot per second per day = 86,400 cubic feet = 1.98 acre-feet

ADDITIONAL ENERGY AND POWER MEASUREMENTS

1 watt (W)	=	1 joule/sec	1 W/m²	=	2.064 cal/cm²/day
1 joule	=	0.239 calorie	1 W/m²	=	61.91 cal/cm²/month
1 calorie	=	4.186 joules	1 W/m²	=	753.4 cal/cm²/year
1 W/m²	=	0.001433 cal/min	100 W/m²	=	75 kcal/cm²/year
697.8 W/m²	=	1 cal/cm²/minute			

Solar constant:
1372 W/m²
2 cal/cm²/minute